DYNAMICS OF BUBBLES, DROPS AND RIGID PARTICLES

FLUID MECHANICS AND ITS APPLICATIONS

Volume 50

Series Editor: R. MOREAU

MADYLAM

Ecole Nationale Supérieure d'Hydraulique de Grenoble

Boîte Postale 95

38402 Saint Martin d'Hères Cedex, France

Aims and Scope of the Series

The purpose of this series is to focus on subjects in which fluid mechanics plays a fundamental role.

As well as the more traditional applications of aeronautics, hydraulics, heat and mass transfer etc., books will be published dealing with topics which are currently in a state of rapid development, such as turbulence, suspensions and multiphase fluids, super and hypersonic flows and numerical modelling techniques.

It is a widely held view that it is the interdisciplinary subjects that will receive intense scientific attention, bringing them to the forefront of technological advancement. Fluids have the ability to transport matter and its properties as well as transmit force, therefore fluid mechanics is a subject that is particulary open to cross fertilisation with other sciences and disciplines of engineering. The subject of fluid mechanics will be highly relevant in domains such as chemical, metallurgical, biological and ecological engineering. This series is particularly open to such new multidisciplinary domains.

The median level of presentation is the first year graduate student. Some texts are monographs defining the current state of a field; others are accessible to final year undergraduates; but essentially the emphasis is on readability and clarity.

Dynamics of Bubbles, Drops and Rigid Particles

by

Z. ZAPRYANOV

Faculty of Mathematics and Informatics,
Sofia University,
Sofia, Bulgaria

and

S. TABAKOVA

Department of Mechanics,
Technical University of Plovdiv,
Plovdiv, Bulgaria

KLUWER ACADEMIC PUBLISHERS

DORDRECHT / BOSTON / LONDON

A C.I.P. Catalogue record for this book is available from the Library of Congress.

ISBN 0-7923-5347-1

Published by Kluwer Academic Publishers,
P.O. Box 17, 3300 AA Dordrecht, The Netherlands.

Sold and distributed in North, Central and South America
by Kluwer Academic Publishers,
101 Philip Drive, Norwell, MA 02061, U.S.A.

In all other countries, sold and distributed
by Kluwer Academic Publishers,
P.O. Box 322, 3300 AH Dordrecht, The Netherlands.

Printed on acid-free paper

Printed in the Netherlands.

In the Loving Memory of our Parents

PREFACE

1. Objective and Scope

Bubbles, drops and rigid particles occur everywhere in life, from valuable industrial operations like gas-liquid contracting, fluidized beds and extraction to such vital natural processes as fermentation, evaporation, and sedimentation. As we become increasingly aware of their fundamental role in industrial and biological systems, we are driven to know more about these fascinating particles. It is no surprise, therefore, that their practical and theoretical implications have aroused great interest among the scientific community and have inspired a growing number of studies and publications.

Over the past ten years advances in the field of small Reynolds numbers flows and their technological and biological applications have given rise to several definitive monographs and textbooks in the area. In addition, the past three decades have witnessed enormous progress in describing quantitatively the behaviour of these particles. However, to the best of our knowledge, there are still no available books that reflect such achievements in the areas of bubble and drop deformation, hydrodynamic interactions of deformable fluid particles at low and moderate Reynolds numbers and hydrodynamic interactions of particles in oscillatory flows. Indeed, only one more book is dedicated entirely to the behaviour of bubbles, drops and rigid particles ["Bubbles, Drops and Particles" by Clift et al. (1978)] and the authors state its limitations clearly in the preface: "We treat only phenomena in which particle-particle interactions are of negligible importance. Hence, direct application of the book is limited to single-particle systems of dilute suspensions."

The main objective of this textbook, therefore, is to close the gap between current knowledge and available publications with a unique emphasis on bubble and drop deformations, hydrodynamic interactions of deformable fluid particles at low and moderate Reynolds numbers, and hydrodynamic interactions of rigid particles in oscillating flows. Consistently employing the point of view of pure fluid mechanics, the book combines a thorough exploration of recent developments on the dynamics of bubbles, drops, and rigid particles with the classical studies in the field. It creates a coherent framework that unifies the subject and makes the available literature more accessible, especially for those wishing to extend their knowledge of particulate systems.

An additional objective of our work is to analyse hydrodynamic interactions between non-deformable and deformable fluid particles when their deformations are finite. We also strive to provide an overview on the latest advances in topics such as sedimentation, diffusion, and effective viscosity of suspensions and emulsions.

The studies involved are mainly theoretical, and in all considered problems we emphasize hydrodynamic analysis. Experimental data are used only in order to verify the obtained solutions. Experimental results, numerical solutions, and reference to topics not covered by the book are noted when they serve to illustrate a concept, result, or limitation of what has been presented.

The language used in the book is intended to be as clear and understandable as possible, and we pay attention to both the mathematics and the physics needed to apprehend various phenomena in fluid mechanics. Where necessary, the physics and mathematics prevail. However, the authors have aimed at reaching a balance between exact presentation, intuitive grasp of new ideas, and creative application of concepts throughout the text.

Because the text incorporates the essential physical, mathematical and physicochemical fundamentals, it is largely self-contained and appropriate for readers in a wide range of disciplines. It provides both a reference monograph for academic and industrial researchers and a textbook for advanced graduate courses in subjects like fluid mechanics, chemical engineering, physics, and applied mathematics. The coverage of the book is rather broad and some of the chapters have a survey character, but the content of the others is given in detail which allows the researcher, or instructor, to select those sections which suit his or her goals. Actually the book is based partly on the lecture notes for advanced undergraduate and graduate students on fluid mechanics that we teach at Sofia university and Technical University of Plovdiv.

2. Limitations and Specifications

Throughout the text, the focus is on laminar flows of incompressible Newtonian fluids. However, serious consideration is also given to viscous-dominated creeping flows. These flows have historically been of interest only in chemical engineering operations but are now developing uses in various technologies where fluid motions or transport processes on small length scale are involved.

Since the term "particle" is a loose one, our study will limit itself to noncolloidal (rigid or fluid) particles which are smaller than 10 cm (10^{-1} m) and larger than 0.1 micron (10^{-7} m). Such a study provides the basis for fundamental industrial operations like floatation in ore processing; particle filtration operations used in many industries; extraction, aggregation and deposition of pulp fibres in paper manufacturing and so on.

When speaking about the behaviour of these types of particles, Batchelor (1976) has suggested the term *microhydrodynamics*. Microhydrodynamics defines a subject that retains its focus on fluid mechanics, keeping its spread into other fields within the range of manageable properties. Since the dynamics of single particles is a sound basis upon which one can build the knowledge of multiparticle systems, both single (rigid or fluid) particles and two or more (rigid or fluid) particles are considered in the book.

The basic flows connected with bubbles, drops and rigid particles treated in the book are what are known as uniform flows. Examples are the flows due to gravitational forces. Stokes' problem (1851) for a falling spherical rigid particle in a viscous fluid and Hadamard and Rybczynski's problem (1911) for a falling spherical viscous drop in a viscous fluid are the classical studies in the area.

This book will also cover a substantial class of flows called the shear fluid flows, which have been studied intensively during the past 30-35 years. These flows have both fundamental and practical significance, and they appear in the chemical and biotechnological industry as a result of the movement of various apparatus mobile elements in fluid medium. Examples of such flows include the Couette flow between two planar or spherical walls, one of which is moving parallel to the other; the Poiseuille flow in a cylindrical tube, caused by a prescribed pressure shear; flow around a critical point, as observed in particle attraction by a collector; extensional flows in the production of fibre polymer materials; and so forth. The results of shear flows investigations are useful in modelling various disperse systems (suspensions, emulsions, etc.), the erythrocytes movement, the bodies movement in traces (behind other bodies) or boundary layers, and other instances.

Simple shear flows are characterised by a macroscopic parameter G, called the coefficient of the velocity profile change. Complicated shear flows can be defined by two, three, or more parameters. The coefficient of the velocity profile change G at a simple flow has a dimension reciprocal to the time. It participates in the Reynolds number expression, $\mathrm{Re}=\dfrac{a^2G}{\nu}$, which is the basic dynamic parameter in shear flows hydrodynamics (where a is the flow length characteristic, and ν -the fluid kinematic viscosity).

In the study of the motion of a single particle or an assemblage of particles with arbitrary shape in a viscous fluid, it is still very difficult to obtain exact or even approximate solutions. In order to construct tractable mathematical models for these problems it is, therefore, necessary to resort to a number of simplifications. A basic hydrodynamic assumption of this book is that all flows, both steady and unsteady, are considered to be laminar. Furthermore, we shall often assume that a flow is sufficiently "slow" and inertial effects need not be taken into account. One criterion which determines the relative importance of inertial and viscous effects is the Reynolds number. Flows in which the Reynolds number is very small (practically zero) are governed by Stokes equations. In the past 30 years the knowledge of such flows and their applications has grown tremendously and our study will attempt to reflect these advances.

A main goal of this book is to examine the interactions between particles and fluids from the point of view of pure fluid mechanics. The evaluation of the external forces and moments acting on the particles in unbounded or bounded fluid flow is possible if the relations between their motion and hydrodynamic force and/or moment are known, which implies that the mobility relation between particles motion must be studied. This statement is only applicable for noncolloidal particles whose size is larger than one micron ($>10^{-6}$ m) or large enough to neglect the molecular processes inside them. These types of particles can be divided depending on the dispersed phase, that is, on the material from which they are formed, namely: rigid particles, which are not deformed considerably under the action of large stresses; drops, where the dispersed phase is assumed to be liquid, and in our case, Newtonian; bubbles formed from gas (usually considered as perfect) or from a void. Drops and bubbles are often referred to as fluid particles.

Particle geometry is also very important when looking for mathematical models and their solutions. Simple particle shapes like spheres and ellipsoids allow the construction of analytical solutions in some cases of importance, examples of which will be given in this book. Complex particle shapes, on the other hand, require special numerical schemes. The presence of a wall or deformable interface also complicate the considered problems and the techniques for their solution.

A set of analytical solution techniques is used throughout the book, in treating the behaviour of simple particle shapes in unbounded flow or in flows bounded by simple container geometry. This set of techniques will be explained where appropriate. Complex geometries are handled by numerical methods that are described as well. In terms of simple particle shapes the most basic examples are spheres, spheroids and ellipsoids. The simplest boundary is the plane wall, but analytical and semi-analytical methods are available only for spherical and cylindrical container geometry. The emphasis is on asymptotic approximation techniques applied to classical problems involving laminar flows of incompressible Newtonian fluids. However, there is also a significant amount of material on viscosity dominated creeping flows. Many problems connected with the dynamics of bubbles, drops and rigid particles are also solved numerically on the basis of the Navier-Stokes equations.

3. Organization of the Book.

The book is divided into four parts and consists of 14 chapters.

In Part I, we introduce the governing equations which, along with several fundamental theorems and laws, form the foundation of the book. Chapter 1 defines the governing equations, which are actually no more than Newton's basic laws of classical mechanics, applied to a body that is deformed as a consequence of its movement. The basic laws are coupled with constitutive laws that specify the relationship between dynamic variables (shear rate, etc.) and force, and between temperature gradients and heat flux. For the sake of brevity this chapter maintains a focus on the relationship of the stress tensor with the surface vector, without entering into detailed explanation of the stress structure. Chapter 2 provides a discussion of fundamental theorems and laws, including the Lorentz reciprocal theorem, Faxen laws for spherical particles, general integral representations with single and double layer potential, and the Helmholtz theorems for minimum energy dissipation at Stokes flows.

Part II is composed of four chapters, all of which are dedicated to the steady dynamics of a single rigid or fluid particle. Chapter 3 introduces the singularity solutions for the problems of a viscous flow past a spherical rigid or fluid particle, the link between the singularity solutions and the Faxen laws, the approximate singularity solutions for the translation of a prolate ellipsoid in uniform viscous fluid, and others. Chapter 4 applies the superposition method of vector harmonic functions to solve problems involving the rotation and translation of a rigid spherical particle in quiescent fluid, and a spherical rigid particle in general shear flow. Chapter 5 explores other methods used to study the flow past single rigid or fluid particles. It also considers fluid particle movement due to thermocapillary convection and fluid flows induced by an electric field, both outside and inside a drop. Chapter 6 addresses the deformation of bubbles and drops in viscous flows for the first time in the book. It examines the small deformations of fluid particles at small and large Reynolds numbers, along with finite deformations at moderate Reynolds numbers where the boundary-fitted coordinates method or the boundary element method is employed.

Part III discusses steady problems involving hydrodynamic interactions between two rigid or fluid particles, boundary effects on the motion of rigid or fluid particles and hydrodynamic interactions in particulate systems. Chapter 7 begins by considering the effects of hydrodynamic interactions between two rigid particles and the resistance and mobility relations in unbounded flow. The most important sections of this chapter deal with small deformations of two fluid particles in steady flow and small deformations of a compound drop moving in a viscous fluid. The method of domain perturbations is applied and bispherical coordinates are exploited. Chapter 8 presents the boundary effects on the motion of rigid or fluid particles near a plane wall or circular tube. This chapter also includes an analysis of small deformations of a drop in the presence of a deformable interface. In Chapters 9 and 10 some contemporary reviews are given on the investigations of flows relative to assemblages of rigid or fluid particles and the effective viscosity of suspensions and emulsions.

The last part of the book, Part IV, discusses the various unsteady effects which complicate the dynamics of bubbles, drops and rigid particles. Chapter 11 begins by discussing the effects of time dependent Stokes flows, including the concepts of added mass and Basset forces, and ends with the coupling between shape and volume oscillations for a gas or vapour bubble. In Chapter 12, the unsteady fundamental solutions, the reciprocal theorem, Faxen laws, and integral representations for unsteady flows are considered. The chapter also describes the numerical solution of the oscillatory flow past an axisymmetric rigid particle, followed by the topics of relaxation and break-up of an initially extended drop, finite deformations of a drop moving through a fluid interface, and time-dependent interactions between two buoyancy-driven deformable drops. Chapter 13 deals with unsteady hydrodynamic interactions between two rigid spherical particles performing short-time translation or oscillating with large frequency in a viscous fluid. In the final Chapter 14 we explore the rise of a drop (bubble) towards a rigid wall, and the finite deformations of two drops or a compound drop in viscous flows at moderate Reynolds numbers. The unsteady Navier-Stokes equations are solved numerically using the finite element method, and the obtained numerical solutions enable us to reveal some inertial effects in the case of drop deformations at moderate Reynolds numbers.

ACKNOWLEDGEMENTS

We would like to express our sincere appreciation to Professor Howard Brenner of Massachusetts Institute of Technology (MIT) and Professor Andrea Acrivos of the Levich Institute for Physicochemical Hydrodynamics at CCNY for many helpful discussions on problems connected with dynamics of bubbles, drops and rigid particles. One of us (ZZ) would like to take this opportunity to thank Professor van Dyke of Stanford University and Professor Darsh Wasan of Ilinois Institute of Technology for their consideration and attention toward him during his specialisations in their Departments in 1973 and 1979. In fact, these specialisations have aroused his eversince growing interest to the problems considered in this book.

We are extremely grateful to our colleagues Ch. Christov, P. Kalitzova-Kurteva, E. Chervenivanova, E. Toshev, Tz. Kotzev, N. Kovacheva, I. Lambova, P. Shopov, P. Minev, I. Bazhlekov, I. Bozduganov and T. Partalin for their particularly prolific and beneficial participation in the studies which made this book possible.

We wish to express also our deep and cordial thanks to Dr. E. Chervenivanova for her hard work as a reviewer, and to Dr. E. Manev and Miss Hollynd Feldman for doing the editing job, and to Mrs. A. Stoyanova, Mr. Ya.. Christov and Mr. I. Kolemanov for the technical assistance in preparing the manuscript and figures in camera-ready form.

Finally, we thank our families for the patience and for not complaining during the long period of preparing this book.

PART I

GOVERNING EQUATIONS AND GENERAL PROPERTIES OF FLUID FLOWS

CHAPTER 1

Governing Equations and Boundary Conditions for Fluid Flows

1.1. Introduction

The objective of this chapter is to give a brief description of the basic equations of fluid mechanics. The main method used to analyse the fluid motion is the construction of phenomenological macroscopic theories, based on experimentally established common relations and hypotheses. The fundamental theory of fluid dynamics is based on the three basic conservation laws of mechanics, of matter, momentum and energy, whose modification for fluid flow will be given in the next section 1.2. The laws that we shall use to obtain these equations are of such a fundamental nature that they cannot be proven in the mathematical sense. They are sufficiently general to be applied to all substances including both rigid bodies and fluids. The truthfulness of these basic laws has been established through scientific evolution over a very long period of time. Through the years these laws, as well as the concepts on which they are based, have undergone serious changes. Although fluid mechanics now has a developed content, the major part of the phenomena treated by fluid mechanics is still not sufficiently understood. The derivation of basic laws is in fact the mathematical formulation of the relations between some physical concepts. The final forms of these laws are the differential equations valid for each point inside the considered region.

Assuming the basic laws applicable for the fluid flows description and utilizing some hypotheses specific for different types of fluids, a closed system or model of partial differential equations is achieved. Appropriate boundary conditions must be further imposed and the obtained boundary problems solved analytically or numerically. The full hydrodynamical investigation needs also an experimental verification of the results. One of the hypotheses for the fluids is that they are not a discrete system of material points, but as a continuous phase without intermolecular spaces, i.e., a material continuum, to every point of which a corresponding value, characterizing the density, is ascribed.

The three basic independent dynamic laws in continuum mechanics and, in particular, in fluid mechanics - the continuity equation, the momentum equation and the internal energy equation - will be given in section 1.2. From the momentum equation several additional laws can be derived: the kinetic energy equation, the angular moment of momentum equation and the vorticity equation.

The fluid motion is fully determined if the velocity **v** of each fluid point is known as a function of time and position. Therefore, velocity is a function of space coordinates and of time. The stress and the rate of strain tensors play a main role in fluid mechanics. The relation between them can vary, depending on the fluid type. When it is linear, the fluids are called Newtonian. These fluids consist mainly of the fluids of common use, including water, glycerol, liquid metals and most of the gases. The non-linear relation between the stress and rate of strain tensors corresponds to the non-Newtonian fluids. Usually these are high polymers, polymer solutions, oil dies, petroleum, blood, biological solutions, etc. As already noted, in this book we shall confine ourselves to Newtonian fluids only.

1.2. Balance Laws for Newtonian Fluids

The three basic conservation laws applied to a fluid phase lead correspondingly to three differential equations valid at every point in the fluid volume. The conservation of mass law transforms into the continuity equation for the fluid phase, the conservation of momentum law yields the momentum equation, and the conservation of energy law: the energy equation.

Continuity Equation. A fluid volume V(t) with surface S is considered (as shown in Fig.1.2.1.) with density ρ, dynamic viscosity μ, velocity **v** and unit normal vector **n**. The mass conservation principle is stated as : "The time rate of change of the mass of a fluid volume is zero". The mass M(t) of V(t) is calculated by integrating the density over the volume

$$\frac{\mathrm{dM}}{\mathrm{dt}}=\frac{\mathrm{d}}{\mathrm{dt}}\int_{\mathrm{V}}\rho\,\mathrm{d}\tau=0\,. \tag{1.2.1}$$

Since the surface S moves with the local fluid velocity **v**, the Leibnitz's theorem (see Panton, 1984) transforms (1.2.1) into

$$\int_{\mathrm{V}}\frac{\partial\rho}{\partial \mathrm{t}}\,\mathrm{d}\tau+\int_{\mathrm{S}}\rho\,\mathbf{v}.\mathbf{n}\,\mathrm{d}\sigma=0\,, \tag{1.2.2}$$

which due to the divergence theorem takes the form of a single volume integral

$$\int_V \left[\frac{\partial \rho}{\partial t} + \nabla.(\rho \mathbf{v})\right] d\tau = 0. \tag{1.2.3}$$

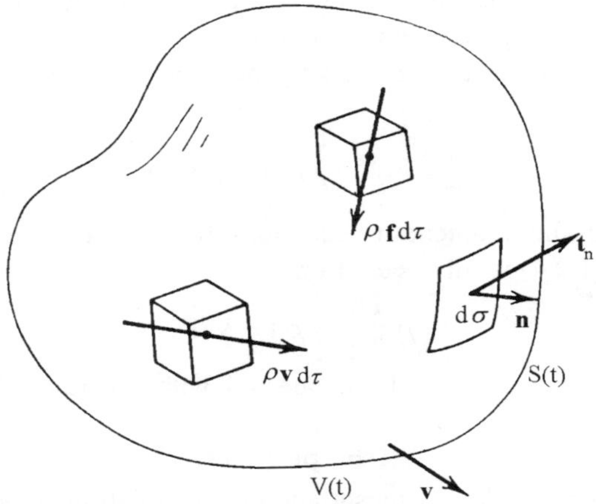

Fig.1.2.1. A sketch of the fluid volume

As the fluid volume is arbitrary and the integrand is a continuous function, equation (1.2.3) is only true when the integrand is zero, i.e.,

$$\frac{\partial \rho}{\partial t} + \nabla.(\rho \mathbf{v}) = 0. \tag{1.2.4}$$

If substantial derivative D/Dt is introduced in (1.2.4), the final form of the continuity equation becomes

$$\frac{D\rho}{Dt} + \rho \nabla.\mathbf{v} = 0. \tag{1.2.5}$$

In the case when some fluid mass dm = $\gamma\, d\tau\, dt$ enters the fluid surface S from some source, the right-hand side of (1.2.4) or (1.2.5) will be equal to γ, where γ is the source density. The steady flow continuity equation is easily obtained from (1.2.4) at $\partial/\partial t = 0$, while at the constant density - from (1.2.5) at D/Dt=0. The latter reads

$$\nabla.\mathbf{v} = 0, \tag{1.2.6}$$

which is the known as the continuity equation for incompressible fluid.

Momentum Equation. The momentum principle states: "The time rate of change of the linear momentum of a fluid volume is equal to the sum of body forces acting on the bulk in the volume and surface forces acting on the boundary surface." Here, the momentum is calculated again by integration of $\rho\,\mathbf{v}$ over the volume V

$$\frac{d}{dt}\int_V \rho \mathbf{v} d\tau = \int_V \rho \mathbf{f} d\tau + \int_S \mathbf{t}_n \, d\sigma , \tag{1.2.7}$$

where **f** is the external body force per unit mass, $\mathbf{t}_n$ - the stress vector as $\mathbf{t}_n = \mathbf{T}.\mathbf{n}$ and **T** - the stress tensor (symmetric, due to the moment of momentum balance). The common fluids obey the Stokes hypothesis that there is no distinction between thermodynamic and mechanical pressure. In the gravity field **f** is denoted by the gravity acceleration **g**.

Applying the Leibnitz's and divergence theorems to (1.2.7), we reach its differential analogue

$$\frac{\partial(\rho \mathbf{v})}{\partial t} + \nabla.(\rho \mathbf{v}\mathbf{v}) = \rho \mathbf{f} + \nabla.\mathbf{T} , \tag{1.2.8}$$

which is known as the momentum equation. It is significantly simplified if the continuity equation (1.2.4) is subtracted from it

$$\rho \frac{D\mathbf{v}}{Dt} = \rho \mathbf{f} + \nabla.\mathbf{T} . \tag{1.2.9}$$

Now the stress tensor can be connected with the rate of strain tensor **E** by Newton's linear law:

$$\mathbf{T} = -p\mathbf{I} + 2\mu \mathbf{E} , \tag{1.2.10}$$

where p is the pressure, **I** - identity tensor and the rate of strain tensor **E** is expressed by the velocity gradients

$$\mathbf{E} = \frac{1}{2}\left[\nabla \mathbf{v} + (\nabla \mathbf{v})^T\right]. \tag{1.2.11}$$

The equation (1.2.10) is the constitutive equation for Newtonian fluids. As mentioned in section 1.1. a lot of industrially important products do not fulfil Newton's viscosity law and are considered as non-Newtonian fluids (see Coleman et al., 1966).

<u>Energy Equation.</u> The first law of thermodynamics states: "The time rate of change of the energy of a fluid volume is equal to sum of the rate of body and surface forces work and the rate of heat transfer from the fluid boundary." The integral form of this statement is:

$$\frac{d}{dt}\int_V \rho \left(e + \frac{1}{2}v^2\right) d\tau = \int_V \rho \mathbf{f}.\mathbf{v} d\tau - \int_S q_n \, d\sigma + \int_S \mathbf{t}_n.\mathbf{v} \, d\sigma , \tag{1.2.12}$$

where e is the absolute thermodynamic internal energy per unit mass, $\frac{1}{2}v^2$ - kinetic energy per unit mass, and q_n - the normal heat flux, usually connected with temperature T via the Fourier's law $\mathbf{q} = -\kappa \nabla T$, where κ is fluid thermal conductivity.

As in the previous two cases, we apply successively the Leibnitz's and divergence theorems to (1.2.12) to get

$$\frac{\partial}{\partial t}\left[\rho\left(e + \frac{1}{2}v^2\right)\right] + \nabla.\left[\rho\left(e + \frac{1}{2}v^2\right)\right] = \rho \mathbf{f}\mathbf{v} + \nabla.(\kappa \nabla T) - \mathbf{v}\nabla p - p\nabla.\mathbf{v} + 2\nabla.(\mu \mathbf{E}.\mathbf{v}) \tag{1.2.13}$$

From this equation we subtract the kinetic energy equation obtained after a scalar product of the momentum equation (1.2.8) with **v** and finally reach the internal energy equation

$$\rho \frac{De}{Dt} = \nabla.(\kappa \nabla T) - p\nabla.\mathbf{v} + 2\mu\, \mathbf{E}{:}\nabla\mathbf{v}\ . \qquad (1.2.14)$$

The last term in (1.2.14) is called viscous dissipation, which is responsible for heat generation due to friction. Its contribution to temperature rise can be neglected under certain conditions as its magnitude is ~ $O(\mu\, U^2/\kappa)$, where U is the fluid velocity scale. In most of the cases treated in this book, we shall assume that fluid processes occur under isothermal conditions.

1.3. Governing Equations in Orthogonal Curvilinear Coordinates

With the postulations made in the previous section 1.2. we are ready to formulate the Navier-Stokes system of differential equations for the motion of an incompressible isothermal Newtonian fluid, combining (1.2.6) and (1.2.9) with the constitutive equations (1.2.10) and (1.2.11):

$$\rho \frac{D\mathbf{v}}{Dt} = \rho\,\mathbf{f} - \nabla p + \mu\, \nabla^2 \mathbf{v}\ , \qquad (1.3.1)$$

$$\nabla.\mathbf{v} = 0\ . \qquad (1.3.2)$$

An analysis of these equations can be found in many classical fluid mechanics books (Lamb, 1945), (Batchelor, 1967), (Landau and Lifshitz, 1959), (Levich, 1962), etc.

Since hydrostatic pressure is caused by the force of gravity, it can be subtracted from the total pressure, that is $p_{dyn} = p - p_{hyd}$. In the following presentations we shall refer to pressure as dynamic pressure, but shall drop the subscript "dyn". Thus (1.3.1) transforms into

$$\rho \frac{D\mathbf{v}}{Dt} = -\nabla p + \mu\, \nabla^2 \mathbf{v}\ . \qquad (1.3.3)$$

If the Reynolds number $Re = \rho UL/\mu$ and the Strouhal number $St = L/U\tau$ are introduced, where U, L and τ are the velocity, length and time scales, respectively, then (1.3.3) can be dimensionalized in two different ways depending on the pressure scale. For moderate and high Re >> 1, the pressure is scaled with the dynamic pressure $\frac{\rho}{2}U^2$.

In dimensionless form (1.3.3) may be rewritten as

$$St\frac{\partial \mathbf{v}}{\partial t} + \mathbf{v}.\nabla\mathbf{v} = -\nabla p + \frac{1}{Re}\nabla^2\mathbf{v}\ , \qquad (1.3.4)$$

where all variables are dimensionless. In the second case, when Re is small (Re << 1), i.e., the viscous forces are dominant, an appropriate pressure scale is $\mu U/L$ and the dimensionless Navier-Stokes equations appear as

$$\beta \frac{\partial \mathbf{v}}{\partial t} + \mathrm{Re}\mathbf{v}.\nabla\mathbf{v} = -\nabla p + \nabla^2\mathbf{v}, \tag{1.3.5}$$

where $\beta = \mathrm{ReSt} = L^2/\nu\tau$ is a frequency parameter.

An approximate solution to the Navier-Stokes equations can be obtained in cases where the Reynolds number, or the ratio of the inertial to viscous forces, is very small. Thus, the inertial effects can be neglected, and the influence of viscosity is considered to be leading. We can assume that the Reynolds number is small (Re << 1) either because the fluid is very viscous ($\mu \to \infty$), or because the inertia or density is very small ($\rho \to 0$). Such flows are frequently called "creeping" flows. This simplification is justified, since many multiparticle systems do involve sufficiently slow motions (small Re). This class of fluid mechanics problems, known as Stokes linear problems, at Reynolds number tending to zero, leads to a number of analytical solutions. If in (1.3.5) Re $\to$ 0 and $\beta \to 0$, since St = O(1), then the steady dimensionless Stokes equations are simply

$$\begin{aligned} -\nabla p + \nabla^2\mathbf{v} &= 0 \\ \nabla.\mathbf{v} &= 0, \end{aligned} \tag{1.3.6}$$

or in dimensional form

$$\begin{aligned} -\nabla p + \mu\nabla^2\mathbf{v} &= 0 \\ \nabla.\mathbf{v} &= 0. \end{aligned} \tag{1.3.7}$$

As a consequence of (1.3.7) the pressure governing equation is found to be

$$\Delta p = 0. \tag{1.3.8}$$

Unsteady creeping flow problems can be governed by (1.3.6), if the quasi-steady assumptions are appropriate, for example in the case of evolution problems with deformable boundary. In more general cases, such problems can be made to fit within the so-called unsteady Stokes equations with retained local time derivative in (1.3.5), when St $\to \infty$ and $\beta = \mathrm{ReSt} = O(1)$

$$\rho \frac{\partial \mathbf{v}}{\partial t} = -\nabla p + \mu\nabla^2\mathbf{v}. \tag{1.3.9}$$

The Stokes flow problems have been studied extensively by many authors. A good review for the work up until the 1970's can be found in (Happel and Brenner, 1973). The paper of Odqvist (1930) builds the fundamentals of the Stokes flow theory with its existence and the uniqueness theorems.

The scalar and vector equations (1.3.2) and (1.3.3) are applicable in an arbitrary coordinate system. The Cartesian coordinate system (x, y, z) is often used in the practice. When solving hydrodynamic problems with complex geometry it is convenient to employ curvilinear orthogonal coordinates instead of the Cartesian ones. For example, when a spherical particle is passed by a fluid flow, it is more suitable to utilise a spherical coordinate system, while for two spherical particles in a fluid flow, a bispherical coordinate system is the more appropriate to use. To this end we shall write the Navier-Stokes equations in curvilinear coordinates.

Let (q_1, q_2, q_3) be curvilinear orthogonal coordinates connected with the Cartesian coordinates (x, y, z) by the vector relation $\mathbf{r} = \mathbf{r}(q_1, q_2, q_3)$, where $\mathbf{r}$ is the radius vector of the considered point. The unit vectors $\mathbf{e}_1$, $\mathbf{e}_2$, $\mathbf{e}_3$ are parallel to the tangents of the corresponding coordinate lines. The first order differential operators of any quantity **A** in the coordinate system (q_1, q_2, q_3) are:

$$\nabla\mathbf{A} \equiv \mathrm{grad}\mathbf{A} \equiv \left(\frac{1}{h_1}\frac{\partial \mathbf{A}}{\partial q_1}, \frac{1}{h_2}\frac{\partial \mathbf{A}}{\partial q_2}, \frac{1}{h_3}\frac{\partial \mathbf{A}}{\partial q_3}\right);$$

$$\nabla.\mathbf{A} \equiv \mathrm{div}\mathbf{A} \equiv \frac{1}{h_1h_2h_3}\left[\frac{\partial}{\partial q_1}(h_2h_3A_1) + \frac{\partial}{\partial q_2}(h_1h_3A_3) + \frac{\partial}{\partial q_3}(h_1h_2A_3)\right];$$

$$\nabla \times \mathbf{A} \equiv \mathrm{rot}\mathbf{A} \equiv \frac{1}{h_1h_2h_3}\begin{vmatrix} h_1\mathbf{e}_1 & h_2\mathbf{e}_2 & h_3\mathbf{e}_3 \\ \frac{\partial}{\partial q_1} & \frac{\partial}{\partial q_2} & \frac{\partial}{\partial q_3} \\ h_1A_1 & h_2A_2 & h_3A_3 \end{vmatrix}.$$

Then the continuity equation (1.3.2) in the system (q_1, q_2, q_3) takes the form

$$\frac{\partial(h_2h_3v_1)}{\partial q_1} + \frac{\partial(h_1h_3v_2)}{\partial q_2} + \frac{\partial(h_1h_2v_3)}{\partial q_3} = 0, \tag{1.3.10}$$

where $h_i = \left|\frac{\partial \mathbf{r}}{\partial q_i}\right|$ are the Lame coefficients.

In cylindrical coordinates (r, φ, z) connected with the Cartesian system by

$$x = r\cos\varphi,\ y = r\sin\varphi,\ z = z, \tag{1.3.11}$$

the Lame coefficients are

$$h_1 = 1,\ h_2 = r,\ h_3 = 1. \tag{1.3.12}$$

Usually, for planar (2D) flows $q_3 \equiv z$ and then $h_3 \equiv 1$. In spherical coordinates (r, θ, φ), given by

$$x = r\sin\theta\cos\varphi,\ y = r\sin\theta\sin\varphi,\ z = r\cos\theta, \tag{1.3.13}$$

the Lame coefficients have the form

$$h_1 = 1,\ h_2 = r,\ h_3 = r\sin\theta. \tag{1.3.14}$$

We have to note that for the axisymmetric case, $q_3 \equiv \varphi$ is the angle about the symmetry axis, and h_3 measures the distance from the axis of symmetry.

In this way we obtain for the Navier-Stokes equations (1.3.3) in curvilinear orthogonal coordinates

$$\begin{aligned}&\frac{\partial v_i}{\partial t} + \frac{v_1}{h_1}\frac{\partial v_i}{\partial q_1} + \frac{v_2}{h_2}\frac{\partial v_i}{\partial q_2} + \frac{v_3}{h_3}\frac{\partial v_i}{\partial q_3} + \frac{v_i}{h_i}\left(\frac{v_1}{h_1}\frac{\partial h_i}{\partial q_1} + \frac{v_2}{h_2}\frac{\partial h_i}{\partial q_2} + \frac{v_3}{h_3}\frac{\partial h_i}{\partial q_3}\right) \\ &- \frac{1}{h_i}\left(\frac{v_1^2}{h_1}\frac{\partial h_1}{\partial q_i} + \frac{v_2^2}{h_2}\frac{\partial h_2}{\partial q_i} + \frac{v_3^2}{h_3}\frac{\partial h_3}{\partial q_i}\right) = -\frac{1}{\rho h_i}\frac{\partial p}{\partial q_i} + \nu(\Delta\mathbf{v})_i,\end{aligned} \tag{1.3.15}$$

where i= 1, 2, 3, $\nu = \mu/\rho$ - kinematic viscosity coefficient and $\Delta\mathbf{v} = \mathrm{grad}(\mathrm{div}\mathbf{v}) - \mathrm{rot}(\mathrm{rot}\mathbf{v})$.

Here we shall note that the second order differential operator Δ of a scalar is the Laplace operator:

$$\Delta \equiv \frac{1}{h_1h_2h_3}\left[\frac{\partial}{\partial q_1}\left(\frac{h_2h_3}{h_1}\frac{\partial}{\partial q_1}\right)+\frac{\partial}{\partial q_2}\left(\frac{h_1h_3}{h_2}\frac{\partial}{\partial q_2}\right)+\frac{\partial}{\partial q_3}\left(\frac{h_1h_2}{h_3}\frac{\partial}{\partial q_3}\right)\right] \quad (1.3.16)$$

From the continuity equation (1.3.10) in the axisymmetric or planar case the stream function Ψ is introduced through its relation with the velocity components

$$v_1 = \frac{1}{h_2h_3}\frac{\partial \Psi}{\partial q_2}, \quad v_2 = -\frac{1}{h_1h_3}\frac{\partial \Psi}{\partial q_1}. \quad (1.3.17)$$

Since the vorticity vector $\mathbf{\Omega}$ is connected with the velocity by $\mathbf{\Omega}$ = rot$\mathbf{v}$, for its non-zero component we have

$$\Omega_3 = -\frac{1}{h_3}E^2\Psi, \quad (1.3.18)$$

where in the axisymmetric case

$$E^2 \equiv \frac{h_3}{h_1h_2}\left[\frac{\partial}{\partial q_1}\left(\frac{h_2}{h_1h_3}\frac{\partial}{\partial q_1}\right)+\frac{\partial}{\partial q_2}\left(\frac{h_1}{h_2h_3}\frac{\partial}{\partial q_2}\right)\right] \quad (1.3.19)$$

and at the planar case E^2 reduces to the 2D (two-dimensional) Laplace operator (1.3.16) with $h_3 = 1$ and $\partial/\partial q_3 = 0$. Then the Navier-Stokes equations given in their ($\mathbf{v}$ - p) formulation (1.3.10) and (1.3.15) may be transformed in (Ψ) formulation, which by use of (1.3.17) reads

$$\frac{\partial}{\partial t}(E^2\Psi) - \frac{1}{h_1h_2h_3}\left[\frac{\partial(\Psi, E^2\Psi)}{\partial(q_1, q_2)}\right] + \frac{2E^2\Psi}{h_1h_2h_3^2}\left[\frac{\partial(\Psi, h_3)}{\partial(q_1, q_2)}\right] = \nu E^4\Psi. \quad (1.3.20)$$

If the problem is axisymmetric or planar, the steady Stokes equations (1.3.6) or (1.3.7) may be rewritten in terms of the stream function Ψ in curvilinear orthogonal coordinates

$$E^4\Psi = 0. \quad (1.3.21)$$

Since the Navier-Stokes equations form a complicated system of partial differential equations, their integration is only possible in a few exceptionally simple cases. On the other hand, the comparison shows that there is a good correspondence between the experimental results and the calculated results, based on the known exact solutions. This fact means that the Navier-Stokes equations describe to a large extent correctly the real Newtonian fluids motion.

1.4. Boundary Conditions

In order to obtain a full model describing the viscous incompressible flow motion for given hydrodynamical problems, the Navier-Stokes equations (1.3.2) and (1.3.3) must be supplemented with the correspondent initial and boundary conditions. In this section we shall discuss only the latter, while leaving the initial conditions to be considered further in the book for the specific unsteady problems (part IV).

Boundary conditions on a rigid particle surface Let a rigid particle be passed by a viscous fluid and let the coordinate system, to which the motion is related, be fixed upon the particle. The particle surface is denoted by S_p and the unit normal vector to it by **n**. In terms of velocity, the no-slip boundary conditions on S_p must be fulfilled

$$\mathbf{v}\big|_{S_p} = 0, \quad \text{i.e.,} \quad v_n\big|_{S_p} = 0, \quad v_{\tau_1}\big|_{S_p} = 0, \quad v_{\tau_2}\big|_{S_p} = 0, \tag{1.4.1}$$

where v_n, v_{τ_1} and v_{τ_2} are the normal and tangential velocity components, respectively. If the particle moves with velocity $\mathbf{U}_p$, then the fluid and particle velocities must be equal on the particle surface

$$\mathbf{v}\big|_{S_p} = \mathbf{U}_p\,. \tag{1.4.2}$$

If the particle is located in an unbounded flow, the fluid flow velocity at infinity tends toward the prescribed velocity $\mathbf{v}^\infty$

$$\mathbf{v} \to \mathbf{v}^\infty \quad \text{as} \quad \mathbf{r} \to \infty\,. \tag{1.4.3}$$

Boundary conditions at the interface S between two immiscible fluids We shall begin with the assumption that the boundary conditions at the interface S between two immiscible fluids include first of all the continuity condition for the velocities **v** and $\hat{\mathbf{v}}$ on both sides of the interface S

$$\mathbf{v} = \hat{\mathbf{v}}\,, \tag{1.4.4}$$

where **v** is the velocity above the interface S and $\hat{\mathbf{v}}$ - under it.

Let us assume that the motion and change of the interface S is defined by the equation

$$F(\mathbf{r}, t) = 0. \tag{1.4.5}$$

Then the unit normal vector **n** to S equals

$$\mathbf{n} = \pm\frac{\nabla F}{|\nabla F|}\,, \tag{1.4.6}$$

where the sign "+" is for the normal **n** directed outward and "-" is for **n** directed inward (see Fig.1.2.1). Then the full derivative of F with respect to time t is given by

$$\frac{\partial F}{\partial t} + \mathbf{v}.\nabla F = 0\,, \tag{1.4.7}$$

where $\mathbf{v} = \hat{\mathbf{v}}$ is the surface points velocity, as well as the fluids velocity from the two sides of S. Combining (1.4.6) and (1.4.7), we can derive the kinematic velocity condition on a deformed fluid interface

$$\mathbf{v}.\mathbf{n} = \hat{\mathbf{v}}.\mathbf{n} = -\frac{\dfrac{\partial F}{\partial t}}{|\nabla F|}\,. \tag{1.4.8}$$

The stress balance on both sides of the interface S reads (Scriven, 1960)

$$\left(\mathbf{T}.\mathbf{n} - \hat{\mathbf{T}}.\mathbf{n}\right) + \nabla_s\sigma - \sigma\mathbf{n}\left(\nabla_s.\mathbf{n}\right) = 0\,, \tag{1.4.9}$$

where σ is the surface tension acting on S and $\nabla_s = (\mathbf{I} - \mathbf{nn}).\nabla$ is the surface gradient operator. If the equality (1.4.9) is projected on **n**, the normal stress balance condition is reached

$$\left(\mathbf{T}.\mathbf{n} - \hat{\mathbf{T}}.\mathbf{n}\right).\mathbf{n} = \sigma\left(\nabla_s.\mathbf{n}\right), \tag{1.4.10}$$

which shows that, due to the surface tension, a normal stress jump exists when passing from one side to the other through S. Utilising the type of the stress tensor **T** given by Newton's viscosity law (1.2.10), the stress balance condition (1.4.10) can be rewritten as

$$\hat{p}_{tot} - p_{tot} + 2\left[\mu\,\mathbf{E}.\mathbf{n} - \hat{\mu}\,\hat{\mathbf{E}}.\mathbf{n}\right].\mathbf{n} = \sigma\left(\nabla_s.\mathbf{n}\right), \tag{1.4.11}$$

where μ and $\hat{\mu}$ are the two fluids viscosities, while p_{tot} and $\hat{p}_{tot}$ are the total pressures of the two fluids, incorporating the dynamic and hydrostatic pressure. The normal divergence $\nabla.\mathbf{n}$ may be expressed by the main curvature radii R_1 and R_2

$$\nabla_s.\mathbf{n} = \frac{1}{R_1} + \frac{1}{R_2}. \tag{1.4.12}$$

For spherical interface S, there is only one principal radius $R_1 = R_2 = R$, and for planar interface, $R_1 = R_2 = \infty$, implying that the pressures on both sides of S are equal.

In the general case two tangential force components exist with respect to S. If the two tangential unit vectors are denoted by $\boldsymbol{\tau}_i$ (i = 1, 2), each of them is perpendicular to **n** and to the other tangential vector, $\boldsymbol{\tau}_1 . \boldsymbol{\tau}_2 = 0$, then from (1.4.9) we get

$$\left(\mathbf{T}.\mathbf{n} - \hat{\mathbf{T}}.\mathbf{n}\right).\boldsymbol{\tau}_i + \left(\nabla_s\sigma\right).\boldsymbol{\tau}_i = 0, \quad \left(i = 1,2\right). \tag{1.4.13}$$

From this equality it follows that when surface tension is constant, tangential stress is continuous over the interface. When the surface tension is non-uniform, there is a jump of the tangential stress. In other words, if $\nabla\sigma \neq 0$, one or both stress components $\mathbf{T}.\mathbf{n}.\boldsymbol{\tau}_i$ and $\hat{\mathbf{T}}.\mathbf{n}.\boldsymbol{\tau}_i$ must be different from zero. It should be noted that $\mathbf{T}.\mathbf{n}$ or $\hat{\mathbf{T}}.\mathbf{n}$ can be non-zero only if the fluid or fluids are moving. In this case the surface tension gradients cause a surface convection, called a Marangoni convection (see section 5.5).

For charged surfaces or interfaces, the stress balance conditions must be modified to include surface charges and electric stresses (Melcher and Taylor, 1969).

In the problems treated further in this book, the spherical fluid particle shape is encountered many times, so now we shall summarise the derived boundary conditions on S_p, assumed fixed, when σ = const. The kinematic conditions (1.4.2) and (1.4.4) lead to

$$\left(\mathbf{v} - \mathbf{U}_p\right).\mathbf{n} = 0 \qquad \text{on } S_p, \tag{1.4.14}$$

$$\left(\hat{\mathbf{v}} - \mathbf{U}_p\right).\mathbf{n} = 0 \qquad \text{on } S_p. \tag{1.4.15}$$

From the equality

$$\mathbf{v} = \mathbf{v}.\mathbf{nn} + \left(\mathbf{I} - \mathbf{nn}\right).\mathbf{v} = v_n\mathbf{n} + v_\tau\boldsymbol{\tau},$$

where **I** is the identity tensor and the expression **I** - **nn** can be regarded as a tangential projector, we can write the tangential velocity and stress boundary conditions (1.4.4) and (1.4.13) as

$$\mathbf{v}.(\mathbf{I}-\mathbf{nn})=\hat{\mathbf{v}}.(\mathbf{I}-\mathbf{nn})\,, \tag{1.4.16}$$

$$(\mathbf{E}.\mathbf{n}).(\mathbf{I}-\mathbf{nn})=\lambda\left(\hat{\mathbf{E}}.\mathbf{n}\right).(\mathbf{I}-\mathbf{nn})\,, \tag{1.4.17}$$

where $\lambda=\dfrac{\hat{\mu}}{\mu}$ is the viscosity ratio of the two fluids.

The normal stress jump condition (1.4.11) combined with (1.4.12) produces the following formula for the spherical interface S_p

$$\hat{p}_{tot}-p_{tot}+2\mu\left[\,\mathbf{E}.\mathbf{n}-\lambda\,\hat{\mathbf{E}}.\mathbf{n}\right].\mathbf{n}=\frac{2\sigma}{R}\,. \tag{1.4.18}$$

CHAPTER 2

Fundamental Theorems and General Properties of Stokes Flows

2.1. Introduction

We shall start Chapter 2 by considering the solution of the creeping flow equations for a point force $\mathbf{F}^{(c)}$ in an unbounded fluid. For this problem we shall obtain a fundamental solution (Green's function) when using the Fourier transforms.

We present a general integral theorem (reminiscent of the Green's theorem from vector calculus) which is known as Lorentz reciprocal theorem, a reciprocal relation between two solutions of the Stokes equations. This integral theorem leads directly to a general integral representation of the creeping flow equations. Indeed, the integral representation is a reciprocal relation between the above mentioned Green's function and the velocity field of interest.

The integral representation in terms of boundary velocity and surface forces, which has been already known to Lorentz (1907), is later used by Odqvist (1930) to obtain solutions of Stokes and continuity equations. Odqvist calls the two terms in the integral representation of Stokes solutions, respectively, the single and double layer. The reason for this terminology is that the single layer potentials are just superpositions of the hydrodynamic potentials, while the double layer potentials are caused by a layer of sources (or sinks) and doublets of point forces (see Ladyzhenskaya, 1969).

The integral representation provides a formal solution in a compact form and it transforms the governing 3D (three-dimensional) partial differential equations into 2D integral equations at the boundary of the fluid domain. The main objective when deriving the integral representation is to show that the general solution of the creeping flow equations for a flow past a particle can be completely expressed as a superposition of surface forces (Stokeslets) on the particle surface S_p. We discuss also the multipole expansion for the disturbance velocity field induced by a particle immersed in an ambient flow. The multipole expansion follows from a Taylor series expansion of the Green's function given in an integral representation about a reference point inside the particle. Away from the particle, the multipole expansion is a universal form for the disturbance produced by the particle motion.

The knowledge of energy dissipation rate of a fluid flow often helps us to obtain useful results without the requirement of detailed solutions of the equations of motion.

The minimum energy dissipation theorem is presented (Helmholtz theorem), which states that the solution of the creeping flow equations show less energy dissipation than any other solenoidal field with the same boundary velocities. The Helmholtz theorem is closely related to the uniqueness proof for the solutions of the Stokes equations.

The Hill and Power's results for the energy dissipation of particulate flows are given in the present Chapter. Hill and Power proved that by reversing the roles of the velocity and stress fields we may derive lower bounds for the energy dissipation rate of the Stokes velocity fields. At the end of Chapter 2 the so-called "inclusion monotonicity statements" of Hill and Power are given.

2.2. Fourier Transforms and Dirac Delta Function

Fourier Transform Definition Some of the important solutions of the Stokes equations are those that describe the flows caused by a point force, by a force dipole, etc. Hence, the flow, caused by a particle moving due to an applied force through an unbounded quiescent fluid, could be modelled using some of these solutions. The same velocity and pressure fields as those due to a concentrated force $\mathbf{F}^{(c)}$ of equal magnitude could be expected sufficiently far from the particle. Using Fourier transforms Ladyzhenskaya (1969) outlined how the point force solution was obtained. In this connection, it is helpful to recall briefly (see Stakgold 1967, 1968, **vol.1, 2**) some information on the Fourier transforms and Dirac delta function.

The Fourier transform of a function f(x) on the real line, $-\infty < x < \infty$, is a function $\hat{f}(k)$ defined from

$$\hat{f}(k) = \int_{-\infty}^{\infty} f(x)\, e^{ikx}\, dx\,, \tag{2.2.1}$$

where the transform variable k is taken real.

For the simple function f(x) = 1 the integral (2.2.1) fails to converge for every k. It can be established, however, that $\hat{f}(k)$ will exist for every real value of k if

$$\int_{-\infty}^{\infty} |f(x)|^2\, dx < \infty\,.$$

If f(x) is piecewise smooth function then we can recover f from $\hat{f}$ by an inversion formula (see Stakgold 1967, 1968, **vol.1, 2**)

$$f(x) = \lim_{\lambda\to\infty} \frac{1}{2\pi} \int_{-\lambda}^{\lambda} \hat{f}(k)\, e^{-ikx}\, dk\,. \tag{2.2.2}$$

For the 3D case f is a function of the position vector

$$\mathbf{r} = x_1\mathbf{e}_1 + x_2\mathbf{e}_2 + x_3\mathbf{e}_3\,,$$

while $\hat{f}$ is a function of

$$\mathbf{k} = k_1\mathbf{e}_1 + k_2\mathbf{e}_2 + k_3\mathbf{e}_3\,.$$

If the Fourier transform of $f(\mathbf{r})$ is

$$\hat{f}(\mathbf{k}) = \iiint_{-\infty}^{\infty} f(\mathbf{r})e^{i\mathbf{k}.\mathbf{r}}\, d\mathbf{r}, \tag{2.2.3}$$

then the inversion formula is

$$f(\mathbf{r}) = \frac{1}{(2\pi)^3}\iiint_{-\infty}^{\infty} \hat{f}(\mathbf{k})e^{-i\mathbf{k}.\mathbf{r}}\, d\mathbf{k} \tag{2.2.4}$$

where $d\mathbf{r} = dx_1dx_2dx_3$, $d\mathbf{k} = dk_1dk_2dk_3$ and $\mathbf{k}.\mathbf{r} = k_1x_1 + k_2x_2 + k_3x_3$.

<u>Dirac Delta Function Definition</u> To determine the velocity and the pressure fields induced into an infinite fluid by a point force $\mathbf{F}^{(c)}$, imposed at the origin of the coordinate system, we use the Dirac delta function. To describe mathematically a concentrated force in a mechanical problem, we can consider the force as an idealised (concentrated) source that is almost localised around the point where the force acts. We assume that such a source is characterised by its "action" on a suitable accessory function. This "function " in our case is the so called Dirac delta function $\delta(x)$, which is defined as follows: "$\delta(x)$ vanishes for $x \neq 0$ and is positively infinite at $x = 0$ in such a way that $\int_{-\infty}^{\infty}\delta(x)\,dx = 1$". This definition is not precise and has to be revised (see Stakgold 1967, 1968, **vol.1, 2**).

The symbolic function $\delta(x)$ may be interpreted as the charge density for a unit charge placed at $x = 0$. Similarly, the symbolic function $\delta(x - x_0)$ represents the charge density for a unit charge placed at $x = x_0$.

The general property of the Dirac delta function is the so-called shifting property.

$$\int_{-\infty}^{\infty} f(x)\,\delta(x - a)\,dx = f(a). \tag{2.2.5}$$

If $a = 0$ we obtain

$$\int_{-\infty}^{\infty} f(x)\,\delta(x)\,dx = f(0). \tag{2.2.6}$$

Therefore, for the particular function $f(x) = x$, we have

$$\int_{-\infty}^{\infty} x\,\delta(x)\,dx = f(0) = 0. \tag{2.2.7}$$

The second property of the Dirac delta function is

$$\delta(cx) = \frac{1}{|c|}\delta(x) \quad (c = \text{const.} \neq 0). \tag{2.2.8}$$

$$\delta(cx) = \frac{1}{|c|}\delta(x) \quad (c = \text{const.} \neq 0). \tag{2.2.8}$$

The following relations could be easily proved: $|c|\delta(cx) = 0$ at $x \neq 0$ and $|c|\delta(cx) = \infty$ at $x = 0$. If $c > 0$, then

$$\int_{-\infty}^{\infty}|c|\,\delta(cx)\,dx = \int_{-\infty}^{\infty}\delta(cx)\,c\,dx = \int_{-\infty}^{\infty}\delta(cx)\,d(cx) = \int_{-\infty}^{\infty}\delta(x_1)\,d(x_1) = 1$$

and $\int_{-\infty}^{\infty}\delta(cx)\,dx = \frac{1}{|c|}$.

Similarly, if $c < 0$, then

$$\int_{-\infty}^{\infty}|c|\,\delta(cx)\,dx = -\int_{-\infty}^{\infty}\delta(cx)\,c\,dx = \int_{\infty}^{-\infty}\delta(cx)\,d(cx) = \int_{-\infty}^{\infty}\delta(x_1)\,d(x_1) = 1$$

and $\int_{-\infty}^{\infty}\delta(cx)\,dx = \frac{1}{|c|}$.

The so-called Heavyside function is defined as follows

$$H(x) = \int_{-\infty}^{\infty}\delta(x)\,dx = \begin{cases} 0 & \text{as } x < 0, \\ 1 & \text{as } x \geq 0. \end{cases} \tag{2.2.9}$$

It is observed that H(x) is a piecewise continuous function, whereas $\delta(x)$ is a symbolic function. In ordinary analysis the derivative of H(x) is 0, except at the point $x = 0$, where it fails to exist.

The third property of the Dirac delta function states that

$$\frac{dH(x)}{dx} = \delta(x). \tag{2.2.10}$$

This property shows that the derivative of a piecewise continuous function could be expressed with the help of the Dirac delta function.

According to the shifting property of delta function (2.2.6), we have

$$\int_{-\infty}^{\infty}\delta(x)\,e^{ikx}\,dx = e^{ik0} = 1. \tag{2.2.11}$$

This equation shows that the number 1 can be considered as an image of the function e^{ikx}. By using the inversion formula (2.2.2), we obtain the following expression for $\delta(x)$

$$\delta(x) = \frac{1}{2\pi}\int_{-\infty}^{\infty}e^{-ikx}\,dk\ . \tag{2.2.12}$$

It is helpful to write down another property of the Dirac delta function, which states that

$$x\,\delta'(x) = -\,\delta(x). \tag{2.2.13}$$

In 3D case the Dirac delta function is defined as follows

$$1)\ \delta(x_1,x_2,x_3) = 0 \quad \text{at } x_1^2 + x_2^2 + x_3^2 \neq 0$$
$$2)\ \delta(x_1,x_2,x_3) = \infty \quad \text{at } x_1^2 + x_2^2 + x_3^2 = 0 \qquad (2.2.14)$$
$$3)\ \iiint_{-\infty}^{\infty} \delta(x_1,x_2,x_3)\delta(x_1,x_2,x_3)\, dx_1dx_2dx_3 = 1.$$

It can be checked that

$$\delta(x_1,x_2,x_3) = \delta(x_1)\delta(x_2)\delta(x_3), \qquad (2.2.15)$$

i.e., the 3D delta function is expressed by means of three 1D delta functions.

Clearly, the 3D Dirac delta function can be written in the form

$$\delta(\mathbf{r}) = \delta(x_1,x_2,x_3) = \frac{1}{(2\pi)^3}\iiint_{-\infty}^{\infty} e^{-i\mathbf{k}.\mathbf{r}}d\mathbf{k}, \qquad (2.2.16)$$

where $\mathbf{k} = k_1\mathbf{e}_1 + k_2\mathbf{e}_2 + k_3\mathbf{e}_3$, $d\mathbf{k} = dk_1dk_2dk_3$.

Stokes Solutions from Stokes Equations. Since the Stokes equations (1.3.7) are linear, one can easily prove that if the velocity and pressure field pair (**v**, p) satisfies these equations, then the derivatives of these quantities satisfy the Stokes equations as well.

There exists another way of obtaining new solutions from a known solution of the Stokes equations. Assume that the pair (**v**, p) satisfies the Stokes equations. Then since $T_{ij} = -p\delta_{ij} + \mu\left(\frac{\partial v_i}{\partial x_j} + \frac{\partial v_j}{\partial x_i}\right)$, we have $\nabla_i(T_{ij}) = -\nabla_j p + \mu\left[\frac{\partial}{\partial x_j}(\nabla_i v_i) + \frac{\partial^2 v_j}{\partial x_i^2}\right]$, or, in tensorial form,

$$\nabla.\mathbf{T} = -\nabla p + \mu\nabla^2\mathbf{v}.$$

Taking the operator $\nabla.$ of the last equation and bearing in mind (1.3.7), we obtain

$$\nabla.(\nabla.\mathbf{T}) = -\nabla.\nabla p + \mu\nabla.(\nabla^2\mathbf{v}) = -\nabla^2 p + \mu\nabla.\left(\frac{\nabla p}{\mu}\right) = 0.$$

Therefore, the pair [$\mathbf{T}(\mathbf{v})$, $\mu\nabla p$] is also a Stokes solution, one order higher in tensorial rank. However, the order can be reduced by taking, for example, dot products with some fixed vector.

2.3. The Green's Function for Stokes Flow

Let us consider the solution of the creeping flow equations:

$$\nabla.\mathbf{T} = \mu\nabla^2\mathbf{v} - \nabla p = -\mathbf{F}^{(c)}\delta(\mathbf{r}), \qquad (2.3.1)$$
$$\nabla.\mathbf{v} = 0. \qquad (2.3.2)$$

for a point force $\mathbf{F}^{(c)}$ applied to the fluid at $\mathbf{r} = \mathbf{0}$. Here the meaning of the first equation is:

1) for $\mathbf{r} \neq \mathbf{0}$, $\quad \nabla.\mathbf{T} = 0$;

2) for any volume V that encloses the point $\mathbf{r} = \mathbf{0}$,

$$\iiint\limits_V \nabla.\mathbf{T}\, d\tau = -\mathbf{F}^{(c)}.$$

In this section we shall derive this fundamental solution via Fourier transform. Since the velocity $\mathbf{v}$ and pressure p tend to zero as $\mathbf{r} \to \infty$, the Fourier transform of (2.3.1) and (2.3.2) yields

$$-\mu k^2 \hat{\mathbf{v}} + i\hat{p}\mathbf{k} = -\mathbf{F}^{(c)}, \tag{2.3.3}$$

$$\mathbf{k}.\hat{\mathbf{v}} = 0. \tag{2.3.4}$$

If we take the inner product of (2.3.3) with $\mathbf{k}$ and use (2.3.4) the result is

$$\hat{p} = i\frac{\mathbf{k}.\mathbf{F}^{(c)}}{k^2}. \tag{2.3.5}$$

Substituting (2.3.5) into (2.3.3) we find that

$$\mu k^2 \hat{\mathbf{v}} + \frac{(\mathbf{k}.\mathbf{F}^{(c)})\mathbf{k}}{k^2} = \mathbf{F}^{(c)},$$

whence

$$\hat{\mathbf{v}} = \frac{1}{\mu k^2}\left[\mathbf{F}^{(c)} - \frac{(\mathbf{k}.\mathbf{F}^{(c)})\mathbf{k}}{k^2}\right]. \tag{2.3.6}$$

By the inversion formula (2.2.4) we obtain for the Fourier transforms

$$p(\mathbf{r}) = \frac{i}{(2\pi)^3}\iiint\limits_{-\infty}^{\infty} \frac{\mathbf{k}.\mathbf{F}^{(c)}}{k^2} e^{-i\mathbf{k}.\mathbf{r}} d\mathbf{k}, \tag{2.3.7}$$

$$\mathbf{v}(\mathbf{r}) = \frac{1}{\mu(2\pi)^3}\iiint\limits_{-\infty}^{\infty} \frac{1}{k^2}\left[\mathbf{F}^{(c)} - \frac{(\mathbf{k}.\mathbf{F}^{(c)})\mathbf{k}}{k^2}\right] e^{-i\mathbf{k}.\mathbf{r}} d\mathbf{k}. \tag{2.3.8}$$

In order to calculate these integrals we shall use the fact that the function $\varphi(r) = \frac{1}{r}$ is a fundamental solution of the equation

$$\nabla^2 \varphi(r) = -4\pi\delta(\mathbf{r}), \tag{2.3.9}$$

i.e.,

$$\nabla^2\left(\frac{1}{r}\right) = -4\pi\delta(\mathbf{r}). \tag{2.3.10}$$

Therefore the following three formulae are valid:

$$\nabla^2\left(\frac{1}{4\pi r}\right) = -\delta(\mathbf{r}) = -\frac{1}{(2\pi)^3}\iiint\limits_{-\infty}^{\infty} e^{-i\mathbf{k}.\mathbf{r}} d\mathbf{k}, \tag{2.3.11}$$

$$\nabla\left(\frac{1}{4\pi r}\right) = -\frac{i}{(2\pi)^3}\iiint\limits_{-\infty}^{\infty} \frac{\mathbf{k}}{k^2} e^{-i\mathbf{k}.\mathbf{r}} d\mathbf{k}, \tag{2.3.12}$$

$$\frac{1}{4\pi r} = \frac{1}{(2\pi)^3}\iiint\limits_{-\infty}^{\infty} \frac{1}{k^2} e^{-i\mathbf{k}.\mathbf{r}} d\mathbf{k}. \tag{2.3.13}$$

If we take the inner product of (2.3.12) with $\mathbf{F}^{(c)}$ the result is

$$\mathbf{F}^{(c)}.\nabla\left(\frac{1}{4\pi r}\right) = -\frac{i}{(2\pi)^3}\iiint_{-\infty}^{\infty}\frac{(\mathbf{F}^{(c)}.\mathbf{k})}{k^2}e^{-i\mathbf{k}.\mathbf{r}}d\mathbf{k}. \tag{2.3.14}$$

Then, from (2.3.7) and (2.3.14) we find

$$p = -\mathbf{F}^{(c)}.\frac{1}{4\pi}\nabla\left(\frac{1}{r}\right) = \frac{\mathbf{F}^{(c)}.\mathbf{r}}{4\pi r^3}. \tag{2.3.15}$$

The final step in completing the solution of the creeping flow equations (2.3.1) and (2.3.2) is to determine **v(r)**. For this purpose, we firsts rewrite the equation (2.3.13) in the form:

$$\frac{\mathbf{F}^{(c)}}{4\pi r} = \frac{1}{(2\pi)^3}\iiint_{-\infty}^{\infty}\frac{\mathbf{F}^{(c)}}{k^2}e^{-i\mathbf{k}.\mathbf{r}}d\mathbf{k}. \tag{2.3.16}$$

Since

$$\nabla^4\left(\frac{r}{8\pi}\right) = -\delta(\mathbf{r}), \tag{2.3.17}$$

it follows that

$$\nabla^4\left(\frac{r}{8\pi}\right) = -\frac{1}{(2\pi)^3}\iiint_{-\infty}^{\infty}e^{-i\mathbf{k}.\mathbf{r}}\,d\mathbf{k}. \tag{2.3.18}$$

Therefore

$$\frac{r}{8\pi} = -\frac{1}{(2\pi)^3}\iiint_{-\infty}^{\infty}\frac{1}{k^4}e^{-i\mathbf{k}.\mathbf{r}}d\mathbf{k} \tag{2.3.19}$$

and

$$\nabla\nabla\left(\frac{r}{8\pi}\right) = \frac{1}{(2\pi)^3}\iiint_{-\infty}^{\infty}\frac{\mathbf{kk}}{k^4}e^{-i\mathbf{k}.\mathbf{r}}d\mathbf{k}. \tag{2.3.20}$$

Then, from (2.3.8), (2.3.16) and (2.3.20) we reach the expression

$$\mathbf{v} = \frac{\mathbf{I}.\mathbf{F}^{(c)}}{4\pi r\mu} - \frac{\mathbf{F}^{(c)}}{\mu}.\nabla\nabla\left(\frac{r}{8\pi}\right). \tag{2.3.21}$$

Since $\nabla r = \dfrac{\mathbf{r}}{r}$ and $\nabla\mathbf{r} = \mathbf{I}$, where **I** is the unit tensor, we have

$$\frac{1}{8\pi}\nabla(\nabla r) = \frac{1}{8\pi}\nabla\left(\frac{\mathbf{r}}{r}\right) = -\frac{1}{8\pi}\frac{\mathbf{rr}}{r^3} + \frac{\mathbf{I}}{8\pi r}. \tag{2.3.22}$$

We shall note that in rectangular Cartesian coordinates with the basis **i**, **j**, **k** the unit tensor **I** has the form $\mathbf{I} = \mathbf{ii} + \mathbf{jj} + \mathbf{kk}$. Using (2.3.21) and (2.3.22) we finally obtain

$$\mathbf{v} = \frac{1}{8\pi\mu}\left(\frac{\mathbf{I}}{r} + \frac{\mathbf{rr}}{r^3}\right).\mathbf{F}^{(c)}. \tag{2.3.23}$$

The multiplier of $\dfrac{\mathbf{F}^{(c)}}{8\pi\mu}$ in the last equation is called Oseen-Burgers tensor **B**, i.e.,

$$B_{ij} = \frac{\delta_{ij}}{r} + \frac{x_i x_j}{r^3} \quad (i, j = 1, 2, 3) \tag{2.3.24}$$

The velocity field (2.3.23) due to $\mathbf{F}^{(c)}$ is called a Stokeslet of strength $\boldsymbol{\alpha} = \dfrac{\mathbf{F}^{(c)}}{8\pi\mu}$.

The expression $\mathbf{G} = \dfrac{\mathbf{B}}{8\pi\mu}$ is known as Green's function. We claim that the fundamental solution consists of the velocity and pressure pair:

$$\mathbf{v} = \frac{\mathbf{B}}{8\pi\mu}.\mathbf{F}^{(c)} \quad , \qquad p(\mathbf{r}) = \frac{\mathbf{P(r)}}{8\pi\mu}.\mathbf{F}^{(c)}. \tag{2.3.25}$$

We shall note that the Oseen-Burgers tensor, **B**, is independent of the fluid properties, i.e., **B** is a purely geometric quantity. The pressure field of the Oseen-Burgers tensor is

$$P_j = 2\mu\frac{x_j}{r^3} + P_j^{\infty} \qquad (j = 1, 2, 3), \tag{2.3.26}$$

where P_j^{∞} is the pressure field at infinity.

For the point force $\mathbf{F}^{(c)}$ applied to the fluid at a point $\mathbf{r} = \boldsymbol{\rho}$ the Stokeslet has the form

$$\mathbf{v} = \frac{1}{8\pi\mu}\left(\frac{\mathbf{I}}{R} + \frac{(\mathbf{r} - \boldsymbol{\rho})(\mathbf{r} - \boldsymbol{\rho})}{R^3}\right).\mathbf{F}^{(c)}, \tag{2.3.27}$$

where $R = |\mathbf{r} - \boldsymbol{\rho}|$ is the distance between a point $\mathbf{r}$ and the point of application of the force, $\boldsymbol{\rho}(\xi_1, \xi_2, \xi_3)$. As the observation point $\mathbf{r}$ approaches the pole $\boldsymbol{\rho}$, the Stokeslet (2.3.27) exhibits a singular behaviour.

An alternative way to obtain the Green's function is by the use of delta and harmonic function properties (Pozrikidis, 1992a). For that purpose we return to equation (2.3.1) and replace the delta function in it with the expression (2.3.10), and the pressure with its equivalent form

$$p(\mathbf{r}) = -\frac{1}{4\pi}\mathbf{F}^{(c)}.\nabla\left(\frac{1}{r}\right), \tag{2.3.28}$$

correspondent to the harmonic properties of the pressure, where $\mathbf{F}^{(c)}$ is applied at the point $\mathbf{r} = \mathbf{0}$. Thus, (2.3.1) is reduced to

$$\mu\nabla^2\mathbf{v} = -\frac{1}{4\pi}\mathbf{F}^{(c)}.\left(\nabla\nabla - \mathbf{I}\nabla^2\right)\left(\frac{1}{r}\right), \tag{2.3.29}$$

which can be rewritten in terms of a scalar function f as

$$\left(\nabla\nabla - \mathbf{I}\nabla^2\right)\left(\nabla^2 f + \frac{1}{4\pi r}\right) = 0, \tag{2.3.30}$$

where f is connected with the velocity by the relation $\mathbf{v}=\frac{1}{\mu}\mathbf{F}^{(c)}.\left(\nabla\nabla-\mathbf{I}\nabla^2\right)f$. The equation (2.3.30) is fulfilled by every solution of Poison's equation $\nabla^2 f=-\frac{1}{4\pi r}$, while f is a solution of the biharmonic equation $\nabla^4 f=\delta(\mathbf{r})$ and $f=-\frac{r}{8\pi}$. As a result we get the previously obtained formula for the velocity (2.3.23), i.e., the Stokeslet or the so-called Oseen-Burgers tensor.

It is convenient to classify the Green's functions into three categories: i) the free-space Green's function for infinite unbounded flow; ii) the Green's function for infinite or semi-infinite flow that is bounded by a rigid wall or by a fluid/fluid interface; iii) the Green's function for internal flow that is completely confined by rigid surfaces. In this section we have considered the Green's function of the first kind. The Green's functions for infinite or semi-infinite bounded flow will be considered in the next chapters.

2.4. Lorentz Reciprocal Theorem. Applications

Useful general relationships could be derived through the application of the point force approach. Many of them stem from the work of Lorentz (1907).

Let us consider two steady, incompressible creeping flows of the same fluid and let $(\mathbf{v},\ \mathbf{T})$ and $(\mathbf{v}^*,\ \mathbf{T}^*)$ be the velocity and the stress fields corresponding to these flows. If they conform to equations (2.3.1) and (2.3.2), then it can be shown that

$$\iint_{S_0}\mathbf{v}^*.(\mathbf{T}.\mathbf{n})\,d\sigma=\iint_{S_0}\mathbf{v}.(\mathbf{T}^*.\mathbf{n})\,d\sigma, \tag{2.4.1}$$

where S_0 is a closed surface or distinct surfaces bounding any fluid volume V and $\mathbf{n}$ is the outward directed unit normal vector to S_0.

Using continuity equation we compute

$$v_i^*\frac{\partial T_{ij}}{\partial x_j}=\frac{\partial}{\partial x_j}\left(v_i^*T_{ij}\right)-T_{ij}\frac{\partial v_i^*}{\partial x_j}=\frac{\partial}{\partial x_j}\left(v_i^*T_{ij}\right) \tag{2.4.2}$$

$$-\left[-p\delta_{ij}+\mu\left(\frac{\partial v_i}{\partial x_j}+\frac{\partial v_j}{\partial x_i}\right)\right]\frac{\partial v_i^*}{\partial x_j}=\frac{\partial}{\partial x_j}\left(v_i^*T_{ij}\right)-\mu\left(\frac{\partial v_i}{\partial x_j}+\frac{\partial v_j}{\partial x_i}\right)\frac{\partial v_i^*}{\partial x_j}.$$

Reversing the roles of the two Stokes flows, we find

$$v_i\frac{\partial T_{ij}^*}{\partial x_j}=\frac{\partial}{\partial x_j}\left(v_i T_{ij}^*\right)-\mu\left(\frac{\partial v_i^*}{\partial x_j}+\frac{\partial v_j^*}{\partial x_i}\right)\frac{\partial v_i}{\partial x_j}. \tag{2.4.3}$$

Next, we subtract (2.4.3) from (2.4.2) to obtain

$$\frac{\partial}{\partial x_j}\left(v_i^*T_{ij}-v_i T_{ij}^*\right)=v_i^*\frac{\partial T_{ij}}{\partial x_j}-v_i\frac{\partial T_{ij}^*}{\partial x_j}. \tag{2.4.4}$$

If there is no singular points in the flows $\mathbf{v}$ and $\mathbf{v}^*$, then the corresponding Stokes equations are

$$\frac{\partial T_{ij}}{\partial x_j} = 0, \qquad \frac{\partial T^*_{ij}}{\partial x_j} = 0 \qquad (i, j = 1, 2, 3). \qquad (2.4.5)$$

It means that the right-hand side of (2.4.4) vanishes and we obtain the remarkable (Lorentz, 1907) reciprocal identity

$$\nabla.(\mathbf{v}^*.\mathbf{T} - \mathbf{v}.\mathbf{T}^*) = 0. \qquad (2.4.6)$$

If both sides of equation (2.4.6) are multiplied by an elementary fluid volume $d\tau$ and integrated over an arbitrary fluid volume V, we obtain

$$\iiint_V \nabla.(\mathbf{v}^*.\mathbf{T})d\tau = \iiint_V \nabla.(\mathbf{v}.\mathbf{T}^*)d\tau. \qquad (2.4.7)$$

Applying Gauss' divergence theorem, the volume integrals in (2.4.7) may be converted to the surface integrals in (2.4.1).

If the flows $\mathbf{v}$ and $\mathbf{v}^*$ are unbounded, then we can select a control volume that is confined by the surface of rigid or fluid particles S_p and a large surface S_∞ extending to infinity, i.e., we shall have $S_0 = S_p + S_\infty$ and

$$\iint_{S_p} \mathbf{v}^*.(\mathbf{T}.\mathbf{n})\, d\sigma + \iint_{S_\infty} \mathbf{v}^*.(\mathbf{T}.\mathbf{n})\, d\sigma = \iint_{S_p} \mathbf{v}.(\mathbf{T}^*.\mathbf{n})\, d\sigma + \iint_{S_\infty} \mathbf{v}.(\mathbf{T}^*.\mathbf{n})\, d\sigma. \quad (2.4.8)$$

We shall note that if the assumptions $\nabla.\mathbf{T} = \mathbf{0}$ and $\nabla.\mathbf{T}^* = \mathbf{0}$ are relaxed and the divergence theorem is applied, then the corresponding result is

$$\iint_{S_0} \mathbf{v}^*.(\mathbf{T}.\mathbf{n})\, d\sigma - \iiint_V \mathbf{v}^*.(\nabla.\mathbf{T})d\tau = \iint_{S_0} \mathbf{v}.(\mathbf{T}^*.\mathbf{n})\, d\sigma - \iiint_V \mathbf{v}.(\nabla.\mathbf{T}^*)d\tau. \quad (2.4.9)$$

Application 1. Now we shall show that Stokes flow "transmits" the total force and torque unchanged from an inner closed surface to an outer enclosing surface.

Let us consider the fluid region between the surfaces S_p and S_∞; let us choose $\mathbf{v}^*$ as the velocity of a rigid body motion, i.e.,

$$\mathbf{v}^* = \mathbf{U}_p + \boldsymbol{\omega}_p \times \mathbf{r}_p, \qquad (2.4.10)$$

where $\mathbf{U}_p$ and $\boldsymbol{\omega}_p$ are the body velocity of translation and rotation, respectively. Since for a rigid body motion $e_{ij}^* = 0$ and $\mathbf{T}^*.\mathbf{n} = -p_0^*.\mathbf{n}$, where p_0^*=const. from (2.4.8), we get the following result:

$$\iint_{S_p} (\mathbf{U}_p + \boldsymbol{\omega}_p \times \mathbf{r}_p).(\mathbf{T}.\mathbf{n})\, d\sigma + \iint_{S_\infty} (\mathbf{U}_p + \boldsymbol{\omega}_p \times \mathbf{r}_p).(\mathbf{T}.\mathbf{n})\, d\sigma =$$

$$\iint_{S_p} \mathbf{v}.(-p_0^*\mathbf{I}).\mathbf{n}\, d\sigma + \iint_{S_\infty} \mathbf{v}.(-p_0^*\mathbf{I}).\mathbf{n}\, d\sigma,$$

or

$$\iint_{S_p} (\mathbf{U}_p + \boldsymbol{\omega}_p \times \mathbf{r}_p).(\mathbf{T}.\mathbf{n})\, d\sigma + \iint_{S_\infty} (\mathbf{U}_p + \boldsymbol{\omega}_p \times \mathbf{r}_p).(\mathbf{T}.\mathbf{n})\, d\sigma = 0.$$

This implies that

$$\mathbf{U}_p \cdot \iint_{S_p} \mathbf{T}.\mathbf{n}\, d\sigma = \mathbf{U}_p \cdot \iint_{S_\infty} \mathbf{T}.\hat{\mathbf{n}}\, d\sigma \tag{2.4.11}$$

and

$$\boldsymbol{\omega}_p \cdot \iint_{S_p} \left[\mathbf{r}_p \times (\mathbf{T}.\mathbf{n})\right] d\sigma = \boldsymbol{\omega}_p \cdot \iint_{S_\infty} \left[\mathbf{r}_p \times (\mathbf{T}.\hat{\mathbf{n}})\right] d\sigma . \tag{2.4.12}$$

Since $\mathbf{U}_p$ and ω_p are arbitrary, we obtain

$$\iint_{S_p} \mathbf{T}.\mathbf{n}\, d\sigma = \iint_{S_\infty} \mathbf{T}.\hat{\mathbf{n}}\, d\sigma$$

and

$$\iint_{S_p} \left[\mathbf{r}_p \times (\mathbf{T}.\mathbf{n})\right] d\sigma = \iint_{S_\infty} \left[\mathbf{r}_p \times (\mathbf{T}.\hat{\mathbf{n}})\right] d\sigma ,$$

where $\hat{\mathbf{n}} = -\mathbf{n}$.

Application 2. As a second application we shall derive Brenner's result (Brenner, 1964b) for the drag on a rigid particle of arbitrary shape immersed in an ambient fluid of velocity $\mathbf{v}^\infty(\mathbf{r})$.

Let us consider two flow fields $(\mathbf{v}^*, \mathbf{T}^*)$ and $(\mathbf{v}^D, \mathbf{T}^D)$, the first of which is induced by a rigid particle of arbitrary shape translating with a steady velocity $\mathbf{v}^* = \mathbf{U} =$ const. and the second is induced by the same particle placed in an arbitrary ambient flow field, $\mathbf{v}^\infty(\mathbf{r})$.

Applying the Lorentz reciprocal theorem (2.4.1) for these two flow fields we get

$$\iint_{S_p} \mathbf{v}^*.(\mathbf{T}^D.\hat{\mathbf{n}})\, d\sigma = \iint_{S_p} \mathbf{v}^D.(\mathbf{T}^*.\hat{\mathbf{n}})\, d\sigma ,$$

or

$$\mathbf{U}. \iint_{S_p} (\mathbf{T}^D.\hat{\mathbf{n}})\, d\sigma = - \iint_{S_p} \mathbf{v}^\infty(\mathbf{r}).(\mathbf{T}^*.\hat{\mathbf{n}})\, d\sigma , \tag{2.4.13}$$

Since $\iint_{S_p} \mathbf{T}^D.\mathbf{n}\, d\sigma = \mathbf{F}$ is the drag on the particle in an arbitrary ambient field which satisfies the equation $\nabla.\mathbf{T}^\infty = \mathbf{0}$, it follows from (2.4.13) that

$$\mathbf{U}.\mathbf{F} = - \iint_{S_p} \mathbf{v}^\infty(\mathbf{r}).(\mathbf{T}^*.\hat{\mathbf{n}})\, d\sigma . \tag{2.4.14}$$

In order to obtain all the three components of **F**, we require a solution for an uniform flow past the same particle along any three mutually perpendicular directions. With a given surface stress tensor for the translation problem, the hydrodynamic drag for the same particle in an arbitrary field, $\mathbf{v}^\infty(\mathbf{r})$, could be computed directly.

2.5. General Representation with Single and Double Layer Potentials

The Lorentz theorem (2.4.9) leads directly to a general representation of the solutions of the creeping flow equations (2.3.1) and (2.3.2). This can be seen if the Oseen-Burgers tensor

$$\mathbf{B}(\mathbf{r}-\boldsymbol{\rho}) = \frac{\mathbf{I}}{R} + \frac{(\mathbf{r}-\boldsymbol{\rho})(\mathbf{r}-\boldsymbol{\rho})}{R^3} \tag{2.5.1}$$

is substituted for $\mathbf{v}^*$ in equation (2.4.9). Here $\boldsymbol{\rho}$ is the pole and $\mathbf{r}$ is the field point, and $\mathbf{v}$ is replaced by the solution of the Stokes equations throughout a region V, for which the integral representation is sought.

The pressure and stress field associated with the flow

$$v_i = \frac{1}{8\pi\mu} B_{ij} F_j^{(c)} \quad , \text{as } B_{ij} = \frac{\delta_{ij}}{R} + \frac{(x_i - \xi_i)(x_j - \xi_j)}{R^3}, \tag{2.5.2}$$

may be presented in the form

$$p(\mathbf{r}) = \frac{1}{8\pi} P_j(\mathbf{r},\boldsymbol{\rho}) F_j^{(c)}, \tag{2.5.3}$$

$$T_{ik}(\mathbf{r}) = \frac{1}{8\pi} K_{ijk}(\mathbf{r},\boldsymbol{\rho}) F_j^{(c)}, \tag{2.5.4}$$

where $\mathbf{K}$ is the stress tensor defined as

$$K_{ijk} = -\delta_{ik} P_j(\mathbf{r},\boldsymbol{\rho}) + \frac{\partial B_{ij}(\mathbf{r},\boldsymbol{\rho})}{\partial x_k} + \frac{\partial B_{kj}(\mathbf{r},\boldsymbol{\rho})}{\partial x_i} \tag{2.5.5}$$

and the pressure vector $\mathbf{P}$ is associated with the Green's function.

Using (2.5.1), (2.5.2), (2.5.4) and (2.5.5) we obtain

$$K_{ijk}(\hat{\mathbf{r}}) = -\frac{6(x_i - \xi_i)(x_j - \xi_j)(x_k - \xi_k)}{R^5}, \tag{2.5.6}$$

where $\hat{\mathbf{r}} = \mathbf{r} - \boldsymbol{\rho}$ and $R = |\hat{\mathbf{r}}|$ (Pozrikidis, 1992a, 1997).

Let us now compute the surface force $\mathbf{F}$ exerted on a fluid sphere of radius R centred at the pole of a point force $\mathbf{F}^{(c)}$. Substituting (2.5.6) into (2.5.4) we find for the surface force

$$f_i(\mathbf{r}) = T_{ik}(\mathbf{r}) n_k(\mathbf{r}) = \frac{1}{8\pi} K_{ijk}(\mathbf{r},\boldsymbol{\rho}) n_k F_j^{(c)} = -\frac{3}{4\pi}\frac{\hat{r}_i \hat{r}_j}{R^4} F_j^{(c)}.$$

Thus the force exerted on the sphere is

$$F_i = \iint_{\text{sphere}} f_i(\mathbf{r}) d\sigma(\mathbf{r}) = -\frac{3}{4\pi}\frac{F_j^{(c)}}{R^4} \iint_{\text{sphere}} \hat{r}_i \hat{r}_j d\sigma(\mathbf{r}). \tag{2.5.7}$$

Applying the divergence theorem to convert the surface integral into a volume integral, we find

$$\iint\limits_{\text{sphere}} \hat{r}_i \hat{r}_j d\sigma(\mathbf{r}) = R \iint\limits_{\text{sphere}} \hat{r}_i n_j d\sigma(\mathbf{r}) = R \iiint\limits_{\text{sphere}} \frac{\partial \hat{r}_i}{\partial \hat{r}_j} d\sigma(\mathbf{r}) = \delta_{ij} \frac{4}{3}\pi R^4 . \tag{2.5.8}$$

Substituting (2.5.8) into (2.5.7) we obtain $\mathbf{F} = -\mathbf{F}^{(c)}$ independently of the sphere radius.

We shall note that the torque with respect to the pole of a point force on any surface that encloses the pole is equal to zero (prove this assertion).

Consider two flows with velocities $\mathbf{v}$ and $\mathbf{v}^*$ of the same fluid of viscosity μ. If $\mathbf{T}$ and $\mathbf{T}^*$ are their correspondent stresses, then from equation (2.4.6) we have

$$\frac{\partial}{\partial x_k}\left(v_i^* T_{ik} - v_i T_{ik}^*\right) = 0 . \tag{2.5.9}$$

Let $\mathbf{v}^*$ is the flow velocity, due to the point force $\mathbf{F}^{(c)}$, acting in the point $\boldsymbol{\rho}^0(\xi_1^0, \xi_2^0, \xi_3^0)$. According to (2.5.2) and (2.5.4) we can write down

$$v_i^* = \frac{1}{8\pi\mu} B_{ij}(\mathbf{r}, \boldsymbol{\rho}^0) F_j^{(c)}, \qquad T_{ik}^* = \frac{1}{8\pi} K_{ijk}(\mathbf{r}, \boldsymbol{\rho}^0) F_j^{(c)} . \tag{2.5.10}$$

From (2.5.9) and (2.5.10), it follows:

$$\frac{\partial}{\partial x_k}\left[B_{ij}(\mathbf{r}, \boldsymbol{\rho}^0) T_{ik} - \mu v_i K_{ijk}(\mathbf{r}, \boldsymbol{\rho}^0)\right] = 0 . \tag{2.5.11}$$

As a control volume of the fluid, we consider the volume V which is enclosed by the surface S_0, as shown in Fig.2.5.1. S_0 consists of the fluid and rigid particles surfaces, as well as of a fluid surface, such that the particles remain inside the fluid.

With respect to the situation of the singular point $\boldsymbol{\rho}^0(\xi_1^0, \xi_2^0, \xi_3^0)$ we shall focus attention on three distinct cases:

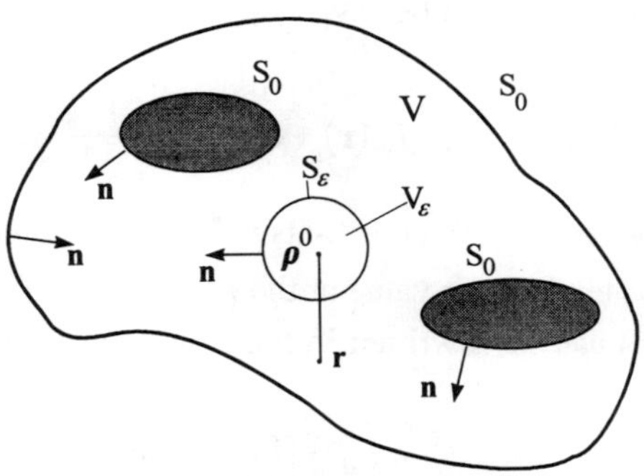

Fig.2.5.1. The control volume V with its boundary S_0.

i) The point $\boldsymbol{\rho}^0$ lies outside the control volume V. Since there is no singularity in the fluid occupying the control volume V, after integrating (2.5.11), we obtain

$$\iiint\limits_{V} \frac{\partial}{\partial x_k}\left[B_{ij}(\mathbf{r}, \boldsymbol{\rho}^0) T_{ik}(\mathbf{r}) - \mu v_i(\mathbf{r}) K_{ijk}(\mathbf{r}, \boldsymbol{\rho}^0)\right] d\tau(\mathbf{r}) = 0 . \tag{2.5.12}$$

Applying the divergence theorem, the volume integral in (2.5.12) is transformed into a surface integral

$$\iint\limits_{S_0}\left[B_{ij}(\mathbf{r},\boldsymbol{\rho}^0)T_{ik}(\mathbf{r})-\mu\,v_i(\mathbf{r})K_{ijk}(\mathbf{r},\boldsymbol{\rho}^0)\right]n_k\,d\sigma(\mathbf{r})=0, \tag{2.5.13}$$

where the normal vector **n** is directed towards the fluid in the control volume V (see Fig.2.5.1).

ii) The point $\boldsymbol{\rho}^0$ lies inside the control volume V. In order to isolate the singular point $\boldsymbol{\rho}^0$ in the flow, a sphere of small radius ε is taken (see Fig.2.5.1), whose centre coincides with the point $\boldsymbol{\rho}^0$ and the fluid inside the volume $V - V_\varepsilon$ is considered, where V_ε is the volume of the sphere of radius ε. If (2.5.11) is integrated in the volume $V - V_\varepsilon$ and the divergence theorem is applied, a result analogous to (2.5.13) is reached

$$\iint\limits_{S_0,\,S_\varepsilon}\left[B_{ij}(\mathbf{r},\boldsymbol{\rho}^0)T_{ik}(\mathbf{r})-\mu\,v_i(\mathbf{r})K_{ijk}(\mathbf{r},\boldsymbol{\rho}^0)\right]n_k\,d\sigma(\mathbf{r})=0, \tag{2.5.14}$$

where S_ε is the surface of a sphere of radius ε.

If ε is tending to zero, then on S_ε the Stokeslet **B** and its associate stress tensor **K**, given by (2.5.1) and (2.5.6), are approximated by the expressions

$$B_{ij}\approx\frac{\delta_{ij}}{\varepsilon}+\frac{\hat{r}_i\hat{r}_j}{\varepsilon^3},\qquad K_{ijk}(\hat{\mathbf{r}})\approx-\frac{6\hat{r}_i\hat{r}_j\hat{r}_k}{\varepsilon^5}, \tag{2.5.15}$$

where $\hat{\mathbf{r}}=\mathbf{r}-\boldsymbol{\rho}^0$. Taking into account that the normal **n** to S_ε and the elementary surface element on S_ε are respectively equal to $\hat{\mathbf{r}}/\varepsilon$ and $\varepsilon^2 d\omega$ (here $d\omega$ is the differential solid angle), we get after substituting these expressions in (2.5.14)

$$\iint\limits_{S_0}\left[B_{ij}(\mathbf{r},\boldsymbol{\rho}^0)T_{ik}(\mathbf{r})-\mu\,v_i(\mathbf{r})K_{ijk}(\mathbf{r},\boldsymbol{\rho}^0)\right]n_k\,d\sigma(\mathbf{r})$$

$$=-\iint\limits_{S_\varepsilon}\left[\left(\delta_{ij}+\frac{\hat{r}_i\hat{r}_j}{\varepsilon^2}\right)T_{ik}(\mathbf{r})+6\mu\,v_i(\mathbf{r})\frac{\hat{r}_i\hat{r}_j\hat{r}_k}{\varepsilon^4}\right]\hat{r}_k\,d\omega. \tag{2.5.16}$$

When $\varepsilon\to0$, the velocity vector $\mathbf{v}(\mathbf{r})$ tends to its value at the point $\boldsymbol{\rho}^0$, i.e., to $\mathbf{v}(\boldsymbol{\rho}^0)$ and the stress tensor $\mathbf{T}(\mathbf{r})$ also tends to its value at the pole $\mathbf{T}(\boldsymbol{\rho}^0)$. Then, the right-hand side of the expression (2.5.16) can be rewritten in the following form, bearing in mind that $\mathbf{r}=\varepsilon\,\mathbf{n}\to0$:

$$-\frac{6\mu}{\varepsilon^4}\,v_i(\boldsymbol{\rho}^0)\iint\limits_{S_\varepsilon}\hat{r}_i\hat{r}_j\,d\sigma(\mathbf{r}), \tag{2.5.17}$$

which transforms into $-\frac{6\mu}{\varepsilon^4}\,v_i(\boldsymbol{\rho}^0)\delta_{ij}\frac{4}{3}\pi\varepsilon^4$ or $-8\pi\mu\,v_j(\boldsymbol{\rho}^0)$ due to (2.5.8). Now, taking in view (2.5.16), (2.5.17) and the last consideration, we can express the velocity in the pole as

$$v_j(\boldsymbol{\rho}^0)=-\frac{1}{8\pi\mu}\iint\limits_{S_0}B_{ij}(\mathbf{r},\boldsymbol{\rho}^0)T_{ik}(\mathbf{r})n_k\,d\sigma(\mathbf{r})+\frac{1}{8\pi}\iint\limits_{S_0}v_i(\mathbf{r})K_{ijk}(\mathbf{r},\boldsymbol{\rho}^0)n_k\,d\sigma(\mathbf{r}). \tag{2.5.18}$$

If the surface stress forces are denoted by **f**, i.e. **f** = **T.n**, then (2.5.18) can be given in the form

$$v_j(\boldsymbol{\rho}^0) = -\frac{1}{8\pi\mu}\iint_{S_0} f_i(\mathbf{r})B_{ij}(\mathbf{r},\boldsymbol{\rho}^0)d\sigma(\mathbf{r}) + \frac{1}{8\pi}\iint_{S_0} v_i(\mathbf{r})K_{ijk}(\mathbf{r},\boldsymbol{\rho}^0)n_k d\sigma(\mathbf{r}).$$

(2.5.19)

This formula expresses the velocity in an arbitrary point from the interior of the flow control volume V by means of the Green's function **B** and its associated stress tensor **K**. The first term in (2.5.19) is called the single-layer potential, and the second one - the double-layer potential.

<u>iii) The point $\boldsymbol{\rho}^0$ lies on S_0.</u> In order to obtain the corresponding velocity expression, the formulae (2.5.13) and (2.5.18) are used with $\boldsymbol{\rho}^0$ tending to S_0 from the inner or outer side of S_0. It turns out that the double-layer potential is a discontinuous function of $\boldsymbol{\rho}^0$ on the surface S_0 , because two different values are obtained when $\boldsymbol{\rho}^0$ tends to S_0 from both sides. If the surface S_0 has a continuously changing normal [S_0 is a Lyapunov's type surface, Jaswon and Symm (1977)] and the velocity **v**(**r**) is a continuous function of **r** (see section 2.7), we have

$$\lim_{\boldsymbol{\rho}^0\to S_0}\iint_{S_0} v_i(\mathbf{r})K_{ijk}(\mathbf{r},\boldsymbol{\rho}^0)n_k d\sigma(\mathbf{r}) = \pm 4\pi v_j(\boldsymbol{\rho}^0) + \int\int_{S_0}^{PV} v_i(\mathbf{r})K_{ijk}(\mathbf{r},\boldsymbol{\rho}^0)n_k d\sigma(\mathbf{r}),$$

(2.5.20)

where the plus sign is taken when the point $\boldsymbol{\rho}^0$ approaches normally S_0 from the side of the fluid, while the minus sign is taken in the reverse case. When $\boldsymbol{\rho}^0$ is on the surface S_0, the current point **r** on S_0 could coincide with $\boldsymbol{\rho}^0$ and the double-layer integral in (2.5.20) becomes improper, therefore its principle value is taken and denoted as the superscript PV. If the limit $\boldsymbol{\rho}^0 \to S_0$ is taken in (2.5.18) from the side of the normal vector **n** direction, as shown in Fig.2.5.1, then

$$v_j(\boldsymbol{\rho}^0) = -\frac{1}{8\pi\mu}\lim_{\boldsymbol{\rho}^0\to S_0}\iint_{S_0} T_{ik}(\mathbf{r},\boldsymbol{\rho}^0)B_{ij}(\mathbf{r},\boldsymbol{\rho}^0)n_k(\mathbf{r})d\sigma(\mathbf{r})$$
$$+\frac{1}{8\pi}\lim_{\boldsymbol{\rho}^0\to S_0}\iint_{S_0} v_i(\mathbf{r})K_{ijk}(\mathbf{r},\boldsymbol{\rho}^0)n_k(\mathbf{r})d\sigma(\mathbf{r}).$$

If the equation (2.5.20) with plus sign is substituted in the above expression, we reach eventually

$$v_j(\boldsymbol{\rho}^0) = -\frac{1}{4\pi\mu}\iint_{S_0} T_{ik}(\mathbf{r},\boldsymbol{\rho}^0)B_{ij}(\mathbf{r},\boldsymbol{\rho}^0)n_k(\mathbf{r})d\sigma(\mathbf{r})$$
$$+\frac{1}{4\pi}\int\int_{S_0}^{PV} v_i(\mathbf{r})K_{ijk}(\mathbf{r},\boldsymbol{\rho}^0)n_k(\mathbf{r})d\sigma(\mathbf{r}). \qquad (2.5.21)$$

An analogous formula is obtained when (2.5.13) and (2.5.20) are taken with minus sign.

Now we consider a particle moving in a unbounded or bounded infinite flow. Suppose that the control volume is enclosed by the surface $S_0 = S_p + S_\infty$, where S_p is a rigid or fluid boundary and S_∞ is a large surface of dimension R_∞^2, which we intend to expand by applying the limit of large $R_\infty \to \infty$. Then if the fluid is quiescent at infinity, the velocity must decay at least as 1/r, whereas the pressure and stress must decay at least as fast as $1/r^2$, where r is a typical distance from S_p. This suggests that as the radius R_∞ of S_∞ tends to infinity, both the single-layer and double-layer potentials over S_∞ tend to zero. As a result the surface of integration S_0 of the boundary integral equation (2.5.21) is reduced only to S_p, i.e.

$$v_j(\boldsymbol{\rho}^0) = -\frac{1}{4\pi\mu}\iint_{S_p} T_{ik}(\mathbf{r},\boldsymbol{\rho}^0)B_{ij}(\mathbf{r},\boldsymbol{\rho}^0)n_k(\mathbf{r})d\sigma(\mathbf{r})$$

$$+\frac{1}{4\pi}\overset{PV}{\iint_{S_p}} v_i(\mathbf{r})K_{ijk}(\mathbf{r},\boldsymbol{\rho}^0)n_k(\mathbf{r})d\sigma(\mathbf{r}). \tag{2.5.22}$$

2.6. Properties of the Pressure and Stress

In this section we shall discuss the properties of the pressure p(**r**) and the stress **T**(**r**), when imposing the requirement for the pressure vector **P**, vorticity vector **Ω** and tensor **K** (defined by (2.5.5)) to tend to zero with $\boldsymbol{\rho}^0 \to \infty$. After substituting (2.5.2), (2.5.3) and (2.5.5) in the equation

$$\mu\nabla^2\mathbf{v} - \nabla p + \mathbf{F}^{(c)}\delta(\mathbf{r}-\boldsymbol{\rho}^0) = 0, \tag{2.6.1}$$

we get the identities

$$\nabla^2 B_{kj}(\mathbf{r},\boldsymbol{\rho}^0) - \frac{\partial P_j(\mathbf{r},\boldsymbol{\rho}^0)}{\partial x_k} = -8\pi\,\delta_{kj}\delta(\mathbf{r}-\boldsymbol{\rho}^0), \tag{2.6.2}$$

or

$$\frac{\partial K_{ijk}(\mathbf{r},\boldsymbol{\rho}^0)}{\partial x_i} = \frac{\partial K_{jki}(\mathbf{r},\boldsymbol{\rho}^0)}{\partial x_i} = -8\pi\,\delta_{kj}\delta(\mathbf{r}-\boldsymbol{\rho}^0). \tag{2.6.3}$$

Integrating (2.6.3) over the fluid volume V enclosed by the smooth surface S_0 and using the divergence theorem, we obtain the expression

$$\iint_{S_0} K_{ijk}(\mathbf{r},\boldsymbol{\rho}^0)n_i(\mathbf{r})d\sigma(\mathbf{r}) = \iint_{S_0} K_{jki}(\mathbf{r},\boldsymbol{\rho}^0)n_i(\mathbf{r})d\sigma(\mathbf{r}) = -\begin{bmatrix}8\pi\\4\pi\\0\end{bmatrix}\delta_{jk}, \tag{2.6.4}$$

where the normal **n** is directed outwards the control volume. The values -8π, -4π and 0 correspond to the cases when the point $\boldsymbol{\rho}^0$ is situated inside, exactly on, and outside the surface S_0. We have to note that when the point $\boldsymbol{\rho}^0$ lies on S_0 the integrals in (2.6.4) are improper, but convergent.

It can easily be proved that the stress tensor **K** associated with the Green's function **B** and the pressure vector **P** correspondent to the Green's function **B** are two fundamental solutions of the Stokes equation (2.6.1), when the flow domain is unbounded or bounded, but infinite (Pozrikidis, 1992a, 1997):

$$v_j(\boldsymbol{\rho}^0) = K_{ijk}(\mathbf{r}, \boldsymbol{\rho}^0) q_{ik}, \tag{2.6.5}$$

$$\mathbf{v}(\boldsymbol{\rho}^0) = \alpha \mathbf{P}(\mathbf{r}, \boldsymbol{\rho}^0). \tag{2.6.6}$$

Here α and **q** are an arbitrary constant scalar and a matrix, respectively, and define two solutions of the creeping flows equations.

Now we shall prove that if S is an arbitrary closed surface and its normal vector is directed outwards, then

$$\iint_S P_j(\mathbf{r}, \boldsymbol{\rho}^0) n_j(\boldsymbol{\rho}^0) d\sigma(\boldsymbol{\rho}^0) = \begin{cases} -8\pi, & \text{if } \mathbf{r} \text{ lies inside S}, \\ -4\pi, & \text{if } \mathbf{r} \text{ lies on S}, \\ 0, & \text{if } \mathbf{r} \text{ lies outside S}. \end{cases} \tag{2.6.7}$$

Indeed, the components of the pressure vector are determined from the equality [see (2.3.26)]

$$P_j(\mathbf{r}, \boldsymbol{\rho}^0) = \frac{2\hat{r}_j}{R^3}, \qquad \text{where } \hat{\mathbf{r}} = \mathbf{r} - \boldsymbol{\rho}^0 \text{ and } R = |\hat{\mathbf{r}}|. \tag{2.6.8}$$

The expression on the right-hand side of (2.6.8) corresponds to a velocity of a flow, induced by a potential source with potential $\Phi = \frac{Q}{4\pi R}$, as $\mathbf{v} = \text{grad}\Phi = -\frac{Q}{4\pi}\frac{\hat{\mathbf{r}}}{R^3}$. This means that the strength of the source correspondent to (2.6.8), is equal to -8π. Since the integral on the left-hand side of (2.6.7) represents the flow rate due to the pressure vector, i.e., due to a source of strength $Q = -8\pi$, then the relation (2.6.7) is proved for the case when **r** is inside the domain bounded by S. In the opposite case, when **r** is outside of S, the domain enclosed by S is free of singularities and the flow rate across S equals zero. When **r** is exactly on the surface S, the left-hand side of (2.6.7) is an improper integral and its principle value equals the average of -8π and 0, i.e., -4π.

From the equality (2.6.5) it follows that

$$v_j(\boldsymbol{\rho}^0) = K_{ijk}(\mathbf{r}, \boldsymbol{\rho}^0) q_{ik} = -K_{ijk}(\boldsymbol{\rho}^0, \mathbf{r}) q_{ik}, \tag{2.6.9}$$

where **q** is a square matrix of third order with constant elements. The velocity field defined by (2.6.9) is a flow due to a singularity called a stresslet with pole at **r**. Let us rewrite the pressure correspondent to the velocity field (2.6.9) in the form

$$p(\boldsymbol{\rho}^0) = \mu \Pi_{ik}(\boldsymbol{\rho}^0, \mathbf{r}) q_{ik}. \tag{2.6.10}$$

Here $\Pi_{ik}(\boldsymbol{\rho}^0, \mathbf{r})$ is named pressure tensor of the flow (2.6.5). We shall establish that there exists a direct connection between the pressure fields correspondent to the velocities given by (2.6.5) and (2.6.6). Taking into account (2.5.5) and the symmetry property of the Green's function **B**, we find that the pressure due to the velocity (2.6.6) equals $\alpha \mu \mathcal{P}(\boldsymbol{\rho}^0, \mathbf{r})$, while the pressure tensor correspondent to the flow (2.6.5) is

$$\Pi_{ik}\left(\boldsymbol{\rho}^0,\mathbf{r}\right)=-\delta_{ik}\mathcal{P}\left(\boldsymbol{\rho}^0,\mathbf{r}\right)+\frac{\partial \mathrm{P}_i\left(\boldsymbol{\rho}^0,\mathbf{r}\right)}{\partial \mathrm{x}_k}+\frac{\partial \mathrm{P}_k\left(\boldsymbol{\rho}^0,\mathbf{r}\right)}{\partial \mathrm{x}_i}, \tag{2.6.11}$$

where $\mathrm{P}_i\left(\boldsymbol{\rho}^0,\mathbf{r}\right)$ are the components of the pressure vector due to the Stokeslet $\mathrm{B}_{ik}\left(\boldsymbol{\rho}^0,\mathbf{r}\right)$. The formulae (2.6.5), (2.6.6), (2.6.10) and (2.6.11) indicate, that the pressure tensor Π_{ik} of the velocity field (2.6.5) is a stress tensor of the velocity field (2.6.6). In this way, for $\Pi_{ik}\left(\boldsymbol{\rho}^0,\mathbf{r}\right)$ we find that

$$\Pi_{ik}\left(\boldsymbol{\rho}^0,\mathbf{r}\right)=4\left(-\frac{\delta_{ik}}{R^3}+3\frac{\hat{r}_i\hat{r}_k}{R^5}\right). \tag{2.6.12}$$

It is known that the single-layer potential represents the flow induced by continuously distributed point forces (Stokeslets) over the surface S_p:

$$\mathrm{v}_i\left(\boldsymbol{\rho}^0\right)=\iint_{S_p}\mathrm{B}_{ij}\left(\boldsymbol{\rho}^0,\mathbf{r}\right)\mathrm{q}_j(\mathbf{r})d\sigma(\mathbf{r}), \tag{2.6.13}$$

where $\mathbf{q}$ is the density of the point forces distribution. The pressure and stress, correspondent to the velocity (2.6.13) are given by the formulae:

$$\mathrm{p}\left(\boldsymbol{\rho}^0\right)=\mu\iint_{S_p}\mathrm{P}_j\left(\boldsymbol{\rho}^0,\mathbf{r}\right)\mathrm{q}_j(\mathbf{r})d\sigma(\mathbf{r}), \tag{2.6.14}$$

$$\mathrm{T}_{ik}\left(\boldsymbol{\rho}^0\right)=\mu\iint_{S_p}\mathrm{K}_{ijk}\left(\boldsymbol{\rho}^0,\mathbf{r}\right)\mathrm{q}_j(\mathbf{r})d\sigma(\mathbf{r}), \tag{2.6.15}$$

where **P** and **K** are the pressure vector and stress tensor, respectively, connected with the Green's function $\mathbf{B}(\boldsymbol{\rho}^0,\mathbf{r})$.

On the other hand, the double-layer potential represent the flow induced by the continuous distribution of the stress tensor, correspondent to the Green's function **B**, i.e.,

$$\mathrm{v}_i\left(\boldsymbol{\rho}^0\right)=\iint_{S_p}\mathrm{q}_j(\mathbf{r})\mathrm{K}_{jik}\left(\boldsymbol{\rho}^0,\mathbf{r}\right)\mathrm{n}_k(\mathbf{r})d\sigma(\mathbf{r}). \tag{2.6.16}$$

The pressure resulting from the velocity (2.6.16) can provisionally be presented by the expression

$$\mathrm{p}\left(\boldsymbol{\rho}^0\right)=\mu\iint_{S_p}\mathrm{q}_i(\mathbf{r})\Pi_{ik}\left(\boldsymbol{\rho}^0,\mathbf{r}\right)\mathrm{n}_k(\mathbf{r})d\sigma(\mathbf{r}), \tag{2.6.17}$$

where $\boldsymbol{\Pi}$ is the pressure tensor due to the stresslet $\mathrm{v}_j=\mathrm{K}_{jik}\mathrm{q}_{ik}$.

2.7. Representation of a Flow Only in Terms of a Single-Layer Potential or Double-Layer Potential

A basic property of the single-layer (2.6.13) is that it creates surface forces $\mathrm{f}_i\left(\boldsymbol{\rho}^0\right)=\mathrm{T}_{ik}\left(\boldsymbol{\rho}^0\right)\mathrm{n}_k\left(\boldsymbol{\rho}^0\right)$ on the surface S_p of the Stokeslets distribution, which have two different values on either side of S_p, i.e., they suffer a jump across S_p. If the normal vector $\mathbf{n}(\boldsymbol{\rho}^0)$ is directed to the external side of S_p, then if $\boldsymbol{\rho}^0 \to S_p$ from outside S_p, we have

$$f_i^+(\boldsymbol{\rho}^0) \equiv T_{ik}^+(\boldsymbol{\rho}^0) n_k(\boldsymbol{\rho}^0) = -4\pi\mu q_i(\boldsymbol{\rho}^0) + \mu n_k(\boldsymbol{\rho}^0) \overset{PV}{\int\int_{S_p}} K_{ijk}(\boldsymbol{\rho}^0, \mathbf{r}) q_j(\mathbf{r}) d\sigma(\mathbf{r}) \tag{2.7.1}$$

and when $\boldsymbol{\rho}^0 \to S_p$ from inside S_p:

$$f_i^-(\boldsymbol{\rho}^0) \equiv T_{ik}^-(\boldsymbol{\rho}^0) n_k(\boldsymbol{\rho}^0) = 4\pi\mu q_i(\boldsymbol{\rho}^0) + \mu n_k(\boldsymbol{\rho}^0) \overset{PV}{\int\int_{S_p}} K_{ijk}(\boldsymbol{\rho}^0, \mathbf{r}) q_j(\mathbf{r}) d\sigma(\mathbf{r}) \tag{2.7.2}$$

The last two terms in (2.7.1) and (2.7.2) express the principle value of the improper integrals for which the point $\boldsymbol{\rho}^0$ must lie exactly on S_p. We shall note that when S_p is a Lyapunov surface (as we have assumed it to be), the principle value of the surface stress force is equal to the average of the surface force values calculated on either side of S_p.

The single-layer potential can be used when solving the Dirichlet problem, as well as the Neumann problem. The Dirichlet problem arises when the velocity over S_p is known and the density $q_i(\boldsymbol{\rho}^0)$ of the Stokeslets distribution over S_p is sought from (2.6.13). Then, a first kind Fredholm integral equation has to be solved in order to define $q_i(\boldsymbol{\rho}^0)$. The Neumann problem corresponds to the case when the surface stress forces acting on external or internal side of S_p are given and the density $q_i(\boldsymbol{\rho}^0)$ of the Stokeslets distribution over S_p must be found from any of (2.7.1) or (2.7.2). Then, a second kind Fredholm integral equation is obtained and its solution represents the unknown density function $q_i(\boldsymbol{\rho}^0)$.

A main property of the double-layer (2.6.15) is the discontinuity of the velocity, defined by it, on the singularities distribution surface. If the normal vector $\mathbf{n}(\boldsymbol{\rho}^0)$ is directed outwards from S_p, then as $\boldsymbol{\rho}^0 \to S_p$ from outside S_p, we have

$$v_i^+(\boldsymbol{\rho}^0) = 4\pi\, q_i(\boldsymbol{\rho}^0) + \overset{PV}{\int\int_{S_p}} q_j(\mathbf{r}) K_{ijk}(\mathbf{r}, \boldsymbol{\rho}^0) n_k(\mathbf{r}) d\sigma(\mathbf{r}), \tag{2.7.3}$$

and as $\boldsymbol{\rho}^0 \to S_p$ from inside S_p:

$$v_i^-(\boldsymbol{\rho}^0) = -4\pi\, q_i(\boldsymbol{\rho}^0) + \overset{PV}{\int\int_{S_p}} q_j(\mathbf{r}) K_{ijk}(\mathbf{r}, \boldsymbol{\rho}^0) n_k(\mathbf{r}) d\sigma(\mathbf{r}), \tag{2.7.4}$$

where the above integrals are taken with their principle values (since they are improper when $\boldsymbol{\rho}^0$ stands exactly on S_p).

If the velocity from one of both sides of the surface S_p is prescribed, i.e., a Dirichlet problem arises, then in order to find the density of the singularities distribution on S_p one of the two Fredholm integral equations of second kind (2.7.3) or (2.7.4) must be solved with respect to $q_i(\boldsymbol{\rho}^0)$.

2.8. The Multipole Expansion of the Velocity of a Creeping Flow Field

Let a rigid particle in an ambient flow field $\mathbf{v}^{\infty}(\mathbf{r})$ move with a velocity $\mathbf{U}_p + \mathrm{w}_p \times \mathbf{r}_p$. If we place point forces on the points $\mathbf{r}$ of the particle surface S_p, then for the disturbance velocity of the flow we have

$$v_j^{D}(\boldsymbol{\rho}^0) = -\frac{1}{8\pi\mu}\iint_{S_p} B_{ji}(\boldsymbol{\rho}^0 - \mathbf{r})T_{ik}(\mathbf{r})n_k(\mathbf{r})\, d\sigma(\mathbf{r}). \tag{2.8.1}$$

It is clear that when we are interested in the points at great distances from the particle, i.e. $|\boldsymbol{\rho}^0| >> |\mathbf{r}|$, we cannot distinguish between the points $\mathbf{r}$ on the surface of the particle and a reference origin O located at a convenient point inside the particle and $\mathbf{B}(\boldsymbol{\rho}^0 - \mathbf{r}) \approx \mathbf{B}(\boldsymbol{\rho}^0)$. Then the equation (2.8.1) takes the form

$$\mathbf{v}(\boldsymbol{\rho}^0) = -\frac{1}{8\pi\mu}\mathbf{B}(\boldsymbol{\rho}^0).\iint_{S_p}\mathbf{T}(\mathbf{r}).\ \mathbf{n}(\mathbf{r})\, d\sigma(\mathbf{r}) = -\frac{1}{8\pi\mu}\mathbf{B}(\boldsymbol{\rho}^0).\mathbf{F}, \tag{2.8.2}$$

where $\mathbf{F} = \iint_{S_p}\mathbf{T}(\mathbf{r}).\ \mathbf{n}(\mathbf{r})\, d\sigma(\mathbf{r})$ is the force exerted by the fluid on the particle. It means that sufficiently far from the particle the disturbance velocity is independent of the details of the particle shape. For any fixed point $M(\xi_1^0, \xi_2^0, \xi_3^0) = M(\boldsymbol{\rho}^0)$, the function

$$B_{ij}(\boldsymbol{\rho}^0 - \mathbf{r}) = \frac{\delta_{ij}}{|\boldsymbol{\rho}^0 - \mathbf{r}|} + \frac{(\xi_i^0 - x_i)(\xi_j^0 - x_j)}{|\boldsymbol{\rho}^0 - \mathbf{r}|^3} \tag{2.8.3}$$

can be expressed in terms of a formal multipole expansion in the form in a generalised Taylor series in $\mathbf{r}(x_1, x_2, x_3)$ about $\mathbf{r} = \mathbf{r}^c(x_1^c, x_2^c, x_3^c)$:

$$B_{ji}(\boldsymbol{\rho}^0 - \mathbf{r}) = B_{ji}(\boldsymbol{\rho}^0, \mathbf{r}^c) - \frac{\partial B_{ij}(\boldsymbol{\rho}^0, \mathbf{r}^c)}{\partial x_k^c}.(x_k - x_k^c) + \ldots, \tag{2.8.4}$$

Thus the asymptotic expansion for the disturbance velocity has the form

$$v_j^D(\boldsymbol{\rho}^0) = -\frac{1}{8\pi\mu}\Bigg[B_{ji}(\boldsymbol{\rho}^0, \mathbf{r}^c)\iint_{S_p} f_i(\mathbf{r})d\sigma(\mathbf{r})$$
$$+\frac{\partial B_{ij}(\boldsymbol{\rho}^0, \mathbf{r}^c)}{\partial x_k^c}\iint_{S_p}(x_k - x_k^c)f_i(\mathbf{r})d\sigma(\mathbf{r}) + \ldots\Bigg], \tag{2.8.5}$$

where $f_i(\mathbf{r}) = T_{ik}(\mathbf{r})n_k(\mathbf{r})$ is the interfacial force, $F_i = \iint_{S_p} f_i(\mathbf{r})d\sigma(\mathbf{r})$ is the force exerted by the fluid on the particle and the expression $\iint_{S_p}(x_k - x_k^c)f_i(\mathbf{r})d\sigma(\mathbf{r})$ is the corresponding first moment of $f_i(\mathbf{r})$, while the subsequent integrals are the higher-order moments of $f_i(\mathbf{r})$.

We have to note that the first term on the right-hand side of the equality (2.8.5) can be considered as a flow induced by a point force, i.e., a Stokeslet with coefficient **F**, the second one as a flow due to a force dipole and the next terms as flows due to higher-order multipoles.

Further, we shall decompose the coefficient of the second term on the right-hand side of (2.8.5) into a symmetric part S_{ij}, an isotropic component, which has again a symmetric and an antisymmetric component R_{ij}, i.e., we shall set

$$\iint_{S_p} \hat{x}_k f_i(\mathbf{r}) d\sigma(\mathbf{r}) = S_{ki} + \frac{1}{3}\delta_{ki} \iint_{S_p} \hat{x}_l f_l(\mathbf{r}) d\sigma(\mathbf{r}) + R_{ki}, \tag{2.8.6}$$

where

$$S_{ki} = \frac{1}{2} \iint_{S_p} \left[\hat{x}_k f_i(\mathbf{r}) + \hat{x}_i f_k(\mathbf{r}) - \frac{2}{3}\delta_{ik} \hat{x}_l f_l(\mathbf{r}) \right] d\sigma(\mathbf{r}), \tag{2.8.7}$$

$$R_{ki} = \frac{1}{2} \iint_{S_p} \left[\hat{x}_k f_i(\mathbf{r}) - \hat{x}_i f_k(\mathbf{r}) \right] d\sigma(\mathbf{r}) \tag{2.8.8}$$

and $\hat{\mathbf{r}} = \mathbf{r} - \mathbf{r}^c$. We note also that, the antisymmetric part of the force dipole R_{ij} can be represented as the hydrodynamic torque exerted on the particle, i.e., if $\mathbf{L} = \iint_{S_p} (\mathbf{r} \times \mathbf{f}) d\sigma(\mathbf{r})$, then $R_{ki} = \frac{1}{2}\varepsilon_{kil} L_l$, where $\boldsymbol{\varepsilon}$ is the alternating tensor*.

Since $\frac{\partial B_{ji}(\boldsymbol{\rho}^0, \mathbf{r}^c)}{\partial x_k^c} = \frac{1}{2}\left(\frac{\partial B_{ji}}{\partial x_k^c} + \frac{\partial B_{jk}}{\partial x_i^c} \right) + \frac{1}{2}\left(\frac{\partial B_{ji}}{\partial x_k^c} - \frac{\partial B_{jk}}{\partial x_i^c} \right)$, the second term on the right-hand side of (2.8.5) is

$$\frac{\partial B_{ji}(\boldsymbol{\rho}^0, \mathbf{r}^c)}{\partial x_k^c} \iint_{S_p} \hat{x}_k f_i(\mathbf{r}) d\sigma(\mathbf{r}) = \frac{1}{2}\left(\frac{\partial B_{ji}}{\partial x_k^c} + \frac{\partial B_{jk}}{\partial x_i^c} \right)(\boldsymbol{\rho}^0, \mathbf{r}^c) S_{ki}$$
$$+ \frac{1}{4}\varepsilon_{kil} \left(\frac{\partial B_{ji}}{\partial x_k^c} - \frac{\partial B_{jk}}{\partial x_i^c} \right)(\boldsymbol{\rho}^0, \mathbf{r}^c) L_l \tag{2.8.9}$$

because $\delta_{ki} \iint_{S_p} \hat{x}_l f_l(\mathbf{r}) d\sigma(\mathbf{r}) = 0$. In order to prove it the divergence theorem is applied, together with the identity $\frac{\partial B_{jl}}{\partial x_l^c} = 0$, for j, l =1, 2, 3.

---

* $\boldsymbol{\varepsilon} = \varepsilon_{ijk} \mathbf{e}_i \mathbf{e}_j \mathbf{e}_k$, where ε_{ijk} equals +1, when the number of inversions in the permutation of the three indices i, j, k is even, and -1, when it is odd. Moreover, if at least two indices are equal, then ε_{ijk} =0.

The expression $\frac{1}{2}\left(\frac{\partial B_{ji}}{\partial x_k^c}+\frac{\partial B_{jk}}{\partial x_i^c}\right)(\boldsymbol{\rho}^0,\mathbf{r}^c)S_{ki}$ represents a flow induced by a symmetric couple of two point force dipoles, which can be split into two more singular flow components: the so called stresslet and the point source (see section 3.4.) The coefficient S_{ki} in this expression is named the stresslet coefficient. The second term of (2.8.9) $\frac{1}{2}\left(\frac{\partial B_{ji}}{\partial x_k^c}-\frac{\partial B_{jk}}{\partial x_i^c}\right)(\boldsymbol{\rho}^0,\mathbf{r}^c)R_{ki}$ corresponds to a flow induced by another singular flow, called a rotlet or couplet [see (3.4.6)]. From (2.8.9) it follows, that when the torque **L** equals zero, the flow far away from the particle is similar to the flow created by a symmetric pair of point force dipoles.

In the case of a fluid particle (drop or bubble) passed by another fluid, the flow representation only by means of the single-layer potential is no more effective, and the expression (2.5.18) must be used instead of (2.8.1). If here again we expand the stress tensor K_{ijl} in Taylor series

$$K_{ijl}(\mathbf{r},\,\boldsymbol{\rho}^0)=K_{ijl}(\mathbf{r}^c,\,\boldsymbol{\rho}^0)+\frac{\partial K_{ijl}(\mathbf{r}^c,\,\boldsymbol{\rho}^0)}{\partial x_k^c}\hat{x}_k+\ldots, \tag{2.8.10}$$

where $\hat{x}_k=x_k-x_k^c$, and exploit (2.8.4) and (2.8.10), we obtain

$$\begin{aligned}v_j^D(\boldsymbol{\rho}^0)=&-\frac{1}{8\pi\mu_1}\left[B_{ji}(\boldsymbol{\rho}^0,\mathbf{r}^c)\iint\limits_{S_p^+}f_i(\mathbf{r})d\sigma(\mathbf{r})+\frac{\partial B_{ji}(\boldsymbol{\rho}^0,\mathbf{r}^c)}{\partial x_k^c}\iint\limits_{S_p^+}\hat{x}_k f_i(\mathbf{r})d\sigma(\mathbf{r})+\ldots\right]\\&+\frac{1}{8\pi}\left[K_{ijl}(\mathbf{r}^c,\,\boldsymbol{\rho}^0)\iint\limits_{S_p}v_i(\mathbf{r})n_l(\mathbf{r})d\sigma(\mathbf{r})+\frac{\partial K_{ijl}(\mathbf{r}^c,\,\boldsymbol{\rho}^0)}{\partial x_k^c}\iint\limits_{S_p}\hat{x}_k v_i(\mathbf{r})n_l(\mathbf{r})d\sigma(\mathbf{r})+\ldots\right]\end{aligned} \tag{2.8.11}$$

In (2.8.11) the fluid flow viscosity is denoted by μ_1, and the superscript + means that the surface stress force is calculated from the particle outer side.

After using (2.5.5), the symmetry condition $B_{ij}(\mathbf{r}^c,\,\boldsymbol{\rho}^0)\boldsymbol{\rho}=B_{ji}(\boldsymbol{\rho}^0,\mathbf{r}^c)$, while preserving only the point force and the point force dipole from the first series in brackets in (2.8.11) and the stresslet **K** from the second series, the expression (2.8.11) is reduced to

$$\begin{aligned}v_j^D(\boldsymbol{\rho}^0)=-\frac{1}{8\pi\mu_1}&\left\{B_{ji}(\boldsymbol{\rho}^0,\mathbf{r}^c)\iint\limits_{S_p^+}f_i(\mathbf{r})d\sigma(\mathbf{r})\right.\\&\left.+\frac{\partial B_{ji}(\boldsymbol{\rho}^0,\mathbf{r}^c)}{\partial x_k^c}\iint\limits_{S_p^+}\left[\hat{x}_k f_i(\mathbf{r})-\mu_1(v_k n_i+v_i n_k)(\mathbf{r})\right]d\sigma(\mathbf{r})+\ldots\right\}\\&-\frac{1}{8\pi}P_j(\mathbf{r}^c,\,\boldsymbol{\rho}^0)\iint\limits_{S_p}v_i(\mathbf{r})n_l(\mathbf{r})d\sigma(\mathbf{r})+\ldots\end{aligned} \tag{2.8.12}$$

It is clear, that the pressure coefficient in the last term of (2.8.12) is proportional to the flow rate $Q = \iint_{S_p} \mathbf{v}.\mathbf{n}d\sigma(\mathbf{r})$ through the particle surface S_p. As we have seen in section 2.6, the pressure vector $\mathbf{P}(\mathbf{r}^c, \boldsymbol{\rho}^0)$ can be considered (except in the inner flows case with $Q = 0$) as a velocity field, induced by a point source with strength -8π and pole $\mathbf{r}^c$.

We have to note that all terms after the last written one are expressed due to the Stokes equations by means of the Laplace operator acting on the Green's function , i.e., by second derivatives of $B_{jk}(\boldsymbol{\rho}^0, \mathbf{r}^c)$

$$\frac{\partial P_j(\mathbf{r}^c, \boldsymbol{\rho}^0)}{\partial x_k^c} = \nabla_c^2 B_{jk}(\boldsymbol{\rho}^0, \mathbf{r}^c), \tag{2.8.13}$$

where ∇_c denotes a differentiation with respect to $\mathbf{r}^c$. Now, we can deduce that if the terms connected with the pressure derivatives from the second series are added to the terms of the first series, the asymptotically disturbed flow $v_j^D(\boldsymbol{\rho}^0)$ can be expressed by a multipole expansion of the point force and a point source.

If now we split the coefficient of the point force dipole (the coefficient of $\frac{\partial B_{ji}(\boldsymbol{\rho}^0, \mathbf{r}^c)}{\partial x_k^c}$) to a symmetric part, isotropic part and antisymmetric part, then for the antisymmetric part we shall have

$$S_{ki} = \frac{1}{2} \iint_{S_p^+} \left[\hat{x}_k f_i(\mathbf{r}) + \hat{x}_i f_k(\mathbf{r}) - \frac{2}{3} \delta_{ik} \hat{x}_l f_l(\mathbf{r}) - 2(v_k n_i + v_i n_k)(\mathbf{r}) \right] d\sigma(\mathbf{r}), \tag{2.8.14}$$

and (2.8.8) for the asymmetric one.

2.9. Faxen Laws for Spherical Rigid Particle

In section 3.2 we shall study the problem of a rigid spherical particle motion with velocity $\mathbf{U}_p$ = const in a viscous fluid and we shall calculate the drag, which the particle suffers from the fluid side. For incompressible fluids this problem is equivalent to the problem of the flow of velocity $\mathbf{v}^\infty = -\mathbf{U}_p$ past a particle. It is well known that the Stokes formula (Stokes, 1851) for drag exerted by a stationary particle on the fluid flow, has the form:

$$\mathbf{F} = 6\pi \mu a \mathbf{v}^\infty, \tag{2.9.1}$$

where a is the particle radius and μ - the fluid viscosity.

Faxen (1924) generalises this relation, finding a formula by means of which we could calculate the particles drag action on a shear flow of velocity $\mathbf{v}^\infty(\mathbf{r})$, if we know the stress vector $\mathbf{t}_n = \mathbf{T}.\mathbf{n}$ on the surface of a rigid particle immersed in a uniform flow.

Since Faxen has derived his relations for the force and torque on a sphere by using of Lamb's general solution, his original approach is not readily extended to any particle shape. Brenner (1964b) first derived the Faxen laws via the Lorentz reciprocal theorem (see Application 2 in section 2.4) and all subsequent works in this area have followed his approach. As a simple illustration of Brenner's approach we shall derive the Faxen's formula for the drag on a sphere by using Lorentz reciprocal theorem.

Let us consider the following two Stokes flows:

1) The flow generated by a rigid spherical particle immersed in a uniform flow of velocity $\mathbf{U}_p$ (this flow is equivalent to the flow induced by the movement of a rigid spherical particle with velocity $-\mathbf{U}_p$ in quiescent fluid);

2) The flow generated by the same particle immersed in a gradient flow of velocity $\mathbf{v}^* = \mathbf{v}^\infty(\mathbf{r})$ (this flow is equivalent to the flow induced by the movement of a rigid spherical particle with velocity $-\mathbf{v}^\infty(\mathbf{r})$ in quiescent fluid).

Since we assume that the Stokes equation holds well for both flows, we have $\nabla.\mathbf{T} = \mathbf{0}$ and $\nabla.\mathbf{T}^* = \mathbf{0}$. In this case the Lorenz theorem is written in the form:

$$\iint_{S_p} \mathbf{v}^*.(\mathbf{T}.\mathbf{n})\, d\sigma = \iint_{S_p} \mathbf{v}.(\mathbf{T}^*.\mathbf{n})\, d\sigma \,. \tag{2.9.2}$$

Substituting the velocities of the two flows in (2.9.2) correspondingly with $\mathbf{v}^* = -\mathbf{v}^\infty(\mathbf{r})$ and $\mathbf{v} = -\mathbf{U}_p$ and changing the signs in the obtained equality, we find:

$$\iint_{S_p} \mathbf{v}^\infty(\mathbf{r}).(\mathbf{T}.\mathbf{n})\, d\sigma = \iint_{S_p} \mathbf{U}_p.(\mathbf{T}^*.\mathbf{n})\, d\sigma \tag{2.9.3}$$

or

$$\iint_{S_p} \mathbf{v}^\infty(\mathbf{r}).(\mathbf{T}.\mathbf{n})\, d\sigma = \mathbf{U}_p.\mathbf{F}^*, \tag{2.9.4}$$

where

$$\mathbf{F}^* = \iint_{S_p} \mathbf{T}^*.\mathbf{n}\, d\sigma \tag{2.9.5}$$

is the sought drag force. In order to calculate $\mathbf{F}^*$ it is sufficient to solve three problems of uniform flow past the rigid spherical particle in three mutually perpendicular directions (i.e., the three coordinate axes). After calculating the forces F_x^*, F_y^* and F_z^* corresponding to these three flows it is easy to find the sought force $\mathbf{F}^* = F_x^*\mathbf{i} + F_y^*\mathbf{j} + F_z^*\mathbf{k}$.

It is well known (Batchelor, 1967) that for the uniform flow past a rigid spherical particle with velocity $\mathbf{U}_p$, we have:

$$\mathbf{T}.\mathbf{n}\,|_{S_p} = \frac{3\mu\mathbf{U}_p}{2a} \,. \tag{2.9.6}$$

Then from (2.9.4) and (2.9.6) we get:

$$\mathbf{U}_p.\mathbf{F}^* = \iint\limits_{S_p} \mathbf{v}^\infty(\mathbf{r}).(\frac{3\mu\,\mathbf{U}_p}{2a})\,d\sigma = \frac{3\mu\,\mathbf{U}_p}{2a}\iint\limits_{S_p}\mathbf{v}^\infty(\mathbf{r})\,d\sigma\,.$$

Since $\mathbf{U}_p$ is arbitrary,

$$\mathbf{F}^* = \frac{3\mu}{2a}\iint\limits_{S_p}\mathbf{v}^\infty(\mathbf{r})\,d\sigma\,. \tag{2.9.7}$$

If we expand $\mathbf{v}^\infty(\mathbf{r})$ in a Taylor's series at the point $\mathbf{r} = \mathbf{0}$

$$\mathbf{v}^\infty(\mathbf{r}) = \mathbf{v}^\infty(\mathbf{0}) + \frac{\mathbf{r}}{1!}.(\nabla\mathbf{v}^\infty(\mathbf{r}))\Big|_{r=0} + \frac{\mathbf{r}\,\mathbf{r}}{2!}:[\nabla(\nabla\mathbf{v}^\infty(\mathbf{r}))]\Big|_{r=0} + \ldots \tag{2.9.8}$$

and substitute the obtained expression in (2.9.7), we reach:

$$\mathbf{F}^* = \frac{3\mu}{2a}4p\,a^2\left[\mathbf{v}^\infty(0) + \frac{a^2}{6}\nabla^2\mathbf{v}^\infty(\mathbf{r})\Big|_{r=0} + \text{const.}\nabla^4\mathbf{v}^\infty(\mathbf{r})\Big|_{r=0} + \ldots\right]. \tag{2.9.9}$$

When deriving (2.9.9) it is taken into account that the integrals of all the odd terms of (2.9.8) are zero because S_p is a sphere.

Since $\iint\limits_{S_p}\mathbf{n}\,\mathbf{n}\,d\sigma = \frac{4}{3}\,\pi\mathbf{I}$, for the third integral in (2.9.7) with substituting $\mathbf{v}^\infty(\mathbf{r})$ from (2.9.8) we have

$$\iint\limits_{S_p}\frac{\mathbf{r}\,\mathbf{r}}{2}:[\nabla(\nabla\mathbf{v}^\infty(\mathbf{r}))]\Big|_{r=0}\,d\sigma = \frac{4}{6}\,\pi a^2\,\nabla^2\mathbf{v}^\infty(\mathbf{r})\Big|_{r=0}\,.$$

Moreover, from the equation $\nabla^2\mathbf{v}^\infty = \frac{1}{\mu}\nabla p$ it follows that $\nabla^4\mathbf{v}^\infty = 0$, as well as $\nabla^{2n}\mathbf{v}^\infty = 0$, at $n \geq 2$. Then from (2.9.9) we obtain the Faxen's formula (Faxen, 1924)

$$\mathbf{F}^* = 6\pi\,\mu\,a\left[\mathbf{v}^\infty(0) + \frac{a^2}{6}\nabla^2\mathbf{v}^\infty(\mathbf{r})\Big|_{r=0}\right]. \tag{2.9.10}$$

From this formula it follows that the drag, which is exerted by a rigid spherical particle in a gradient flow of velocity $\mathbf{v}^\infty(\mathbf{r})$ could be found without solving the corresponding boundary problem for the velocity field and pressure. This formula is a generalisation of the Stokes formula, because if $\mathbf{v}^\infty(\mathbf{r})$ = const then we get (2.9.1) from (2.9.10). Faxen derives also the corresponding formula for the torque, which is exerted by the rigid spherical particle in a flow of angular velocity $\mathbf{\Omega}^\infty(\mathbf{r}) = \frac{1}{2}\nabla\times\mathbf{v}^\infty(\mathbf{r})$:

$$\mathbf{M}^* = 4\pi\,\mu\,a^3\nabla\times\mathbf{v}^\infty(\mathbf{r})\Big|_{r=0} = 8\pi\,\mu\,a^3\mathbf{\Omega}^\infty(\mathbf{r})\Big|_{r=0}. \tag{2.9.11}$$

In section 3.6 we shall show that if the particle rotates with angular velocity $\boldsymbol{\omega}$, then the torque formula will be transformed into

$$\mathbf{M}^* = 8\pi\mu\,a^3[\mathbf{\Omega}^\infty(\mathbf{r}) - \boldsymbol{\omega}]\Big|_{r=0}. \tag{2.9.12}$$

Batchelor and Green (1972a) have derived an analogous relation for the stresslet $\mathbf{S}^*$ for a rigid sphere in a strain flow $\mathbf{E}^\infty$:

$$\mathbf{S}^* = \frac{20}{3}\pi\mu a^3\left(1+\frac{1}{10}a^2\nabla^2\right)\mathbf{E}^\infty(\mathbf{r})\Big|_{r=0}. \tag{2.9.13}$$

As shown by Felderhof (1977a, b) and Rallison (1977) these relations can be used in the multipole expansion of the hydrodynamic functions for the interaction of two or more particles.

2.10. Energy Dissipation. Uniqueness. Hill and Power Drag Statements.

Theorem 1. If for a given fluid motion the deformation velocity tensor e_{ij} is equal to zero, then the fluid is moving as a rigid body.

Therefore, we have to prove that, if $e_{ij} = 0$, (i, j = 1, 2, 3) the fluid velocity $\mathbf{v}$ consists of translation $\mathbf{U}$ and rotation $\boldsymbol{\omega}$ [see (Kim and Karrila, 1991)]. In fact, from $e_{ij} = 0$ it follows

$$\frac{\partial v_i}{\partial x_j} = -\frac{\partial v_j}{\partial x_i} \qquad (i, j=1, 2, 3).$$

After differentiating this equality with respect to x_k, we obtain:

$$\frac{\partial^2 v_i}{\partial x_j \partial x_k} = -\frac{\partial^2 v_j}{\partial x_i \partial x_k}, \qquad \left(v_{i,jk} = -v_{j,ik}\right). \tag{2.10.1}$$

This equality is true for every j. Therefore, it is also true for j = i, i.e.,

$$\frac{\partial^2 v_i}{\partial x_i \partial x_k} = -\frac{\partial^2 v_i}{\partial x_i \partial x_k} \tag{2.10.2}$$

Here there is no summation on the repeating index.

From (2.10.2) it follows, that the velocity components can be linear functions of the three independent variables x_1, x_2, x_3. By this, the statement is proved, because at the fluid motion as a rigid body, the velocity components are linear functions of x_1, x_2, x_3:

$$\mathbf{v} = \mathbf{U} + \boldsymbol{\omega}\,x\,\mathbf{r}$$

or

$$\mathbf{v} = \mathbf{U} + \begin{vmatrix} \mathbf{i} & \mathbf{j} & \mathbf{k} \\ \omega_1 & \omega_2 & \omega_3 \\ x_1 & x_2 & x_3 \end{vmatrix} = \mathbf{U} + (\omega_2 x_3 - \omega_3 x_2)\mathbf{i} + (\omega_3 x_1 - \omega_1 x_3)\mathbf{j} + (\omega_1 x_2 - \omega_2 x_1)\mathbf{k}.$$

Theorem 2. For the energy dissipation velocity E_k the following expressions are valid:

$$E_k = \iiint_{V_0} \Phi_v d\tau = \iiint_{V_0} 2\mu\, E:\nabla \mathbf{v} d\tau = \iiint_{V_0} 2\mu\, e_{ij} \frac{\partial v_j}{\partial x_i} d\tau = \iiint_{V_0} 2\mu\, e_{ij} e_{ij} d\tau = \iiint_{V_0} T_{ij} e_{ij} d\tau .$$

Since for Newtonian fluids we have: (2.10.3)

$$T_{ij} = -p\delta_{ij} + 2\mu\, e_{ij}, \quad \text{where } e_{ij} = \frac{1}{2}\left(\frac{\partial v_i}{\partial x_j} + \frac{\partial v_j}{\partial x_i}\right).$$

Then

$$T_{ij} e_{ij} = -p\delta_{ij} e_{ij} + 2\mu\, e_{ij} e_{ij} = -p e_{ii} + 2\mu\, e_{ij} e_{ij} = 2\mu\, e_{ij} e_{ij},$$

because of continuity equation $e_{ii} = \nabla . \mathbf{v} = 0$.

On the other hand

$$e_{ij} e_{ij} = \frac{1}{2}\left(\frac{\partial v_i}{\partial x_j} + \frac{\partial v_j}{\partial x_i}\right) e_{ij} = \frac{1}{2}\frac{\partial v_i}{\partial x_j} e_{ij} + \frac{1}{2}\frac{\partial v_j}{\partial x_i} e_{ij}. \tag{2.10.4}$$

Taking into account that $e_{ij} = e_{ji}$ and that in the first addend of the last expression of (2.10.4) the indices i and j are "mute", then we can change their positions. In such a way we get $e_{ij} e_{ij} = e_{ij} \dfrac{\partial v_j}{\partial x_i}$ and the proof is completed.

<u>Theorem 3.</u> The energy dissipation of Stokes flow is less than the energy dissipation of every other viscous flow, if the two flows satisfy the same boundary conditions.

It is given, that $\nabla . \mathbf{v} = 0$, $\nabla . \mathbf{v}^* = 0$ and $\mu \nabla^2 \mathbf{v} - \nabla p = 0$, where $\mathbf{v}$ and p are the velocity and pressure of the Stokes flow, respectively, and $\mathbf{v}^*$ is the velocity of an arbitrary viscous flow. We shall prove, that

$$E_k^* \geq E_k^{st}. \tag{2.10.5}$$

First of all, we shall prove the identity:

$$A = \iiint_{V_0} \left(e_{ij}^* - e_{ij} \right) e_{ij} d\tau = 0 .$$

Let us make the transformations:

$$\left(e_{ij}^* - e_{ij} \right) e_{ij} = \left[\frac{1}{2}\left(\frac{\partial v_i^*}{\partial x_j} + \frac{\partial v_j^*}{\partial x_i}\right) - \frac{1}{2}\left(\frac{\partial v_i}{\partial x_j} + \frac{\partial v_j}{\partial x_i}\right) \right] e_{ij}$$

$$= \frac{1}{2}\left(\frac{\partial v_i^*}{\partial x_j} - \frac{\partial v_i}{\partial x_j}\right) e_{ij} + \frac{1}{2}\left(\frac{\partial v_j^*}{\partial x_i} - \frac{\partial v_j}{\partial x_i}\right) e_{ij} = \left(\frac{\partial v_i^*}{\partial x_j} - \frac{\partial v_i}{\partial x_j}\right) e_{ij}$$

$$= \frac{\partial}{\partial x_j}\left[\left(v_i^* - v_i \right) e_{ij} \right] - \left(v_i^* - v_i \right) \frac{\partial e_{ij}}{\partial x_j} .$$

From the divergence theorem

$$A=\iint_{S_0}\left(v_i^*-v_i\right)e_{ij}n_j d\sigma-\iiint_{V_0}\left(v_i^*-v_i\right)\frac{1}{2}\left[\frac{\partial^2 v_i}{\partial x_i \partial x_j}+\frac{\partial^2 v_j}{\partial x_i \partial x_j}\right]d\tau$$

and since $\dfrac{\partial^2 v_j}{\partial x_i \partial x_j}=\dfrac{\partial}{\partial x_i}\left(\dfrac{\partial v_j}{\partial x_j}\right)=0$, we obtain:

$$-\left(v_i^*-v_i\right)\frac{\partial^2 v_i}{\partial x_j \partial x_j}=-\frac{1}{\mu}\left(v_i^*-v_i\right)\frac{\partial p}{\partial x_i}.$$

Therefore

$$A=\iiint_{V_0}\left(e_{ij}^*-e_{ij}\right)e_{ij}d\tau=-\frac{1}{\mu}\iiint_{V_0}\left(v_i^*-v_i\right)\frac{\partial p}{\partial x_i}d\tau$$

$$=-\frac{1}{\mu}\iiint_{V_0}\frac{\partial}{\partial x_i}\left[\left(v_i^*-v_i\right)p\right]d\tau+\frac{1}{\mu}\iiint_{V_0}p\frac{\partial}{\partial x_i}\left(v_i^*-v_i\right)d\tau=0.$$

Thus it is proved that

$$\iiint_{V_0}\left(e_{ij}^*-e_{ij}\right)e_{ij}d\tau=0. \tag{2.10.6}$$

It is given that 1) both $\mathbf{v}^*$ and $\mathbf{v}$ are solenoidal fields, i.e., $\nabla.\mathbf{v}^*=\nabla.\mathbf{v}=0$; 2) the field $\mathbf{v}$ satisfies the Stokes equation. We have to prove (2.10.5).

We start from the proved equality (2.10.6) to obtain:

$$E_k^*=\iiint_{V_0}2\mu e_{ij}^*e_{ij}^*d\tau=2\mu\iiint_{V_0}\left(e_{ij}^*e_{ij}^*-2e_{ij}^*e_{ij}+2e_{ij}e_{ij}\right)d\tau$$

$$=2\mu\iiint_{V_0}\left[e_{ij}e_{ij}+\left(e_{ij}^*-e_{ij}\right)\left(e_{ij}^*-e_{ij}\right)\right]d\tau=2\mu\iiint_{V_0}e_{ij}e_{ij}d\tau$$

$$+2\mu\iiint_{V_0}\left(e_{ij}^*-e_{ij}\right)\left(e_{ij}^*-e_{ij}\right)d\tau.$$

Therefore

$E_k^*=E_k+\alpha^2$, where $\alpha^2=2\mu\iiint_{V_0}\left(e_{ij}^*-e_{ij}\right)\left(e_{ij}^*-e_{ij}\right)d\tau$, i.e., $E_k^*\geq E_k^{st}$.

In Stokes flow the drag of an object is fairly insensitive to the exact shape. Hill and Power (1956) prove the following so-called "inclusion monotonicity statements" valid for Stokes flows:

a) If a body of surface S_2 could be contained within a body of surface S_1, then the drag F_1 on S_1 is greater than or equal to the drag F_2 on S_2 (the velocity of translation and external fluid boundary being the same);
b) The drag F_1 on a given body moving translationally through a fluid in a stationary rigid container C_1 is greater than or equal to the drag F_2 relevant to the same body, but in another container C_2, completely enclosing C_1;
c) The drag on a body S tends to increase in the presence of other bodies S_1. S_2, …, S_n, whether fixed in position or free to move without restraint;

d) The apparent viscosity of a fluid of solid suspension is a monotonic increasing function of the volume concentration of solute.

Therefore, the drag of an object must be larger than the drag of any inscribed figure, but smaller than that of any circumscribed figure. A sphere circumscribed around a body has a larger drag than the body itself, though it has no sharp corners; while a sphere inscribed within the body has a smaller drag than the latter.

PART II.

STEADY FLOWS. HYDRODYNAMICS OF A SINGLE RIGID OR FLUID PARTICLE

CHAPTER 3.

Application of the Singularity Method for a Single Rigid or Fluid Particle

3.1. Introduction

The linearity of the creeping flow equations allows the creation of a class of solution methods that is readily applied to various type of hydrodynamic problems. Solutions of problems involving applications of internal concentrated forces are of fundamental interest for fluid mechanics. Apart from being Green's functions, these fundamental solutions can be used in constructing the new solutions of more complicated and physically realizable problems. The construction of such solutions is known as a singularity method. This method is employed not only for obtaining solutions which are not known so far, but also to recover already known solutions in a very simple fashion. If the boundary shapes of the problem under consideration are simple, then the analytical solution can be achieved via the internal distributions of force and force multipole singularities. Unfortunately, a systematic procedure for choosing the singularities necessary to solve any particular problem does not exist yet. One relies on a combination of intuition and a knowledge of the properties of the basic singularities to make an appropriate guess. It is clear that suggestions about the symmetry of the considered flow have an important significance.

The fundamental singularity solution of Stokes and continuity equations for an infinite incompressible viscous fluid is available in the works of Oseen, (1927), Burgers, (1938a) and the text-book of Lamb (1945). Some other references for singularity solutions of hydrodynamic problems may be found in the interesting papers of Chwang and Wu Yao-Tsu (1974, 1975, 1976), Chwang (1975), Johnson and Wu Yao-Tsu (1979) and Huang and Chwang (1986), and many others.

In the present chapter we shall briefly consider the application of the singularity method in solving some hydrodynamic problems of single bubbles, drops and rigid particles; the link between the singularity solutions and Faxen's laws, etc.

In this connection it is useful to present some of the derivatives of the Oseen-Burgers tensor (Stokeslet)

$$B_{ij} = \frac{\delta_{ij}}{r} + \frac{x_i x_j}{r^3}, \qquad r = |\mathbf{r}| = \sqrt{x_i x_i} \tag{3.1.1}$$

and the degenerate quadrupole $\nabla^2 B_{ij}$:

i) the first derivative of the Oseen-Burgers tensor (Stokes dipole)

$$\nabla B_{ij} = \frac{\partial B_{ij}}{\partial x_k} = B_{ij,k} = \frac{1}{r^3}\left(-\delta_{ij}x_k + \delta_{jk}x_i + \delta_{ki}x_j\right) - \frac{3x_i x_j x_k}{r^5}, \tag{3.1.2}$$

ii) The second derivative of the Oseen-Burgers tensor

$$\frac{\partial^2 B_{ij}}{\partial x_k \partial x_k} = B_{ij,kk} = -\frac{\delta_{ij}}{r^3} + \frac{3\delta_{ij}}{r^3} + \frac{2}{r^3}\delta_{ik}\delta_{jk}$$

$$-\frac{3}{r^5}\left(\delta_{jk}x_i + \delta_{ik}x_j\right)x_k - \frac{3}{r^5}\left(\delta_{ik}x_j x_k + \delta_{jk}x_k x_i + x_i x_j\right) + \frac{15}{r^5}x_i x_j, \tag{3.1.3}$$

iii) the degenerate quadrupole

$$\nabla^2 B_{ij} = \nabla.\left(\nabla B_{ij}\right) = B_{ij,ii} = \frac{2\delta_{ij}}{r^3} - \frac{6x_i x_j}{r^5}, \tag{3.1.4}$$

iv) the derivative of the degenerate quadrupole (the degenerate octupole)

$$\nabla\left(\nabla^2 B_{ij}\right) = \nabla^2 B_{ij,k} = -\frac{6}{r^5}\left(\delta_{ij}x_k + \delta_{jk}x_i + \delta_{ik}x_j\right) + \frac{30}{r^7}x_i x_j x_k \tag{3.1.5}$$

(here we shall note that in order to obtain (3.1.2)-(3.1.5) the following formulae have been used:

$$\delta_{ij}x_i = x_j, \quad \frac{\partial x_i}{\partial x_j} = \delta_{ij}, \quad x_i x_i = r^2, \quad \frac{\partial r}{\partial x_i} = \frac{x_i}{r} \;).$$

For the general particle shape, the multipole expansion representation (obtained in section 2.8) requires an infinite number of terms. If the particle shape is simple, the multipole expansion representation may contain only a finite number of terms. For the case of a spherical particle, for example, the multipole expansion contains (as we shall see in section 3.2) only two terms. For other simple shapes like ellipsoid, a truncated expansion in the lower order singularities is the only possible, providing the singularities to be distributed over a region. The other example is the solution for very elongated, slender particles where the integral representation (obtained in section 2.5) can approximately be reduced to a line distribution of point forces along the centreline of the particle.

For interior flows (like the flow inside a drop) the velocity field of the Stokeson

$$v_i = H_{ij}U_j = \left(2r^2\delta_{ij} - x_i x_j\right)U_j, \text{ i.e., } \mathbf{v} = \mathbf{H(r).U} \tag{3.1.6}$$

is used. Since the Stokeson is linear with respect to a constant vector **U** (as we shall see in section 3.3), it enters into the solution for a translating drop. In view of the fact that the Stokeson is quadratic with respect to r, its gradient is linear with respect to r.

Other interior solutions are roton and stresson which are equal to the symmetric and antisymmetric derivatives of the Stokeson, respectively. Since the roton and the stresson correspond to a rigid body rotation and a constant rate-of-strain field (which are not typical for fluid flows), they are used rarely.

3.2. Flow Due to the Translation or Rotation of a Rigid Spherical Particle

The theoretical examination of the motion of bodies or particles in viscous fluids at small values of the Reynolds number starts in 1851 with the remarkable work of Stokes (1851), in which the famous law for the force, exerted by a sphere in a uniform viscous flow with velocity **U**, is derived:

$$\mathbf{F} = 6\pi\,\mu\,\mathrm{a}\mathbf{U}, \tag{3.2.1}$$

where μ is the dynamic viscosity coefficient, and a is the sphere radius.

Beside the numerous theoretical studies, some experiments have been conducted both with spherical particles (Maxworthy, 1965) as well as nonspherical particles (Becker, 1959), (McNown and Malaika, 1950), (Willmarth et al., 1964). Several types of nonspherically shaped particles: plane (disks), prolate (cylindrical), rounded, sharp, etc., were mainly investigated.

Kirchoff (1876) approximately solved the problem of the slow rotation of a rigid sphere of radius a in a viscous fluid around an axis passing through its centre. For the torque, which acts on the particle, he obtained the formula:

$$\mathbf{M} = -8\pi\,\mu\,\mathrm{a}^3\boldsymbol{\Omega}, \tag{3.2.2}$$

where $\boldsymbol{\Omega}$ is the angular velocity of the sphere.

The slow translation of a rigid spherical particle of radius a through a quiescent viscous fluid induces a flow which can be found by means of the singularity method. Since this flow produces a net force on the spherical particle, we require a Stokeslet $\mathbf{B}.\,\mathbf{F}^{(c)}$ located at the sphere centre in order to construct a solution via internal singularities. However, the Stokeslet is most often accompanied by a degenerate quadrupole. Thus, we suggest the construction of a solution as a superposition of a Stokeslet and a degenerate quadrupole both located at the centre of the spherical particle, i.e.,

$$\mathbf{v}(\mathbf{r}) = \mathbf{p}.\mathbf{B}(\mathbf{r}) + \mathbf{q}.\ \nabla^2\mathbf{B}(\mathbf{r})\,. \tag{3.2.3}$$

Since each term in this expansion satisfies the creeping flow equations (2.3.1) and (2.3.2), further we shall try to determine the unknown vectors **p** and **q** from the following boundary conditions

$$\mathbf{v} = \mathbf{U} \qquad \text{at } \mathrm{r} = \mathrm{a}, \tag{3.2.4}$$

$$\mathbf{v} \to 0 \qquad \text{as } \mathrm{r} \to \infty, \tag{3.2.5}$$

where **U** is the velocity of the particle.

In fact, since $\mathbf{B}(\mathbf{r})$ and $\nabla^2\mathbf{B}(\mathbf{r})$ tend to zero as $r \to \infty$, the boundary condition (3.2.5) is fulfilled automatically. Then from the uniqueness theorem (proved in section 2.7) it follows that the solution of the considered problem is found.

Introducing the explicit forms of the singularities in (3.2.3), we obtain

$$\mathbf{v}(\mathbf{r}) = \mathbf{p}.\left(\frac{\mathbf{I}}{r}+\frac{\mathbf{rr}}{r^3}\right)+\mathbf{q}.\left(\frac{2\mathbf{I}}{r^3}-\frac{6\mathbf{rr}}{r^5}\right). \tag{3.2.6}$$

If $\mathbf{n}$ is the unit normal vector to the spherical particle surface, we have $\mathbf{r} = a\,\mathbf{n}$ at $r = a$ and the boundary condition (3.2.4) gives

$$\mathbf{U} = \mathbf{p}.\left(\frac{\mathbf{I}}{a}+\frac{\mathbf{nn}}{a}\right)+\mathbf{q}.\left(\frac{2\mathbf{I}}{a^3}-\frac{6\mathbf{nn}}{a^3}\right)$$

or

$$a^3\mathbf{U} = a^2\mathbf{p} + 2\mathbf{q} + \left(a^2\mathbf{p} - 6\mathbf{q}\right).\mathbf{nn}. \tag{3.2.7}$$

Since the problem considered is linear, it is reasonable to assume that the unknown vectors $\mathbf{p}$ and $\mathbf{q}$ in the equation (3.2.3) are expressed linearly via the particle velocity $\mathbf{U}$, i.e.,

$$\mathbf{p} = C_0\mathbf{U}, \qquad \mathbf{q} = C_2\mathbf{U}, \tag{3.2.8}$$

where C_0 and C_2 are unknown constants.

With (3.2.8) the equation (3.2.7) becomes

$$\left(-a^3 + C_0a^2 + 2C_2\right)\mathbf{U} + \left(C_0a^2 - 6C_2\right)\mathbf{U}.\mathbf{nn} = 0$$

or

$$\left(-a^3 + C_0a^2 + 2C_2\right)\mathbf{U} + \left(C_0a^2 - 6C_2\right)U_n\mathbf{n} = 0. \tag{3.2.9}$$

Taking into account that $\mathbf{U}$ and $\mathbf{n}$ are independent vectors, we obtain the following system for the constants C_0 and C_2:

$$\left|\begin{array}{l} C_0a^2 + 2C_2 = a^3 \\ C_0a^2 - 6C_2 = 0 \end{array}\right.$$

Hence, $C_0 = \dfrac{3}{4}a$ and $C_2 = \dfrac{a^3}{8}$, and (3.2.3) becomes

$$\mathbf{v}(\mathbf{r}) = \frac{3a}{4}\mathbf{U}.\mathbf{B}(\mathbf{r}) + \frac{a^3}{8}\mathbf{U}.\ \nabla^2\mathbf{B}(\mathbf{r}) = 6\pi\mu a\mathbf{U}.\left(1+\frac{a^2}{6}\nabla^2\right)\frac{\mathbf{B}(\mathbf{r})}{8\pi\mu}. \tag{3.2.10}$$

It is easy to show that this expression is identical to the standard result in spherical coordinates given in elementary books on fluid mechanics.

Therefore, the translating spherical particle in Stokes flow requires a degenerate quadrupole, in addition to a Stokeslet with coefficient $6\pi\mu a\mathbf{U}$. Of special interest is the fact that we have derived the Stokes law, $\mathbf{F} = 6\pi\mu a\mathbf{U}$ for the force on the spherical particle undergoing steady translation, without explicit computation of the surface stress vector $\mathbf{t}_n = \mathbf{T}.\mathbf{n}|_{r=a}$. Here we have used the statement from section 2.9 that solutions expressed as a multipole expansion yield quantities of interest, such as the hydrodynamic force, in a straightforward fashion.

Now let us consider the flow round a spherical particle of radius a rotating with angular velocity $\boldsymbol{\omega}$ through an unbounded quiescent viscous fluid. If we take a Cartesian coordinate system with origin that coincides with the sphere centre then the boundary conditions of the considered problem are as follows

$$\mathbf{v} \to 0 \qquad \text{as } r \to \infty \tag{3.2.11}$$

$$\mathbf{v} = \boldsymbol{\omega} \times a\mathbf{e}_r \quad \text{at } r = a \tag{3.2.12}$$

where $\mathbf{e}_r$ is a unit vector in direction $\mathbf{r}$.

Equation (3.2.12) suggests that the produced flow may be represented merely in terms of a rotlet (couplet) with strength $C_1\,\boldsymbol{\omega}$, located at the centre of the sphere, i.e.,

$$\mathbf{v} = C_1 \boldsymbol{\omega} \times \frac{\mathbf{r}}{r^3}. \tag{3.2.13}$$

With the boundary condition (3.2.12) the velocity field (3.2.13) gives

$$\boldsymbol{\omega} \times a\mathbf{e}_r = C_1 \boldsymbol{\omega} \times \frac{a\mathbf{e}_r}{a^3}$$

whence, $C_1 = a^3$. Therefore,

$$\mathbf{v} = \frac{a^3}{r^3}(\boldsymbol{\omega} \times \mathbf{r}). \tag{3.2.14}$$

Using (3.2.14) we can calculate the torque acting on the particle

$$\mathbf{M} = \iint_S \mathbf{r} \times (\mathbf{T}.\mathbf{n})\, d\sigma = 8\pi \mu a^3 \boldsymbol{\omega}. \tag{3.2.15}$$

3.3. A Translating Spherical Liquid Drop in Viscous Quiescent Fluid

One of the most important solutions in the fluid particle hydrodynamics is the solution derived independently of one another by Hadamard (1911) and Rybczynski (1911). They considered the droplet passed by another viscous fluid as a fluid sphere with constant surface tension on the interface. For the force exerted on the drop in the viscous fluid flow of velocity **U** they obtained the formula:

$$\mathbf{F} = 6\pi a \mu \mathbf{U} \frac{2\mu + 3\hat{\mu}}{3(\mu + \hat{\mu})}, \tag{3.3.1}$$

where $\hat{\mu}$ is the fluid viscosity inside the droplet, and μ is the fluid viscosity outside it.

It is interesting to note, that the Hadamard-Rybczynski theory predicts terminal velocity of a spherical drop in a creeping flow by up to 50% higher than that of a rigid sphere of the same size and density. Moreover, the experimental investigations show that for some drops the measured drag is closer to the calculated by the Stokes formula (3.2.1) and corresponding to rigid spheres rather than to the Hadamard-Rybczynski law

(3.3.1). It follows from these results, that for the above cases the internal circulation in the drop has a small influence on the drag value.

Boussinesq (1913a, b) explained the lack of the influence of inner circulation in the drop with the presence of an interfacial monolayer on the fluid sphere surface acting as a viscous membrane. In addition to the surface tension the constitutive equations proposed by Boussinesq involve another parameter, referred to as surface viscosity μ_s. On the basis of this new "Newtonian surface fluid model" Boussinesq derived an exact solution of the Stokes equation for the viscous fluid flow similar to the Hadamard-Rybczynski solution involving the surface viscosity as a new physical quantity. For the drop force Boussinesq got the formula:

$$\mathbf{F} = 6\pi \, \mathrm{a} \, \mu \, \mathbf{U} \frac{\mu_s + \mathrm{a}(2\mu + 3\hat{\mu})}{\mu_s + 3\mathrm{a}(\mu + \hat{\mu})}. \tag{3.3.2}$$

Although this solution gives values of the spherical drops drag closer to the experimental ones, the differences still remain.

Frumkin and Levich (1947) gave another explanation. According to them surfactants accumulate on the interface, which change the surface tension action. When the drop moves in the fluid the surfactants tend to occupy its rear end where the surface tension tends to zero and the interface conducts itself as a "rigid" boundary. According to Frumkin and Levich the created surface tension gradient on the interface and the tangential stress retard the drop motion.

An important tendency of the investigations of viscous flows with interfaces is the creation of thin films on their surface. When the bulk phase in a fluid flow contains surfactants, the composite structure of the region in vicinity to the interface is completely different from that pertaining to the bulk. Recent studies have proven that the hydrodynamic films on the interface between air and different types of petroleum result from the surfactants action rather than from oxidation.

The rheological properties of films created by surfactants on fluid interfaces are examined by many authors. The effect of the insoluble surfactants on the bulk phase dynamics is described by the coefficient of surface shear viscosity ε. For the first time in the scientific literature Boussinesq (1913a, b) proposed a 2D analogous of the 3D Navier-Stokes equations for a spherical fluid surface. A generalisation of this model for the case of arbitrary fluid surfaces is done by Scriven (1960) and Slattery (1964). Besides the coefficient of surface shear viscosity ε they introduce the so-called coefficient of surface dilatational viscosity k.

As soon as the surface viscosity is introduced (with its both components) as a characteristics of the fluid way of flow along the corresponding interfaces, necessity of methods for its experimental measurement arise. First measurements of the surface viscosity were conducted by Wilson and Ries (1923) by means of an oscillating circular disk on the fluid surface. Langmuir and Schaefer (1937) also proposed a surface viscosity measurement method. The methods developed till 1961 are reviewed by Davies and Rideal (1961). None of these methods however, is based on a strictly proved mathematical formula connecting the fluid velocity on the interface with the surface viscosity.

Burton and Mannhemer (1967) developed a surface viscosity measurement method for Newtonian fluids and they called their apparatus "deep channel viscometer". Later, Pintar et al. (1971) established that this apparatus can be successfully used for surface viscosity measurement of various polymer solutions. Goodrich and Allen (1972) and Shail (1978) studied the dynamics of stationary rotating disk viscometer. The apparatus proposed by them consists of a thin disk between the surfactants film and the fluid next under it. The torque necessary to support the rotation is measured and compared with the theoretically obtained torque due to the steady Stokes equations.

Since this method turns also not to be sufficiently sensitive, Brenner (1962) has suggested the idea to place the disk not on the fluid interface but in the bulk, in order to avoid the incorrect evaluation of the torque of the displaced fluid. Similar is the method of Shail (1979), in which the stationary rotating disk is immersed in the bulk on a given distance from the interface. Other investigations of fluid flows with surfactant covered interfaces are carried out by Shail and Gooden (1981).

Important results about the influence of the surfactants on the delay of sedimenting fluid particles of reaching their terminal velocity are obtained by Levich (1981) and Newman (1967). Saville (1973) obtained basic formulae for the drag coefficient and terminal velocity of a falling fluid particle in a viscous fluid from which the Harper's (Harper, 1972) formulae for large surfactants concentrations and Saville's (Saville, 1973) formulae for small surfactants concentrations were derived as special cases. These studies supplement the Savic's model (Savic, 1953) related to such surfactants concentrations, accumulating on the rear end of the fluid particle surface and completely changing the well-known flow pattern given by Hadamard and Rybczynski. The surfactants effects on the terminal velocity of sedimenting drops are examined by Levan and Newman (1976) , too. Agrawal and Wasan (1979) analysed the influence of the surface shear viscosity and surface dilatational viscosity, assuming that the surface tension gradient is due to the surfactants, which are adsorbed on the interface along the linearized isotherm. A more general solution of the same problem (without use of the linearized isotherm) is obtained by Levan (1981).

Let us consider a spherical drop moving with velocity $\mathbf{U}$ in a viscous quiescent fluid. We suppose that the fluids both outside and inside the drop are immiscible and that the surface tension, σ, at the interface is sufficiently high to keep the drop approximately spherical against any deforming effect of viscous forces. It is also assumed that the Reynolds number of the motion within the drop is small compared to unity, as it is with the motion outside the drop. The two fluid motions are described by the equations (2.3.1) and (2.3.2) with different values of the viscosity coefficient - μ and $\hat{\mu}$ (here the caret indicates a quantity related to the internal fluid and its motion).

We choose the origin of the coordinate system to be at the instantaneous centre position of the drop with radius a. The velocity $\mathbf{v}$ and the difference $p - p_\infty$ must vanish at infinity, while $\hat{\mathbf{v}}$ and $\hat{p} - \hat{p}_0$ are finite everywhere within the fluid particle.

The boundary conditions at the interface of the droplet, r = a, are (1.4.14) and (1.4.15) for the normal component of the velocity vectors, (1.4.16) for the tangential velocity vectors and (1.4.17) for the tangential components of the stress vectors, where $\lambda = \dfrac{\hat{\mu}}{\mu}$ is the ratio of the drop and solvent viscosities.

Cursory inspection of the menu of the available singularities results in the following selections: a Stokeslet and a degenerate quadrupole outside the drop and a Stokeson and a uniform field **U** inside the drop. Then the velocity fields outside and inside the drop can be written respectively as

$$\mathbf{v} = \frac{3a}{4}\mathbf{U}.\left(C_0 + C_2\, a^2\, \nabla^2\right)\mathbf{B}(\mathbf{r}), \tag{3.3.3}$$

$$\hat{\mathbf{v}} = D_0\mathbf{U} + D_2\, a^{-2}\, \mathbf{U}.\,\mathbf{H}(\mathbf{r}), \tag{3.3.4}$$

where the four unknown constants C_0, C_2, D_0 and D_2 are determined from the boundary conditions at the drop interface.

Taking into account formulae (3.1.1), (3.1.4) from (3.3.3), we obtain

$$v_i = C_0 \frac{3a}{4}\, B_{ij}\, U_j \; + C_2 \frac{3a^3}{4}\; \nabla^2 B_{ij}\, U_j \tag{3.3.5}$$

$$= C_0 \frac{3a}{4}\left(\frac{\delta_{ij}}{r} + \frac{x_i x_j}{r^3}\right) U_j \; + C_2 \frac{3a^3}{4}\left(\frac{2\delta_{ij}}{r^3} - \frac{6 x_i x_j}{r^5}\right) U_j .$$

Since at r = a we have $x_i = a\, n_i$, then (3.3.3) gives

$$v_i\big|_{r=a} = \frac{3}{2}\, U_i \left(\frac{C_0}{2} + C_2\right) + \frac{3}{2}\, n_i n_j U_j \left(\frac{C_0}{2} - 3C_2\right) \tag{3.3.6}$$

and thus

$$v_i n_i\big|_{r=a} = \frac{3}{2} n_i U_i \left(\frac{C_0}{2} + C_2\right) + \frac{3}{2}\, n_i n_i n_j U_j \left(\frac{C_0}{2} - 3C_2\right) \tag{3.3.7}$$

$$= n_i U_i \left(\frac{3}{2} C_0 - 3C_2\right).$$

Therefore from the first kinematic condition (1.4.14) it is found

$$3C_0 - 6C_2 = 2 . \tag{3.3.8}$$

Returning to (3.3.4) and using (3.1.6) we obtain

$$\hat{v}_i = D_0 U_i + D_2\, a^{-2} \left(2r^2 \delta_{ij} - x_i x_j\right) U_j = \left(D_0 + 2\frac{r^2}{a^2} D_2\right) U_i - D_2 \frac{x_i x_j}{a^2} U_j \tag{3.3.9}$$

and thus

$$\hat{v}_i n_i\big|_{r=a} = \left(D_0 + D_2\right) n_i U_i . \tag{3.3.10}$$

It follows from the second kinematic condition (1.4.15) that

$$D_0 + D_2 = 1 . \tag{3.3.11}$$

The condition of continuity of the velocity at the surface of the drop (1.4.16) requires to accomplish the following computations

$$[(\mathbf{I}-\mathbf{nn}).\mathbf{v}]_i\Big|_{r=a} = [(\mathbf{v}-\mathbf{nn.v})]_i\Big|_{r=a} = \left(\frac{3}{4}C_0 + \frac{3}{2}C_2\right)\left(U_i - n_i n_j U_j\right)$$

and

$$[(\mathbf{I}-\mathbf{nn}).\hat{\mathbf{v}}]_i\Big|_{r=a} = [(\hat{\mathbf{v}}-\mathbf{nn}.\hat{\mathbf{v}})]_i\Big|_{r=a} = \left(D_0 + 2D_2\right)\left(U_i - n_i n_j U_j\right).$$

Substituting these expressions into (1.4.16) yields

$$3C_0 + 6C_2 = 4D_0 + 8D_2. \tag{3.3.12}$$

In order to apply the boundary condition (1.4.17) we have to calculate the rate of strain of stress tensors $\mathbf{E} = \left\{e_{ij}\right\}$ and $\hat{\mathbf{E}} = \hat{e}_{ij}$, and from (3.3.3) and (3.3.4) it follows that

$$e_{ij}\Big|_{r=a} = \frac{1}{2}\left(\frac{\partial v_i}{\partial x_j} + \frac{\partial v_j}{\partial x_i}\right)\Bigg|_{r=a} = \frac{3}{8}aC_0\left(B_{ik,j} + B_{jk,i}\right)U_k\Big|_{r=a} \tag{3.3.13}$$

$$+\frac{3}{8}a^3 C_2\left(\nabla^2 B_{ik,j} + \nabla^2 B_{jk,i}\right)U_k\Big|_{r=a},$$

$$\hat{e}_{ij}\Big|_{r=a} = \frac{1}{2}\left(\frac{\partial \hat{v}_i}{\partial x_j} + \frac{\partial \hat{v}_j}{\partial x_i}\right)\Bigg|_{r=a} = \frac{D_2}{2a}\left(3n_i U_j + 3n_j U_i - 2\delta_{ij} n_k U_k\right). \tag{3.3.14}$$

Using (3.1.5) it is easy to obtain the formula

$$\left(\nabla^2 B_{ik,j} + \nabla^2 B_{jk,i}\right)\Big|_{r=a} = \frac{12}{a^4}\left(\delta_{jk} n_i + \delta_{ki} n_j + \delta_{ij} n_k\right) - \frac{60}{a^4} n_i n_j n_k$$

and thus

$$e_{ij}\Big|_{r=a} = \frac{3}{4a}C_0 n_k U_k\left(\delta_{ij} - 3n_i n_j\right) - \frac{9}{2a}C_2 n_k U_k\left(\delta_{ij} - 5n_i n_j\right) - \frac{9}{2a}C_2\left(n_j U_i + n_i U_j\right)$$

and

$$e_{ij} n_i n_j\Big|_{r=a} = -\frac{3}{a} n_k U_k\left(\frac{1}{2}C_0 - 3C_2\right).$$

Hence

$$[(\mathbf{I}-\mathbf{nn}).(\mathbf{E.n})]_i\Big|_{r=a} = [\mathbf{E.n} - \mathbf{nn}.(\mathbf{E.n})]_i\Big|_{r=a} = -\frac{9}{2a}C_2 U_i + \frac{9}{2a}C_2 n_i n_k U_k \tag{3.3.15}$$

$$= \frac{9}{2a}C_2\left(n_i n_k U_k - U_i\right),$$

$$\left[(\mathbf{I}-\mathbf{nn}).\left(\hat{\mathbf{E}}.\mathbf{n}\right)\right]_i\Big|_{r=a} = \left[\hat{\mathbf{E}}.\mathbf{n} - \mathbf{nn}.\left(\hat{\mathbf{E}}.\mathbf{n}\right)\right]_i\Big|_{r=a} = \frac{D_2}{2a}\left(n_i n_k U_k + 3U_i\right) \tag{3.3.16}$$

$$-\frac{2D_2}{a} n_i n_k U_k = -\frac{3}{2a}D_2\left(n_i n_k U_k - U_i\right).$$

Substituting (3.3.15) and (3.3.16) into (1.4.17) we find the fourth equation for the constants C_0, C_2, D_0 and D_2:

$$3C_2 + \lambda D_2 = 0. \tag{3.3.17}$$

Solving the system (3.3.8), (3.3.11), (3.3.12) and (3.3.17) for the coefficients C_0, C_2, D_0 and D_2 we find (Zapryanov and Markov, 1994):

$$C_0 = \frac{2+3\lambda}{3(1+\lambda)}, \quad C_2 = \frac{\lambda}{6(1+\lambda)}, \quad D_0 = \frac{3+2\lambda}{2(1+\lambda)}, \quad D_2 = -\frac{1}{2(1+\lambda)} .$$

Consequently the equations (3.3.3) and (3.3.4) acquire the form

$$\mathbf{v} = 2\pi\,\mu\, a\,\frac{2+3\lambda}{(1+\lambda)}\left[1+\frac{\lambda\, a^2\nabla^2}{2(2+3\lambda)}\right]\mathbf{U}.\,\frac{\mathbf{B(r)}}{8\,\pi\,\mu}, \tag{3.3.18}$$

$$\hat{\mathbf{v}} = \frac{3+2\lambda}{2(1+\lambda)}\,\mathbf{U} - \frac{1}{2a^2(1+\lambda)}\,\mathbf{U}.\,\mathrm{H}(\mathbf{r}). \tag{3.3.19}$$

A notable feature of (3.3.18) is that in the limiting case of $\lambda \to \infty$ it assumes for a rigid particle the form given previously in (3.2.10), while in the limiting case of $\lambda \to 0$ the degenerate quadrupole vanishes and the Stokeslet provides alone the exact solution for a translating bubble.

In order to calculate the force on the fluid particle we have to integrate the surface stress vector over the drop surface

$$\mathbf{F} = \iint_S (\mathbf{T.n})\, d\sigma = -2\pi\,\mu\, a\,\mathbf{U}\left[\frac{2+3\lambda}{(1+\lambda)}\right]. \tag{3.3.20}$$

In the limiting case, $\lambda \to \infty$, this expression for the drag becomes $\mathbf{F} = -6\pi\,\mu\, a\,\mathbf{U}$, which is simply the Stokes law for the drag on a rigid spherical particle. In the limit $\lambda \to 0$, the expression (3.3.20) becomes

$$\mathbf{F} = -4\pi\,\mu\, a\,\mathbf{U}, \tag{3.3.21}$$

which is the drag on a spherical bubble at Re << 1.

3.4. Other Free Singularities of Stokes Flow. The Potential Sources, Stresslet and Rotlet

We shall start with the singularities caused by a point force in the point with radius - vector $\boldsymbol{\rho}\,(\xi_1,\ \xi_2,\ \xi_3)$. The first order singularity is the Stokeslet induced velocity and its corresponding pressure tensor, presented by (3.1.1) and (2.6.8), respectively. Higher order singularities will be constructed when taking the different order derivatives of the Stokeslet (3.1.1) and its pressure tensor (2.6.8). From their first derivatives with respect to $\boldsymbol{\rho}$ the new velocity field and pressure can be written in the form:

$$v_i = B_{ijk}\, d_{jk} = B^D_{ijk}\, d_{jk}, \qquad p = \mu\, P_{jk}\, d_{jk}, \tag{3.4.1}$$

where $\mathbf{d}$ is a second order matrix with constant elements,

$$B^D_{ijk} = \left[\nabla_\rho \mathbf{B}(\mathbf{r}-\boldsymbol{\rho})\right]_{ijk} = \frac{1}{R^3}\left(\delta_{ij}\hat{x}_k - \delta_{jk}\hat{x}_i - \delta_{ki}\hat{x}_j\right) + \frac{3\hat{x}_i\hat{x}_j\hat{x}_k}{R^5}, \tag{3.4.2}$$

$$P_{jk} = \left[\nabla_\rho\left(\frac{\mathbf{r}-\boldsymbol{\rho}}{R^3}\right)\right]_{jk} = -2\frac{\delta_{jk}}{R^3} + 6\frac{\hat{x}_j\hat{x}_k}{R^5},$$

and $\hat{\mathbf{r}} = \mathbf{r} - \boldsymbol{\rho}$, $R = |\mathbf{r} - \boldsymbol{\rho}|$.

The first of these formulae participating in the velocity expression is called a Stokeslet doublet. The corresponding to this flow stress is easily found

$$T_{il} = \mu K_{ijkl} d_{jk}, \tag{3.4.3}$$

where $K_{ijkl} = -6\left[\nabla_\rho\left(\frac{(\mathbf{r}-\boldsymbol{\rho})(\mathbf{r}-\boldsymbol{\rho})(\mathbf{r}-\boldsymbol{\rho})}{R^5}\right)\right]_{ijkl}$

$$= \frac{6}{R^5}\left(\delta_{il}\hat{x}_j\hat{x}_k + \delta_{jl}\hat{x}_i\hat{x}_k + \delta_{kl}\hat{x}_i\hat{x}_j\right) - 30\frac{\hat{x}_i\hat{x}_j\hat{x}_k\hat{x}_l}{R^7}.$$

If we denote the symmetric part of the constant matrix **d** by **s** and the antisymmetric part by **a**, then

$$\mathbf{s} = \frac{\mathbf{d}+\mathbf{d}^T}{2}, \qquad \mathbf{a} = \frac{\mathbf{d}-\mathbf{d}^T}{2}, \tag{3.4.4}$$

where the superscript T stands for the transposed matrix. By the use of the alternating tensor $\boldsymbol{\varepsilon}$ the antisymmetric part **a** can be given as

$$a_{jk} = -\frac{1}{2}\varepsilon_{jkm}L_m, \qquad L_m = -\varepsilon_{mjk}a_{jk}. \tag{3.4.5}$$

Finally, the velocity induced by the Stokeslet doublet expressed on (3.4.1) can be rewritten in the form

$$v_i = \left(-\frac{\delta_{jk}\hat{x}_i}{R^3} + 3\frac{\hat{x}_i\hat{x}_j\hat{x}_k}{R^5}\right)s_{jk} + \frac{1}{R^3}\left(\delta_{ij}\hat{x}_k - \delta_{ik}\hat{x}_j\right)a_{jk} = B^s_{ijk}s_{jk} + B^a_{im}L_m, \tag{3.4.6}$$

where the parts in brackets are denoted with B^s_{ijk} and B^a_{im}.

The flow having a singularity of the type $B^a_{im} = \varepsilon_{imk}\frac{\hat{x}_k}{R^3} = \frac{1}{2}\varepsilon_{mkj}\frac{\partial B_{ij}}{\partial \xi_k}$ is known as the rotlet or couplet. Examining carefully the symmetric part of (3.4.6), we note that the expression $-\frac{\delta_{jk}\hat{x}_i}{R^3}$ can be regarded as a velocity induced by a point source with potential $\Phi = \frac{1}{R}$, while the second term under the brackets defines a new singularity flow called the stresslet (see section 2.5), i.e.

$$v_i^{STR} = B_{ijk}^{STR}s_{jk}, \tag{3.4.7}$$

where

$$B_{ijk}^{STR} = 3\frac{\hat{x}_i\hat{x}_j\hat{x}_k}{R^5}. \tag{3.4.8}$$

It follows from (2.5.6) and (3.4.8), that the stresslet coincides with the stress induced by the Stokeslet with an accuracy up to a proportionality constant.

If in a Stokes flow the pressure is assumed constant, then its velocity will be a harmonic function, i.e., $\nabla^2\mathbf{v}=0$. The potential $\Phi=-\frac{1}{4\pi R}$ corresponds to a point source in the point $\boldsymbol{\rho}$. Then the induced velocity by the source will be $\mathbf{v} = s_1\boldsymbol{\Sigma}$, where $\Sigma_i=(\mathrm{grad}\Phi)_i=\frac{\hat{x}_i}{R^3}$ and $s_1=\frac{1}{4\pi}$. The first three derivatives of the point source with respect to the pole $\boldsymbol{\rho}$ are the potential dipole, the potential quadrupole and the potential octupole, correspondingly. The potential dipole velocity can be represented in the form $v_i = D_{ij}b_j$, where **b** is a constant vector and

$$D_{ij}=(\nabla_\rho\boldsymbol{\Sigma})_{ij}=-\frac{\delta_{ij}}{R^3}+3\frac{\hat{x}_i\hat{x}_j}{R^5}. \tag{3.4.9}$$

If we carefully inspect the formulae obtained in section 3.1, we can notice that the potential dipole velocity coincides up to a constant with the degenerate quadrupole given by (3.1.4), i.e., $D_{ij}=-\frac{1}{2}\nabla^2B_{ij}$. Therefore, in our further analysis we may use one instead of both singularities types.

Similarly, if we put $v_i=Q_{ijk}q_{jk}$ for the potential quadrupole velocity, then

$$Q_{ijk}=(\nabla_\rho\mathbf{D})_{ijk}=-\frac{3}{R^5}(\delta_{ij}\hat{x}_k+\delta_{ik}\hat{x}_j+\delta_{jk}\hat{x}_i)+15\frac{\hat{x}_i\hat{x}_j\hat{x}_k}{R^7}. \tag{3.4.10}$$

Let us now see how the considered singularities influence the general characteristics of the flow, such as the force, torque and the stresslet, and the flow rate through a closed surface S_0, whose interior domain contains some singularities.

The flow rate Q passing through S_0, with a point source or stresslet in its interior, is obtained by the formulae:

$$Q=4\pi s_1 \qquad \text{or} \qquad Q=4\pi s_{ii}. \tag{3.4.11}$$

The flow rates for the remaining singularities are equal to zero.

The only non-zero force exerted on S_0 by the flow is that induced by a Stokeslet with coefficient $8\pi\mu\mathbf{F}^{(c)}$, i.e.

$$\mathbf{F}=-8\pi\mu\mathbf{F}^{(c)}. \tag{3.4.12}$$

The torque due to all the singularities is zero except for the rotlet

$$\mathbf{M}=-8\pi\mu\mathbf{L}, \tag{3.4.13}$$

and for the Stokeslet doublet

$$\mathbf{M}=8\pi\mu\,\boldsymbol{\varepsilon}:\mathbf{d}. \tag{3.4.14}$$

The coefficient of the stresslet **S** is related only to the point force dipole and its value is

$$\mathbf{S}=-8\pi\mu\,\mathbf{s}. \tag{3.4.15}$$

In the spirit of the basic ideas of the singularity method and of the considerations done up to this moment, we can model a given flow using discrete or continuous distributed singularities of the type:

$$v_i(\mathbf{r}) = \iint_{S_\Sigma} \Sigma_i(\mathbf{r}, \boldsymbol{\rho})\, s_1 d\sigma(\rho) + \iint_{S_D} D_{ij}(\mathbf{r}, \boldsymbol{\rho})\, b_j d\sigma(\boldsymbol{\rho}) \qquad (3.4.16)$$

$$+ \iint_{S_Q} Q_{ijk}(\mathbf{r}, \boldsymbol{\rho})\, q_{jk} d\sigma(\boldsymbol{\rho}) + ... + \iint_{S_B} B_{ij}(\mathbf{r}, \boldsymbol{\rho})\, g_j d\sigma(\boldsymbol{\rho}) + \iint_{S_B^D} B_{ijk}^D(\mathbf{r}, \boldsymbol{\rho})\, d_{jk} d\sigma(\boldsymbol{\rho}) + ...,$$

where the surfaces S_Σ, S_D, S_Q, ... on which the singularities are distributed outside of the flow domain and $\mathbf{q}$ is a constant elements matrix. In the first group of integrals in the right-hand side of (3.4.16) point sources, point source dipoles, point source quadrupoles, etc. are included, while point forces, point force dipoles, etc. are in the second group. All the terms in (3.4.16) satisfy the Stokes equation and the continuity equation in the flow domain and on its boundaries, as well.

Taking into account that the point source dipoles, as well as the source singularities of higher-order can be expressed by means of the Laplacian or derivatives of the Green's function Laplacian, we can conclude that all the singularities of the first series of (3.4.16), except the first source term, can be included in the second series of the same formula.

The flow rate, due to the velocity given by (3.4.16), through a closed surface S_0 containing point sources and stresslets in its interior is

$$Q = 4\pi \iint_{S_0} (s_1 + s_{ii})\, d\sigma \qquad (3.4.17).$$

The force exerted on the surface S_0, which encloses Stokeslets, is equal to

$$\mathbf{F} = -8\pi\mu \iint_{S_0} \mathbf{F}^{(c)}\, d\sigma . \qquad (3.4.18)$$

3.5. A Rigid Spherical Particle in an Arbitrary Linear Shear Flow

Let an arbitrary linear shear flow pass a rigid spherical particle with undisturbed velocity at infinity $\mathbf{v}^\infty = \mathbf{A}.\mathbf{r}$, where $\mathbf{A}$ is a square matrix with constant elements and Trace($\mathbf{A}$) = 0. We should note again that there does not exist yet a systematic procedure for choosing the singularities necessary to solve any particulate problem. In order to make an educated guess, we are forced to rely on a combination of our intuition and the knowledge of the properties of the basic singularities. After looking carefully at the menu of the above cited (section 3.4.) singularities, we chose the Stokes dipole and the potential quadrupole with poles at the particle centre, in terms of which to represent the considered flow:

$$v_i(\mathbf{r}) = B_{ijk}^D(\mathbf{r}, \boldsymbol{\rho})\, d_{jk} + Q_{ijk}(\mathbf{r}, \boldsymbol{\rho})\, q_{jk}, \qquad (3.5.1)$$

where $B_{ijk}^D(\mathbf{r}, \boldsymbol{\rho})$ and $Q_{ijk}(\mathbf{r}, \boldsymbol{\rho})$ are given by (3.4.2) and (3.4.10), respectively.

Applying the boundary condition $\mathbf{v} = -\mathbf{A}.\mathbf{r}$ at r = a, where a is the particle radius, we obtain

$$\mathbf{d} = -\frac{5}{a^2}\mathbf{q}, \qquad \mathbf{A} = \frac{2}{a^5}\left[4\mathbf{q} - \mathbf{q}^T - \mathbf{I}\ \mathrm{Trace}(\mathbf{q})\right]. \tag{3.5.2}$$

Since Trace(**A**) = 0, then Trace(**q**) = 0, too. In order to calculate **q**, let us decompose both **A** and **q** into their symmetrical

$$\mathbf{A}^s = \frac{\mathbf{A} + \mathbf{A}^T}{2}, \qquad \mathbf{q}^s = \frac{\mathbf{q} + \mathbf{q}^T}{2},$$

and their antisymmetrical

$$\mathbf{A}^a = \frac{\mathbf{A} - \mathbf{A}^T}{2}, \qquad \mathbf{q}^a = \frac{\mathbf{q} - \mathbf{q}^T}{2}$$

components. Then

$$\mathbf{A}^s + \mathbf{A}^a = \frac{6\mathbf{q}^s}{a^5} + \frac{10\mathbf{q}^a}{a^5}. \tag{3.5.3}$$

Therefore

$$\mathbf{q}^s = \frac{a^5}{6}\mathbf{A}^s, \qquad \mathbf{q}^a = \frac{a^5}{10}\mathbf{A}^a. \tag{3.5.4}$$

Since $\mathbf{A}^s - \mathbf{A}^a = \mathbf{A}^T$, we have

$$\mathbf{q} = \frac{a^5}{30}\left(4\mathbf{A} + \mathbf{A}^T\right). \tag{3.5.5}$$

Applying (3.5.2) we can calculate the coefficient of the Stokeslet

$$\mathbf{d} = -\frac{a^3}{6}\left(4\mathbf{A} + \mathbf{A}^T\right). \tag{3.5.6}$$

Let us now decompose the Stokeslet dipole coefficient into a symmetrical component **s** and antisymmetrical component **a**:

$$\mathbf{s} = \frac{\mathbf{d} + \mathbf{d}^T}{2} = -\frac{5a^3}{12}\left(\mathbf{A} + \mathbf{A}^T\right), \quad \mathbf{a} = \frac{\mathbf{d} - \mathbf{d}^T}{2} = -\frac{a^3}{4}\left(\mathbf{A} - \mathbf{A}^T\right). \tag{3.5.7}$$

Using (3.4.5) the coefficient of the rotlet due to the Stokeslet dipole is found to be

$$L_m = -\varepsilon_{mjk} a_{jk} = \frac{1}{2} a^3 \varepsilon_{mjk} A_{jk}, \tag{3.5.8}$$

while the coefficient of the stresslet is obtained from (3.4.15)

$$\mathbf{S} = -8\pi\mu\ \mathbf{s} = \frac{10}{3}\pi\mu a^3\left(\mathbf{A} + \mathbf{A}^T\right). \tag{3.5.9}$$

The total force exerted on the sphere is equal to zero, but the torque given by (3.4.14) in this case will be

$$M_m = 8\pi\mu\,\varepsilon_{mjk} a_{jk} = -4\pi\mu a^3\ \varepsilon_{mjk} A_{jk}. \tag{3.5.10}$$

If the flow at infinity is purely straining, then **A** is a symmetric matrix, i.e., $\mathbf{A}^T = \mathbf{A} = \mathbf{E}$ and

$$\mathbf{q} = \frac{a^5}{6}\mathbf{E}, \qquad \mathbf{d} = \mathbf{s} = -\frac{5}{6}a^3\mathbf{E}, \qquad \mathbf{a} = \mathbf{L} = \mathbf{0}, \qquad \mathbf{S} = \frac{20}{3}\pi\mu a^3\mathbf{E}. \quad (3.5.11)$$

It is worth noting here, that the torque is also zero, i.e., the purely straining flow does not create any force or torque on the rigid sphere.

3.6. Faxen Laws for Rigid or Fluid Particles

The Faxen's investigations (see section 2.9) are generalised also for particles of arbitrary shape (Brenner, 1964b). On the basis of the considered multipole expansion of the velocity of the flow past a particle of arbitrary shape more general formulae of the type (2.9.10), (2.9.11) and (2.9.13) could be derived here. The problem that has to be solved is to find a method to calculate the generalised moments and the multipole expansion of the velocity of the disturbed flow, if the velocity of the undisturbed flow $\mathbf{v}^{\infty}(\mathbf{r})$ and the particle velocity $\mathbf{U}_p$ are given, i.e., to find the coefficients of this expansion. Such a method contributes again the reciprocal theorem for two appropriately chosen flows.

Hinch (1977) and Kim (1985b) have established a remarkable connection between the singularity solutions for the velocity and Faxen's laws for the force and torque in steady Stokes flows. They have shown the existence of a functional similarity between them. Kim (1985b) has demonstrated that the singularity representations for uniform stream, linear and higher-order flow yield Faxen's laws for force, torque, stresslet and higher-order moments in a direct manner. Furthermore, this duality appears even in two-phase problems, such as Stokes flow past a viscous drop suspended in different flows. Kim and Lu (1987) have shown explicitly that this duality originates from the Faxen relations being always the result of an integration involving generalized functions associated with the distribution of singularities of the exterior solution of the conjugate boundary value problem.

Rigid particle of arbitrary shape. In order to derive the generalized Faxen's formulae by use of the Lorentz theorem, we have to consider two flows $(\mathbf{v},\mathbf{T})$ and $(\mathbf{v}^*,\mathbf{T}^*)$, for the first of which the Stokes equation $\nabla.\mathbf{T} = \mathbf{0}$, and for the second the generalized Stokes equation $\nabla.\mathbf{T}^* = -\mathbf{F}^{(c)}\delta(\mathbf{r}-\mathbf{r}_1)$ are valid. We suppose that the first flow is generated by a slow translational motion of a solid particle in quiescent viscous fluid, while the second one is generated by the action of a concentrated force $\mathbf{F}^{(c)}$ in a point of a radius-vector $\mathbf{r}_1$ when the same particle is stationary and situated in the fluid flow (see Fig.3.6.1). Then from (2.3.25) or (2.3.27) we have for the second flow velocity $\mathbf{v}^*$:

$$\mathbf{v}^* = \frac{1}{8\pi\mu}\mathbf{F}^{(c)}.\mathbf{B}(\mathbf{r}-\mathbf{r}_1), \qquad (3.6.1)$$

where $\mathbf{r}_1$ is outside of the particle, while on it $\mathbf{v}^*|_{S_p} = 0$.

Let us write down the Lorentz theorem in its most general formulation as given by (2.4.9):

$$\iint_{S_0} \mathbf{v}^*.(\mathbf{T}.\hat{\mathbf{n}})\, d\sigma + \iiint_V \mathbf{v}^*.(\nabla.\mathbf{T}) d\tau = \iint_{S_0} \mathbf{v}.(\mathbf{T}^*.\hat{\mathbf{n}})\, d\sigma + \iiint_V \mathbf{v}.(\nabla.\mathbf{T}^*) d\tau \quad (3.6.2)$$

where V is the fluid volume bounded by the particle surface and a large surface at infinity, i.e., $S_0 = S_p + S_\infty$. Here the sign of the first terms of both sides of this equality is changed due to the use of the normal $\hat{\mathbf{n}}$ instead of the normal $\mathbf{n}$, i.e., the outer normal to S_0 is used, which is directed towards the fluid. In all further applications of the reciprocal theorem the velocity and stress fields decay at infinity and instead of S_0 the particle boundary S_p is supposed in (3.6.2).

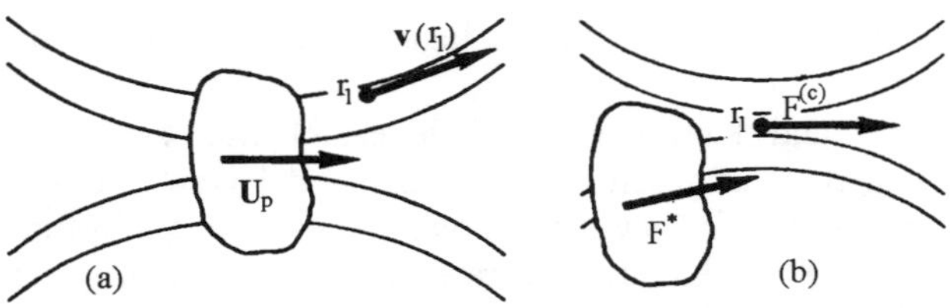

Fig.3.6.1. Two Stokes flows around: a) a translating particle; b) a fixed particle near a Stokeslet

Taking into account the conditions $\mathbf{v}^*|_{S_p} = 0$ and $\nabla.\mathbf{T} = \mathbf{0}$, $\nabla.\mathbf{T}^* = -\mathbf{F}^{(c)}\delta(\mathbf{r}-\mathbf{r}_1)$ in the volume V, from (3.6.2) we obtain:

$$\mathbf{U}_p.\iint_{S_p}(\mathbf{T}^*.\hat{\mathbf{n}})\, d\sigma - \iiint_V \mathbf{v}.\mathbf{F}^{(c)}\delta(\mathbf{r}-\mathbf{r}_1) d\tau = 0$$

or

$$\mathbf{U}_p.\mathbf{F}^* - \mathbf{v}(\mathbf{r}_1).\mathbf{F}^{(c)} = 0, \quad (3.6.3)$$

where $\mathbf{F}^*$ is the drag force caused by the velocity (3.6.1).

Due to the linearity of the Stokes equation and continuity equation, the velocity $\mathbf{v}(\mathbf{r}_1)$ participating in the equality (3.6.3) depends linearly on the translational particle velocity $\mathbf{U}_p$. Suppose that the velocity field of the first flow is constructed by the singularity method in the form:

$$\mathbf{v}(\mathbf{r}) = \mathbf{U}_p.\boldsymbol{\Phi}\left[\frac{\mathbf{B}(\mathbf{r}-\boldsymbol{\rho})}{8\pi\mu}\right], \quad (3.6.4)$$

where $\mathbf{\Phi}$ is a linear differential, integral or integro-differential functional, and the radius- vector $\boldsymbol{\rho}$ refers to points (point), lying inside the moving particle, over which the singularities are distributed. When the particle is spherical the velocity expression is given by the formula (3.2.10) and therefore the form of the functional is $\mathbf{\Phi} \equiv 6\pi\mu a\left(1+\frac{a^2}{6}\nabla^2\right)$ [see (2.9.10)]. In the general case we have

$$\mathbf{U}_p.\mathbf{F}^* - \mathbf{U}_p.\mathbf{\Phi}\left[\frac{\mathbf{F}^{(c)}.\mathbf{B}(\mathbf{r}_1-\boldsymbol{\rho})}{8\pi\mu}\right] = 0 .$$

Since $\mathbf{U}_p$ is arbitrary

$$F_i^* = \mathbf{\Phi}\left[\frac{F_j^{(c)}B_{ji}(\mathbf{r}_1-\boldsymbol{\rho})}{8\pi\mu}\right]. \tag{3.6.5}$$

However, the expression $F_j^{(c)}B_{ji}(\mathbf{r}_1-\boldsymbol{\rho}) = F_j^{(c)}B_{ij}(\boldsymbol{\rho}-\mathbf{r}_1)$ is calculated on the system of images $\mathbf{r} = \boldsymbol{\rho}$, lying inside the particle and describes the flow passing the particle. This means that the drag $\mathbf{F}^*$ is expressed by the operator $\mathbf{\Phi}$ taking part in the singular solution for the velocity of the particle in the flow

$$\mathbf{F}^* = \mathbf{\Phi}\left[\mathbf{v}^\infty(\boldsymbol{\rho})\right]. \tag{3.6.6}$$

Usually the singularities inside the particle are situated in its centre, i.e. $\boldsymbol{\rho} = 0$ in the Faxen 's law (3.6.6). The obtained results show that in order to find the drag when a fluid flow of velocity $\mathbf{v}^\infty$ passes a particle, first of all it is necessary to obtain the operator acting on the Oseen-Burgers tensor via the singularity method for which $\mathbf{v}^\infty(\boldsymbol{\rho})$ has to be substituted. This is true because the function $\dfrac{\mathbf{F}^{(c)}.\mathbf{B}(\mathbf{r}_1-\boldsymbol{\rho})}{8\pi\mu}$, calculated in the image system, is actually the flow velocity, i.e., $\mathbf{v}^\infty(\boldsymbol{\rho})$ for the fixed particle. In this respect, we are going to give two more examples. The first one concerns the Faxen law for the torque on a rigid particle.

Similarly to the particle translation formula (3.6.4), the disturbance velocity $\mathbf{v}$ due to particle rotation in a quiescent fluid can be expressed by means of the singularity distributions of Green's functions as

$$\mathbf{v}(\mathbf{r}) = \boldsymbol{\omega}.\mathbf{\Phi}\left[\frac{\mathrm{B}(\mathbf{r}-\boldsymbol{\rho})}{8\pi\mu}\right], \tag{3.6.7}$$

where $\boldsymbol{\omega}$ is the particle angular velocity of rotation.

From the singularity solution for a spherical particle rotating with angular velocity $\boldsymbol{\omega}$ (3.2.14) in a stationary fluid, the velocity $\mathbf{v}$ can be expressed by means of the Green's function, i.e., by the rotlet B_{im}^a as it was shown on (3.4.6) with the coefficient of the rotlet given on (3.5.8):

$$\mathrm{v_i} = 4\pi\mu \mathrm{a}^3 \varepsilon_{\mathrm{jkl}}\omega_{\mathrm{l}} \frac{\partial}{\partial \xi_{\mathrm{k}}} \frac{\mathrm{B_{ij}}(\mathbf{r}-\boldsymbol{\rho})}{8\pi\mu} = \mathrm{a}^3\omega_{\mathrm{l}}\varepsilon_{\mathrm{ikl}}\frac{\hat{\mathrm{x}}_{\mathrm{k}}}{\mathrm{R}^3}. \tag{3.6.8}$$

Proceeding as for the drag formula derivation, we can get the torque exerted on the particle when it remains stationary in an ambient flow of velocity $\mathbf{v}^{\infty}$

$$\mathbf{M}^* = \mathbf{\Phi}[\mathbf{v}^{\infty}(\boldsymbol{\rho})]. \tag{3.6.9}$$

So that due to (3.6.8) the torque on a spherical particle is:

$$\mathrm{M}_{\mathrm{l}}^* = 4\pi\mu \mathrm{a}^3 \varepsilon_{\mathrm{jkl}} \left.\frac{\partial \mathrm{v}_{\mathrm{j}}^{\infty}}{\partial \xi_{\mathrm{k}}}\right|_{\mathrm{r}=0} = 8\pi\mu \mathrm{a}^3 \Omega_{\mathrm{l}}^{\infty}\Big|_{\mathrm{r}=0}, \tag{3.6.10}$$

as $\mathbf{\Omega}^{\infty}(\mathbf{r}) = \frac{1}{2}\nabla \mathrm{x} \mathbf{v}^{\infty}(\mathbf{r})$ and functional $\mathbf{\Phi} \equiv 4\pi\mu \mathrm{a}^3 \nabla\times$. The formula derived for the torque is exactly the Faxen's law for the torque (2.9.11). When the spherical particle rotates with angular velocity $\boldsymbol{\omega}$ in the ambient flow $\mathbf{v}^{\infty}$, a superposition is applied to get (2.9.12).

The second example concerns the stresslet on a rigid particle. As mentioned by Batchelor (1974), the stresslet is the key suspension mechanical quantity in the expression for the effective viscosity of the suspensions and emulsions [see Chapter 10)] The Faxen law for the stresslet on a rigid spherical particle was first derived by Batchelor and Green (1972a).

If we assume that the particle is stationary in a purely straining flow $\mathbf{v}^{\infty} = \mathbf{E.r}$, then its velocity can be represented as

$$\mathbf{v}(\mathbf{r}) = \mathbf{E}{:}\mathbf{\Phi}\left[\frac{\mathrm{B}(\mathbf{r}-\boldsymbol{\rho})}{8\pi\mu}\right], \tag{3.6.11}$$

with $\mathbf{E} = \{\mathrm{e_{ij}}\}$ - a symmetric traceless tensor.

As we know, the velocity components for a spherical particle in a purely straining field are obtained from the formula (3.5.1) when having in view (3.5.11)

$$\mathrm{v_i} - \mathrm{e_{ij}x_j} = -\frac{20}{3}\pi\mu \mathrm{a}^3 \mathrm{e_{jk}}\left(1+\frac{\mathrm{a}^2}{10}\nabla^2\right)\frac{\partial}{\partial \xi_{\mathrm{k}}}\left(\frac{\mathrm{B_{ij}}}{8\pi\mu}\right). \tag{3.6.12}$$

As in the previous two cases, we will get the stresslet coefficient for a stationary particle in the purely straining flow $\mathbf{v}^{\infty}$ represented by the Faxen relation

$$\mathbf{S}^* = -\frac{1}{2}\left\{\mathbf{\Phi}[\mathbf{v}^{\infty}(\boldsymbol{\rho})] + \mathbf{\Phi}^{\mathrm{T}}[\mathbf{v}^{\infty}(\boldsymbol{\rho})]\right\}, \tag{3.6.13}$$

where the superscript T denotes the transposed operator $\mathbf{\Phi}$ with respect to the first two indices.

Here again we reach to the stresslet Faxen formula for a stationary spherical particle in a purely straining flow as presented by (2.9.13)

$$\mathrm{S}_{\mathrm{jk}}^* = \frac{20}{3}\pi\mu \mathrm{a}^3\left(1+\frac{\mathrm{a}^2}{10}\nabla^2\right)\mathrm{e}_{\mathrm{jk}}^{\infty}(\mathbf{r})\Big|_{\mathbf{r}=0}, \tag{3.6.14}$$

because $\boldsymbol{\rho} = 0$ and $e_{jk}^{\infty} = \frac{1}{2}\left(\frac{\partial v_j^{\infty}}{\partial \xi_k} + \frac{\partial v_k^{\infty}}{\partial \xi_j}\right)$. The correspondent operator $\boldsymbol{\Phi}$ is $\boldsymbol{\Phi} = -\frac{20}{3}\pi\mu a^3\left(1+\frac{a^2}{10}\nabla^2\right)\nabla$.

Fluid particle of arbitrary shape. Now we are going to discuss the more general case when the particle is fluid. We have to answer the question how the generalized moments of the multipole expansion of the disturbed fluid flow $\mathbf{v}^{D}$ can be computed with known velocity of the undisturbed flow $\mathbf{v}^{\infty}(\mathbf{r})$ and translational velocity of the fluid particle $\mathbf{U}_p$.

In order to prove the Faxen's theorem for a fluid particle of arbitrary shape we have to use again the general formulation of the Lorentz theorem (3.6.2) concerning two fluid flows $(\mathbf{v},\mathbf{T})$ and $(\mathbf{v}^*,\mathbf{T}^*)$ for the same particle.

At the first flow the fluid particle moves with velocity $\mathbf{U}_p$. This is a Stokes flow without any singularities and therefore $\nabla\cdot\mathbf{T} = \mathbf{0}$. If we take a point inside the flow with radius-vector $\mathbf{r}_1$, then its velocity will be $\mathbf{v}(\mathbf{r}_1)$ (see Fig.3.6.1).

At the second flow the fluid particle is motionless, but a concentrated force $\mathbf{F}^{(c)}$ acts in the point with radius-vector $\mathbf{r}_1$, which generates a flow with velocity of the type (3.6.1). In this case the flow has a singularity and because of that it is described by the generalized Stokes equation $\nabla\cdot\mathbf{T}^* = -\mathbf{F}^{(c)}\delta(\mathbf{r}-\mathbf{r}_1)$.

For these two flows the Lorentz theorem has the form:

$$\iint_{S_0}\mathbf{v}^*.(\mathbf{T}.\hat{\mathbf{n}})\,d\sigma = \iint_{S_0}\mathbf{v}.(\mathbf{T}^*.\hat{\mathbf{n}})\,d\sigma - \iiint_{V}\mathbf{v}.\mathbf{F}^{(c)}\delta(\mathbf{r}-\mathbf{r}_1)d\tau, \tag{3.6.15}$$

where $S_0 = S_{\infty} + S_p$ and $S_0 = \partial V$.

The kinematics conditions on the fluid particle surface for the first flow outside and inside the fluid particle, respectively, are:

$$\left(\mathbf{v}^{(o)} - \mathbf{U}_p\right).\hat{\mathbf{n}} = 0; \qquad \left(\mathbf{v}^{(i)} - \mathbf{U}_p\right).\hat{\mathbf{n}} = 0, \tag{3.6.16}$$

and for the second flow

$$\mathbf{v}^{*(o)}.\hat{\mathbf{n}} = 0; \qquad \mathbf{v}^{*(i)}.\hat{\mathbf{n}} = 0. \tag{3.6.17}$$

Since the considered volume V and its boundary S_0 are occupying the particle exterior, and due to the assumption that velocities decay at infinity, both integrals over S_{∞} tend to zero and the equation (3.6.15) is simplified to:

$$\iint_{S_p}\mathbf{v}^{*(o)}.(\mathbf{T}^{(o)}.\hat{\mathbf{n}})\,d\sigma + \mathbf{F}^{(c)}\cdot\mathbf{v}(\mathbf{r}_1) = \iint_{S_p}\mathbf{v}^{(o)}.(\mathbf{T}^{*(o)}.\hat{\mathbf{n}})\,d\sigma,$$

or

$$-\iint\limits_{S_p} \mathbf{v}^{*(o)}.(\mathbf{T}^{(o)}.\mathbf{n})\, d\sigma + \mathbf{F}^{(c)}\cdot\mathbf{v}(\mathbf{r}_1) = -\iint\limits_{S_p} \mathbf{v}^{(o)}.(\mathbf{T}^{*(o)}.\mathbf{n})\, d\sigma, \tag{3.6.18}$$

where $\mathbf{n} = -\hat{\mathbf{n}}$.

If we subtract the expression $\mathbf{U}_p \cdot \mathbf{F}^*$, where $\mathbf{F}^* = \iint\limits_{S_p} \mathbf{T}^{*(o)}.\hat{\mathbf{n}}\, d\sigma$, from both sides of the equation (3.6.18), then it takes the form:

$$-\mathbf{U}_p\cdot\mathbf{F}^* + \mathbf{F}^{(c)}\cdot\mathbf{v}(\mathbf{r}_1) = -\iint\limits_{S_p}\left(\mathbf{v}^{(o)} - \mathbf{U}_p\right).(\mathbf{T}^{*(o)}.\mathbf{n})\, d\sigma + \iint\limits_{S_p} \mathbf{v}^{*(o)}.(\mathbf{T}^{(o)}.\mathbf{n})\, d\sigma \tag{3.6.19}$$

We have to note that if the Lorentz theorem is applied for both flows inside the fluid particle, we reach to:

$$\iint\limits_{S_p}\left(\mathbf{v}^{(i)} - \mathbf{U}_p\right).(\mathbf{T}^{*(i)}.\mathbf{n})\, d\sigma = \iint\limits_{S_p} \mathbf{v}^{*(i)}.(\mathbf{T}^{(i)}.\mathbf{n})\, d\sigma. \tag{3.6.20}$$

Since $\mathbf{T}.\mathbf{n} = t_{nn}\mathbf{n} + t_{n\tau}\boldsymbol{\tau}$ and $\mathbf{T}^*.\mathbf{n} = t^*_{nn}\mathbf{n} + t^*_{n\tau}\boldsymbol{\tau}$, then

$$\left(\mathbf{v} - \mathbf{U}_p\right)\cdot\left(\mathbf{T}^*.\mathbf{n}\right) = \left(\mathbf{v} - \mathbf{U}_p\right)\cdot t^*_{nn}\mathbf{n} + \left(\mathbf{v} - \mathbf{U}_p\right)\cdot t^*_{n\tau}\boldsymbol{\tau}, \mathbf{v}^*\cdot(\mathbf{T}.\mathbf{n}) = \mathbf{v}^*\cdot t_{nn}\mathbf{n} + \mathbf{v}^*\cdot t_{n\tau}\boldsymbol{\tau}.$$

According to the kinematic conditions (3.6.16) and (3.6.17) the first terms on the right-hand sides of the last two equations are equal to zero. Therefore, only the tangential components of the stress vectors of both flows remain in (3.6.20). However, the continuity of the tangential velocity vectors and the continuity of the tangential components of the stress vectors is fulfilled on the fluid particle surface, i.e., $\mathbf{v}^{*(o)}.\boldsymbol{\tau} = \mathbf{v}^{*(i)}.\boldsymbol{\tau}$ and $t^{*(o)}_{n\tau} = t^{*(i)}_{n\tau}$ on S_p. Then from (3.6.20) it follows that the left-hand side of (3.6.19) is equal to zero, i.e., the following equality holds:

$$-\mathbf{U}_p\cdot\mathbf{F}^* + \mathbf{F}^{(c)}\cdot\mathbf{v}(\mathbf{r}_1) = 0. \tag{3.6.21}$$

As the velocity $\mathbf{v}(\mathbf{r})$ is known from the first flow (3.6.4), i.e.,

$$\mathbf{F}^{(c)}\cdot\mathbf{v}(\mathbf{r}) = \mathbf{U}_p.\boldsymbol{\Phi}\left[\frac{\mathbf{F}^{(c)}.\mathbf{B}(\mathbf{r}-\boldsymbol{\rho})}{8\pi\mu}\right],$$

and according to (3.6.21) the force $\mathbf{F}^*$ for the second flow satisfies the equation

$$-\mathbf{U}_p.\mathbf{F}^* + \mathbf{U}_p.\boldsymbol{\Phi}\left[\frac{\mathbf{F}^{(c)}.\mathbf{B}(\mathbf{r}_1-\boldsymbol{\rho})}{8\pi\mu}\right] = 0.$$

The velocity of the second flow passing the fluid particle is $\mathbf{v}^\infty(\boldsymbol{\rho}) = \dfrac{\mathbf{F}^{(c)}.\mathbf{B}(\mathbf{r}-\boldsymbol{\rho})}{8\pi\mu}$.

Therefore, the force $\mathbf{F}^*$ of the second flow is expressed by means of the same operator $\boldsymbol{\Phi}$ as the first flow velocity **v**, but now it is acting on the velocity $\mathbf{v}^\infty(\boldsymbol{\rho})$, i.e.,

$$\mathbf{F}^* = \boldsymbol{\Phi}\left[\mathbf{v}^\infty(\boldsymbol{\rho})\right], \tag{3.6.22}$$

which is known as the drag force Faxen's law for a fluid particle and its functional form is the same as for a rigid particle (3.6.6).

For a spherical drop the singularity velocity solution is given by the formula (3.3.18) and the operator $\mathbf{\Phi}$ has the form

$$\mathbf{\Phi} = 2\pi\,\mu\,\mathrm{a}\left(\frac{2+3\lambda}{1+\lambda}\right)\left(1+\frac{\lambda\,\mathrm{a}^2}{2(2+3\lambda)}\nabla^2\right). \qquad (3.6.23)$$

This result was first shown by Hetsroni and Haber (1970).

The Faxen's law for the stresslet on a drop was first presented by Rallison (1978b). To avoid the lengthy derivation, here we shall give only its final form. Utilising the fundamental definition of the stresslet S^* (Kim and Lu, 1987)

$$\mathbf{S}^* = \frac{1}{2}\iint_{S_p}\left[\mathbf{r}\left(\mathbf{T}^*.\mathbf{n}\right)+\left(\mathbf{T}^*.\mathbf{n}\right)\mathbf{r}-2\mu\left(\mathbf{v}^*\mathbf{n}+\mathbf{n}\mathbf{v}^*\right)\right]d\sigma, \qquad (3.6.24)$$

where $\mathbf{S}^*$ participates in the general expression

$$\mathbf{E}^\infty:\mathbf{S}^* = \frac{1}{2}\mathbf{v}^{\mathrm{D}}\left(\mathbf{r}_1\right).\mathbf{F}^{(c)}, \qquad (3.6.25)$$

the stresslet formula is found as $\mathbf{S}^* = -\frac{1}{2}\left\{\mathbf{\Phi}\left[\mathbf{v}^\infty(\boldsymbol{\rho})\right]+\mathbf{\Phi}^{\mathrm{T}}\left[\mathbf{v}^\infty(\boldsymbol{\rho})\right]\right\}$, which is absolutely identical with the rigid particle stresslet formula (3.6.13).

Taking into account the spherical drop velocity formula (see section 6.2)

$$\mathbf{v}^{\mathrm{D}} = 2\pi\,\mu\,\mathrm{a}^3\mathbf{E}^\infty:\nabla\left[\frac{2(5\lambda+2)}{3(\lambda+1)}+\frac{\lambda\,\mathrm{a}^2}{3(\lambda+1)}\nabla^2\right]\frac{\mathbf{B}(\mathbf{r})}{8\pi\,\mu} = \mathbf{E}^\infty:\mathbf{\Phi}\left[\frac{\mathbf{B}(\mathbf{r}-\boldsymbol{\rho})}{8\pi\,\mu}\right] \qquad (3.6.26)$$

and the stresslet generalized Faxen's formula, we get the following stresslet law of Faxen's laws type

$$S^*_{jk} = \frac{4}{3}\pi\,\mu\,\mathrm{a}^3\left(\frac{2+5\lambda}{1+\lambda}\right)\left[1+\frac{\lambda\,\mathrm{a}^2\nabla^2}{2(2+5\lambda)}\right]e^\infty_{jk}(\mathbf{r})\Big|_{\mathbf{r}=0}. \qquad (3.6.27)$$

In the limit $\lambda \to \propto$ the upper formula reduces to the rigid spherical particle stresslet given by (3.6.14).

3.7. A Translating Prolate Spheroid in a Uniform Viscous Flow

In the previous sections we have met with the creeping flow problems treated by the singularity method. This method is quite suitable also for non-spherical bodies, because it does not need to exploit any other coordinate system different from the Cartesian, cylindrical or spherical. The problems of prolate spheroids and other non-spherical bodies solved by internal singularities distributions are discussed in several papers by Chwang and Wu Yao-Tsu (1974, 1975) and Kim (1986).

Here we shall consider the Stokes flow solution for a prolate spheroid in a uniform flow. In Cartesian coordinates $\mathbf{r}$ = (x, y, z) the spheroid surface is described by the equation

$$\frac{x^2}{a^2}+\frac{s^2}{b^2}=1, \tag{3.7.1}$$

where $s^2=y^2+z^2$ and $a \geq b$ are the semimajor and semiminor axes, respectively.

The undisturbed flow $\mathbf{v}^\infty$ is taken in the plane (x, y), in which the ellipse, rotating around its semimajor axis, is lying:

$$\mathbf{v}^\infty = U_1\mathbf{i}+U_2\mathbf{j} \qquad \text{at } |\mathbf{r}| \to \infty.$$

The semimajor axis value a is taken as a characteristic length scale, while the velocity U_1 - as a characteristic velocity scale. Then for the dimensionless undisturbed velocity

$$\mathbf{v}^\infty = \mathbf{i}+\frac{U_2}{U_1}\mathbf{j} \qquad \text{at } |\mathbf{r}| \to \infty. \tag{3.7.2}$$

The Stokes flow solutions for non-spherical bodies are usually sought as distributions of singularities on the bodies' surfaces as discussed in section 3.5., while for prolate spheroids we shall utilise a line distribution of singularities between the two foci, along the axis of symmetry. The dimensionless spheroid foci are defined by $x = \pm c$, where

$$c=\left(1-\frac{b^2}{a^2}\right)^{1/2} \tag{3.7.3}$$

and $e = c$ is the spheroid eccentricity with $0 \leq e < 1$. As shown in section 3.5., in order to have a net force on the body, Stokeslets must be distributed along an internal line (for the spheroid this is the axial line between the two foci). As for the solid sphere case (see section 3.2.), in order to satisfy the zero velocity boundary condition on the spheroid, a superposition of the Stokeslets and potential dipoles is usually proposed:

$$\mathbf{v}=\mathbf{v}^\infty-\int_{-c}^{c}\left[\alpha_1 v_x^B(\mathbf{r}-\boldsymbol{\rho})\mathbf{i}+\alpha_2 v_y^B(\mathbf{r}-\boldsymbol{\rho})\mathbf{j}\right]d\xi_1 \tag{3.7.4}$$

$$+\int_{-c}^{c}\left(c^2-\xi_1^2\right)\left[\beta_1 v_x^D(\mathbf{r}-\boldsymbol{\rho})\mathbf{i}+\beta_2 v_y^D(\mathbf{r}-\boldsymbol{\rho})\mathbf{j}\right]d\xi_1$$

and

$$p=-\int_{-c}^{c}\left[\alpha_1 p^B(\mathbf{r}-\boldsymbol{\rho})\mathbf{i}+\alpha_2 p^B(\mathbf{r}-\boldsymbol{\rho})\mathbf{j}\right]d\xi_1, \tag{3.7.5}$$

where $\boldsymbol{\rho}=\xi_1\mathbf{i}$, the indices B and D denote the Stokeslet and potential dipole components, respectively given by the Oseen-Burgers tensor (3.1.1), (2.3.25), (3.4.9) and $p^D \equiv 0$.

The Stokes flow velocity assumed in the form (3.7.4) satisfies the infinity boundary condition (3.7.2). The no-slip boundary condition on spheroid surface gives additional conditions for the constants α_1, α_2, β_1 and β_2. After performing the integration in (3.7.4), $\mathbf{v}$ is expressed by

$$\mathbf{v} = \mathbf{v}^{\infty} - (2\alpha_1\mathbf{i} + \alpha_2\mathbf{j})\gamma_1 - (\alpha_1 s\mathbf{i}_s + \alpha_2 y\mathbf{i})\left(\frac{1}{R_2} - \frac{1}{R_1}\right) \tag{3.7.6}$$

$$+ (\alpha_1 s\mathbf{i} - \alpha_2 y\mathbf{i}_s)s\gamma_3 + \nabla\left[-2\beta_1\gamma_2 + \beta_2 y\left(\frac{x-c}{s^2}R_1 - \frac{x+c}{s^2}R_2 + \gamma_1\right)\right],$$

where $\mathbf{i}_s = \frac{y}{s}\mathbf{j} + \frac{z}{s}\mathbf{k}$ is the radial unit vector in the (y, z) plane,

$$R_1 = \left[(x+c)^2 + s^2\right]^{1/2},\ R_2 = \left[(x-c)^2 + s^2\right]^{1/2},\ \gamma_1 = \ln\frac{R_2 - (x-c)}{R_1 - (x+c)},\ \gamma_2 = R_2 - R_1 + x\gamma_1,$$

$$\gamma_3 = \frac{1}{s^2}\left(\frac{x+c}{R_1} - \frac{x-c}{R_2}\right).$$

Taking into account that on the spheroid surface the above subsidiary formulae change to

$$s^2 = (1-e^2)(1-x^2),\ R_1 = 1+ex,\ R_2 = 1-ex,\ \gamma_1 = \ln\frac{1+e}{1-e} = \lambda_e,$$

$$\gamma_3 = \frac{2e}{(1-e^2)(1-e^2x^2)},$$

we substitute them in the expression for $\mathbf{v}$ (3.7.6) evaluated on the spheroid surface (3.7.1). The no-slip boundary condition is satisfied, if the unknown coefficients are given by

$$\alpha_1 = \frac{2\beta_1 e^2}{(1-e^2)} = e^2\left[-2e + (1+e^2)\lambda_e\right]^{-1}, \tag{3.7.7}$$

$$\alpha_2 = \frac{2\beta_2 e^2}{(1-e^2)} = 2\frac{U_2}{U_1}e^2\left[2e + (3e^2-1)\lambda_e\right]^{-1}.$$

The force exerted on the spheroid by the fluid flow is only due to the Stokeslets distributed along the central line of the spheroid and its dimensionless form is

$$\mathbf{F} = 8\pi\int_{-c}^{c}\left[\alpha_1\mathbf{i} + \alpha_2\mathbf{j}\right]dx, \tag{3.7.8}$$

which can be rewritten with respect to (3.7.7) as

$$\mathbf{F} = 6\pi\left[C_{F_1}\mathbf{i} + \frac{U_2}{U_1}C_{F_2}\mathbf{j}\right], \tag{3.7.9}$$

where

$$C_{F_1} = \frac{8}{3}e^3\left[-2e + (1+e^2)\ln\frac{1+e}{1-e}\right]^{-1} \tag{3.7.10}$$

and

$$C_{F_2} = \frac{16}{3}e^3\left[2e + (3e^2-1)\ln\frac{1+e}{1-e}\right]^{-1} \tag{3.7.11}$$

are the drag coefficients.

Then the dimensional drag force is

$$\mathbf{F} = 6\pi\mu a\left[U_1 C_{F_1}\mathbf{i} + U_2 C_{F_2}\mathbf{j}\right], \tag{3.7.12}$$

which reduces to the Stokes law for the limiting case of a rigid sphere, i.e., $e \to 0$ and $C_{F_1} = C_{F_2} = 1$.

3.8. Approximate Solutions of Stokes Equations via Slender-Body Theory

As discussed in the previous several sections, the singularity superposition method is very powerful for simple flows, while for more complicated flows there is no receipt what type of singularities are necessary and how they should be distributed on the body surface or inside the body, or somewhere else in the flow region. Therefore, this difficulty reduces the applicability spectrum of the singularity method. In this section we shall deal with the creeping flow problem around a slender-body with arbitrary cross section that is very small compared to the body length. As in the prolate spheroid case, here again the Stokeslets distribution along the centreline is proposed.

We assume that the distance between the body surface and the centreline is very small of order ε, where $\varepsilon \ll 1$ is a small parameter. A detailed explanation will be given when considering the slender rigid particle sketch in Fig.3.8.1. The axis x_1 of the coordinate system $Ox_1x_2x_3$ is directed along the particle length. Let the particle length be equal to $2l$. In the case of a rotational particle the cross section perpendicular to the axis x_1 has a circular shape, while in the general case it is arbitrary. The distance between the surface points and the axis x_1 is denoted by s, where $s^2 = x_2{}^2 + x_3{}^2$. For the different surface points s is a function of x_1, as well as of the azimuthal angle θ with respect to the axis x_1. Taking into account that the particle is long and thin, for the particle surface equation we have in dimensionless form:

$$s = \varepsilon\, l\varphi(x_1,\ \theta), \tag{3.8.1}$$

where $\varphi(x_1,\ \theta)$ is a dimensionless function. If for example, the particle is a prolate spheroid $\varepsilon = \frac{b}{a}$, where a and b are the semimajor and semiminor axes of the spheroid. In this case the function φ is independent of θ.

As shown in section 2.5., it stems from formula (2.5.12) that the flows past a rigid particle can be represented by a single layer potential alone. We shall rewrite it in the dimensionless form

$$\mathbf{v}(\mathbf{r}) = \mathbf{v}^{\infty}(\mathbf{r}) - \frac{1}{8\pi}\iint_{S_p}\left[\frac{\mathbf{I}}{R} + \frac{(\mathbf{r}-\boldsymbol{\rho})(\mathbf{r}-\boldsymbol{\rho})}{R^3}\right].\mathbf{t}\, d\sigma(\boldsymbol{\rho}), \tag{3.8.2}$$

where S_p is the particle surface, the stress vector $\mathbf{t} = \mathbf{T}.\ \mathbf{n}$ is the searched distribution density of the Stokeslets on the particle surface, $\mathbf{v}^{\infty}(\mathbf{r})$ is the undisturbed flow velocity at infinity, and $R = |\mathbf{r} - \boldsymbol{\rho}|$ is the length of the vector $\mathbf{r} - \boldsymbol{\rho}$. Here the vector $\mathbf{r}$ is the radius-vector of an arbitrary flow point, while $\boldsymbol{\rho}$ is the radius-vector of a particle surface point.

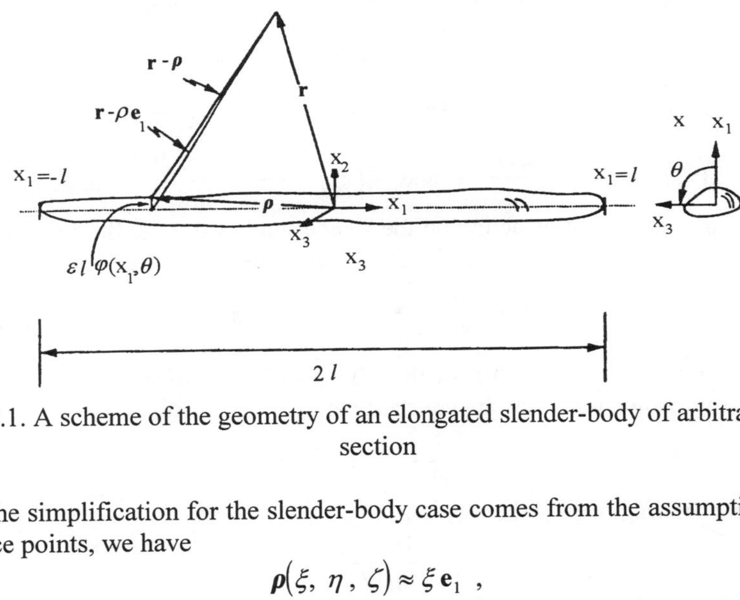

Fig.3.8.1. A scheme of the geometry of an elongated slender-body of arbitrary cross section

The simplification for the slender-body case comes from the assumption that for the surface points, we have

$$\boldsymbol{\rho}(\xi,\ \eta,\ \zeta) \approx \xi\, \mathbf{e}_1\ , \tag{3.8.3}$$

where $-1 \leq \xi \leq 1$. Then $\mathbf{r} - \boldsymbol{\rho} \approx \mathbf{r} - \xi\, \mathbf{e}_1 = (x_i - \xi\, \delta_{i1})\mathbf{e}_i$ and instead of (3.8.2) the approximation expression is

$$v_i(\mathbf{r}) = v_i^{\infty}(\mathbf{r}) - \frac{1}{8\pi}\int_{-1}^{1}\left[\frac{\delta_{ij}}{R_1} + \frac{(x_i - \xi\,\delta_{i1})(x_j - \xi\,\delta_{j1})}{R_1^3}\right] f_j(\xi)\, d\xi\ , \tag{3.8.4}$$

where $R_1 = \sqrt{(x_1 - \xi)^2 + x_2^2 + x_3^2}$, and $f_j(\xi)$ is the integral of the local surface stress vector $t_j(\xi)$ around the closed cross-section perimeter of the particle at fixed $x_1 = \xi$.

We have to note that the error in the approximate formula (3.8.4) with respect to the exact one (3.8.2) depends on the geometry of the particle ends; for rounded ends Cox (1970, 1971) proved that the error is of the order of $\varepsilon^2 \ln \varepsilon$.

From the no-slip boundary condition on the particle surface $\mathbf{v}|_{S_p} = 0$, we get the integral equations defining the unknown functions $f_j(\xi)$ (j = 1, 2, 3), which are connected with the density distribution of the Stokeslet on S_p:

$$v_i^{\infty}(\mathbf{r}) = \frac{1}{8\pi} \int_{-1}^{1} \left[\frac{\delta_{ij}}{R_1} + \frac{(x_i - \xi \delta_{i1})(x_j - \xi \delta_{j1})}{R_1^3} \right] f_j(\xi)\, d\xi, \tag{3.8.5}$$

where $\mathbf{r} \in S_p$.

Applying the limiting procedure $\varepsilon \to 0$ to both integrals in (3.8.5) and after some calculations we reach to the approximation of v_i^{∞} on the centreline:

$$v_i^{\infty}(x_1,0,0) = -\frac{\ln \varepsilon}{4\pi} \left[f_i(x_1) + \delta_{i1} f_1(x_1) \right] + O(1). \tag{3.8.6}$$

The unknown functions $f_i(x_1)$ can be determined if the value of $v_i^{\infty}(x_1,0,0)$ is known:

$$f_1(x_1) = -\frac{2\pi v_1^{\infty}(x_1,0,0)}{\ln \varepsilon} \quad \text{and} \quad f_i(x_1) = -\frac{4\pi v_i^{\infty}(x_1,0,0)}{\ln \varepsilon} \quad \text{at } i = 2, 3. \tag{3.8.7}$$

The dimensionless drag force acting on the slender-body from the fluid side is obtained in a similar manner as in section 3.7. for the prolate spheroid case (3.7.8):

$$\mathbf{F} = \int_{-1}^{1} \left[f_1 \mathbf{e}_1 + f_2 \mathbf{e}_2 + f_3 \mathbf{e}_3 \right] dx_1 . \tag{3.8.8}$$

For constant flow velocity at infinity, parallel to the slender-body axis, i.e., $\mathbf{v}^{\infty} = \mathbf{e}_1$, the unknown functions are constant too:

$$f_1 = -\frac{2\pi}{\ln \varepsilon}, \qquad f_2 = f_3 = 0 . \tag{3.8.9}$$

The dimensional drag force is then

$$\mathbf{F} = \frac{4\pi \mu l U}{\ln 1/\varepsilon} \mathbf{e}_1, \tag{3.8.10}$$

where U is the characteristic velocity, e.g., the magnitude of the dimensional velocity at infinity.

Furthermore, if the flow velocity at infinity, perpendicular to the slender-body axis, i.e., $\mathbf{v}^{\infty} = \mathbf{e}_1$, then

$$f_2 = -\frac{4\pi}{\ln \varepsilon}, \qquad f_1 = f_3 = 0 \tag{3.8.11}$$

and the dimensional drag is

$$\mathbf{F} = \frac{8\pi \mu l U}{\ln 1/\varepsilon} \mathbf{e}_1 . \tag{3.8.12}$$

It is evident from (3.8.10) and (3.8.12) that the drag force for the perpendicular flow motion with respect to the slender-body axis is twice bigger than the drag force for the parallel flow motion. Moreover, this result is confirmed by the formulae (3.7.10) - (3.7.12) obtained for the prolate spheroid in the limiting case of $b/a \to 0$ ($e \to 1$).

Hancock (1953) and Gray and Hancock (1955) have represented the flow due to a moving flagellum in terms of distributions of point forces and potential dipoles. A comprehensive account and extensions of their solutions are given by Lighthill (1975).

3.9. Image System for the Stokeslet near a Rigid Plane or Fluid/Fluid Interface

In the present section the Green's function for a semi-infinite flow bounded by a rigid wall or by a fluid/fluid interface will be examined. It appears that this problem, i.e., the problem of the fluid motion due to a point force in the presence of a wall or interface can successfully approach the problem of a flow generated by the motion of a rigid sphere near a wall or interface, when the sphere radius is much smaller than the distance between its centre and the wall or interface.

The treatment of the stated rigid wall problem started with the pioneering work of Lorentz (1907), who used it as an example of his reciprocal theorem. He obtains the general solution of the Stokes equation for a sphere approaching a plane wall with the appropriate boundary conditions on the wall in terms of velocity components and pressure, which corresponds exactly to the image of the free-space point force solution, i.e., Stokeslet, with respect to the wall. First, we shall present the generalized Lorentz' solution for arbitrary motion of a sphere near a plane wall in the form proposed by Blake (1971), who shows that the correspondent Green's function $\mathbf{G}^{W}$ consists of a Stokeslet located at the point $\boldsymbol{\rho} = (-h, 0, 0)$ together with some image singularities acting in the image point $\boldsymbol{\rho}^{IM}(h, 0, 0)$ of the Stokeslet with respect to the wall: a Stokeslet $\mathbf{B}$ (2.3.24), a potential dipole $\mathbf{D}$ (3.4.9) and a Stokeslet dipole $\mathbf{B}^{D}$ (3.4.2), i.e.,

$$\mathbf{G}^{W}(\mathbf{r},\boldsymbol{\rho}) = \mathbf{B}(\hat{\mathbf{r}}) - \mathbf{B}(\hat{\mathbf{r}}^{IM}) \pm 2h^2\mathbf{D}(\hat{\mathbf{r}}^{IM}) \pm 2h\mathbf{e}_1.\mathbf{B}^{D}(\hat{\mathbf{r}}^{IM}), \qquad (3.9.1)$$

where $\mathbf{r} = (x_1, x_2, x_3)$ is the observation point, $\hat{\mathbf{r}} = \mathbf{r} - \boldsymbol{\rho}$, $\hat{\mathbf{r}}^{IM} = \mathbf{r} - \boldsymbol{\rho}^{IM}$. The upper sign of $\pm$ corresponds to the second index of all tensors in (3.9.1) equal to 1, while the lower to 2 and 3. Here the wall coincides with the x_2x_3 plane and is given by the equation $x_1 = 0$, and $\mathbf{e}_1$ is the unit vector perpendicular to the wall (see Fig.3.9.1). The velocity field due to the wall Green's function $\mathbf{G}^{W}$ creates a pressure field vector $\mathbf{p}^{W}$, which is a superposition of the pressure fields related to each singularity in (3.9.1)

$$\mathbf{p}^{W}(\mathbf{r},\boldsymbol{\rho}) = \mathbf{p}(\hat{\mathbf{r}}) - \mathbf{p}(\hat{\mathbf{r}}^{IM}) \pm 2h\mathbf{P}(\hat{\mathbf{r}}^{IM}).\mathbf{e}_1, \qquad (3.9.2)$$

where $\mathbf{p}$ is the pressure field induced by the Stokeslet and given by (2.3.26), while $\mathbf{P}$ corresponds to the Stokeslet doublet (3.4.2) and the potential dipole, as noted in section 3.4, gives no contribution to the pressure field.

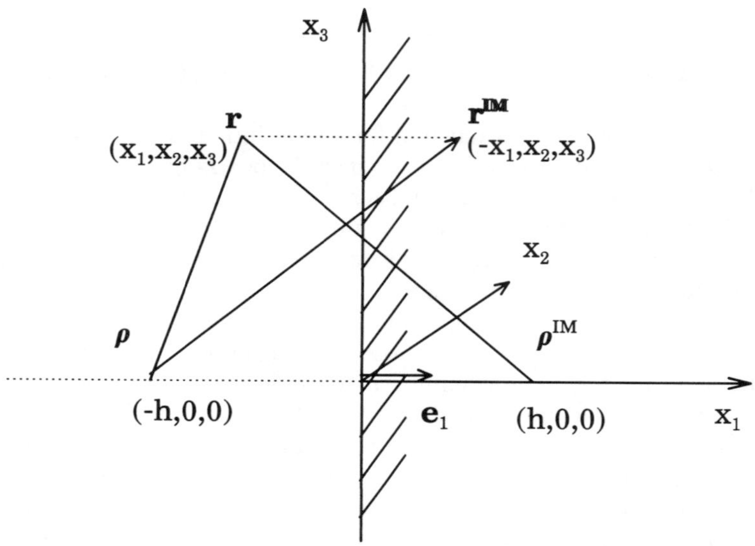

Fig.3.9.1. Schematic representation of the image system of a Stokeslet near a plane wall or interface.

Exploiting the symmetry property of the free-space Green's function, the expressions (3.9.1) and (3.9.2) can be rewritten in a form similar to the original form given by Lorentz and applied by other authors (Kim and Karrila, 1991, Ch. 12), (Lee et al., 1979). The theory developed in the latter work will be of use for the fluid/fluid interface problem, but as a particular case the rigid wall solution is obtained. Then, if the observation point image is denoted by $\mathbf{r}^{IM} = (-x_1, x_2, x_3)$, as shown on Fig.3.9.1., the alternative solution is:

$$\mathbf{G}^{W}(\mathbf{r},\boldsymbol{\rho}) = \mathbf{B}(\hat{\mathbf{r}}) - \mathbf{B}(\hat{\mathbf{r}}^{IM}) + h^2\nabla\mathbf{p}(\hat{\mathbf{r}}^{IM}) \pm 2h\nabla(\mathbf{B}(\hat{\mathbf{r}}^{IM}).\mathbf{e}_1) \tag{3.9.3}$$

$$= \mathbf{B}(\hat{\boldsymbol{\rho}}) + \hat{\mathbf{B}}(\hat{\boldsymbol{\rho}}^{\,IM});$$

$$\mathbf{p}^{W}(\mathbf{r},\boldsymbol{\rho}) = \mathbf{p}(\hat{\mathbf{r}}) - \mathbf{p}(\hat{\mathbf{r}}^{IM}) \mp 2h\nabla\mathbf{p}(\hat{\mathbf{r}}^{IM}).\mathbf{e}_1 = \mathbf{p}(\hat{\boldsymbol{\rho}}) + \hat{\mathbf{p}}(\hat{\boldsymbol{\rho}}^{IM}), \tag{3.9.4}$$

where $\hat{\boldsymbol{\rho}} = \boldsymbol{\rho} - \mathbf{r}$, $\hat{\boldsymbol{\rho}}^{\,IM} = \boldsymbol{\rho} - \mathbf{r}^{IM}$ and

$$\hat{\mathbf{B}}(\hat{\boldsymbol{\rho}}^{\,IM}) = \mathcal{L}_v(\mathbf{B}^*), \quad \hat{\mathbf{p}}(\hat{\boldsymbol{\rho}}^{IM}) = \mathcal{L}_p(\mathbf{p}^*). \tag{3.9.5}$$

The operators $\mathcal{L}_v$ and $\mathcal{L}_p$ are the respective Lorentz' operators for velocity and pressure given by

$$\mathcal{L}_v(\mathbf{B}^*) = -\mathbf{J}.\mathbf{B}^*(\hat{\boldsymbol{\rho}}^{IM}) - 2x_1\nabla\left[\left(\mathbf{B}^*(\hat{\boldsymbol{\rho}}^{IM}).\mathbf{e}_1\right)\right] + x_1^2\nabla^2\left[\mathbf{B}^*(\hat{\boldsymbol{\rho}}^{IM})\right], \tag{3.9.6}$$
$$\mathcal{L}_p(\mathbf{p}^*) = \mathbf{p}^*(\hat{\boldsymbol{\rho}}^{IM}) + 2x_1\nabla\mathbf{p}^*(\hat{\boldsymbol{\rho}}^{IM}).\mathbf{e}_1 - 4\nabla\left[\left(\mathbf{B}^*(\hat{\boldsymbol{\rho}}^{IM}).\mathbf{e}_1\right)\right].\mathbf{e}_1 .$$

In the upper expressions the operator $\mathbf{J} = \mathbf{I} - 2\mathbf{e}_1\mathbf{e}_1$ is called the mirror reflection operator as $\mathbf{I}$ is the identity tensor. Then, the pair ($\mathbf{B}^*$, $\mathbf{p}^*$) is the reflected image solution of ($\mathbf{B}$, $\mathbf{p}$), when the mirror reflection operator is applied to it, i.e.,

$$\mathbf{B}^* = \mathbf{J}.\mathbf{B}(\hat{\boldsymbol{\rho}}^{IM}) \text{ and } \mathbf{p}^* = \mathbf{p}(\hat{\boldsymbol{\rho}}^{IM}). \tag{3.9.7}$$

Using the notation $R^{IM} = \left[(h - x_1)^2 + x_2^2 + x_3^2\right]^{1/2}$ and the formulae for the Stokeslet (2.3.24) and its associated pressure vector (2.3.26), the wall Green's function $\mathbf{G}^W$ and the wall pressure vector $\mathbf{p}^W$ finally become:

$$\mathbf{G}^W(\mathbf{r}, \boldsymbol{\rho}) = \mathbf{B}(\hat{\boldsymbol{\rho}}) - \mathbf{B}(\hat{\boldsymbol{\rho}}^{IM}) + 2\frac{x_1}{(R^{IM})^3}\mathbf{J}.\left[\hat{\boldsymbol{\rho}}^{IM}\mathbf{e}_1 + h\mathbf{I} - \mathbf{e}_1\hat{\boldsymbol{\rho}}^{IM} - \frac{3h}{(R^{IM})^2}\hat{\boldsymbol{\rho}}^{IM}\hat{\boldsymbol{\rho}}^{IM}\right],$$
$$\mathbf{p}^W(\mathbf{r}, \boldsymbol{\rho}) = \mathbf{p}(\hat{\boldsymbol{\rho}}) + \mathbf{p}(\hat{\boldsymbol{\rho}}^{IM}) + 4\left[-\frac{\mathbf{e}_1 x_1}{(R^{IM})^3} + \frac{\hat{\boldsymbol{\rho}}^{IM}}{(R^{IM})^3} + \frac{3h(\hat{\boldsymbol{\rho}}^{IM})_1}{(R^{IM})^5}\hat{\boldsymbol{\rho}}^{IM}\right]. \tag{3.9.8}$$

The Lorentz' operator can be applied to each solution of the Stokes equation, for it transforms it again into a Stokes equation solution. The same property is possessed by the mirror reflection operator.

Following Pozrikidis (1992a), we shall present the correspondent stress field due to the wall Green's function:

$$\mathbf{K}^W(\mathbf{r}, \boldsymbol{\rho}) = \mathbf{k}(\hat{\mathbf{r}}) - \mathbf{k}(\hat{\mathbf{r}}^{IM}) \pm 2h^2\mathbf{K}(\hat{\mathbf{r}}^{IM}) \pm 2h\left[\mathbf{e}_1.\mathbf{K}^D(\hat{\mathbf{r}}^{IM})\right]', \tag{3.9.9}$$

where the sign ' denotes a transposed matrix with respect of the second and third indices and $\mathbf{k}$ is the Stokeslet stress field (2.5.6), $\mathbf{K}^D$ is the Stokeslet doublet stress field (3.4.3) and $\mathbf{K}$ is the stress field of potential dipole, i.e.,

$$K_{ijk}(\hat{\mathbf{r}}^{IM}) = 6\left(\frac{\delta_{ik}\hat{x}_j^{IM} + \delta_{ij}\hat{x}_k^{IM} + \delta_{jk}\hat{x}_i^{IM}}{(R^{IM})^5} - 5\frac{\hat{x}_i^{IM}\hat{x}_j^{IM}\hat{x}_k^{IM}}{(R^{IM})^7}\right). \tag{3.9.10}$$

When the singular point of the Stokeslet tends to be located on the wall, the wall Green's function becomes zero. Now, in this limit, we shall give some assessment to $\mathbf{G}^W$ upon expanding it into Taylor series about $\mathbf{r}$. As a result, after some algebra, we yield:

$$\mathbf{G}^W(\mathbf{r}, \boldsymbol{\rho}) = 2h\left[\frac{\partial\mathbf{B}(\mathbf{r})}{\partial x_1} \pm \mathbf{e}_1.\mathbf{B}^D(\mathbf{r})\right] \pm 2h^2\left[\mathbf{D}(\mathbf{r}) - \frac{\partial(\mathbf{e}_1.\mathbf{B}^D(\mathbf{r}))}{\partial x_1}\right] + O(h^3), \tag{3.9.11}$$

or

$$G_{i1}^{W}(\mathbf{r},\boldsymbol{\rho}) = -6h^2 x_1\left(2\frac{x_i}{R^5} + \delta_{i1}\frac{x_1}{R^5} - 5\frac{x_1^2 x_i}{R^7}\right) \text{ and} \tag{3.9.12}$$

$$G_{ik}^{W}(\mathbf{r},\boldsymbol{\rho}) = -12\,h\frac{x_1 x_i x_k}{R^5} \quad \text{at } k = 2, 3, \tag{3.9.13}$$

which is obtained, when the singularities are substituted by their correspondents. From the two latter expressions we can conclude that the wall Green's function behaves as a stresslet [see (3.4.8)] and decays as R^{-2}, when the point force is parallel to the wall, and as a stresslet doublet decaying as R^{-3}, when the point force is perpendicular to the wall. These results can be successfully applied to the rigid sphere translating (parallel or perpendicular) near to a wall, as will be discussed in section 8.2.

In the rest of this section, we shall discuss the more general case of a point force near a fluid/fluid interface. Aderogba (1976) has given a solution of the Stokes equation for the same problem, but without satisfying the normal velocity condition on the interface. In this respect his solution does not conserve a steady state in the interface and is only valid at the initial moment, when the Stokeslet is placed. Later on, Lee et al. (1979) have solved successfully the two-fluid system problem and generalize the Lorentz' reciprocal theorem, giving a lemma for Stokes solutions which obeys the full system of boundary conditions excluding the normal stress balance on the plane interface.

For two immiscible fluids, named fluid 1 and fluid 2, with viscosities ratio $\lambda = \mu_1/\mu_2$, in contact at the plane z = 0, the Lee's reflection lemma states that if $\mathbf{v}$ and p satisfy the Stokes equations for an infinite fluid 2, then the corresponding velocity and pressure fields for the semi-infinite fluids 1 and 2 are given by

$$\mathbf{v}_1 = \frac{1}{1+\lambda}\left[\mathbf{v} - \mathcal{L}_v(\mathbf{v})\right], \quad p_1 = \frac{\lambda}{1+\lambda}\left[p - \mathcal{L}_p(p)\right] \text{ and} \tag{3.9.14}$$

$$\mathbf{v}_2 = \left(\mathbf{v}+\mathbf{v}^*\right) - \frac{\lambda}{1+\lambda}\left[\mathbf{v}^* - \mathcal{L}_v\left(\mathbf{v}^*\right)\right], \; p_2 = \left(p+p^*\right) - \frac{\lambda}{1+\lambda}\left[p^* - \mathcal{L}_p\left(p^*\right)\right], \tag{3.9.15}$$

and also satisfy the Stokes equations with the boundary conditions for zero normal velocities and continuity of velocity and tangential stress on the interface

$$\mathbf{v}_1, \mathbf{v}_2 \to 0 \text{ at infinity}, \qquad \mathbf{v}_2 = \mathbf{U} \text{ on the particle surface}, \tag{3.9.16}$$

$$\mathbf{v}_1 = \mathbf{v}_2,\ \mathbf{v}_1 .\mathbf{n} = \mathbf{v}_2 .\mathbf{n}, = 0, \quad [[\mathbf{t}.\mathbf{n}.\mathbf{T}]]_1^2 = 0 \quad \text{on the interface}, \tag{3.9.17}$$

where $\mathbf{U}$ is the particle velocity.

In the expressions (3.9.14) and (3.9.15) the operators $\mathcal{L}_V$ and $\mathcal{L}_p$, and the asterisks * are the Lorentz' operators and the mirror reflection operator as defined by (3.9.6) and (3.9.7). This lemma is proved by different means by Lee (1979) and Kim and Karrila (1991, Ch. 12). When $\lambda \to \infty$, equation (3.9.15) yields the Lorentz' rigid plane problem solution for the fluid near the wall, while the infinite fluid solution ($\mathbf{v}$, p) results for the distant flow.

Afterwards this lemma serves to obtain the Stokes solution for the arbitrary motion of a particle modelled as a point force, near a plane interface at the assumption of negligible interface deformations. The velocity and pressure due to a point force $\mathbf{F}^{(c)}$ in a unbounded fluid is discussed in detail in the previous chapters of the book and introduced through (2.3.25). If the strength of the Stokeslet is $\boldsymbol{\alpha} = \frac{\mathbf{F}^{(c)}}{8\pi\mu_2}$, then $\mathbf{v} = \boldsymbol{\alpha}.\mathbf{B}$ and $p = \boldsymbol{\alpha}.\mathbf{p}$. Similar to the rigid wall case, the point force is situated in the pole (-h, 0, 0) and then (3.9.14) and (3.9.15) transform into:

$$\mathbf{v}_1 = \frac{1}{1+\lambda}(\mathbf{I}+\mathbf{J}).\mathbf{B}(\hat{\boldsymbol{\rho}}).\boldsymbol{\alpha} + \frac{2x_1}{(1+\lambda)R^3}\left[\mathbf{e}_1\hat{\boldsymbol{\rho}}_1 + h\mathbf{I} - \hat{\boldsymbol{\rho}}\mathbf{e}_1 - \frac{3h}{R^2}\hat{\boldsymbol{\rho}}\hat{\boldsymbol{\rho}}\right].\boldsymbol{\alpha}\,; \qquad (3.9.18)$$

$$p_1 = -\frac{4\lambda}{(1+\lambda)R^3}\left[x_1\mathbf{e}_1 + \hat{\boldsymbol{\rho}} + \frac{3h}{R^2}\hat{\boldsymbol{\rho}}\hat{\rho}_1\right].\boldsymbol{\alpha}\,;$$

and

$$\mathbf{v}_2 = \mathbf{v}(\hat{\boldsymbol{\rho}}) + \left(\frac{1}{1+\lambda}\mathbf{J} - \frac{\lambda}{1+\lambda}\mathbf{I}\right).\mathbf{B}(\hat{\boldsymbol{\rho}}^{IM}).\boldsymbol{\alpha} \qquad (3.9.19)$$

$$+\frac{2\lambda\, x_1}{(1+\lambda)(R^{IM})^3}\mathbf{J}.\left[\hat{\boldsymbol{\rho}}^{IM}\mathbf{e}_1 + h\mathbf{I} - \mathbf{e}_1\hat{\boldsymbol{\rho}}^{IM} - \frac{3h}{(R^{IM})^2}\hat{\boldsymbol{\rho}}^{IM}\hat{\boldsymbol{\rho}}^{IM}\right].\boldsymbol{\alpha}\,;$$

$$p_2 = \mathbf{p}(\hat{\boldsymbol{\rho}}) - 2\frac{\hat{\boldsymbol{\rho}}^{IM}}{(R^{IM})^3}.\boldsymbol{\alpha} + 4\frac{\lambda}{1+\lambda}\left[-\frac{\mathbf{e}_1 x_1}{(R^{IM})^3} + \frac{\hat{\boldsymbol{\rho}}^{IM}}{(R^{IM})^3} + \frac{3h\hat{\rho}_1^{IM}}{(R^{IM})^5}\hat{\boldsymbol{\rho}}^{IM}\right].\boldsymbol{\alpha},$$

where $R = \left[(h+x_1)^2 + x_2^2 + x_3^2\right]^{1/2}$. If $\lambda \to \propto$ the upper expressions reduce to (3.9.8) for the pair ($\mathbf{v}_2$, p_2), while $\mathbf{v}_1$ vanishes.

As a result of (3.9.18) and (3.9.19) we can obtain the normal stress jump, which occurs to be independent on the viscosity ratio λ (Lee et al., 1979). In the case when the sphere translates perpendicularly towards the fluid/fluid interface, i.e., the point force is directed normally to the interface $\boldsymbol{\alpha} = \alpha\,\mathbf{e}_1$, then the normal stress jump is:

$$[[\mathbf{n}.\mathbf{n}.\mathbf{T}]]_1^2 = -12\frac{\alpha h^3}{R_0^5}, \qquad (3.9.20)$$

where $R_0 = \left[h^2 + x_2^2 + x_3^2\right]^{1/2}$, and when translation is parallel to the interface, $\boldsymbol{\alpha} = \alpha\,\mathbf{e}_i$,

$$[[\mathbf{n}.\mathbf{n}.\mathbf{T}]]_1^2 = -12\frac{\alpha x_i h^2}{R_0^5}, \quad (i = 2, 3). \qquad (3.9.21)$$

It is evident from the above expressions that the normal stress jump becomes larger as the point force approaches the interface, and then the interface deformation must be taken into account.

The presented lemma can be used to predict the resistance and mobility functions of arbitrary shaped bodies (Lee et al., 1979), (Yang and Leal, 1983, 1984). In the limit of small deformations the boundary condition on the interface has to be linearized (Yang and Leal, 1990). Some numerical algorithms can also be based on this lemma when treating the interface finite deformations (Yang and Leal, 1983).

CHAPTER 4.

Solutions via Superposition of Vector Harmonic Functions

4.1. Introduction

Together with the general Lamb's solution (Lamb, 1945) for the Stokes equations in the 3D case, the so called "method of the vector harmonic functions" is sometimes applied (Kynch, 1956; Happel and Brenner; 1973; Leal, 1992). It is based on some general assumptions and ideas related to every equality between physical quantities, expressing any law or principle of the type:

$$\mathrm{X} = \mathrm{Y}, \tag{4.1.1}$$

where X and Y are scalars, vectors or tensors of arbitrary rank.

For the quantities X and Y in the equality (4.1.1) the following properties hold:

i) They have the same physical dimensions. If **X** has the dimension of force, then **Y** has the dimension of force, too; if X has the dimension of mass, then Y has also the dimension of mass, and so on.

ii) They are tensors of the same rank, i.e., if X is a scalar, it follows that Y is a scalar; if **X** is a vector, **Y** is a vector, etc.

iii) If **X** is a true vector, then **Y** also is a true vector; if **X** is a pseudo vector, **Y** is a pseudo vector, too, and so on.

iv) The quantities X and Y have the same symmetry, i.e., if for example the tensor **X** is antisymmetric, the tensor **Y** is also antisymmetric, etc.

We shall recall that true tensor (vector) quantities are those, which are invariant according to the coordinate system change from a right- to a left-handed type or vice versa. On the other hand, pseudo tensor (pseudo vector) quantities are those, which are not invariant when changing the type of the coordinate system. We are going to give some examples illustrating true and pseudo tensor quantities:

a) The cross product $\mathbf{a} \times \mathbf{b}$ of two true vectors **a** and **b** is a pseudo vector, because the cross product is not invariant when changing the type of the coordinate system;

b) The moment **M** of the force **F** is a pseudo tensor $\mathbf{M} = \mathbf{r} \times \mathbf{F}$;

c) The vorticity vector $\boldsymbol{\omega} = \frac{1}{2}\operatorname{rot}\mathbf{v} = \frac{1}{2}\nabla \times \mathbf{v}$ is a pseudo vector;

d) The dot product of a true vector $\mathbf{r}$ and the third-rank unit alternator pseudo tensor $\boldsymbol{\varepsilon}$ is a second-rank pseudo tensor;

e) The dot product of two pseudo tensor quantities (e.g. $\mathbf{v} = \mathbf{A}.\boldsymbol{\omega}$, where $\mathbf{A}$ is a pseudo tensor and $\boldsymbol{\omega}$ - pseudo vector) gives a true scalar, vector or tensor.

When applying the method of the vector harmonic functions to describe the creeping flow solution, usually the disturbed pressure $p' = p - p_\infty$, where p_∞ is a constant pressure at infinity, is taken into account. The disturbed pressure satisfies also the Laplace equation, which means that it is a harmonic function.

It is well known, that the general solution of the Stokes equation

$$\mu \nabla^2 \mathbf{v} = \nabla p' \tag{4.1.2}$$

could be expressed in the dimensional form (Happel and Brenner, 1973, Chapter 9):

$$\mathbf{v} = \frac{p'}{2\mu}\mathbf{r} + \mathbf{v}^{(H)}, \tag{4.1.3}$$

where $\mathbf{v}^{(H)}$ is a harmonic vector function, i.e.,

$$\nabla^2 \mathbf{v}^{(H)} = 0. \tag{4.1.4}$$

We shall show that (4.1.3) satisfies the Stokes equation (4.1.2), if $\mathbf{v}^{(H)}$ obeys the Laplace equation (4.1.4).

First of all, let us calculate $\nabla^2 \mathbf{v}$:

$$\nabla \mathbf{v} = \nabla\left(\frac{p'}{2\mu}\mathbf{r}\right) + \nabla \mathbf{v}^{(H)} = \frac{p'}{2\mu}\nabla \mathbf{r} + \frac{1}{2\mu}\nabla p' \mathbf{r} + \nabla \mathbf{v}^{(H)} \tag{4.1.5}$$

$$= \frac{p'}{2\mu}\mathbf{I} + \frac{1}{2\mu}\nabla p' \mathbf{r} + \nabla \mathbf{v}^{(H)}.$$

Applying the differential operator div $= \nabla.$ to both sides of (4.1.5)

$$\nabla.(\nabla \mathbf{v}) = \nabla^2 \mathbf{v} = \nabla.\left(\frac{p'}{2\mu}\mathbf{I}\right) + \nabla.\left(\frac{1}{2\mu}\nabla p' \mathbf{r}\right) + \nabla.\left(\nabla \mathbf{v}^{(H)}\right)$$

and using the formula $\nabla.(\mathbf{ab}) = \mathbf{b}(\nabla.\mathbf{a}) + (\mathbf{a}.\nabla)\mathbf{b}$, we get:

$$\nabla^2 \mathbf{v} = \frac{\nabla p'}{2\mu} + \frac{1}{2\mu}\mathbf{r}\nabla^2 p' + \frac{1}{2\mu}\nabla p'.\nabla \mathbf{r} + \mathbf{0} = \frac{1}{\mu}\nabla p'.$$

This proves our statement.

Beside the Stokes equation, the velocity (4.1.3) must satisfy the continuity equation, that is:

$$\nabla.\left(\frac{p'}{2\mu}\mathbf{r} + \mathbf{v}^{(H)}\right) = 0.$$

Because of the fact that $\nabla.\mathbf{r} = 3$, the following condition for the function $\mathbf{v}^{(H)}$ remains:

$$\nabla.\mathbf{v}^{(H)} = -\frac{1}{2\mu}\left(3p' + \mathbf{r}.\nabla p'\right). \tag{4.1.6}$$

The equality (4.1.3) gives an opportunity to reduce the search for the Stokes equation solution to a search of appropriate solutions for the Laplace equation, i.e., to the use of the so called vector harmonic functions, which for outer problems are given by the formula:

$$\Phi_{-(n+1)} = \frac{(-1)^n}{1.3.5...(2n-1)} \cdot \frac{\partial^n}{\partial x_i \partial x_j \partial x_k ...}\left(\frac{1}{r}\right), \tag{4.1.7}$$

where $r = |\mathbf{r}| = |x_1\mathbf{i} + x_2\mathbf{j} + x_3\mathbf{k}|$ and $n = 0, 1, 2, 3, ...$

The first several decaying vector harmonic functions have the form:

$$\frac{1}{r}, \quad \frac{\mathbf{r}}{r^3}, \quad \frac{\mathbf{rr}}{r^5} - \frac{\mathbf{I}}{3r^3}, \quad \frac{\mathbf{rrr}}{r^7} - \frac{\mathbf{rI} + \mathbf{Ir} + (\mathbf{Ir})^T}{5r^5}, \text{ etc.} \tag{4.1.8}$$

For the sake of completeness we shall write down also the growing vector harmonic functions, which are used when solving inner problems. They are derived from the formula:

$$r^{2n+1}\Phi_{-(n+1)} = \frac{(-1)^n r^{2n+1}}{1.3.5...(2n-1)} \cdot \frac{\partial^n}{\partial x_i \partial x_j \partial x_k ...}\left(\frac{1}{r}\right) \tag{4.1.9}$$

and have the form:

$$1, \quad \mathbf{r}, \quad \mathbf{rr} - \frac{r^2\mathbf{I}}{3}, \quad \mathbf{rrr} - \frac{r^2\left[\mathbf{rI} + \mathbf{Ir} + (\mathbf{Ir})^T\right]}{5}, \text{ etc.} \tag{4.1.10}$$

In the present chapter the method of harmonic functions is applied for solving some problems for the translation and rotation of spherical rigid or fluid particles in a quiescent viscous fluid.

4.2. The Rotation and Translation of a Rigid Spherical Particle in a Quiescent Fluid

Let a solid spherical particle of radius a performs a slow rotation in a viscous fluid with angular velocity $\mathbf{\Omega}$. Consider a spherical coordinate system (r, θ, φ), with the origin at the particle centre (Fig.4.2.1). Then for the angular velocity at the sphere rotation around the axis Oz, we have $\Omega = \dot{\varphi} = \dfrac{d\varphi}{dt}$, and for the linear velocity $\mathbf{v}_\varphi\big|_{r=a} = a\,\Omega \sin\theta\ \mathbf{e}_\varphi$.

Since the angular velocity vector $\mathbf{\Omega}$ is directed towards the axis Oz and $\mathbf{e}_z \times \mathbf{e}_r = \sin\theta\ \mathbf{e}_\varphi$, then

$$\mathbf{v}_\varphi\big|_{r=a} = a\,\Omega \sin\theta\ \mathbf{e}_\varphi = a\,\Omega\left(\mathbf{e}_z \times \mathbf{e}_r\right) = \Omega\,\mathbf{e}_z \times a\,\mathbf{e}_r = \mathbf{\Omega} \times a\,\mathbf{e}_r. \tag{4.2.1}$$

When applying the vector harmonic functions, the calculations always start with the determination of the disturbed pressure p'. The pressure p' is a true scalar, which satisfies the Laplace equation and is decaying when $r \to \infty$. From this follows that p' is a linear function of the angular velocity Ω, depending only on the radius-vector $\mathbf{r}$ and can be expressed by the vector harmonic functions (4.1.8).

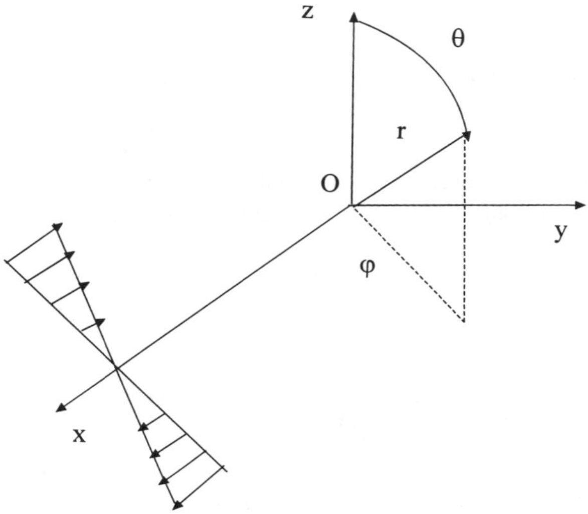

Fig.4.2.1.Spherical coordinate system in shear flow.

The only linear combination of $\boldsymbol{\Omega}$ and the vector harmonic functions (4.1.8) with respect to $\boldsymbol{\Omega}$, which gives a scalar, is the dot product of the vectors $\boldsymbol{\Omega}$ and $\frac{\mathbf{r}}{r^3}$, i.e.,

$$p' = C_1 \boldsymbol{\Omega}.\frac{\mathbf{r}}{r^3}. \qquad (4.2.2)$$

This is true, because the dot product of the vector $\boldsymbol{\Omega}$ and the other tensors from (4.1.8) yields a vector or a tensor, and the cross product of $\boldsymbol{\Omega}$ and the other tensor functions (except the first one) also yields a vector or a tensor, but not a scalar.

In the equality (4.2.2) $\boldsymbol{\Omega}$ is a pseudo vector, while **r** is a true vector. That is why the dot product of $\boldsymbol{\Omega}$ and $\frac{\mathbf{r}}{r^3}$ is a pseudo scalar. Because of the fact that p' is a true scalar, the equality (4.2.2) is only possible when $C_1 = 0$. Consequently, $p' = 0$ in this problem. Then from the general solution (4.1.3) we find

$$\mathbf{v} = \mathbf{v}^{(H)}, \qquad (4.2.3)$$

i.e., the velocity solution is given only by an appropriate harmonic function $\mathbf{v}^{(H)}$.

In the case under consideration, we are looking for a function $\mathbf{v}^{(H)}$, which is: i) a true vector; ii) decaying when $r \to \infty$; iii) linearly depending on the vector $\boldsymbol{\Omega}$.

The only combination of $\mathbf{\Omega}$ and the decaying vector functions (4.1.8) satisfying the upper conditions is given by the expression $\mathbf{\Omega}\times\frac{\mathbf{r}}{r^3}$, i.e.,

$$\mathbf{v}^{(H)} = C_2\mathbf{\Omega}\times\frac{\mathbf{r}}{r^3}, \tag{4.2.4}$$

where C_2 is a constant. The quantity $C_2\mathbf{\Omega}\times\frac{\mathbf{r}}{r^3}$ is a true vector because $\mathbf{\Omega}$ is a pseudo vector and the cross product is not invariant when changing the type of the coordinate system.

We must note here, that the dot product of $\mathbf{\Omega}$ with the third tensor function $\frac{\mathbf{rr}}{r^5}-\frac{\mathbf{I}}{3r^3}$ will give a pseudo vector and, for this reason, it is inappropriate when constructing the true vector $\mathbf{v}^{(H)}$.

The constant C_2 is to be determined from the boundary condition on the sphere surface (4.2.1). In fact, from $\mathbf{v}_\varphi\big|_{r=a} = \mathbf{\Omega}\times a\,\mathbf{e}_r$ and (4.2.4), we find $C_2 = a^3$. Therefore, the velocity of the fluid flow induced by the rotating sphere has the form:

$$\mathbf{v} = \mathbf{v}^{(H)} = \mathbf{\Omega}\times\frac{a^3}{r^3}\mathbf{r}. \tag{4.2.5}$$

Let us now consider the problem of the uniform viscous flow past a solid spherical particle of radius a. Here, as in the previous case, we are considering a spherical coordinate system (r, θ, φ) with the origin at the particle centre. If the velocity of the flow passing the particle is $\mathbf{U}$, then the velocity of the disturbed flow $\mathbf{v}'$ is equal to $\mathbf{v}-\mathbf{U}$. Consequently, $\mathbf{v}'$ satisfies the equations:

$$\mu\nabla^2\mathbf{v}' = \nabla p', \qquad \nabla.\mathbf{v}' = 0, \tag{4.2.6}$$

and the boundary conditions:

a) on the particle surface $r = a$, $\mathbf{v}' = -\mathbf{U}$;

b) at infinity $r\to\infty$, $\mathbf{v}'\to 0$.

In order to solve the problem, we are starting with the pressure p'. It is a true scalar, satisfying the Laplace equation and linearly depending on the velocity $\mathbf{U}$ of the flow passing the particle. The only appropriate linear combination of $\mathbf{U}$ and the vector harmonic functions (4.1.8) is $\mathbf{U}.\frac{\mathbf{r}}{r^3}$, i.e.,

$$p' = C_1\mathbf{U}.\frac{\mathbf{r}}{r^3}, \tag{4.2.7}$$

where C_1 is an arbitrary constant.

Then from (4.1.3) it follows, that the disturbed fluid velocity has the form:

$$\mathbf{v}' = \frac{\mathbf{r}}{2\mu}\left(C_1\mathbf{U}.\frac{\mathbf{r}}{r^3}\right) + \mathbf{v}^{(H)}. \tag{4.2.8}$$

When seeking $\mathbf{v}^{(H)}$, we have to take into account, that $\mathbf{v}^{(H)}$ is a true vector, which depends linearly on **U** and is a decaying harmonic function. Now the given velocity **U** can be combined with the first and the third harmonic functions from (4.1.8): $\mathbf{U}\frac{1}{r}$ and $\left(\frac{\mathbf{rr}}{r^5}-\frac{\mathbf{I}}{3r^3}\right).\mathbf{U}$. Therefore, the vector harmonic function $\mathbf{v}^{(H)}$ can be sought into the form:

$$\mathbf{v}^{(H)} = \delta\frac{\mathbf{U}}{r} + \gamma\left(\frac{\mathbf{rr}}{r^5}-\frac{\mathbf{I}}{3r^3}\right).\mathbf{U}, \tag{4.2.9}$$

where δ and γ are unknown constants.

After substituting (4.2.9) into the limiting equality (4.1.6), we reach $\delta = \frac{C_1}{2\mu}$, i.e.,

$$\mathbf{v}' = \frac{1}{2\mu}C_1\mathbf{U}.\left(\frac{\mathbf{rr}}{r^3}\right) + \gamma\frac{\mathbf{rr}}{r^5}.\mathbf{U} - \gamma\frac{\mathbf{U}}{3r^3} + \frac{C_1}{2\mu}\frac{\mathbf{U}}{r} = \frac{(\mathbf{U}.\mathbf{r})\mathbf{r}}{r^3}\left(\frac{C_1}{2\mu}+\frac{\gamma}{r^2}\right) + \frac{\mathbf{U}}{r}\left(\frac{C_1}{2\mu}-\frac{\gamma}{3r^2}\right). \tag{4.2.10}$$

The boundary condition for the disturbance vanishing at infinity is fulfilled automatically, because we have used decaying vector harmonic functions when constructing $\mathbf{v}^{(H)}$.

From the boundary condition on the particle surface $\mathbf{v}' = -\mathbf{U}$ at r = a, we find

$$-\mathbf{U} = \frac{(a\,\mathbf{e}_r.\mathbf{U})\,a\,\mathbf{e}_r}{a^3}\left(\frac{C_1}{2\mu}+\frac{\gamma}{a^2}\right) + \frac{\mathbf{U}}{a}\left(\frac{C_1}{2\mu}-\frac{\gamma}{3a^2}\right)$$

or

$$\frac{\mathbf{U}}{a}\left(a+\frac{C_1}{2\mu}-\frac{\gamma}{3a^2}\right) + \frac{(\mathbf{e}_r.\mathbf{U})\,\mathbf{e}_r}{a}\left(\frac{C_1}{2\mu}+\frac{\gamma}{a^2}\right) = 0. \tag{4.2.11}$$

Since in the vector equality (4.2.11) the vectors **U** and $\mathbf{e}_r$ are not collinear, their coefficients have to be zero, i.e.,

$$\left|\begin{array}{l} 1+\dfrac{C_1}{2\mu a}-\dfrac{\gamma}{3a^3}=0, \\ \dfrac{C_1}{2\mu}+\dfrac{\gamma}{a^2}=0. \end{array}\right. \tag{4.2.12}$$

To find the system constants C_1 and γ, we shall multiply the second equation by $\frac{1}{3a}$ and add it to the first one. Hence we get $1+\frac{C_1}{\mu a}\left(\frac{1}{2}+\frac{1}{6}\right)=0$, and the constant C_1 is defined as

$$C_1 = -\frac{3\mu a}{2}. \tag{4.2.13}$$

The other unknown constant γ is found from the second equation of the system (4.2.12)

$$-\frac{3\mu\,a}{4\mu}+\frac{\gamma}{a^2}=0,$$

from where it follows $\gamma=\dfrac{3a^3}{4}$.

Therefore

$$\mathbf{v}'=\frac{p'}{2\mu}\mathbf{r}+\mathbf{v}^{(H)}=\frac{p'}{2\mu}\mathbf{r}+\delta\frac{\mathbf{U}}{r}+\gamma\left(\frac{\mathbf{rr}}{r^5}-\frac{\mathbf{I}}{3r^3}\right).\mathbf{U}$$

$$=-\left(\frac{3a}{4r}+\frac{a^3}{4r^3}\right)\mathbf{U}-\left(\frac{3a}{4r^3}-\frac{3a^3}{4r^5}\right)(\mathbf{U}.\mathbf{r})\mathbf{r}\,. \tag{4.2.14}$$

Since $\mathbf{v}=\mathbf{U}+\mathbf{v}'$, we finally get for the flow velocity induced by the translational motion of a spherical particle with velocity $\mathbf{U}$ immersed in the fluid

$$\mathbf{v}=\mathbf{U}-\left(\frac{3a}{4r}+\frac{a^3}{4r^3}\right)\mathbf{U}-\left(\frac{3a}{4r^3}-\frac{3a^3}{4r^5}\right)(\mathbf{U}.\mathbf{r})\mathbf{r}, \tag{4.2.15}$$

which coincides with the Stokes solution of the same problem. The corresponding pressure distribution is:

$$p=-\frac{3\mu\,a}{2}\frac{\mathbf{U}.\mathbf{r}}{r^3}\,. \tag{4.2.16}$$

4.3. A Spherical Rigid Particle in General Linear Flow

In this section we shall consider the problem of the arbitrary linear shear flow past a solid spherical particle

$$\mathbf{v}^\infty=\boldsymbol{\Gamma}^\infty.\mathbf{r}=\mathbf{E}^\infty.\mathbf{r}+\boldsymbol{\Omega}^\infty.\mathbf{r}, \tag{4.3.1}$$

where $\mathbf{E}^\infty=\text{grad}\,\mathbf{v}^\infty$ is the symmetric rate of strain tensor and

$$e_{ii}^\infty=\frac{\partial v_1^\infty}{\partial x_1}+\frac{\partial v_2^\infty}{\partial x_2}+\frac{\partial v_3^\infty}{\partial x_3}=0, \tag{4.3.2}$$

while $\boldsymbol{\Omega}^\infty$ is the antisymmetric vorticity tensor and $\Omega_{ij}^\infty=\frac{1}{2}\varepsilon_{kji}\,\omega_k$, where ε_{kji} are the components of the third-rank unit alternator pseudo tensor and $\boldsymbol{\omega}$ is the vorticity vector.

A spherical coordinate system with the origin at the particle mass centre is again utilised. Replacing $\mathbf{v}=\mathbf{v}^\infty+\mathbf{v}'$ from (1.4.2) and (1.4.3) we obtain the boundary conditions for the disturbance velocity $\mathbf{v}'$

a) on the particle: $\mathbf{v}'\big|_{r=a}=-\mathbf{E}^\infty.a\mathbf{e}_r-\frac{1}{2}(\boldsymbol{\omega}\times a\mathbf{e}_r)$;

b) at infinity: $\mathbf{v}'\to\mathbf{0}$ at $\mathbf{r}\to\infty$.

In this case we start again with the pressure p', which is a true scalar, a harmonic decaying function at infinity and depending linearly on the tensors $\mathbf{E}^\infty$ and $\mathbf{\Omega}^\infty$. Since the first and the second vector function in (4.1.8) cannot be combined linearly with the tensors $\mathbf{E}^\infty$ and $\mathbf{\Omega}^\infty$ in a way to get a scalar as a result, then the third tensor function $\frac{\mathbf{rr}}{r^5}-\frac{\mathbf{I}}{3r^3}$ is considered. The inner products of this function respectively with $\mathbf{E}^\infty$ and $\mathbf{\Omega}^\infty$ are the scalars

$$\mathbf{E}^\infty:\left(\frac{\mathbf{rr}}{r^5}-\frac{\mathbf{I}}{3r^3}\right) \qquad \text{and } \mathbf{\Omega}^\infty:\left(\frac{\mathbf{rr}}{r^5}-\frac{\mathbf{I}}{3r^3}\right).$$

Since the second one is a pseudo scalar, we shall seek the pressure p' in the form:

$$p' = C_1\mathbf{E}^\infty:\left(\frac{\mathbf{rr}}{r^5}-\frac{\mathbf{I}}{3r^3}\right), \tag{4.3.3}$$

where C_1 is a unknown constant.

Further we have to find the vector harmonic function $\mathbf{v}^{(H)}$, which is a true vector and depends linearly on $\mathbf{E}^\infty$ and $\mathbf{\Omega}^\infty$. In fact, we have to look for some multiplication of $\mathbf{E}^\infty$ and $\mathbf{\Omega}^\infty$ with the tensor functions $\frac{1}{r}$, $\frac{\mathbf{r}}{r^3}$, $\frac{\mathbf{rr}}{r^5}-\frac{\mathbf{I}}{3r^3}$, $\frac{\mathbf{rrr}}{r^7}-\frac{\mathbf{rI}+\mathbf{Ir}+(\mathbf{Ir})^T}{5r^5}$, etc. in order to get true (not pseudo) vectors.

It is clear that this could be achieved by the dot product of $\mathbf{E}^\infty$ and $\frac{\mathbf{r}}{r^3}$. If $\mathbf{E}^\infty$ is multiplied by the fourth tensor function, i.e., if the inner product is formed $\mathbf{E}^\infty:\left(\frac{\mathbf{rrr}}{r^7}-\frac{\mathbf{rI}+\mathbf{Ir}+(\mathbf{Ir})^T}{5r^5}\right)$ and the vectors $\boldsymbol{\omega}$ and $\frac{\mathbf{r}}{r^3}$ are multiplied, i.e., if the cross product $\boldsymbol{\omega}\times\frac{\mathbf{r}}{r^3}$ is formed, then in both cases a true vector will be obtained.

In other words, for the vector harmonic function we have

$$\mathbf{v}^{(H)} = C_2\mathbf{E}^\infty.\mathbf{r} + C_3\mathbf{E}^\infty:\left(\frac{\mathbf{rrr}}{r^7}-\frac{\mathbf{rI}+\mathbf{Ir}+(\mathbf{Ir})^T}{5r^5}\right)+C_4\boldsymbol{\omega}\times\frac{\mathbf{r}}{r^3}, \tag{4.3.4}$$

where C_2, C_3 and C_4 are arbitrary constants.

Deriving $\nabla p'$ from (4.3.3) and $\nabla\mathbf{v}^{(H)}$ from (4.3.4) and substituting these expressions in (4.1.6) we arrive at $C_2 = 0$.

Since $\operatorname{tr}\mathbf{E}^\infty = e_{11}+e_{22}+e_{33} = 0$, then $\mathbf{E}^\infty:\mathbf{I} = 0$ and

$$p' = C_1\mathbf{E}^\infty:\left(\frac{\mathbf{rr}}{r^5}-\frac{\mathbf{I}}{3r^3}\right) = C_1\mathbf{E}^\infty:\frac{\mathbf{rr}}{r^5} = C_1\frac{\mathbf{r}.\mathbf{E}^\infty.\mathbf{r}}{r^5}.$$

Hence $\mathbf{E}^\infty:\mathbf{rI} = \mathbf{E}^\infty:\mathbf{Ir} = \mathbf{E}^\infty.\mathbf{r}$ and $\mathbf{E}^\infty:(\mathbf{Ir})^T = e^\infty_{ii}x_k = 0$, the harmonic vector function is

$$\mathbf{v}^{(\mathrm{H})} = \mathrm{C}_3 \frac{\mathbf{r}.\mathbf{E}^\infty.\mathbf{r}}{\mathrm{r}^7}\mathbf{r} - \mathrm{C}_3 \frac{2}{5\mathrm{r}^5}\mathbf{E}^\infty.\mathbf{r} + \mathrm{C}_4 \boldsymbol{\omega} \times \frac{\mathbf{r}}{\mathrm{r}^3}$$

Then $\mathbf{v} = \mathbf{E}^\infty.\mathbf{r} + \frac{1}{2}\left(\boldsymbol{\omega} \times \frac{\mathbf{r}}{\mathrm{r}^3}\right) + \mathbf{v}'$, where due to (4.1.3)

$$\mathbf{v}' = \frac{\mathrm{p}'}{2\mu}\mathbf{r} + \mathbf{v}^{(\mathrm{H})} = \frac{\mathrm{C}_1}{2\mu}\frac{\mathbf{r}.\mathbf{E}^\infty.\mathbf{r}}{\mathrm{r}^5}\mathbf{r} + \mathrm{C}_3 \frac{\mathbf{r}.\mathbf{E}^\infty.\mathbf{r}}{\mathrm{r}^7}\mathbf{r} - \mathrm{C}_3 \frac{2}{5\mathrm{r}^5}\mathbf{E}^\infty.\mathbf{r} + \mathrm{C}_4 \boldsymbol{\omega} \times \frac{\mathbf{r}}{\mathrm{r}^3}$$

$$= \left(\mathbf{r}.\mathbf{E}^\infty.\mathbf{r}\right)\mathbf{r}\left(\frac{\mathrm{C}_1}{2\mu \mathrm{r}^5} + \frac{\mathrm{C}_3}{\mathrm{r}^7}\right) + \mathbf{E}^\infty.\mathbf{r}\left(\frac{-2\mathrm{C}_3}{5\mathrm{r}^5}\right) + \mathrm{C}_4\left(\boldsymbol{\omega} \times \frac{\mathbf{r}}{\mathrm{r}^3}\right). \tag{4.3.5}$$

In order to determine the constants C_1, C_3 and C_4 we shall apply the boundary condition on the particle surface

$$\mathbf{v}'\big|_{\mathrm{r=a}} = -\mathbf{E}^\infty.\mathrm{a}\mathbf{e}_\mathrm{r} - \frac{1}{2}\left(\boldsymbol{\omega} \times \mathrm{a}\mathbf{e}_\mathrm{r}\right). \tag{4.3.6}$$

From (4.3.5) and (4.3.6) it follows that

$$-\mathbf{E}^\infty.\mathrm{a}\mathbf{e}_\mathrm{r} - \frac{1}{2}\left(\boldsymbol{\omega} \times \mathrm{a}\mathbf{e}_\mathrm{r}\right)$$

$$= \mathrm{a}\mathbf{e}_\mathrm{r}\left(\mathrm{a}\mathbf{e}_\mathrm{r}.\mathbf{E}^\infty.\mathrm{a}\mathbf{e}_\mathrm{r}\right)\left(\frac{\mathrm{C}_1}{2\mu \mathrm{a}^5} + \frac{\mathrm{C}_3}{\mathrm{a}^7}\right) + \mathbf{E}^\infty.\mathrm{a}\mathbf{e}_\mathrm{r}\left(\frac{-2\mathrm{C}_3}{5\mathrm{a}^5}\right) + \mathrm{C}_4\left(\boldsymbol{\omega} \times \frac{\mathrm{a}\mathbf{e}_\mathrm{r}}{\mathrm{a}^3}\right)$$

This equality is fulfilled when the coefficients of both sides are respectively equal. Hence,

1) $\quad \dfrac{\mathrm{C}_4}{\mathrm{a}^2} = -\dfrac{\mathrm{a}}{2} \quad \Rightarrow \quad \mathrm{C}_4 = -\dfrac{\mathrm{a}^3}{2}$;

2) $\quad \dfrac{-2\mathrm{C}_3}{5\mathrm{a}^4} = -\mathrm{a} \quad \Rightarrow \quad \mathrm{C}_3 = \dfrac{5\mathrm{a}^5}{2}$;

3) $\quad \dfrac{\mathrm{C}_1}{2\mu \mathrm{a}^2} + \dfrac{\mathrm{C}_3}{\mathrm{a}^4} = 0$.

From the last equation we find $\mathrm{C}_1 = -5\mu \mathrm{a}^3$.

Therefore, the velocity field of the flow induced by an arbitrary linear shear flow of velocity $\mathbf{E}^\infty.\mathbf{r} + \boldsymbol{\Omega}^\infty.\mathbf{r}$ passing a solid spherical particle has the form

$$\mathbf{v} = \mathbf{E}^\infty.\mathbf{r}\left(1 - \frac{\mathrm{a}^5}{\mathrm{r}^5}\right) + \frac{1}{2}\left(\boldsymbol{\omega} \times \frac{\mathbf{r}}{\mathrm{r}^3}\right)\left(1 - \frac{\mathrm{a}^3}{\mathrm{r}^3}\right) - \mathbf{r}\left(\mathbf{r}.\mathbf{E}^\infty.\mathbf{r}\right)\left(\frac{5\mathrm{a}^3}{2\,\mathrm{r}^5} - \frac{5\mathrm{a}^5}{2\mathrm{r}^7}\right), \tag{4.3.7}$$

and the pressure

$$\mathrm{p} = -5\mu\left(\mathbf{r}.\mathbf{E}^\infty.\mathbf{r}\right)\left(\frac{\mathrm{a}^3}{\mathrm{r}^3}\right) + \mathrm{p}_\infty. \tag{4.3.8}$$

Now let us apply the obtained results to a special fluid flow, called a simple shear flow, with undisturbed velocity of

$$\mathbf{v}^\infty = \mathrm{Gy}\mathbf{i}. \tag{4.3.9}$$

When calculating the components of the rate of strain tensor $\mathbf{E}^{\infty}$ and the vorticity vector $\boldsymbol{\omega}$, we get

$$\mathbf{E}^{\infty} = \begin{pmatrix} 0 & \frac{G}{2} & 0 \\ \frac{G}{2} & 0 & 0 \\ 0 & 0 & 0 \end{pmatrix}, \quad \boldsymbol{\omega} = -G\mathbf{k}. \tag{4.3.10}$$

It is convenient to represent the found exact solution (4.3.7) of the Stokes equation in spherical coordinates (Fig.4.2.1). For this end we shall use the general relations between Cartesian coordinates (x, y, z) and spherical coordinates (r, θ, φ) given by (1.3.13) and the equations

$$\begin{aligned} \mathbf{i} &= \sin\theta\cos\varphi\,\mathbf{e}_r + \cos\theta\cos\varphi\,\mathbf{e}_\theta - \sin\varphi\,\mathbf{e}_\phi, \\ \mathbf{j} &= \sin\theta\sin\varphi\,\mathbf{e}_r + \cos\theta\sin\varphi\,\mathbf{e}_\theta + \cos\varphi\,\mathbf{e}_\phi, \\ \mathbf{k} &= \cos\theta\,\mathbf{e}_r + \cos(90+\theta)\mathbf{e}_\theta = \cos\theta\,\mathbf{e}_r - \sin\theta\,\mathbf{e}_\theta. \end{aligned} \tag{4.3.11}$$

Then from (1.3.13), (4.3.11) and (4.3.7), we obtain consecutively

$$\mathbf{r} = x\mathbf{i} + y\mathbf{j} + z\mathbf{k} = r(\sin^2\theta - 2\cos^2\theta)\mathbf{e}_r + 3r\sin\theta\cos\theta\,\mathbf{e}_\theta \tag{4.3.12}$$

$$\mathbf{E}^{\infty}.\mathbf{r} = \begin{pmatrix} 0 & \frac{G}{2} & 0 \\ \frac{G}{2} & 0 & 0 \\ 0 & 0 & 0 \end{pmatrix}.(x\mathbf{i} + y\mathbf{j} + z\mathbf{k}) = \frac{G}{2}(y\mathbf{i} + x\mathbf{j}), \tag{4.3.13}$$

$$\frac{1}{2}\boldsymbol{\omega} \times \mathbf{r} = \frac{1}{2}\begin{pmatrix} \mathbf{i} & \mathbf{j} & \mathbf{k} \\ 0 & 0 & -G \\ x & y & z \end{pmatrix} = -\frac{G}{2}(x\mathbf{j} - y\mathbf{i}), \tag{4.3.14}$$

$$\mathbf{r}(\mathbf{r}.\mathbf{E}^{\infty}.\mathbf{r}) = \mathbf{r}(x\mathbf{i} + y\mathbf{j} + z\mathbf{k}).\begin{pmatrix} 0 & \frac{G}{2} & 0 \\ \frac{G}{2} & 0 & 0 \\ 0 & 0 & 0 \end{pmatrix}.(x\mathbf{i} + y\mathbf{j} + z\mathbf{k}) \tag{4.3.15}$$

$$= \mathbf{r}(x\mathbf{i} + y\mathbf{j} + z\mathbf{k}).\left(\frac{G}{2}y\mathbf{i} + \frac{G}{2}x\mathbf{j}\right) = \frac{G}{2}(xy + xy)\mathbf{r} = Gxy(x\mathbf{i} + y\mathbf{j} + z\mathbf{k})$$

$$\mathbf{v} = \frac{G}{2}(y\mathbf{i} + x\mathbf{j})\left(1 - \frac{a^5}{r^5}\right) - \frac{G}{2}(x\mathbf{j} - y\mathbf{i})\left(1 - \frac{a^3}{r^3}\right) - \frac{5}{2}Gxy(x\mathbf{i} + y\mathbf{j} + z\mathbf{k})\left(\frac{a^3}{r^5} - \frac{a^5}{r^7}\right). \tag{4.3.16}$$

In spherical coordinates the upper formula for the fluid velocity around an irrotational, motionless rigid spherical particle is rewritten as

$$\mathbf{v} = \mathrm{Gr}\ \sin^2\theta \sin\varphi \cos\varphi \left(1 - \frac{5}{2}\frac{a^3}{r^3} + \frac{3}{2}\frac{a^5}{r^5}\right)\mathbf{e}_r \tag{4.3.17}$$

$$+ \mathrm{Gr}\ \sin\theta\cos\theta\sin\varphi\cos\varphi\left(1 - \frac{a^5}{r^5}\right)\mathbf{e}_\theta + \mathrm{Gr}\ \sin\theta\left(-\sin^2\varphi + \frac{a^3}{2r^3} - \frac{\cos 2\varphi}{2}\frac{a^5}{r^5}\right)\mathbf{e}_\varphi$$

The streamlines of this flow projected on the plane xy are given on Fig.4.3.1.

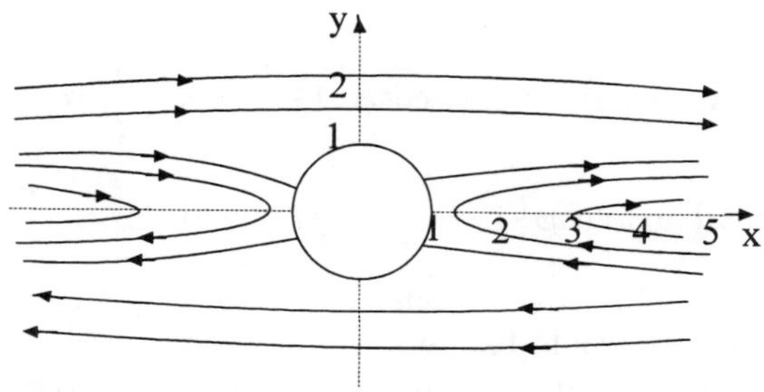

Fig.4.3.1. Streamlines pattern due to a sphere in a simple shear flow (from Leal, 1992)

Now we have at our disposal both the velocity field of a particle rotating with velocity $\boldsymbol{\Omega}$ in a quiescent fluid (see 4.2.5) and the velocity field induced by the simple shear flow passing an irrotational solid spherical particle (see 4.3.17). Thus we can solve the problem of the simple shear flow passing a spherical particle which is rotating with a certain angular velocity $\boldsymbol{\Omega}$ under the action of the flow. For this purpose it is sufficient to apply the superposition principal for the two flows. If the spherical particle rotates under the influence of the simple shear flow, then the angular velocity $\boldsymbol{\Omega}$ must have the same direction as the vorticity vector $\boldsymbol{\omega}$, i.e.

$$\boldsymbol{\Omega} = -\Omega\,\mathbf{k}. \tag{4.3.18}$$

Then, the formula (4.2.5) will have the form

$$\mathbf{v} = \left(-\Omega\,x\mathbf{j} - \Omega\,y\mathbf{i}\right)\frac{a^3}{r^3}. \tag{4.3.19}$$

Substitution of the corresponding expressions from (1.3.13) and (4.3.11) for **i**, **j**, x and y transforms the velocity formula (4.3.19) into

$$\mathbf{v}' = -\Omega\left(\frac{a}{r}\right)^3 r\ \sin\theta\,\mathbf{e}_\varphi. \tag{4.3.20}$$

Then, in order to find the velocity field around a spherical particle that rotates under the action of the flow, the velocities (4.3.17) and (4.3.20) have to be summed. After calculating the particle moment corresponding to the sum of these two flows, we obtain

$$\mathbf{M} = -8\pi a^3 \mu \left(\frac{G}{2} - \Omega \right) \mathbf{k} .$$

When the particle is left to rotate freely under the influence of the simple shear flow, the global moment due to the action of the considered two flows has to be equal to zero ($\mathbf{M} = \mathbf{0}$), which is only possible when

$$\Omega = \frac{G}{2} . \qquad (4.3.21)$$

Taking into account (4.3.21) we get for the sum of the velocities (4.3.17) and (4.3.20):

$$\mathbf{v} = \mathrm{Gr}\ \sin^2\theta \sin\varphi \cos\varphi \left(1 - \frac{5}{2}\frac{a^3}{r^3} + \frac{3}{2}\frac{a^5}{r^5} \right) \mathbf{e}_r \qquad (4.3.22)$$

$$+ \mathrm{Gr}\ \sin\theta \cos\theta \sin\varphi \cos\varphi \left(1 - \frac{a^5}{r^5} \right) \mathbf{e}_\theta - \mathrm{Gr}\ \sin\theta \left(\sin^2\varphi + \frac{\cos 2\varphi}{2}\frac{a^5}{r^5} \right) \mathbf{e}_\varphi$$

The streamlines of this flow projected on the plane xy are given on Fig.4.3.2.

As a conclusion we would like to mention that the presented method of vector harmonic functions is extremely simple and rational. However, the constants determination from the boundary conditions when applying this method is sometimes very difficult. This type of difficulties can be resolved, due to the fact that the general form of the solution is the same for the whole class of problems of a given family. Therefore, very often it is possible to use one of the simple flows of the considered family to determine the constants of a more complex flow. Sometimes the application of the vector harmonic functions method is useful to be combined with the other classical methods and especially with the Lamb's method (Lamb, 1945) for the general solution of the Stokes equations.

Fig.4.3.2. Streamlines pattern due to a rotating sphere in a simple shear flow (from Leal, 1992).

CHAPTER 5

Other Methods to Study the Flow Past Single Rigid or Fluid Particles

5.1. Introduction

In the present chapter we consider the steady flow past a single rigid or fluid particle at small, moderate and large Reynolds numbers. To this end we use analytical, numerical and asymptotic methods.

In many hydrodynamic problems solved in Stokes approximation, the variables in the corresponding equations can be separated. With respect to the geometry of the considered hydrodynamic problem, the most suitable coordinate system is to be chosen and the corresponding equations written in it. Generally, the most suitable coordinate system for a particular problem is that, at which the surface of the particle in the fluid flow coincides with a given coordinate surface. In other words, if for example a problem is examined, the particle surface in the considered coordinate system (ξ,η,ζ) must have an equation $\xi=\xi_1=\text{const.}$ or $\eta=\eta_1=\text{const.}$.

When the spectrum of the solved problem is discrete, the obtained analytical solution is written in the form of infinite series, and when the spectrum is continuous, the solution can be given in an integral form. In this chapter we shall discuss some of the fundamental hydrodynamic problems for a single particle solved by different authors by means of this method.

The theoretical examination of the motion of bodies or particles in viscous fluids at zero Reynolds number starts in 1851 with the remarkable work of Stokes (1851), in which the famous law for the force on a sphere in a uniform viscous flow with velocity U, is given by (2.9.1).

A substantial class of flows studied very intensively during the last 15-20 years is the shear fluid flows which are of fundamental as well as of practical importance. The shear fluid flows arise in the chemical and biotechnological industry at the movement of the various apparatuses walls in fluid medium. Examples for such flows are: the Couette flow between two planar or spherical walls, one of which is moving in direction parallel to the other; Poiseuille flow in a cylindrical tube, caused by a prescribed pressure gradient; flow around a critical point, for example observed at the particle attraction by a collector; extensional flows at the production of fibber polymer materials, etc. The results of the shear flows investigations are used in modeling various disperse systems (suspensions, emulsions, etc.), the erythrocytes movement, the bodies movement in the traces (behind other bodies) or boundary layers and others.

The simple shear flows are characterised by a macroscopic parameter G, called a coefficient of the velocity profile change. The complicated shear flows can be defined by two, three, etc. parameters. The coefficient of the velocity profile change G at the simple shear flows has a dimension of reciprocal time. It participates in the Reynolds number expression $Re = \frac{a^2 G}{\nu}$, which is the basic dynamical parameter in the shear flows hydrodynamics (a is the flow length characteristics, and ν -the fluid kinematic viscosity).

The first fundamental results in the study of the shear flows are obtained by Einstein (1906, 1911), who calculating the effective viscosity of a suspension of equal spherical particles solved the problem for the hydrodynamic interaction of a simple shear flow and a neutrally suspended in it spherical particle, i.e., a particle to which the gravity force or other external forces are not acting. In this case only the shear flow acts on the particle, which creates non-zero stresses on its surface (see Chapter 10).

The Einstein theory was generalized by Jeffery (1922b) for a suspension of ellipsoidal particles, for which the orientation to the shear flow plane had a significant role. Because of the ellipsoid different geometry with respect to the sphere, the Jeffery's investigations were applied in the rheology of anisotropic disperse and polymer systems, the fluid crystals, the flows with double refraction and others.

The Jeffery's studies, done on the basis of the Stokes equations, show, that at the flow past a rotational ellipsoid by a simple shear flow, there is no tendency of its axis to be situated in some preferential situation to the undisturbed fluid flow. This is the Jeffery's model defect. The ellipsoid moment, calculated in the Stokes approximation, is equal to zero independently of the ellipsoid orientation with respect to the flow direction. A similar defect of the Stokes equations is also observed (Lorentz, 1907) at the eccentrically situated spherical particle gravitational fall inside a vertical circular cylinder. At its motion the sphere does not experience any force, which can push it to migrate in radial direction simultaneously with its motion downwards. This fact contradicts the experimental observations (Segre and Silberberg, 1962a,b).

A general solution of the Stokes equations for spherical geometries of the particles (rigid or fluid) was derived by Lamb (1945). A general theory applicable to a single particle of arbitrary shape has been developed by Brenner (1963, 1964a, b, c). Interesting results for the flow past particles of spherical and arbitrary shape are obtained by Brenner and Cox (1963) and many others. A full summary of these results is given in the Happel and Brenner's book (1973).

5.2. Eigenfunction Expansion for Axisymmetric Flows in Spherical Coordinates

In this section we shall find the general solution of the biharmonic equation

$$E^4\Psi = 0, \tag{5.2.1}$$

written in spherical coordinates (r, θ, φ), which describes the fluid flow in Stokes approximation. Putting $\cos\theta = \beta$, we have for the operator E^2

$$E^2 \equiv \frac{\partial^2}{\partial r^2} + \frac{\sin\theta}{r^2}\frac{\partial}{\partial\theta}\left(\frac{1}{\sin\theta}\frac{\partial}{\partial\theta}\right) = \frac{\partial^2}{\partial r^2} + \frac{1-\beta^2}{r^2}\frac{\partial^2}{\partial\beta^2}. \tag{5.2.2}$$

The solution of (5.2.1) could be found with the replacement

$$E^2\Psi = f(r,\beta), \tag{5.2.3}$$

because

$$E^2 f(r,\beta) = E^4\Psi = 0. \tag{5.2.4}$$

Therefore the following equation has to be solved first:

$$E^2 f(r,\beta) \equiv \frac{\partial^2 f}{\partial r^2} + \frac{1-\beta^2}{r^2}\frac{\partial^2 f}{\partial\beta^2} = 0 \tag{5.2.5}$$

and after that equation (5.2.3) is solved, as the found solution of the function $f(r,\beta)$ is replaced in its right-hand side.

The solution of equation (5.2.5) will be sought by the variables separation method:

$$f(r,\beta) = F(r)\Phi(\beta). \tag{5.2.6}$$

Substituting (5.2.6) in (5.2.5), we find that the functions F(r) and $\Phi(\beta)$ satisfy the equation

$$F''(r)\Phi(\beta) + \frac{1-\beta^2}{r^2}F(r)\Phi''(\beta) = 0. \tag{5.2.7}$$

From (5.2.7) it follows

$$\frac{r^2}{F(r)}\frac{d^2F}{dr^2} = -\frac{1-\beta^2}{\Phi(\beta)}\frac{d^2\Phi}{d\beta^2} = \text{const.} \tag{5.2.8}$$

Without loosing the generality, the constant in (5.2.8) is denoted by n(n+1) and therefore the functions F(r) and $\Phi(\beta)$ have to satisfy the equations

$$r^2\frac{d^2F(r)}{dr^2} - n(n+1)F(r) = 0, \tag{5.2.9}$$

$$\left(1-\beta^2\right)\frac{d^2\Phi(\beta)}{d\beta^2} + n(n+1)\Phi(\beta) = 0 \tag{5.2.10}$$

As the equation (5.2.9) has two partial integrals r^{n+1} and r^{-n}, then its general solution has the form:

$$F(r) = a_n r^{n+1} + b_n r^{-n}, \tag{5.2.11}$$

where a_n and b_n are arbitrary constants.

Differentiating the equation (5.2.10) with respect to β and putting $\Phi'(\beta) = Y(\beta)$, the Legendre equation is obtained

$$\frac{d}{d\beta}\left[\left(1-\beta^2\right)\frac{dY(\beta)}{d\beta}\right] + n(n+1)Y(\beta) = 0.$$

This equation is satisfied by the Legendre polynomials of I-st kind $P_n(\beta)$, which are defined by the formula

$$P_n(\beta) = \frac{1}{2^n n!}\frac{d^n}{d\beta^n}\left(\beta^2 - 1\right)^n. \tag{5.2.12}$$

The first several Legendre polynomials are correspondingly equal to:

$$P_0(\beta) = 1,\ P_1(\beta) = \beta,\ P_2(\beta) = \frac{1}{2}\left(3\beta^2 - 1\right), \text{ etc.} \tag{5.2.13}$$

Equation (5.2.10) is also satisfied by the Legendre functions of II-nd kind, which have a logarithmic singularity at $\beta = \pm 1$. Neglecting this second solution (because of its singularity at $\beta = \pm 1$), for all integer values of n we have $Y(\beta) = P_n(\beta)$. In order to satisfy the symmetry condition $\Phi(\beta) = 0$ at $\beta = \pm 1$, which assures the axisymmetry of Ψ, we substitute

$$\Phi(\beta) = \int_{-1}^{\beta} P_n(\beta)\, d\beta, \tag{5.2.14}$$

for all $P_n(\beta)$ except for $n = 0$.

In order to prove that $\int_{-1}^{1} P_n(\beta)\, d\beta = 0$, the integrand is written in the form $\int_{-1}^{1} P_0(\beta) P_n(\beta)\, d\beta$. From the orthonormality condition of the Legendre polynomials

$$\int_{-1}^{1} P_n(\beta) P_m(\beta)\, d\beta = \begin{cases} 0 & \text{at } n \neq m \\ \dfrac{2}{2n+1} & \text{at } n = m \end{cases} \tag{5.2.15}$$

we reach to $\Phi(1) = 0$.

The numbers $\lambda_n = n(n+1)$ are eigenvalues, and the functions $\Phi(\beta) = \int_{1}^{\beta} P_n(\beta)\, d\beta = Q_n(\beta)$, (where $Q_0(\beta) = \beta - 1$, $Q_1(\beta) = \frac{1}{2}\left(\beta^2 - 1\right)$, $Q_2(\beta) = \frac{1}{2}\left(\beta^3 - \beta\right)$, etc.) are eigenfunctions of equation (5.2.10). Therefore, for $f(r,\beta) = F(r)\Phi(\beta)$ we have

$$f(r,\beta)=\sum_{n=1}^{\infty}\left(a_n r^{n+1}+b_n r^{-n}\right)Q_n(\beta). \tag{5.2.16}$$

To find the function $\Psi(r,\beta)$ we have to solve the equation

$$E^2\Psi=\sum_{n=1}^{\infty}\left(a_n r^{n+1}+b_n r^{-n}\right)Q_n(\beta) \tag{5.2.17}$$

at arbitrary values of the constants a_n and b_n . Since the equations $E^2\Psi=0$ and $E^2 f=0$ are identical, the solution of Ψ_{hom} is sought in the type:

$$\Psi_{hom}=\sum_{n=1}^{\infty}\left(B_n r^{n+1}+D_n r^{-n}\right)Q_n(\beta), \tag{5.2.18}$$

while for the partial integral of (5.2.17):

$$\Psi_{part}=r^{\lambda}Q_n(\beta). \tag{5.2.19}$$

Due to the property of the functions $Q_n(\beta)$:

$$\left(1-\beta^2\right)\frac{d^2Q_n(\beta)}{d\beta^2}=-n(n+1)Q_n(\beta)$$

we get

$$E^2\left[r^{\lambda}Q_n(\beta)\right]=\alpha r^{\lambda-2}Q_n(\beta), \tag{5.2.20}$$

where α is a constant dependent on λ and n.

It follows from (5.2.19) and (5.2.20) that equation (5.2.17) has a partial integral of the form:

$$\Psi_{part}=\sum_{n=1}^{\infty}\left(A_n r^{n+3}+C_n r^{2-n}\right)Q_n(\beta), \tag{5.2.21}$$

where A_n and C_n are arbitrary constants. Then from (5.2.18) and (5.2.21) we obtain the following general solution of the biharmonic equation in spherical coordinates

$$\Psi(r,\beta)=\sum_{n=1}^{\infty}\left(A_n r^{n+3}+B_n r^{n+1}+C_n r^{2-n}+D_n r^{-n}\right)Q_n(\beta), \tag{5.2.22}$$

where A_n, B_n, C_n and D_n are arbitrary constants to be determined by the boundary condition of any creeping flow problem.

The force F_z that the passing flow, given by (5.2.22), exerts on an axisymmetric particle, whose mass centre coincides with the coordinate system origin, has an interesting property: it depends only on the constant C_1:

$$F_z=4\pi\mu d_c C_1 U_c, \tag{5.2.23}$$

where d_c and U_c are the characteristic length and velocity scales, respectively, and z is the direction of the symmetry axis.

In order to prove this statement, we shall note that the following formula is valid when dealing with the dimensionless variables

$$\mathbf{F} = \mu \, d_c U_c \iint_{S_p} (\mathbf{T}.\mathbf{n}) \, d\sigma . \tag{5.2.24}$$

Moreover, according to the divergence theorem, for the Stokes flows we have

$$0 = \iiint_V (\nabla.\mathbf{T}) \, d\tau = \iint_{S_p} (\mathbf{T}.\mathbf{n}) \, d\sigma - \iint_{S^*} (\mathbf{T}.\mathbf{n}^*) \, d\sigma , \tag{5.2.25}$$

where V is the flow volume between the particle surface S_p and another surface S^*, which envelops the particle. The vectors $\mathbf{n}$ and $\mathbf{n}^*$ are the outer normals to the surfaces S_p and S^*. Then from (5.2.25) it follows

$$\iint_{S_p} (\mathbf{T}.\mathbf{n}) \, d\sigma = \iint_{S^*} (\mathbf{T}.\mathbf{n}^*) \, d\sigma ,$$

which gives us an opportunity to change the complicated form of S_p with another, simpler surface S^* and thus to facilitate the calculation of the integral in (5.2.24). A sphere having a centre at the mass centre of a particle of arbitrary surface S_p and a radius of unity can be an example of a suitable surface S^*.

Then the following identity holds

$$\mathbf{F}.\mathbf{e}_z = \mu \, d_c U_c \int_0^{2\pi} \left[\int_0^{\pi} \mathbf{e}_z.(\mathbf{T}.\mathbf{n}^*) \sin\theta \, d\theta \right] d\varphi . \tag{5.2.26}$$

In the considered axisymmetric case we have

$$\mathbf{T}.\mathbf{n}^* = \mathbf{T}.\mathbf{e}_r = T_{rr} \mathbf{e}_r + T_{r\theta} \, \mathbf{e}_\theta , \tag{5.2.27}$$

where

$$T_{rr} = -p + 2 \frac{\partial v_r}{\partial r}, \qquad T_{r\theta} = r \frac{\partial}{\partial r} \left(\frac{v_\theta}{r} \right) + \frac{1}{r} \frac{\partial v_r}{\partial \theta} .$$

By the use of formulae (5.2.22), (1.3.14) and (1.3.17) we calculate the expressions for T_{rr} and $T_{r\theta}$ and substitute (5.2.25) in (5.2.24). After carrying out the integration in (5.2.24), we finally obtain the formula (5.2.23).

5.3. Uniform Flow Past an Arbitrary Axisymmetric Rigid Particle

Let an arbitrary axisymmetric rigid particle is passed by a flow with velocity U_c. If the axis Oz is directed to the particle axis of symmetry (Fig.5.3.1.), a spherical coordinate system (r, θ, φ), fixed with the particle, is introduced. In order to find the boundary conditions at $r \to \infty$, we have to take into account that

$$\mathbf{e}_z = \mathbf{e}_r \cos\theta + \mathbf{e}_\theta \cos(90 + \theta) = \cos\theta \, \mathbf{e}_r - \sqrt{1 - \cos^2\theta} \, \mathbf{e}_\theta . \tag{5.3.1}$$

Then at $r \to \infty$, putting $\cos\theta = \beta$, we obtain

$$v_r = U_c \beta , \tag{5.3.2}$$

$$v_\theta = -U_c\left(1-\beta^2\right)^{1/2}. \tag{5.3.3}$$

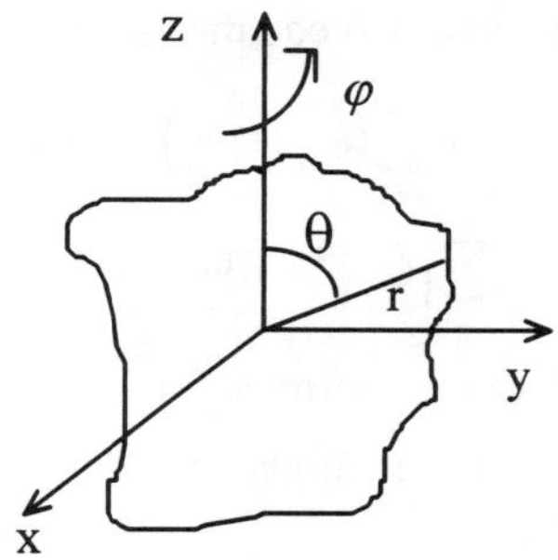

Fig.5.3.1. Spherical coordinate system.

From the relations between the dimensionless velocity components and the dimensionless stream function (1.3.14) and (1.3.17), we have

$$v_r = \frac{1}{r^2 \sin\theta}\frac{\partial\Psi}{\partial\theta}, \qquad v_\theta = -\frac{1}{r\sin\theta}\frac{\partial\Psi}{\partial r}. \tag{5.3.4}$$

Then the boundary conditions (5.3.2) and (5.3.3) for the dimensionless stream function Ψ are changed:

$$\frac{\partial\Psi}{\partial r} = r\left(1-\beta^2\right), \qquad \frac{\partial\Psi}{\partial\theta} = -\beta r^2, \text{ at } r \to \infty. \tag{5.3.5}$$

After integrating (5.3.5), we reach to

$$\Psi \to \frac{1}{2}r^2\left(1-\beta^2\right) \qquad \text{at } r\to\infty,$$

or

$$\Psi \to -r^2 Q_1(\beta) \qquad \text{at } r\to\infty. \tag{5.3.6}$$

Using the general solution of the stream function Stokes equation (5.2.22), we have

$$\Psi(r,\beta) = \sum_{n=1}^{\infty}\left(A_n r^{n+3} + B_n r^{n+1}\right)Q_n(\beta) + \sum_{n=1}^{\infty}\left(C_n r^{2-n} + D_n r^{-n}\right)Q_n(\beta). \tag{5.3.7}$$

Taking into account the boundary condition (5.3.6) we obtain that $A_n = 0$ at $n \ge 1$, $B_1 = -1$ and $B_n = 0$ at $n \ge 2$. Therefore

$$\Psi(r,\beta) = -r^2 Q_1(\beta) + \sum_{n=1}^{\infty}\left(C_n r^{2-n} + D_n r^{-n}\right)Q_n(\beta). \tag{5.3.8}$$

The constants C_n and D_n are defined after applying the boundary conditions on the particle surface. If the particle is spherical, the velocity boundary conditions on its surface are

$$v_r = v_\theta = 0 \qquad \text{at } r = 1, \tag{5.3.9}$$

correspondent to the stream function boundary conditions

$$\Psi = \frac{\partial \Psi}{\partial r} = 0 \quad \text{at } r = 1. \tag{5.3.10}$$

These two boundary conditions give two equations for the coefficients C_n and D_n :

$$\left| \begin{aligned} -Q_1 + \sum_{n=1}^{\infty} (C_n + D_n) Q_n(\beta) = 0 \\ -2Q_1 + \sum_{n=1}^{\infty} [C_n (2-n) - nD_n] Q_n(\beta) = 0 \end{aligned} \right.$$

Because the functions $Q_n(\beta)$ form a full system of orthogonal functions, the coefficients C_n and D_n satisfy the equations:

i) at $n = 1$

$$\left| \begin{aligned} -1 + C_1 = \frac{3}{2} + D_1 = 0 \\ -2 + C_1 - D_1 = 0 \end{aligned} \right.$$

ii) at $n \geq 2$

$$\left| \begin{aligned} C_n + D_n = 0 \\ C_n (2-n) - nD_n = 0 \end{aligned} \right.$$

From these two systems we find $C_1 = \frac{3}{2}$, $D_1 = -\frac{1}{2}$ and $C_n = D_n = 0$ at $n \geq 2$. Therefore for the stream function we have

$$\Psi = -\left(r^2 - \frac{3}{2} r + \frac{1}{2r} \right) Q_1(\beta). \tag{5.3.11}$$

Then using $C_1 = \frac{3}{2}$ for a spherical particle, from (5.2.23) we get the famous Stokes formula (2.9.1)

$$F_z = 6\pi \mu a U_c, \tag{5.3.12}$$

with characteristic scales a and U_c.

5.4. A Rigid Spherical Particle in an Axisymmetric Extensional Flow

Beside considering the uniform flow past rigid particles, which occurs at sedimentation, in a lot of applications it is necessary to know the fluid velocity field around the rigid particles passed by a non-uniform flow. If the particles density is equal to the density of the surrounding fluid, then it is said that the particles are neutrally suspended in the fluid. In that case the only reason for the fluid motion around the particles is the passing non-uniform flow, which is undisturbed at infinity but interacts with the particles in the vicinity.

Of special interest to the study of the so called slow linear gradient flows past rigid particles is the case when the latter are very small with respect to the characteristic linear scale L of the flow. In this case we can approximate the undisturbed velocity field at the particle vicinity by the Taylor's formula, i.e.

$$\mathbf{v}'^{\infty} = \mathbf{v}'_0 + (\nabla'\mathbf{v}')_0 . \mathbf{r}' + \ldots, \tag{5.4.1}$$

where $(\nabla\mathbf{v})_0$ is the velocity gradient, calculated at the particle mass centre, and $\mathbf{v}'_0$ is the velocity at its mass centre. With respect to the coordinate system moving together with the particle $\mathbf{v}'_0 = 0$ and

$$\mathbf{v}^{\infty} = (\nabla\mathbf{v})_0 . \mathbf{r} + \ldots \tag{5.4.2}$$

The flow given by (5.4.2) can be represented as a combination of purely deformational and purely rotational flow

$$\mathbf{v}^{\infty} = \mathbf{E}^{\infty} . \mathbf{r} + \mathbf{\Omega}^{\infty} . \mathbf{r} + \ldots, \tag{5.4.3}$$

where $\mathbf{E}^{\infty}$ and $\mathbf{\Omega}^{\infty}$ are the tensors of the deformation rate and vorticity.

From both flows: purely deformational and purely rotational, the first one is more complicated. Here we shall consider the flow past a rigid particle of radius a in an extensional axisymmetrical flow

$$\mathbf{v}^{\infty} = -\mathbf{E}^{\infty} . (x\mathbf{i} + y\mathbf{j} + z\mathbf{k}), \tag{5.4.4}$$

where the deformation rate tensor components are given by the matrix

$$\mathbf{E}^{\infty} = \mathbf{E}_0 \begin{pmatrix} 1 & 0 & 0 \\ 0 & 1 & 0 \\ 0 & 0 & -2 \end{pmatrix}.$$

For $\mathbf{E}_0 > 0$ the flow is directed outward the particle, along its axis of symmetry. Besides this flow, there exists another flow, directed towards the plane perpendicular to this axis. This flow is named uniaxial extensional flow. For $\mathbf{E}_0 < 0$ the fluid flow is in the opposite direction and is named biaxial extensional flow. In both cases the flow is axisymmetrical and we can use the stream function Ψ obtained in section 5.2. to solve the posed problem

$$\Psi(r,\beta) = \sum_{n=1}^{\infty} \left(A_n r^{n+3} + B_n r^{n+1} + C_n r^{2-n} + D_n r^{-n}\right) Q_n(\beta). \tag{5.4.5}$$

Since for spherical particles it is convenient to explore a spherical coordinate system (r, θ, φ), we shall transform (5.4.4) in spherical coordinates using the general relations with Cartesian coordinates (1.3.13) and (4.3.11). Thus we reach to

$$\mathbf{v}^{\infty} = -\mathbf{E}_0 r(1 - 3\cos^2\theta)\mathbf{e}_r - 3\mathbf{E}_0 r \sin\theta\cos\theta\, \mathbf{e}_{\theta} . \tag{5.4.6}$$

From (5.4.4) it follows that the quantity $\mathbf{E}_0$ has dimension of T^{-1} and

$$v_c = a\mathbf{E}_0 \tag{5.4.7}$$

can be chosen as a typical velocity scale of the flow, where a is the spherical rigid particle radius.

Thus the dimensionless velocity at infinity will be

$$\mathbf{v}^{\infty} = -r(1 - 3\cos^2\theta)\mathbf{e}_r - 3r\sin\theta\cos\theta\, \mathbf{e}_{\theta} ,$$

or

$$\mathbf{v}^{\infty} = -r\left(1-3\beta^2\right)\mathbf{e}_r - 3r\beta\sqrt{1-\beta^2}\,\mathbf{e}_\theta\,, \qquad \text{at } r\to\infty, \tag{5.4.8}$$

where $\beta = \cos\theta$.

From $\mathbf{v}_r^{\ \infty} = -r\left(1-3\beta^2\right)$ and $\mathbf{v}_\theta^{\ \infty} = -3r\beta\sqrt{1-\beta^2}$, it follows that

$$\Psi^{\infty} = r^3\beta\left(1-\beta^2\right) = -2r^3 Q_2(\beta) \qquad \text{at } r\to\infty, \tag{5.4.9}$$

because $Q_2(\beta) = -\dfrac{1}{2}\beta\left(\beta^2-1\right)$.

From the boundary condition (5.4.9) and (5.4.5) we derive that

$$A_n = 0 \quad \text{at } n \ge 1, \qquad B_1 = 0,\ B_2 = -2,\ B_n = 0 \quad \text{at } n \ge 3. \tag{5.4.10}$$

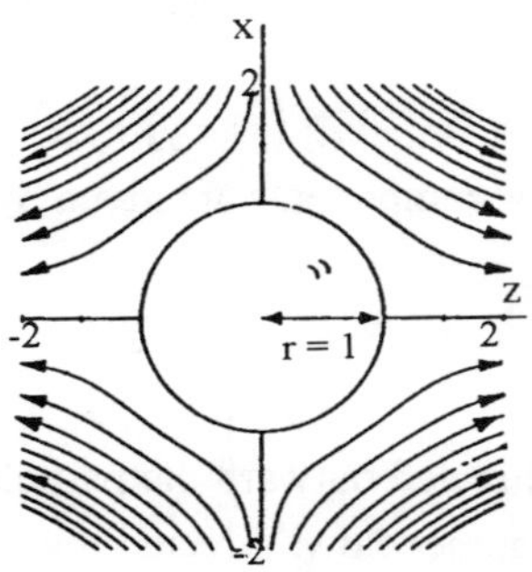

Fig.5.4.1. Streamlines pattern around a rigid sphere in extensional flow (from Leal, 1992).

The remaining constants in (5.4.5) will be determined by the boundary conditions on the particle surface

$$\Psi = \frac{\partial\Psi}{\partial r} = 0 \quad \text{at } r = 1. \tag{5.4.11}$$

The following equations are derived by the equalities (5.4.5), (5.4.10) and (5.4.11)

$$0 = \Psi = -2Q_2(\beta) + \sum_{n=1}^{\infty}\left(C_n + D_n\right)Q_n(\beta),$$

$$0 = \frac{\partial\Psi}{\partial r} = -6Q_2(\beta) + \sum_{n=1}^{\infty}\left[(2-n)C_n - nD_n\right]Q_n(\beta),$$

from which the result is $C_n = D_n = 0$ at $n \ne 2$ and

$$\left|\begin{array}{l} -2 + C_2 + D_2 = 0 \\ -6 + (2\text{-}2)C_2 - 2D_2 = 0 \end{array}\right. \tag{5.4.12}$$

Upon solving this system we get $D_2 = -3$ and $C_2 = 5$. Therefore, the sought solution is

$$\Psi(r,\beta) = -2r^3 Q_2(\beta) + 5Q_2(\beta) - 3r^{-2}Q_2(\beta) = 2\left(-r^3 + \frac{5}{2} - \frac{3}{2}\frac{1}{r^2}\right)Q_2(\beta). \tag{5.4.13}$$

The streamlines flow pattern corresponding to the obtained solution for the flow past a rigid spherical particle is given in Fig.5.4.1.

Since the coefficient C_1 of (5.4.13) is equal to zero, from (5.2.23) it follows that no drag force is exerted by the sphere. Moreover, the sign of E_0 does not influence the stream function form (5.4.13) (since the stream function is dimensionless and E_0^{-1} is characteristic time) and it is valid for both uniaxial and biaxial flows.

5.5. Spherical Bubble Motion due to Thermocapillary Convection

The thermocapillary (Marangoni) flows are known for long time, but the number of experimental and theoretical studies until recently has been very small. Hershey (1939) made some experiments to show that the temperature gradient in direction to the interface creates a steady flow of the fluids on either side of the interface. Block (1956) confirmed experimentally that the motion in the Benard's cells is a result of the surface tension change due to the temperature gradient. The first experimental and theoretical investigation, in which the thermocapillary bubble drift has been discovered, was carried out by Young et al. (1959). In their experiment, a vertical temperature gradient creates a thermocapillary force balanced by the gravity force and the bubble rests motionless instead of to raise upwards. This fully confirms the theoretically made predictions. Later, Papazian and Wilcox (1976) made an unsuccessful Marangoni experiment in a weak gravity field because of some technical difficulties.

Another reason for surface tension change on the particles surface S is the temperature variation on S. The fluids motion caused by the non-uniform temperature distribution on the interface is called thermocapillary convection (Subramanian., 1987). Nowadays, the thermocapillary convection is used for the creation of new production technologies of high quality materials for the space industry in microgravity conditions. An important application of the thermocapillary droplets movement is the creation of a mechanism for gas bubbles removal at high quality glass and other fine materials production.

In this section we shall consider the thermocapillary movement of a spherical bubble in gravity field (see Fig.5.5.1.) in a simplified formulation (without solving the Laplace equation for the temperature on the bubble surface) for the problem studied by Young et al. (1959). Suppose that a gas bubble is in a viscous fluid with a temperature gradient field acting on it, so that the fluid temperature decreases upwards.

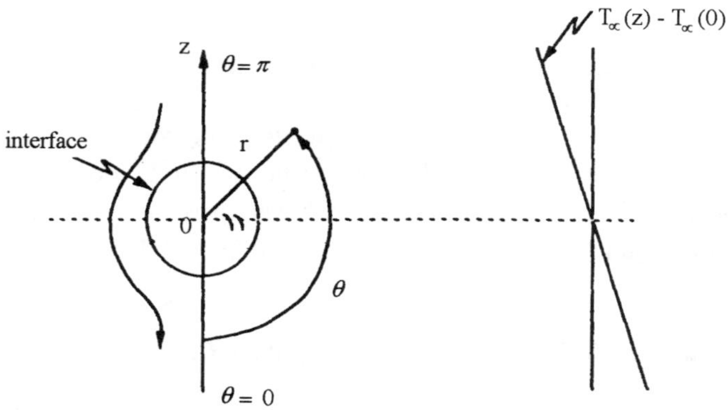

Fig.5.5.1. A sketch of the bubble geometry and temperature distribution for thermocapillary motion of a gas bubble.

The derivative $\frac{\partial \sigma}{\partial T}$ in the expression

$$\frac{\partial \sigma}{\partial \theta} = \frac{\partial \sigma}{\partial T} \cdot \frac{\partial T}{\partial \theta} \tag{5.5.2}$$

decreases with the temperature growth and therefore

$$\frac{\partial \sigma}{\partial T} = -\gamma < 0 \qquad \text{(if } \gamma > 0\text{).} \tag{5.5.3}$$

We suppose that the temperature field far away from the bubble is given in the form

$$T_\infty(z) = T_\infty(0) - \alpha z, \tag{5.5.4}$$

where $T_\infty(0)$ is the temperature on a horizontal plane z = 0 passing through the spherical bubble centre. Assuming the absence of viscous dissipation and negligible convection with respect to the molecular heat transfer, the temperature distribution in the fluid is described by the Laplace equation

$$\nabla^2 T = 0. \tag{5.5.5}$$

The boundary conditions for the temperature problem are (5.5.4) at $r \to \infty$, and at $r = 1$:

$$k_1 \left.\frac{\partial T}{\partial r}\right|_{r=1+\varepsilon} = k_2 \left.\frac{\partial T}{\partial r}\right|_{r=1-\varepsilon}, \tag{5.5.6}$$

where k_1 and k_2 are the conductivity coefficients of the fluid inside and outside the particle, respectively, and $\varepsilon \ll 1$. In our case $k_1 \ll k_2$ and (5.5.6) is simply reduced to

$\frac{\partial \mathrm{T}}{\partial \mathrm{r}}=0$. The function $\mathrm{T}_\infty(\mathrm{z})$, given by (5.5.4), occurs to be the exact solution of (5.5.5), when $\mathrm{k}_1=\mathrm{k}_2$ is assumed. However, this is absolutely correct for the region away from the bubble and approximately correct in the vicinity of the bubble.

Now, let us write down the dynamic problem boundary conditions as the spherical coordinate system has its origin at the particle centre. The dimensionless boundary conditions for the velocity and stress (1.4.3), (1.4.14) and (1.4.9) have the form

$$\mathbf{v} \to -\mathbf{e}_z \qquad \text{at } |\mathbf{r}| \to \infty, \tag{5.5.7}$$

$$\mathrm{v}_\mathrm{r}=0 \qquad \text{at } \mathrm{r}=1, \tag{5.5.8}$$

$$\mathrm{T}_{\mathrm{r}\theta}+\frac{1}{\mu \mathrm{U}}\frac{\partial \sigma}{\partial \theta}=0 \qquad \text{at } \mathrm{r}=1, \tag{5.5.9}$$

where U is the characteristic flow velocity.

From $\mathrm{z}=\mathrm{r}\cos(180-\theta)=-\mathrm{r}\cos\theta$ and (5.5.4), it follows that

$$\mathrm{T}_\infty(\mathrm{z})=\mathrm{T}_\infty(0)+\alpha\,\mathrm{r}\cos\theta\ . \tag{5.5.10}$$

If we substitute $\cos\theta=\beta$ and take into account that

$$\left.\frac{\partial \mathrm{T}}{\partial \theta}\right|_{\mathrm{r}=1}=-\alpha\,\mathrm{r}\sin\theta\big|_{\mathrm{r}=1}=-\alpha\sin\theta$$

we obtain

$$\frac{\partial \sigma}{\partial \theta}=\frac{\partial \sigma}{\partial \mathrm{T}}\cdot\frac{\partial \mathrm{T}}{\partial \theta}=-\gamma(-\alpha\sin\theta)=\gamma\,\alpha\sin\theta=\gamma\,\alpha\left(1-\beta^2\right)^{1/2}. \tag{5.5.11}$$

Then the tangential stress condition (5.5.9) takes the form

$$\mathrm{T}_{\mathrm{r}\theta}+\frac{1}{\mu \mathrm{U}}\gamma\,\alpha\left(1-\beta^2\right)^{1/2}=0\,. \tag{5.5.12}$$

Since the studied problem is axisymmetrical and in terms of the boundary condition (5.5.7) we can use the stream function obtained in section 5.3.

$$\Psi(\mathrm{r},\beta)=-\mathrm{r}^2\mathrm{Q}_1(\beta)+\sum_{\mathrm{n}=1}^{\infty}\left(\mathrm{C}_\mathrm{n}\mathrm{r}^{2-\mathrm{n}}+\mathrm{D}_\mathrm{n}\mathrm{r}^{-\mathrm{n}}\right)\mathrm{Q}_\mathrm{n}(\beta), \tag{5.5.13}$$

where $\mathrm{P}_\mathrm{n}(\beta)$ and $\mathrm{Q}_\mathrm{n}(\beta)$ are given in section 5.2.

The coefficients C_n and D_n will be determined by the boundary conditions. The condition (5.5.8) following from (5.3.4) is equivalent to the identity $\Psi=0$ at $\mathrm{r}=1$. Therefore,

$$0=-\mathrm{Q}_1(\beta)+\sum_{\mathrm{n}=1}^{\infty}\left(\mathrm{C}_\mathrm{n}+\mathrm{D}_\mathrm{n}\right)\mathrm{Q}_\mathrm{n}(\beta). \tag{5.5.14}$$

From the boundary conditions (5.5.12) and taking into account that $\mathrm{v}_\theta=-\frac{1}{\mathrm{r}\sin\theta}\frac{\partial \Psi}{\partial \mathrm{r}}=-\frac{1}{\mathrm{r}\sqrt{1-\beta^2}}\frac{\partial \Psi}{\partial \mathrm{r}}$ and $\left.\mathrm{T}_{\mathrm{r}\theta}\right|_{\mathrm{r}=1}=\mu\,\mathrm{r}\left.\frac{\partial}{\partial \mathrm{r}}\left(\frac{\mathrm{v}_\theta}{\mathrm{r}}\right)\right|_{\mathrm{r}=1}$, we reach to

$$-2Q_1(\beta)+2C_1Q_1(\beta)-4D_1Q_1(\beta) \qquad (5.5.15)$$
$$+\sum_{n=2}^{\infty}\left[(2-n)(n+1)C_n-n(n+3)D_n\right]Q_n(\beta)=\frac{2}{\mu U}\gamma\,\alpha Q_1(\beta),$$

while the boundary condition at infinity (5.5.7) is identically fulfilled by Ψ given with (5.5.13).

At n = 1 a system of two equations for the two unknown quantities C_1 and D_1 is derived from (5.5.14) and (5.5.15)

$$\left|\begin{aligned} C_1+D_1&=1\\ C_1-2D_1&=1+\frac{\gamma\,\alpha}{\mu U}\end{aligned}\right. \qquad (5.5.16)$$

which gives $C_1=1+\frac{\gamma\,\alpha}{3\mu U}$ and $D_1=-\frac{\gamma\,\alpha}{3\mu U}$.

At $n\ge 2$ a homogeneous system is obtained

$$\left|\begin{aligned} C_n+D_n&=0\\ (2-n)(n+1)C_n-n(n+3)D_n&=0\end{aligned}\right. \qquad (5.5.17)$$

whose determinant is not equal to zero and thus $C_n=D_n=0$.

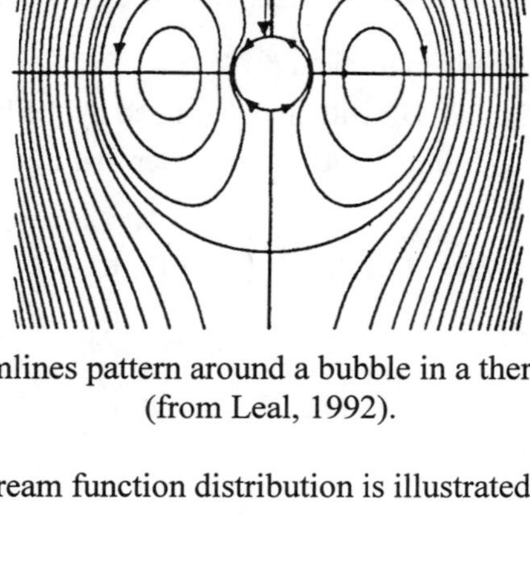

Fig.5.5.2. Streamlines pattern around a bubble in a thermocapillary flow (from Leal, 1992).

The fluid flow stream function distribution is illustrated in Fig.5.5.2.

From (5.2.23) the drag force exerted by the bubble at its motion in the fluid is

$$\mathbf{F} = \iint_S (\mathbf{T}.\mathbf{n})\, d\sigma = 4\pi\,\mu\, a\mathbf{U}\left(1+\frac{\gamma\,\alpha}{3\mu\,U}\right), \qquad (5.5.18)$$

where $\mathbf{U} = (0,0,U)$ is the bubble velocity vector.

The terminal velocity U is obtained after balancing the gravity force and the drag force (5.5.18), namely

$$\frac{4}{3}\pi\, a^3 \rho\, g = 4\pi\,\mu\, aU\left(1+\frac{\gamma\,\alpha}{3\mu\,U}\right),$$

from where

$$U = \frac{1}{3}\frac{a^2\rho g}{\mu}\left(1-\frac{\gamma\,\alpha}{a^2\rho\, g}\right).$$

Then the temperature gradient correspondent to a motionless bubble, i.e., zero drift velocity, is $\alpha = \dfrac{a^2\rho\, g}{\gamma}$, while the corresponding result of Young et al. (1959) differs only by a factor of 2/3.

An interesting theoretical analysis of the thermocapillary drift of a deformable gas bubble in a fluid under the action of a constant temperature gradient is done by Bratuhin (1975). By the perturbation method he obtained the terminal velocity of the bubble in a series expansion of the Marangoni number $Mn = \dfrac{\sigma\,\beta\, a|\nabla T|}{\rho\,\nu^2}$, where $|\nabla T|$ is the absolute value of the constant temperature gradient, $\beta =$ const. >0 and a is the radius of the spherical fluid particle of equivalent volume. From this solution, as a particular case at $Mn = 0$, the solution of Young et al. (1959) can be derived.

The accuracy of the mathematical model of the spherical drop motion under the action of gravity in a non-uniform temperature field is studied in (Rivkind and Sigovtzev, 1980) and the velocity and temperature field at different Marangoni numbers are given, as well. It is shown that the thermocapillary convection strongly influences the drop shape deformation. It is also established that the drift velocity increases with the decrease of the Prandtl number at fixed Marangoni number. The Reynolds number dependence on the Marangoni number at different Prandtl numbers is investigated. The problem for the fluid particles motion under the action of temperature gradient, when the Reynolds number and the temperature gradient are independently posed, represents a separate interest, too.

Using the method of the matched asymptotic expansions, Subramanian (1981) studied the slow motion of a gas bubble under a temperature gradient.

More detailed investigations of the single bubble drift in a unbounded fluid are performed by Subramanian (1983, 1985) and Bauer (1984). It is worth mentioning here the work of Thompson et al. (1980) in which the steady Navier-Stokes equations and the heat transfer equation are used to obtain the bubble deformation, the temperature distribution and the terminal velocity of a bubble motion under the action of a given temperature gradient.

5.6. Fluid Flows Outside and Inside a Drop, Induced by an Electric Field

As outlined in (Melcher and Taylor, 1969), if for a given fluid volume the electric permittivity ε is constant and its bulk charge density q is zero (electric conduction prevails over the electric charge convection in the bulk, i.e., the electric Reynolds number Re_e is zero), then the fluid is not coupled to the electric field in the bulk. If the electrical conductivity R is constant, the electric field intensity $E = -\mathrm{grad}\phi$ is a solenoidal vector and its potential ϕ satisfies the Laplace's equation, i.e., $\Delta\phi = 0$. In such cases the electromechanical coupling could arise at the fluid interfaces.

This physical situation can be regarded in the following example (Taylor, 1966). If an electrified fluid drop of radius a, constant electric permittivity $\hat{\varepsilon}$ and electrical conductivity $\hat{R}$ is suspended in another electrified fluid of constant electric permittivity ε and electrical conductivity R, on which a homogeneous electric field with intensity $\mathbf{E}_0$ is acting, then an electric charge is created on the fluid drop interface. The interaction of the electric fields with the accumulated electric charge on the interface gives rise to normal and shear electric stresses on it. While the normal electric stress leads (depending on the interfacial tension σ value) to smaller or larger deformation of the drop, then the electric shear stress induces fluid flows both inside and outside the drop.

We shall start with the electrical problem, since it is not influenced by the hydrodynamics. Following the model of Melcher and Taylor (1969) and the above assumptions, the only unknown quantities which have to be determined are the electric field intensities inside and outside the drop:

$$\mathbf{E} = -\mathrm{grad}\phi \ , \ \ \hat{\mathbf{E}} = -\mathrm{grad}\hat{\phi} \ . \tag{5.6.1}$$

There are four boundary conditions to be satisfied by the electric field intensities:

(i^e) no singularities at the drop origin;

(ii^e) uniform electric field at infinity, $\mathbf{E} \to \mathbf{E}_0$;

(iii^e) tangential electric field is continuous at the interface, $\mathbf{n}\mathrm{x}[\mathbf{E}] = 0$;

(iv^e) normal conduction current is continuous at the interface, $\mathbf{n}.[R\mathbf{E}] = 0$.

Here the square brackets [.] denote the jump across the interface of the function inside of them.

The formulated electric problem (5.6.1) with the boundary conditions (i^e) - (iv^e) is electrostatic and its potentials ϕ and $\hat{\phi}$ must satisfy the Laplace equation [see (1.3.16) and (1.3.14)].

In this section we assume that the drop preserves its spherical shape and the normal stress balance (of the hydrodynamic problem) on the interface is dropped. A spherical coordinate system (r, θ, φ) is adopted with an origin in the drop centre. Due to the axis symmetry, the solution of Laplace equation for both potentials is sought by the separation of the variables method in the form:

$$\phi(\mathrm{r},\theta) = \mathrm{X}(\mathrm{r}).\,\mathrm{Y}(\theta)\,. \tag{5.6.2}$$

The first function X(r) is a polynomial of r, while the second - Y(θ) is the first kind Legendre polynomial, i.e.,

$$\phi(\mathrm{r},\theta) = \sum_{n=0}^{\infty}\left(\mathrm{A}_n \mathrm{r}^n + \mathrm{B}_n \mathrm{r}^{-n-1}\right)\mathrm{P}_n(\cos\theta)\,. \tag{5.6.3}$$

From the boundary condition (ii^e), which implies $\phi \to -\mathrm{E}_0 \mathrm{r}\cos\theta$ when $\mathrm{r}\to\infty$, it follows that the potential $\phi(\mathrm{r},\theta)$ contains the function $\cos\theta$. Then the electric potential outside the drop (5.6.3) will be:

$$\phi(\mathrm{r},\theta) = \left(\mathrm{A}_1 \mathrm{r} + \mathrm{B}_1 \mathrm{r}^{-2}\right)\cos\theta\,. \tag{5.6.4}$$

and inside the drop:

$$\hat{\phi}(\mathrm{r},\theta) = \left(\hat{\mathrm{A}}_1 \mathrm{r} + \hat{\mathrm{B}}_1 \mathrm{r}^{-2}\right)\cos\theta\,. \tag{5.6.5}$$

Due to the boundary condition (i^e), it is evident that $\hat{\mathrm{B}}_1 = 0$ and due to (ii^e): $\mathrm{A}_1 = -\mathrm{E}_0$. From the boundary condition (iii^e) $\left[\mathrm{E}_\theta\right]_{\mathrm{r}=\mathrm{a}} = 0$, which means that $\phi(\mathrm{a},\theta) = \hat{\phi}(\mathrm{a},\theta)$, we have

$$-\mathrm{E}_0 \mathrm{a} + \mathrm{B}_1 \mathrm{a}^{-2} = \hat{\mathrm{A}}_1 \mathrm{a}\,, \tag{5.6.6}$$

and from (iv^e) $\left[R\mathrm{E}_\mathrm{r}\right]_{\mathrm{r}=\mathrm{a}} = 0$, written in terms of the potentials $R\left.\frac{\partial\phi}{\partial \mathrm{r}}\right|_{\mathrm{r}=\mathrm{a}} = \hat{R}\left.\frac{\partial\hat{\phi}}{\partial \mathrm{r}}\right|_{\mathrm{r}=\mathrm{a}}$, we find the latter relation for the unknown coefficients:

$$R\left(-\mathrm{E}_0 - 2\mathrm{B}_1 \mathrm{a}^{-3}\right) = \hat{R}\hat{\mathrm{A}}_1\,. \tag{5.6.7}$$

Finally, we get for the two potentials the expressions:

$$\phi(\mathrm{r},\theta) = -\mathrm{E}_0\left(\mathrm{r} + \frac{\mathrm{a}^3(1-\lambda)}{(\lambda+2)}\frac{1}{\mathrm{r}^2}\right)\cos\theta \tag{5.6.8}$$

and

$$\hat{\phi}(\mathrm{r},\theta) = -\frac{3\mathrm{E}_0}{(\lambda+2)}\mathrm{r}\cos\theta\,, \tag{5.6.9}$$

where $\lambda = \dfrac{\hat{R}}{R}$.

The components of the electric surface stress tensor, as given by Melcher and Taylor (1969), are acting on the two sides of the interface. Their jump across the interface contributes to the normal and tangential electro-mechanical stress balance:

$$\mathrm{T}_\mathrm{r}^\mathrm{e} = \left[\mathrm{T}_{\mathrm{rr}}^\mathrm{e}\right]_{\mathrm{r}=\mathrm{a}} = \frac{1}{2}\left\{\left[\varepsilon \mathrm{E}_\mathrm{r}^2\right]_{\mathrm{r}=\mathrm{a}} - \left[\varepsilon \mathrm{E}_\theta^2\right]_{\mathrm{r}=\mathrm{a}}\right\} \tag{5.6.10$_1$}$$

$$\mathrm{T}_\theta^\mathrm{e} = \left[\mathrm{T}_{\mathrm{r}\theta}^\mathrm{e}\right]_{\mathrm{r}=\mathrm{a}} = \left[\varepsilon \mathrm{E}_\mathrm{r}\mathrm{E}_\theta\right]_{\mathrm{r}=\mathrm{a}} \tag{5.6.10$_2$}$$

and causes the fluid flow inside and outside the drop. It can be observed that $T_r^e \sim \cos^2\theta$ and $T_\theta^e \sim \cos\theta\sin\theta$, which will be helpful when solving the fluid dynamics problem.

In the framework of the Stokes approximation of the hydrodynamic problem, it is clear that its flow fields, expressed by means of the stream function Ψ for the outer flow and by $\hat{\Psi}$ for the inner flow, have the form (5.2.22). The unknown coefficients will be determined by the boundary conditions:

(i^m) no singularities at the drop origin;

(ii^m) zero velocity vector at infinity;

(iii^m) zero normal velocities on the interface: $\Psi|_{r=a} = \hat{\Psi}|_{r=a} = 0$;

(iv^m) continuous tangential velocities on the interface: $\left.\frac{\partial\Psi}{\partial r}\right|_{r=a} = \left.\frac{\partial\hat{\Psi}}{\partial r}\right|_{r=a}$;

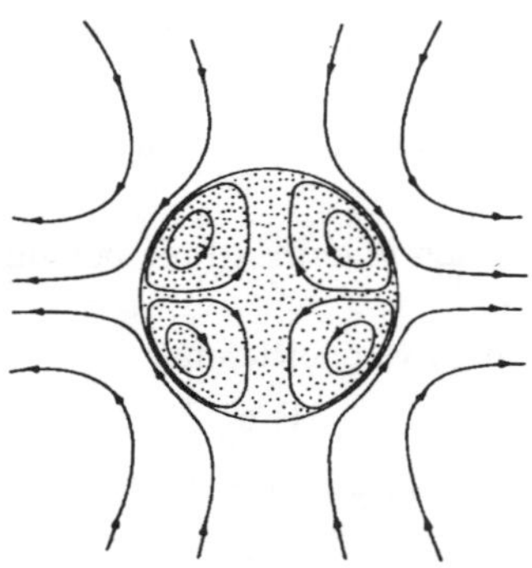

Fig.5.6.1. Streamlines pattern around a drop in an electrical field acting in the direction of the fluid flow (from Melcher and Taylor, 1969).

(v^m) tangential stress balance $T_\theta^e + \left[T_{r\theta}^m\right]_{r=a} = 0$.

From the boundary condition (v^m) written in terms of the stream functions

$$\left[\varepsilon E_r E_\theta\right]_{r=a} - \left[\mu r \frac{\partial}{\partial r}\left(\frac{1}{r^2\sin\theta}\frac{\partial\Psi}{\partial r}\right)\right]_{r=a} = 0, \tag{5.6.11}$$

it follows that they are proportional to $\sin^2\theta\cos\theta$. Turning back to (5.2.22), the only Gegenbauer polynomial having any contribution to the stream functions expressions is at n = 2, i.e., $Q_2(\cos\theta) = -0.5\sin^2\theta\cos\theta$. Then, taking into account the first two boundary conditions, the stream function for both regions will have the form

$$\Psi = \left(Ca^2 + Da^4r^{-2}\right)\sin^2\theta\cos\theta\,, \qquad (5.6.12_1)$$

$$\hat{\Psi} = \left(\hat{A}a^{-3}r^5 + \hat{B}a^{-1}r^3\right)\sin^2\theta\,\cos\theta\,. \tag{5.6.12$_2$}$$

The third and fourth boundary conditions, (iii^m) and (iv^m), give the following three relations for the unknown coefficients

$$C + D = 0, \tag{5.6.13$_1$}$$

$$\hat{A} + \hat{B} = 0, \tag{5.6.13$_2$}$$

$$-2D = 5\hat{A} + 3\hat{B}\,. \tag{5.6.13$_3$}$$

With the boundary condition expressed by equation (5.6.11) and taking into account (5.6.1), (5.6.8) and (5.6.9) we obtain a closed algebraic system for the unknown coefficients:

$$\frac{10}{a}\left(\mu D - \hat{\mu}\hat{A}\right) = -\frac{9E_0^{\,2}(\lambda\varepsilon - \hat{\varepsilon})}{(\lambda+2)^2}\,. \tag{5.6.13$_4$}$$

Finally, with the determined coefficients $-C = D = -\hat{A} = \hat{B} = U$, the stream functions outside and inside the fluid drop are respectively:

$$\Psi = -Ua^2\left(1 - a^2r^{-2}\right)\sin^2\theta\,\cos\theta\,, \tag{5.6.14}$$

$$\hat{\Psi} = -Ua^{-1}\left(a^{-2}r^5 - r^3\right)\sin^2\theta\,\cos\theta\,, \tag{5.6.15}$$

where $U = -\dfrac{9\hat{\varepsilon}E_0^{\,2}a(\lambda S - 1)}{10(\lambda+2)^2(\mu+\hat{\mu})}$ is the maximum surface velocity $v_\theta\big|_{r=a} = U\sin 2\theta$ at $\theta = \dfrac{\pi}{4}$, which can be used as a characteristic velocity and $S = \dfrac{\varepsilon}{\hat{\varepsilon}}$.

For the case of $\lambda S < 1$ (corresponding to a highly insulating drop) the direction of the flow induced by the field is sketched in Fig.5.6.1.

5.7. Lamb's General Solution

The solution of Laplace equation for an arbitrary function Φ in spherical coordinates (r, θ, φ) is a spherical harmonic of the type (Lamb, 1945):

$$\Phi_n(r,\theta,\varphi) = r^n S_n(\theta,\varphi), \tag{5.7.1}$$

where n = 0, ±1, ±2, ... and S_n is a surface harmonic of degree n. If it is a finite and uniform function of φ, then it can be expanded into series of cos(mφ) and sin(mφ). If, on the other hand, this expansion is valid everywhere on the sphere, i.e., $\varphi \in [0, 2\pi]$, then on the ground of Fourier's theorem we can assume that m is an integer. The differential equation satisfied by these harmonics is

$$\frac{d}{d\beta}\left\{\left(1-\beta^2\right)\frac{dS_n}{d\beta}\right\} + \left\{n(n+1) - \frac{m^2}{\left(1-\beta^2\right)}\right\}S_n = 0 \tag{5.7.2}$$

where $\beta = \cos\theta$ and $\beta \in [-1, 1]$. The finite solutions of this equation correspond to the first kind associated Legendre functions $P_n^m(\beta)$ of degree n and order m connected with the Legendre polynomials $P_n(\beta)$ is given by (5.2.12):

$$P_n^m(\beta) = (-1)^m (1-\beta^2)^{\frac{m}{2}} \frac{d^m}{d\beta^m} P_n(\beta). \tag{5.7.3}$$

Finally for S_n we have:

$$S_n = \sum_{m=0}^{n} (A_{nm} \cos m\varphi + B_{nm} \sin m\varphi) P_n^m(\beta). \tag{5.7.4}$$

Since the pressure is a harmonic function, it can be represented as a summation of spherical harmonic functions

$$p = \sum_{n=-\infty}^{\infty} p_n, \tag{5.7.5}$$

where

$$p_n = r^n \sum_{m=0}^{n} (a_{nm} \cos m\varphi + a_{nm}^* \sin m\varphi) P_n^m(\beta). \tag{5.7.6}$$

The Stokes equation (2.3.1) is more complicated than the Laplace's equation and its solution given by Lamb (1945) is expressed by three harmonic functions p, ϕ, κ of the type (5.7.5). Analogously to p_n, we can write down the solutions ϕ_n and κ_n:

$$\phi_n = r^n \sum_{m=0}^{n} (b_{nm} \cos m\varphi + b_{nm}^* \sin m\varphi) P_n^m(\beta), \tag{5.7.7}$$

$$\kappa_n = r^n \sum_{m=0}^{n} (c_{nm} \cos m\varphi + c_{nm}^* \sin m\varphi) P_n^m(\beta). \tag{5.7.8}$$

If the problem is axisymmetrical, the spherical harmonic functions are simplified, because they are independent on φ, i.e., m = 0 in (5.7.6)-(5.7.8):

$$p_n = a_n r^n P_n(\beta), \qquad \phi_n = b_n r^n P_n(\beta), \qquad \kappa_n = c_n r^n P_n(\beta).$$

Then Lamb's solution is:

$$\mathbf{v} = \sum_{\substack{n=-\infty \\ n\neq -1}}^{\infty} \left[\frac{(n+3) r^2 \nabla p_n}{2\mu(n+1)(2n+3)} - \frac{n \mathbf{r} p_n}{\mu(n+1)(2n+3)} \right] + \sum_{n=-\infty}^{\infty} \left[\nabla \phi_n + \nabla \times (\mathbf{r} \kappa_n) \right]. \tag{5.7.9}$$

We shall prove that its first two terms give a particular integral of the Stokes equation $\mu \nabla^2 \mathbf{v}_p = \nabla p$ and the remaining terms represent a homogeneous solution of the homogeneous equation $\nabla^2 \mathbf{v}_h = 0$.

Let us put $\mathbf{v}_p = \alpha_n r^2 \nabla p_n + \beta_n \mathbf{r} p_n$, where α_n and β_n are unknown constants and substitute in the continuity equation $\nabla . \mathbf{v} = 0$ and Stokes equation, from where we get two algebraic equations for the two coefficients:

$$\left| \begin{aligned} &2n\alpha_n + (n+3)\beta_n = 0, \\ &2(2n+1)\alpha_n + 2\beta_n = 1. \end{aligned} \right.$$

As a result we obtain the values of the unknown coefficients:

$$\alpha_n = -\frac{n+3}{2n}\beta_n, \qquad \beta_n = -\frac{n}{(n+1)(2n+3)},$$

which confirms the form of the first two terms in Lamb's solution (5.7.9).

Further on, we shall show that the third and fourth terms of (5.7.9) satisfy the homogenous Laplace equation for the velocity and the continuity equation. Indeed, after changing the places of the differential operators and using the property of the harmonic functions, we get:

$$\nabla^2 \mathbf{v}_h = \nabla^2(\nabla\phi_n) = \nabla(\nabla^2\phi_n) = 0, \ \nabla.\mathbf{v}_h = \nabla.\nabla\phi_n = \nabla^2\phi_n = 0;$$

$$\nabla^2 \mathbf{v}_h = \nabla^2(\nabla\times \mathbf{r}\kappa_n) = \nabla\times\left[\nabla^2(\mathbf{r}\kappa_n)\right] = \nabla\times\left[\nabla\nabla.(\mathbf{r}\kappa_n) - \nabla\times\nabla\times(\mathbf{r}\kappa_n)\right] =$$

$$-\nabla\times\nabla\times\left[\kappa_n\nabla\times\mathbf{r} + \nabla\kappa_n\times\mathbf{r}\right] = -2\nabla\times\nabla\kappa_n = 0, \qquad \nabla.\mathbf{v}_h = \nabla.(\nabla\times\mathbf{r}\kappa_n) = 0.$$

Formula (5.7.9), thus proved, regards both outer and inner problems. We shall rewrite it in separate forms, suitable for outer or inner problems, respectively.

When treating inner problems the terms containing negative powers of r have to be dropped from the expressions (5.7.6)-(5.7.8). Then the general Lamb's solution (5.7.9), noting that $\mathbf{v} = 0$ at $n = 0$, takes the form:

$$\mathbf{v} = \sum_{n=1}^{\infty}\left[\frac{(n+3)r^2\nabla p_n}{2\mu(n+1)(2n+3)} - \frac{n\mathbf{r}p_n}{\mu(n+1)(2n+3)}\right] + \sum_{n=1}^{\infty}\left[\nabla\phi_n + \nabla\times(\mathbf{r}\kappa_n)\right]. \quad (5.7.10)$$

For outer problems with flow velocity created by the body disturbances in a quiescent fluid at infinity, only the negative powers of r have to remain in (5.7.9). That is why we have to write the sums from $-\infty$ to -2, but we change n with -n - 1, because we prefer to operate with positive numbers n. Then the coefficients α_n and β_n change to:

$$\frac{n+3}{(n+1)(2n+3)} \to -\frac{n-2}{n(2n-1)} \quad \text{and} \quad -\frac{n}{(n+1)(2n+3)} \to \frac{n+1}{n(2n-1)}.$$

Therefore the Lamb's solution for outer problems has the form:

$$\mathbf{v} = \sum_{n=1}^{\infty}\left[-\frac{(n-2)r^2\nabla p_{-n-1}}{2\mu n(2n-1)} + \frac{(n+1)\mathbf{r}p_{-n-1}}{\mu n(2n-1)}\right] + \sum_{n=1}^{\infty}\left[\nabla\phi_{-n-1} + \nabla\times(\mathbf{r}\kappa_{-n-1})\right], \quad (5.7.11)$$

where $p_{-n-1} = r^{-n-1}\sum_{m=0}^{n}(a_{nm}\cos m\varphi + a^*_{nm}\sin m\varphi)P_n^m(\beta)$, and for ϕ_{-n-1} and κ_{-n-1} we have similar expressions.

Another solution method for spherical coordinates has been developed by Schmitz and Felderhof (1982a, b). This method is closely related to Lamb's solution and Kim and Karrila called it "the adjoint method" [see Kim and Karrila (1992)].

5.8. Inertial Effects for a Flow at Small Reynolds Numbers

In 1851, Stokes (1851) established that, at infinitely small Reynolds number, the problem for the viscous fluid flow around a circular cylinder had no solution. This fact is renowned as the Stokes paradox. It occurs that, neglecting the nonlinear terms in the Navier-Stokes equations leads to serious discrepancies in the equations away from the cylinder. Revealing the reasons of the Stokes paradox, Oseen (1911) preserved one of the convective terms in the Navier-Stokes equations and in this way succeeded to solve them. Thirty eight years after Stokes had published his results, Whitehead (1889) made an unsuccessful effort to obtain higher-order approximations to the flow past a sphere in a slow uniform stream of viscous incompressible fluid. In order to find the viscous terms in the Navier-Stokes equations he proposed an iterative procedure, calculating the inertia terms by using lower-order approximations. The attempt to expand the solution in powers of Reynolds number leads to a situation in which it is impossible to satisfy the boundary conditions at infinity for all terms except the leading one. This mathematical problem occurs to be common for all uniform flows past finite bodies and is referred as the "Whitehead paradox". In fact, the perturbation represented by a small Reynolds number has a singularity at infinity and the Stokes and Whitehead's solutions are not uniformly valid everywhere in the problem region. This paradox was later resolved by Oseen (1910), who proposed a method of strained coordinates scaled by a function of the Reynolds number in order to transform the infinite region far from the body into a finite one. He reached to the well-known Oseen equation, which provides a uniformly valid first approximation for the velocity and its derivatives. As a result, Oseen obtained the following formula for the force, exerted by a sphere in a uniform viscous flow of velocity U:

$$F = 6\pi \mu aU \left[1 + \frac{3}{8} \mathrm{Re} + O(\mathrm{Re}) \right]. \tag{5.8.1}$$

Creating the matching of the asymptotic expansions method, Kaplun (1957), Lagestrom and Cole (1962) and Van Dyke (1964) contributed to solving in a more efficient manner the flow problems past a cylinder and sphere, as well as other problems, which include small but finite values of Reynolds number. In the works of Kaplun and Lagestrom (1957) and Proudman and Pearson (1957) a more accurate formula for the force exerted by a sphere in a uniform viscous flow is derived:

$$F = 6\pi \mu aU \left[1 + \frac{3}{8} \mathrm{Re} + \frac{9}{40} \mathrm{Re}^2 \ln \mathrm{Re} + O\left(\mathrm{Re}^2\right) \right]. \tag{5.8.2}$$

These authors showed that Oseen solution is valid up to accuracy o(1), but not to accuracy o(Re). Although formulae (5.8.1) and (5.8.2) coincide up to accuracy o(Re), their coincidence is by chance. In fact, the velocity distribution found by Oseen does not coincide to accuracy o(Re) with one found from the Navier-Stokes equations by the matching of the asymptotic expansions method. The appearance of a transcendental term in the correctly derived formula (5.8.2) corrects the wrong formula for the force of the sphere obtained by Goldstein (1965) with accuracy to $O(\mathrm{Re}^5)$.

Applying the method of matched asymptotic expansions, Breach (1961) solved the problem of the flow past a prolate or oblate spheroid (rotational ellipsoid). Due to the linearity of the Stokes equations, the full flow passing a spheroid with arbitrary velocity **U** can be obtained as a sum of the flow in the direction of the rotational axis with velocity $U\cos\theta$ and the flow perpendicular to the same axis with velocity $U\sin\theta$. Here the angle θ is the angle between the flow direction and the rotational axis direction. By the method proposed by Breach, the force and moment acting on a sphere, which translates and simultaneously rotates in a viscous fluid was calculated with accuracy to O(Re) by Rubinow and Keller (1961). For the force they obtained:

$$\mathbf{F} = -6\pi\mu a\mathbf{U}\left(1+\frac{3}{8}\mathrm{Re}\right)+\mathrm{Re}\frac{\pi\mu a^2}{U}(\mathbf{\Omega x U})+\mu a\mathbf{U}o(\mathrm{Re}), \qquad (5.8.3)$$

and for the moment:

$$\mathbf{M} = -8\pi\mu a\Omega\left[1+O(\mathrm{Re})\right]. \qquad (5.8.4)$$

From (5.8.3) it follows, that the sphere exerts a drag, directed reversely to the velocity

$$\mathbf{F}_D = -6\pi\mu a\mathbf{U}\left[1+\frac{3}{8}\mathrm{Re}+o(\mathrm{Re})\right]$$

and a lift force, perpendicular to it

$$\mathbf{F}_L = \pi\rho a^3(\mathbf{\Omega x U})\left[1+o(\mathrm{Re})\right]. \qquad (5.8.5)$$

Let us examine in details the derivation of the higher order approximations for the flow field of a solid sphere in a uniform viscous flow. The Stokes and Oseen approximations are based on neglecting fully or partially the nonlinear convective terms in the Navier-Stokes equations. This simplification is only valid for small Reynolds numbers. The consideration of the second and higher order effects, connected with the convective terms, is interesting from theoretical, as well as practical, point of view.

The analysis of the first order Stokes and Oseen approximation shows, that the nonlinear terms cannot be treated by means of the series expansions on the powers of Re. In fact, there is no solution of the problem for the steady flow past a cylinder in the Stokes approximation, while in Oseen approximation the obtained solution contains a logarithm of Re. Similar difficulties can be encountered when seeking higher order approximations for the sphere problem.

The boundary element method has been applied for small Reynolds numbers flows by Bush and Tanner (1983), the nonlinear inertia terms being treated iteratively as forcing functions.

The fundamental works of Kaplun and Lagestrom (1957) and Proudman and Pearson (1957) gave rise to a lot of studies devoted to the inertial effects of low Reynolds numbers flows of particulate systems and especially to those of shear flows. Saffman (1965) has shown that a sphere in a uniform simple shear flow experiences a lift force perpendicular to the fluid motion as a result of the inertial effects. Harper and Chang (1968) have generalized Saffman's analysis for an arbitrary 3D body and have obtained a lift tensor relative to the Stokes translational dyadic.

Here, we shall focus on a slightly different problem, as presented by Lin et al. (1970b). Let us consider a neutrally buoyant spherical particle of radius a immersed in an infinite incompressible Newtonian fluid, which away from the particle moves as a simple shear flow. Our aim is to analyse the effect of the fluid inertia on the velocity and pressure distribution near the particle. This analysis is directed towards modification of the well-known creeping flow solution (Landau and Lifshitz, 1959) due to inertia and rheological behaviour of a dilute suspension of spheres at small, but non-zero, Reynolds numbers.

The flow field is described with respect to a fixed coordinate system with origin in the sphere centre. The dimensional fluid velocity at infinity is taken in the form $\mathbf{v}'^{\infty} = \mathrm{G}y'\mathbf{e}_x$, where $\mathbf{e}_x$ is a unit vector in x' direction and G is the velocity gradient of the undisturbed flow. On the condition of free suspension no net force or torque acts on the sphere and rotation $\mathbf{\Omega}'$ around the centre of the sphere is imposed which will be found in the course of solution. The fluid velocity $\mathbf{v}'$ has to satisfy the steady Navier-Stokes equations and the continuity equation. The boundary conditions are the no-slip conditions on the particle surface and $\mathbf{v}' \to \mathbf{v}'^{\infty}$; $p' \to p'^{\infty}$ = const. at $|\mathbf{r}'| \to \infty$. Introducing the following dimensionless variables $\mathbf{r} = \mathbf{r}'/a$, $\mathbf{v} = \mathbf{v}'/a\mathrm{G}$, $p = p'/\mu\,\mathrm{G}$, $p^{\infty} = p'^{\infty}/\mu\,\mathrm{G}$ and $\mathbf{\Omega} = \mathbf{\Omega}'/\mathrm{G}$ (μ is the fluid dynamic viscosity) we can write down the governing equations

$$-\nabla p + \nabla^2 \mathbf{v} = \mathrm{Re}\;\mathbf{v}.\nabla\mathbf{v}, \qquad (5.8.6)$$

$$\nabla.\mathbf{v} = 0,$$

with the corresponding boundary conditions

$$\mathbf{v} = \Omega \times \mathbf{r} \qquad \text{at } |\mathbf{r}| = 1, \qquad (5.8.7_1)$$

$$\mathbf{v} \to y\mathbf{e}_x \text{ and } p \to p^{\infty} \text{ as } |\mathbf{r}| \to \infty, \qquad (5.8.7_2)$$

where $\mathrm{Re} = a^2\mathrm{G}\rho/\mu$ is the shear Reynolds number and ρ is the fluid density.

Here we shall use the method of inner and outer expansions of the unknown solution as it was done by Proudman and Pearson (1957). The inner region is assumed to cover the particle vicinity at $r = |\mathbf{r}| = O(1)$ and the inner solution expansions to have the form:

$$\{\mathbf{v},\, p,\, \mathbf{\Omega}\} = \sum_{n=0}^{\infty} f_n(\mathrm{Re})\{\mathbf{v}_n,\, p_n,\, \mathbf{\Omega}_n\}, \qquad (5.8.8)$$

where

$$\lim_{\mathrm{Re}\to 0}\left[f_{n+1}(\mathrm{Re})/f_n(\mathrm{Re})\right] = 0 \text{ and } f_0(\mathrm{Re}) = 1. \qquad (5.8.9)$$

Therefore, the solution with index 0 corresponds to the Stokes creeping flow solution. Each term of the expansion (5.8.8) has to satisfy (5.8.6) and the no-slip condition $(5.8.7_1)$. The condition at infinity $(5.8.7_2)$ is replaced with a suitable matching condition with the outer solution expansions.

After substituting (5.8.8) into (5.8.6) and $(5.8.7_1)$ and equating the same order of magnitude in Re, we reach to the equations:

$$-\nabla p_n + \nabla^2 \mathbf{v}_n = \sum_{\substack{l=0 \\ (\mathrm{Re}\, f_l f_m = f_n)}} \sum_{m=0} \left(\mathbf{v}_l . \nabla \mathbf{v}_m\right), \tag{5.8.10}$$

$$\nabla . \mathbf{v}_n = 0, \tag{5.8.11}$$

$$\mathbf{v}_n = \boldsymbol{\Omega}_n \times \mathbf{r} \qquad \text{at } r = 1. \tag{5.8.12}$$

Due to the linear character of the above equations, their solutions can be sought as a sum of homogeneous and particular terms:

$$\mathbf{v}_n = \mathbf{v}_{nh} + \mathbf{v}_{np}, \qquad p_n = p_{nh} + p_{np}, \tag{5.8.13}$$

where

$$-\nabla p_{nh} + \nabla^2 \mathbf{v}_{nh} = 0, \quad \nabla . \mathbf{v}_{nh} = 0 \tag{5.8.14}$$

with $\mathbf{v}_{nh} = \boldsymbol{\Omega}_n \times \mathbf{r}$ at $r = 1$ and

$$-\nabla p_{np} + \nabla^2 \mathbf{v}_{np} = -g_n, \qquad \nabla . \mathbf{v}_{np} = 0, \tag{5.8.15}$$

where $g_n = -\sum_l \sum_m \left(\mathbf{v}_l . \nabla \mathbf{v}_m\right)$ and with the boundary condition $\mathbf{v}_{np} = 0$ at $r = 1$.

The general solution of (5.8.14) can be expressed in spherical coordinates as the Lamb's general solution (Lamb, 1945) proposed in section 5.7, i.e., (5.7.5) for the pressure and (5.7.9) for the velocity.

The outer region includes the flow field far from the particle and there an appropriate change of variables $\tilde{\mathbf{r}} = \mathrm{Re}^{1/2}\, \mathbf{r}$, $\tilde{\mathbf{v}} = \mathrm{Re}^{1/2}\, \mathbf{v}$ and $\tilde{p} = p$ is applied in order to exclude the Reynolds number from the governing equations:

$$-\tilde{\nabla}\tilde{p} + \tilde{\nabla}^2 \tilde{\mathbf{v}} = \tilde{\mathbf{v}} . \tilde{\nabla}\tilde{\mathbf{v}}, \quad \tilde{\nabla} . \tilde{\mathbf{v}} = 0. \tag{5.8.16}$$

The outer Oseen expansion is supposed to be of the form:

$$\{\tilde{\mathbf{v}}, \tilde{p}\} = \sum_{n=0}^{\infty} F_n(\mathrm{Re}) \{\tilde{\mathbf{v}}_n, \tilde{p}_n\}, \tag{5.8.17}$$

with

$$\lim_{\mathrm{Re}\to 0} \left[F_{n+1}(\mathrm{Re}) / F_n(\mathrm{Re})\right] = 0 \text{ and } F_0(\mathrm{Re}) = 1. \tag{5.8.18}$$

All the terms of the outer expansion must satisfy the equation (5.8.16) and approach a uniform shear flow at infinity $\tilde{r} = |\tilde{\mathbf{r}}| \to \infty$. Instead of the no-slip boundary condition on the sphere surface, a matching condition with the inner (Stokes) expansion (5.8.8) is imposed.

Proceeding as with the inner expansions, the coefficients before the equal powers of Re in (5.8.16) give:

$$-\tilde{\nabla}\tilde{p}_n + \tilde{\nabla}^2 \tilde{\mathbf{v}}_n - \tilde{y}\frac{\partial \tilde{\mathbf{v}}_n}{\partial \tilde{x}} - \tilde{\mathbf{v}}_{ny} \mathbf{e}_x = \sum_{\substack{l=1 \\ (F_l F_m = F_n)}} \sum_{m=1} \left(\tilde{\mathbf{v}}_l . \tilde{\nabla}\tilde{\mathbf{v}}_m\right), \tag{5.8.19}$$

$$\tilde{\nabla} . \tilde{\mathbf{v}}_n = 0$$

The boundary conditions at infinity are such that $\tilde{\mathbf{v}}_0 = \tilde{y}\mathbf{e}_x$ and $\tilde{p}_0 = p^{\infty}$, while $\tilde{\mathbf{v}}_n = 0$ and $\tilde{p}_n = 0$ for n>0.

The matching principle consists in the limiting procedure for both regions (Van Dyke, 1964; Cole, 1968)

$$\mathrm{Re}^{\frac{1}{2}} \lim_{r\to\infty} \mathbf{v}(\mathbf{r}) = \lim_{\tilde{r}\to 0} \tilde{\mathbf{v}}(\tilde{\mathbf{r}}) \qquad \text{and} \qquad \lim_{r\to\infty} p(\mathbf{r}) = \lim_{\tilde{r}\to 0} \tilde{p}(\tilde{\mathbf{r}}). \tag{5.8.20}$$

In the limit Re $\to 0$ the expansions for the velocity field (5.8.8) and (5.8.17) reduce to $\mathbf{v} \to \mathbf{v}_0$ and $\tilde{\mathbf{v}} \to \tilde{\mathbf{v}}_0$. The condition of free suspension (zero net force and torque on the sphere) leads to $\mathbf{\Omega}_0 = -\frac{1}{2}\mathbf{e}_z$. Then the zero order solution in the inner and outer regions are:

$$\mathbf{v}_0 = \mathbf{e}_x\left[y - \frac{y}{2r^5} - \frac{5x^2y}{2r^5}\left(1-\frac{1}{r^2}\right)\right] + \mathbf{e}_y\left[-\frac{x}{2r^5} - \frac{5xy^2}{2r^5}\left(1-\frac{1}{r^2}\right)\right] + \mathbf{e}_z\left[-\frac{5xyz}{2r^5}\left(1-\frac{1}{r^2}\right)\right] \tag{5.8.21$_1$}$$

$$p_0 = p_\infty - \frac{5xy}{2r^5}, \qquad \tilde{\mathbf{v}}_0 = \tilde{y}\mathbf{e}_x, \qquad \tilde{p}_0 = \tilde{p}_\infty. \tag{5.8.21$_2$}$$

If (5.8.21$_1$) is expressed in terms of the outer coordinates, then its contribution to $\tilde{\mathbf{v}}$ is:

$$\tilde{y}\mathbf{e}_x + \mathrm{Re}^{\frac{1}{2}}\left(-\frac{5\tilde{x}\tilde{y}\tilde{\mathbf{r}}}{2\tilde{r}^5}\right) + O\left(\mathrm{Re}^{\frac{5}{2}}\right), \tag{5.8.22}$$

where the first term is exactly $\tilde{\mathbf{v}}_0$ and the second one gives

$$F_1(\mathrm{Re}) = \mathrm{Re}^{\frac{1}{2}}. \tag{5.8.23}$$

Further, it turns out that $\mathbf{\Omega}_1 = 0$ and $f_1(\mathrm{Re}) = \mathrm{Re}$. The first order velocity and pressure terms in the inner region expansions are the particular solutions of (5.8.10) - (5.8.11) at n = 1. The corresponding terms of the outer expansions are found by the application of a Fourier transformation procedure similar to the one used by Saffman (1965), who has modified a method introduced by Childress (1964). This approach is essential for determining the unknown constants in the next terms of the inner expansions.

The inner solution is used afterwards to derive the constitutive equation up to terms $O(\mathrm{Re}^{\frac{3}{2}})$ of dilute suspension of neutrally suspended rigid spheres uniformly distributed in an incompressible Newtonian fluid of viscosity μ (see Chapter 10).

5.9. Spherical Rigid or Fluid Particle in a Uniform Flow at Moderate or Large Reynolds Numbers

Rigid Particle. The Stokes approximation is also unsuitable for the cases where the particles are passed by flows at medium Reynolds numbers. Viscous flows at medium Reynolds numbers can be observed in many technological processes. In these cases the hydrodynamic interaction of the particles is modelled by the full Navier-Stokes equations, because the inertia fluid effects are significant and cannot be neglected. Due to the equations nonlinearity, solving such problems encounters difficulties and finding analytical solutions is very hard or even impossible. The moderate Reynolds number flows are those where both pressure and viscous forces are equally important for the flow acceleration over the greater part of the flow field. In some cases their competition produces complicated and intricate flow patterns (Van Dyke, 1982).

The numerical methods leading to computer realisation and simulation appears as a natural universal tool for predicting and confirming the flow pattern at moderate Reynolds numbers. The main numerical methods, used for this class of problems are: the finite differences method, which strongly depends on the problem geometry and eventually leads to a marked complexity of the differential equations when applied to an arbitrary geometry; the finite elements method, whose complexity is independent of the geometry; and the boundary elements method, giving the solution only at the boundary nodes. The last method is suitable when interested in the results only on the boundaries and is directed towards further integration of these results for finding forces, moments and drag or lift coefficients.

In the recent years the modern rapid computers have created suitable conditions for solving numerically many problems of non-linear fluid mechanics. By means of different numerical methods problems connected with the flow around one or two spherical or non-spherical particles are solved. Numerical solutions of the steady flow past a circular cylinder are given by Thom (1933) at $Re = 10$ and 20, by Kawaguti (1953) - at $Re = 40$ and by Allen and Southwell (1955) - at $Re = 1 - 100$. Dennis and Chang (1969) obtained the flow field due to an elliptic cylinder in a uniform flow for $Re = 40$. Fornberg (1980), using a numerical scheme of high accuracy, solved the steady flow problem around a circular cylinder for Re up to 300. The experimental studies show that the steady flow becomes unstable for Re bigger than 100. There exist a lot of numerical solutions of the sphere problem (Jenson, 1959; Rimon and Cheng, 1969). Masliyah and Epstein (1970) gave a numerical solution of the flow problem around an oblate or prolate spheroid. They analysed the cases when the ratio of the smaller to bigger axis equals 0.2, 0.5, 0.9 and 1 and obtained the velocity and pressure distribution, and the drag coefficient as a function of Reynolds number.

In this section we shall focus on the problem of a single rigid sphere immersed in a uniform viscous flow. The great number of numerical results and experimental data allows us a detailed analysis of the flow pattern with the increase of Reynolds number. When $Re < 10$ the flow is not separated from the spherical surface but certain fore-aft asymmetry of the flow can already be detected, in contrast to the Stokes flow pattern. The Reynolds number in the range $10 < Re < 65$ is characterised with the existence of two symmetrical closed stable eddies at the rear region of the sphere with respect to the flow direction. With the increase of Re the vortices stretch and the separation point shifts from the rear point. The separation angle measured from the rear stagnation point is θ_s and for $Re = 10$ it has a zero value, while at $Re = 200$ $\theta_s = 72^0$. This change can be expressed by the function (Clift et al., 1978)

$$\theta_s = 42.5\left(\ln\frac{\mathrm{Re}}{10}\right)^{0.483} \quad \text{at } 10 < \mathrm{Re} < 200. \tag{5.9.1}$$

If Re > 65 the flow in the wake looses its stability and becomes unsteady. In the range 65 < Re < 200 a vorticity shedding begins in the wake behind the sphere, as large vortices form periodically and move downstream. The wake gradually obtains a turbulent character at 200 <Re < $1.5\text{x}10^5$ and for the separation angle we have

$$\theta_s = 102 - 213(\mathrm{Re})^{-0.37}. \tag{5.9.2}$$

For Re > 1500 the boundary layer theory is successfully applied (Schlichting, 1964), but it is only correct in the region ahead of flow separation, i.e., in the sphere front region. Separation is predicted at $\theta_s = 71.4^0$ which corresponds to Re = 200, while for very high Reynolds numbers connected with more realistic boundary layer conditions, it occurs at $\theta_s = 109^0$.

For the drag coefficient as a function of Re there exist many empirical or semi-empirical formulae, but they do not cover the entire spectrum of Re. For example (Clift et al., 1978):

$$c_D = \frac{12}{\mathrm{Re}}\left(1 + 0.241\,\mathrm{Re}^{0.687}\right) \quad \text{for Re} < 400, \tag{5.9.3}$$

In a wider range of Reynolds numbers a more complex approximate formula can be used (Clift and Gauvin, 1970)

$$c_D = \frac{12}{\mathrm{Re}}\left(1 + 0.241\,\mathrm{Re}^{0.687}\right) + 0.42\left(1 + 1.902\text{x}10^4\,\mathrm{Re}^{-1.16}\right)^{-1} \quad \text{at Re}<1.5\text{x}10^5 \tag{5.9.4}$$

More precise examinations of sphere drag are given in (Nguyen, 1973):

$$c_D = \begin{cases} 28.18 - 5.3\lg\mathrm{Re} & \text{at } 1.7\text{x}10^5 \le \mathrm{Re} \le 2\text{x}10^5, \\ 0.1\lg\mathrm{Re} - 0.46 & \text{at } 2\text{x}10^5 < \mathrm{Re} \le 5\text{x}10^5, \\ 0.19 - 4\text{x}10^4\,\mathrm{Re}^{-1} & \text{at } 5\text{x}10^5 < \mathrm{Re}. \end{cases} \tag{5.9.5}$$

Further on, we shall give the classical boundary layer solution for a rigid sphere in a uniform viscous flow. From the Navier-Stokes equation for the region outside the boundary layer, in terms of the stream function Ψ (1.3.20), it is clear that the approximation of O(1) must satisfy the following boundary value problem:

$$E^2\Psi = 0, \tag{5.9.6}$$

with

$$\Psi = 0 \quad \text{at } r = 1, \tag{5.9.7}$$

$$\Psi \to \frac{1}{2} r^2 \sin^2\theta \quad \text{as } r \to \infty. \tag{5.9.8}$$

This problem is equivalent to the 2D potential flow problem and its solution is:

$$\Psi = -\frac{1}{2}\sin^2\theta\left(r^2 - \frac{1}{r}\right). \tag{5.9.9}$$

Then, the velocity components are:

$$\mathrm{v_r} = -\cos\theta\left(1 - \frac{1}{\mathrm{r}^3}\right), \qquad (5.9.10)$$

$$\mathrm{v}_\theta = \sin\theta\left(1 + \frac{1}{2\mathrm{r}^3}\right). \qquad (5.9.11)$$

It is evident that the normal velocity is zero at the sphere surface r = 1, while the tangential velocity does not vanish at r = 1, i.e., the first order outer solution does not satisfy the no-slip boundary condition on the sphere. This causes the necessity of introducing a thin boundary layer adjacent to the sphere surface where viscous forces are significant. The scaling of the normal velocity and radial coordinate in the boundary layer is as follows:

$$\mathrm{v_r} = \frac{1}{\sqrt{\mathrm{Re}}}\mathrm{V} \qquad \text{and} \qquad \mathrm{r} - 1 = \frac{\mathrm{Y}}{\sqrt{\mathrm{Re}}}. \qquad (5.9.12)$$

Since Re $\to \propto$, from the continuity equation in spherical coordinates (1.3.2) and (5.9.12), we obtain

$$\frac{\partial \mathrm{V}}{\partial \mathrm{Y}} + \frac{1}{\sin\theta}\frac{\partial}{\partial\theta}\left(\mathrm{v}_\theta \sin\theta\right) = \mathrm{O}(\mathrm{Re}^{-\frac{1}{2}}) \to 0. \qquad (5.9.13)$$

In analogous manner the tangential component of the momentum equation gives

$$\frac{\partial^2 \mathrm{v}_\theta}{\partial \mathrm{Y}^2} = \frac{\partial \mathrm{p}}{\partial\theta} + \mathrm{V}\frac{\partial \mathrm{v}_\theta}{\partial \mathrm{Y}} + \mathrm{v}_\theta\frac{\partial \mathrm{v}_\theta}{\partial\theta} + \mathrm{O}(\mathrm{Re}^{-\frac{1}{2}}), \qquad (5.9.14)$$

while its normal component is simply

$$\frac{\partial \mathrm{p}}{\partial \mathrm{Y}} = \mathrm{O}(\mathrm{Re}^{-\frac{1}{2}}) \to 0. \qquad (5.9.15)$$

The leading terms in the boundary layer approximations must satisfy (5.9.13) - (5.9.15) with the boundary conditions :

$$\mathrm{v}_\theta = \mathrm{V} = 0 \qquad \text{at } \mathrm{Y} = 0 \qquad (5.9.16)$$

and the matching condition

$$\mathrm{v}_\theta \to \frac{3}{2}\sin\theta \quad \text{as } \mathrm{Y} \to \propto. \qquad (5.9.17)$$

As the pressure in the boundary layer is a function only of *6* , then from the Bernoulli equation applied for the potential flow in the outer region with velocities given by (5.9.10) and (5.9.11) and differentiated with respect to *6* , in the boundary layer we have:

$$\frac{\partial \mathrm{p}}{\partial\theta} = \frac{9}{8}\sin 2\theta. \qquad (5.9.18)$$

Schlichting (1964) has shown that the boundary layer equations (5.9.13) - (5.9.15) can successfully be applied for the case of an arbitrary axisymmetrical body, whose geometry is defined with r(x) instead of sin*6* for the sphere case. The solution of this problem can be found as a power expansion on x, respectively on *6* , similar to the Blasius series (Schlichting, 1964), i.e.:

$$v_\theta = v_{\theta 1}\, x\, f_1'(\eta) + 2v_{\theta 3}\, x^3\, f_3'(\eta) + 3v_{\theta 5}\, x^5\, f_5'(\eta) + O(x^7)\,. \qquad (5.9.19)$$

where $\eta = Y\sqrt{2v_{\theta 1}}$ and $v_{\theta n}$ are the coefficients in the power expansion in x of the boundary value $v_{\theta e} = \lim\limits_{r\to 1} v_\theta$ from the potential flow (5.9.11):

$$v_{\theta e} = v_{\theta 1}\, x + v_{\theta 3}\, x^3 + v_{\theta 5}\, x^5 + \ldots \qquad (5.9.20)$$

Because of the symmetry the geometry function r(x) can also be expressed by the odd power series of x:

$$r = r_1\, x + r_3\, x^3 + r_5\, x^5 + \ldots \qquad (5.9.21)$$

Frössling (1940) has shown that after suitable substitutions the problem (5.9.13)-(5.9.18) can be reduced to a problem for the unknown coefficients in (5.9.19)-(5.9.21). As a results for the sphere case, these coefficients are:

$$r_1 = 1,\ r_3 = -\frac{1}{6}, \ldots;\ v_{\theta 1} = \frac{3}{2},\ v_{\theta 2} = -\frac{1}{4}, \ldots. \qquad (5.9.22)$$

The existence of a separation point means that the normal gradient of v_θ on the sphere surface is zero:

$$\left.\frac{\partial v_\theta}{\partial Y}\right|_{Y=0} = \frac{3\sqrt{3}}{2}\left(1 - 0.3925x^2 + \ldots\right)0.9277 = 0\,. \qquad (5.9.23)$$

If in the approximation only two terms are taken, then the separation point is at $x_s = 1.596$ or at $\theta_s = 91.5^0$ (as x = 0 is at the front stagnation point). Schlichting gives a more accurate value when four terms are present and $x_s = 1.913$ or $\theta_s = 109.6^0$. These results are very close to the corresponding results for a circular cylinder.

Fluid Particle. It is experimentally established (Clift et al., 1978) that the fluid drops and bubbles remain almost spherical for Reynolds numbers up to 300, if the surface tension is not very high. The disturbed fluid flow around the fluid sphere is quite similar to the corresponding flow around a rigid sphere when the ratio of internal and external viscosity $\lambda = \dfrac{\hat{\mu}}{\mu}$ is high, or when the surfactants are "solidifying" the interface as explained in section 5.5. Separations occurs at rather high Reynolds numbers - Re ≈ 50 (the Reynolds numbers are defined as $Re = \dfrac{\rho U a}{\mu}$ and $\hat{Re} = \dfrac{\hat{\rho} U a}{\hat{\mu}}$). In the range $Re \le 50$ the numerical methods are the most universal tool to solve the flow problem outside and inside the fluid sphere at different Reynolds numbers and viscosity ratios. The results obtained by them are discussed in (Clift et al., 1978). The inner fluid circulation is significantly more intensive than that described by the Hadamard-Rybczynski solution. The boundary velocity of the drop increases very fast with the increase of Reynolds number even for very viscous drops.

The error distribution method (Galerkin method) is highly suitable when analytical solutions are aimed. For small Re the following approximation formula of the drag coefficient (Clift et al., 1978) is successfully applied:

$$c_D = \frac{1.83(738\lambda^2 + 2142\lambda + 1080)}{(60 + 29\lambda)(4 + 3\lambda)} Re^{-0.74} \quad \text{at } 2 < Re < 50. \qquad (5.9.24)$$

Since the boundary layer around a drop or bubble is considerably thinner than that around a rigid sphere, then the boundary layer methods can be applied for flows in the range 50 < Re < 300. On the basis of these methods the drag coefficient has been calculated by many authors, a list of which can be found (Clift et al., 1978). However, we shall point out only the work of Harper and Moore (1968), in which a formula has been derived for Re >> 1:

$$c_D = \frac{24}{Re}\left[1 + \frac{3}{2}\lambda + \frac{(2+3\lambda)^2}{Re^{0.5}}\left(B_1 + B_2 \ln Re\right)\right]. \qquad (5.9.25)$$

The values of B_1 and B_2 are given in the next table (Table 5.9.1), where $\gamma = \frac{\hat{\rho}}{\rho}$ is the density ratio of the fluid inside and fluid outside the particle.

Table 5.9.1

$\lambda\gamma$	25	4.0	1.0	0.25	0.04	0
B_1	-0.429	-0.457	-0.460	-0.446	-0.434	-0.391
B_2	0.00202	0.0062	0.0100	0.0113	0.00842	0

The limit $\lambda\gamma \to 0$ of (5.9.25) gives the bubble drag coefficient as found by Moore (1963), and the first term of (5.9.25) is exactly the Levich formula (Levich, 1949) $c_D = \frac{24}{Re}$. The latter formula is obtained also by another method of a direct integration of the normal stress over the bubble surface by Kang and Leal (1988a, b). These authors show that the drag coefficient up to $O(Re^{-1})$ depends only on the vorticity on the bubble surface and on the completely irrotational fluid flow.

Following Leal (1992) we shall present in details the boundary layer solution for a spherical bubble in a uniform viscous flow. This example is chosen for simplicity, because there is no fluid motion to be examined inside the bubble, i.e., the problem is uncoupled, contrary to the general fluid drop problem where the coupling of inside flow with the outside flow is necessary on the interface. Moreover, we shall be able to see directly that the boundary layer existence is connected with the vorticity generated at the body (solid or fluid) surface.

On the solid sphere the boundary conditions are the no-slip conditions, while on the bubble surface they are replaced by the zero tangential stress condition. The impossibility for the potential flow to satisfy these no-slip conditions on the spherical interface makes necessary to introduce a boundary layer for the solid sphere. Here we expect again to encounter a similar difficulty, as the potential flow will yield non-zero shear stress on the bubble surface. Therefore, we shall treat the bubble problem again in the framework of the boundary layer theory.

For the flow past a solid body, the mechanism of generating vorticity is the velocity gradient created to satisfy the no-slip surface boundary condition and the vorticity is of the order $O(\mathrm{Re}^{\frac{1}{2}})$. On the other hand, for an interface the no-slip condition is dropped, the tangential velocity is non-zero and the vorticity is produced by the fluid particles rotation along the surface curvature. The dimensionless tangential shear stress for a spherical bubble must be zero on its surface, i.e.,

$$\tau_{r\theta}\Big|_{r=1} = \frac{\partial v_\theta}{\partial r} - \frac{v_\theta}{r} = 0 . \qquad (5.9.26)$$

Then the dimensionless vorticity ω can be expressed as:

$$\omega\Big|_{r=1} = \frac{2v_\theta}{r}\Bigg|_{r=1} = 2v_\theta . \qquad (5.9.27)$$

Further, we shall show that the strength of the boundary layer expressed as velocity gradient or vorticity magnitude is quite different for the no-slip and slip surfaces.

As for the solid sphere case, a spherical coordinate system with its origin in the bubble centre is imposed. The bubble is assumed spherical, when the Reynolds number is not small, but is in the limits discussed at the beginning of this section and the bubble radius is taken as a characteristic length. The dimensionless Navier-Stokes equations are (1.3.5) written in spherical coordinates (1.3.13) with the boundary conditions

$$v_r = 0 \quad \text{at } r = 1, \qquad (5.9.28)$$

$$\tau_{r\theta} = \frac{\partial v_\theta}{\partial r} - \frac{v_\theta}{r} = 0 \quad \text{at } r = 1, \qquad (5.9.29)$$

and

$$v_r \to -\cos\theta,\ v_\theta \to \sin\theta \qquad \text{as } r \to \infty \qquad (5.9.30)$$

If $\mathrm{Re} \to \infty$, the leading order terms of the solution will be the same as the potential flow solution of the solid sphere problem (5.9.10) and (5.9.11). The no-slip condition change with (5.9.29) has no influence on the potential flow solution, as it satisfies only the kinematic condition (5.9.28). It occurs that the zero shear stress condition cannot be fulfilled by the outer region solution. In addition, an appropriate boundary layer region must be introduced; its solution will correct this discrepancy on the bubble surface.

In the inner boundary layer region a scaling is applied given by (5.9.12). The rescaled equations of motion preserving terms of order O(1) and $O(\mathrm{Re}^{-1/2})$ are shortly given by (5.9.13) - (5.9.15) with the boundary conditions (5.9.28) and (5.9.29) expressed in the inner variables

$$V = 0 \quad \text{at } Y = 0 \qquad (5.9.31)$$

and

$$\frac{\partial v_\theta}{\partial Y} - \frac{v_\theta}{\sqrt{\mathrm{Re}}} + O\left(\mathrm{Re}^{-1}\right) = 0 \qquad \text{at } Y = 0. \qquad (5.9.32)$$

The matching with the outer potential flow (5.9.10) and (5.9.11) must be also imposed on the inner solution, i.e.:

$$\lim_{Y\to\infty} v_\theta = \lim_{r\to 1}\left(1+\frac{1}{2r^3}\right)\sin\theta = \frac{3}{2}\sin\theta - \frac{3Y}{2\sqrt{Re}}\sin\theta + O\left(Re^{-1}\right), \tag{5.9.33}$$

$$\lim_{Y\to\infty}\frac{V}{\sqrt{Re}} = -\lim_{r\to 1}\left(1-\frac{1}{r^3}\right)\cos\theta = -\frac{3Y}{\sqrt{Re}}\cos\theta + O\left(Re^{-1}\right). \tag{5.9.34}$$

The form of the above expressions suggests to us to seek the inner boundary layer solutions in similar forms as power expansions of $Re^{-1/2}$:

$$v_\theta = v_{\theta 0} + \frac{1}{\sqrt{Re}}v_{\theta 1} + \ldots, \qquad V = V_0 + \frac{1}{\sqrt{Re}}V_1 + \ldots, \tag{5.9.35}$$

$$p = p_0 + \frac{1}{\sqrt{Re}}p_1 + \ldots.$$

The governing equations for the zero order terms are exactly the same as for the solid sphere case (5.9.13) - (5.9.15). The boundary conditions (5.9.31) - (5.9.33) give, respectively:

$$V_0 = 0 \text{ and } \frac{\partial v_{\theta 0}}{\partial Y} = 0 \qquad \text{at } Y = 0 \tag{5.9.36}$$

and

$$v_{\theta 0} \to \frac{3}{2}\sin\theta \text{ as } Y \to \infty. \tag{5.9.37}$$

The solutions of (5.9.13) - (5.9.15) with (5.9.36) and (5.9.37) are exactly the first terms of (5.9.33) and (5.9.34), i.e.:

$$v_{\theta 0} = \frac{3}{2}\sin\theta \quad \text{and} \quad V_0 = -3Y\cos\theta. \tag{5.9.38}$$

It turns out that this is the leading term of the inviscid solution, while the correction to it from the boundary layer zero shear stress is only of order $O(Re^{-1/2})$. This is well seen if the potential flow solution (5.9.10) and (5.9.11) is expressed in inner variables, as given by the matching procedure (5.9.33) and (5.9.34)

$$v_\theta \to \frac{3}{2}\sin\theta - \frac{3Y}{2\sqrt{Re}}\sin\theta + O\left(Re^{-1}\right) \tag{5.9.39}$$

and

$$v_r\sqrt{Re} = V \to -3Y\cos\theta + 6Y^2\cos\theta\frac{1}{\sqrt{Re}} + O\left(Re^{-1}\right),$$

whose first terms are exactly (5.9.38). Here, it is worth to remind that the corresponding correction from the no-slip solid surface condition from the boundary layer was of order $O(Re^{-1/2})$.

The next terms in the asymptotic expansions for the solution in the outer region will be of order $O(Re^{-1})$, as it is seen from the Navier-Stokes equation (1.3.20). However, the inner solution correction is of order $O(Re^{-1/2})$ and it must satisfy the following boundary value problem, derived from (5.9.13) - (5.9.15), (5.9.31) - (5.9.32), when taking into account the form of the zero order solution (5.9.38):

$$\frac{\partial V_1}{\partial Y}+\frac{1}{\sin\theta}\frac{\partial}{\partial\theta}\left(v_{\theta 1}\sin\theta\right)=0, \tag{5.9.40}$$

$$\frac{\partial^2 v_{\theta 1}}{\partial Y^2}=V_0\frac{\partial v_{\theta 1}}{\partial Y}+v_{\theta 1}\frac{\partial v_{\theta 0}}{\partial\theta}+v_{\theta 0}\frac{\partial v_{\theta 1}}{\partial\theta}, \tag{5.9.41}$$

and

$$\frac{\partial p_1}{\partial\theta}=\frac{9}{2}Y\sin\theta\cos\theta \tag{5.9.42}$$

with

$$V_1=0 \quad \text{and} \quad \frac{\partial v_{\theta 1}}{\partial Y}=v_{\theta 0} \quad \text{at } Y=0 \tag{5.9.43}$$

and

$$v_{\theta 1}\to-\frac{3}{2}Y\sin\theta \qquad \text{as } Y\to\infty. \tag{5.9.44}$$

The formulated problem is linear. Since (5.9.41) is independent of V_1, it can be solved first, to find $v_{\theta 1}$, while V_1 can be obtained afterwards from the continuity equation (5.9.40). Here we shall give the final analytical form of the tangential velocity correction $v_{\theta 1}$:

$$v_{\theta 1}=-\frac{3}{2}Y\sin\theta+3g(\theta)\sin\theta f(\eta), \tag{5.9.45}$$

where $g(\theta)$ is given by

$$g(\theta)=\frac{\sqrt{8/3}}{\sin^2\theta}\left(\frac{2}{3}-\cos\theta+\frac{\cos^3\theta}{3}\right)^{1/2} \tag{5.9.46}$$

and $f(\eta)$ is

$$f(\eta)=-\frac{1}{\sqrt{\pi}}\exp(-\eta^2)+\eta\,\mathrm{erfc}(\eta) \tag{5.9.47}$$

with $\mathrm{erfc}(\eta)\equiv\frac{2}{\sqrt{\pi}}\int_\eta^\infty\exp(-t^2)\,dt$ - the complementary error function (Abramowitz and Stegun, 1965).

Finally, we can write down the inner region solution for $v_{\theta 1}$ of order $O(Re^{-1})$:

$$v_\theta=\frac{3}{2}\sin\theta+\frac{1}{\sqrt{Re}}\left[-\frac{3}{2}Y\sin\theta+3g(\theta)\sin\theta\ f(\eta)\right]+O(Re^{-1}), \tag{5.9.48}$$

whose first two terms come directly from the potential flow solution, namely, (5.9.11). The $O(Re^{-1/2})$ viscous correction is found in a closed form, in contrast to the solid sphere case where approximate or numerical solutions are necessary. The reason for this fact is the linear character of the governing equation (5.9.41) for the bubble case or, more general, for a zero shear stress surface.

5.10. Rigid Particle of Arbitrary Shape in a Uniform Flow

Interesting results for the flow past bodies of arbitrary shape are obtained by Brenner (1963, 1964a) and Brenner and Cox (1963). A full summary of these results can be found in Brenner (1972b) and in the Happel and Brenner's book (Happel and Brenner, 1973).

In this section we shall give some typical examples for the drag force experienced by nonspherically shaped solid particles in a unbounded Stokes flow and in flows of high Reynolds numbers. Prior to it, we shall show how the drag force and its moment can be expressed for arbitrary shaped particle in a Stokes flow.

In order to determine the interaction between the fluid and the particles it is necessary to define the so called resistance tensor and mobility tensor. When the forces **F** and their moments **M**, acting on the particles of the suspension, are given and the particles relative motion in the fluid is sought, then we say that we deal with the particles mobility problem. On the other hand, if the particles relative motion is known, as well as the forces **F** and their moments **M** experienced by the particles are sought, then the drag problem is said to be treated.

Denote the translational velocity of a given particle in a unbounded fluid with **U** and its rotational velocity with $\boldsymbol{\omega}$. The translational and rotational velocities of the undisturbed fluid flow at infinity are expressed as $\mathbf{v}^\infty$ and $\boldsymbol{\Omega}^\infty$, respectively. Then the velocity $\mathbf{v}^D(\mathbf{r})$ of the disturbed fluid motion on the particle surface S_p is

$$\mathbf{v}^D(\mathbf{r}) = \mathbf{U} - \mathbf{v}^\infty + (\boldsymbol{\omega} - \boldsymbol{\Omega}^\infty) \times \mathbf{r} - \mathbf{E}^\infty . \mathbf{r}, \qquad (5.10.1)$$

where $\mathbf{E}^\infty$ is the deformation rate tensor of the undisturbed flow at infinity. If the relative particle motion has only a translational velocity $\mathbf{U} - \mathbf{v}^\infty$, then due to linearity of the Stokes equations the relation between the drag **F** and the velocity $\mathbf{U} - \mathbf{v}^\infty$ is:

$$\mathbf{F} = \mu \mathbf{A}(\mathbf{v}^\infty - \mathbf{U}), \qquad (5.10.2)$$

where **A** is a second rank tensor depending on the moving particle shape, and μ is the fluid viscosity.

When the velocity $\mathbf{v}^D(\mathbf{r})$ of the disturbed flow on the particle surface has the form (5.10.1), formula (5.10.2) is generalized by the linear relation between the moments and flow parameters:

$$\begin{pmatrix} \mathbf{F} \\ \mathbf{M} \\ \mathbf{S} \end{pmatrix} = \mu \begin{pmatrix} \mathbf{A} & \tilde{\mathbf{B}} & \tilde{\mathbf{G}} \\ \mathbf{B} & \mathbf{C} & \tilde{\mathbf{H}} \\ \mathbf{G} & \mathbf{H} & \mathbf{T} \end{pmatrix} \begin{pmatrix} \mathbf{v}^\infty - \mathbf{U} \\ \boldsymbol{\Omega}^\infty - \boldsymbol{\omega} \\ \mathbf{E}^\infty \end{pmatrix}. \qquad (5.10.3)$$

In this equation **A, B, C** are second rank tensors, **G** and **H** are third rank tensors, **T** is a fourth rank tensor and **S** is the particle Stokeslet, a second rank tensor. The transposed matrices and tensors are denoted with the mark ~. The quadratic matrix in (5.10.3) is called resistance matrix.

The rules for obtaining inner tensor products follow from the multiplication rules for the matrix elements of (5.10.3), i.e.:

$$F_i = \mu A_{ij}\left(U_j^\infty - U_j\right) + \mu \widetilde{B}_{ij}\left(\Omega_j^\infty - \omega_j\right) + \mu \widetilde{G}_{ijk} E_{jk}^\infty, \tag{5.10.4}$$

$$S_{ij} = \mu G_{ijk}\left(U_k^\infty - U_k\right) + \mu H_{ijk}\left(\Omega_k^\infty - \omega_k\right) + \mu T_{ijkl} E_{kl}^\infty.$$

For the particles mobility problem we have the equality

$$\begin{pmatrix} \mathbf{v}^\infty - \mathbf{U} \\ \boldsymbol{\Omega}^\infty - \boldsymbol{\omega} \\ \mu^{-1}\mathbf{S} \end{pmatrix} = \begin{pmatrix} \mathbf{a} & \widetilde{\mathbf{b}} & \widetilde{\mathbf{g}} \\ \mathbf{b} & \mathbf{c} & \widetilde{\mathbf{h}} \\ \mathbf{g} & \mathbf{h} & \mathbf{m} \end{pmatrix} \begin{pmatrix} \mu^{-1}\mathbf{F} \\ \mu^{-1}\mathbf{M} \\ \mathbf{E}^\infty \end{pmatrix}. \tag{5.10.5}$$

In this equation **a**, **b**, **c** are second rank tensors, **g**, **h** are tensors of third rank, and **m** is a tensor of fourth rank. The quadratic matrix in (5.10.5) is called mobility matrix.

We have to note that if the Stokeslet **S** is excluded from the analysis, then there exists a relationship between the two mentioned matrices. Indeed, from

$$\begin{pmatrix} \mathbf{F} \\ \mathbf{M} \end{pmatrix} = \mu \begin{pmatrix} \mathbf{A} & \widetilde{\mathbf{B}} \\ \mathbf{B} & \mathbf{C} \end{pmatrix} \begin{pmatrix} \mathbf{v}^\infty - \mathbf{U} \\ \boldsymbol{\Omega}^\infty - \boldsymbol{\omega} \end{pmatrix} + \mu \begin{pmatrix} \widetilde{\mathbf{G}} \\ \widetilde{\mathbf{H}} \end{pmatrix} \mathbf{E}^\infty$$

$$= \mu \begin{pmatrix} \mathbf{A} & \widetilde{\mathbf{B}} \\ \mathbf{B} & \mathbf{C} \end{pmatrix} \left[\begin{pmatrix} \mathbf{a} & \widetilde{\mathbf{b}} \\ \mathbf{b} & \mathbf{c} \end{pmatrix} \begin{pmatrix} \mu^{-1}\mathbf{F} \\ \mu^{-1}\mathbf{M} \end{pmatrix} + \begin{pmatrix} \widetilde{\mathbf{g}} \\ \widetilde{\mathbf{h}} \end{pmatrix} \mathbf{E}^\infty \right] + \mu \begin{pmatrix} \widetilde{\mathbf{G}} \\ \widetilde{\mathbf{H}} \end{pmatrix} \mathbf{E}^\infty$$

the next equalities follow:

$$\begin{pmatrix} \mathbf{A} & \widetilde{\mathbf{B}} \\ \mathbf{B} & \mathbf{C} \end{pmatrix} \begin{pmatrix} \mathbf{a} & \widetilde{\mathbf{b}} \\ \mathbf{b} & \mathbf{c} \end{pmatrix} = \begin{pmatrix} \delta & 0 \\ 0 & \delta \end{pmatrix}, \quad \begin{pmatrix} \mathbf{A} & \widetilde{\mathbf{B}} \\ \mathbf{B} & \mathbf{C} \end{pmatrix} \begin{pmatrix} \widetilde{\mathbf{g}} \\ \widetilde{\mathbf{h}} \end{pmatrix} + \begin{pmatrix} \widetilde{\mathbf{G}} \\ \widetilde{\mathbf{H}} \end{pmatrix} = \begin{pmatrix} 0 \\ 0 \end{pmatrix}.$$

Therefore,

$$\begin{pmatrix} \mathbf{a} & \widetilde{\mathbf{b}} \\ \mathbf{b} & \mathbf{c} \end{pmatrix} = \begin{pmatrix} \mathbf{A} & \widetilde{\mathbf{B}} \\ \mathbf{B} & \mathbf{C} \end{pmatrix}^{-1}, \quad \begin{pmatrix} \widetilde{\mathbf{g}} \\ \widetilde{\mathbf{h}} \end{pmatrix} = -\begin{pmatrix} \mathbf{A} & \widetilde{\mathbf{B}} \\ \mathbf{B} & \mathbf{C} \end{pmatrix}^{-1} \begin{pmatrix} \widetilde{\mathbf{G}} \\ \widetilde{\mathbf{H}} \end{pmatrix}.$$

After some transformations, we get

$$\mathbf{a} = \left(\mathbf{A} - \widetilde{\mathbf{B}}\mathbf{C}^{-1}\mathbf{B}\right)^{-1}, \quad \mathbf{b} = -\mathbf{C}^{-1}\left(\mathbf{A} - \widetilde{\mathbf{B}}\mathbf{C}^{-1}\mathbf{B}\right)^{-1}, \tag{5.10.6}$$

$$\mathbf{c} = \left(\mathbf{C} - \mathbf{B}\mathbf{A}^{-1}\widetilde{\mathbf{B}}\right)^{-1}.$$

The axisymmetric problem for the uniform flow past an ellipsoidal particle in Stokes approximation has an exact analytical solution. Here we shall confine ourselves to the results given by Happel and Brenner (1973).

If an oblate ellipsoid of revolution with semimajor axis a and semiminor axis b is passed by an uniform Stokes flow of velocity U and fluid viscosity μ, then the force exerted by the fluid on the ellipsoid is:

$$F = \frac{8\pi \mu U \sqrt{a^2 - b^2}}{\sigma_0 - \left(\sigma_0^2 - 1\right)\operatorname{arcctg}\sigma_0}, \tag{5.10.7}$$

where $\sigma_0 = \left[(a/b)^2 - 1\right]^{-1/2}$.

At $\sigma_0 \to 0$ the oblate ellipsoid degenerates into a flat infinitely thin disk of radius a. The force acting on a disk, moving perpendicularly to its plane with velocity U in a quiescent fluid, is given by

$$F = 16\,\mu\,aU\,, \tag{5.10.8}$$

which is less than the force acting on a sphere of the same radius (for a sphere $F = 6\pi\,\mu\,aU$).

The drag force exerted on a prolate ellipsoid of revolution with semiaxes a > b in a uniform Stokes flow of velocity U and viscosity μ is calculated by the formula

$$F = \frac{8\pi\,\mu\,U\sqrt{a^2 - b^2}}{\left(\sigma_0^2 + 1\right)\operatorname{arccth}\sigma_0 - \sigma_0}, \tag{5.10.9}$$

where $\sigma_0 = \left[1 - \left(b/a\right)^2\right]^{-1/2}$.

If a >> b the prolate ellipsoid degenerates into a needle-shaped rod. In this case the force acting on a needle of length a and radius b moving in direction parallel to its axis with velocity U has the form

$$F = \frac{4\pi\,\mu\,aU}{\ln\left(a/b\right) + 0.193}. \tag{5.10.10}$$

If we compare the results for the forces acting on oblate and prolate ellipsoids with the similar expressions for a sphere of equivalent equatorial radius, we can give the generalized formula:

$$F = 6\pi\,\mu\,lUK\left(\frac{b}{a}\right), \tag{5.10.11}$$

where l = a for an oblate ellipsoid and l = b for a prolate ellipsoid. The numerical values of the correction coefficient K to the Stokes formula are given for different ratios b/a in (Happel and Brenner, 1973).

Now we shall discuss the low Reynolds numbers flow past bodies of revolution of arbitrary shape. The body axis is assumed to be inclined with angle α with respect to the flow direction at infinity, as shown in Fig.5.10.1. The unit vector **i** directed along the flow can be represented as the sum $\mathbf{i} = \boldsymbol{\tau}\cos\alpha + \mathbf{n}\sin\alpha$, where $\boldsymbol{\tau}$ is the unit vector along the body axis of revolution and **n** is a unit vector lying in the plane of revolution. In Stokes approximation for the drag force concerning the general case we can obtain (Happel and Brenner, 1973):

$$\mathbf{F} = \boldsymbol{\tau}\,F_{II}\cos\alpha + \mathbf{n}F_{\perp}\sin\alpha, \tag{5.10.12}$$

where F_{II} and $F_{\perp}$ are the drag force values in the case of parallel flow ($\alpha = 0$) and of perpendicular flow ($\alpha = \pi/2$) with respect to the body axis of revolution.

The value of the force projected on the flow direction is equal to the dot product

$$\mathbf{F}.\mathbf{i} = F_{II}\cos^2\alpha + F_{\perp}\sin^2\alpha. \tag{5.10.13}$$

Then the calculation of the drag force on an arbitrary shaped body of revolution in a uniform Stokes flow is reduced to the determination of two particular spatial situations.

The parallel and perpendicular drag F_{II} and $F_{\perp}$ can be determined theoretically or experimentally. Here, we shall present expressions for F_{II} and $F_{\perp}$ as referred by Happel and Brenner (1973):

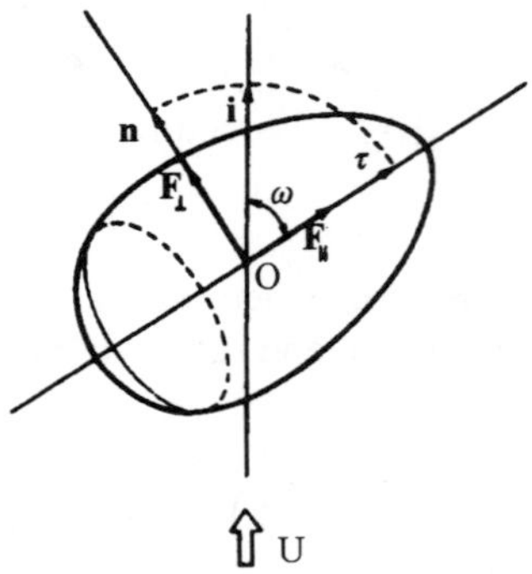

Fig.5.10.1. A body of revolution of arbitrary shape in an uniform flow

i) for a thin circular disk of radius a:

$$F_{II} = 16\mu aU, \; F_{\perp} = \frac{32}{3}\mu aU. \tag{5.10.14}$$

ii) For oblate ellipsoid of revolution with semi-axes a > b:

$$F_{II} = 3.77(4a + b), \qquad F_{\perp} = 3.77(3a + 2b). \tag{5.10.15}$$

iii) For prolate ellipsoid of revolution with semi-axes b > a:

$$F_{II} = 3.77(a + 4b), \qquad F_{\perp} = 3.77(2a + 3b). \tag{5.10.16}$$

As the Stokes flow past a particle of arbitrary shape is without separation, the streamlines are not closed, they come from infinity, pass around the body smoothly and go again to infinity. However, at high Reynolds numbers the flow separates from the body surface. This leads to a vorticity region and a wake after the body. The length of this region depends on the Reynolds number and on the body shape. In Fig.5.10.2. the experimental and numerical data of the relative wake length L_w/d (d - diameter of the equatorial section) are given as function of Reynolds number for different axisymmetric bodies: sphere, disk and ellipsoids with different ratios E of the axial length to equatorial.

The drag force coefficient depends on the Reynolds number. At $Re \to 0$ it corresponds to Stokes regime, while at $Re \to \infty$ it corresponds to Newtonian regime which is characterised by a constant drag coefficient c_D.

The flow passing a disk in its axial direction is given as a limiting case of the axisymmetrical bodies of small prolongation in Clift et al., (1978). The drag coefficients are obtained for the whole spectrum of Reynolds number calculated on the radius:

$$c_D = 10.2\,\mathrm{Re}^{-1}(1+0.318\mathrm{Re}) \quad \text{at } \mathrm{Re} \le 0.005, \qquad (5.10.17)$$

$$c_D = 10.2\,\mathrm{Re}^{-1}\left(1+10^{s}\right) \quad \text{at } 0.005 < \mathrm{Re} \le 0.75,$$

$$c_D = 10.2\,\mathrm{Re}^{-1}\left(1+0.239\mathrm{Re}^{0.792}\right) \text{ at } 0.75 < \mathrm{Re} \le 66.5,$$

$$c_D = 1.17 \quad \text{at } \mathrm{Re} > 66.5,$$

where $s = -0.61 + 0{,}906 \lg \mathrm{Re} - 0.025(\lg \mathrm{Re})^2$.

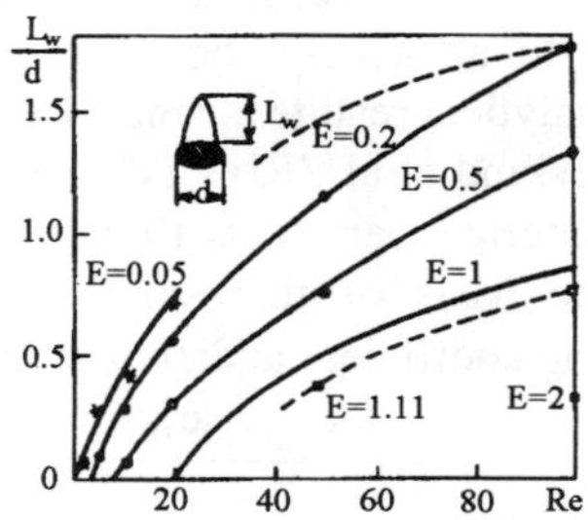

Fig.5.10.2. Relative wake length as a function of Reynolds number

5.11. Numerical Solutions for Flows Past Rigid or Fluid Spherical Particles at Arbitrary Reynolds Numbers

The numerous hydrodynamic investigations, performed up to now, show that analytical (exact or approximate) solutions for flows around drops, bubbles or rigid particles are only possible in some particular cases. In the preceding chapters of this book we considered some exact or approximate analytical solutions of viscous flows passing single fluid or rigid particles. In the present section we shall make a brief review of the numerical models of the flows around arbitrary rigid and spherical fluid particles at arbitrary Reynolds numbers and we shall focus on the numerical solution, given by Youngren and Acrivos (1975), of the problem of an arbitrary shaped rigid particle in viscous flow at zero-valued Reynolds numbers.

The first numerical solutions of the flow problem over rigid spherical particle for a wide range of Reynolds number values are proposed by Dennis and Walker (1964), Hamielec et al. (1967) and LeClair et al. (1970). The same problem is also studied experimentally in Maxworthy (1965) and Beard and Pruppacher (1969), and comparisons with the numerical results are made.

The first attempts for finding a numerical solution of the spherical fluid particle problem at moderate Reynolds numbers are not completely successful. A typical example in this respect is the paper of Nakano and Tien (1967), in which the Galerkin method is exploited. This solution does not accurately describe the considered flow, because it does not contain enough terms and the trial functions are not suitably chosen to converge to the well-known solution of Hadamard-Rybczynski for zero Reynolds number.

LeClair et al. (1972) are the first to apply the finite-difference method techniques for the steady problem of water drop(s) in air. Their numerical results are in good agreement with their own experimental measurements. Brabston and Keller (1975) use a hybrid method including the finite-difference numerical approximation and the series-truncation technique to solve the flow passing a gas bubble. They obtain result which correlate well with those due to an asymptotic analysis at large Reynolds numbers (Re > 40) and with the exact analytical results at small Reynolds numbers (Re < 0.5). By the finite-difference method Rivkind et al. (Rivkind et al., 1976) solve the problem for the flow outside and inside a spherical particle at Reynolds numbers of the outer flow in the range 0.5 to 200. On the basis of this solution they reach to the following approximate formula for the drag coefficient at different viscosities ratio λ:

$$c_D = \frac{\lambda c_{D_\infty} + c_{D_0}}{\lambda + 1},$$

where c_{D_∞} is the drag coefficient for a spherical rigid particle ($\lambda \to \infty$) and c_{D_0} - the drag coefficient for a spherical gas bubble. The numerical results for the drag coefficient at small Reynolds numbers, obtained in a similar work by Rivkind and Ryskin (1976), are in good correlation with the analytical results for this type of flow.

Oliver and Chung (1985) treat the same problem at small Reynolds by the above mentioned method: a combination of the finite-difference numerical method and the series-truncation technique. Further, in a subsequent paper (Oliver and Chung, 1987), they modify their method by replacing the finite-difference numerical method with a cubic finite-element scheme and apply it again in combination with the series-truncation for fluid sphere problem but at moderate Reynolds numbers. This new techniques occurs to be quite effective when solving strongly nonlinear flow problems. The performed comparisons with other numerical solutions and experimental data show good agreement.

The paper of Youngren and Acrivos (1975) plays a crucial role in the development of the numerical methods for analysing problems of fluid flow around arbitrary rigid particles in Stokes approach. There the problem for finding a solution of the Stokes equations together with the continuity equation (1.3.6) is considered at boundary conditions

$$\mathbf{v}(\mathbf{r}) = -\mathbf{U}(\mathbf{r}), \quad \text{for } \mathbf{r} \in S_p, \tag{5.11.1}$$

$$\mathbf{v}(\mathbf{r}), p(\mathbf{r}) \to 0 \text{ at } |\mathbf{r}| \to \infty, \tag{5.11.2}$$

where S_p is the particle surface, while the flow region is denoted by Ω.

The solution of such a type of problems can be reduced to the solution of a system of integral equations, which is known from the works of Odqvist (1930) and Ladyzhenskaya (1969) and has been thoroughly investigated in Chapters 2 and 3 of the present book. The problem of Youngren and Acrivos is the numerical realisation of the same method for a flow around rigid particles of arbitrary shape.

Using the fundamental representation of the creeping flow velocity as a sum of a single-layer potential and double-layer potential (2.5.18), which is transformed into (2.5.22) when the singular points are lying on the particle surface and afterwards applying to it the boundary conditions (5.11.1) and (5.11.2), the following linear integral equations of first kind for the surface stress force or Stokeslets density $\mathbf{f}(\boldsymbol{\rho}) = \mathbf{T}(\boldsymbol{\rho}).\mathbf{n}(\boldsymbol{\rho})$ on the surface S_p are yielded:

$$-\mathbf{U}(\mathbf{r}) = \frac{3}{2\pi} \overset{PV}{\iint_{S_p}} \mathbf{U}(\boldsymbol{\rho}) \frac{(\mathbf{r}-\boldsymbol{\rho})(\mathbf{r}-\boldsymbol{\rho})(\mathbf{r}-\boldsymbol{\rho})}{R^5} \mathbf{n}(\boldsymbol{\rho}) d\sigma(\boldsymbol{\rho}) \tag{5.11.3}$$

$$+\frac{1}{4\pi\mu} \iint_{S_p} \mathbf{f}(\boldsymbol{\rho}) \left[\frac{\mathbf{I}}{R} + \frac{(\mathbf{r}-\boldsymbol{\rho})(\mathbf{r}-\boldsymbol{\rho})}{R^3} \right] d\sigma(\boldsymbol{\rho}),$$

where $\mathbf{r} \in S_p$, $R = |\mathbf{r} - \boldsymbol{\rho}|$ and $\mathbf{n}$ is the unit normal vector towards the particle surface.

On the basis of the analysis of Odqvist (1930) and Ladyzhenskaya (1969), Youngren and Acrivos (1975) prove, that if S_p is a Lyapunov surface, the formulated boundary problem has an unique solution for every $\mathbf{U}(\mathbf{r})$, and its solution is reduced to finding the functions $\mathbf{f}$ from (5.11.3). The uniqueness of the solution of (5.11.3) will be proved, if it is proved that the homogeneous system

$$0 = \frac{1}{8\pi\mu} \iint_{S_p} \boldsymbol{\phi}(\boldsymbol{\rho}) \left[\frac{\mathbf{I}}{R} + \frac{(\mathbf{r}-\boldsymbol{\rho})(\mathbf{r}-\boldsymbol{\rho})}{R^3} \right] d\sigma(\boldsymbol{\rho}), \qquad \mathbf{r} \in S_p, \tag{5.11.4}$$

has only a trivial solution.

For this reason the velocity field due to (2.6.13) corresponding to the solution $\boldsymbol{\phi}$ of (5.11.4) is considered

$$\mathbf{w}(\mathbf{r}) = \frac{1}{8\pi\mu} \iint_{S_p} \boldsymbol{\phi}(\boldsymbol{\rho}) \left[\frac{\mathbf{I}}{R} + \frac{(\mathbf{r}-\boldsymbol{\rho})(\mathbf{r}-\boldsymbol{\rho})}{R^3} \right] d\sigma(\boldsymbol{\rho}), \qquad \mathbf{r} \in \Omega, \tag{5.11.5}$$

as well as its stress field according to (2.5.17)

$$\mathbf{T}(\mathbf{w}) = -\frac{3}{4\pi} \overset{PV}{\iint_{S_p}} \boldsymbol{\phi}(\boldsymbol{\rho}) \frac{(\mathbf{r}-\boldsymbol{\rho})(\mathbf{r}-\boldsymbol{\rho})(\mathbf{r}-\boldsymbol{\rho})}{R^5} d\sigma(\boldsymbol{\rho}), \quad \mathbf{r} \in \Omega. \tag{5.11.6}$$

Since $\mathbf{T}$ suffers a jump across the surface S_p, then using (2.7.1) and changing the normal direction, we shall have

$$\lim_{\mathbf{r}\to\mathbf{r}_0} \mathbf{T}(\mathbf{w}).\mathbf{n}(\mathbf{r}) = \frac{1}{2}\boldsymbol{\phi}(\mathbf{r}_0) + \mathbf{T}^{PV}(\mathbf{w})\big|_{\mathbf{r}=\mathbf{r}_0}.\mathbf{n}(\mathbf{r}_0), \quad \mathbf{r}\in\Omega,\ \mathbf{r}_0\in S_p, \tag{5.11.7}$$

where the superscript PV means the principal value of the surface force on S_p.

Since the single-layer potential determined by (5.11.4) is continuous over the flow domain Ω, which includes also S_p, then $\mathbf{w}(\mathbf{r}) = 0$ on S_p. Moreover, from (5.11.5) it is clear, that $\mathbf{w}(\mathbf{r}) \to 0$ as $\mathbf{r} \to \infty$. As $\mathbf{w}$ and the correspondent pressure p, defined also by the single-layer potential (2.5.16) with density $\boldsymbol{\phi}$, satisfy the Stokes flow equations and $\mathbf{w}$ is equal to zero on both boundaries: infinity and particle surface, then the uniqueness of the Stokes flow solution (Odqvist, 1930) means that $\mathbf{w}$ and its associate stress $\mathbf{T}$ must be equal to zero everywhere in Ω. Then due to (5.11.7) $\boldsymbol{\phi}$ is obliged to be zero, too. Thus the uniqueness of (5.11.3) is proved.

In the mentioned paper Youngren and Acrivos treat the axisymmetric and three-dimensional flow around spheroids and cylinders of finite length. They solve the equations (5.11.3) numerically by the method of Krylov-Bogoliubov (Kantorovich and Krylov, 1958) when transforming the integral equations into a linear system of algebraic equations. The body surface S_p is divided into N elements Δ_m (m = 1, 2, .., N), which are sufficiently small to assume, with some approximation, $\mathbf{f}(\mathbf{r})$ constants on each element and equal to their values at the central point of the element. The integral equations (5.11.3) are satisfied on N discrete points $\mathbf{r}^m$ at the centre of each element and in this manner the integral equations are reduced to 3N linear algebraic equations with 3N unknowns $\mathbf{f}(\mathbf{r}^m)$ for the 3D case and to 2N equations with 2N unknowns for the axisymmetric case:

$$\sum_{j=1}^{N} \mathbf{A}^j(\mathbf{r}^m)\mathbf{f}(\mathbf{r}^j) = \mathbf{g}(\mathbf{r}^m), \qquad m = 1, 2, ..., N. \tag{5.11.8}$$

In (5.11.8) the coefficients $\mathbf{A}^j$ are connected with the integrals in (5.11.3) in front of the unknowns on the surface elements Δ_j:

$$\mathbf{A}^j(\mathbf{r}^m) = \frac{1}{4\pi\mu}\iint_{S_p}\mathbf{f}(\boldsymbol{\rho})\left[\frac{\mathbf{I}}{R^m} + \frac{(\mathbf{r}^m-\boldsymbol{\rho})(\mathbf{r}^m-\boldsymbol{\rho})}{R^{m3}}\right]d\sigma(\boldsymbol{\rho}), \tag{5.11.9}$$

while $\mathbf{g}$ are expressions obtained from the terms in (5.11.3), which do not contain the unknowns:

$$\mathbf{g}(\mathbf{r}^m) = -\mathbf{U}(\mathbf{r}^m) - \frac{3}{2\pi}\iint_{S_p}^{PV}\frac{(\mathbf{r}^m-\boldsymbol{\rho})(\mathbf{r}^m-\boldsymbol{\rho})(\mathbf{r}^m-\boldsymbol{\rho})}{R^{m^5}}\mathbf{n}(\boldsymbol{\rho})\mathbf{U}(\boldsymbol{\rho})d\sigma(\boldsymbol{\rho}). \tag{5.11.10}$$

After calculating the local stress forces on S_p the force $\mathbf{F}$ and the torque $\mathbf{M}$, acting on the particle surface, are found as a superposition of their counterparts on the all N elements.

An evident advantage of this method, named the boundary elements method, is that it reduces the 3D boundary-value problems to 2D, and 2D to 1D problems.

It turns out that the computation of the integrals included in $\mathbf{A}^j$ and $\mathbf{g}$ is the most time consuming. The linear algebraic system (5.11.8) is usually completely dense and it is solved by the standard Gaussian elimination techniques. The solution uniqueness of (5.11.3) does not necessarily mean that its discrete approximation (5.11.8) has also a unique solution but for the cases studied in (Youngren and Acrivos, 1975) this is true.

Another problem arises due to the singular behaviour of some of the integrals in $\mathbf{A}^j$ and $\mathbf{g}$, which implies a prior analytical integration in the elements with singularities. Moreover, the first kind integral equations (5.11.3) are ill conditioned, i.e., small errors in $\mathbf{A}^j$ and $\mathbf{g}$ lead to large errors in the solution $\mathbf{f}$, but this problem for the considered cases is less serious than usual and very accurate results are obtained.

An important characteristics of the method used by Youngren and Acrivos is that it can be applied successfully for the flow problems around clusters of rigid particles. However, at the particles surface discretization a large number of unknowns is required. For example, when solving the problem of a simple shear flow passing spheroids with small aspect ratio, Youngren and Acrivos used 144 surface elements to get results with average relative accuracy up to 1%. For more particles the number of unknowns increases significantly, which decreases the method efficiency. The finite-difference and the finite-elements methods can also be exploited but they are hardly applicable for infinite domains, since the disturbances in the fluid caused by the particles decay very slowly and the computational domains must be very large.

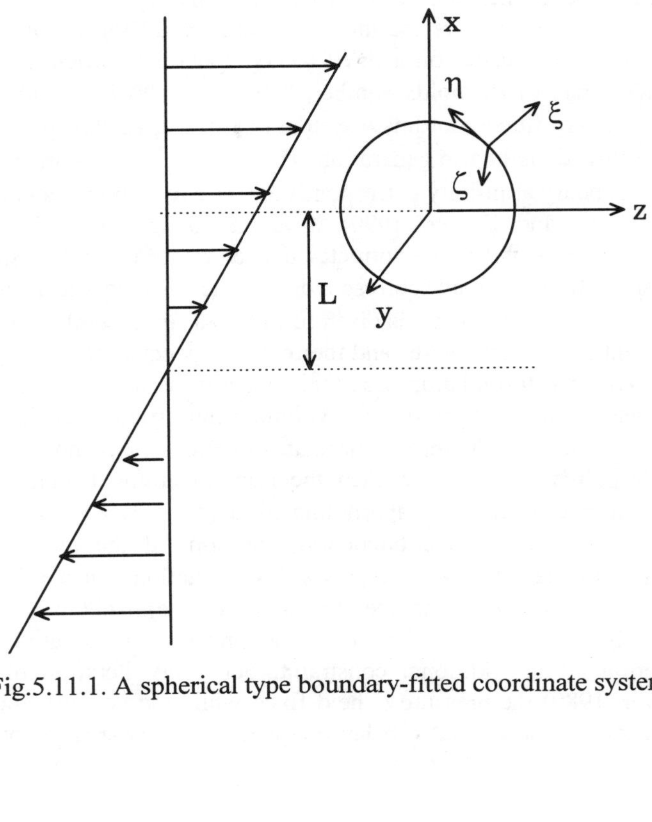

Fig.5.11.1. A spherical type boundary-fitted coordinate system

Aiming to improve the accuracy of calculations, better approximating formulae for the unknowns of each surface element have been proposed recently, based on linear or quadratic interpolation formulae and able to overcome some boundary discontinuity for non Lyapunov surfaces, e.g. a spherical bubble rise from a hole in a plane wall (Yan et al., 1987) and the 3D motion of a sphere approaching a circular orifice (Dagan et al., 1988).

Some global expansions of **f** are proposed using eigenfunctions (such as Legendre functions or trigonometric functions) in order to approximate better the integral equations by algebraic equations for the coefficients of these walls [Yan et al., (1987), Hsu and Ganatos, (1989)]. These improved numerical algorithms are applied mainly for bounded flows, where greater accuracy is necessary.

The theoretical works done on flows around particles at finite Reynolds numbers are mainly for axisymmetric outer flows, which reduces significantly the computational time. The requirement for a spherical body shape is mostly used [Masliyah, (1970), Woo, (1971), Brabston and Keller, (1975), LeClair et al., (1972), Rivkind and Ryskin, (1976), Oliver and Chung, (1987), Fornberg, (1988)] as well as that for small deviations from sphericity [Ryskin and Leal, (1984a, b), Christov and Volkov, (1985), Dandy and Leal, (1989)]. All these papers deal with a unbounded domain and uniform motion. If shear flow is introduced, then the problem becomes 3D and it is very consuming to study it at finite Reynolds numbers. Besides this, Dandy and Dwyer (1990) study the lift and drag force and heat transfer on a fixed heated spherical particle in a linear shear flow over a wide range of Reynolds number, $0.1 \leq Re \leq 100$ for small shear rates. A spherical type boundary-fitted coordinate system (ξ , η, ζ), as shown in Fig.5.11.1., is exploited for the flow domain grid generation. The particle surface corresponds to $\xi = 0$, while the outer boundary at infinity corresponds to $\xi = 1$, which in the numerical model presented in (Dandy, and Dwyer, 1990) is about 25 sphere radii. The boundary conditions on this outer boundary are corrected to account for its finite distance location. Then the physical flow domain corresponds to the mathematical model domain $0 \leq \xi, \eta, \zeta \leq 1$, as the model domain is discretized into a finite number of volumes. The full time-dependent 3D Navier-Stokes and thermal energy equation are approximated in terms of a finite-volume formulation, i.e., their integral variants are computed in each element by evaluating each integrand at the volume centre (node) and multiplying it by the volume of the element. If the three coordinate directions ξ , η and ζ are divided into M_ξ, M_η and M_ζ points, respectively, then the four governing balance equations for momentum and thermal energy are approximated at $(M_\xi - 2) \times (M_\eta - 2) \times (M_\zeta - 2)$ interior volume elements plus the boundary conditions at the remaining boundary elements, i.e., a complete set of $4 \times M_\xi \times M_\eta \times M_\zeta$ equations for $5 \times M_\xi \times M_\eta \times M_\zeta$ unknowns (3 velocity components, temperature and pressure) is obtained. Therefore, the problem is not fully determined and the set of equations is solved iteratively, leaving the continuity equation as an additional constraint. For every iteration of the iterative algorithm (Dwyer, 1989) the pressure is held fixed while the velocity and temperature are calculated from the approximated balance equations, afterwards the pressure due to the current velocity and temperature fields is calculated; with this new value the pressure enters the next iteration, etc. The proposed algorithm for fitting the pressure is similar to the artificial compressibility method (Chorin, 1967, 1968).

CHAPTER 6

Deformations of a Single Fluid Particle in a Viscous Flow

6.1. Introduction

Beside the subject of bubbles and drops as an area of intense study for many years, the attention is naturally focused on the problem of fluid particle deformation. One of the earliest analysis in this field goes as far back as the 1930's when Taylor (1934) treated the deformation of drops in linear shear and hyperbolic flows by means of the domain perturbation techniques. In most studies in this area, the matter of drop deformation was found to require special attention because of its well-known influence on the dynamics of such objects and on some processes like extraction, for example. Other areas of such studies application are the emulsion formation, where one fluid phase must be dispersed in another fluid, and the design of efficient mixing devices. The problems of oil recovery, where oil drops must be displaced by water in porous rock, are also related to the drop deformations.

In this chapter we shall present the basic results for the deformation of a single bubble and drop in a viscous flow. The behaviour of fluid particles suspended in shear flows differs strongly from their behaviour in uniform flows at infinity. As it will be shown in the following sections, the shape of a single spherical droplet remains unaltered for a uniform flow in Stokes approximation, while it is not true for a shear flow. At small values of the shear velocity G, the droplet shape in Stokes approximation is not spherical, but deviates weakly from this shape. At large values of G, however, large deformations take place, at which the fluid particle becomes unstable and usually it splits into two or more smaller fluid particles.

We have to note, that although Stokes equations (1.3.6) and the corresponding boundary conditions (1.4.14), (1.4.15), (1.4.16) and (1.4.17) are linear, as the droplet shape is determined from the normal stress balance condition (1.4.18) on the droplet surface, the problem becomes non-linear. This can be seen from the condition (1.4.18) given in dimensionless form:

$$\mathbf{n}.\left[\mathbf{T}.\mathbf{n}-\lambda\,\hat{\mathbf{T}}.\mathbf{n}\right]=\frac{1}{\mathrm{Ca}}\left(\frac{1}{\mathrm{R}_1}+\frac{1}{\mathrm{R}_2}\right), \qquad (6.1.1)$$

where $\lambda = \hat{\mu}/\mu$ is the viscosity ratio of the fluid inside and outside the drop, $\mathrm{Ca} = \mathrm{U}\mu/\sigma$ - the capillary number and R_1 and R_2 are the principal radii of the droplet curvature. The condition (6.1.1), due to the velocity gradient, transforms into a non-linear equation for the droplet shape with **r** as the surface position vector and a as the "equivalent" spherical droplet radius,

$$f\left(\frac{\mathbf{r}}{a}, \mathrm{Ca}, \lambda\right) = 0, \tag{6.1.2}$$

which makes the considered problem non-linear, too. This is the reason why exact solutions of shear viscous flows around non-spherical fluid particles have not been found till now, in contrast to the rigid particles case.

Since simultaneous solution of the flow fields and the equation of the interface is extremely difficult, an iterative procedure (called a domain perturbation technique) was adopted by Taylor (1934). First, the drop is postulated to be spherical and the flow fields are determined by using the boundary conditions of continuity of the tangential velocity vectors, vanishing normal components of the velocity vectors and continuity of the tangential components of the stress vectors inside and outside the spherical drop. Further, the function describing the deviation of the droplet from sphericity is determined using the relation between the outside and inside values of the normal components of the stress vectors. The newly determined interface may then be used for calculating the flow fields of the second iteration and so on.

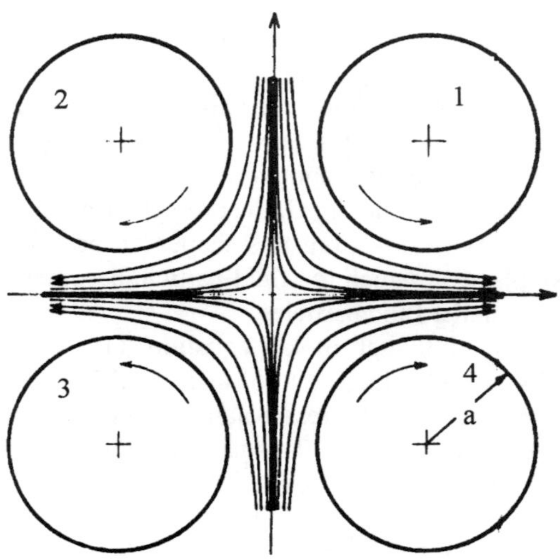

Fig.6.1.1. A top view scheme of four-roll mill.

Some experiments on deformation of fluid particles subjected to extensional flow, at very low Reynolds number, were performed by Taylor (1934). In his experiments a neutrally buoyant single drop was placed in a four-roller mill (see Fig.6.1.1.) filled with golden syrup (a highly viscous fluid) and then the effect of increasing flow velocity on the drop shape was observed. Therefore, the velocity gradient is responsible for the drop deformation and its subsequent burst. The problem becomes more complicated when both shear and gravity forces act on the drop and up to our knowledge remains still unsolved. The value of the shear rate, at which the steady drop shape ceases to exist and leads to the drop burst, is called a critical shear rate G_{cr} and the related with it capillary number: a critical capillary number Ca_{cr}. In the works of Taylor (1934) and Rumscheidt and Mason (1961) it is shown that at $G < G_{cr}$, the drop accepts a steady shape, while at $G > G_{cr}$ it starts to elongate transiently. The latter state is known as "drop breakup". Actually, the drop stretches and thins, obtaining a thread-like form. It is possible that for sufficiently small drop radii capillary waves become strong, leading to drop fragmentation into a large number of smaller droplets. In the experimental work of Grace (1971) this fragmentation is achieved by means of a gradual decrease of the shear rate. More details about the time dependent effects of the drop deformation and burst will be found in the subsequent Chapter 12 of the present book.

In the case of simple shear flows, there exist circumstances, e.g., at $\lambda > 1$, for which the drop assumes some limiting unchangeable shape, nearly spherical, even when G increases. For this kind of flows no drop breakup occurs as schematically illustrated in Fig.6.1.2.

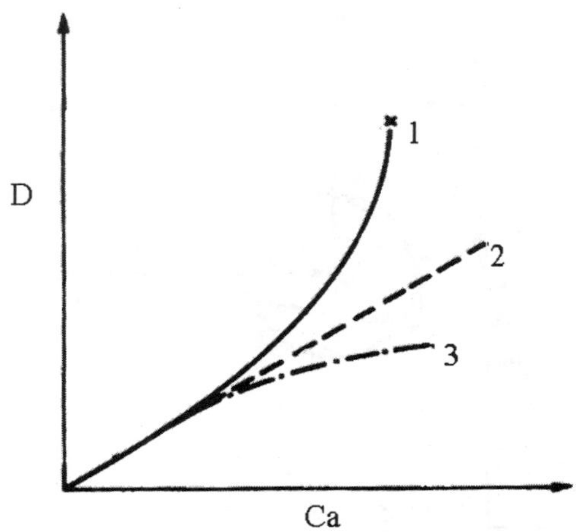

Fig.6.1.2. Experimentally found relations for the drop deformation: 1 - drop breakup at the point **x**; 2 - linear theory; 3 - drop limiting shape (from Grace, 1971).

6.2. Small Deformations of a Drop in Shear Flows at Zero Reynolds Number. Drop Breakup

Using the Stokes equations Taylor (1932, 1934) obtains the first fundamental results for the drop behaviour in a shear Couette flow and hyperbolic flow. Satisfying the boundary conditions of velocity and tangential stress continuity on the fixed drop spherical surface, he finds the velocity and the normal stresses in the flow inside and outside the drop. If the shear G is weak, i.e., the capillary number $Ca = Ga\mu/\sigma << 1$, at fixed λ, or if the fluid inside the drop is more viscous than the surrounding one with $\lambda >> 1$ at fixed Ca, then the drop shape is nearly spherical or its deformation from sphericity is small. Taylor considers the two possibilities in his studies. In the special case of a simple shear flow with undisturbed velocity $\mathbf{v}^{\infty} = Gx_2\mathbf{i}$, Taylor finds the shape of the fluid particle for a weak shear flow, as an asymptotic series in the small parameter $\varepsilon = Ca << 1$,

$$r = a\left(1 + \frac{2D}{a^2}x_1x_2\right) + O\left(Ca^2\right), \tag{6.2.1}$$

where r is the value of the drop surface position vector $\mathbf{r}$ and D is the deformation parameter characterizing the degree of deformation

$$D_{Taylor} = D = \frac{L-b}{L+b} = Ca\frac{16+19\lambda}{16+16\lambda} << 1. \tag{6.2.2}$$

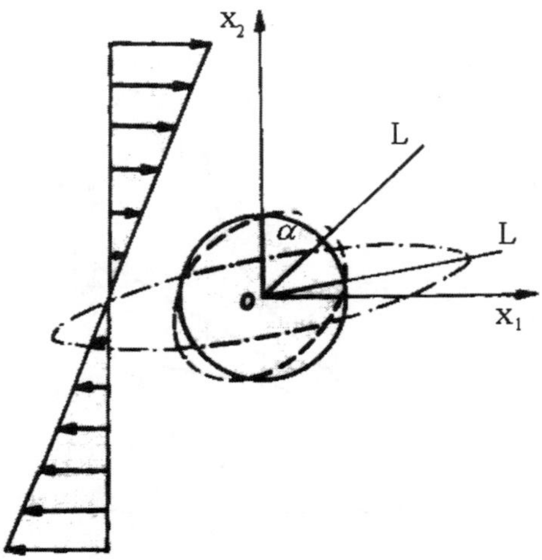

Fig.6.2.1. Deformation stages of a drop in a simple shear flow with the increase of G .

Here b = a(1 - D) is the minor deformed axis and L = a(1 + D) is the major one. Formula (6.2.1) expresses the shape of a rotational ellipsoid with axes oriented along the principal strain axes of the undisturbed flow. The orientation of the ellipsoid is set by the angle α between the ellipsoid major axis L and the coordinate axis Ox_2 as shown in Fig.6.2.1., which is obtained experimentally by Taylor (1934) in a four-roller mill and Rumscheidt and Mason (1961). For weak shear flows Ca << 1 and $\alpha = 45^0$, i.e., the drop is elongated along the principal extension direction and shrunk in normal direction. When increasing Ca the drop equilibrium shape becomes elongated and α tends to zero. It is established that for Ca > 0.2 significant deviations from the theoretically determined elliptical form appear. Moreover, at Ca = 0.5, the viscous forces, leading to the drop shape change exceed the surface tension forces tending to preserve the existing drop shape and it breaks up, loosing its stability. At these values of Ca, the drop shape dependence on the surface tension becomes nonlinear. Rumscheidt and Mason note that this dependence immediately before the drop breakup accepts different forms with respect to the different types of shear flows. Taylor's result has been improved by the theoretical work of Chaffey and Brenner (1967), who have studied the deformations of a drop suspended in a Couette flow and have shown that the drop shape is not spherical if the second-order approximation is taken into account. The latter model is in excellent agreement with the experimental results of Rumscheidt and Mason.

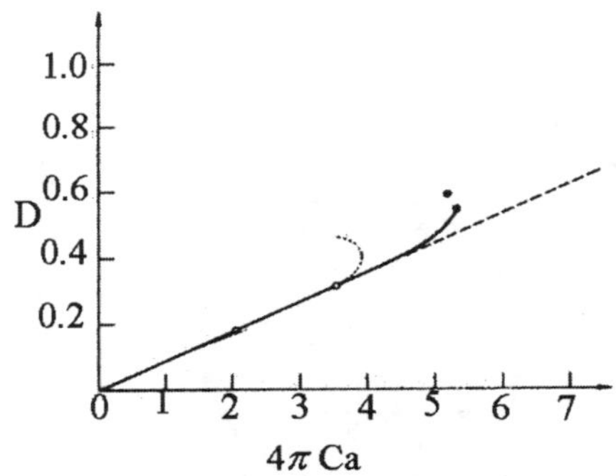

Fig.6.2.2. Drop deformation at $\lambda = 1$ according to: (●) numerically determined point of breakup and numerical results (———) of Rallison (1981); (O) experiments of Rumscheidt and Mason (1961); (------) linear theory of Taylor (1932); (……) non-linear theory of Barthes-Biesel and Acrivos (1973b).

Since the shape function (6.2.2) is linear in Ca and then in the shear rate G, no drop burst due to it can be expected. A higher-order analysis in Ca is performed by Barthes-Biesel and Acrivos (1973b) and the drop shape is represented of order $O(Ca^2)$.

In the limit of still small deformations at $\lambda = 1$, in Fig.6.2.2. the drop deformation rate D is plotted versus Ca based on the experimental observations of Rumscheidt and Mason (1961), the numerical computations of Rallison (1981) by the boundary integral technique, the linear theory of Taylor (1932) and the non-linear analysis of Barthes-Biesel and Acrivos. Although Taylor's linear theory predicts well the deformation rate, it does not give a criterion to determine Ca_{cr} or critical shear rate G_{cr} at which the droplet breakup appears. However, from the non-linear theory it is possible to reach a critical capillary number Ca_{cr} but it overestimates the drop breakup and is less accurate than the linear Taylor's theory (6.2.2). The numerical results of Rallison are in excellent agreement with the experimental ones up to the breakup point.

As pointed by Rallison (1984), small deformations may also occur with highly viscous drops, i.e., $\lambda >> 1$, and not very strong shear flow, $Ca \leq 1$. In this case the drop deformation is sought as asymptotic series on the small parameter $\varepsilon = 1/\lambda$. If, on the other hand, the fluid has sufficient vorticity, the drop can revolve as a rigid body. As noted by Taylor (1934), the presence of vorticity enhances the drop shape stability. He shows that at $\lambda > 4$ the shear rate G does not affect the drop deformation and the drop remains nearly spherical, regardless of higher shear rates. However, for a planar extensional flow, which has no vorticity the drop breakup is possible for every viscosity ratio λ.

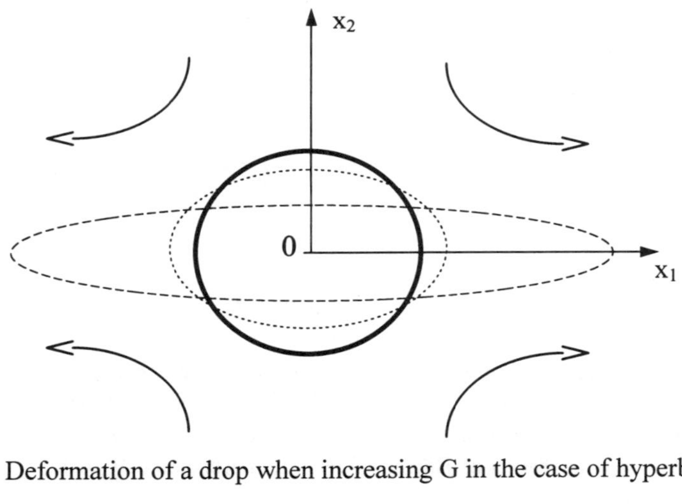

Fig.6.2.3. Deformation of a drop when increasing G in the case of hyperbolic flow.

The experimentally registered deformations of a drop in hyperbolic flow $\mathbf{v}^{\infty} = Gx_1\mathbf{i} - Gx_2\mathbf{j}$ in a four-roll mill by Taylor (1934) and Rumscheidt and Mason (1961) are shown in Fig.6.2.3. At small values of Ca and all λ the stable steady drop shape is elliptical. At higher Ca and $\lambda > 0.2$, the shapes are still rounded, but at Ca > Ca_c, the drop deforms into a long thin thread that afterwards breaks into a number of small droplets. Cox (1969) theoretically investigates an unsteady shear and hyperbolic flow impulsively started from rest, in which a fluid particle is submerged. It is shown that for both flows after a long period of time the drop reaches its steady shape, which differs from the initially spherical one by

$$D = D_{Taylor}\left[1+\left(\frac{19Ca\lambda}{20}\right)^2\right]^{-1/2} \text{ and } \alpha = \frac{\pi}{4}+\frac{1}{2}\tan^{-1}\left(\frac{19Ca\lambda}{20}\right) \tag{6.2.3}$$

for the shear flow, and by

$$D = 2D_{Taylor},\ \alpha = 0 \tag{6.2.4}$$

for the hyperbolic flow. It is interesting to note that the deformations (6.2.3) remain small when either $\lambda >> 1$ or Ca << 1, or both together are valid, while the deformations (6.2.4) are small only when Ca << 1. From (6.2.3) the Taylor's formula (6.2.2) is obtained at $\lambda = O(1)$ and Ca << 1. If, however, the interfacial tension is zero, i.e., Ca $\to\infty$ and $\lambda >> 1$, the deformation occurs as an undamped periodic oscillation of the drop:

$$D \cong \frac{5}{2\lambda}\sin\frac{t}{2},\ \alpha = \frac{\pi + t}{4}. \tag{6.2.5}$$

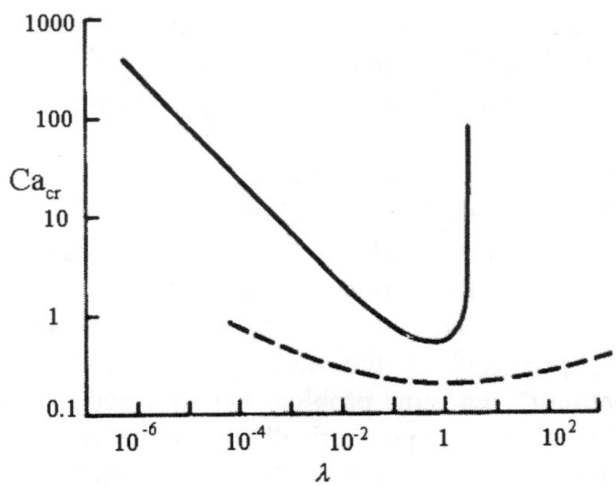

Fig.6.2.4. Experimentally determined relations between the viscosity ratio λ and the critical values of shear rate G (i.e., Ca) in the case of: (-------) hyperbolic flow and (______) simple shear flow (from Grace, 1971).

The time effects on deformation and burst are studied experimentally by Torza et al. (1972) and Grace (1971) for both shear and hyperbolic flow. The values of the critical capillary number versus λ are illustrated in Fig.6.2.4. with the observation data of Grace.

Another expression for the orientation angle of a droplet in shear flow is given by Cerf (1951) and afterwards corrected by Roscoe (1967):

$$\alpha = \frac{\pi}{4} + \mathrm{D}_{\mathrm{Taylor}} \frac{3+2\lambda}{5}. \tag{6.2.6}$$

Recently, Uijttewaal et al. (1993) have considered the full 3D drop deformation problem in the presence and in the absence of a wall in a linear shear flow by the boundary integral method. For the numerical implementation of this method the droplet surface is covered by a grid of boundary elements, triangles with collocation points at their vertices. The final steady droplet shape is reached by an iterative procedure with initially spherical droplet shape. The problem parameters $\lambda = 1$ and Ca ≤ 0.5 do not permit transient effects, or droplet breakup. The numerical analysis yields a drop deformation orientation that is in good agreement with the corrected results of Cerf and with the experiments of de Bruin (1989).

Bentley and Leal (1986) by means of computer controlled four-roll mill have studied experimentally the droplet deformation in the 2D linear flow $\mathbf{v}^{\infty} = \mathbf{\Gamma}.\mathbf{x}$, where

$$\mathbf{\Gamma} = \frac{\mathrm{G}}{2}\begin{pmatrix} 1+\alpha & 1-\alpha & 0 \\ -1+\alpha & -1-\alpha & 0 \\ 0 & 0 & 0 \end{pmatrix} \tag{6.2.7}$$

and G is the shear rate and α is the flow type parameter. Different flows in a wide range of $\alpha \in [0.2, 1]$ have been considered by Bentley and Leal (1986) in order to predict the deformation parameter D as a function of the capillary number Ca. In contrast to previous investigators, they show that the critical capillary number $\mathrm{Ca_{cr}}$ remains almost constant when $\lambda > 5$, which confirms the early second-order small deformation theory of Barthes-Biesel and Acrivos and the boundary integral numerical results of Rallison (1981).

The drop surface deformation and the critical capillary number for drop break up are highly influenced by surfactants present in the flow. Flumerfelt (1980) considers the Cox's problem (Cox, 1969) for small Ca, but with more general stress boundary conditions on the drop surface accounting for a variable surface tension and shear and the presence of dilatation surface viscosity. As a result the droplet shape deviates significantly from that predicted by the linear Taylor's theory. Simultaneously, an experimental investigation of the same problem is performed by Phillips et al. (1980). Stone and Leal (1990) when studying numerically the finite drop deformations derive an analytical expression for the small deformations of a spherical droplet with uniformly distributed surfactants at small surface Peclet numbers

$$D \approx \frac{3CaB}{4+CaB}, \text{ with } B = \frac{5}{4} \cdot \frac{(16+19\lambda)+4\beta\delta/(1-\beta)}{10(1+\lambda)+2\beta\delta/(1-\beta)}, \tag{6.2.8}$$

where β is the dimensionless interface tension gradient due to the surfactant and δ is the ratio between the surface Peclet number and the capillary number Ca. If $\beta \to 0$ or $\delta \to 0$ (6.2.8) is reduced to the Taylor's deformation formula (6.2.2) for uniaxial extensional flow. As discussed in section 3.3., the surfactants accumulate on the drop rear end and the surface tension decreases to zero. Thus, larger surface deformations are necessary in order to balance the viscous forces.

6.3. Small Deformations of a Drop in Uniform Viscous Flow at Small Reynolds Numbers

A first attempt to determine the drop shape in an uniform flow is made by Saito (1913), but the obtained solution is not correct, as the problem singularity is not taken into account. The same problem is solved successfully in the classical work of Taylor and Acrivos (1964) utilising the stationary Navier-Stokes equations at small Reynolds numbers and the method of matched asymptotic expansions. Some additional extensions of this solution are performed in Brignell (1973) and Pan and Acrivos (1968). Here we shall present briefly the Taylor and Acrivos' solution method and results.

Let us formulate the problem determining the small deformations of a drop of equivalent radius a in a slow steady viscous flow of velocity U. Since the problem is axisymmetric, the dimensionless form of the Navier-Stokes equations (1.3.20) in terms of the stream functions Ψ and $\hat{\Psi}$ outside and inside the drop, correspondingly, is written in spherical coordinates (r, θ):

$$\frac{1}{Re} E^4\Psi = \frac{1}{r^2}\frac{\partial(\Psi, E^2\Psi)}{\partial(r,\beta)} + \frac{2}{r^2}E^2\Psi\left[\frac{\beta}{1-\beta^2}\frac{\partial\Psi}{\partial r} + \frac{1}{r}\frac{\partial\Psi}{\partial\beta}\right], \tag{6.3.1}$$

$$\frac{\lambda}{\gamma\, Re} E^4\hat{\Psi} = \frac{1}{r^2}\frac{\partial(\hat{\Psi}, E^2\hat{\Psi})}{\partial(r,\beta)} + \frac{2}{r^2}E^2\hat{\Psi}\left[\frac{\beta}{1-\beta^2}\frac{\partial\hat{\Psi}}{\partial r} + \frac{1}{r}\frac{\partial\hat{\Psi}}{\partial\beta}\right]. \tag{6.3.2}$$

Here the differential operator E^2 is given by (5.2.2), $\beta = \cos\theta$, $\lambda = \hat{\mu}/\mu$, $\gamma = \hat{\rho}/\rho$ and $Re = Ua\rho/\mu$.

The drop surface equation in dimensionless form is given by

$$R(\beta) = 1 + H(\beta), \qquad \text{where } \max_{\theta} |H(\beta)| < 1. \tag{6.3.3}$$

If the unit normal and tangential vectors of the drop surface are denoted by $\mathbf{n}$ and τ, respectively, then the boundary conditions on the drop surface $r = R(\beta)$ are: zero normal velocities on either side of the surface (1.4.14) and (1.4.15); continuity of the tangential velocities (1.4.16); continuity of the tangential components of the stress vectors (1.4.17).

The dimensionless normal stress jump on the drop surface, in view of (1.4.11) and (1.4.12), is given by:

$$\mathbf{n}.\left[\mathbf{T}.\mathbf{n}-\hat{\mathbf{T}}.\mathbf{n}\right]=\frac{1}{\mathrm{We}}\left(\frac{1}{\mathrm{R}_1}+\frac{1}{\mathrm{R}_2}\right). \qquad (6.3.4)$$

Here the stress is dimensionalized by the dynamic pressure $\rho\,U^2$ and the Weber number is defined as We = ReCa= $\rho\, aU^2/\sigma$. However, for Stokes flow the stress scale is the viscous stress μ U/a and there, instead of We, the capillary number Ca stands as in (6.1.1). The uniform velocity boundary condition at infinity expressed by the stream function in spherical coordinates is

$$\Psi=\frac{1}{2}\mathrm{r}^2\left(1-\beta^2\right) \qquad \text{for } \mathrm{r}\to\infty. \qquad (6.3.5)$$

Since small drop deformations are only considered, the following relations are true up to second degree accuracy of H(β):

$$\frac{1}{\mathrm{R}_1}+\frac{1}{\mathrm{R}_2}\approx 2-2\mathrm{H}-\frac{\mathrm{d}}{\mathrm{d}\beta}\left[\left(1-\beta^2\right)\frac{\mathrm{dH}}{\mathrm{d}\beta}\right], \qquad (6.3.6)$$

$$\frac{1}{\mathrm{r}}\frac{\mathrm{dr}}{\mathrm{d}\theta}\approx-\left(1-\beta^2\right)^{1/2}\frac{\mathrm{dH}}{\mathrm{d}\beta}. \qquad (6.3.7)$$

In the first stage of the full problem (6.3.1) - (6.3.5) equations (6.3.1) and (6.3.2) are solved at the assumption that the drop has a spherical shape and the boundary condition (6.3.4) is neglected. Since Re << 1 the stream functions are expressed as asymptotic series expansions on the small parameter Re:

$$\Psi=\Psi_0+\mathrm{Re}\Psi_1+\ldots, \quad \text{for } \mathrm{r}\geq 1+\mathrm{H}, \qquad (6.3.8)$$

$$\hat{\Psi}=\hat{\Psi}_0+\mathrm{Re}\hat{\Psi}_1+\ldots, \quad \text{for } 0\leq \mathrm{r}\leq 1+\mathrm{H}, \qquad (6.3.9)$$

where (6.3.8) holds only in the Stokes region and must be matched (Proudman and Pearson, 1957) with the corresponding stream function solution of the outer region r >> 1. The zero-order terms of (6.3.8) and (6.3.9) correspond to the Stokes problem solution of Hadamard (1911) and Rybczynski (1911) (see section 3.3), for which the spherical drop shape remains unchanged. The higher order approximations of the stream functions are then expected to give some contribution to the drop surface deformation. Taylor and Acrivos (1964) confine themselves only to the first-order solutions outside and inside the drop.

In the second stage the drop shape is determined by the normal stress balance condition (6.3.4). As mentioned before for the zero-order solutions, the latter condition transforms into

$$\frac{\mathrm{d}}{\mathrm{d}\beta}\left[\left(1-\beta^2\right)\frac{\mathrm{dH}}{\mathrm{d}\beta}\right]+2\mathrm{H}=2-\mathrm{We}\,\Pi.$$

It is satisfied for Π = 2/We and H(β) ≡ 0 (see Taylor and Acrivos, 1964), which implies that the drop spherical shape is preserved and the system origin coincides with the drop centre, as given further by the conditions (6.3.11) and (6.3.12). Thus, the deformation is calculated only to the first-order stream functions Ψ_1 and $\hat{\Psi}_1$. In view of (6.3.6) and (6.3.7) the condition (6.3.4) is simplified as

$$\frac{d}{d\beta}\left[\left(1-\beta^2\right)\frac{dH}{d\beta}\right]+2H=-\mathrm{We}\,\mathbf{n}.\left[\mathbf{T}_1.\mathbf{n}-\lambda\,\hat{\mathbf{T}}_1.\mathbf{n}\right], \qquad (6.3.8\,\mathrm{a})$$

where $\mathbf{T}_1$ and $\hat{\mathbf{T}}_1$ are the first-order stresses.

Since the right-hand side of the upper equation is already known, it can be expanded in the Legendre polynomials of first kind:

$$-\mathrm{We}\,\mathbf{n}.\left[\mathbf{T}_1.\mathbf{n}-\hat{\mathbf{T}}_1.\mathbf{n}\right]=\gamma_0+\sum_{n=1}^{\infty}\gamma_n P_n(\beta).$$

Then the solution of (6.3.8a) can be sought in the form

$$H(\beta)=\alpha\,P_1(\beta)+\frac{\gamma_0}{2}+\sum_{n=2}^{\infty}\frac{\gamma_n P_n(\beta)}{(2+n)(1-n)}, \qquad (6.3.10)$$

where the condition of a finite drop deformation at $\beta=\pm 1$ is taken into account.

From the fluid incompressibility it follows that, the disturbance of the spherical shape at the drop deformation must satisfy the volume preservation condition:

$$\frac{4}{3}\pi=\frac{2}{3}\pi\int_{-1}^{1}\left[1+H(\beta)\right]^3 d\beta,$$

which is rewritten up to first-order terms as

$$\int_{-1}^{1}H(\beta)d\beta=0. \qquad (6.3.11)$$

The coordinate system origin coincides with the drop centre when H(β) obeys the condition:

$$\int_{-1}^{1}\left[1+H(\beta)\right]^4 \beta d\beta=0,$$

which for the first-order terms transforms into

$$\int_{-1}^{1}H(\beta)\beta d\beta=0. \qquad (6.3.12)$$

From the last two linearized conditions, it follows that $\gamma_0=0$ and $\alpha=0$, respectively. The remaining coefficients in (6.3.10) can also be determined and finally the drop shape is expressed by an expansion in We << 1, as We = O(Re^2):

$$r=1-\alpha_1 \mathrm{We} P_2(\beta)-\alpha_2\,\mathrm{We}^2/\mathrm{Re}\,P_3(\beta)+\ldots, \qquad (6.3.13)$$

$$\alpha_1=\frac{1}{4(1+\lambda)^3}\left\{\left(\frac{81}{80}\lambda^3+\frac{57}{20}\lambda^2+\frac{103}{40}\lambda+\frac{3}{4}\right)-\frac{(\gamma-1)(1+\lambda)}{12}\right\},\quad \alpha_2=\frac{3\alpha_1(10+11\lambda)}{70(1+\lambda)}.$$

Therefore, it can be seen from (6.3.13) that with the increase of Weber number, due to the second term of (6.3.13), the spherical drop will first deform into an oblate spheroid whose minor axis is parallel to the undisturbed flow direction and then, due to the third term of (6.3.13), will turn into a spherical cap. This result is confirmed by the experiments of Haberman and Morton (1953).

The drag force coefficient on the deformed drop yields

$$\frac{c_D \operatorname{Re}}{2\pi} = \frac{3\lambda+2}{1+\lambda} + \frac{\operatorname{Re}}{8}\left(\frac{3\lambda+2}{1+\lambda}\right)^2 + \frac{1}{40}\left(\frac{3\lambda+2}{1+\lambda}\right)^3 \operatorname{Re}^2 \ln \operatorname{Re} + \frac{\alpha_1 \operatorname{We}}{5(1+\lambda)^2}\left(3\lambda^2 - \lambda + 8\right) + \dots \tag{6.3.14}$$

The last term in the upper expression represents the inertial effect.

As discussed in section 3.3., the surfactants should change the drag and the terminal velocity on a spherical drop (Frumkin and Levich, 1947), but should not affect the spherical drop shape, since the inertia is neglected. However, Taylor and Acrivos do not expect any significant changes in the drop deformation (6.3.13) if surfactants are present in the flow, as the deformation is a direct consequence of the hydrodynamic forces and the static surface tension.

6.4. Small Deformations of a Drop in a General Linear Flow at Zero Reynolds Numbers

In section 3.3 we solved the problem of a spherical drop of radius a in a linear flow at zero Reynolds number. It is also of interest to determine the small deformations of a drop in a general linear flow at zero Reynolds number.

We assume that at large distances from the fluid particle the fluid performs a general linear flow

$$\mathbf{v}^\infty = G\left(\mathbf{E}^\infty + \mathbf{\Omega}^\infty\right).\mathbf{r}' \quad , \mathrm{r}' \to \infty , \tag{6.4.1}$$

where $\mathbf{r}'$ is a position vector, e_{ij}^∞ = const., Ω_{ij}^∞ = const. (i, j = 1, 2, 3) are the dimensionless rate of strain and vorticity tensors, respectively and G represents the magnitude of $\nabla \mathbf{v}$. It is known (see sections 6.2. and 6.3.) that a drop is deformed in almost any viscous flow, other than the uniform (steady) translation through a stationary fluid. With characteristic velocity $v_c = G a$ the dimensionless form of the equation (6.4.1) is

$$\mathbf{v}^\infty = \left(\mathbf{E}^\infty + \mathbf{\Omega}^\infty\right).\mathbf{r} \quad , \mathrm{r} \to \infty . \tag{6.4.2}$$

We assume that the density of the fluid inside the drop is equal to that outside it, but the viscosities of the two fluids $\hat{\mu}$ and μ are different (here the physical parameters pertaining to the interior of the drop will be distinguished from the corresponding exterior parameters by a caret). Further, we assume that the surface tension σ, is constant on the drop surface and the capillary number $\mathrm{Ca} = \frac{\mu v_c}{\sigma}$ is small, i.e., $\mathrm{Ca} \ll 1$.

The magnitude of the drop deformation depends on the capillary number and the ratio of internal to external viscosity: $\lambda = \dfrac{\hat{\mu}}{\mu}$. For Ca << 1, very small deviations from the spherical shape are possible. Taking into account this fact, we express the drop shape in the form

$$F(\mathbf{r},t) \equiv r - [1 + \text{Ca}\, f(\mathbf{r})] \equiv 0. \tag{6.4.3}$$

So we suppose that the deviation from sphericity is contained in the function $f(\mathbf{r})$ and that the magnitude of this deviation is proportional to Ca. It is also assumed that the Reynolds number of the motion within the drop is small compared to unity, like that of the motion outside the drop.

Generally, when the motion inside and outside the drop is analysed, the shape of the drop has to be determined. We should emphasise that the shape of a neutrally buoyant immiscible liquid drop immersed in a continuous liquid undergoing shear is not governed solely by the bulk and interfacial properties of the two phases, but also depends on the rate of shear. Usually, the determination of the drop surface equation constitutes a non-linear problem, because the unknown shape has to be defined together with the solution of the equation of motion inside and outside the drop which, on the other hand, is a linear Stokes equation. As a consequence of this non-linearity, the droplet shape has not yet been found in its general form but rather for small deviations from the spherical form only. For small values of the capillary number Ca the boundary conditions on the drop surface could be linearized with respect to the boundary conditions for an exactly spherical drop and we shall see that for the above problem an approximate analytic solution could be obtained. The two fluid motions are described by the equations (2.3.1) and (2.3.2) with different value of the viscosity - $\hat{\mu}$ and μ. We choose the origin of the coordinate system to be at the instantaneous position of the drop centre. The boundary conditions in dimensionless form at the interface of the drop are given by (1.4.14) and (1.4.15) with $\mathbf{U}_p = \mathbf{0}$, (1.4.16), (1.4.17) and (1.4.11). The latter condition expresses the relation between the outside and inside values of the normal components of the stress vectors and its dimensionless form reads

$$\frac{\hat{p} - p}{\mu} + 2\left(\mathbf{E}.\mathbf{n} - \lambda\, \hat{\mathbf{E}}.\mathbf{n}\right).\mathbf{n} = \frac{\nabla_s.\mathbf{n}}{\text{Ca}}, \tag{6.4.4}$$

where ∇_s is acting over the drop surface, $\mathbf{n}$ is its outer normal and p/μ and $\hat{p}/\mu$ are the dimensionless pressures outside and inside the drop, correspondingly. In addition, there is a boundary condition at infinity, namely (6.4.2), and the requirement that the solution is finite everywhere.

Applying the method of domain perturbations, we first postulate that the drop is spherical with radius a, and the fields inside and outside it are solved, using only the equations (6.4.2) and (1.4.14)-(1.4.17). Later the function $f(\mathbf{r})$ describing the deviation of the droplet from sphericity is determined using the boundary condition (6.4.4). Therefore, the solution presented herein should be considered as a first approximation to a much more complex problem.

If we consider the disturbance of the flow $\mathbf{v}'$ due to the drop, then the total flow is $\mathbf{v} = \mathbf{v}^\infty + \mathbf{v}'$ and $\mathbf{v}' \to \mathbf{0}$, as $r \to \infty$. Inspecting the functional form of the various singularities presented in section 3.1, we decide to represent the disturbance velocity field outside of the drop in terms of a Stokes dipole and a degenerate octupole

$$\mathbf{v}' = 2\pi\mu a^3 (\mathbf{E}^\infty.\nabla).(C_1 + C_2 a^2 \nabla^2)\frac{\mathbf{B}(\mathbf{r})}{8\pi\mu}, \tag{6.4.5}$$

where C_1 and C_2 are unknown constants.

Further on, we shall prove that in order to model correctly the flow inside the drop, we have to use the velocity field

$$\hat{\mathbf{v}} = d_1\, \mathbf{E}^\infty.\mathbf{r} + d_2\ \boldsymbol{\omega}^\infty \times \mathbf{r} + d_3\left[5\, r^2(\mathbf{E}^\infty.\mathbf{r}) - 2\,\mathbf{r}(\mathbf{r}.\mathbf{E}^\infty.\mathbf{r})\right], \tag{6.4.6}$$

where d_1, d_2 and d_3 are specified by the boundary conditions.

Since the Stokeson $\mathbf{H} = 2r^2\mathbf{I} - \mathbf{rr}$ is quadratic in r, the gradient of $\mathbf{H}$ is a third-order tensor that is linear on r. It turned out that the symmetric and antisymmetric derivatives (which are known as roton and stresson) can be used as "building materials" in the construction of the interior flow field for a spherical drop in the linear field (6.4.2). That is so because roton and stresson simply correspond to a rigid-body rotation and a constant rate-of-strain field. In (6.4.6) we also use a cubic field that is the less obvious portion of the solution. It should be emphasised that to avoid any singularity at the origin of the coordinate system, we have to use increasing harmonics inside the drop.

The cubic field could be obtained by the appropriate linear combination of $\mathbf{E}^\infty:\mathbf{rrr}$ and $r^2\mathbf{E}^\infty.\mathbf{r}$. If we seek the velocity field $\hat{\mathbf{v}}$ inside the drop in the form

$$\hat{\mathbf{v}} = d_1\, \mathbf{E}^\infty.\mathbf{r} + d_2\boldsymbol{\omega}^\infty \times \mathbf{r} + \hat{\mathbf{v}}'$$

then $\hat{\mathbf{v}}' = A\,\mathbf{E}^\infty:\mathbf{rrr} + B\, r^2\mathbf{E}^\infty.\mathbf{r}$, where A and B are unknown constants. From the equation of continuity (2.3.2) it is easy to show that $A = -\frac{2}{5}B$.

Therefore we can write down

$$\hat{\mathbf{v}}' = d_3\left(5\, r^2\mathbf{E}^\infty.\mathbf{r} - 2\,\mathbf{E}^\infty:\mathbf{rrr}\right).$$

Further we shall calculate the constants C_1 , C_2, d_1, d_2 and d_3 from the boundary conditions (1.4.14)-(1.4.17).

Taking into account formulae (3.1.2) and (3.1.5), we obtain from (6.4.5)

$$v_i = e_{ki}{}^\infty x_k + \frac{1}{2}\varepsilon_{kji}\omega_k{}^\infty x_j - \frac{3a^3}{4}C_1\, e_{kj}{}^\infty \frac{x_i x_j x_k}{r^5} - 3C_2 a^5\, e_{ij}{}^\infty \frac{x_j}{r^5} + \frac{15}{2}C_2\, a^5 e_{kj}{}^\infty \frac{x_i x_j x_k}{r^7}. \tag{6.4.7}$$

Since at $r = a = 1$ we have $x_i = r\, n_i = n_i$, then (6.4.7) gives

$$v_i = e_{ki}{}^\infty n_k(1 - 3C_2) + e_{kj}{}^\infty n_i n_j n_k\left(-\frac{3}{4}C_1 + \frac{15}{2}C_2\right) + \frac{1}{2}\varepsilon_{kji}\omega_k{}^\infty n_j\ . \tag{6.4.8}$$

Therefore,

$$\mathbf{v.n} = v_i n_i = e_{ki}{}^{\infty} n_i n_k \left(1 + \frac{9}{2} C_2 - \frac{3}{4} C_1\right). \tag{6.4.9}$$

From the normal velocity component inside the drop we get an expression similar to (6.4.6)

$$\hat{v}_i = d_1\, e_{ki}{}^{\infty} n_k + d_2\, \varepsilon_{kji} \omega_k{}^{\infty} n_j + 5\, d_3\, e_{ki}{}^{\infty} n_k - 2\, d_3\, e_{kj}{}^{\infty} n_j n_k n_i\ , \tag{6.4.10}$$

$$\hat{\mathbf{v}}.\mathbf{n} = \hat{v}_i n_i = e_{ki}{}^{\infty} n_k n_i \left(d_1 + 3\, d_3\right). \tag{6.4.11}$$

According to (1.4.14), (1.4.15), (6.4.9) and (6.4.11), we find for the unknown coefficients the following two relations:

$$3\, C_1 - 18\, C_2 = 4, \tag{6.4.12}$$

$$d_1 + 3\, d_3 = 0. \tag{6.4.13}$$

Applying the boundary condition (1.4.16) we derive the expression $d_2 = \frac{1}{2}$ and reach to the third relation:

$$1 - 3C_2 = d_1 + 5\, d_3. \tag{6.4.14}$$

For the boundary condition (1.4.17) we need the quantities:

$$e_{ip} n_p n_i \Big|_{r=1} = e_{kj}{}^{\infty} n_j n_k \left(\frac{3}{2} C_1 - 18 C_2 + 1\right), \tag{6.4.15}$$

$$\hat{e}_{ip} n_p n_i \Big|_{r=1} = e_{kj}{}^{\infty} n_j n_k \left(d_1 + 9\, d_3\right). \tag{6.4.16}$$

Reworking both sides of (1.4.17), we finally get the fourth relation for the solution coefficients:

$$1 - \frac{3}{4} C_1 + 12 C_2 = \lambda\left(d_1 + 8\, d_3\right). \tag{6.4.17}$$

In this way, we have four equations [namely (6.4.12), (6.4.13), (6.4.14) and (6.4.17)] for four unknown constants C_1, C_2, d_1 and d_3. Solving this system, we get

$$C_1 = \frac{2(5\lambda + 2)}{3(\lambda + 1)},\ C_2 = \frac{\lambda}{3(\lambda + 1)},\ d_1 = -\frac{3}{2(\lambda + 1)},\ d_3 = \frac{1}{2(\lambda + 1)}, \tag{6.4.18}$$

which are then replaced in (6.4.4) and (6.4.5):

$$\mathbf{v} = \mathbf{E}^{\infty}.\mathbf{r} + \frac{1}{2}\, \boldsymbol{\omega}^{\infty} \times \mathbf{r} + 2\pi\mu a^3 \left(\mathbf{E}^{\infty}.\nabla\right)\left[\frac{2(5\lambda + 2)}{3(\lambda + 1)} + \frac{\lambda a^2}{3(\lambda + 1)} \nabla^2\right] \frac{\mathbf{B}(\mathbf{r})}{8\pi\mu},$$

$$\hat{\mathbf{v}} = -\frac{3}{2(\lambda + 1)} \mathbf{E}^{\infty}.\mathbf{r} + \frac{1}{2}\, \boldsymbol{\omega}^{\infty} \times \mathbf{r} + \frac{1}{2(\lambda + 1)}\left[5r^2\left(\mathbf{E}^{\infty}.\mathbf{r}\right) - 2\mathbf{E}^{\infty}{:}\mathbf{rrr}\right].$$

The obtained solution can be used now to calculate the deformation of the drop from (6.4.4) for small values of the capillary number Ca. If $\mathbf{e}_r$ is the unit vector in the radial direction of a spherical coordinate system, then the first approximation of the unit normal vector **n** for small Ca is $\mathbf{e}_r$. According to the definition of **n** written in terms of F, and to the equation $\nabla r = \frac{\mathbf{r}}{r} = \mathbf{e}_r$, it follows that

$$\mathbf{n} \equiv \frac{\nabla F}{|\nabla F|} = \frac{\nabla r - Ca\nabla f}{|\nabla r - Ca\nabla f|} = \frac{\mathbf{e}_r - Ca\nabla f}{\sqrt{1 + Ca\left[Ca(\nabla f)^2 - 2(\mathbf{e}_r.\nabla f)\right]}}. \qquad (6.4.19)$$

It is important to note that

$$\mathbf{n} = \mathbf{e}_r - Ca\nabla f + 2Ca(\mathbf{e}_r\mathbf{e}_r.\nabla f) + O(Ca^2), \qquad (6.4.20)$$

$$\nabla_s.\mathbf{n} = \nabla.\mathbf{e}_r - Ca\nabla^2 f + 2Ca\nabla.(\mathbf{e}_r\mathbf{e}_r.\nabla f) + O(Ca^2), \qquad (6.4.21)$$

$$\nabla.\mathbf{e}_r = \frac{\partial}{\partial x_i}\left(\frac{x_i}{r}\right) = \frac{\delta_{ii}}{r} - \frac{x_i x_i}{r^3} = \frac{3}{r} - \frac{1}{r} = \frac{2}{r}. \qquad (6.4.22)$$

It follows from (6.4.20)-(6.4.22) that the surface curvature $\nabla_s.\mathbf{n}$ can be expressed as

$$\nabla_s.\mathbf{n} = \frac{2}{r} - Ca\nabla^2 f + Ca\frac{2\mathbf{r}.\nabla f}{r^2} + O(Ca^2) \qquad (6.4.23)$$

$$= 2(1 - Ca\, f) - Ca\nabla^2 f + 2Ca\,\mathbf{r}.\nabla f + O(Ca^2) = 2 - Ca(2\,f - 2\mathbf{r}.\nabla f + \nabla^2 f) + O(Ca^2).$$

Then the boundary condition (6.4.4) is transformed into:

$$Ca\frac{\hat{p} - p}{\mu} + 2Ca\left\{\mathbf{E}.\left[\mathbf{e}_r - Ca\nabla f + O(Ca^2)\right] - \lambda\,\hat{\mathbf{E}}.\left[\mathbf{e}_r - Ca\nabla f + O(Ca^2)\right]\right\}$$
$$\times\left[\mathbf{e}_r - Ca\nabla f + O(Ca^2)\right] = 2 - Ca(2\,f - 2\mathbf{r}.\nabla f + \nabla^2 f) + O(Ca^2), \qquad (6.4.24)$$

and up to order $O(Ca^2)$ it takes the form:

$$\frac{\hat{p} - p}{\mu} + 2(\mathbf{E}.\mathbf{e}_r - \lambda\,\hat{\mathbf{E}}.\mathbf{e}_r).\mathbf{e}_r = \frac{1}{Ca}\left[2 - Ca(2\,f - 2\mathbf{r}.\nabla f + \nabla^2 f)\right]. \qquad (6.4.25)$$

The shape function f($\mathbf{r}$) is a true scalar and is linearly related to the variables $\mathbf{v}$, $\hat{\mathbf{v}}$, and $(\mathbf{E}.\mathbf{n} - \lambda\,\hat{\mathbf{E}}.\mathbf{n})$. Therefore, f($\mathbf{r}$) can be expressed in an invariant form as a linear function of $\mathbf{E}^\infty$ (or $\boldsymbol{\Omega}^\infty$), i.e.,

$$f(\mathbf{r}) = b\mathbf{r}.\mathbf{E}^\infty.\mathbf{r}, \qquad (6.4.26)$$

where b is unknown constant.

From (6.4.23) and (6.4.26) it follows that the surface curvature $\nabla_s.\mathbf{n}$ is equal to

$$\nabla_s.\mathbf{n} = 2 + 4(\mathbf{r}.\mathbf{E}^\infty.\mathbf{r})bCa + O(Ca^2), \qquad (6.4.27)$$

and thus (6.4.24) becomes

$$\frac{\hat{p} - p}{\mu} + 2(\mathbf{E}.\mathbf{e}_r - \lambda\,\hat{\mathbf{E}}.\mathbf{e}_r).\mathbf{e}_r = \frac{1}{Ca}\left[2 + 4(\mathbf{r}.\mathbf{E}^\infty.\mathbf{r})bCa\right]. \qquad (6.4.28)$$

In order to apply the boundary condition (6.4.28), we have to calculate the pressures p and $\hat{p}$ from the Stokes equations inside and outside the drop. After some algebra we reach to the result for the velocity gradient and, consequently, for the pressure gradient inside the drop:

$$\frac{\partial \hat{v}'_i}{\partial x_j} = 5d_3 e_{ij}{}^\infty x_p x_p + 10 d_3 e_{ki}{}^\infty x_k x_j - 2d_3 e_{kl}{}^\infty x_k x_l d_{ij} - 4d_3 e_{kj}{}^\infty x_k x_i, \qquad (6.4.29)$$

$$\left(\nabla^2\hat{v}'\right)_i = 42 d_3 e_{ki}^{\infty} x_k , \tag{6.4.30}$$

$$\left(\nabla\hat{p}\right)_i = \hat{\mu}\left(\nabla^2\hat{v}'\right)_i = 42\,\hat{\mu}\, d_3 e_{ki}^{\infty} x_k . \tag{6.4.31}$$

Integrating equation (6.4.31) with respect to x_i we find that

$$\hat{p} = 21\,\hat{\mu}\, d_3 e_{ki}^{\infty} x_k x_i + \hat{\mu}\,\hat{p}_0 = 21\,\hat{\mu}\, d_3 \mathbf{E}^{\infty}\!:\!\mathbf{rr} + \hat{\mu}\,\hat{p}_0 , \tag{6.4.32}$$

where $\hat{p}_0 = \text{const}$.

Similarly, we obtain the pressure outside the drop, namely (Zapryanov and Markov, 1994):

$$p = -\frac{3a^3}{2}\,\mu C_1 r^{-5} \mathbf{E}^{\infty}\!:\!\mathbf{rr} + \mu\, p_0 , \tag{6.4.33}$$

where $p_0 = \text{const}$.

We observe that pressure field inside and outside the droplet can be calculated from the Stokes equations up to a constant. The constant p_0 involved in the outside field is determined from the known pressure for the droplet. As we shall see, the constant $\hat{p}_0$ involved in the interior pressure field can be determined from the boundary condition (6.4.4) for the normal components of the stress vectors. Using the equations (6.4.15), (6.4.16), (6.4.28), (6.4.32) and (6.4.33) we obtain an equation for the constants p_0, $\hat{p}_0$ and b:

$$-\mu\, p_0 + \hat{\mu}\,\hat{p}_0 + 21\,\hat{\mu}\, d_3 \mathbf{E}^{\infty}\!:\!\mathbf{rr} + \frac{3}{2}\mu C_1 \mathbf{E}^{\infty}\!:\!\mathbf{rr}$$

$$+2\mu\left[1+\frac{3}{2}C_1 - 18C_2 - \lambda\left(d_1 + 9d_3\right)\right]\mathbf{E}^{\infty}\!:\!\mathbf{rr} = \mu\left(\frac{2}{Ca} + 4b\mathbf{E}^{\infty}\!:\!\mathbf{rr}\right), \tag{6.4.34}$$

which reduces to

$$\lambda\,\hat{p}_0 - p_0 = \frac{2}{Ca}$$

and

$$\frac{3}{4}C_1 + \frac{21}{2}\lambda\, d_3 + 1 + \frac{3}{2}C_1 - 18C_2 - \lambda\left(d_1 + 9d_3\right) = 2b . \tag{6.4.35}$$

From the latter equation we have

$$b = \frac{19\lambda + 16}{8(\lambda + 1)} ,$$

and thus the drop shape is determined as

$$r = 1 + Ca\frac{19\lambda + 16}{8(\lambda + 1)}\left(\mathbf{r}.\mathbf{E}^{\infty}.\mathbf{r}\right). \tag{6.4.36}$$

To illustrate the result reached in (6.4.36), we shall note that for a simple shear flow $\left(v_1^{\infty} = x_2, v_2^{\infty} = 0, v_3^{\infty} = 0\right)$

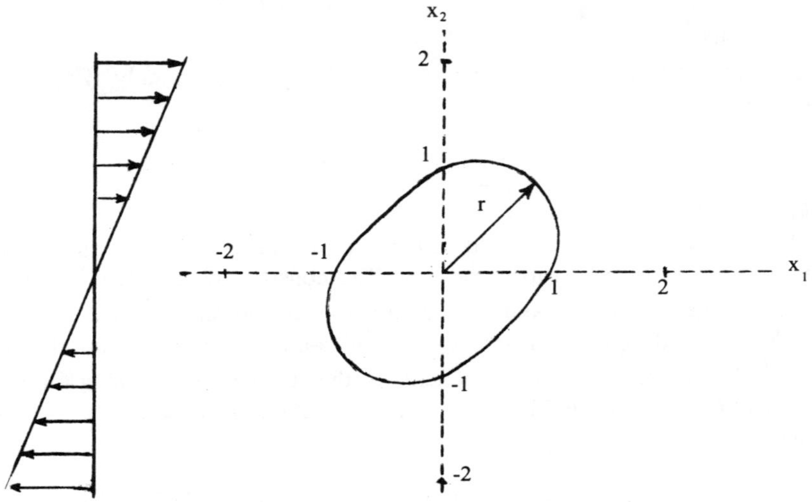

Fig.6.4.1.a) Small deformation of a drop in a simple shear flow (from Leal, 1992).

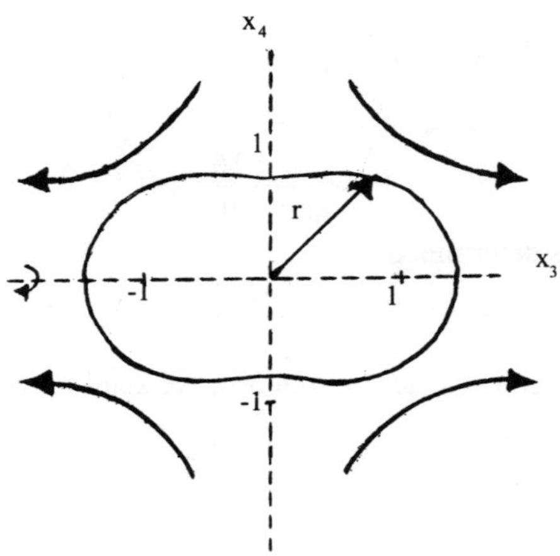

Fig.6.4.1.b) Small deformation of a drop in an extensional flow (from Leal, 1992).

$$\mathbf{E}^{\infty} = \begin{pmatrix} 0 & \frac{1}{2} & 0 \\ \frac{1}{2} & 0 & 0 \\ 0 & 0 & 0 \end{pmatrix} \quad \text{and} \quad r = 1 + \mathrm{Ca}\, x_1 x_2 \frac{19\lambda + 16}{8(\lambda + 1)},$$

whereas for an extensional flow $\left(v_1^{\infty} = -\frac{1}{2} x_1, v_2^{\infty} = -\frac{1}{2} x_2, v_3^{\infty} = x_3 \right)$

$$\mathbf{E}^{\infty} = \begin{pmatrix} -\frac{1}{2} & 0 & 0 \\ 0 & -\frac{1}{2} & 0 \\ 0 & 0 & 1 \end{pmatrix} \quad \text{and} \quad r = 1 + \mathrm{Ca} \frac{19\lambda + 16}{8(\lambda + 1)} \left(x_3^{\,2} - \frac{1}{2} x_1^{\,2} - \frac{1}{2} x_2^{\,2} \right).$$

These two flows are sketched in Fig.6.4.1. One can see that in all cases for the limit Ca << 1, although the deformation is small, the slight difference in the shape of the two flows shows that the extensional flow is more efficient at stretching deformable particles than the simple shear flow.

6.5. Small Deformations of a Fluid Particle at High Reynolds Numbers

This section is an extension of section 5.9., in which we dealt with the spherical drop and bubble dynamics at high Reynolds numbers. Since the fluid drop deformation problem, due to the internal Hill circulation, is more complicated than the bubble problem, we shall start with the shape determination of a bubble rising in liquid.

The rising velocity as a function of the bubble size for different liquids is measured experimentally by Haberman and Morton (1954), while Saffman (1956) and Hartunian and Sears (1957) study experimentally and theoretically the path instability at bubble rising. All these authors show that the physical liquid parameters are very important for the bubble behaviour through the Morton number $M = g\mu^4/\rho\sigma^3$. The bubble deformation at high Reynolds numbers is expected to be small if the Morton number is low. If the relation $\mathrm{Re} \geq 0.55\, M^{-1/5}$ is present, then the bubble shape deviation from sphericity is more than 5%. Then for liquids such as pure water with $M \approx O(10^{-11})$ the bubble deformations are significant at Re > 100, while for oil $M \approx O(10^{-2})$ and the deformations must be taken into account at small Re.

When deriving the velocity distribution (5.9.48) together with (5.9.38) and (5.9.40) in the boundary layer around the bubble, the bubble shape was assumed spherical and the normal stress balance condition was neglected. Similarly to the analysis performed in the previous section 6.3., we shall assume small bubble shape deformation from sphericity, which implies that the hydrodynamic and shape determination problems are uncoupled. Then the results for the velocity and pressure fields found in section 5.9. will be utilised to obtain the weakly deformed bubble shape, assumed to be still axisymmetric and presented in dimensionless form as:

$$r = 1 + \varepsilon f(\theta) + \ldots, \tag{6.5.1}$$

where ε is a small parameter to be determined throughout the solution and $f(\theta)$ is the deviation from sphericity function.

Taking in view that the bubble density and viscosity are negligible with respect to those of the ambient fluid, then the normal stress component inside the bubble is reduced to its pressure with reverse sign, i.e., $\mathbf{n}.\hat{\mathbf{T}}.\mathbf{n} = -p_{bubble}$. Substituting this expression in (6.3.4), the simplified dimensionless normal stress condition on the bubble surface $r = r(\theta)$ yields:

$$\mathbf{n}.\mathbf{T}.\mathbf{n} + p_{bubble} = \frac{\nabla_s.\mathbf{n}}{We}, \tag{6.5.2}$$

where $\nabla_s.\mathbf{n}$ must be calculated with (6.5.1) and is obtained to be

$$\nabla_s.\mathbf{n} = 2 - \varepsilon\left\{2f + \frac{\partial}{\partial\beta}\left[\left(1-\beta^2\right)\frac{\partial f}{\partial\beta}\right] + O\left(\varepsilon^2\right)\right\}, \tag{6.5.3}$$

with $\beta = \cos\theta$. The normal stress component outside the bubble in its dimensionless form is, correspondingly

$$\mathbf{n}.\mathbf{T}.\mathbf{n} = \tau_{rr} = -p + \frac{1}{Re}\left(2\frac{\partial v_r}{\partial r}\right), \tag{6.5.4}$$

which evaluated from the boundary layer on the bubble surface (see section 5.9.) is reduced to $-p_0(\theta) + O(Re^{-1/2})$. The zero-order pressure field $p_0(\theta)$ is calculated from the steady Euler equations (obtained formally from the Navier-Stokes equations (1.3.4) at $Re \to \infty$ and $St \to 0$), resulting from the potential solution (5.9.11) and (5.9.12):

$$p_0 = p_\infty + \frac{1}{2} - \frac{9}{8}\sin^2\theta + p_{static}, \tag{6.5.5}$$

where p_∞ is a constant pressure at infinity and $p_{static} = (-9/Re)\cos\theta$ is the dimensionless hydrostatic pressure on the bubble surface calculated from the balance of drag, based on the Levich result $12\pi\mu Ua$ as a first term of (5.9.25) in its dimensional form, and the buoyancy force.

Substituting (6.5.3), (6.5.4) and (6.5.5) in the boundary condition (6.5.2) and leaving only the terms up $O(Re^{-1/2})$ and $O(\varepsilon^2)$, one finally gets:

$$\left[p_{bubble} - p_\infty - \frac{1}{2} + \frac{9}{8}\sin^2\theta\right] = \frac{1}{We}\left[2 - \varepsilon\left\{2f + \frac{\partial}{\partial\beta}\left[\left(1-\beta^2\right)\frac{\partial f}{\partial\beta}\right]\right\}\right]. \tag{6.5.6}$$

Since $f = O(1)$, then the small parameter must be equal to the Weber number, i.e., $\varepsilon \equiv We$, which implies that in order to preserve the shape near to the spherical one, it is necessary $We \ll 1$. If the left hand side of the upper condition (6.5.6) is expanded in the orthogonal Legendre polynomials of first kind $P_n(\beta)$, then this condition can be written as a differential equation for the shape deformation function f:

$$\frac{\partial}{\partial \beta}\left[\left(1-\beta^2\right)\frac{\partial f}{\partial \beta}\right]+2f=\frac{3}{4}P_2(\beta)+\left[\frac{2}{We}-p_{bubble}+p_\infty-\frac{1}{4}\right]P_0(\beta). \qquad (6.5.7)$$

Analogously to the small Reynolds numbers deformation problem, the conservation conditions (6.3.11) and (6.3.12) are imposed on f. In conclusion, we reach to the following expressions for the bubble pressure

$$p_{bubble}=p_\infty+\frac{2}{We}+\frac{1}{4}, \qquad (6.5.8)$$

and for the deformation function

$$f=-\frac{3}{16}P_2(\beta). \qquad (6.5.9)$$

Then the bubble shape (6.5.1) transforms into the function

$$r=1-We\frac{3}{16}P_2(\beta)+O\left(We^2\right), \qquad (6.5.10)$$

which represents an oblate spheroid of revolution. This fact is registered experimentally by the photographs of Haberman and Morton (1953), from which Hartunian and Sears (1957) determine that the ratio l of the spheroid major axis diameter to the minor axis diameter parallel to the flow trajectories is 2. The deformed shape formula (6.5.10) has been derived by Moore (1959) and the drop deformation ratio is l = 1+(9/32)We. The comparison with (6.3.13) shows that the drop deformation at low Reynolds numbers l = 1+(5/16)We is slightly larger than that at high Reynolds numbers, namely by only 10%.

In section 5.9. the drag coefficient of spherical drops has been derived for large Reynolds numbers (5.9.25). Its first term 24/Re is obtained by Levich (1949) for the bubble rise in a viscous fluid. On the basis of the boundary layer theory Chao (1962) improves Levich's result which, in fact, because of some erroneous assumptions, is not correct. In a subsequent paper Moore (1963) obtains a drag coefficient including a contribution up to o($Re^{-3/2}$)

$$c_D=\frac{24}{Re}\left[1-\frac{1.563}{Re^{1/2}}+O\left(Re^{-5/6}\right)\right], \qquad (6.5.11)$$

that is a particular case of the more general drag coefficient of a drop (5.9.25). The comparison with the experiments of Haberman and Morton (1953) show that (6.5.11) is closer to the them than Levich's drag. Since the applied model imposes a zero tangential stress on the bubble surface, no flow separation from the bubble is predicted by the discussed theory. However, Hartunian and Sears (1957) observed a closed wake with a vortex ring inside the rear of the rising bubble and supposed that the phenomenon was due to impurities.

Generalising the boundary layer solution for oblate spheroidal bubble Moore (1965) calculated its drag

$$c_D=\frac{24}{Re}G(l)\left[1+\frac{H(l)}{Re^{1/2}}+\ldots\right], \qquad (6.5.12)$$

where G(l) is given by

$$G(l) = \frac{1}{3} l^{4/3} \left(l^2 - 1\right)^{3/2} \frac{\left(l^2 - 1\right)^{1/2} - \left(2 - l^2\right) \sec l^{-1}}{\left[l^2 \sec l^{-1} - \left(l^2 - 1\right)^{1/2}\right]^2} \tag{6.5.13}$$

and H(l) are tabulated in (Moore, 1965). The deformation ratio l can be computed from its approximate relation with We, given by Moore:

$$W(l) = 2l^{-4/3}\left(l^3 + l - 2\right)\left[l^2 \sec l^{-1} - \left(l^2 - 1\right)^{1/2}\right]^2 \left(l^2 - 1\right)^{-3}, \tag{6.5.14}$$

which for $l < 2$ is similar to the exact relation of Benjamin (1987) and El Sawi (1974). Again on the potential theory basis Miksis et al. (1981) calculate numerically a bubble shape different from the oblate spheroid. Using a numerical method Ryskin and Leal (1984b) find the drag and bubble deformation as functions of We at different values of Reynolds number. They show that, regardless of Weber number, at high Re the drag coefficient differs from that of Moore and the fore-aft symmetry of the bubble shape does not exist.

Another important feature of the bubble rise process is its rise velocity. The experimental rise velocity values of Haberman and Morton (1954) and the theoretical ones of Moore differ significantly for small bubbles and negligibly for large bubbles. The reason is the presence of surfactants which influences more the small bubbles.

Recent measurements are performed by Duineveld (1995) for the velocity and shape deformation of rising bubbles of equivalent radii $0.33 < a < 1$ mm in "hyper clean" water, supposed to be surfactant free. For small bubbles $a < 0.6$ mm the agreement with the Moore's theory (Moore, 1965) is excellent, while for larger bubbles there is weak deviation from the theoretical prediction due to the inaccurate models used, e.g., the supposed fore-aft symmetry is not present for large bubbles. Duineveld observes bubble front flattening with respect to its rear, which is qualitatively confirmed by the numerical results of Ryskin and Leal. The critical Weber number We_{cr} indicating the beginning of path instability is obtained to be 3.2, close to that in the Hartunian and Sears' experiments (Hartunian and Sears, 1957). The critical We depends on Morton number and at lower M, such as that of water, We_{cr} is smaller. This is a reasonable explanation for the contradiction with other authors indications for $We_{cr} = 5$ (Tsuge and Hibino, 1977).

At We > 10 the large bubbles lose their fore-aft symmetry and begin to look like spherical or ellipsoidal segments, called "spherical caps", for which the following model confirmed by the flow visualisation is adopted. The bubble is approximated by a spherical segment, as shown in Fig.6.5.1., with wake half-angle $0 \le \theta \le \theta_w$, where θ is the angular coordinate with origin at the front stagnation point. The remaining part of the sphere is occupied by a toroidal vortex in such a way that the outer flow passes a full sphere. The outer flow is usually assumed potential.

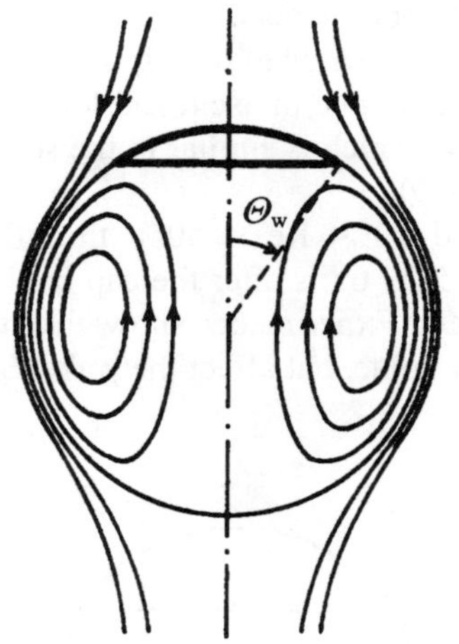

Fig.6.5.1. The structure of the flow around a drop, shaped as a spherical segment .

This interesting phenomenon has been studied experimentally and theoretically by many authors, e.g., Davies and Taylor (1950), Collins (1966), Maxworthy (1967), Maneri and Zuber (1974), Hnat and Buckmaster (1976). A good review on the topic can be found in Wegener and Parlange (1973).

Maxworthy shows that after the cap bubble there is a turbulent wake of width comparable with the bubble size. At high Morton number M the bubble motion is steady and stable (Hartunian and Sears, 1957). If $M > 1.4 \times 10^{-8}$, the bubble rise in any liquid is stable. Assuming laminar wake and neglecting the bubble effect on it, Parlange (1969) proposes a model, which substitutes the circulating wake with a circulating Hill's spherical vortex, called "wake sphere". Thus, he is able to calculate the drag coefficient in terms of Re based on the wake dimension

$$c_D = 6\left[\frac{20}{\mathrm{Re}}\left(1+\frac{0.14\ln\mathrm{Re}-6.6}{\mathrm{Re}^{1/2}}\right)\right]^{1/2}. \tag{6.5.15}$$

The terminal velocity of a large bubble can be determined from the proposed model (Clift et al., 1978):

$$U_r \approx \frac{2}{3}\sqrt{ag}\,, \tag{6.5.16}$$

where a is the curvature radius of the spherical segment. This formula is obtained by Davies and Taylor and describes very well the experimental data. The following semi-empirical formula can be used for the wake angle:

$$\theta_w = 50 + 190\exp\left(-0.62\mathrm{Re}^{0.4}\right). \tag{6.5.17}$$

For Re > 150 the wake angle $6_w \approx 50°$.

The turbulent wake after the spherical cap bubbles is studied by Yang and Levine (1992) with two different models: stagnant wake model and vertical wake model, since the rise velocity is determined only from the bubble tip and the wake form is not significant. The fluid is assumed inviscid and for the 2D and 3D case the Green's function formulation is applied, which is similar to the solution methods for a bubble in a tube (Kessler and Levine, 1989).

At high M > 0.2 and 10 < Re < 50 a thin fluid film, named "skirt", as schematically shown in Fig.6.5.2., trails after the cap bubbles (Davenport et al., 1967), (Guthrie and Bradshaw, 1969). Experiments show that the skirts do not effectively influence the terminal velocity value, but affect the wake. Still the "skirt" problem is not very well understood.

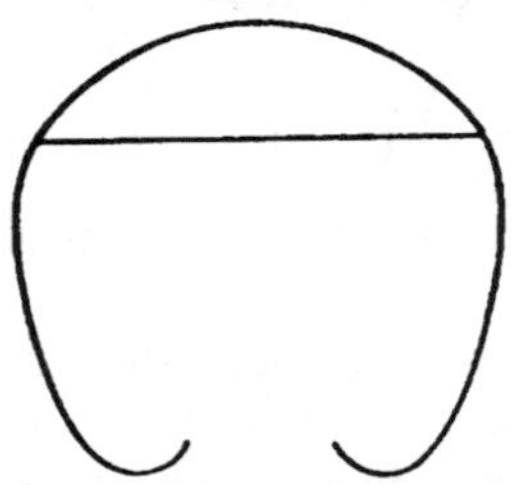

Fig.6.5.2. A sketch of the cross section of a spherical cap bubble with a skirt (from Harper, 1972).

At the end of this section we shall give some results for drop deformation at high Reynolds numbers, when the drops have different behaviour from the bubbles in contrast to the small Reynolds numbers case. The internal circulation cannot be neglected as for the bubble as well as the internal boundary layer. Here we shall mention the work of Parlange (1970), where the spherical drop motion is analysed applying the matched asymptotic expansions for the two boundary layers outside and inside the drop and the Hill's vortex solution. An alternative, but simpler than Harper and Moore's expression (5.9.25) for the drag coefficient is found, which gives the same numerical values

$$c_D \cong \frac{24}{\mathrm{Re}}\left(1+\frac{3\lambda}{2}\right)\left[1-\frac{4}{5(\pi\,\mathrm{Re})^{1/2}}\left(1+\frac{3\lambda}{2}\right)\left(6\sqrt{3}+5\sqrt{2}-14\right)\right]. \qquad (6.5.18)$$

At very high Re the drops cannot be spherical and the experiments with drops and bubbles show similar effects (Winnikov and Chao, 1966). Falling drops have front surface flattened, as sketched in Fig.6.5.3 (Harper, 1972), which is reversed for bubbles. This effect can be explained by the high dynamic pressure at the front stagnation point and the small density difference between inner and outer fluids making the rear spherical. For large Morton numbers and large We skirts formation with dimples in the rear stagnation regions is reported by Shoemaker and Marc de Chazal (1969) for rising drops and for falling drops (Thomson and Newall, 1885).

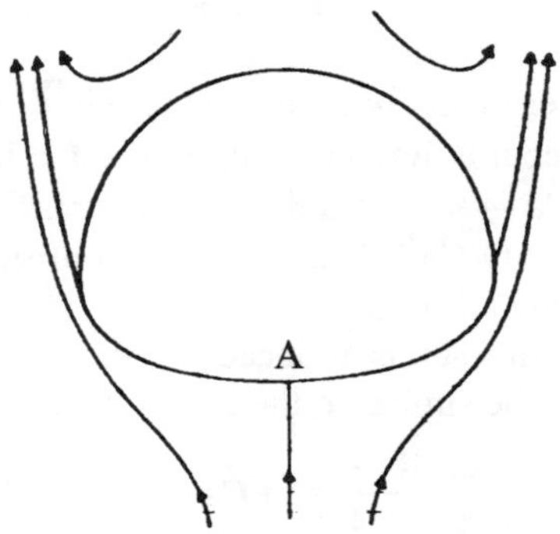

Fig.6.5.3. A sketch of the flow around a falling drop (from Harper, 1972).

6.6. Slender-Body Theory for Drops at Low Viscosity. Inertial Effects

As it has been mentioned in section 6.2. at high velocity gradients G and small viscosity ratios $\lambda \ll 1$ the steady drop shape elongates and becomes thin with pointed ends, as shown in Figs.6.2.1 and 6.2.3. The slender body theory is first applied by Taylor (1964) for the theoretical determination of the large drop deformations in Stokes flows. Inspecting the simple case of a droplet in a purely straining axisymmetric flow, he establishes that it bursts if the condition $Ca > 0.148\lambda^{-1/6}$ is fulfilled, where $Ca = G\mu a^2/\sigma$ is the capillary number. An improvement of Taylor's theory can be found in some later works, e.g., Buckmaster (1972, 1973), Acrivos and Lo (1978), Hinch (1980), Hinch and Acrivos (1979, 1980), a survey of which will be given below. As good reviews on the topic the papers of Acrivos (1983), Rallison (1984) and Stone (1994) must be mentioned.

We shall start with the axisymmetric pure straining flow case when the droplet has a circular cross section and fixed axis of symmetry. In cylindrical coordinates, with axis z along the symmetry axis, the droplet shape is given by $r = R(z)$, where $z \in [-L, L]$, $r \in [-b, b]$ and L and b are the lengths of its major and minor semi-axes, respectively. The schematic illustration of the drop position is represented on Fig.6.6.1. If the radius of an equivalent spherical droplet a is chosen as a characteristic length, then $l = L/a$ is the dimensionless half-length and $\varepsilon = b/L$ is the slenderness parameter. Then, the dimensionless velocity field of the undisturbed flow is written as

$$v_z^\infty = \mathrm{Ca}z, \quad v_r^\infty = -\frac{\mathrm{Car}}{2}. \tag{6.6.1}$$

When applying the slender body theory to analyse large droplet deformations, the flow disturbances due to the droplet presence are substituted by continuously distributed singularities along the droplet axis, instead of over its surface. The central line position and the singularities strength are unknown and R(z) = 0 is assumed at the end points $z = \pm l$.

Since the largest experimentally registered steady droplets shapes are elongated and thin, (i.e., $\varepsilon \ll 1$) and occur at low viscosity ratios and high shear rates, this model is only valid for $\lambda \to 0$ and $Ca \to \infty$. From the normal stress balance on the interface the estimates $\varepsilon = O(Ca^{-3})$ and $l = O(Ca^2)$ follow, which confirms our assumptions for the geometry of the highly deformed droplet.

The flow inside the droplet is induced by the outside purely straining flow $v_z = Caz$ and is considered to be similar to the Poiseuille flow

$$v_z = \frac{1}{\lambda}\frac{dp}{dz}\frac{R^2 - r^2}{4} + Caz \,, \quad \text{for } r \le R, \tag{6.6.2}$$

where $\frac{dp}{dz}$ is the internal pressure gradient. Buckmaster (1973) shows that the pressure inside the drop is given by the formula

$$p_{int}(z) = 8Ca\lambda \int_0^z \frac{s\,ds}{R^2(s)} + p_0 \,, \tag{6.6.3}$$

where p_0 is the pressure at the centre of the droplet.

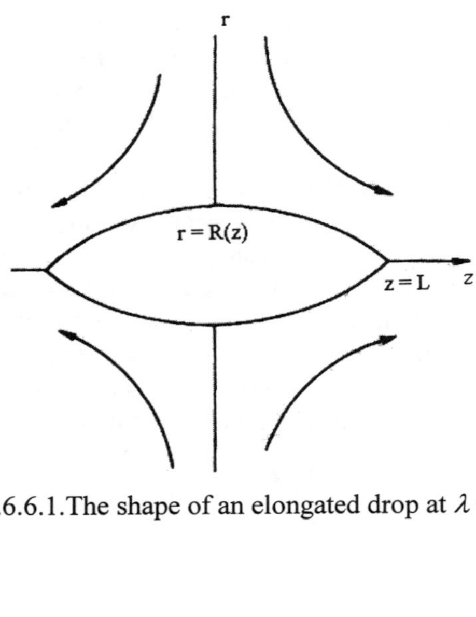

Fig.6.6.1.The shape of an elongated drop at $\lambda \ll 1$.

The exterior flow is obtained as the flow due to the line distribution of Stokeslets and potential sources [see (3.4.16) and section 3.8.], located on the major axis of the deformed droplet, is added to the undisturbed flow and the corresponding stream function acquires the form:

$$\psi = \frac{Ca}{2} r^2 z + \int_{-1}^{1} \frac{f(\xi) r^2}{\left[r^2 + (z-\xi)^2\right]^{1/2}} d\xi - \int_{-1}^{1} \frac{g(\xi)(z-\xi)}{\left[r^2 + (z-\xi)^2\right]^{1/2}} d\xi, \tag{6.6.4}$$

where f and g are unknown source strengths with $\int_{-1}^{1} g(\xi) d\xi = 0$, because the droplet is not a net source. Assuming that the order of f and g is $O(\varepsilon^2)$ then the exterior stream function near the droplet at $r = O(R) = O(\varepsilon)$ yields:

$$\psi = \frac{Ca}{2} R^2 z - \int_{-1}^{1} \frac{g(\xi)(z-\xi)}{|z-\xi|} d\xi. \tag{6.6.5}$$

After applying the streamline boundary condition for the droplet surface, i.e., $\psi = 0$, the source strength is determined as $g \sim \frac{Ca}{4} \frac{d}{dz}(zR^2)$. An analogous analysis of the shear stress boundary condition leads to another relation between f and g. On the other hand, the normal stress balance condition will present an equation for the remaining unknown: the droplet shape R. The external normal stress of order $O(\varepsilon)$ acting on the interface is only due to its viscous part, since the potential source creates only a constant pressure

$$\tau_{nn}\big|_{ext} \sim -Ca - \frac{4g}{R^2}, \tag{6.6.6}$$

Since the droplet viscosity is very small, the internal normal stress is only due to the pressure (6.6.3). Hence, the normal stress balance has the form:

$$CazR'(z) = 4Ca\lambda R(z) \int_0^z \frac{s\, ds}{R^2(s)} + \frac{1}{2} p_0 R(z) - CaR(z) - \frac{1}{2}, \tag{6.6.7}$$

which is the differential equation for the droplet shape R(z). When deriving this equation it has been assumed that the main contribution to the droplet mean curvature comes from the curvature of the cross section normal to z axis, i.e., 1/R(z). The pressure p_0 at the droplet centre is determined from the volume conservation condition $\int_{-1}^{1} \pi R^2(\xi) d\xi = \frac{4}{3}\pi$.

The solution of (6.6.7) at $\lambda = 0$ - for gas bubbles or inviscid drops represents a family of solutions depending on a real positive parameter n (Buckmaster, 1972):

$$R(z) = \frac{1}{2nCa}\left[1 - \left(\frac{z}{l}\right)^n\right], \tag{6.6.8}$$

where $l = \frac{4}{3}(n+1)(2n+1)Ca^2$ and $p_0 = 2Ca(n+1)$.

It is interesting to note that Taylor (1964) has intuitively selected the parameter n = 2. Later Buckmaster discusses that n must be an even integer for having smooth analytic steady shapes. Similar ideas can be found also in Acrivos and Lo (1978), where the choice of n = 2 corresponds to the only stable equilibrium steady droplet resulting from the small perturbation stability analysis. Moreover, observing the shape evolution with time, Hinch (1980) shows that, if larger disturbances are initially applied to the system the stability is lost and the droplet extends infinitely or bursts. The droplet shape (6.6.8) with n = 2 is confirmed numerically by Youngren and Acrivos (1976) as a special case for $Ca \to \infty$.

However, the slightly viscous drop situation $\lambda \neq 0$, due to the pressure gradient inside the drop, is more complicated. Examining (6.6.7), we observe that $\lambda = O(\varepsilon^2)$ and from the relation between ε and Ca we have $Ca\lambda^{1/6} = O(1)$, which in an explicit form given below will be a criterion for the critical capillary number depending on the viscosity ratio. According to Buckmaster (1972) and Acrivos and Lo (1978) the droplet shape differs from (6.6.8) insignificantly:

$$R(z) = \frac{\sqrt{5}}{2l}\left[1 - \left(\frac{z}{l}\right)^2\right]. \tag{6.6.9}$$

The critical capillary number corresponding to the droplet burst can be obtained from the equality (Taylor, 1964), (Buckmaster, 1973), (Acrivos and Lo, 1978):

$$Ca\lambda^{1/6} = \frac{1}{\sqrt{20}}\frac{\alpha^{1/2}}{1+\frac{4}{5}\alpha^3}, \tag{6.6.10}$$

where $\alpha = l\,\lambda^{1/3}$. This equation proves not to have a steady stable solution for capillary numbers larger than its critical value $Ca_{cr} = 0.148\lambda^{-1/6}$, which corresponds to $\alpha = 0.63$. The relation (6.6.10) is plotted in Fig.6.6.2. and the stability curve is shown on as AB with point B = (0.148, 0.63), breakup point. The section BD is the unstable solution branch and has no physical meaning, as noted by Acrivos and Lo. For reference with the inviscid drop case the curve AC is presented; since it has no maximum, a finite Ca_{cr} does not theoretically exist. Reminding that $l = O(Ca^2)$, which denotes increasing the strain rate G, the droplet elongates and, hence, it follows from (6.6.9) that the cross-section radius decreases. As a result we can resume that the pressure gradient inside the droplet increases with the increase of G and when G becomes greater than G_{cr}, the droplet is not able to sustain the enlarged pressure gradient and bursts. We have to point out again, that the droplet breakup is due rather to the absence of a steady solution of the problem than to an unstable equilibrium state.

The boundary integral numerical technique developed by Youngren and Acrivos (1975) and presented in section 5.11. is applied to the problem of a gas bubble in an extensional flow. For large surface tension values, i.e., $0.13 > Ca > 0.067$, their results for bubble shape, deformation parameter D and maximum radius are in excellent agreement with the analytical results of the linear O(Ca) (Frankel and Acrivos, 1970) and non-linear $O(Ca^2)$ theory (Barthes-Biesel and Acrivos, 1973b), while for large $Ca > 0.25$, as noted before in this section, are extrapolated to (6.6.8) at $n = 2$.

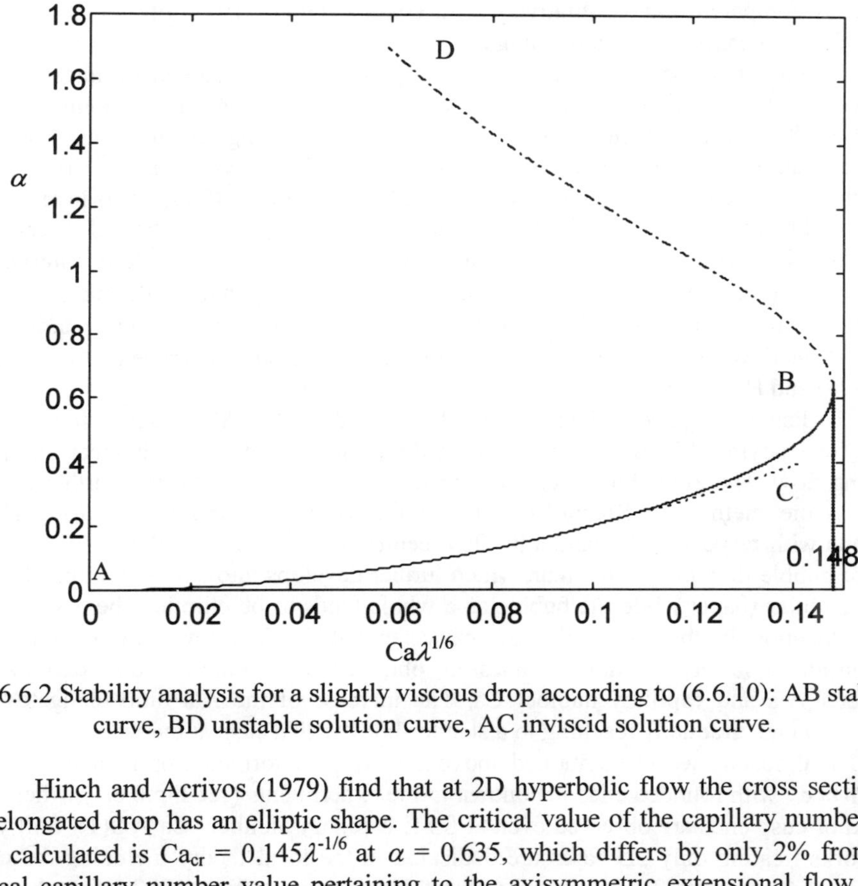

Fig.6.6.2 Stability analysis for a slightly viscous drop according to (6.6.10): AB stability curve, BD unstable solution curve, AC inviscid solution curve.

Hinch and Acrivos (1979) find that at 2D hyperbolic flow the cross section of the elongated drop has an elliptic shape. The critical value of the capillary number that they calculated is $Ca_{cr} = 0.145\lambda^{-1/6}$ at $\alpha = 0.635$, which differs by only 2% from the critical capillary number value pertaining to the axisymmetric extensional flow case. The comparison with the experimentally measured critical capillary number $Ca_{cr} = 0.12\lambda^{-0.16}$ by Grace (1971) and by Yu (1974) shows a qualitative agreement.

In a subsequent paper Hinch and Acrivos (1980) consider a simple shear flow case, which differs significantly from the previously analysed pure straining flows, due to the non-zero vorticity in the undisturbed flow field. They show that the drop axis is inclined towards the flow direction on an angle that decreases with the drop elongation and is of the slenderness order $O(\varepsilon)$. Further they find that a drop steady shape exists for all shear rates. However, at $Ca_{cr} = 0.0541\lambda^{-2/3}$ the equilibrium becomes unstable with respect to small and large disturbances and it can be thought that, contrary to the other straining flows, the critical state of the shear flow is connected with the process instability. It turns out that if the shear rate is abruptly increased from a certain subcritical value to another, drop breakup may be expected. This makes the analysis, as well as the experiments, extremely difficult to be accurate and thus the critical capillary number is compared only qualitatively with Grace's (Grace, 1971) observation for the onset of drop breaking by tip streaming or by fracture.

The deformations and breakup of slightly viscous drop in a general linear flow is studied in Khakhar and Ottino (1986) with the assumption of no pointed drops formation. The proposed model includes both drop stretching and rotation. The drop shape evolution is similar to that of Buckmaster (1973). A comparison is performed with the experimental readings of Grace (1971) and Bentley (1985), who investigates the drop breakup in planar flows generated by a computer controlled four-roller apparatus. Khakhar and Ottino obtain the following generalized breakup criteria: for nearly extensional flow $Ca_{cr} \approx 0.148\phi^{-1/2}\lambda^{-1/6}$, where ϕ is a parameter defining the flow type $\phi = 1$ - pure extensional flow, $\phi = 0$ - simple shear and $\phi = -1$ - pure rotation; for simple shear flow $Ca_{cr} \approx 0.0501\lambda^{-2/3}$. These results are the same as those of Acrivos and Lo (1978) and Hinch and Acrivos (1979), but for axisymmetric flow.

Utilising complex variables methods Richardson (1968) creates a free-surface model for studying 2D bubble behaviour in different gradient flows: a shear and pure straining flow. The 2D bubble shape deformations in a parabolic flow is further treated by the same method in Richardson (1973) for symmetric and asymmetric bubble positions with respect to the parabolic flow centre. In the first case the increase of Ca leads to bubble flattening at the rear, which further develops into a re-entrant cavity of a nearly circular shape, while the bubble as a whole tends to be elliptic. The asymmetric bubble location in the parabolic flow causes a bubble drift towards its symmetric position and a variety of bubble shapes are obtained for a wide range of eccentricity parameter A/a and capillary number Ca. The increase of the A/a ratio elongates the bubble rear end, gradually reaching to a slender-body with a cusp at the rear.

In almost all works mentioned above the largely deformed drop is modelled as a slender-body with rounded ends in contrast to the experiments (see Fig.6.6.1), where the pointed or cusp ends are observed even at finite capillary numbers Ca = O(1). Still very few works treating only 2D cases are dedicated to the local theory including the cusp ends (Buckmaster (1973), Sherwood (1981), Jeong and Moffatt (1992), Joseph et al. (1991) and Antanovskii (1994, 1996)]. On the basis of the slender-body theory Sherwood (1981) has attempted to investigate the tip streaming, described by Taylor (1934) as a streaming of small droplets ejected from the ends of an elongated inviscid or slightly viscous drop, by including non-linear cubic terms in the undisturbed linear extensional flow. In a recent paper Antanovskii (1996) also treats the Taylor's four-roller mill problem of an inviscid drop in a 2D hyperbolic flow by the matched asymptotic expansions method for the outer undisturbed flow and complex variable techniques for the inner flow around the drop. At strong deformation of the drop surface, two cusps are observed to develop at the stagnation points where the inner flow is convergent. It is shown that if the drop size exceeds a critical value, a unique solution exists for every strain rate. In reverse, multiple solutions result for a wide range of the capillary number, which indicates a hysteresial drop shape response.

A new insight of this complicated phenomenon is given by the experiments of de Bruin (1989, 1993). He argues that the tip streaming is caused by the surfactants non-uniform distribution over the droplet surface, with larger concentration at the narrower droplet parts, making its ends rigid. Then the viscous stresses break off these rigid narrow ends. The viscosity ratio necessary for tip streaming occurrence is $\lambda < 0.1$ with capillary numbers even lower than the critical. The surfactants presence in most cases means a tip streaming generation. Experiments with viscoelastic drops of some polymer material are performed by Milliken and Leal (1991) and Milliken et al. (1993). They observe tip streaming mainly as a consequence of the polymer behaving as a surfactant. Up to our knowledge, following Stone (1994), there are still no theoretical or numerical results supporting these experimental data.

In the reviewed papers both Reynolds numbers, of the outer flow and the drop, are assumed sufficiently small for the Stokes approach to hold. However, with the drop elongation the drop Reynolds number changes and the inertia effects may become significant. In order to introduce the inertia in the model Acrivos and Lo (1978) define the dimensionless strain rate at the inviscid drop case as

$$G_{Re} \equiv \frac{G\mu a}{\sigma}\left(\frac{\rho a \sigma}{\mu^2}\right)^{1/5} = Ca^{4/5}\,Re^{1/5}\left(\frac{a}{L}\right)^{2/5}, \tag{6.6.11}$$

where $Re = \rho\, GL^2/\mu$ and the dimensionless drop extension:

$$L_{Re} \equiv \frac{L}{a}\left(\frac{\rho a \sigma}{\mu^2}\right)^{2/5} = Ca^{-2/5}\,Re^{2/5}\left(\frac{L}{a}\right)^{3/5} = \frac{L}{a}Ca^{-2}G_{Re}^2\,. \tag{6.6.12}$$

Similarly to the zero Reynolds number analysis, the drop shape is obtained analytically as a family of curves from which only one corresponds to the drop stable steady deformation. Here the breakup point is given as $\left(G_{Re}\right)_{cr} = 0.284$ and $\left(L_{Re}\right)_{cr} = 2.37$ and there is no steady solution beyond the limit $\left(G_{Re}\right)_{cr}$. The corresponding slenderness ratio is $0.6268\left(\frac{\rho a \sigma}{\mu^2}\right)$, which must be less than 0.1 to permit applying the slender body theory. Then the Reynolds number is $Re \approx 0.065 \ll 1$, while if it is calculated on the drop half-length L, becomes 1.6, which means $Re = O(1)$. Therefore, the inertia must be taken into account for elongated deformed drops. In their approach Acrivos and Lo do not account for the inertia inside a weakly viscous drop. An extension to their model, meant to include the inertia effects inside the drop, is proposed by Brady and Acrivos (1982). When the parameter $A = \left(\frac{\rho a \sigma}{\mu^2}\right)\lambda^{5/6} \to 0$, the critical capillary number is the same as for the zero Re case, i.e., $Ca_{cr} = 0.148\lambda^{-1/6}$, and the inertia effects are negligible.

For large A the inertia outside the drop is dominating and $(\text{Ca}\lambda^{1/6}\text{A}^{1/5})_{cr} = 0.284$. Only in the intermediate region of A the inertia inside the drop plays a stabilising role opposing the outside flow inertia. With respect to this we can conclude that the internal inertia has a very weak effect on the drop behaviour and it may be neglected in the practical purpose models dealing with the drop deformation and breakup at non-zero Reynolds numbers.

As a last reference we shall cite the paper of Miksis (1981) treating the shape deformation of a bubble placed in an axisymmetric shear flow of an inviscid incompressible fluid. The problem is solved numerically in terms of flow potential and bubble shape function. Two branches of solutions are obtained, depending on a dimensionless pressure gradient parameter. The bubble shape depends on the Weber number; a critical value is found beyond which no stable bubble shape exists.

6.7. Boundary Integral Formulation of the Problem of Drop Deformation in an Extensional Flow. Uniqueness of the Solution

The deformed drop shapes different from spherical and slender-bodies can be obtained only via numerical solutions. Often the equilibrium drops shapes assume unusual forms, which can be obtained when solving the transient drop shape evolution problem. Sometimes these problems lead to drops burst.

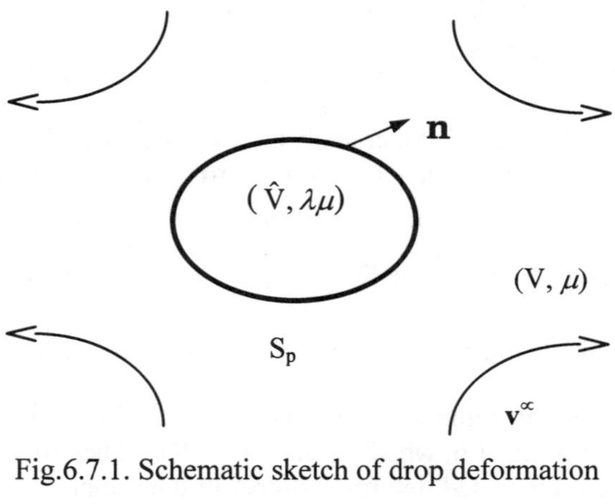

Fig.6.7.1. Schematic sketch of drop deformation

Except for the flows passing rigid particles (Youngren and Acrivos, 1975) the boundary elements method is also implemented by the same authors for the case of a bubble in an extensional flow (Youngren and Acrivos, 1976). The full problem of a fluid drop in an extensional Stokes flow is solved numerically by means of the boundary element method by Rallison and Acrivos (1978). Following Pozrikidis (1992a) in this section we shall present the boundary integral formulation of the drop deformation problem of Rallison and Acrivos.

Let a drop is passed by an extensional fluid flow of velocity $\mathbf{v}^{\infty}(\mathbf{r})$ and viscosity μ (Fig.6.7.1). The viscosity and volume of the fluid inside the drop is denoted respectively by $\lambda\mu$ and $\hat{V}$. For the flow outside the drop, i.e. $\mathbf{r}\in V$, the Stokes flow equations (1.3.6) with the appropriate boundary conditions (1.4.14), (1.4.15), (1.4.16), (1.4.17) and (1.4.18) must hold. In order to obtain the integral representation of the disturbed flow velocity outside the drop, we shall make use of (2.5.18)

$$v_j(\boldsymbol{\rho}^0) = -\frac{1}{8\pi\mu}\iint_{S_p} B_{ij}(\mathbf{r},\boldsymbol{\rho}^0)T_{ik}(\mathbf{r})n_k(\mathbf{r})d\sigma(\mathbf{r}) + \frac{1}{8\pi}\iint_{S_p} v_i(\mathbf{r})K_{ijk}(\mathbf{r},\boldsymbol{\rho}^0)n_k(\mathbf{r})d\sigma(\mathbf{r}), \tag{6.7.1}$$

$\mathbf{T}$ is the stress tensor induced by the disturbance flow. Applying the Lorentz theorem (2.5.13) for the fluid inside the drop, we obtain

$$\iint_{S_p} B_{ij}(\mathbf{r},\boldsymbol{\rho}^0)\hat{T}_{ik}(\mathbf{r})n_k(\mathbf{r})d\sigma(\mathbf{r}) - \lambda\mu\iint_{S_p}\hat{v}_i(\mathbf{r})K_{ijk}(\mathbf{r},\boldsymbol{\rho}^0)n_k(\mathbf{r})d\sigma(\mathbf{r}) = 0. \tag{6.7.2}$$

From (6.7.1), (6.7.2) and $\mathbf{v}(\mathbf{r}) = \hat{\mathbf{v}}(\mathbf{r})$ on S_p, it follows that

$$v_j(\boldsymbol{\rho}^0) = -\frac{1}{8\pi\mu}\iint_{S_p} B_{ij}(\mathbf{r},\boldsymbol{\rho}^0)\Delta f_i(\mathbf{r})d\sigma(\mathbf{r}) + \frac{1-\lambda}{8\pi}\iint_{S_p} v_i(\mathbf{r})K_{ijk}(\mathbf{r},\boldsymbol{\rho}^0)n_k(\mathbf{r})d\sigma(\mathbf{r}) \tag{6.7.3}$$

for outside flow $\boldsymbol{\rho}^0\in V$, where $\Delta\mathbf{f} = \mathbf{f} - \hat{\mathbf{f}} = (\mathbf{T}-\hat{\mathbf{T}}).\mathbf{n}$. Then according to (2.6.14) and (2.6.17) the pressure outside the drop

$$p(\boldsymbol{\rho}^0) = -\frac{1}{8\pi}\iint_{S_p} P_i(\boldsymbol{\rho}^0,\mathbf{r})\Delta f_i(\mathbf{r})d\sigma(\mathbf{r}) + \frac{1-\lambda}{8\pi}\mu\iint_{S_p} v_i(\mathbf{r})\Pi_{ij}(\boldsymbol{\rho}^0,\mathbf{r})n_k(\mathbf{r})d\sigma(\mathbf{r}). \tag{6.7.4}$$

In order to represent the flow inside the drop in an integral form, we shall apply again the integral equation (2.5.18)

$$\hat{v}_j(\boldsymbol{\rho}^0) = \frac{1}{8\pi\lambda\mu}\iint_{S_p} B_{ij}(\mathbf{r},\boldsymbol{\rho}^0)\hat{f}_i(\mathbf{r})d\sigma(\mathbf{r}) - \frac{1}{8\pi}\iint_{S_p} v_i(\mathbf{r})K_{ijk}(\mathbf{r},\boldsymbol{\rho}^0)n_k(\mathbf{r})d\sigma(\mathbf{r}), \tag{6.7.5}$$

where $\boldsymbol{\rho}^0\in\hat{V}$. After that, on the basis of Lorentz reciprocal theorem (2.5.13) we reach the expression for the exterior disturbance flow at $\boldsymbol{\rho}^0\in\hat{V}$

$$\frac{1}{8\pi\lambda\mu}\iint_{S_p} B_{ij}(\mathbf{r},\boldsymbol{\rho}^0)f_i(\mathbf{r})d\sigma(\mathbf{r}) - \frac{1}{8\pi\lambda}\iint_{S_p} v_i(\mathbf{r})K_{ijk}(\mathbf{r},\boldsymbol{\rho}^0)n_k(\mathbf{r})d\sigma(\mathbf{r}) = 0. \tag{6.7.6}$$

Subtracting (6.7.6) from (6.7.5), we finally have

$$\hat{v}_j(\boldsymbol{\rho}^0) = -\frac{1}{8\pi\lambda\mu}\iint_{S_p} B_{ij}(\mathbf{r},\boldsymbol{\rho}^0)\Delta f_i(\mathbf{r})d\sigma(\mathbf{r}) + \frac{1-\lambda}{8\pi\lambda}\iint_{S_p} v_i(\mathbf{r})K_{ijk}(\mathbf{r},\boldsymbol{\rho}^0)n_k(\mathbf{r})d\sigma(\mathbf{r}) \tag{6.7.7}$$

The pressure inside the drop is defined from the formula (6.7.4), when dividing by λ and taking into account that $\boldsymbol{\rho}^0\in\hat{V}$.Further, if in the equalities (6.7.3) and (6.7.7) the left- and right-hand limits are taken, when $\boldsymbol{\rho}^0\to S_p$ as $\boldsymbol{\rho}^0\in V$ and $\boldsymbol{\rho}^0\to S_p$ as $\boldsymbol{\rho}^0\in\hat{V}$, respectively, they transform into a single equation for the drop surface velocity $\mathbf{v}(\boldsymbol{\rho}^0)$ at $\boldsymbol{\rho}^0 \in S_p$

$$v_j(\boldsymbol{\rho}^0) = -\frac{1}{4\pi\mu(\lambda+1)} \iint_{S_p} B_{ij}(\mathbf{r},\boldsymbol{\rho}^0)\Delta f_i(\mathbf{r})d\sigma(\mathbf{r}) + \frac{k}{4\pi} \overset{PV}{\int\int_{S_p}} v_i(\mathbf{r})K_{ijk}(\mathbf{r},\boldsymbol{\rho}^0)n_k(\mathbf{r})d\sigma(\mathbf{r}), \tag{6.7.8}$$

as the expressions (2.7.3) and (2.7.4) are taken in view and $k = (1-\lambda)/(1+\lambda)$. From (6.7.8) it follows, that if we know the surface force jump $\Delta\mathbf{f}$ we obtain a second order integral equation for the surface velocity $\mathbf{v}(\boldsymbol{\rho}^0)$. After determining the velocity $\mathbf{v}(\boldsymbol{\rho}^0)$ at $\boldsymbol{\rho}^0 \in S_p$, the velocity and pressure exterior and interior of the drop is obtained from formulae (6.7.3), (6.7.4) and (6.7.7). Here we must mention that, at equal fluids viscosities, $\lambda = 1$ and $k = 0$, the velocities and pressure expressions (6.7.3), (6.7.4), (6.7.7) and (6.7.8) are simplified to single-layer potentials with known density $\Delta\mathbf{f}$.

If we want to get the corresponding expressions for the real flow field consisting of an incident flow $\mathbf{v}^\infty$ and disturbance flow, we apply the Lorentz reciprocal theorem (2.5.13) for $\mathbf{v}^\infty$ inside the drop, since $\mathbf{v}^\infty$ is continuous everywhere in $V \cup \hat{V}$ and $\mathbf{f}^\infty$ suffers a jump $\Delta\mathbf{f}^\infty$ across the drop surface S_p

$$\frac{1}{8\pi\lambda\mu} \iint_{S_p} B_{ij}(\mathbf{r},\boldsymbol{\rho}^0)\hat{f}_i^\infty(\mathbf{r})d\sigma(\mathbf{r}) - \frac{1}{8\pi} \iint_{S_p} v_i^\infty(\mathbf{r})K_{ijk}(\mathbf{r},\boldsymbol{\rho}^0)n_k(\mathbf{r})d\sigma(\mathbf{r}) = 0, \tag{6.7.9}$$

where at $\boldsymbol{\rho}^0 \in V$. Then, the summation of (6.7.9) with the expression for the disturbance flow (6.7.3) correspondent to $\mathbf{v} - \mathbf{v}^\infty$, yields the equation of the velocity outside the drop

$$v_j(\boldsymbol{\rho}^0) = v_i^\infty(\boldsymbol{\rho}^0) - \frac{1}{8\pi\mu} \iint_{S_p} B_{ij}(\mathbf{r},\boldsymbol{\rho}^0)\Delta f_i(\mathbf{r})d\sigma(\mathbf{r}) + \frac{1-\lambda}{8\pi} \iint_{S_p} v_i(\mathbf{r})K_{ijk}(\mathbf{r},\boldsymbol{\rho}^0)n_k(\mathbf{r})d\sigma(\mathbf{r}) \tag{6.7.10}$$

where $\hat{\mathbf{f}}^\infty = \dfrac{\lambda}{(1-\lambda)}\Delta\mathbf{f}^\infty$ has been taken into account. For the velocity inside the drop an equation similar to (6.7.10) is obtained, as the right-hand side is multiplied by $1/\lambda$. For the surface velocity $\mathbf{v}(\boldsymbol{\rho}^0)$ at $\boldsymbol{\rho}^0 \in S_p$, the respective equation is:

$$\begin{aligned} v_j(\boldsymbol{\rho}^0) = {} & \frac{2}{(\lambda+1)} v_j^\infty(\boldsymbol{\rho}^0) - \frac{1}{4\pi\mu(\lambda+1)} \iint_{S_p} B_{ij}(\mathbf{r},\boldsymbol{\rho}^0)\Delta f_i(\mathbf{r})d\sigma(\mathbf{r}) \\ & + \frac{k}{4\pi} \overset{PV}{\int\int_{S_p}} v_i(\mathbf{r})K_{ijk}(\mathbf{r},\boldsymbol{\rho}^0)n_k(\mathbf{r})d\sigma(\mathbf{r}) \end{aligned} \tag{6.7.11}$$

In the problem considered by Rallison and Acrivos the incident flow is the extensional flow $\mathbf{v}^\infty = \mathbf{E}^\infty.\mathbf{r}$.

In order to investigate the properties of the integral equation (6.7.8), we shall first study its homogeneous correspondent

$$\mathrm{v}_j(\boldsymbol{\rho}^0) = \frac{k}{4\pi} \overset{PV}{\iint\limits_{S_p}} \mathrm{v}_i(\mathbf{r}) K_{ijk}(\mathbf{r}, \boldsymbol{\rho}^0) n_k(\mathbf{r}) d\sigma(\mathbf{r}) . \tag{6.7.12}$$

Actually, it has only real eigenvalues $|k| \geq 1$, from which $k = 1$, i.e., $\lambda = 0$, is a single eigenvalue, while $k = -1$, i.e., $\lambda \to \infty$, is a multiple eigenvalue relative to six independent eigensolutions, each one denoting a different mode of rigid body motion (Ladyzhenskaya, 1969). The first pole physically corresponds to the bubble case, and the second: to the rigid body case. Therefore, the equation (6.7.8) will have a unique solution for every λ different from zero or infinity and this solution may be calculated via some iterative procedure. Moreover, when $\Delta\mathbf{f} = 0$ the single-layer integral is not present in (6.7.8) and this equation is simplified to the homogeneous equation for the velocity $\mathbf{v}$ (6.7.12). Further, we shall show that at the absence of incident flow and finite λ, the homogenous equation has only a trivial solution, i.e., no fluid flow is induced by the drop. For homoviscous fluids at $\lambda = 1$, the double-layer integral vanishes and (6.7.8) transforms into an equation valid everywhere in domain outside and inside the drop and from which the velocity is directly obtained.

In order to prove that (6.7.8) has an unique solution for any $\Delta\mathbf{f}$ and $\lambda \in (0, \infty)$, according to the Fredholm's alternative it is sufficient to show that the homogenous equation (6.7.12) has only a trivial solution. Instead of this statement Power (1987) proves that its adjoint homogeneous equation does not possess a pole at $\lambda \in (0, \infty)$

$$\phi_j(\boldsymbol{\rho}^0) = \frac{k}{4\pi} \overset{PV}{\iint\limits_{S_p}} \phi_i(\mathbf{r}) K_{ijk}(\boldsymbol{\rho}^0, \mathbf{r}) n_k(\mathbf{r}) d\sigma(\mathbf{r}) . \tag{6.7.13}$$

He considers the single-layer potential, which represents a Stokes flow solution

$$V_j(\boldsymbol{\rho}^0) = -\frac{1}{8\pi} \iint\limits_{S_p} B_{ij}(\mathbf{r}, \boldsymbol{\rho}^0) \phi_i(\mathbf{r}) d\sigma(\mathbf{r}) \tag{6.7.14}$$

with $\boldsymbol{\phi}$ standing for a non-trivial solution of (6.7.12). Then the surface normal stress suffers a jump on S_p given by

$$\hat{f}_i(\mathbf{V}) = \frac{1}{2} \phi_i(\boldsymbol{\rho}^0) - \frac{1}{8\pi} \iint\limits_{S_p} K_{ijk}(\boldsymbol{\rho}^0, \mathbf{r}) n_k(\mathbf{r}) \phi_i(\mathbf{r}) d\sigma(\mathbf{r}),$$

$$f_i(\mathbf{V}) = -\frac{1}{2} \phi_i(\boldsymbol{\rho}^0) - \frac{1}{8\pi} \iint\limits_{S_p} K_{ijk}(\boldsymbol{\rho}^0, \mathbf{r}) n_k(\mathbf{r}) \phi_i(\mathbf{r}) d\sigma(\mathbf{r}).$$

After some calculations (6.7.13) can be presented in the form

$$(1 + k) \iint\limits_{S_p} \hat{V}_i(\mathbf{r}) \hat{f}_i(\mathbf{V}(\mathbf{r})) d\sigma(\mathbf{r}) = (1 - k) \iint\limits_{S_p} V_i(\mathbf{r}) f_i(\mathbf{V}(\mathbf{r})) d\sigma(\mathbf{r}) . \tag{6.7.15}$$

For the Stokes flow described by (6.7.14) the following relations hold (Ladyzhenskaya, 1969):

$$\iint\limits_{S_p} \hat{V}_i(\mathbf{r}) \hat{f}_i(\mathbf{V}(\mathbf{r})) d\sigma(\mathbf{r}) = \frac{1}{2} \iiint\limits_{\hat{V}} \left(\frac{\partial V_j}{\partial x_i} + \frac{\partial V_i}{\partial x_j} \right)^2 d\mathbf{r} > 0, \tag{6.7.16}$$

$$\iint_{S_p} V_i(\mathbf{r}) f_i(\mathbf{V}(\mathbf{r})) d\sigma(\mathbf{r}) = -\frac{1}{2} \iiint_V \left(\frac{\partial V_j}{\partial x_i} + \frac{\partial V_i}{\partial x_j} \right)^2 d\mathbf{r} < 0,$$

and then the only possibility for the integrals in (6.7.15) is to be identically zero as $-1 < k < 1$. On the other hand from (6.7.16) this implies that

$$\frac{\partial V_j}{\partial x_i} + \frac{\partial V_i}{\partial x_j} = 0 \quad \text{for all } \mathbf{r} \in V \cup \hat{V}, \tag{6.7.17}$$

which corresponds to the motion of the fluid as a rigid body. Hence, for the drop deformation problem $\mathbf{V}$ as well as $\boldsymbol{\phi}$ must be zero, which proves that the homogeneous equation (6.7.12) has only a trivial solution and the original equation (6.7.8) has an unique solution for the range of the parameter $k \in (-1, 1)$ or $\lambda \in (0, \infty)$.

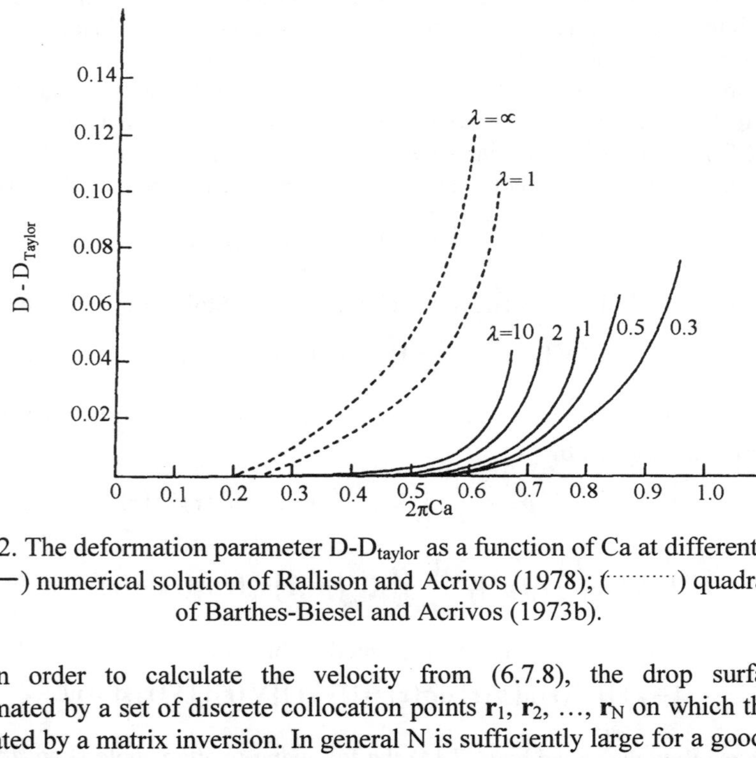

Fig.6.7.2. The deformation parameter $D\text{-}D_{taylor}$ as a function of Ca at different values of λ: (———) numerical solution of Rallison and Acrivos (1978); (··········) quadratic theory of Barthes-Biesel and Acrivos (1973b).

In order to calculate the velocity from (6.7.8), the drop surface S_p is approximated by a set of discrete collocation points $\mathbf{r}_1, \mathbf{r}_2, \ldots, \mathbf{r}_N$ on which the velocity is evaluated by a matrix inversion. In general N is sufficiently large for a good approach to the drop shape, e.g., if N = 100 a matrix 100x100 must be inverted. The results of Rallison and Acrivos (1978), for example, are obtained at N = 20. In the case $\lambda = 1$ no matrix inversion is necessary, which permits nonaxisymmetric drop shapes to be studied without large computer time losses (Rallison, 1981). We have to point out that the shape of the deformed drop surface S_p is also an unknown function, which couples the problem. If the velocity is decomposed into normal and tangential parts and taking into account that the normal component vanishes on S_p, then only the tangential velocity components remain unknown, together with the drop shape function. There are different ways to solve the vector equation or the system of three scalar equations (6.7.8) with the included boundary conditions for the three unknowns: the Newton-Raphson method directly applied to it; iterative determination of the shape function starting with an appropriate initial guess [Youngren and Acrivos (1976), Pozrikidis (1988)].

In an alternative approach to the drop deformation problem some of the boundary conditions are not included into the integral equations and left as additional constraints for the shape function determination. Generally the stress conditions serve for the purpose, i.e., the normal stress method (normal stress constraint) or the shear stress method (shear stress constraint). The normal stress method is appropriate for small capillary numbers Ca, while the shear stress method: for large Ca [Reinelt (1987), Pozrikidis (1988)].

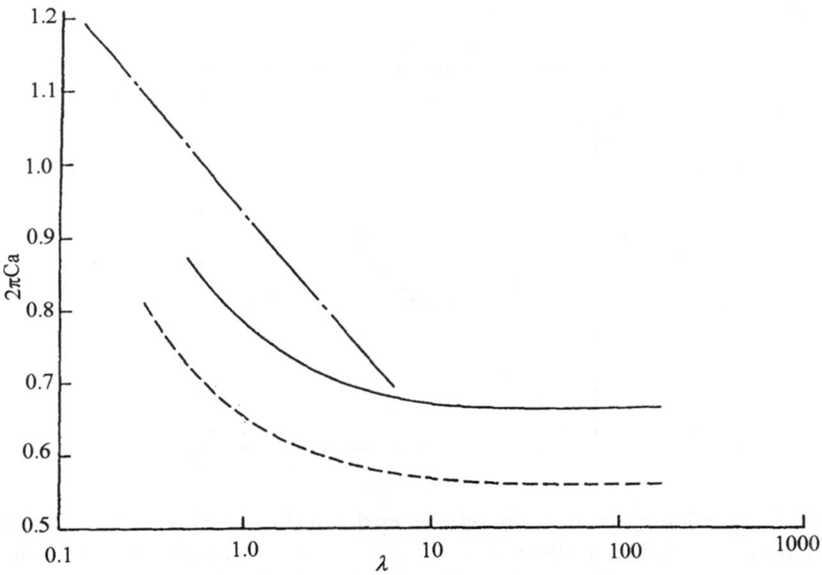

Fig.6.7.3. Critical value of Ca as function of λ: (——) numerical solution of Rallison and Acrivos (1978); (⋯⋯) quadratic theory of Barthes-Biesel and Acrivos (1973b); (– · – ·) slender body theory of Taylor (1964)

Here we shall present some of the results of Rallison and Acrivos (1978) for the distortions of an initially spherical drop to prolate spheroid drop and will discuss the behaviour of the deformation parameter D and critical capillary number Ca_{cr}. The critical capillary number indicates for the drop break up, i..e., the drop either attains a steady shape or extends on its longer axis without restriction.

In Fig.6.7.2. the deviation of the deformation parameter $D\text{-}D_{Taylor}$ from the linear theory of Taylor (6.2.2) is illustrated for $0 < Ca < 0.16$ and $0.3 \leq \lambda \leq 10$. It is shown that this deviation is sufficiently small and allows to use the linear theory results when predicting the uniaxial drop deformation. However, the non-linear second order theory of Barthes-Biesel and Acrivos (1973b) gives worse approximation than the linear theory.

In Fig.6.7.3. the critical capillary number Ca_{cr}, corresponding to the limit of the steady shape existence is plotted as a function of λ. The numerical results are in good agreement with the slender body asymptote of Taylor (1964) $Ca_{cr} = 0.148\lambda^{-1/6}$ at $\lambda \to 0$ and the stability analysis of Acrivos and Lo, (1978). Moreover, the numerical prediction is also confirmed (up to 80%) by the finite λ prediction of Barthes-Biesel and Acrivos, (1973b).

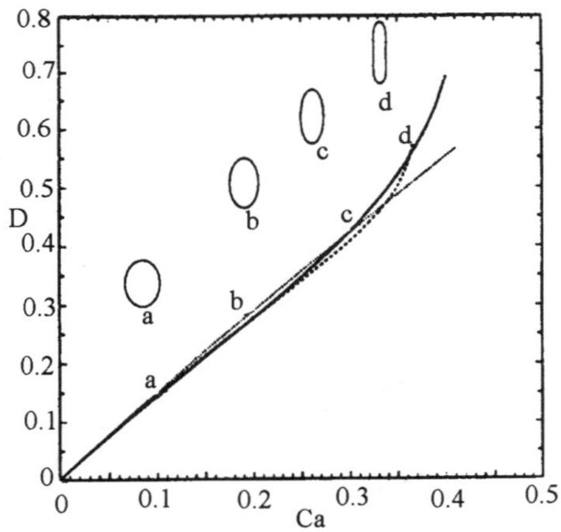

Fig.6.7.4. Drop deformation in a biaxial extensional flow, $\lambda=1$: (———) , numerical solution of Stone and Leal (1989c); (-----) and (...........) O(Ca) and O(Ca^2) results after Barthes-Biesel and Acrivos (1973b).

The drop deformation in a biaxial extensional flow $\mathbf{v}^{\infty} = Gx_1\mathbf{i} + Gx_2\mathbf{j} - 2Gx_3\mathbf{k}$ for zero Reynolds numbers and small deformations is studied by Frankel and Acrivos (1970) and Barthes-Biesel and Acrivos (1973b) via perturbation analysis. The finite drop deformations and break-up are calculated by Stone and Leal (1989c) using the boundary integral method. They find that the critical capillary number at which the drop starts a continuous extension in radial direction is $Ca_{cr} = 0.41$, i.e., no steady drop shapes exist for Ca > Ca_{cr}. A comparison between their results and the small deformation theory results of Barthes-Biesel and Acrivos are presented in Fig.6.7.4. Contrary to the drop deformed shapes in uniaxial flow which are prolate, here the deformed shapes are oblate. For capillary number values close to Ca_{cr} the drop shape is almost flat in its midsection but no dimple is observed, as Reynolds number is still zero.

6.8. Buoyancy-Driven Motion of a Deformable Fluid Particle through a Quiescent Liquid at Intermediate Reynolds Numbers

Due to the difficulties connected with the finite drop deformations at non-zero Reynolds number, the flow field and drop surface shape computations can be performed numerically by means of finite-difference method or finite-element method. One of the first works by the latter method is presented by Tsukada et al. (1984) for the buoyancy-driven fluid particle deformation. The finite-difference method has been successfully applied by Ryskin and Leal (1984a,b) and Christov and Volkov (1985) for inviscid bubble finite deformations. The numerical method of Ryskin and Leal (1983) for free boundary problems, originally created for bubble in a fluid, is extended to the case of a viscous fluid particle moving steadily under the gravity action in another viscous fluid by Dandy and Leal (1989). They give a detailed analysis of the flow field for a variety of the four dimensionless parameters: Reynolds number $Re = 2\rho aU/\mu$, Weber number $We = 2\rho aU^2/\sigma$, viscosity ratio $\lambda = \hat{\mu}/\mu$ of the fluid inside and outside the droplet and the corresponding density ratio $\gamma = \hat{\rho}/\rho$, where U is the flow velocity at infinity, a is the radius of an equivalent droplet of the same volume and σ is surface tension. The droplet is assumed axisymmetric and cylindrical coordinates are introduced (z, r, ϕ). As said before, the numerical technique of Ryskin and Leal, consisting in boundary-fitted orthogonal curvilinear coordinates (ξ, η) connected with the droplet surface, is employed. Then the mapping functions z(ξ, η) and r(ξ, η) are constructed numerically inside and outside the droplet. Due to this mapping the drop surface is defined by $\xi = 1$ and infinity by $\xi = 0$. The authors use the strong constraint method outside the droplet (Ryskin and Leal, 1983, 1984a,b), i.e., the solution of Laplace covariant equations for z and r in a unit square (ξ, η)

$$\frac{\partial}{\partial \xi}\left(f\frac{\partial z}{\partial \xi}\right) + \frac{\partial}{\partial \eta}\left(\frac{1}{f}\frac{\partial z}{\partial \eta}\right) = 0$$
$$\frac{\partial}{\partial \xi}\left(f\frac{\partial r}{\partial \xi}\right) + \frac{\partial}{\partial \eta}\left(\frac{1}{f}\frac{\partial r}{\partial \eta}\right) = 0 \qquad (6.8.1)$$

where f(ξ, η) is the distortion function, representing the ratio of the scale factors h_η/h_ξ, with $h_\eta = \left[z_\eta^2 + r_\eta^2\right]^{1/2}$ and $h_\xi = \left[z_\xi^2 + r_\xi^2\right]^{1/2}$ for the coordinate mapping. In the strong constraint method of Ryskin and Leal it is freely determined to control the mesh properties. Since at large Re and We the flow structure becomes complicated at the droplet rear, the mesh density is managed via this distortion function, which is chosen (Ryskin and Leal, 1984a,b) as $f(\xi,\eta) = \alpha\pi\xi(1-\alpha\cos\pi\eta)$ with $0 \le \alpha < 1$. In order to match meshes both outside and inside the droplet at the droplet surface, a weak constraint method is applied to generate the mesh inside the droplet, such as $\xi = 0$ at the origin (0, 0).

Following Ryskin and Leal (1984a, b) Dandy and Leal utilise the conformal mapping for the outer domain

$$z + ir = \frac{1}{z^* - ir^*}, \tag{6.8.2}$$

to transform the infinite domain into a finite in the new coordinates (z^*, r^*), which are afterwards mapped to a unit square in (ξ, η) plane.

In the boundary-fitted curvilinear coordinates the velocity components are expressed by the stream function ψ as

$$v_\xi = -\frac{1}{rh_\eta}\frac{\partial\psi}{\partial\eta}, \quad v_\eta = \frac{1}{rh_\xi}\frac{\partial\psi}{\partial\xi}. \tag{6.8.3}$$

Then the full Navier-Stokes equations in stream function ψ - vorticity ω formulation for both domains appear as

$$L^2(\omega r) - \frac{Re}{2}\frac{1}{h_\xi h_\eta}\left[\frac{\partial\psi}{\partial\xi}\frac{\partial}{\partial\eta}\left(\frac{\omega}{r}\right) - \frac{\partial\psi}{\partial\eta}\frac{\partial}{\partial\xi}\left(\frac{\omega}{r}\right)\right] = 0,$$

$$L^2\psi + \omega = 0, \tag{6.8.4}$$

$$\text{where } L^2 \equiv \frac{1}{h_\xi h_\eta}\left[\frac{\partial}{\partial\xi}\left(\frac{1}{fr}\frac{\partial}{\partial\xi}\right) + \frac{\partial}{\partial\eta}\left(\frac{f}{r}\frac{\partial}{\partial\eta}\right)\right]$$

We have to note here that the equations inside the droplet are the same as (6.8.4) with the exception of the Reynolds number, which for the inner flow is $\hat{Re} = \lambda\, Re$.

The boundary conditions of (6.8.4) for ψ - ω in (ξ, η) coordinates are:
on the axis of symmetry

$$\omega, \psi^*, \hat{\omega}, \hat{\psi} = 0 \quad \text{at} \quad \eta = 0,\ 1, \qquad \text{where } \psi^* = \psi - \frac{1}{2}r^2\left(1-\xi^2\right); \tag{6.8.5a}$$

at infinity

$$\psi^*, \omega = 0 \quad \text{at} \quad \xi = 0; \tag{6.8.5b}$$

at the droplet centre

$$\hat{\psi}, \hat{\omega} = 0 \quad \text{at} \quad \xi = 0; \tag{6.8.5c}$$

the no-slip velocity condition at droplet surface

$$\hat{\psi} = \psi = 0, \quad v_\eta = \hat{v}_\eta \quad \text{at} \quad \xi = 1, \tag{6.8.5d}$$

the tangential stress

$$[\lambda\,\hat{\omega} - \omega]_{\xi=1} = 2k_1 v_\eta(\lambda - 1) \quad \text{at} \quad \xi = 1 \tag{6.8.5e}$$

and the normal stress balance at droplet surface

$$\hat{p}-p+\frac{8}{\mathrm{Re}}\left[\frac{\lambda}{\hat{r}\hat{h}_\eta}\frac{\partial}{\partial\eta}\left(\hat{r}\hat{v}_\eta\right)-\frac{\lambda}{rh_\eta}\frac{\partial}{\partial\eta}\left(rv_\eta\right)\right]=\frac{4}{\mathrm{We}}\left[k_1+k_2\right]_{\xi=1}\quad\text{at}\quad\xi=1,\quad(6.8.5f)$$

where the pressure difference

$$\hat{p}-p=-\frac{3}{4}C_D z-\gamma\hat{v}_\eta^2+v_\eta^2+\frac{4}{\mathrm{Re}}\left[\lambda\int\frac{\hat{f}}{\hat{r}}\frac{\partial}{\partial\xi}(\hat{r}\hat{\omega})d\eta-\int\frac{f}{r}\frac{\partial}{\partial\xi}(r\omega)d\eta\right]+C_1\Big|_{\xi=1}$$

is calculated with the hydrostatic pressure taken into account, the first term on the right-hand side of the upper expression, while the remaining part refers to the dynamic pressures. The constant C_1 is obtained from the volume conservation constraint. The normal curvatures k_1 and k_2 are respectively

$$k_1=\frac{1}{h_\eta^3}\left(\frac{\partial z}{\partial\eta}\frac{\partial^2 r}{\partial\eta^2}-\frac{\partial r}{\partial\eta}\frac{\partial^2 z}{\partial\eta^2}\right)_{\xi=1},\quad k_2=-\frac{1}{rh_\xi}\frac{\partial r}{\partial\xi}\bigg|_{\xi=1}.$$

The system (6.8.4)-(6.8.5) is solved by Dandy and Leal by the ADI method of Peacemann and Rachford (1955). An artificial time dependence is introduced in (6.8.4) as an iteration parameter, which checks for the steady solution convergence. Instead of solving the two sets (6.8.4) and (6.8.1) simultaneously at every fictitious time step, as each of them is coupled, they are solved successively. At each time step first the droplet shape is chosen and the orthogonal coordinate system connected with is generated, after which the equations (6.8.4) are solved with the boundary conditions (6.8.5a)-(6.8.5e). The remaining normal stress condition (6.8.5f) is used to adjust the new droplet shape for the next time step. It turns out that this simple approach provides the numerical solution stability and efficiency.

Dandy and Leal perform numerical experiments for a wide range of the four problem parameters, i.e., $0.005 \le \mathrm{Re} \le 250$, $0.005 \le \mathrm{We} \le 14$, $0.001 \le \lambda \le 1000$ and $0.001 \le \gamma \le 1000$, which are difficult to be executed in laboratory. The observations due to their results show that at small $\mathrm{Re} < 1$, the droplet shapes change from spherical to spherical cap, and at large $\mathrm{Re} > 100$, the droplets progress into disk shaped with the increase of We. Moreover, an eddy behind the droplet is found to be disjoint from the droplet surface and its size and distance from the drop depends on the problem dimensionless parameters. For small values of Re and We their results are analogous to those of Ryskin and Leal (1984b) for gas bubble rise in a quiescent fluid under the gravity action, where $1 \le \mathrm{Re} \le 200$ and $\mathrm{We} \le 20$. Ryskin and Leal show that at $\mathrm{Re} \le 20$ and $\mathrm{We} \approx O(15)$, the bubble shape becomes independent of We in conformity with the experimental conclusions of Bhaga and Weber (1981). For the bubbles case, the separation occurrence and size depends strongly on Re and We, but for $\mathrm{Re} \ge O(100)$ this dependence is very weak. The computed bubble deformation is confirmed by the experimental pictures of Hnat and Buckmaster (1976).

In addition Ryskin and Leal (1984c) present also results for a bubble in an uniaxial extensional flow with Reynolds numbers in the range $0.1 \le \mathrm{Re} \le 100$. It is shown that for relatively low $\mathrm{Re} \ge O(10)$ the bubble bursts without reaching the experimentally predicted very elongated shapes. It also turns out that the performed calculations for a potential (instead of viscous) flow outside the bubble are satisfactory when Re ≥ O(100). The onset of bubble break-up at small Re as shown by Acrivos and Lo (1978) is at higher We than the obtained by Ryskin and Leal. The solution at Re = 0.1 agrees with the Stokes flow solution of Youngren and Acrivos (1976) via the boundary-integral method. In Fig.6.8.1. the bubble surface shapes for different Re and We are illustrated according to Ryskin and Leal. From the shape evolution, it is clear that with the further increase of We, the bubble will become unstable and burst into two parts, as said in section 6.6. This situation corresponds to We_{cr}, which for Re = 10 is $0.9 < We_{cr} < 1$ and for Re = 100 is $2.1 < We_{cr} < 2.2$. It is interesting to note that for the shapes presented on Fig.6.8.1. there is not any flow separation.

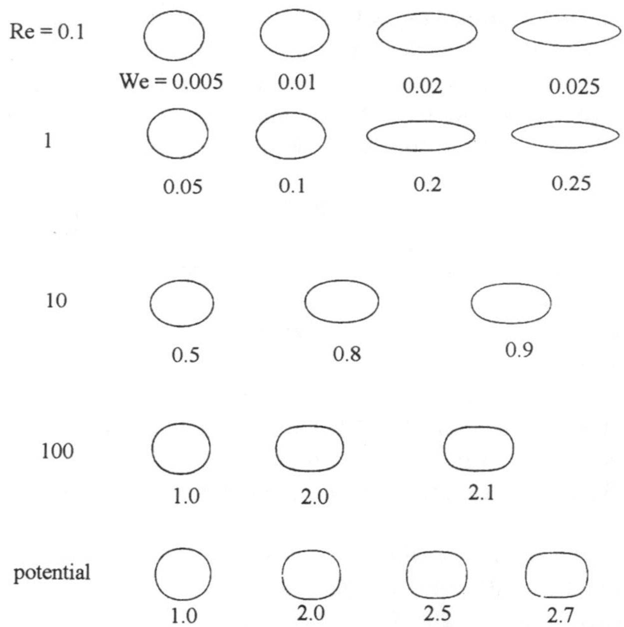

Fig.6.8.1. Shapes of a bubble in uniaxial extensional flow after Ryskin and Leal (1984c) for different Re and We.

The bubble deformations at finite Reynolds numbers biaxial flow case is studied by Kang and Leal (1989) by the numerical scheme of Ryskin and Leal (1984a). The computed bubbles shapes for different Re and We are shown in Fig.6.8.2. The difference with the shapes correspondent to uniaxial flow (Fig.6.8.1.) is that the deformed shape is oblate and that steady shapes are obtained at the same Re for much larger We. Thus the steady bubble shape subjected to biaxial flow can have a dimpled oblate configuration with negative curvature, while the corresponding waisted prolate shape with negative curvature in axial direction cannot be a steady configuration for the uniaxial flow case. A further comparison between the biaxial flow solutions for Re = 0 of Kang and Leal (1989) and of Frankel and Acrivos (1970), on the one hand, and the uniaxial flow solutions of Youngren and Acrivos (1976) and of Ryskin and Leal (1984c), on the other hand, is presented in Fig.6.8.3. for the deformation parameter D as a function of Ca.

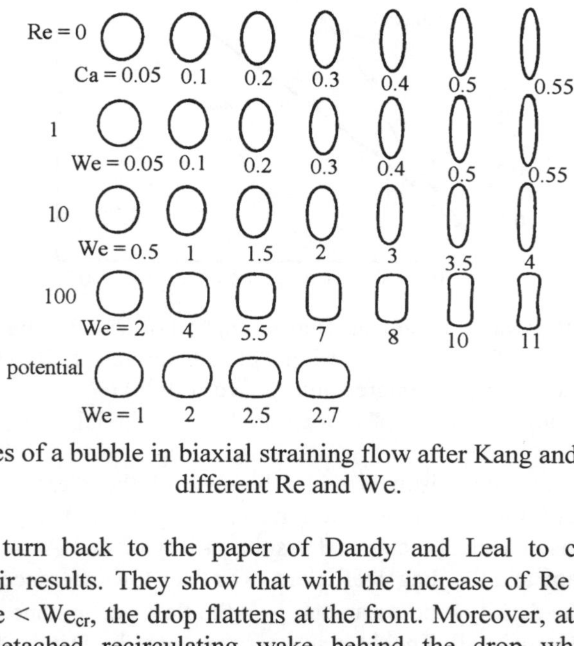

Fig.6.8.2. Shapes of a bubble in biaxial straining flow after Kang and Leal (1989) for different Re and We.

Now we turn back to the paper of Dandy and Leal to continue with the discussion of their results. They show that with the increase of Re ≤1 00, at $\lambda = 4$, $\gamma = 0.909$ and $We < We_{cr}$, the drop flattens at the front. Moreover, at Re = 60 and 100 there exists a detached recirculating wake behind the drop which is calculated numerically by Rivkind and Ryskin (1976). Unfortunately, there are still no experimental observations of such a disjoint recirculation zone, as recently reported by Bozzi et al. (1997). However, the wake is attached for bubbles as registered by Ryskin and Leal (1984a, b) and Miksis et al. (1981) and for rigid particles by Taneda (1956) and Rimon and Cheng (1969). After the analysis of Dandy and Leal for a wide range of problem parameters it is supposed that the disjoint wake formation is caused by the internal flow. Even for very high $\lambda = 100$ and 1000, the wake is still detached, although that at $\lambda \to \infty$ the droplet transforms into a rigid particle and the wake must stick to the particle surface. Then the limit $\lambda \to \infty$ may be regarded as singular with respect to the wake attachment. For higher Reynolds numbers $60 \geq Re \geq 350$ at We = 1 and $\lambda = 2.5$,

$\gamma = 0.909$ the size of the eddy in the wake decreases and gradually disappears and the vortex inside the drop takes the form of Hill's vortex. Therefore, the results of Harper and Moore (1968) can be regarded for very high Reynolds numbers, $Re \to \infty$.

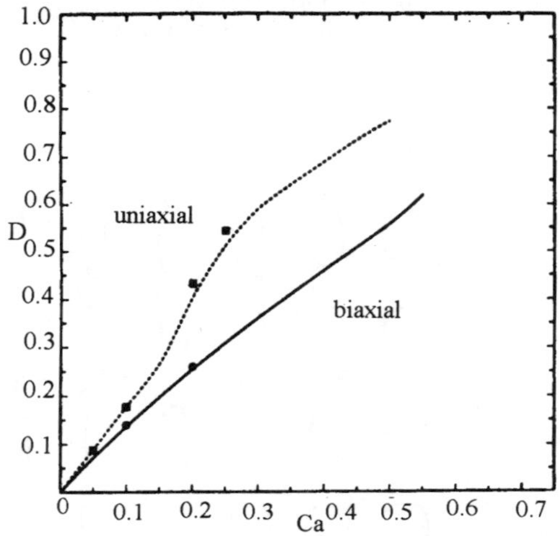

Fig.6.8.3. The deformation parameter D as a function of Ca after the results: of Kang and Leal (1989) at Re = 0 (——) and of Frankel and Acrivos (1970) (●) for a bubble in a biaxial straining flow; of Youngren and Acrivos (1976) (········) and of Ryskin and Leal (1984c) at Re = 0.1 (■) for a bubble in a uniaxial straining flow.

The small Re comparison is easier because of the plenty of theoretical and experimental works for a drop deformation in a creeping flow. Dandy and Leal present a very good comparison with the classical results of Taylor and Acrivos (1964) for the deformed drop shape and the drag coefficient (less than 1% difference) at Re, We << 1. The influence of the density ratio parameter γ on the flow performance is very weak and we shall not consider it at all. The drop shapes, computed by Dandy and Leal when the problem parameters are in the limits $0.5 \le Re \le 300$ and $0.5 \le We \le 15$, are in qualitative agreement with the correlations of Clift et al. (1978) and some experimental data.

PART III.

STEADY FLOWS. HYDRODYNAMIC INTERACTIONS BETWEEN RIGID OR FLUID PARTICLES

CHAPTER 7.

Hydrodynamic Interactions between Two Rigid or Fluid Particles

7.1. Introduction

In this chapter we shall consider the behaviour of two rigid or fluid particles moving through a viscous fluid under the action of external forces, typically the gravity force. When two rigid or fluid particles suspended in a viscous fluid approach one another, the motion of each one of them is influenced by the other, even in the absence of direct interparticle interactions, such as van-der-Waals and electrostatic forces. The velocity field caused by the motion of one of these particles is transmitted through the fluid medium and influences the motion and, therefore, the hydrodynamic force, torque and stresslet on the other one. This indirect particles interaction is known as a hydrodynamic interaction. The magnitude of the hydrodynamic interaction between rigid or fluid particles depends on the following variables: (a) their shapes and sizes; (b) the distance between them; (c) their orientation with respect to one another; (d) their individual orientation relative to the active external forces; (e) their velocities and spins relative to the fluid motion at infinity.

By virtue of the linearity of Stokes equations and boundary conditions that relate the fluid and particles velocities, the slow flow problems may be separated into two sorts: (i) the axisymmetric translational motion along the line of centres and (ii) asymmetric translational motion along the line of centres and rotational motion around an axis perpendicular to the line of centres.

The method of reflections, first developed by Smoluchowski (1911) and used systematically for two spherical particles by Happel and Brenner (1973), is an iterative approximation technique. For the first reflection, perturbations resulting from the velocity field, due to one (rigid or fluid) of the particles, reflected from the boundary of the second (rigid or fluid) particle are used to correct the zero-order velocity field of the second particle calculated in the absence of the first one. Then, the nth-order reflection is the reflection required to satisfy the no-slip boundary conditions at the surface of each particle caused by the disturbance field of the (n-1)th-order reflection of the other particle. It has been shown that the solution obtained by this technique converges to the exact solution for the two spherical particles problem (Luke, 1989).

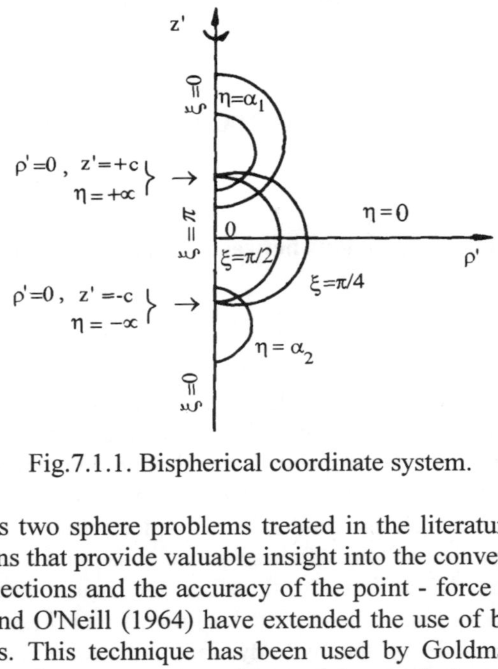

Fig.7.1.1. Bispherical coordinate system.

The numerous two sphere problems treated in the literature form an important class of exact solutions that provide valuable insight into the convergence characteristics of the method of reflections and the accuracy of the point - force approximation. Dean and O'Neill (1963) and O'Neill (1964) have extended the use of bipolar coordinates to asymmetric problems. This technique has been used by Goldman et al. (1966) and Wakiya (1967) to solve the asymmetric problem of the translation of two equal spherical particles along the axes perpendicular to their line of centres. Thus, the utilisation of the bispherical coordinates (ξ , η , φ) connected with the cylindrical coordinates $(\mathrm{r}', \varphi, \mathrm{z}')$ through the relations

$$\mathrm{z}' = \frac{\mathrm{c}\ \mathrm{sh}\eta}{\mathrm{ch}\eta - \cos\xi}, \quad \rho' = \frac{\mathrm{c}\ \sin\xi}{\mathrm{ch}\eta - \cos\xi}, \tag{7.1.1}$$

($0 \le \xi \le \pi$, $-\infty < \eta < \infty$, $0 \le \varphi < 2\pi$ and $c > 0$ is the focal distance) proves to be appropriate to solving a lot of problems of two particles or particle and interface interactions. For example, the two spheres can be defined by $\eta = \alpha_1 < 0$ ($a = c|\mathrm{cosech}\ \alpha_1|$) and $\eta = \alpha_2 > 0$ ($b = c|\mathrm{cosech}\ \alpha_2|$) as shown in Fig.7.1.1.

Through the use of bispherical coordinates Davis (1971) and O'Neill and Majumdar (1970) obtained an exact solution to the Stokes equations for the motion of two unequal-sized spherical particles arbitrary oriented with respect to a shear field. Lin et al. (1970b) computed the relative trajectories of two equal-size spherical particles freely suspended in a simple shear flow which is of particular value to the study of suspension rheology.

A systematic study of the hydrodynamic interaction of two freely moving spherical particles in shear viscous flow is performed by Batchelor and Green (1972a, b). The results of Davis (1971) and Lin et al. (1970a, b) are obtained as a particular case from their results.

Jeffrey and Onishi (1984a, b) and Davis and Hill (1992) derived the far-field expansions for the coefficients of the mobility tensor for two rigid spherical particles in shear flow. Kim and Mifflin (1985) considered also the interactions between two spherical particles in shear flow that provide the basis for the rheology of semi-dilute suspensions (see chapter 10).

Darabaner and Mason (1967) calculate also the trajectories of two particles passed by a shear flow. The comparison of the theoretical with the experimental data shows a good conformity. These results are used by some other authors when defining the effective viscosity of suspensions of spherical particles with volume concentration up to 1 or 2%.

The flow field and the hydrodynamic interaction forces between two rigid spherical particles passed by an axisymmetrical parabolic viscous flow is found by Kalitzova-Kurteva and Zapryanov (1989).

The solution of the Stokes equation in bispherical coordinates for the fluid motion in the vicinity of spherical particles in linear shear flow (giving an account of the van der Waals and electrostatic forces) has been discussed by Curtis and Hocking (1970). As in the above mentioned case, the rate of convergence of their solution (in the form of series) decreases with smaller clearance between the particles. In fact the series diverge if the particles are in contact. That is why through the use of tangent-sphere coordinates O'Neill (1969) has obtained an exact solution for the special case of two equal-size spherical particles in contact, settling with equal velocities perpendicular to their line of centres. The creeping motion of the two arbitrary-size touching spherical particles in a linear shear field has been discussed by Nir and Acrivos (1973). The explicit results covering the entire range of size ratios are presented for: (i) the forces and torques on aggregates; (ii) the hydrodynamic forces on the individual particles; (iii) the contribution of the pair to the bulk stress of a dilute suspension.

So far most studies of the pair hydrodynamic interaction have been based on the Stokes equations. However, they cannot fully explain interesting features of the interaction that can be observed in the experiments. For example, in the case when two spherical particles of equal size are settling vertically one above the other in an unbounded fluid, the difference between the forces on the leading and trailing spheres cannot be explained through an analysis based on the Stokes equation. In order to study such properties we need to take into account the inertial effects. Vasseur and Cox (1977) and Kaneda and Ishii (1982) have investigated the hydrodynamic interaction of two spherical particles moving in unbounded fluid at small but finite Reynolds numbers through the method of matched asymptotic expansions. To study such a problem one has to consider at least two cases: case (i), when the particles are sufficiently separated so that each of them is located in the outer region of expansion of the other; and case (ii), when they are sufficiently close to one another, so that each of them is located in the inner region of expansion of the other. Case (i) was studied by Vasseur and Cox (1977), while case (ii) was studied by Kaneda and Ishii (1982). Some properties of the force, not expected from the Stokes equation, are revealed.

In the experiments of Taneda (1979) it is established, that when a flow passes two spheres at small Reynolds numbers, there is a flow separation from the particles surface. The first investigation, in which it is established also theoretically that there exists a flow separation from the two spheres, although that the model stems from the Stokes equations, belongs to Davis et al. (1976).

The first attempts to study numerically the hydrodynamic interaction of two rigid spherical particles in a uniform viscous flow are done by Ivanov and Rivkind (1982). In these works flow field patterns are obtained for Reynolds numbers up to 40 for spheres of equal and unequal radii. The problem of uniform viscous flow past two spherical particles at Reynolds numbers up to 100 is investigated by Zapryanov and Toshev (1985, 1986). In these papers the relations of the drag forces exerted by the spheres with the distance between the spheres and the Reynolds numbers are found. Moreover, in the same papers, the convective heat-mass transfer around the two spheres is studied. The obtained results are in good correspondence with the experimental results of Rowe and Henwood (1961).

The analytical analysis of the flow past two spheroids in Stokes approximation is performed in Kim (1985a). The same problem is solved numerically at moderate Reynolds numbers by Kotzev and Zapryanov (1989).

We have seen in Chapter 6 that if a net external force (for example, gravity) acts on a drop and makes it move with a nonzero velocity relative to the fluid, then the solution, correspondent to the zero - Reynolds number case, has the remarkable feature that the drop remains undeformed for all values of the capillary number and preserves its spherical form (Taylor and Acrivos, 1964). Zapryanov and Chervenivanova (1983), and Chervenivanova and Zapryanov (1985, 1989) have obtained zero - Reynolds number solutions for the problems of two separate drops and a compound drop moving steadily in quiescent fluid. The most interesting result of their analysis is that in the two problems both drops deform, unlike a single sedimenting drop which remains exactly spherical at zero Reynolds numbers. They have shown that in the case of a compound drop the outer drop deforms into a prolate spheroid while the inner drop deforms into an oblate spheroid. Chervenivanova et al. (1994) have found the small deformations of two separate drops induced by an electric field in a viscous fluid.

7.2. Hydrodynamic Interactions between Two Rigid Particles

The hydrodynamic interaction of two spherical particles is investigated first by Smoluchowski (1911). He supposes, that the particles move slowly in a viscous fluid parallel to the axis of their centres. Proposing the reflection method (Happel and Brenner, 1973) he solves approximately this problem only in the case of large distances between the particles.

An exact analytical solution of the axisymmetrical problem in Stokes approximation for the settling of two rigid spherical particles with equal velocities in a direction parallel to their line of centres is obtained by Stimson and Jeffery (1926) when applying a bispherical coordinate system. This solution is used to evaluate the accuracy of the approximation methods at the study of the hydrodynamic interaction of two particles at more complicated geometrical or hydrodynamical conditions.

The Stimson and Jeffery's solution is based on the stream function determination in bispherical coordinates, yielding the forces supporting the two particles motion. When the particles are not far away from each other, they exert a drag which can not be derived from the Stokes drag formula (2.9.1), referring to a flow past a single sphere. Therefore, Stimson and Jeffery introduce a correction coefficient Φ to obtain a new drag formula for the two equal spheres problem

$$\mathbf{F} = 6\pi\,\mu\,a\mathbf{U}\Phi\,, \tag{7.2.1}$$

where a is the radius of each sphere and

$$\Phi = \frac{4}{3}\operatorname{sh}\alpha\sum_{n=1}^{\infty}\frac{n(n+1)}{(2n-1)(2n+3)}\left[1-\frac{4\operatorname{sh}^2(n+1/2)\alpha-(2n+1)^2\operatorname{sh}^2\alpha}{2\operatorname{sh}(2n+1)\alpha+(2n+1)\operatorname{sh}2\alpha}\right],$$

while $\alpha = \operatorname{arch}\left(\frac{l}{2a}\right) > 0$ defines the surface of one of the spheres and l is the distance between the spheres centres.

We have to note that due to the Stokes equations reversibility, the fluid flows structure analyzed by these equations is symmetric and there is no tendency to variation in the interparticle distance. Of course the particles motion direction is of a significant importance.

In order to illustrate the idea of the reflection method, let us suppose that we have two rigid spherical particles having radii a and b, that move in an unbounded fluid with velocities $\mathbf{U}_a$ and $\mathbf{U}_b$. The line of centres is assumed to form an angle θ with the horizontal line. Thus, at $\theta = 0$, the line of centres is horizontal, while at $\theta = \pi/2$ the line of centres is vertical (see Fig.7.2.1). It is convenient to chose one axis of the reference coordinates system, e.g. z, along the line of centres. If we suppose that the particles move under the influence of a body force, then it is necessary to specify only one additional coordinate since the particles will move in a plane, e.g. the plane xz.

The leading order approximation to a velocity field in the vicinity of particle of radius a is thus the creeping flow solution $\mathbf{v}_a^{(1)}$, which vanishes at infinity and satisfies the boundary condition

$$\mathbf{v}_a^{(1)} = \mathbf{U}_a \qquad \text{on } S_a. \tag{7.2.2}$$

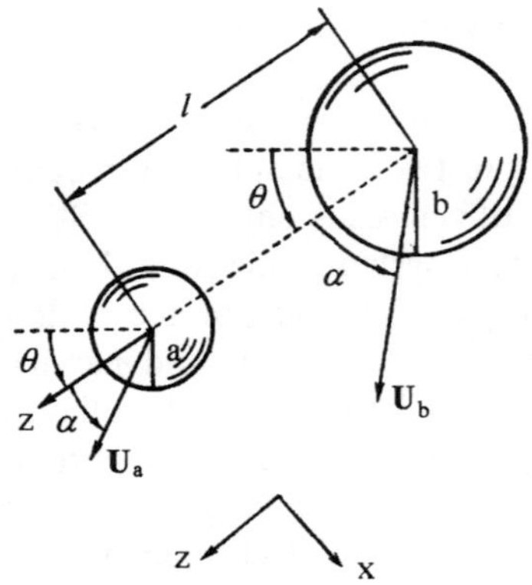

Fig.7.2.1. Settling of two unequal spheres in viscous fluid.

The next boundary conditions that we require to be satisfied are the following

$$\mathbf{v}_b^{(2)} = \mathbf{U}_b - \mathbf{v}_a^{(1)} \qquad \text{on } S_b, \tag{7.2.3}$$

$$\mathbf{v}_a^{(3)} = -\mathbf{v}_b^{(2)} \qquad \text{on } S_a, \tag{7.2.4}$$

$$\mathbf{v}_b^{(4)} = -\mathbf{v}_a^{(3)} \qquad \text{on } S_b, \text{ etc.} \tag{7.2.5}$$

Clearly, this procedure can be continued infinitely and the velocity field will be approximated as

$$\mathbf{v} = \mathbf{v}_a^{(1)} + \mathbf{v}_b^{(2)} + \mathbf{v}_a^{(3)} + \mathbf{v}_b^{(4)} + \ldots \tag{7.2.6}$$

The initial field $\mathbf{v}_a^{(1)}$ will correspond to the settling of the first particle (of radius a) in an unbounded fluid. Since the distance between the two particles is assumed relatively large with respect to their diameters, then in the absence of the second particle (of radius b) we may compute the translational effect of the first particle by supposing that it generates the same field as that which would be produced by a point force (a Stokeslet) situated at the centre of the first particle. Therefore, owing to (2.3.27) and (2.3.28), the velocity and pressure fields caused by the first particle are:

$$\mathbf{v}_a^{(1)} = -\frac{\mathbf{F}_a^{(1)}}{8\pi\mu}\cdot\left(\frac{\mathbf{I}}{r} + \frac{\mathbf{rr}}{r^3}\right) \tag{7.2.7}$$

and

$$p_a^{(1)} = -\frac{\mathbf{F}_a^{(1)}}{4\pi}\cdot\frac{\mathbf{r}}{r^3}, \tag{7.2.8}$$

where **r** denotes the position vector of any point in the fluid relative to the centre of first particle. To the degree of approximation to which (7.2.7) and (7.2.8) are valid, the initial field is independent of the shape of first particle, being determined entirely by its resistance $\mathbf{F}_a^{(1)}$. Taking into account that (Happel and Brenner, 1973)

$$\mathbf{F}_a^{(1)} = -\mu K_a \left(U_{ax}\mathbf{i} + U_{az}\mathbf{k} \right), \tag{7.2.9}$$

where $K_a = 6\pi$ a, we obtain

$$\mathbf{v}_a^{(1)} = \frac{K_a U_{ax}}{8\pi}\left(\frac{\mathbf{i}}{r} + \frac{x\mathbf{r}}{r^3} \right) + \frac{K_a U_{az}}{8\pi}\left(\frac{\mathbf{k}}{r} + \frac{z\mathbf{r}}{r^3} \right). \tag{7.2.10}$$

Since the position vector of the centre of second particle is $\mathbf{r} = l\mathbf{k}$, the value of $\mathbf{v}_a^{(1)}$ on S_b is approximately equal to

$$\mathbf{v}_a^{(1)}\Big|_{S_b} = \frac{K_a}{8\pi l}\left(U_{ax}\mathbf{i} + 2U_{az}\mathbf{k} \right). \tag{7.2.11}$$

Then the force acting on the second particle is

$$\mathbf{F}_b^{(2)} = -\mu K_b \left[\mathbf{U}_b - \mathbf{v}_a^{(1)}\Big|_{S_b} \right] = -\mu K_b \left(U_{bx} - \frac{K_a U_{ax}}{8\pi l} \right)\mathbf{i} - \mu K_b \left(U_{bz} - \frac{K_a U_{az}}{4\pi l} \right)\mathbf{k}, \tag{7.2.12}$$

where $K_b = 6\pi$ b.

Now, using the force $\mathbf{F}_b^{(2)}$, we can calculate the velocity field caused by a Stokeslet acting at the centre of second particle and it has a similar form as (7.2.7). In this case we have to take the origin of the coordinate system at the centre of second particle and then the value of $\mathbf{v}_b^{(2)}$ at S_a will be

$$\mathbf{v}_b^{(2)}\Big|_{S_a} = \frac{K_b}{8\pi l}\left(U_{bx} - \frac{K_a U_{ax}}{8\pi l} \right)\mathbf{i} + \frac{K_b}{4\pi l}\left(U_{bz} - \frac{K_a U_{az}}{4\pi l} \right)\mathbf{k}, \tag{7.2.13}$$

as the position vector of the first particle centre in the new coordinate system is $\mathbf{r} = -l\mathbf{k}$.

Similarly, for the force contribution $\mathbf{F}_a^{(3)} = \mu K_a \mathbf{v}_b^{(2)}\Big|_{S_a}$ we get

$$\mathbf{F}_a^{(3)} = \frac{\mu K_a K_b}{8\pi l}\left(U_{bx} - \frac{K_a U_{ax}}{8\pi l} \right)\mathbf{i} + \frac{\mu K_a K_b}{4\pi l}\left(U_{bz} - \frac{K_a U_{az}}{4\pi l} \right)\mathbf{k}. \tag{7.2.14}$$

In a similar manner we find that

$$\mathbf{F}_a^{(5)} = \mu K_a \left(\frac{K_a}{8\pi l} \right)\left(\frac{K_b}{8\pi l} \right)^2 \left(U_{bx} - \frac{K_a U_{ax}}{8\pi l} \right)\mathbf{i} + \mu K_a \left(\frac{K_a}{4\pi l} \right)\left(\frac{K_b}{4\pi l} \right)^2 \left(U_{bz} - \frac{K_a U_{az}}{4\pi l} \right)\mathbf{k}. \tag{7.2.15}$$

After adding all the forces due to first particle $\mathbf{F}_a = \mathbf{F}_a^{(1)} + \mathbf{F}_a^{(3)} + \mathbf{F}_a^{(5)} + \ldots$, eventually we reach to the expression

$$\frac{\mathbf{F}_a}{\mu K_a} = -\frac{U_{ax} - \left(K_b U_{bx}/8\pi l \right)}{1 - \left(K_a K_b \right)/(8\pi l)^2}\mathbf{i} - \frac{U_{az} - \left(K_b U_{bz}/4\pi l \right)}{1 - \left(K_a K_b \right)/(4\pi l)^2}\mathbf{k}. \tag{7.2.16}$$

If we interchange the subscripts b and a in this equation, we shall obtain the force $\mathbf{F}_b$ exerted on second particle

$$\frac{\mathbf{F}_b}{\mu K_b} = -\frac{U_{bx} - \left(K_a U_{ax}/8\pi l \right)}{1 - \left(K_a K_b \right)/(8\pi l)^2}\mathbf{i} - \frac{U_{bz} - \left(K_a U_{az}/4\pi l \right)}{1 - \left(K_a K_b \right)/(4\pi l)^2}\mathbf{k}. \tag{7.2.17}$$

If the particles are settling under the action of gravity, then $\mathbf{F}_a$ and $\mathbf{F}_b$ are known values, depending on the particles shape, size and density as well as the surrounding fluid density. Then, from the four equations (7.2.16) and (7.2.17) we can determine U_{ax}, U_{az}, U_{bx} and U_{bz}. As an example we shall discuss the case of two equal spheres of radius a and (7.2.16) is transformed into

$$-\frac{\mathbf{F}_a}{6\pi\mu a} = \frac{U_x}{1+(3/4)(a/l)}\mathbf{i} + \frac{U_z}{1+(3/2)(a/l)}\mathbf{k}. \quad (7.2.18)$$

Since the forces on the two spheres are equal, they will move parallel to each other with the same velocity, which may be directed at some angle α with respect to the line of spheres centres, while the spheres distance l remains unchanged. If the spheres velocity $\mathbf{U}$ is given, as well as the angle α, then the force acting on the sphere from the fluid side can be split into two components: longitudinal F_U and transverse F_D:

$$F_U = F_z \cos\alpha + F_x \sin\alpha, \quad (7.2.19)$$

where $F_x = -\frac{6\pi\mu aU}{1+(3/4)(a/l)}\sin\alpha$ and $F_z = -\frac{6\pi\mu aU}{1+(3/2)(a/l)}\cos\alpha$, U is the absolute value of $\mathbf{U}$.

As a result we get the following expressions for F_U and F_D:

$$F_U = -6\pi\mu aU\left\{\frac{1}{1+(3/4)(a/l)} - \left[\frac{1}{1+(3/4)(a/l)} - \frac{1}{1+(3/2)(a/l)}\right]\cos^2\alpha\right\}, \quad (7.2.20)$$

$$F_D = -6\pi\mu aU\sin\alpha\cos\alpha\left[\frac{1}{1+(3/2)(a/l)} - \frac{1}{1+(3/4)(a/l)}\right].$$

Now, if we want to calculate the velocities of two equal particles settling under the action of gravity, it is quite suitable to split the velocity into two parts: vertical U_F and horizontal U_H (drift):

$$U_F = -\frac{F}{6\pi\mu a}\left[1 + \frac{3a}{4l}\left(1+\cos^2\beta\right)\right], \quad U_H = -\frac{F}{6\pi\mu a}\frac{3a}{4l}\sin\beta\cos\beta, \quad (7.2.21)$$

where β is the angle between the spheres line of centre and the gravity force vector and F is the value of the gravity force $\mathbf{F}$. From the derived formula, it is clear that the drift force exists only if $\beta \neq 0$ or $\pi/2$, i.e., if the particles do not fall one after the other along their line of centres or perpendicularly to it.

When more accurate results are necessary than the expressions just derived for the forces (7.2.16), (7.2.17), it is possible to apply the method of reflections, but exploiting the general solution of the two particles boundary value problem. In this case for each particle the general solution is applied with arbitrary boundary conditions on its surface. Since the discussed method is applicable for solving the problem of two spheres arbitrarily situated with respect to the coordinate system, the solution of this problem can be constructed as a combination of two particular solutions for: (i) spheres moving parallel to their line of centres; (ii) spheres moving perpendicular to their line of centres.

Instead of the point force field (7.2.7) and (7.2.8), the Lamb's general solution for a single sphere problem (5.7.11) is used as a general solution for the creeping velocity and pressure. As a result, when the spheres are falling parallel to their line of centres, i.e., in z direction, the drag force exerted on first sphere is:

$$\mathbf{F}_a = \mathbf{k}6\pi\mu a\left\{U_{az}\left[1+\frac{9ab}{4l^2}+\frac{3}{4l^4}\left(-2a^3b+\frac{27}{4}a^2b^2+3ab^3\right)\right]\right. \tag{7.2.22}$$

$$\left.+U_{bz}\left[-\frac{3b}{2l}+\frac{1}{2l^3}\left(a^2b-\frac{27}{4}ab^2+b^3\right)-\frac{9}{4l^5}\left(a^3b^2+\frac{27}{8}a^2b^3+ab^4\right)\right]\right\}=\mathbf{k}6\pi\mu aT_1$$

If the direction of spheres fall is taken as positive direction on z, then the sign of $\mathbf{F}_a$ will be negative. The force $\mathbf{F}_b$ exerted on second sphere is obtained from (7.2.22) after interchanging the symbols b and a everywhere in the expression. Using slightly different method, Faxen (1925) found the same result for the drag force as (7.2.22) when the spheres move parallel to their line of centres.

On the basis of the reflection method Wakiya (1957) also considers the case of two spherical particles of different radii moving perpendicularly to their line of centres (see Fig.7.2.1. at $\theta = \pi/2$). For two particles translating without rotation he obtains for the drag force:

$$\mathbf{F}_a = \mathbf{k}6\pi\mu a\left\{U_{az}\left[1+\frac{9ab}{16l^2}+\frac{3}{8l^4}\left(a^3b+\frac{27}{32}a^2b^2+3ab^3\right)\right]\right. \tag{7.2.23}$$

$$\left.-U_{bz}\left[\frac{3b}{4l}+\frac{1}{4l^3}\left(a^2b+\frac{27}{16}ab^2+b^3\right)+\frac{63}{64l^5}\left(a^3b^2+\frac{27}{112}a^2b^3+ab^4\right)\right]\right\}=\mathbf{k}6\pi\mu aT_2$$

Since in the discussed case the particles cannot rotate, then torque must act on them:

$$\mathbf{T}_a = \mathbf{j}8\pi\mu a^3\left\{U_{by}\left[\frac{3b}{4l^2}+\frac{27ab^2}{64l^4}+\frac{9}{32l^6}\left(3a^3b^2+\frac{27}{32}a^2b^3+2ab^4\right)\right]\right. \tag{7.2.24}$$

$$\left.-U_{ay}\left[\frac{9ab}{16l^3}+\frac{3}{16l^5}\left(a^3b+\frac{27}{16}a^2b^2+3ab^3\right)\right]\right\}.$$

The torque acting on second particle has equal value but opposite direction to $\mathbf{T}_a$, i.e., opposite sign. Its value is obtained as in the previous case for the drag force after interchanging the symbols a and b.

For spherical particles that move perpendicularly to their line of centres and are free to rotate, for the drag force on first particle Wakiya (1957) gives the expression:

$$\mathbf{F}_a = \mathbf{k}6\pi\mu a\left\{U_{az}\left[1+\frac{9ab}{16l^2}+\frac{3}{8l^4}\left(a^3b+\frac{27}{32}a^2b^2+ab^3\right)\right]\right. \tag{7.2.25}$$

$$\left.-U_{bz}\left[\frac{3b}{4l}+\frac{1}{4l^3}\left(a^2b+\frac{27}{16}ab^2+b^3\right)+\frac{27}{64l^5}\left(a^3b^2+\frac{9}{16}a^2b^3+ab^4\right)\right]\right\}=\mathbf{k}6\pi\mu aT_3$$

and for the rotational velocity

$$\omega_a = \mathbf{j}\left\{U_{by}\left[\frac{3b}{4l^2}+\frac{27ab^2}{64l^4}+\frac{9}{32l^6}\left(a^3b^2+\frac{27}{32}a^2b^3+ab^4\right)\right]\right. \tag{7.2.26}$$

$$\left. -U_{ay}\left[\frac{9ab}{16l^3}+\frac{3}{16l^5}\left(a^3b+\frac{27}{16}a^2b^2+ab^3\right)\right]\right\}.$$

The obtained results for the parallel and perpendicular motions of two spheres with respect to their line of centres can simply be applied for the derivation of some general relations. Therefore, instead of (7.2.16) for two spheres devoid of rotation, we can write

$$\frac{\mathbf{F}_a}{6\pi\mu a} = \mathbf{i}T_1 + \mathbf{k}T_2, \tag{7.2.27}$$

where T_1 and T_2 are given by (7.2.22) and (7.2.23), respectively.

Analogously, for freely rotating spheres, the drag force on first sphere is:

$$\frac{\mathbf{F}_a}{6\pi\mu a} = \mathbf{i}T_1 + \mathbf{k}T_3, \tag{7.2.28}$$

where T_3 is determined by (7.2.25).

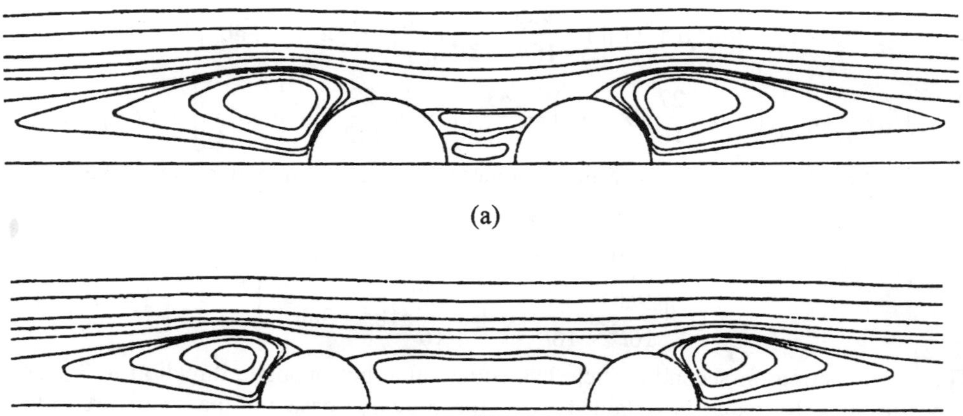

(b)

Fig.7.2.2. Streamlines pattern around two equal spheres in shear flow: a) d/a = 1; b) d/a = 4

The hydrodynamic interaction problem of a steady axisymmetric Poiseuille flow passing two rigid spherical particles at zero-Reynolds number is considered by Kalitzova-Kurteva and Zapryanov (1989). The streamlines pattern strongly depends on the distance between the particles. It is interesting to note, as shown in Fig.7.2.2., that at moderate distance between two identical particles there are more than one vortex in direction transverse to the flow, while the vortex is single in longitudinal direction.

Moreover, with the increase of the distance between the particles, the vortex in longitudinal direction splits into two vortices, tending to be similar to the flow pattern around a single particle in the Poiseuille flow.

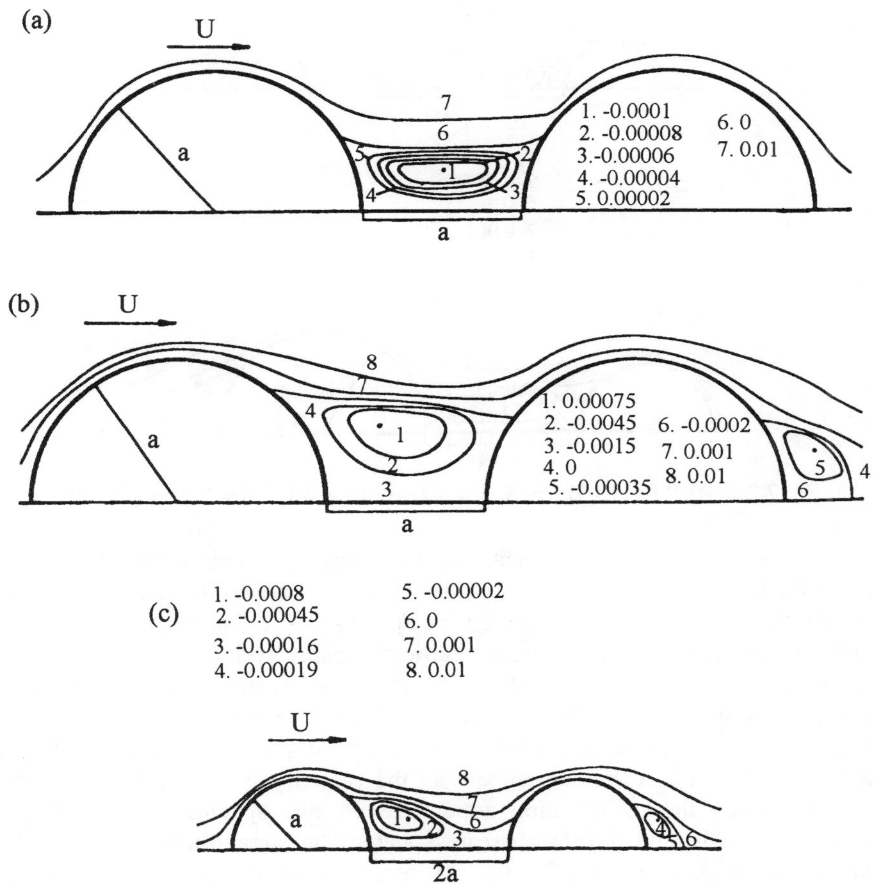

Fig.7.2.3. Streamlines pattern around two spheres: a) Re = 1, d/a = b/a = 1; b) Re = 40, d/a = b/a = 1; c) Re = 40, d/a = 2, b/a = 1.

The axisymmetric problem of hydrodynamic interactions between two rigid spherical particles passed by a steady viscous flow at moderate Reynolds numbers has been studied by Zapryanov and Toshev (1985, 1986). The problem has been solved numerically by an extension of the method of fractional steps (Yanenko, 1971), which yields good results in a wide range from low to reasonably high Reynolds numbers.

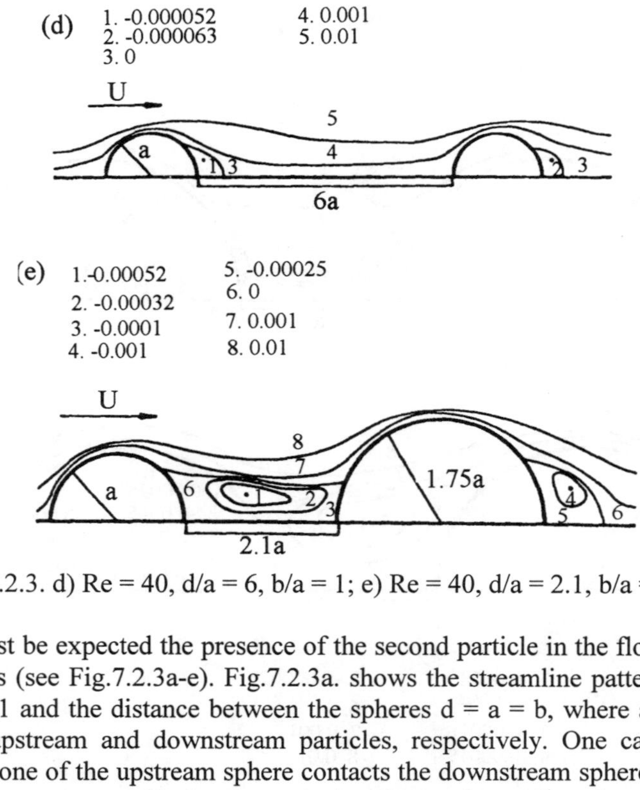

Fig.7.2.3. d) Re = 40, d/a = 6, b/a = 1; e) Re = 40, d/a = 2.1, b/a =1.75.

As must be expected the presence of the second particle in the flow gives rise to closed vortices (see Fig.7.2.3a-e). Fig.7.2.3a. shows the streamline pattern at Reynolds number Re = 1 and the distance between the spheres d = a = b, where a and b are the radii of the upstream and downstream particles, respectively. One can see that the recirculation zone of the upstream sphere contacts the downstream sphere. This result is in qualitative agreement with the theoretical and experimental works of Stimson and Jeffrey (1926) and Schlichting (1964), performed for pairs of spheres under much different conditions of Reynolds numbers. In the case of Re = 40 and d = a = b (see Fig.7.2.3b), beside the recirculation zone between two spheres, there is a second recirculation zone behind the downstream sphere. The streamlines for Re = 40 and d = 2a = 2b are presented in Fig.7.2.3c. It is interesting to note that there is a tendency for the first recirculation zone to recede from the rear particle. As it is shown in Fig.7.2.3d. (Re = 40, d = 6a = 6b) when the distance between the spheres is sufficiently large the recirculation zone of the upstream sphere does not contact the downstream sphere and the flow pattern resembles the flow past a single sphere. The influence of the radii ratio on the hydrodynamic interaction between two spheres is given in Fig.7.2.3e., where b = 1.75a, d = 2a and Re = 40. Fig.7.2.4a. shows the dependence of the drag forces F_a and F_b, acting on two equal spheres, on the dimensionless distance between the spheres d/a. The obtained numerical results show that the drag force acting on each of the spheres is smaller than the same force experienced on a single sphere and the drag on the front sphere is greater than on the rear one. As must be expected, when the distance d/a tends to infinity, there is no interaction between them. When d/a = 20, our results are in good quantitative agreement with the experiments presented by Schlichting (1964) for the drag on a single sphere. In Fig.7.2.4b. the drag force is plotted against Re for d = a = b.

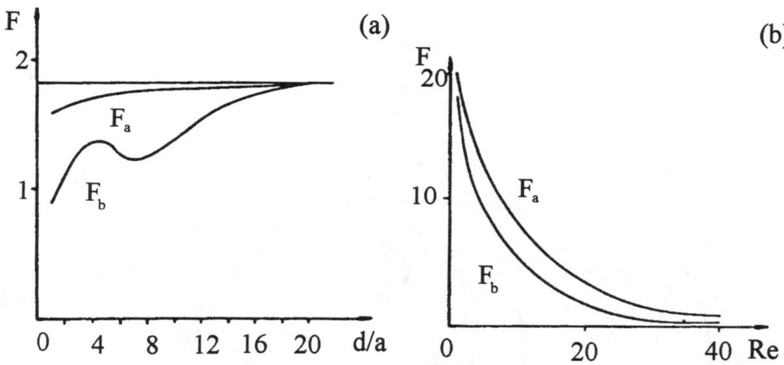

Fig.7.2.4. The drag force F_a and F_b on spheres of equal radii a = b as function of: a) d/a at Re = 40; b) Re at d/a = 1.

The method of numerical matching (see Alam et al., 1981) has been extended to solve the problem of hydrodynamic interaction between two equal oblate spheroids in tandem arrangement at moderate Reynolds numbers by Kotzev and Zapryanov (1989) and Kotzev et al. (1989). It has been founded that when the Reynolds number is low, the interactions between the spheroids fade away rapidly if the spheroids are placed at a large distance. If the Reynolds number is moderate, the interactions are strong even when the separation distance between the particles is not too small.

For moderate Reynolds numbers and not so large separation (d = 1.38a, where a is the large semi-axis), two shear layers separated from the upstream spheroid stick onto the downstream one and there exists a second intensive recirculation zone behind the downstream spheroid. (see Fig.7.2.5a,b). When the separation distance between the two particles is not small, the recirculation zone of the upstream spheroid does not contact the downstream one and the flow pattern resembles the flow past an isolated spheroid.

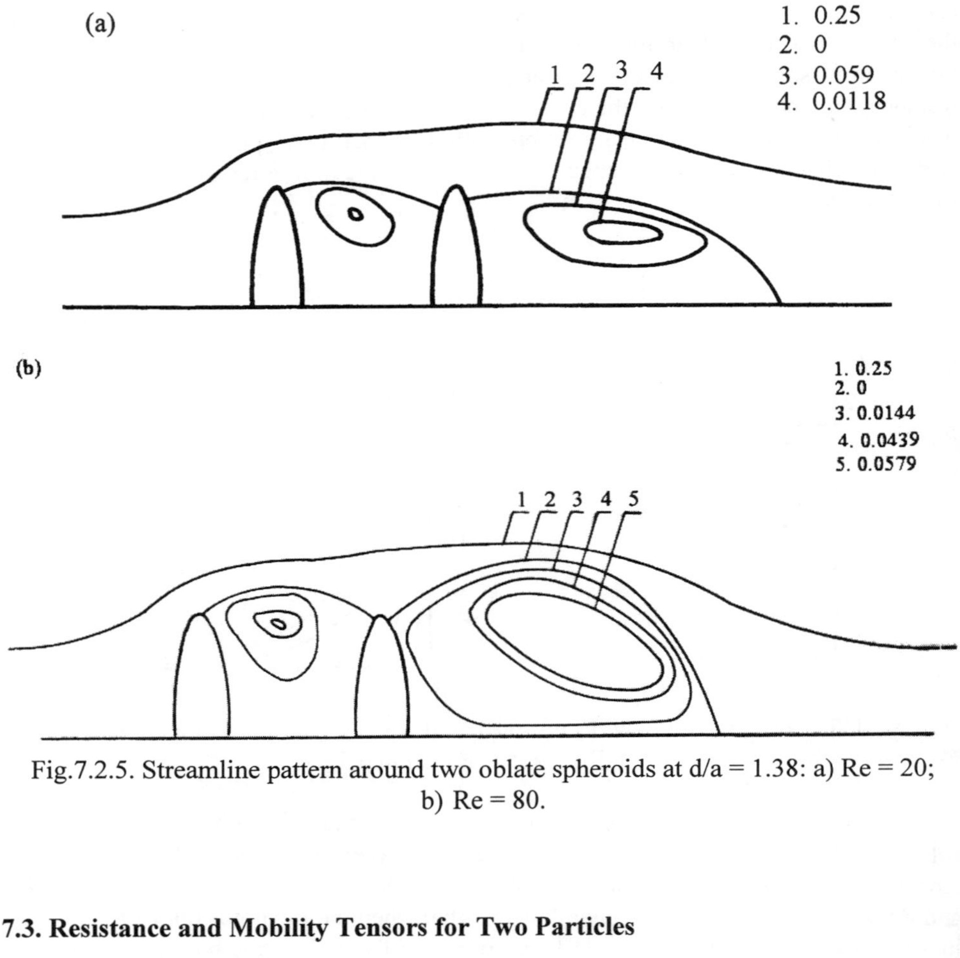

Fig.7.2.5. Streamline pattern around two oblate spheroids at d/a = 1.38: a) Re = 20; b) Re = 80.

7.3. Resistance and Mobility Tensors for Two Particles

In 5.10 we have defined the resistance tensor and mobility tensor of a single particle. In the present chapter we shall generalise these concepts [Brenner and O'Neill (1972); Kim and Mifflin (1985)] for the case of two particles.

Let us denote with $\mathbf{F}_k$, $\mathbf{M}_k$ and $\mathbf{S}_k$ (k=1, 2) respectively the forces, torques and stresslets exerted by the fluid on the two particles. Suppose, that the particles are passed by a flow

$$\mathbf{v}^{\infty} = \mathbf{U}^{\infty} + \mathbf{\Omega}^{\infty} \times \mathbf{r} + \mathbf{E}^{\infty}.\mathbf{r} \tag{7.3.1}$$

and they move with velocities

$$\mathbf{U}_k + \boldsymbol{\omega}_k \times (\mathbf{r} - \mathbf{r}_k), \tag{7.3.2}$$

where k =1, 2 and $\mathbf{r}_k$ is the radius vector of each particle centre.

For the resistance problem, we assume that the particles motion is known and $\mathbf{F}_k$, $\mathbf{M}_k$ and $\mathbf{S}_k$ (k=1, 2) must be found. Similarly to the single particle case, we can write

$$\begin{pmatrix} \mathbf{F}_1 \\ \mathbf{F}_2 \\ \mathbf{M}_1 \\ \mathbf{M}_2 \\ \mathbf{S}_1 \\ \mathbf{S}_2 \end{pmatrix} = \mu \begin{pmatrix} \mathbf{A}_{11} & \mathbf{A}_{12} & \widetilde{\mathbf{B}}_{11} & \widetilde{\mathbf{B}}_{12} & \widetilde{\mathbf{G}}_{11} & \widetilde{\mathbf{G}}_{12} \\ \mathbf{A}_{21} & \mathbf{A}_{22} & \widetilde{\mathbf{B}}_{21} & \widetilde{\mathbf{B}}_{22} & \widetilde{\mathbf{G}}_{21} & \widetilde{\mathbf{G}}_{22} \\ \mathbf{B}_{11} & \mathbf{B}_{12} & \mathbf{C}_{11} & \mathbf{C}_{12} & \widetilde{\mathbf{H}}_{11} & \widetilde{\mathbf{H}}_{12} \\ \mathbf{B}_{21} & \mathbf{B}_{22} & \mathbf{C}_{21} & \mathbf{C}_{22} & \widetilde{\mathbf{H}}_{21} & \widetilde{\mathbf{H}}_{22} \\ \mathbf{G}_{11} & \mathbf{G}_{12} & \mathbf{H}_{11} & \mathbf{H}_{12} & \mathbf{T}_{11} & \mathbf{T}_{12} \\ \mathbf{G}_{21} & \mathbf{G}_{22} & \mathbf{H}_{21} & \mathbf{H}_{22} & \mathbf{T}_{21} & \mathbf{T}_{22} \end{pmatrix} \begin{pmatrix} \mathbf{U}^\infty - \mathbf{U}_1 \\ \mathbf{U}^\infty - \mathbf{U}_2 \\ \mathbf{\Omega}^\infty - \boldsymbol{\omega}_1 \\ \mathbf{\Omega}^\infty - \boldsymbol{\omega}_2 \\ \mathbf{E}^\infty \\ \mathbf{E}^\infty \end{pmatrix}. \tag{7.3.3}$$

As with the single particle, here the tensors **A**, **B** and **C** are of second rank tensors, **G** and **H** are third rank tensors, **T** is a fourth rank tensor. With the mark ~ the transposed matrices of the corresponding tensors are denoted.

Often the opposite problem rises, in which the forces $\mathbf{F}_k$, the torques $\mathbf{M}_k$ and the velocities of the flow at infinity $\mathbf{U}^\infty$ are known, while the motion and the stresslets of the two particles must be determined. Profiting from the considered problem linearity, we can write down

$$\begin{pmatrix} \mathbf{U}^\infty - \mathbf{U}_1 \\ \mathbf{U}^\infty - \mathbf{U}_2 \\ \mathbf{\Omega}^\infty - \boldsymbol{\omega}_1 \\ \mathbf{\Omega}^\infty - \boldsymbol{\omega}_2 \\ \mu^{-1}\mathbf{S}_1 \\ \mu^{-1}\mathbf{S}_2 \end{pmatrix} = \begin{pmatrix} \mathbf{a}_{11} & \mathbf{a}_{12} & \widetilde{\mathbf{b}}_{11} & \widetilde{\mathbf{b}}_{12} & \widetilde{\mathbf{g}}_{11} & \widetilde{\mathbf{g}}_{12} \\ \mathbf{a}_{21} & \mathbf{a}_{22} & \widetilde{\mathbf{b}}_{21} & \widetilde{\mathbf{b}}_{22} & \widetilde{\mathbf{g}}_{21} & \widetilde{\mathbf{g}}_{22} \\ \mathbf{b}_{11} & \mathbf{b}_{12} & \mathbf{c}_{11} & \mathbf{c}_{12} & \widetilde{\mathbf{h}}_{11} & \widetilde{\mathbf{h}}_{12} \\ \mathbf{b}_{21} & \mathbf{b}_{22} & \mathbf{c}_{21} & \mathbf{c}_{22} & \widetilde{\mathbf{h}}_{21} & \widetilde{\mathbf{h}}_{22} \\ \mathbf{g}_{11} & \mathbf{g}_{12} & \mathbf{h}_{11} & \mathbf{h}_{12} & \mathbf{m}_{11} & \mathbf{m}_{12} \\ \mathbf{g}_{21} & \mathbf{g}_{22} & \mathbf{h}_{21} & \mathbf{h}_{22} & \mathbf{m}_{21} & \mathbf{m}_{22} \end{pmatrix} \begin{pmatrix} \mu^{-1}\mathbf{F}_1 \\ \mu^{-1}\mathbf{F}_2 \\ \mu^{-1}\mathbf{M}_1 \\ \mu^{-1}\mathbf{M}_2 \\ \mathbf{E}^\infty \\ \mathbf{E}^\infty \end{pmatrix} \tag{7.3.4}$$

Here, again **a**, **b**, **c** are second rank tensors, **g**, **h** are tensors of third rank, and **m** is a tensor of fourth rank. The quadratic matrix in (7.3.4) is called mobility matrix.

If the components of the resistance problem tensors are known, then the tensors connected with the mobility problem can be found from the equalities:

$$\begin{pmatrix} \mathbf{a}_{11} & \mathbf{a}_{12} & \widetilde{\mathbf{b}}_{11} & \widetilde{\mathbf{b}}_{12} \\ \mathbf{a}_{21} & \mathbf{a}_{22} & \widetilde{\mathbf{b}}_{21} & \widetilde{\mathbf{b}}_{22} \\ \mathbf{b}_{11} & \mathbf{b}_{12} & \mathbf{c}_{11} & \mathbf{c}_{12} \\ \mathbf{b}_{21} & \mathbf{b}_{22} & \mathbf{c}_{21} & \mathbf{c}_{22} \end{pmatrix} = \begin{pmatrix} \mathbf{A}_{11} & \mathbf{A}_{12} & \widetilde{\mathbf{B}}_{11} & \widetilde{\mathbf{B}}_{12} \\ \mathbf{A}_{21} & \mathbf{A}_{22} & \widetilde{\mathbf{B}}_{21} & \widetilde{\mathbf{B}}_{22} \\ \mathbf{B}_{11} & \mathbf{B}_{12} & \mathbf{C}_{11} & \mathbf{C}_{12} \\ \mathbf{B}_{21} & \mathbf{B}_{22} & \mathbf{C}_{21} & \mathbf{C}_{22} \end{pmatrix}^{-1}, \tag{7.3.5}$$

$$\begin{pmatrix} \mathbf{g}_{11} & \mathbf{g}_{12} & \mathbf{h}_{11} & \mathbf{h}_{12} \\ \mathbf{g}_{21} & \mathbf{g}_{22} & \mathbf{h}_{21} & \mathbf{h}_{22} \end{pmatrix} = \begin{pmatrix} \mathbf{G}_{11} & \mathbf{G}_{12} & \mathbf{H}_{11} & \mathbf{H}_{12} \\ \mathbf{G}_{21} & \mathbf{G}_{22} & \mathbf{H}_{21} & \mathbf{H}_{22} \end{pmatrix} \begin{pmatrix} \mathbf{a}_{11} & \mathbf{a}_{12} & \widetilde{\mathbf{b}}_{11} & \widetilde{\mathbf{b}}_{12} \\ \mathbf{a}_{21} & \mathbf{a}_{22} & \widetilde{\mathbf{b}}_{21} & \widetilde{\mathbf{b}}_{22} \\ \mathbf{b}_{11} & \mathbf{b}_{12} & \mathbf{c}_{11} & \mathbf{c}_{12} \\ \mathbf{b}_{21} & \mathbf{b}_{22} & \mathbf{c}_{21} & \mathbf{c}_{22} \end{pmatrix}$$

$$\begin{pmatrix} \mathbf{m}_{11} + \mathbf{m}_{12} \\ \mathbf{m}_{21} + \mathbf{m}_{22} \end{pmatrix} = \begin{pmatrix} \mathbf{T}_{11} + \mathbf{T}_{12} \\ \mathbf{T}_{21} + \mathbf{T}_{22} \end{pmatrix} + \begin{pmatrix} \mathbf{G}_{11} & \mathbf{G}_{12} & \mathbf{H}_{11} & \mathbf{H}_{12} \\ \mathbf{G}_{21} & \mathbf{G}_{22} & \mathbf{H}_{21} & \mathbf{H}_{22} \end{pmatrix} \begin{pmatrix} \tilde{\mathbf{g}}_{11} + \tilde{\mathbf{g}}_{12} \\ \tilde{\mathbf{g}}_{21} + \tilde{\mathbf{g}}_{22} \\ \tilde{\mathbf{h}}_{11} + \tilde{\mathbf{h}}_{12} \\ \tilde{\mathbf{h}}_{21} + \tilde{\mathbf{h}}_{22} \end{pmatrix}.$$

Unfortunately, the only exact analytical values for the resistance tensors in a multiparticle system are those related to the two spheres case and here, we shall present two examples of two spherical particles in Stokes flows.

As a first example, we choose the uniform motion of two spherical particles in a quiescent fluid. If no torques are acting on the particles, then due to the linearity of Stokes equations, we can write for the velocity of the two particles (Batchelor, 1976a)

$$\mathbf{U}_i = \mathbf{a}_{i1}\mathbf{F}_1 + \mathbf{a}_{i2}\mathbf{F}_2, \qquad (i = 1, 2). \tag{7.3.6}$$

The values of the components of the mobility tensor $\mathbf{a}_{ij}$, characterizing the influence of the j-th particle on the i-th one, depend on the distance between them (see Fig.7.3.1.). Taking into account that the relative motion of the two particles and the motion of their mass centre are defined by the equalities

$$\frac{d\boldsymbol{\rho}}{dt} = \mathbf{U}_2 - \mathbf{U}_1, \qquad \frac{d\mathbf{r}_c}{dt} = \frac{\mathbf{U}_2 + \mathbf{U}_1}{2} \tag{7.3.7}$$

we can write down the formulae

$$\frac{d\boldsymbol{\rho}}{dt} = (\mathbf{a}_{21} - \mathbf{a}_{11})\mathbf{F}_1 + (\mathbf{a}_{22} - \mathbf{a}_{12})\mathbf{F}_2, \tag{7.3.8}$$

$$\frac{d\mathbf{r}_c}{dt} = \frac{(\mathbf{a}_{21} + \mathbf{a}_{11})}{2}\mathbf{F}_1 + \frac{(\mathbf{a}_{22} + \mathbf{a}_{12})}{2}\mathbf{F}_2.$$

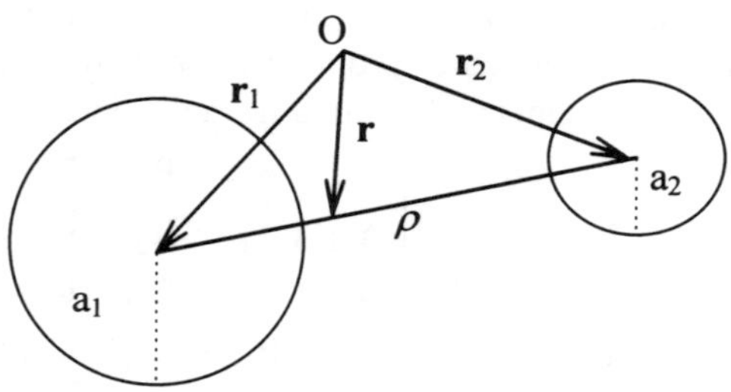

Fig.7.3.1. Geometry of two spheres in a quiescent fluid.

From physical considerations it is clear, that when the distance between the particles tends to infinity, the hydrodynamic interaction between the particles disappears, i.e., at $i \neq j$

$$\lim_{\rho \to \infty} \mathbf{a}_{ij} = 0, \tag{7.3.9}$$

and at $i = j$

$$\lim_{\rho\to\infty} \mathbf{a}_{ii} = \frac{\mathbf{I}}{6\pi\mu\,a_i}. \tag{7.3.10}$$

From the obtained results for the dynamics of two spherical particles (Batchelor, 1976a), it follows that the mobility tensor $\mathbf{a}_{ij}$ can be represented by the formula

$$\mathbf{a}_{ij} = \frac{1}{3\pi\mu\,(a_1+a_2)}\left[x_{ij}(\boldsymbol{\rho})\frac{\boldsymbol{\rho\rho}}{\rho^2} + y_{ij}(\boldsymbol{\rho})\left(\mathbf{I} - \frac{\boldsymbol{\rho\rho}}{\rho^2}\right)\right], \tag{7.3.11}$$

where some evaluations of the coefficients x_{ij} and y_{ij} will be given further.
Right here we show that, because of (7.3.9) and (7.3.10), x_{ij} and y_{ij} have the following properties:

$$\lim_{\rho\to\infty} x_{ij}(\boldsymbol{\rho}) = \lim_{\rho\to\infty} y_{ij}(\boldsymbol{\rho}) = 0 \qquad \text{at } i \neq j, \tag{7.3.12}$$

$$\lim_{\rho\to\infty} x_{ii}(\boldsymbol{\rho}) = \lim_{\rho\to\infty} y_{ii}(\boldsymbol{\rho}) = 1 \qquad \text{at } i = j. \tag{7.3.13}$$

If the problem is simplified, considering two equal spheres ($a_1 = a_2 = a$), then $\mathbf{a}_{11} = \mathbf{a}_{22}$ and $\mathbf{a}_{12} = \mathbf{a}_{21}$. When a force is exerted by the fluid only on the first sphere: $\mathbf{F}_1 = \mathbf{F}$ and $\mathbf{F}_2 = 0$, then at very large distances between the spheres the first one will move with velocity $\mathbf{U}_1 = \dfrac{\mathbf{F}}{6\pi\mu\,a_1}$ and will generate around itself a fluid flow with velocity (3.2.13) and (4.2.14):

$$\mathbf{v}_1 = \frac{3a}{4\rho}\left(1 + \frac{a^2}{3\rho^2}\right)\mathbf{U}_1 + \frac{3a}{4\rho}\left(1 - \frac{a^2}{\rho^2}\right)\frac{\boldsymbol{\rho}.\mathbf{U}_1\boldsymbol{\rho}}{\rho^2}. \tag{7.3.14}$$

According to the Faxen's law for the drag, the velocity v_1 of the disturbed flow around the first particle induces a drag force F^*, given by (2.9.10), which creates a motion of the second particle with velocity

$$\mathbf{U}_2 = \mathbf{v}_1(\boldsymbol{\rho}) + \frac{a_2^2}{6}\nabla^2\mathbf{v}_1(\boldsymbol{\rho}), \tag{7.3.15}$$

calculated at the centre of second sphere.

We have to note, also, that the velocity $\mathbf{v}_1(\boldsymbol{\rho})$ acts on the second particle as a force dipole with intensity

$$\mathbf{S}_2 = \frac{20}{3}\pi\mu\,a_2^3\left(1 + \frac{1}{10}a^2\nabla^2\right)\mathbf{E}_1(\boldsymbol{\rho}), \tag{7.3.16}$$

where $\mathbf{E}_1(\boldsymbol{\rho})$ is the flow rate of strain tensor generated by $\mathbf{v}_1(\boldsymbol{\rho})$. On the other hand, the force dipole with intensity $\mathbf{S}_1$ generates a flow around the second particle with velocity

$$\mathbf{v}_2 = \mathbf{S}_2 : \frac{\nabla\mathbf{B}}{8\pi\mu}, \tag{7.3.17}$$

where $\mathbf{B}$ is the Oseen-Burgers tensor (3.1.1).
This flow changes the velocity $\mathbf{U}_1$ of the first particle and for it we have

$$\mathbf{U}_1 = \frac{\mathbf{F}}{6\pi\mu\,a_1} + \mathbf{v}_2(\boldsymbol{\rho}). \tag{7.3.18}$$

The evaluations of x_{ij} and y_{ij} (i, j =1, 2) at big ρ are obtained when comparing the equalities (7.3.15), (7.3.18) and (7.3.11). The final results for the coefficients x_{ij} and y_{ij} at big values of $\xi = \dfrac{2\rho}{a_1 + a_2}$ are:

$$x_{11} = 1 - \frac{60\eta^3}{(1+\eta)^4 \xi^4} + O(\xi^{-6}), \quad x_{12} = \frac{3}{2\xi} - \frac{2(1+\eta^2)}{(1+\eta)^2 \xi^3} + O(\xi^{-7}), \qquad (7.3.19)$$

$$y_{11} = 1 + O(\xi^{-6}), \qquad y_{12} = \frac{3}{4\xi} + \frac{(1+\eta^2)}{(1+\eta)^2 \xi^3} + O(\xi^{-7}),$$

where $\eta = \dfrac{a_2}{a_1}$. We have to note that at $\eta \to 0$, the influence of the second particle on the first one disappears.

When analysing the resistance problem, we have

$$\mathbf{F}_i = \mathbf{A}_{i1}\mathbf{U}_1 + \mathbf{A}_{i2}\mathbf{U}_2, \qquad (7.3.20)$$

where $\mathbf{A}_{ij}$ (i, j = 1, 2) are the components of the resistance tensors of the two interacting particles. Comparing (7.3.6) with (7.3.20), we find

$$\mathbf{A}_{ii} = \left(\mathbf{a}_{ii} - \mathbf{a}_{ij}.\mathbf{a}_{jj}^{-1}.\mathbf{a}_{ji}\right)^{-1}, \quad \mathbf{A}_{ij} = \left(\mathbf{a}_{ji} - \mathbf{a}_{jj}.\mathbf{a}_{ji}^{-1}.\mathbf{a}_{ii}\right)^{-1} \text{ for } i \neq j. \qquad (7.3.21)$$

From these equalities the scalar coefficients of the resistance tensor can be defined as functions of the mobility tensor coefficients x_{ij} and y_{ij}:

$$\mathbf{A}_{ij} = 3\pi\mu(a_1 + a_2)\left[X_{ij}(\boldsymbol{\rho})\frac{\boldsymbol{\rho\rho}}{\rho^2} + Y_{ij}(\boldsymbol{\rho})\left(\mathbf{I} - \frac{\boldsymbol{\rho\rho}}{\rho^2}\right)\right] \qquad (7.3.22)$$

The second example concerns the two spherical particles motion in shear flow. Let the first particle moves under the action of an external force **F** with linear velocity $\mathbf{U}_1$, while the second particle moves with linear velocity $\mathbf{U}_2$, due to the fluid disturbances created by the first particle motion (Fig.7.3.1). Batchelor (1982) and Batchelor and Wen (1982) derived the following formulae for the translational linear velocities of the two particles, moving in a quiescent, unbounded viscous fluid

$$\mathbf{U}_1 = \frac{\mathbf{F}}{6\pi\mu a_1}\left[x_{11}(\boldsymbol{\rho})\frac{\boldsymbol{\rho\rho}}{\rho^2} + y_{11}(\boldsymbol{\rho})\left(\mathbf{I} - \frac{\boldsymbol{\rho\rho}}{\rho^2}\right)\right], \qquad (7.3.23)$$

$$\mathbf{U}_2 = \frac{\mathbf{F}}{6\pi\mu(a_1 + a_2)}\left[x_{12}(\boldsymbol{\rho})\frac{\boldsymbol{\rho\rho}}{\rho^2} + y_{12}(\boldsymbol{\rho})\left(\mathbf{I} - \frac{\boldsymbol{\rho\rho}}{\rho^2}\right)\right], \qquad (7.3.24)$$

where $\boldsymbol{\rho}$ is the vector connecting the two particles centres and $\rho = |\boldsymbol{\rho}|$. At large values of the dimensionless length $\xi = \dfrac{2\rho}{a_1 + a_2}$ and at $\eta = \dfrac{a_2}{a_1} = O(1)$, Jeffrey and Onishi (1984a, b) and Davis and Hill (1992) derived the following far-field expansions for the coefficients of the mobility tensor

$$1-\mathrm{y}_{11}=\frac{68\eta^5}{(1+\eta)^6\xi^6}+\frac{32\eta^5(10-9\eta^2+9\eta^4)}{(1+\eta)^8\xi^8}+\frac{192\eta^5(35-18\eta^2+6\eta^4)}{(1+\eta)^{10}\xi^{10}}+O(\xi^{-12}) \quad (7.3.25)$$

$$\mathrm{y}_{11}-\mathrm{x}_{11}=\frac{60\eta^3}{(1+\eta)^4\xi^4}+\frac{60\eta^3(8-\eta^2)}{(1+\eta)^6\xi^6}+\frac{32\eta^3(20-123\eta^2+9\eta^4)}{(1+\eta)^8\xi^8}$$
$$+\frac{64\eta^2(175+1500\eta-426\eta^2+18\eta^4)}{(1+\eta)^{10}\xi^{10}}+O(\xi^{-12}),$$

$$\mathrm{x}_{11}-\frac{2\mathrm{x}_{12}}{1+\eta}=1-\frac{3}{(1+\eta)\xi}+\frac{4(1+\eta^2)}{(1+\eta)^3\xi^3}-\frac{60\eta^3}{(1+\eta)^4\xi^4}+\frac{32\eta^3(15-4\eta^2)}{(1+\eta)^6\xi^6}-\frac{2400\eta^3}{(1+\eta)^7\xi^7}$$
$$-\frac{192\eta^3(5-22\eta^2+3\eta^4)}{(1+\eta)^8\xi^8}+\frac{1920\eta^3(1+\eta^2)}{(1+\eta)^9\xi^9}-\frac{256\eta^5(70-375\eta-120\eta^2+9\eta^4)}{(1+\eta)^{10}\xi^{10}}$$
$$-\frac{1536\eta^3(10-151\eta^2+10\eta^4)}{(1+\eta)^{11}\xi^{11}}+O(\xi^{-12}),$$

$$\mathrm{y}_{11}-\frac{2\mathrm{y}_{12}}{1+\eta}=1-\frac{3}{2(1+\eta)\xi}-\frac{2(1+\eta^2)}{(1+\eta)^3\xi^3}-\frac{68\eta^5}{(1+\eta)^6\xi^6}-\frac{32\eta^3(10-9\eta^2+9\eta^4)}{(1+\eta)^8\xi^8}$$
$$-\frac{192\eta^5(35-18\eta^2+6\eta^4)}{(1+\eta)^{10}\xi^{10}}-\frac{16\eta^3(560-553\eta^2+560\eta^4)}{(1+\eta)^{11}\xi^{11}}+O(\xi^{-12}).$$

Similar to these formulae are obtained by Fuentes et al. (1988) for the case $\eta >> 1$, when a small sphere falls through a suspension of large particles. The other limiting case of a large particle in a suspension of small spheres $\eta << 1$ is discussed by Batchelor (1982) and Fuentes et al. (1988).

7.4. Two Rigid Particles Near Contact

In order to calculate the effective viscosity coefficient of strongly concentrated suspensions it is necessary to consider two fundamental cases for the relative motion of two close particles: (i) a spherical particle passing a second stationary spherical particle perpendicularly to their line of centres generates a "shearing" flow of the fluid between them; (ii) a spherical particle approaching a second stationary one along their line of centres generates the so-called "squeezing" flow in the gap between the particles. For both problems the lubrication theory is applied in the subsequent presentation and the force and torque will be shown to possess different types of singularities for the two cases, dependent on the relative gap width.

<u>Shearing Motion of Rigid Particles.</u> In the first case we consider a sphere of radius a (Fig.7.4.1.) moving with velocity $\mathbf{U}_a$ with respect to a stationary sphere of radius $b = \beta a$

in x direction perpendicular to the common axis of the spheres coinciding with z direction. The width of the gap between the two spheres $a\varepsilon$ is small compared with the moving sphere radius, i.e., $\varepsilon << 1$ is a small parameter.

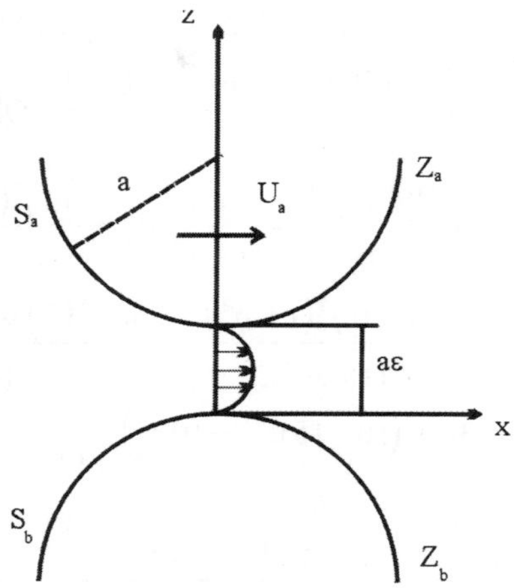

Fig.7.4.1. A rigid sphere translating perpendicularly to the common axis with respect to a stationary rigid sphere.

Since the problem is 3D, a cylindrical coordinate system (r, θ, z) connected with the stationary sphere is introduced. Then the continuity and Stokes equations are given by (1.3.6) and the boundary conditions for the velocity components on the spheres surfaces are:

$$v_r = U_a \cos\theta, \qquad v_\theta = -U_a \sin\theta, \qquad v_z = 0 \text{ on first sphere,} \tag{7.4.1}$$

$$v_r = v_\theta = v_z = 0 \text{ on second sphere.} \tag{7.4.2}$$

If the unknown velocity components and pressure are sought as Fourier series with respect to the angle θ, then due to (7.4.1), it can be easily shown that they are of the type:

$$v_r = U_a U(r,z)\cos\theta, \quad v_\theta = U_a V(r,z)\sin\theta, \tag{7.4.3}$$

$$v_z = U_a W(r,z)\cos\theta, \; p = \frac{\mu U_a}{a} P(r,z)\cos\theta .$$

Performing the substitution (7.4.3) in (1.3.6) we reach to the following equations for the new unknown functions independent of θ:

$$\frac{\partial U}{\partial r} + \frac{U+V}{r} + \frac{\partial W}{\partial z} = 0, \qquad \frac{1}{a}\frac{\partial P}{\partial r} = L_0^2 U - \frac{2}{r^2}(U+V), \tag{7.4.4}$$

$$-\frac{1}{a}\frac{P}{r} = L_0^2 V - \frac{2}{r^2}(U+V), \quad \frac{\partial P}{\partial z} = L_1^2 W,$$

where the operator $L^2{}_m$ is given by

$$L_m^2 \equiv \frac{\partial^2}{\partial r^2} + \frac{1}{r}\frac{\partial}{\partial r} - \frac{m^2}{r^2} + \frac{\partial^2}{\partial z^2}. \tag{7.4.5}$$

The solution of the upper equations will be sought in an "inner region" corresponding to the region near the spheres contact and in an "outer region" corresponding to the remaining space outside the particles.

For the "inner region", where the gap width between the two spheres is very small, of the order of $O(\varepsilon)$, "stretched" coordinates of the order of $O(1)$ will be introduced. The spheres surface near the origin of the coordinate system ($r \ll a$), as shown in Fig.7.4.1., can be approximated as

$$z_a = a(1+\varepsilon) - \sqrt{a^2 - r^2} \approx a\varepsilon + \frac{r^2}{2a}, \tag{7.4.6}$$

$$z_b = -\beta a + \sqrt{\beta^2 a^2 - r^2} \approx -\frac{r^2}{2\beta a}.$$

Then, if the "stretched" coordinates in both directions are supposed to be

$$Z = \frac{z}{a\varepsilon}, \qquad R = \frac{r}{a\sqrt{\varepsilon}}, \tag{7.4.7}$$

the spheres surfaces expressions (7.4.6) become

$$Z_a = 1 + \frac{1}{2}R^2 + O(\varepsilon),\ Z_b = -\frac{1}{2\beta}R^2 + O(\varepsilon). \tag{7.4.8}$$

Assuming the "inner" solution to be represented as power series of $\varepsilon^{-1/2}$, from the boundary conditions (7.4.1), (7.4.2) and from the governing equations (7.4.4), the asymptotic expansions follow

$$U(R,Z) = U_0(R,Z) + \varepsilon U_1(R,Z) + \ldots, \tag{7.4.9}$$

$$V(R,Z) = V_0(R,Z) + \varepsilon V_1(R,Z) + \ldots,$$

$$W(R,Z) = \varepsilon^{1/2} W_0(R,Z) + \varepsilon^{3/2} W_1(R,Z) + \ldots,$$

$$P(R,Z) = \varepsilon^{-3/2} P_0(R,Z) + \varepsilon^{-1/2} P_1(R,Z) + \ldots.$$

In order to obtain the correspondent equations to the first approximation (with indices 0), the upper expansions (7.4.9) are substituted in the equations (7.4.4) written in the "stretched" coordinates. As a result we find equations relevant to the lubrication theory (Kim and Karrila, 1991)

$$\frac{\partial U_0}{\partial R} + \frac{U_0 + V_0}{R} + \frac{\partial W_0}{\partial Z} = 0,\ \frac{\partial P_0}{\partial R} = \frac{\partial^2 U_0}{\partial Z^2},\ -\frac{P_0}{R} = \frac{\partial^2 V_0}{\partial Z^2},\ \frac{\partial P_0}{\partial Z} = 0, \tag{7.4.10}$$

with boundary conditions

$$U_0(R,Z_a) = -V_0(R,Z_a) = 1,\ W_0(R,Z_a) = 0, \tag{7.4.11}$$

$$U_0(R,Z_b) = V_0(R,Z_b) = W_0(R,Z_b) = 0.$$

The solutions of (7.4.10) and (7.4.11) are given by the expressions

$$U_0 = (Z - Z_b)\left[-\frac{1}{2}P_0'(Z_a - Z) + \frac{1}{H}\right],\quad V_0 = (Z - Z_b)\left[\frac{1}{2R}P_0(Z_a - Z) - \frac{1}{H}\right], \tag{7.4.12}$$

where $H(R) = Z_a - Z_b = 1 + \frac{1+\beta}{2\beta} R^2 + O(\varepsilon)$ is the gap width in the "stretched" coordinates and P'_0 is the first derivative of P_0 with respect to R.

If the first of the equations (7.4.10) is integrated from Z_b to Z_a and the boundary conditions (7.4.11) are taken into account, then the Reynolds equation of the lubrication theory is obtained

$$R^2 P_0'' + \left[R + 3\left(1 + \frac{1}{\beta}\right)\frac{R^3}{H} \right] P_0' - P_0 = -6\left(1 - \frac{1}{\beta}\right)\frac{R^3}{H^3}. \tag{7.4.13}$$

Passing to a spherical coordinate system allows us to find the approximate expressions for the force and torque on each of the spheres. When matching procedure of the "inner" solutions with the "outer" solutions is applied (O'Neill and Stewartson, 1967), the force and torque coefficients on first sphere are found in asymptotic expansions of the small parameter ε (Jeffrey and Onishi, 1984b):

$$\frac{F_x}{6\pi\mu a U_a} = -\frac{4(2+\beta+2\beta^2)}{15(1+\beta)^3}\ln\frac{1}{\varepsilon} + A(\beta) \tag{7.4.14}$$

$$-\frac{4(16 - 45\beta + 58\beta^2 - 45\beta^3 + 16\beta^4)}{375(1+\beta)^4}\varepsilon\ln\frac{1}{\varepsilon} + O(\varepsilon),$$

$$\frac{T_y}{8\pi\mu a^2 U_a} = \frac{(4+\beta)}{10(1+\beta)^2}\ln\frac{1}{\varepsilon} + B(\beta) + \frac{(32 - 33\beta + 83\beta^2 + 43\beta^3)}{250(1+\beta)^3}\varepsilon\ln\frac{1}{\varepsilon} + O(\varepsilon) \tag{7.4.15}$$

The terms $A(\beta)$ and $B(\beta)$ are found after the matching of the "inner" with the "outer" solution, but they may be obtained by a curve fit of numerical results for small ε.

Similar results are obtained when the first sphere rotates around y axis with velocity ω_a. Then, following Jeffrey and Onishi (1984b), the force coefficient $\frac{F_x}{8\pi\mu a^2\omega_a}$ is given exactly by the right-hand side of (7.4.15), while the torque coefficient is

$$\frac{T_y}{8\pi\mu a^3 U_a} = -\frac{2\beta}{5(1+\beta)}\ln\frac{1}{\varepsilon} + C(\beta) - \frac{2(8+6\beta+33\beta^2)}{125(1+\beta)^2}\varepsilon\ln\frac{1}{\varepsilon} + O(\varepsilon). \tag{7.4.16}$$

Squeezing Motion of Rigid Particles. The second example that we shall consider treats the so-called "squeezing" flow in the gap between the two spherical particles. As in the previous case the first sphere moves with velocity $\mathbf{U}_a$ towards the second stationary sphere of radius $b = \beta a$ in (-z) direction coincident with the spheres line of centres (see Fig.7.4.2). Here, again the width of the gap between the two spheres $a\varepsilon$ is small compared with the moving sphere radius, i.e., $\varepsilon \ll 1$. Since the problem is axisymmetrical, the stream function ψ connected with the velocity components in cylindrical coordinates is introduced

$$v_r = U_a \frac{1}{r}\frac{\partial \psi}{\partial z}, \qquad v_z = -U_a \frac{1}{r}\frac{\partial \psi}{\partial r}, \qquad v_\theta = 0. \tag{7.4.17}$$

The flow governing equation in terms of the stream function reads (1.3.21)

$$E^2\left(E^2 \psi\right) = 0, \tag{7.4.18}$$

where $E^2 \equiv \frac{\partial^2}{\partial r^2} - \frac{1}{r}\frac{\partial}{\partial r} + \frac{\partial^2}{\partial z^2}$.

The boundary conditions on the spherical surfaces are, respectively:

$$\psi = \frac{1}{2} r^2, \qquad \frac{\partial \psi}{\partial z} = 0 \qquad \text{on first sphere,} \tag{7.4.19}$$

$$\psi = \frac{\partial \psi}{\partial z} = 0 \quad \text{on second sphere.}$$

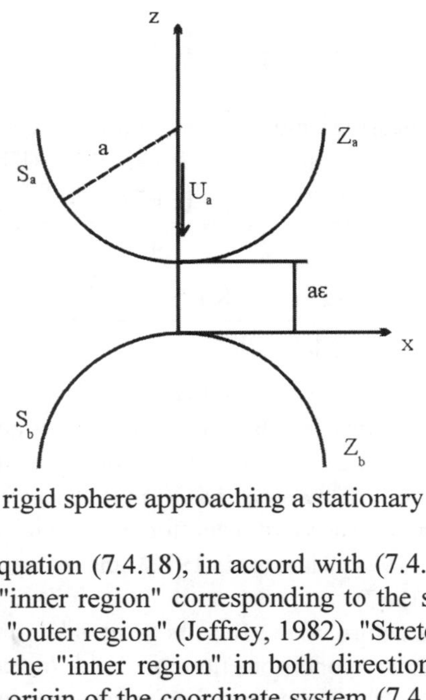

Fig.7.4.2. A rigid sphere approaching a stationary rigid sphere.

The solution of equation (7.4.18), in accord with (7.4.19), will be sought, as in the previous case, in an "inner region" corresponding to the squeezing region near the spheres contact and in an "outer region" (Jeffrey, 1982). "Stretched" coordinates of O(1) order are introduced for the "inner region" in both directions with (7.4.7). Then the spheres surfaces near the origin of the coordinate system (7.4.6) is approximated in the "stretched" coordinates by (7.4.8). Taking into account the first condition of (7.4.19), we reach to the conclusion that $\psi \sim O(\varepsilon)$ in the "inner" region, i.e.,

$$a^{-2}\psi(R,Z) = \varepsilon\, \psi_0(R,Z) + \varepsilon^2\, \psi_1(R,Z) + \ldots. \tag{7.4.20}$$

Now, replacing the initial coordinates with their "stretched" correspondents into equation (7.4.18) and ψ by its expansion on ε, we get an infinite system of partial differential equations for the successive stream functions approximations:

$$\frac{\partial^4 \psi_0}{\partial Z^4} = 0, \tag{7.4.21}$$

$$\frac{\partial^4 \psi_1}{\partial Z^4} = -2\left(\frac{\partial^2}{\partial R^2} - \frac{1}{R}\frac{\partial}{\partial R}\right)\frac{\partial^2 \psi_0}{\partial Z^2}, \ldots,$$

which are the lubrication theory equations in terms of the stream function. The boundary conditions for the first order approximation ψ_0 are exactly (7.4.19), while for the further approximations ψ_i ($i \geq 1$), they are zeros on both surfaces.

Integrating the first equation of (7.4.21), we obtain a polynomial of third degree in Z

$$\psi_0(R,Z) = A_0(R)Z^3 + B_0(R)Z^2 + C_0(R)Z + D_0(R) \tag{7.4.22}$$

with coefficients determined from the boundary conditions (7.4.19)

$$A_0 = -\frac{R^2}{H^3}, \quad B_0 = \frac{3}{2}\frac{R^2}{H^3}(Z_a + Z_b), \quad C_0 = -3\frac{R^2}{H^3}Z_a Z_b, \tag{7.4.23}$$

$$D_0 = \frac{1}{2}\frac{R^2}{H^3}(3Z_a - Z_b)Z_b^2,$$

where $H(R) = Z_a - Z_b$.

Similarly to the "shearing" motion case, the force coefficient on first sphere is calculated approximately as a function of ε (Jeffrey, 1982), (Kim and Karrila, 1991):

$$\frac{F_x}{6\pi\mu a U_a} = -\frac{\beta^2}{(1+\beta)^2}\varepsilon^{-1} - \frac{(1+7\beta+\beta^2)}{5(1+\beta)^3}\ln\frac{1}{\varepsilon} + K(\beta) \tag{7.4.24}$$

$$-\frac{(1+18\beta-29\beta^2+18\beta^3+\beta^4)}{21(1+\beta)^4}\varepsilon\ln\frac{1}{\varepsilon} + O(\varepsilon).$$

$K(\beta)$ may be determined after applying a matching procedure with the "outer" solution, or by a curve fit of the numerical results at small ε. The singularity of the leading term in the "squeezing" motion force expression is of the order of $O(\varepsilon^{-1})$, which is higher than that of the "shearing" motion of the order of $O(\ln\varepsilon^{-1})$. These results are consistent with the exact solutions in bispherical coordinates for the special cases of spheres of equal radii, i.e., $\beta = 1$ (Green, 1971; Hansford, 1970) and of sphere approaching a plane $\beta \to \infty$ (Cox and Brenner, 1967).

If the molecular structure of the fluid is to be taken into account, as in the gas problems, the no-slip boundary conditions no more hold. In such approaches a modified boundary condition is used, which accounts for the non-zero tangential fluid velocity on the body surface:

$$v_t = \frac{C_m l}{\mu}\tau_{nt},$$

where v_t is the tangential fluid velocity, τ_{nt} is shear stress, l is the mean free path and C_m is the slip coefficient, whose value depends on the force law. Extending the work of Cooley and O'Neill (1969) of two spheres in contact moving along their line of centres for the slip flow conditions, Williams (1997) uses a tangential coordinate system (ξ, η, ϕ) connected with the cylindrical coordinate system (z, ρ, ϕ) by the relation:

$$z = \frac{\xi}{\xi^2 + \eta^2}, \quad \rho = \frac{\eta}{\xi^2 + \eta^2}, \tag{7.4.25}$$

In this coordinate system the tangent spheres of radii a and b with centres on the z axis (see Fig.7.4.3) at z = -a and z = b are defined by the coordinate surfaces $\xi = -\frac{1}{2a}$ and $\xi = \frac{1}{2b}$. Then, for the touching spheres, called bisphere, after solving the Stokes equation in tangential coordinates with slip boundary condition, Williams obtains a generalised formula for the drag force:

$$F = 6\pi\,\mu\,aU(1.2902819)\left[1 - 1.1676677\,\delta + \ldots\right], \tag{7.4.26}$$

with $\delta = \frac{C_m l}{a}$. The first term of (7.4.26) at $\delta = 0$ coincides with result of Cooley and O'Neill for no-slip tangential velocity condition.

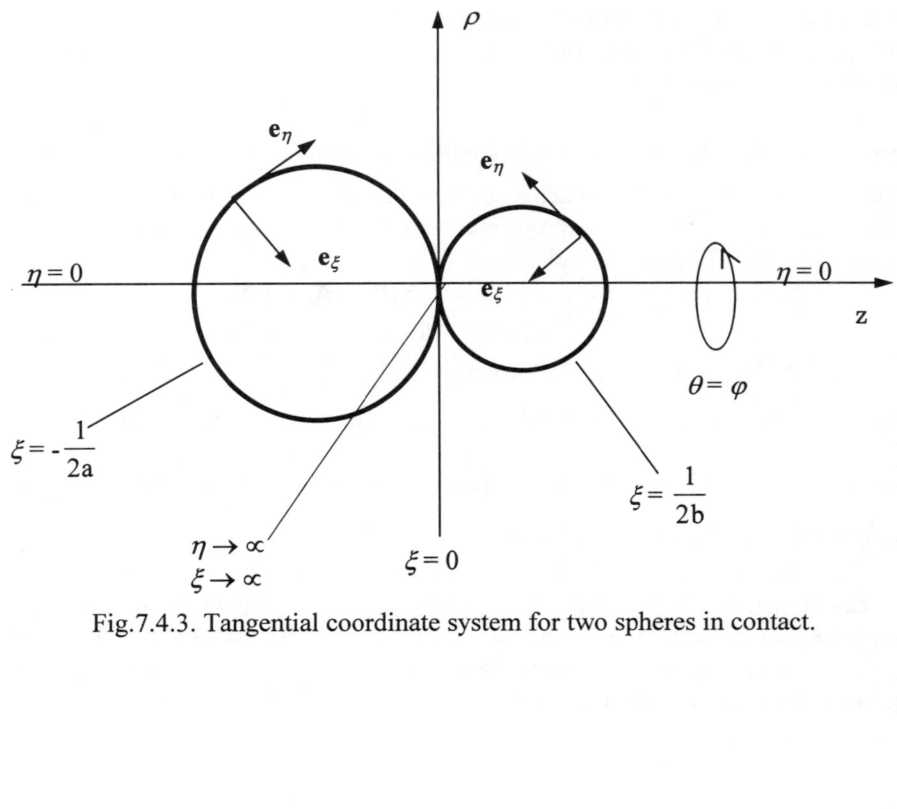

Fig.7.4.3. Tangential coordinate system for two spheres in contact.

7.5. Hydrodynamic Interactions between Two Spherical Fluid Particles

Occurrence of bubbles and drops is common in many industrial and biological systems and processes such as liquid-liquid extraction, flotation, fermentation, distillation, etc. That is why in recent years an increasing amount of fundamental and applied research has been directed to cases involving single bubbles and drops and swarms of bubbles and drops participating in various continuous moving phases. In particular, a problem of interest is that of the motion of two droplets submerged in an unbounded arbitrary velocity field flow. Its significance lies in the fact that it is a necessary first step of the solutions relevant to emulsions.

The theoretical studies of the spherical drops settling along their line of centres in a quiescent fluid started with the exact solution of Bart (1968), using a similar procedure as that of Stimson and Jeffery (1926) for solid spheres. Further, different mathematical approaches have been applied when treating the general case of different drops' velocities, radii and viscosities. The bispherical coordinates method, analogous to that of Stimson-Jeffery, was used by Wacholder and Weihs (1972), when investigating the sedimentation problem of two equal spherical fluid particles in a direction parallel to their central line. By the same method Rushton and Davies (1970, 1973), and independently of them Haber et al. (1973), give exact solutions of the two arbitrary spherical drops (bubbles) problem in quasisteady Stokes approximation for the induced flow field and for corrections to the Hadamard - Rybczynski drag force, as well. In a form equivalent to (7.2.1), the drag forces on the drops with radii a and b and velocities U_1 and U_2, respectively are:

$$F_\alpha = 6\pi\mu a U_1 \Phi_\alpha, \qquad F_\beta = 6\pi\mu b U_2 \Phi_\beta, \tag{7.5.1}$$

where Φ_α and Φ_β (Haber et al., 1973) are the correction factors to the Stokes law with $\alpha > 0$ and $\beta < 0$ defining the spheres surfaces in bispherical coordinates, respectively. For the special case of two equal drops moving towards each other with equal velocities the drags given by (7.5.1) are identical, i.e.,

$$\Phi_\alpha = \frac{4}{3}\mathrm{sh}\alpha \sum_{n=1}^{\infty} \frac{n(n+1)}{(2n-1)(2n+3)} \frac{\Delta_1 + \lambda\Delta_2}{\Delta_3 + \lambda\Delta_4}, \tag{7.5.2}$$

where $\Delta_1 = 2\{(2n+1)\mathrm{sh}2\alpha + 2\mathrm{ch}2\alpha - 2\exp[-(2n+1)\alpha]\}$,

$$\Delta_2 = \{(2n+1)^2\mathrm{ch}2\alpha + 2(2n+1)\mathrm{sh}2\alpha - (2n-1)(2n+3) + 4\exp[-(2n+1)\alpha]\}$$

$\Delta_3 = 8\mathrm{sh}(n-1/2)\alpha\,\mathrm{sh}(n+3/2)\alpha$, $\Delta_4 = 2[2\mathrm{sh}(2n+1)\alpha - (2n+1)\mathrm{sh}2\alpha]$ and $\lambda = \dfrac{\hat{\mu}}{\mu}$ is the ratio of the drop and solvent viscosities.

The drag force given by (7.5.1) is the most general one and some other special cases can be derived from it, e.g., at $\beta = 0$ (correspondent to $b \to \infty$) the Bart solution (Bart, 1968) is obtained; at $\lambda \to \infty$ and a = b we have the Stimson and Jeffery's solution (7.2.1); and when one of the drops vanishes, i.e., $\beta \to \infty$ (correspondent to b = 0), the single drop solution of Hadamard-Rybczynski (3.3.1) is obtained as a limiting case.

Due to the poor convergence of the series in (7.5.1), only asymptotic evaluations are derived by Beshkov et al. (1978) for the drag forces on two fluid drops approaching one another when the separation between them is very large or very small.

A summary of the results of the drag coefficients and interaction between two droplets moving along their central line or between a single droplet in the proximity of a plane interface is presented also by Rushton and Davies (1978). A similar problem is treated by Fuentes et al. (1988) but using the imaging techniques.

An approximate solution for the problem of two drops moving arbitrarily is presented by Hetsroni and Haber (1978). The flow field inside and outside the drops is calculated via the reflection method and the drag forces on the droplets and their terminal velocities are obtained when the drops are falling in an unbounded quiescent fluid in a gravitational field and when the drops are submerged in a Couette flow.

In all these results the drag force is in the form of an infinite series which diverges when the gap between the drops decreases, i.e., the drag force becomes unbounded. Similarly to the problem discussed in section 7.4., we shall present some results concerning both fundamental cases for the relative motion of two close drops, i.e., "squeezing" flow for the axisymmetric case and "shearing" flow for the 3D case.

In the axisymmetric case, when utilizing the lubrication theory for the flow in the gap between the drops interfaces and the boundary integral theory for the flow inside the drops, Davis et al. (1989) and in a subsequent paper Barnocky and Davis (1989) found asymptotic and numerical results for the hydrodynamic lubrication resistance force correspondent to different drops viscosities and gap widths. Here we shall present the main ones.

When two spherical drops of radii a_1 and a_2 approach one another with relative velocity $W = U_1 - U_2$ in an immiscible fluid of viscosity μ (see Fig.7.5.1.) the gap width $h = a\varepsilon$ is small compared with the reduced radius $a = a_1 a_2/(a_1 + a_2)$, i.e., $\varepsilon << 1$ is a small parameter. For the case of equal drops viscosities $\lambda\mu$, Davis et al. introduce a dimensionless parameter $m \equiv \lambda^{-1}\varepsilon^{-1/2}$, to express the ratio between the tangential velocity due to tangential stress and the pressure driven velocity. When $m << 1$ the drops interfaces are almost immobile, while at $m >> 1$ they are fully mobile. Then the parameter m is called an interface mobility parameter. For small m the drops interfaces perform as nearly rigid surfaces and the lubrication force is calculated approximately

$$F = 6\pi\mu a \frac{W}{\varepsilon}\left[1 - 1.31m + 1.78m^2 - 2.46m^3 + 3.44m^4 - 4.83m^5 + O\left(m^6\right)\right] \quad (7.5.3)$$

and is inversely proportional to the minimum gap width, i.e., $F \approx O(\varepsilon^{-1})$.

In the opposite case of large m, correspondent to fully mobile drops and the force is found to be inversely proportional to the square root of the minimum gap width:

$$F = 16.5\lambda\mu a \frac{W}{\varepsilon^{1/2}}. \quad (7.5.4)$$

Here the singularity is $O(\varepsilon^{-1/2})$ instead of $O(\varepsilon^{-1})$ for solid spheres (7.4.24) and the drops can touch in a finite time when pushed with constant force . This result is of importance for drop collision and coalescence.

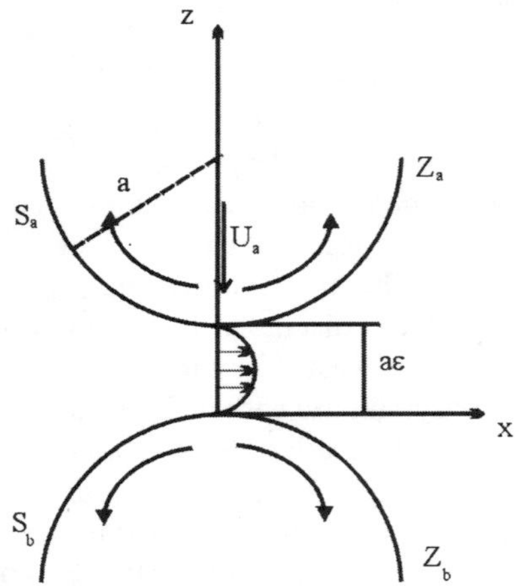

Fig.7.5.1. A fluid drop approaching another stationary fluid drop.

For drops with partially mobile interfaces, m = O(1), only numerical results for the lubrication resistance force are obtained, which can be approximated by a Padé type approximation with accuracy up to 1% or 2% for all values of m

$$F = 6\pi\mu a \frac{W}{\varepsilon}\left[\frac{1+0.38m}{1+1.69m+0.43m^2}\right]. \tag{7.5.5}$$

Further, this expression shows a good agreement with the exact results of Haber et al. (1973) at small gap widths for different viscosity ratios λ. However, for the gas bubbles case, the lubrication force is no more dominant and the near-contact results obtained by Davis et al. (1989) can not be used.

Beshkov et al. (1978) arrived to similar formulae for the drag forces on two fluid drops approaching each other when the separation between them is very small at very large drop viscosities and moderate viscosities.

In the general case of drops with different viscosities $\lambda_1\mu$ and $\lambda_2\mu$, Barnocky and Davis (1989) give two dimensionless parameters m_1 and m_2, respectively, whose ratio $m_1/m_2 = \lambda_1/\lambda_2$ is the relative viscosity between the two drops and measures approximately the ratios between the interfacial velocities of the drops and the radial velocity of the fluid in the gap. Therefore, the different combinations of the interfacial parameters m_1 and m_2 correspond to differently shaped parabolic velocity profile in the gap. If $m_2 \to 0$ the first drop approaches a rigid sphere or plate, while at $m_2 \to \infty$ it approaches a gas bubble or free surface. Some of these cases will be discussed in sections 8.2 and 8.3.

The results for the lubrication force obtained by Barnocky and Davis are derived by the same techniques by Davis et al. (1989). These near-contact results of the lubrication theory combined with the boundary integral method are in good agreement with the exact solutions of Haber et al., who used the bispherical coordinates method.

Utilizing the bispherical coordinates techniques Davis (1978) has shown that, when considering two drops at rest in a uniform flow along their line of centres, no separation from the drops interfaces occurs in contrast to the solid spheres case. However, for oil drops in water radii ratio up to 7, free toroidal eddies in a reversed flow in the gap between the drops exist.

As it was denoted, the lubrication theory is not suitable to approximate the flow in the gap between two touching bubbles. For this case it is necessary to derive the drag force results directly from the exact solution in bispherical coordinates of Haber et al. with some special attention to the series when the gap width becomes small. Following Kim and Karrila (1991) (see also Harper, 1997) we can show that the drag of two identical approaching bubbles exhibit a weaker singularity $O(\ln(\varepsilon^{-1}))$ than the presented result (7.5.4) for fully mobile drops $O(\varepsilon^{-1/2})$:

$$F = 6\pi\mu a U\left[\frac{1}{3}\ln\left(\varepsilon^{-1}\right) + \frac{2}{3}\left(\gamma + \ln 2\right) + o(1)\right], \tag{7.5.6}$$

where $\gamma = 0.577216\ldots$ is Euler's constant and U is the each bubble velocity.

The "shearing" flow case corresponds to the transverse translations of drops, where the lubrication theory is no more applicable, in spite of the small gaps width. The boundary collocation method (Kim and Karrila, 1991) gives results which are in correspondence with those obtained by an asymptotic analysis of the bispherical coordinates method by Zinchenko (1978, 1981, 1982, 1984a, 1984b).

Another analytical method providing the examination of a near-contact flow is proposed by Reed and Morrison (1974), who analyse the translational axisymmetric motion of two touching undeformable fluid spheres by use of a tangential coordinate system [see section 7.4., formula (7.4.25)]. These investigations, as well as those previously described in this chapter, refer to isothermal viscous flows in which the particles motion is governed by the buoyancy migration.

In the presence of an ambient temperature gradient the thermocapillary migration is dominating the drops motion. Utilizing the tangential coordinates method, Loewenberg and Davis (1993) describe the axisymmetric thermocapillary-driven motion of nearly touching non-conducting droplets (tangent fluid spheres) at small Reynolds and Marangoni numbers. They give a parallel analysis of a buoyancy-driven motion to show the similarities and differences in both cases. The numerical results for the pairwise migration velocity for the buoyancy-driven motion agree with those of Reed and Morrison.

The approach of Stimson and Jeffery to the solution of the Stokes equations is applied by Meyyappan et al. (1981a) for the thermocapillary migration of a gas bubble perpendicularly to an infinite interface in the absence of gravity. The bubble shape is assumed spherical and the infinite interface is considered planar. In a subsequent work Meyyappan et al. (1983) find an exact solution in a quasisteady Stokes approximation of the axisymmetric problem for two spherical bubbles motion along their line of centres induced by a constant temperature gradient acting in the same direction. An extension of this problem for the 3D case, when the bubbles central line and the direction of the temperature gradient cross at an arbitrary angle, is presented by Meyyappan and Subramanian (1984). The thermocapillary migration of two concentric fluid droplets under the action of temperature gradient is studied by Shankar et al. (1981). The case of an eccentrically placed bubble inside a drop at zero gravity is investigated by Shankar and Subramanian (1983). The motion of a spherical gas bubble inside a spherical container, when their centres coincide is examined by Mok and Kim (1987). The bubble motion is caused by gravity and a given vertical temperature gradient. With the considered model the authors obtain a general formula for the terminal velocity of the bubble motion in the container.

In connection with the study of double and triple emulsions, in the recent years the interest towards the investigation of the compound liquid-liquid and liquid-gas drops hydrodynamics has grown. The compound drops are divided (Florence and Whitehill, 1981) into three main types "A", "B" and "C". Of type "A" are these drops, which contain one large internal drop (bubble), several small internal drops (bubbles) of type "B" and a great number of internal drops (bubble) of type "C".

The study of the compound multiple drops starts with the work of Chambers and Kopac (1937) connected with the coalescence of live cell and drops of different oils. In some technological processes in medicine mixtures are used, in which some liquid or gas phase moves inside another liquid. Such mixture is observed at the artificial blood oxygenation (Li and Asher, 1973), simultaneous injection of gas and liquid through a nozzle into another liquid (Hayakawa and Shigeta, 1974), vaporisation of some volatile liquid from a less volatile liquid (Tochitani et al., 1977), rise of gas bubbles through layers of two immiscible liquids in contact (Mori, 1978), etc.

Solutions in a closed form for the problem of a quasisteady Stokes flow past drops (bubbles) partially coated with thin films are obtained by Johnson and Sadhal (1983). The translatory motion of a compound spherical drop in a viscous fluid is studied in Stokes approximation in (Sadhal and Oguz, 1985) by use of bispherical coordinates method. The exact solution is found for the case of eccentric spherical drops, when the inner drop moves relatively to the outer together with the motion of the system in the continuous phase. The performed investigations of compound drops, discussed in the review of the fluid mechanics of compound multiphase drops and bubbles (Johnson and Sadhal, 1985), refer only to fluid particles of fixed form. An extension for drops or bubbles with caps of arbitrary size is presented by Lerner and Harper (1991). They solved this problem as Stokes problem for two identical bubbles moving parallel to their line of centres in quiescent fluid and afterwards applied the inversion theorem of Harper (1983). The same method is applicable to three or more bubbles, as well as to a bubble approaching a plane interface. Lerner and Harper used two methods to calculate the boundary value problem: the reflection method, suitable for multi-bubbles case, and the eigenfunctions expansion for the stream function in bispherical coordinates, which can provide solution inside and outside of the two drops. They obtained that the interaction between the bubbles is governed by the surfactants action. If the distance between the bubbles is large or the cap size is small, in order to preserve equal drag on both bubbles the lower bubble has a higher concentration of surfactants and will rise slower than the upper bubble.

Vuong and Sadhal (1989) consider a vapour bubble partially engulfed by its own liquid when as a whole the liquid-vapour compound drop rises in a second continuous liquid phase (see Fig.7.5.2). The mechanism of evaporation is governed by the heat transfer between the liquid drop and the ambient liquid. Although the process is transient, the drop and bubble size and shape change in time, it is possible to be observed as quasi-steady in the restriction of small drop size, highly viscous continuous phase and fast viscous diffusion. Then the Stokes flow model is applicable for the compound drop translation and growth in the quiescent surrounding fluid. The surface tension on each of the interfaces is assumed to be sufficiently large to preserve the spherical shape of the surfaces. This allows the authors to use toroidal coordinate system (Happel and Brenner, 1973) to solve analytically the Stokes equation in both liquid phases. As a result the maximum drag force is achieved when the liquid and vapour volumes are almost equal.

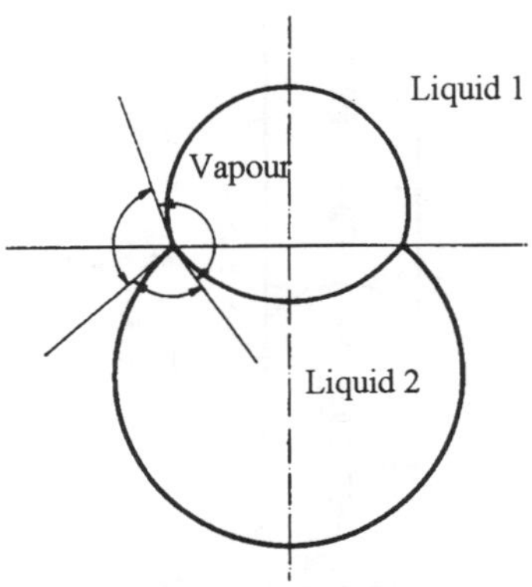

Fig.7.5.2. A partially engulfed liquid-vapour compound drop.

In the membrane extraction the main aim is to separate a given substance from the liquid, where it is situated. This is carried out by means of two liquid phases, called donor and receptor and a liquid membrane, through which the extract particle can diffuse. Moreover, the membrane liquid must be immiscible with the other two liquids. Through a liquid membrane, for example, a dilute copper solution can be extracted (Martin and Davies, 1976). In some column extractors (see Fig.7.5.3) the receptor phase 3 is dispersed in the form of drops, which passing through the membrane phase 1 are coated with a thin film of it. The created in this way compound multiphase drops continue their motion in the donor phase 2 to extract from it by means of diffusion the extraction component. The work of this type extractors depends on the buoyancy velocity of the compound drops. Therefore, the calculation of the terminal velocity and drag force acting on the compound multiphase drops has a significant importance for the design of these apparatuses. Since the presence of a membrane layer influences the correspondent viscous fluid flow, the terminal velocity and drag force cannot be calculated neither from the laws of Stokes and Hadamard - Rybczynski.

In order to analyse and evaluate approximately the terminal velocity and drag, Rushton and Davies (1983) consider a three phase system, consisting of a given number of spherical compound two phase drops (phase 3 + phase 1) moving in the continuous phase 2, in which the extract is dissolved (Fig.7.5.4). To simplify the calculations they assume that the membrane layer (phase 1) is concentric to the receptor spherical nucleus. A slow flow of the incompressible viscous fluid is considered to be induced by the buoyancy of the compound two phase spherical drop with terminal velocity U_{ter}.

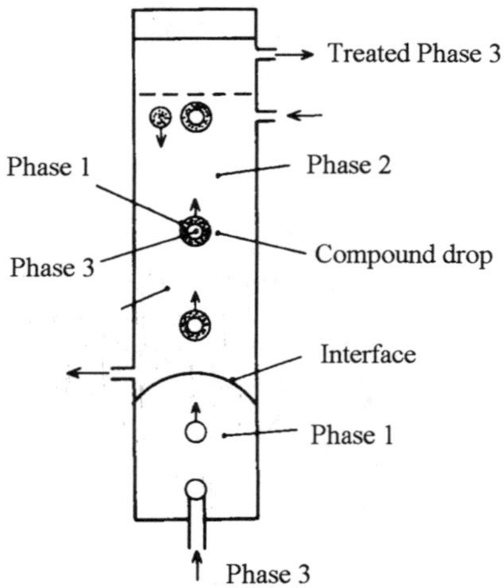

Fig.7.5.3. A column extractor (from Rushton and Davis, 1983).

The compound spherical drop with radius a is supposed to be composed of a spherical nucleus of radius $a\varepsilon$, where $0 < \varepsilon < 1$, coated by a liquid spherical membrane with thickness $a(1 - \varepsilon)$.

On the basis of the Stokes equations with the correspondent boundary conditions on the two interfaces and at infinity, Rushton and Davies obtain the following formula for the drag force due to friction:

$$F = 6\pi \mu_2 a U_{ter} \Phi , \tag{7.5.7}$$

where $\Phi = \frac{2}{3}\frac{\mu_{32} + 6\mu_{12}^2 G(\varepsilon) + \mu_{12}(2 + 3\mu_{32})F(\varepsilon)}{\mu_{32} + 4\mu_{12}^2 G(\varepsilon) + 2\mu_{12}(1 + \mu_{32})F(\varepsilon)}$, $\mu_{12} = \frac{\mu_1}{\mu_2}$, $\mu_{32} = \frac{\mu_3}{\mu_2}$,

$F(\varepsilon) = \frac{(1+\varepsilon)(2\varepsilon^2 + \varepsilon + 2)}{(1-\varepsilon)(4\varepsilon^2 + 7\varepsilon + 4)}$, $G(\varepsilon) = \frac{1-\varepsilon^5}{(1-\varepsilon)^3(4\varepsilon^2 + 7\varepsilon + 4)}$. The functions $F(\varepsilon)$ and $G(\varepsilon)$ are monotonously increasing in the interval (0, 1) and $F(0) = 1/2$, $G(0) = 1/4$ and $F(\varepsilon) \to \propto$, $G(\varepsilon) \to \propto$ as $\varepsilon \to 1$. The coefficient Φ is the correction multiplier in the Stokes drag formula and it is found from the relation $U_{ter}\Phi = U_s$, where U_s is the Stokes velocity of sedimentation. From the expression of Φ, the following relations are obtained:

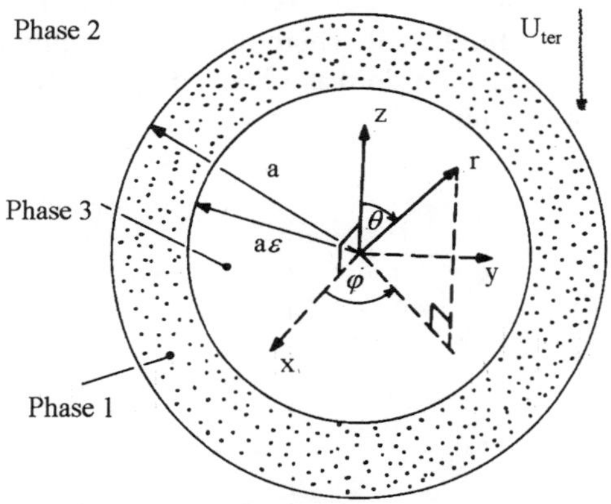

Fig.7.5.4. A compound two-phase drop.

(i) if $\mu_3 >> \mu_1$ and $\mu_3 >> \mu_2$, then

$$\Phi \to \frac{2}{3}\frac{1 + 3\mu_{12}F(\varepsilon)}{1 + 2\mu_{12}F(\varepsilon)}, \tag{7.5.8}$$

which corresponds to the correction coefficient of a rigid particle coated with a liquid membrane;

(ii) if $\mu_1 >> \mu_2$ and $\mu_1 >> \mu_3$, then the liquid membrane is solidified, i.e., $\Phi \to 1$ and $U_{ter} \to U_s$, and the terminal velocity is calculated from the Stokes drag force law using the equivalent density of the nucleus and membrane coat;

(iii) if $\varepsilon \to 0$ and μ_{12}, μ_{32} are arbitrary, the correction coefficient Φ tends to the correction coefficient Φ_f, correspondent to the Hadamard - Rybczynski's solution for a single drop with viscosity μ_1 moving in an infinite fluid of viscosity μ_2, i.e.,

$$\Phi \to \frac{2+3\mu_{12}}{3(1+\mu_{12})} = \Phi_f ; \qquad (7.5.9)$$

(iv) if $\varepsilon \to 1$, i.e., when the membrane is very thin, we have

$$\lim_{\varepsilon\to 1}\frac{F(\varepsilon)}{G(\varepsilon)} = 0 = \lim_{\varepsilon\to 1}\frac{1}{G(\varepsilon)} \text{ and } \Phi \to 1. \qquad (7.5.10)$$

Therefore, when the drop of fluid 3 is coated by a very thin membrane of fluid 1, it moves in the fluid 2 with the Stokes velocity of a rigid particle and not with the velocity of the equivalent fluid drop. This is an unexpected result, which can be explained by the motion direction of the particles pertaining to the three phases (see Fig.7.5.5). The streamlines direction in the receptor phase 3 is opposite to the streamlines direction of a single drop, moving in a viscous fluid. At $\varepsilon \approx 1$ the velocity $(v_\theta)_3$ is opposite to $(v_\theta)_2$ and thus at $\varepsilon \to 1$ the only possible value of v_θ on the interface between phase 3 and phase 2 is $v_\theta = 0$. In this way, we conclude that at $\varepsilon \to 1$ the fluid 3 in the nucleus is stationary and is coated by a stationary fluid shell. Therefore, for very thin concentric membranes the drag coefficient of a compound drop is obtained by the Stokes law instead of the Hadamard - Rybczynski's formula.

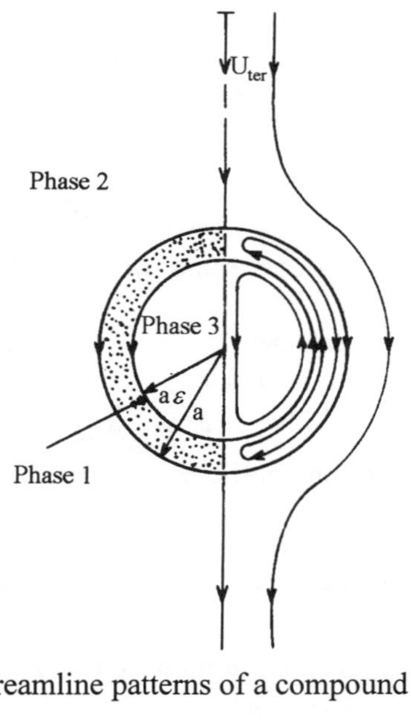

Fig.7.5.5. Streamline patterns of a compound drop motion.

In practice, thin membrane is used, in which the volume of the membrane liquid phase is minimal and as the mass transfer is performed by diffusion across the membrane, the mass flow of the extract substance is large.

7.6. Calculation of the Pressure and the Curvature of a Fluid-Fluid Interface in Bispherical Coordinates

In sections 7.7 and 7.8 we are going to consider small deformations of two drops and a compound drop at zero - Reynolds numbers. It is convenient to do this in bispherical coordinates. That is why in this section we shall work out some relations (formulae) for the pressure and curvature of a fluid/fluid interface in bispherical coordinates.

In the considered problems the following typical parameters are chosen: as a typical geometrical length, r_1, the radius of first fluid particle; as a typical velocity, U_1, the same particle velocity; as a typical pressure in each of the regions $\frac{\mu_3 U_1}{r_1}$, where μ_3 is the dynamic viscosity coefficient of the continuous phase. Therefore the quasisteady condition in the concerned problems is $\mathrm{Re} = \frac{\rho_3 U_1 r_1}{\mu_3} < \varepsilon$, where ρ_3 is the density of the continuous phase and $\varepsilon = \frac{\varepsilon'}{r_1}$ is the dimensionless minimal distance between the interfaces.

The main necessary condition for the existence of the interface S states that the normal components of the velocity v_n and $\hat{v}_n$ on either side of the interface have to be equal to the normal velocity u_n of the interface itself, i.e.,

$$v_n = \hat{v}_n = u_n \qquad \text{on S.} \tag{7.6.1}$$

Numerous experiments confirm that the fluid particles on either side of the interface S do not slip, i.e., for the tangential velocity components on either side of the interface it is necessary that:

$$v_t = \hat{v}_t \qquad \text{on S.} \tag{7.6.2}$$

Beside the velocity conditions (7.6.1) and (7.6.2), there exist stress balance conditions on the interfaces. If the surface viscosity is not taken into account, the normal stress balance (1.4.10) leads to

$$T_{nn} - \hat{T}_{nn} = \sigma\left(\frac{1}{R_1} + \frac{1}{R_2}\right) \qquad \text{on S,} \tag{7.6.3}$$

where R_1 and R_2 are the principal radii of the interface curvature.

For the tangential stresses at constant surface tension σ and ignoring the surface viscosity we find

$$T_{nt} = \hat{T}_{nt} \qquad \text{on S.} \tag{7.6.4}$$

In order to apply the condition (7.6.3), we have to calculate the curvature $\frac{1}{R_1}+\frac{1}{R_2}$ in bispherical coordinates. In tensor form the curvature could be expressed as:

$$\frac{1}{R_1}+\frac{1}{R_2}=a^{\alpha\beta}\, b_{\alpha\beta}\,, \tag{7.6.5}$$

where $a^{\alpha\beta}$ and $b_{\alpha\beta}$ are the contravariant components of the first metric tensor of the surface and the covariant components of the second metric tensor of the surface, respectively (Aris, 1989).

If $y^i = y^i(u^1,u^2)$, (i=1, 2, 3) are the Cartesian coordinates of the surface S in the Euclidean space, and u^1 and u^2 are the curvilinear coordinates of an arbitrary point lying on the interface surface, then it follows that

$$a_{\alpha\beta}=\sum_{i=1}^{3}\frac{\partial y^i}{\partial u^\alpha}\frac{\partial y^i}{\partial u^\beta}\,, \qquad a_{\alpha\beta}\, a^{\beta\gamma}=\delta_\alpha^\gamma\,, \tag{7.6.6}$$

$$b_{\alpha\beta}=\frac{\partial^2 y^i}{\partial u^\alpha\,\partial u^\beta}\, n_i\,,\; n_i=\frac{1}{2}\varepsilon^{lm}\varepsilon_{ijk}\frac{\partial y^j}{\partial u^l}\,\frac{\partial y^k}{\partial u^m}\,, \tag{7.6.7}$$

where ε_{ijk} (i, j, k =1, 2, 3) are the components of the third-rank unit alternator pseudo tensor, while ε^{lm} (l, m =1, 2) are the contravariant components of the second-rank unit alternator pseudo tensor.

In bispherical coordinate system the coordinate surfaces $\eta=\alpha_1$ and $\eta=\alpha_2$ at $\alpha_1\neq\alpha_2$ are not overlapping spheres (see Fig.7.1.1) whose centres lie on the axis Oz and $\alpha_2=-\mathrm{cth}^{-1}\frac{d_1}{\bar{c}}$, $\alpha_1=\mathrm{cth}^{-1}\frac{d_2}{\bar{c}}$, where d_i (i=1, 2) is the dimensionless distance between the centres of the spheres and the plane z=0, and $\bar{c}=c/r_1>0$ is the dimensionless focal distance.

In this section we confine ourselves solely to small deformations of the interface, i.e., its shape slightly differs from the coordinate surface $\eta_0=$ const. Then

$$\eta^H(\xi)=\eta_0+H(\cos\xi)\,, \tag{7.6.8}$$

where $\max\limits_{\xi}\left|H(\beta)\right|<1$, ($\beta=\cos\xi$). Moreover, it is assumed that the function $H(\beta)$ and its derivatives are sufficiently small, and their squares can be neglected.

In Cartesian coordinates the surface has the parametrical representation:

$$y^1=\bar{c}\frac{\sin\xi\cos\varphi}{\mathrm{ch}\eta^H-\cos\xi}\,,\qquad y^2=\bar{c}\frac{\sin\xi\sin\varphi}{\mathrm{ch}\eta^H-\cos\xi}\,,\qquad y^3=\bar{c}\frac{\mathrm{sh}\eta^H}{\mathrm{ch}\eta^H-\cos\xi}\,.$$

Introducing the notations $H_\xi=\frac{\partial H}{\partial\xi}$ and $H_{\xi\xi}=\frac{\partial^2 H}{\partial\xi^2}$, we find from (7.6.6) and (7.6.7)

$$a_{\xi\xi}\approx\frac{\bar{c}^2}{\left(\mathrm{ch}\eta^H-\cos\xi\right)^2}\,,\; a_{\varphi\varphi}=\frac{\bar{c}^2\sin^2\xi}{\left(\mathrm{ch}\eta^H-\cos\xi\right)^2}\,,\; a_{\varphi\xi}=0\,, \tag{7.6.9}$$

$$a = a_{\xi\xi}\, a_{\varphi\varphi} - a_{\varphi\xi}^2 \approx \frac{\bar{c}^4 \sin^2\xi}{\left(\mathrm{ch}\eta^{H} - \cos\xi\right)^4}, \tag{7.6.10}$$

$$b_{\xi\xi} = \frac{1}{\sqrt{a}}\left[\frac{\partial^2 y^1}{\partial\xi^2}\frac{\partial y^2}{\partial\varphi}\frac{\partial y^3}{\partial\xi} - \frac{\partial^2 y^2}{\partial\xi^2}\frac{\partial y^1}{\partial\varphi}\frac{\partial y^3}{\partial\xi} - \frac{\partial^2 y^3}{\partial\xi^2}\left(\frac{\partial y^2}{\partial\varphi}\frac{\partial y^1}{\partial\xi} - \frac{\partial y^1}{\partial\varphi}\frac{\partial y^2}{\partial\xi}\right)\right]$$

$$\approx -\frac{\bar{c}}{\left(\mathrm{ch}\eta^{H} - \cos\xi\right)^2}\left[-H_{\xi\xi}\left(\mathrm{ch}\eta^{H} - \cos\xi\right) + \left(H_{\xi}\sin\xi - \mathrm{sh}\eta^{H}\right)\right] \tag{7.6.11}$$

$$b_{\varphi\varphi} = \frac{1}{\sqrt{a}}\left[\frac{\partial^2 y^1}{\partial\varphi^2}\frac{\partial y^2}{\partial\varphi}\frac{\partial y^3}{\partial\xi} - \frac{\partial^2 y^2}{\partial\varphi^2}\frac{\partial y^1}{\partial\varphi}\frac{\partial y^3}{\partial\xi}\right] \tag{7.6.12}$$

$$\approx -\frac{\bar{c}\sin\xi}{\left(\mathrm{ch}\eta^{H} - \cos\xi\right)^2}\left[H_{\xi} - H_{\xi}\,\mathrm{ch}\eta^{H}\cos\xi - \mathrm{sh}\eta^{H}\sin\xi\right]$$

For the deformed surface curvature we have from (7.6.5) and (7.6.8)-(7.6.12):

$$\frac{1}{R_1} + \frac{1}{R_2} \approx -\frac{1}{\bar{c}}\left[-H_{\xi\xi}\left(\mathrm{ch}\eta^{H} - \cos\xi\right) + H_{\xi}\sin\xi - \mathrm{sh}\eta^{H}\right] \tag{7.6.13}$$

$$-\frac{1}{\bar{c}\sin\xi}\left[H_{\xi} - H_{\xi}\,\mathrm{ch}\eta^{H}\cos\xi - \mathrm{sh}\eta^{H}\sin\xi\right].$$

Taking into account that the deformations are small, i.e.,

$$\mathrm{ch}\eta^{H} \approx \mathrm{ch}\eta_0 + H(\xi)\,\mathrm{sh}\eta_0,$$

$$\mathrm{sh}\eta^{H} \approx \mathrm{sh}\eta_0 + H(\xi)\,\mathrm{ch}\eta_0,$$

and ignoring the second powers of the function $H(\xi)$ in (7.6.13), we reach to:

$$\frac{1}{R_1} + \frac{1}{R_2} = \pm\frac{1}{\bar{c}}\left\{\frac{\left(\mathrm{ch}\eta_0 - \cos\xi\right)^3}{\sin\xi}\frac{\partial}{\partial\xi}\left[\frac{\sin\xi}{\left(\mathrm{ch}\eta_0 - \cos\xi\right)^2}\frac{\partial H}{\partial\xi}\right] + 2H(\xi)\,\mathrm{ch}\eta_0 + 2\mathrm{sh}\eta_0\right\}. \tag{7.6.14}$$

Here the upper sign corresponds to $\eta_0 > 0$ and the lower one to $\eta_0 < 0$

The equations of motion for each of the phases in the quasisteady formulation of the problem under investigation expressed in terms of the stream function ψ have the form

$$E^2\left(E^2\psi_j^{(q)}\right) = 0, \tag{7.6.15}$$

where $E^2 \equiv \dfrac{\mathrm{ch}\eta - \beta}{\bar{c}^2}\left\{\dfrac{\partial}{\partial\eta}\left[\left(\mathrm{ch}\eta - \beta\right)\dfrac{\partial}{\partial\eta}\right] + \left(1-\beta^2\right)\dfrac{\partial}{\partial\beta}\left[\left(\mathrm{ch}\eta - \beta\right)\dfrac{\partial}{\partial\beta}\right]\right\}.$

The upper index q in (7.6.15) indicates the phase type: dispersed (q = i) or continuous (q = e). At q = i the index j has values of 1 or 2 and shows to which of the dispersion phases it refers, while at q = e, the index j equals 3 or 4 showing to which of the continuous phases it corresponds. When dealing with whatever region or interface the indexes are omitted.

The general solution of (7.6.15) as given by Stimson and Jeffery (1926) is:

$$\psi = (\mathrm{ch}\eta - \beta)^{-\frac{3}{2}} \sum_{n=1}^{\infty} U_n(\eta) V_n(\beta), \tag{7.6.16}$$

where $U_n(\eta) = A_n \mathrm{ch}\left(n - \frac{1}{2}\right)\eta + B_n \mathrm{sh}\left(n - \frac{1}{2}\right)\eta + C_n \mathrm{ch}\left(n + \frac{3}{2}\right)\eta + D_n \mathrm{sh}\left(n + \frac{3}{2}\right)\eta$

and $V_n(\beta)$ are functions connected with Legendre polynomials via the relation

$$V_n(\beta) = P_{n-1}(\beta) - P_{n+1}(\beta).$$

The constants A_n, B_n, C_n and D_n are determined from the boundary conditions (7.6.1), (7.6.2) and (7.6.4), that are rewritten as:

$$\left[(\mathrm{ch}\eta - \beta)^{\frac{3}{2}} \psi^{(e)}\right]_{\eta=\alpha_j} = -\frac{1}{2} V_j \bar{c}^2 (1-\beta^2)\left[(\mathrm{ch}\eta - \beta)^{-\frac{1}{2}}\right]_{\eta=\alpha_j}, \tag{7.6.17}$$

$$\left[(\mathrm{ch}\eta - \beta)^{\frac{3}{2}} \psi^{(i)}\right]_{\eta=\alpha_j} = -\frac{1}{2} V_j \bar{c}^2 (1-\beta^2)\left[(\mathrm{ch}\eta - \beta)^{-\frac{1}{2}}\right]_{\eta=\alpha_j}, \tag{7.6.18}$$

$$\frac{\partial \psi^{(e)}}{\partial \eta} = \frac{\partial \psi^{(i)}}{\partial \eta} \qquad \text{at } \eta = \alpha_j, \tag{7.6.19}$$

$$\sum_{n=0}^{\infty}\left[\frac{\partial^2}{\partial \eta^2} U_n^{(e)}\right]_{\eta=\alpha_j} V_n(\beta) = \lambda_j \sum_{n=0}^{\infty}\left[\frac{\partial^2}{\partial \eta^2} U_n^{(i)}\right]_{\eta=\alpha_j} V_n(\beta)$$
$$-\frac{1}{2}(1-\lambda_j) V_j \bar{c}^2 (1-\beta^2)\left[\frac{\partial^2}{\partial \eta^2}(\mathrm{ch}\eta - \beta)^{-\frac{1}{2}}\right]_{\eta=\alpha_j} \tag{7.6.20}$$

where $\lambda_j = \frac{\mu_j}{\mu^{(e)}}$ is the ratio between the viscosity of the disperse phase and the viscosity of the continuous phase and

$$V_j = \begin{cases} \pm\dfrac{U_2}{U_1} & \text{at } \eta = \alpha_1 \\ \pm 1 & \text{at } \eta = \alpha_2 \end{cases}$$

is the dimensionless velocity of the correspondent interface.

Here, the stream function ψ given by (7.6.16) is defined in different regions from the boundary conditions without using the normal stresses jump condition (7.6.3). From this condition we shall determine the shape of the correspondent interface as a first approximation of an iterative process valid for small deformations of interfaces.

The dimensionless form of equation (7.6.3) is:

$$T^{(e)}_{\eta\eta} - T^{(i)}_{\eta\eta} = \frac{1}{Ca}\left(\frac{1}{R_1} + \frac{1}{R_2}\right), \qquad (7.6.21)$$

where $Ca = \dfrac{\mu U}{\sigma}$ is the capillary number.

In order to find the normal stresses $T^{(e)}_{\eta\eta}$ and $T^{(i)}_{\eta\eta}$, acting on a given interface, it is necessary to determine the pressure p inside and outside the disperse phase, because

$$T^{(e)}_{\eta\eta} = -p^{(e)} - 2\frac{ch\eta - \beta}{\bar{c}}\left\{\frac{\partial}{\partial\eta}\left[\frac{(ch\eta-\beta)^2}{\bar{c}^2}\frac{\partial\psi^{(e)}}{\partial\beta}\right] - \frac{ch\eta-\beta}{\bar{c}^2}\frac{\partial\psi^{(e)}}{\partial\eta}\right\},$$

$$T^{(i)}_{\eta\eta} = -p^{(i)} - 2\lambda\frac{ch\eta - \beta}{\bar{c}}\left\{\frac{\partial}{\partial\eta}\left[\frac{(ch\eta-\beta)^2}{\bar{c}^2}\frac{\partial\psi^{(i)}}{\partial\beta}\right] - \frac{ch\eta-\beta}{\bar{c}^2}\frac{\partial\psi^{(i)}}{\partial\eta}\right\},$$

where $p^{(e)}$ and $p^{(i)}$ are dimensionless pressures outside and inside the fluid particles.

At slow viscous flows in quasisteady Stokes approximation the dimensionless equations for the pressure p and the stream function ψ could be given as:

$$\frac{\partial p^{(e)}}{\partial\eta} = -\frac{ch\eta-\beta}{\bar{c}}\frac{\partial\left(E^2\psi^{(e)}\right)}{\partial\beta};\ \frac{\partial p^{(e)}}{\partial\beta} = -\frac{ch\eta-\beta}{\bar{c}\,(1-\beta^2)}\frac{\partial\left(E^2\psi^{(e)}\right)}{\partial\eta}, \qquad (7.6.22)$$

$$\frac{\partial p^{(i)}}{\partial\eta} = -\frac{ch\eta-\beta}{\bar{c}}\lambda\frac{\partial\left(E^2\psi^{(i)}\right)}{\partial\beta};\ \frac{\partial p^{(i)}}{\partial\beta} = -\frac{ch\eta-\beta}{\bar{c}\,(1-\beta^2)}\lambda\frac{\partial\left(E^2\psi^{(i)}\right)}{\partial\eta}. \qquad (7.6.23)$$

It is impossible to integrate directly these equations in bispherical coordinates when implementing the stream function (7.6.16) and it will be achieved in different manner.

From (7.6.22) and (7.6.23) it follows that the pressure satisfies the Laplace's equation:

$$\Delta p = 0,$$

whose general solution in bispherical coordinates has the form:

$$p^{(e)} = \frac{\sqrt{ch\eta-\beta}}{\bar{c}^3}\sum_{n=0}^{\infty}\left[\alpha^{(e)}_n ch\left(n+\frac{1}{2}\right)\eta + \beta^{(e)}_n sh\left(n+\frac{1}{2}\right)\eta\right]P_n(\beta) + \Pi^{(e)}, \qquad (7.6.24)$$

$$p^{(i)} = \frac{\sqrt{ch\eta-\beta}}{\bar{c}^3}\lambda\sum_{n=0}^{\infty}\left[\alpha^{(i)}_n ch\left(n+\frac{1}{2}\right)\eta + \beta^{(i)}_n sh\left(n+\frac{1}{2}\right)\eta\right]P_n(\beta) + \Pi^{(i)}. \qquad (7.6.25)$$

Here $\Pi^{(e)}$ and $\Pi^{(i)}$ are arbitrary constants, The coefficients $\alpha^{(e)}_n$, $\beta^{(e)}_n$, $\alpha^{(i)}_n$ and $\beta^{(i)}_n$ are to be determined by making use of equations (7.6.22) and (7.6.23).

From (7.6.16) we have

$$E^2\psi = \frac{\sqrt{\mathrm{ch}\eta-\beta}}{\bar{c}^2}\sum_{n=1}^{\infty}\left\{V_n(\beta)\left[\frac{d^2U_n}{d\eta^2}-\frac{2\mathrm{sh}\eta}{\mathrm{ch}\eta-\beta}\frac{dU_n}{d\eta}\right.\right. \tag{7.6.26}$$

$$\left.\left.+\frac{3}{4}\frac{\mathrm{ch}\eta+3\beta}{\mathrm{ch}\eta-\beta}U_n\right]+\left(1-\beta^2\right)U_n(\eta)\left[\frac{d^2V_n}{d\beta^2}+\frac{2}{\mathrm{ch}\eta-\beta}\frac{dV_n}{d\beta}\right]\right\}.$$

In order to separate the variables η and β in (7.6.26) the following recurrence relations could be used:

$$\left(1-\beta^2\right)\frac{dV_n}{d\beta}=\frac{n(n-1)}{2n-1}V_{n-1}-\frac{(n+1)(n+2)}{2n+3}V_{n+1},\qquad \beta V_n=\frac{(n-1)}{2n-1}V_{n-1}+\frac{n+2}{2n+3}V_{n+1},$$

$$\left(1-\beta^2\right)\frac{d^2V_n}{d\beta^2}+n(n+1)V_n=0.$$

Then after some lengthy algebra (7.6.26) could be written in the form

$$E^2\psi=\frac{1}{\bar{c}^2\sqrt{\mathrm{ch}\eta-\beta}}\sum_{n=1}^{\infty}\left[a_n\mathrm{ch}\left(n+\frac{1}{2}\right)\eta+b_n\mathrm{sh}\left(n+\frac{1}{2}\right)\eta\right]V_n, \tag{7.6.27}$$

where

$$a_n=-(2n-1)A_n+(2n+3)C_n+\frac{2n(2n+3)}{2n+1}A_{n+1}-\frac{2(n+1)(2n-1)}{2n+1}C_{n-1},$$

$$b_n=-(2n-1)B_n+(2n+3)D_n+\frac{2n(2n+3)}{2n+1}B_{n+1}-\frac{2(n+1)(2n-1)}{2n+1}D_{n-1}.$$

Substituting (7.6.24) and (7.6.25) into (7.6.22) and (7.6.23) and making use of (7.6.27) we find that

$$\alpha_n=\sum_{m=1}^{n-1}\frac{2m+1}{m(m+1)}b_m+\frac{2n+1}{n}b_n+\alpha_0, \tag{7.6.28}$$

$$\beta_n=\sum_{m=1}^{n-1}\frac{2m+1}{m(m+1)}a_m+\frac{2n+1}{n}a_n+\beta_0.$$

The coefficients $\alpha_0^{(e)}$ and $\beta_0^{(e)}$ are obtained from the condition of finite pressure at infinity

$$\alpha_0^{(e)}=-\lim_{n\to\infty}\left[\sum_{m=1}^{n-1}\frac{2m+1}{m(m+1)}b_m^{(e)}+\frac{2n+1}{n}b_n^{(e)}\right], \tag{7.6.29}$$

$$\beta_0^{(e)}=-\lim_{n\to\infty}\left[\sum_{m=1}^{n-1}\frac{2m+1}{m(m+1)}a_m^{(e)}+\frac{2n+1}{n}a_n^{(e)}\right].$$

From the boundary conditions of the problem it is easy to deduce that $p^{(i)}$ inside the fluid particle is finite as $\eta\to\pm\infty$, i.e.,

$$p^{(i)}=\frac{\sqrt{\mathrm{ch}\eta-\beta}}{\bar{c}^3}\lambda\sum_{n=1}^{\infty}\left[\sum_{k=1}^{n-1}\frac{2k+1}{k(k+1)}b_k^{(i)}+\frac{2n+1}{n}b_n^{(i)}\right]\exp\left[\pm\left(n+\frac{1}{2}\right)\eta\right]P_n(\beta)+\Pi^{(i)} \tag{7.6.30}$$

The sign "+" corresponds to $\eta < 0$ and the sign "-" corresponds to $\eta > 0$. If we take into account the gravitational body force, we reach to

$$p^{(i)} = \frac{\sqrt{\mathrm{ch}\eta - \beta}}{\bar{c}^3}\lambda\sum_{n=1}^{\infty}\alpha_n^{(i)}\exp\left[\pm\left(n+\frac{1}{2}\right)\eta\right]P_n(\beta) - \frac{gr_1\gamma}{U_1^2}\bar{c}\frac{\mathrm{sh}\eta}{\mathrm{ch}\eta - \beta} + \Pi^{(i)}, \quad (7.6.31)$$

$$p^{(e)} = \frac{\sqrt{\mathrm{ch}\eta - \beta}}{\bar{c}^3}\sum_{n=0}^{\infty}\left[\alpha_n^{(e)}\mathrm{ch}\left(n+\frac{1}{2}\right)\eta + \beta_n^{(e)}\mathrm{sh}\left(n+\frac{1}{2}\right)\eta\right]P_n(\beta) - \frac{gr_1}{U_1^2}\bar{c}\frac{\mathrm{sh}\eta}{\mathrm{ch}\eta - \beta} + \Pi^{(e)}, \quad (7.6.32)$$

where $\gamma = \dfrac{\rho^{(i)}}{\rho^{(e)}}$ [for more details see Chervenivanova (1985)].

7.7. Small Deformation of Two Moving Fluid Particles in a Viscous Fluid

In this section we consider two spherical drops (bubbles) moving in an unbounded quiescent viscous fluid with velocities U_1 and U_2, respectively, along their line of centres as given by Zapryanov and Chervenivanova (1983), and Chervenivanova and Zapryanov (1985). The following assumptions are made:

i) The tangential and normal components of the velocity vectors inside and outside the fluid particles are continuous on the interfaces, i.e.,

$$\mathbf{v}^{(e)}.\boldsymbol{\tau}^{0} = \mathbf{v}^{(i)}.\boldsymbol{\tau}^{0}, \quad (7.7.1)$$

$$\mathbf{v}^{(e)}.\mathbf{n}^{0} = \mathbf{U}_j.\mathbf{n}^{0}, \quad (j=1, 2) \quad (7.7.2)$$

$$\mathbf{v}^{(i)}.\mathbf{n}^{0} = \mathbf{U}_j.\mathbf{n}^{0}, \quad (7.7.3)$$

where $\mathbf{n}^0$ and $\boldsymbol{\tau}^0$ are unit vectors directed perpendicularly and tangentially to the interfaces and the subscripts (e) and (i) indicate the fluid outside and inside the particles.

ii) The boundary conditions (7.6.3) and (7.6.4) for the normal and tangential stresses are fulfilled on both interfaces.

iii) Far from the fluid particles the fluid velocity vanishes

$$\mathbf{v}^{(e)} \to 0. \quad (7.7.4)$$

iv) The velocity vector is finite inside the drops (bubbles)

$$\mathbf{v}^{(i)} = O(1). \quad (7.7.5)$$

Making use of (7.6.16) for the stream function in the different regions we have:

a) in the continuous phase 3

$$\psi^{(e)} = (\mathrm{ch}\eta - \beta)^{-\frac{3}{2}}\sum_{n=1}^{\infty}\left[A_n\mathrm{ch}\left(n-\frac{1}{2}\right)\eta + B_n\mathrm{sh}\left(n-\frac{1}{2}\right)\eta + C_n\mathrm{ch}\left(n+\frac{3}{2}\right)\eta + D_n\mathrm{sh}\left(n+\frac{3}{2}\right)\eta\right]V_n(\beta)$$

b) in the disperse phase 1

$$\psi_1^{(i)} = (\mathrm{ch}\eta - \beta)^{-\frac{3}{2}} \sum_{n=1}^{\infty} \left[c_n^1 \exp\left(-n + \frac{1}{2}\right)\eta + d_n^1 \exp\left(-n - \frac{3}{2}\right)\eta \right] V_n(\beta)$$

c) in the disperse phase 2

$$\psi_2^{(i)} = (\mathrm{ch}\eta - \beta)^{-\frac{3}{2}} \sum_{n=1}^{\infty} \left[c_n^2 \exp\left(n - \frac{1}{2}\right)\eta + d_n^2 \exp\left(n + \frac{3}{2}\right)\eta \right] V_n(\beta).$$

Taking into account the boundary conditions given above (without the normal stresses condition) we get eight equations to find numerically the unknown coefficients A_n, B_n, C_n, D_n, c_n^1, d_n^1, c_n^2 and d_n^2. The fluid particles deformations will be determined from the boundary condition (7.6.3.) applied on both interfaces. In order to calculate the difference $T_{\eta\eta}^{(e)} - T_{\eta\eta}^{(i)}$, we use the expressions for $\psi^{(e)}$ and $\psi^{(i)}$ just found and (7.6.32) and (7.6.31) for $p^{(e)}$ and $p^{(i)}$. After some calculations we obtain

$$T_{\eta\eta}^{(e)} - T_{\eta\eta}^{(i)}\Big|_{\eta=\alpha_{2,1}} = \frac{\sqrt{\mathrm{ch}\eta - \beta}}{\bar{c}^3} \left\{ \sum_{n=0}^{\infty} \left[\left(\lambda\, \alpha_n^{(i)} \mp \frac{g r_1}{U_1^2} \bar{c}^4 (1-\gamma)\sqrt{2}(2n+1) \right.\right.\right. \tag{7.7.6}$$

$$\left. \mp \frac{3V_j \bar{c}^2 (1-\lambda)(2n+1)(n^2+n-1)}{2\sqrt{2}(2n-1)(2n+3)} \right) \exp\left[\pm \left(n + \frac{1}{2} \right)\eta \right] - (1-\lambda)\left((2n+3)\frac{dU_{n+1}^{(e)}}{d\eta} \right.$$

$$\left. \mp \frac{3V_j \bar{c}^2 (n+1)(n+2)}{2\sqrt{2}(2n+3)} \right) \exp\left[\pm \left(n + \frac{5}{2} \right)\eta \right] - \alpha_n^{(e)} \mathrm{ch}\left(n + \frac{1}{2} \right)\eta - \beta_n^{(e)} \mathrm{sh}\left(n + \frac{1}{2} \right)\eta \Bigg] P_n(\beta)$$

$$+ (1-\lambda)\left\{ \sum_{n=1}^{\infty} (2n+1)\left(U_n^{(e)} \mathrm{sh}\eta + 2\frac{dU_n^{(e)}}{d\eta} \mathrm{ch}\eta \right) P_n(\beta) \right.$$

$$\left. + \sum_{n=2}^{\infty} \left(-(2n-1)\frac{dU_{n-1}^{(e)}}{d\eta} \pm \frac{3V_j \bar{c}^2 n(n-1)}{2\sqrt{2}(2n-1)} \exp\left[\pm \left(n - \frac{3}{2} \right)\eta \right] \right) P_n(\beta) \right\}$$

$$\left. - \left(\Pi^{(e)} - \lambda\, \Pi^{(i)} \right) \bar{c}^3 \sqrt{2} \sum_{n=0}^{\infty} \exp\left[\pm \left(n + \frac{1}{2} \right)\eta \right] P_n(\beta) \right\} \Bigg|_{\eta=\alpha_{2,1}}.$$

The shape of the two drops is sought as follows:

$$\eta_1^H(\beta) = \eta_1 + H_1(\beta), \quad \eta_2^H(\beta) = \eta_2 + H_2(\beta), \quad \max_{\xi} |H(\beta)| < 1. \tag{7.7.7}$$

When applying the expression for the curvature (7.6.14) on any interface $\eta = \eta_0$ ($\eta_0 = \alpha_1$ or α_2) and from (7.6.21) we obtain the equation for the function $H(\beta)$:

$$\frac{\bar{c}^2}{Ca}\left\{\left(ch\eta_0-\beta\right)^3\frac{d}{d\beta}\left[\frac{\left(1-\beta^2\right)}{\left(ch\eta_0-\beta\right)^2}\frac{dH}{d\beta}\right]+2Hch\eta_0\right\}= \tag{7.7.8}$$

$$\sqrt{ch\eta-\beta}\left\{\sum_{n=0}^{\infty}\left[\left(\lambda\,\alpha_n^{(i)}\mp\frac{3\,\bar{c}^4}{2\sqrt{2}\pi\,\tau_0^3}F_D(2n+1)-\frac{3V_j\bar{c}^2(1-\lambda)(2n+1)\left(n^2+n-1\right)}{2\sqrt{2}(2n-1)(2n+3)}\right)\right.\right.$$

$$\exp\left[\pm\left(n+\frac{1}{2}\right)\eta\right]-(1-\lambda)\left((2n+3)\frac{dU_{n+1}^{(e)}}{d\eta}\mp\frac{3V_j\bar{c}^2(n+1)(n+2)}{2\sqrt{2}(2n+3)}\right)\exp\left[\pm\left(n+\frac{5}{2}\right)\eta\right]$$

$$\left.-\alpha_n^{(e)}ch\left(n+\frac{1}{2}\right)\eta-\beta_n^{(e)}sh\left(n+\frac{1}{2}\right)\eta\right]P_n(\beta)+(1-\lambda)\left\{\sum_{n=1}^{\infty}(2n+1)\left(U_n^{(e)}sh\eta\right.\right.$$

$$\left.\left.+2\frac{dU_n^{(e)}}{d\eta}ch\eta\right)P_n(\beta)+\sum_{n=2}^{\infty}\left(-(2n-1)\frac{dU_{n-1}^{(e)}}{d\eta}\pm\frac{3V_j\bar{c}^2n(n-1)}{2\sqrt{2}(2n-1)}\exp\left[\pm\left(n-\frac{3}{2}\right)\eta\right]\right)P_n(\beta)\right\}$$

$$\left.\left.-\left(\Pi^{(e)}-\lambda\,\Pi^{(i)}+\frac{2sh\eta}{\bar{c}\,Ca}\right)\bar{c}^3\sqrt{2}\sum_{n=0}^{\infty}\exp\left[\pm\left(n+\frac{1}{2}\right)\eta\right]P_n(\beta)\right\}\right|_{\eta=\eta_0}$$

In this equation it is taken into account that the drops move with constant velocity under the action of their weights, i.e.,

$$\frac{gr_1\left(1-\gamma\right)}{U_1^2}=-\frac{3}{4\pi\,\tau_0^3}F_D$$

where τ_0 is the dimensionless radius of the fluid sphere, i.e.,

$$\tau_0=\begin{cases}\tau_1\\ \tau_2\end{cases},\quad\left(\tau_j=\frac{r_j}{r_1},\ j=1,2\right).$$

The drag force F_D is dimensionalized with $r_1\mu_3U_1$. According to Stimson and Jeffery (1926)

$$F_D=\frac{2\sqrt{2}\pi}{\bar{c}}\sum_{n=1}^{\infty}(2n+1)\left(A_n\pm B_n+C_n\pm D_n\right),$$

where the upper sign corresponds to the case of $\eta_0>0$ and the lower sign to $\eta_0<0$.

In addition to the ordinary differential equation (7.7.8) the unknown function $H(\beta)$ has to satisfy the following two integral conditions:

$$\int_{-1}^{1}\frac{H(\beta)\,d\beta}{\left(ch\eta_0-\beta\right)^3}=0 \qquad \text{at } \max_{\beta}\left|H(\beta)\right|<<1, \tag{7.7.9}$$

which expresses the constant volume condition of the fluid particles during the deformation and

$$\int_{-1}^{1}\frac{H(\beta)\,d\beta}{\left(ch\eta_0-\beta\right)^4}=0 \qquad \text{at } \max_{\beta}\left|H(\beta)\right|<<1, \tag{7.7.10}$$

showing that the mass centre of the drops is not displaced by the deformation.

The general solution of the homogeneous equation

$$\left(\mathrm{ch}\eta_0-\beta\right)^3\frac{d}{d\beta}\left[\frac{\left(1-\beta^2\right)}{\left(\mathrm{ch}\eta_0-\beta\right)^2}\frac{dH}{d\beta}\right]+2H\,\mathrm{ch}\eta_0=0 \qquad (7.7.8')$$

has the from

$$A\left(1-\beta\,\mathrm{ch}\eta_0\right)+\hat{A}\left[\frac{1}{2}\left(\beta-\frac{1}{\mathrm{ch}\eta_0}\right)\ln\frac{1+\beta}{1-\beta}+\left(\frac{1}{\mathrm{ch}^2\eta_0}-1\right)\right],$$

where A and $\hat{A}$ are arbitrary constants. While the solution has to be finite at the points $\beta=\pm1$, the constant $\hat{A}$ equals 0 .

Hence, the solution of (7.7.8) could be written as:

$$H(\beta)=\frac{Ca}{\bar{c}^2}\left(\mathrm{ch}\eta_0-\beta\right)^{\frac{3}{2}}\left[A\frac{\left(1-\beta\,\mathrm{ch}\eta_0\right)}{\left(\mathrm{ch}\eta_0-\beta\right)^{\frac{3}{2}}}+\sum_{n=0}^{\infty}H_nP_n(\beta)\right], \qquad (7.7.11)$$

where

$$H_n=H_n^{(1)}+\Pi\,H_n^{(2)}, \qquad \Pi=\left(\Pi^{(e)}-\lambda\,\Pi^{(i)}+\frac{2\mathrm{sh}\eta_0}{\bar{c}\,Ca}\right)c^3\sqrt{2}\,.$$

We will note that the sum

$$\frac{Ca}{\bar{c}^2}\left(\mathrm{ch}\eta_0-\beta\right)^{\frac{3}{2}}\sum_{n=0}^{\infty}H_nP_n(\beta)$$

is a particular solution of (7.7.8). The constants A and Π have to be determined from the conditions (7.7.9) and (7.7.10). Substituting (7.7.11) into (7.7.9) and (7.7.10) we obtain the following equations for A and Π:

$$\sum_{n=0}^{\infty}\left(H_n^{(1)}+\Pi\,H_n^{(2)}\right)\exp\left[\pm\left(n+\frac{1}{2}\right)\eta_0\right]=0, \qquad (7.7.12)$$

$$\sqrt{2}\sum_{n=0}^{\infty}(2n+1)\left(H_n^{(1)}+\Pi\,H_n^{(2)}\right)\exp\left[\pm\left(n+\frac{1}{2}\right)\eta_0\right]=-A\,. \qquad (7.7.13)$$

If the coefficients $H_n^{(1)}$ and $H_n^{(2)}$ are known, then the constants A and Π could be defined easily from (7.7.12) and (7.7.13). When (7.7.11) is substituted in (7.7.8) and the coefficients before the Legendre polynomials are equated, the following recurrence equations for the unknown coefficients $\left\{H_n\right\}_{n=0}$ are derived:

i) for n=0

$$-\frac{1}{2}H_2+\mathrm{ch}\eta_0\,H_1-\left(\frac{5}{2}-2\mathrm{ch}^2\eta_0\right)H_0=\left[\lambda\,\alpha_0^{(i)}\pm\frac{3\,\bar{c}^4}{2\sqrt{2}\pi\,\tau_0^3}F_D+\Pi\right]\exp\left(\pm\frac{\eta_0}{2}\right)-\alpha_0^{(e)}\mathrm{ch}\frac{\eta_0}{2}$$
$$-\beta_0^{(e)}\mathrm{sh}\frac{\eta_0}{2}+(1-\lambda)\left[-3\frac{dU_1^{(e)}}{d\eta}\pm\frac{V_j\bar{c}^2}{\sqrt{2}}\left(\exp\left(\pm\frac{5\eta_0}{2}\right)-\exp\left(\pm\frac{\eta_0}{2}\right)\right)\right] \quad (7.7.14)$$

ii) for n=1

$$-\frac{3}{2}H_3+4\mathrm{ch}\eta_0\,H_2-\frac{7}{2}H_1+\mathrm{ch}\eta_0H_0=\left(\lambda\,\alpha_1^{(i)}\pm\frac{9\,\bar{c}^4}{2\sqrt{2}\pi\,\tau_0^3}F_D+\Pi\right)\exp\left(\pm\frac{3\eta_0}{2}\right)$$
$$-\alpha_1^{(e)}\mathrm{ch}\frac{3\eta_0}{2}-\beta_1^{(e)}\mathrm{sh}\frac{3\eta_0}{2}+(1-\lambda)\left[3\left(\mathrm{sh}\eta_0U_1^{(e)}+2\mathrm{ch}\eta_0\frac{dU_1^{(e)}}{d\eta}\right)\right.$$
$$\left.-5\frac{dU_2^{(e)}}{d\eta}\pm\frac{18V_j\bar{c}^2}{10\sqrt{2}}\left(\exp\left(\pm\frac{7\eta_0}{2}\right)-\exp\left(\pm\frac{3\eta_0}{2}\right)\right)\right]. \quad (7.7.15)$$

iii) for n=2

$$-3H_4+9\mathrm{ch}\eta_0H_3-\left(\frac{11}{2}+4\mathrm{ch}^2\eta_0\right)H_2+4\mathrm{ch}\eta_0H_1-\frac{1}{2}H_0=\left(\lambda\,\alpha_2^{(i)}\pm\frac{5\,\bar{c}^4}{2\sqrt{2}\pi\,\tau_0^3}F_D+\Pi\right)$$
$$\times\exp\left(\pm\frac{5\eta_0}{2}\right)-\alpha_2^{(e)}\mathrm{ch}\frac{5\eta_0}{2}-\beta_2^{(e)}\mathrm{sh}\frac{5\eta_0}{2}+(1-\lambda)\left\{5\left(\mathrm{sh}\eta_0U_2^{(e)}+2\mathrm{ch}\eta_0\frac{dU_2^{(e)}}{d\eta}\right)\right.$$
$$\left.-5\frac{dU_1^{(e)}}{d\eta}-7\frac{dU_3^{(e)}}{d\eta}\pm\frac{3V_j\bar{c}^2}{\sqrt{2}}\left[\frac{6}{7}\exp\left(\pm\frac{7\eta_0}{2}\right)-\frac{5}{7}\exp\left(\pm\frac{5\eta_0}{2}\right)+\exp\left(\pm\frac{\eta_0}{2}\right)\right]\right\} \quad (7.7.16)$$

iv) for n≥3

$$-\frac{(n+1)(n+2)}{4}H_{n+2}+(n+1)^2\mathrm{ch}\eta_0H_{n+1}-\left[\frac{n^2+(n+1)^2+9}{4}+(n-1)(n+2)\mathrm{ch}^2\eta_0\right]H_n$$
$$+n^2\mathrm{ch}\eta_0H_{n-1}-\frac{n(n-1)}{4}H_{n-2}=\left(\lambda\,\alpha_n^{(i)}\pm\frac{2\,\bar{c}^4}{2\sqrt{2}\pi\,\tau_0^3}(2n+1)F_D+\Pi\right)\exp\left[\pm\left(n+\frac{1}{2}\right)\eta_0\right]$$
$$-\alpha_n^{(e)}\mathrm{ch}\left(n+\frac{1}{2}\right)\eta_0-\beta_n^{(e)}\mathrm{sh}\left(n+\frac{1}{2}\right)\eta_0+(1-\lambda)\left\{(2n+1)\left(\mathrm{sh}\eta_0U_n^{(e)}+2\mathrm{ch}\eta_0\frac{dU_n^{(e)}}{d\eta}\right)\right.$$
$$-(2n-1)\frac{dU_{n-1}^{(e)}}{d\eta}-(2n+3)\frac{dU_{n+1}^{(e)}}{d\eta}\pm\frac{3V_j\bar{c}^2}{2\sqrt{2}}\left[\frac{n(n-1)}{2n-1}\exp\left[\pm\left(n-\frac{3}{2}\right)\eta_0\right]\right.$$
$$\left.\left.-\frac{2(2n+1)(n^2+n-1)}{(2n-1)(2n+3)}\exp\left[\pm\left(n+\frac{1}{2}\right)\eta_0\right]+\frac{(n+1)(n+2)}{(2n+3)}\exp\left[\pm\left(n+\frac{5}{2}\right)\eta_0\right]\right]\right\} \quad (7.7.17)$$

For the sake of brevity let us denote with $F\psi$ the sum of the terms in the right-hand sides of (7.7.14)-(7.7.17) which depend on ψ and with $F\Pi$ the sum of the terms containing the unknown constant Π. Then the recurrence equations (7.7.14)-(7.7.17) could be written as:

$$-\frac{1}{2}H_2+\mathrm{ch}\eta_0\, H_1-\left(\frac{5}{2}-2\mathrm{ch}^2\eta_0\right)H_0=F\psi_0+F\Pi_0, \qquad (7.7.18)$$

$$-\frac{3}{2}H_3+4\mathrm{ch}\eta_0\, H_2-\frac{7}{2}H_1+\mathrm{ch}\eta_0 H_0=F\psi_1+F\Pi_1, \qquad (7.7.19)$$

$$-3H_4+9\mathrm{ch}\eta_0 H_3-\left(\frac{11}{2}+4\mathrm{ch}^2\eta_0\right)H_2+4\mathrm{ch}\eta_0 H_1-\frac{1}{2}H_0=F\psi_2+F\Pi_2, \qquad (7.7.20)$$

$$-\frac{(n+1)(n+2)}{4}H_{n+2}+(n+1)^2\mathrm{ch}\eta_0 H_{n+1}-\left[\frac{n^2+(n+1)^2+9}{4}+(n-1)(n+2)\mathrm{ch}^2\eta_0\right]H_n$$

$$+n^2\mathrm{ch}\eta_0 H_{n-1}-\frac{n(n-1)}{4}H_{n-2}=F\psi_n+F\Pi_n. \qquad (7.7.21)$$

Since the function $H(\beta)$ is limited at $\beta = 1$, it follows that $\{H_n\}_{n=0}^{\infty}\to 0$ at $n\to\infty$. If the recurrence equation (7.7.21) is considered, its right-hand side is of the order $O\left[\exp\left(-n|\eta_0|\right)\right]$. Dividing (7.7.21) by $(n+1)(n+2)$ we reach to the following equation valid for sufficiently large $n\geq N$:

$$-\frac{1}{4}H_{n+2}+\mathrm{ch}\eta_0 H_{n+1}-\left(\frac{1}{2}+\mathrm{ch}^2\eta_0\right)H_n+\mathrm{ch}\eta_0 H_{n-1}-\frac{1}{4}H_{n-2}=0. \qquad (7.7.22)$$

In the course of the solution it is convenient to split the infinite system (7.7.18)-(7.7.21) into two parts:

1) finite system for n = 0, 1, 2, ..., N;
2) infinite system for n = N+1, N+2, ...

Using the condition $\{H_n\}_{n=N+1}^{\infty}\to 0$ at $n\to\infty$ the equation (7.7.22) can be solved. The coefficients H_{n-1}, H_n, H_{n+1} and H_{n+2} are expressed by means of two constants a and b. Substituting H_{n-1}, H_n, H_{n+1} and H_{n+2} into (7.7.18)-(7.7.21) we can calculate the constants a, b and the coefficients H_0, H_1, ..., H_{N-2}. The system of equations (7.7.22) has constant coefficients and its solution could be expressed by the solution of the equation:

$$x^4-4\mathrm{ch}\eta_0\, x^3+\left(2+4\mathrm{ch}^2\eta_0\right)x^2-4\mathrm{ch}\eta_0\, x+1=0. \qquad (7.7.23)$$

When solving this equation we find

$$x_{1,2}=\exp(\eta_0) \text{ and } \quad x_{3,4}=\exp(-\eta_0).$$

Consequently, the solution of the system (7.7.22) has the form

$$H_n=(a+b\,n)\exp\left(-n|\eta_0|\right)+(c+d\,n)\exp\left(n|\eta_0|\right) \qquad \text{at } n\geq N-1. \qquad (7.7.24)$$

Then, in view of the condition $\lim_{n\to\infty}H_n=0$, we obtain d = c = 0. The coefficients found from (7.7.24) are substituted in (7.7.21) and H_{N-2} are calculated as functions of a and b. Proceeding in this manner all the coefficients H_{N-3}, H_{N-4}, ..., H_2, H_1, H_0 could be expressed from (7.7.21) and (7.7.20) by a and b. Furthermore, these coefficients are replaced in (7.7.19) and (7.7.18) and an algebraic system of two equations is obtained, which proves to be linear-dependent. Since only a particular solution is necessary, we can simply put a = 0.

On the right-hand sides of the recurrence equations (7.7.18)-(7.7.21) the constant Π participates on first power. Hence, according to the described algorithm we can find $\left\{H_n^{(1)}\right\}_{n=0}^{\infty}$ and $\left\{H_n^{(2)}\right\}_{n=0}^{\infty}$, and afterwards find $\left\{H_n\right\}_{n=0}^{\infty}$ as a linear function of Π:

$$H_n = H_n^{(1)} + \Pi\, H_n^{(2)}.$$

The constant Π is determined from equation (7.7.12).

It is noteworthy that if the exact solution of (7.7.23) is compared with the solution of the homogeneous system (7.7.18)-(7.7.21) (zero right-hand sides), a coincidence of up to 10^{-6} is obtained. This shows that the proposed method is very accurate.

Further some numerical results concerning the deformation of the fluid particles will be given in order to illustrate the model described. In colloid chemistry the case of drops (of same or different fluids) approaching one another under the action of buoyancy are often considered. The value of the ratio l_d (l_{d1} and l_{d2} for the respective drops) of the vertical and horizontal deformation with respect to unity shows whether the deformation corresponds to elongation ($l_d > 1$) or to flattening ($l_d < 1$). Then, as it is shown in Fig.7.7.1., the fluid particles are flattened ($l_{d1} = l_{d2} = 0.67$) and between them an initial stage of "dimpling" takes place (see Frankel and Mysels, 1962; Platikanov, 1964; Scheludko, 1967).

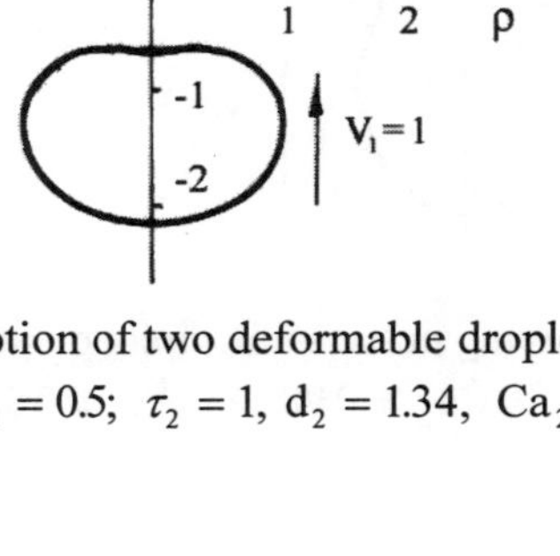

Fig.7.7.1. A relative motion of two deformable droplets for $\tau_1 = 1$, $d_1 = 1.34$, $Ca_1 = 0.28$, $\lambda_1 = 0.5$; $\tau_2 = 1$, $d_2 = 1.34$, $Ca_2 = 0.3$, $\lambda_2 = 1$.

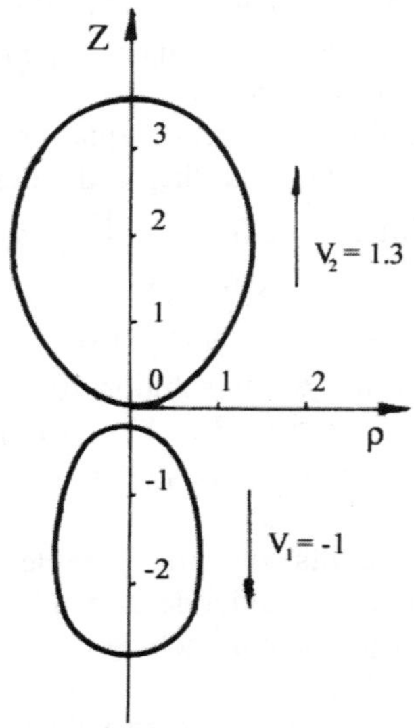

Fig.7.7.2. A relative motion of two deformable droplets for
$\tau_1 = 1$, $d_1 = 1.55$, $Ca_1 = 0.36$, $\lambda_1 = 1.5$; $\tau_2 = 1.5$, $d_2 = 1.9$, $Ca_2 = 0.24$, $\lambda_2 = 0$.

Here, the drops viscosities ratios with respect to the continuous phase viscosity are denoted by $\lambda_1 = \mu_1/\mu_3$, $\lambda_2 = \mu_2/\mu_3$ and the dimensionless distances from the drops centres to the plane z = 0 are indicated by d_1 and d_2, respectively.

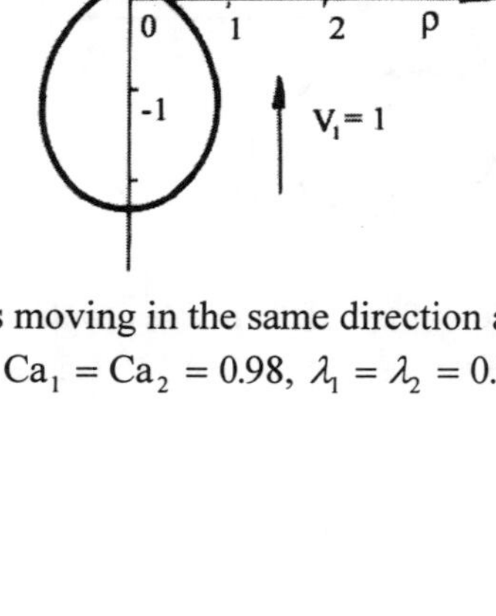

Fig.7.7.3. Two liquid drops moving in the same direction at $\tau_1 = \tau_2 = 1$, $d_1 = d_2 = 1.2$, $Ca_1 = Ca_2 = 0.98$, $\lambda_1 = \lambda_2 = 0.5$.

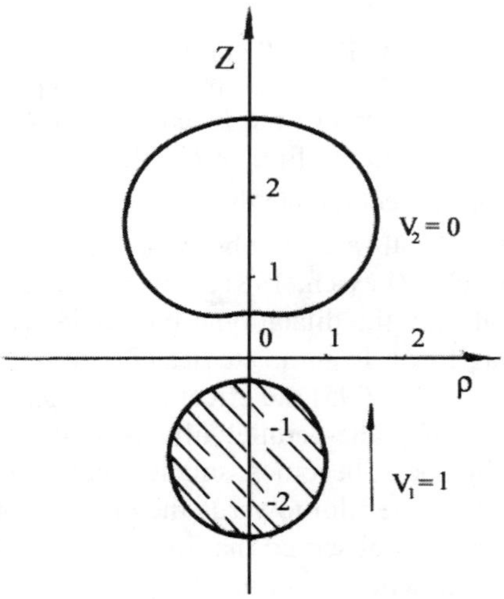

Fig.7.7.4. Solid spherical particle approaching a deformable drop for $\tau_1 = 1$, $d_1 = 1.5$; $\tau_2 = 1.5$, $d_2 = 1.7$, $Ca_2 = 0.16$, $\lambda_2 = 1.5$.

In Fig.7.7.2. the shape deformation of a bubble and drop receding from each other is plotted. The two particles interaction leads to "sucking" and elongation ($l_{d1} = 1.65$, $l_{d2} = 1.3$).

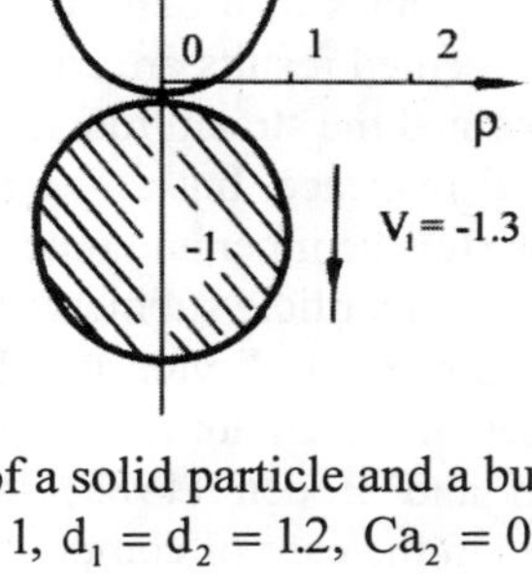

Fig.7.7.5. A relative motion of a solid particle and a bubble in opposite directions for $\tau_1 = \tau_2 = 1$, $d_1 = d_2 = 1.2$, $Ca_2 = 0.1$, $\lambda_2 = 0$.

If two fluid particles of the same fluid travel after one another, they are deformed differently (see Fig.7.7.3.): the leading drop is flattened ($l_{d2} = 0.7$) and an initial stage of "dimple" formation takes place, while the second drop elongates ($l_{d1} = 1.4$), due to the "sucking" effect of the first one.

The particular case of a solid particle approaching or receding from a deformable fluid particle is of a special interest. Fig.7.7.4. presents the approach of a solid sphere to a quiescent drop, while Fig.7.7.5. shows the receding of a solid sphere from a bubble moving in the opposite direction. In the first case a flattening ($l_{d2} = 0.75$) and "dimple" formation is observed and in the second one there is an elongation ($l_{d2} = 1.33$).

The effects of interfacial and Gibbs viscosity essentially affect the drag coefficient of the fluid particles (Levich, 1981; Edwards et al., 1991; He et al., 1991). These authors establish that only the dilatational interfacial viscosity and not the shear viscosity influences the drag force. In some studies (Zapryanov et al., 1983; Tambe and Sharma, 1991; Kralchevsky et al., 1995) the effect of the interfacial viscosity on the rate of thinning and the lifetime of plane-parallel film is studied. This effect is found to decrease when the film thickness becomes smaller and/or the film radius becomes larger. The mass transfers from the film towards the droplets, and the reverse, from the droplets towards the film and it is observed that in both cases, the diffusion fluxes lead to stabilisation of the film (Kralchevsky et al., 1997). In some cases destabilisation of the film is observed which can be attributed to the kinetic instability (Danov et al., 1988). The latter manifests itself through the growth of capillary waves at the interface, which eventually can lead to film rupture.

7.8. Small Deformations of a Compound Drop Moving in a Viscous Fluid

The problems of behaviour and deformation of the compound liquid-liquid and liquid-gas drops surrounded by a third fluid are important in many fields of technology. The study of the deformation of a drop radially moving in a spherical container filled with a viscous fluid at Stokes approximation (Chervenivanova and Zapryanov, 1987) can be regarded as an introductory work on compound, drops, although only two phases are considered. Following the subsequent paper of Chervenivanova and Zapryanov (1989) we shall here consider the shape changes of a deformable compound multiphase drop moving in a viscous fluid. The presented results are continuation of the investigation of Sadhal and Oguz (1985), who obtained the stream function in bispherical coordinates, as well as the drag on each spherical interface, but at the assumption that the compound drop conserves its spherical shape (for the inner and outer drop).

We shall recall first that a fluid particle submerged in an uniform viscous flow in Stokes approximation tends to be spherical (Taylor and Acrivos, 1964). Let us consider a compound drop, the fluid phases of which are moving in an infinite viscous medium with constant velocities. Brunn and Roden (1985) have proved that there is no deformation of the fluid interfaces in the case of compound concentric multiphase drops in an uniform flow in Stokes approximation. That is why we consider the deformation only of compound eccentric multiphase drops (for the case of concentric compound multiphase drops in shear Stokes flow see Stone and Leal, 1990).

We shall note that the deformation of the fluid interfaces depends essentially on the position of the inner droplet within the outer drop. If the minimum separation distance, ε , between the two fluid-fluid interfaces of the compound drop is not small, then their deformation is very weak. As the minimum separation distance decreases, the fluid-fluid interface interactions begin to influence gradually the shape of the inner and outer drops.

The method of finding the deformation of fluid interfaces in bispherical coordinates is the same as that worked out by Chervenivanova (1985) and applied to the deformation of two external to one another drops, presented in the previous sections 7.6. and 7.7. In view of this, here we shall consider only some results.

In order to investigate the deformation of a droplet of radius r_1 (before deformation) shifting radially inside a spherical drop of radius r_2 (before deformation), which is moving in a homogeneous, incompressible, viscous fluid (see Fig.7.8.1) we shall consider two cases: (i) the fluid phases of the compound drop are moving with constant velocities; (ii) the motion of the fluid phases is due to a constant body force.

The density and dynamic viscosity of the continuous phase are denoted by ρ_2 and μ_2 , while the densities and dynamic viscosities of the dispersed phases are expressed by ρ_1, μ_1 and ρ_3, μ_3 , respectively. The velocity of the inner drop is indicated by U_1 and the velocity of the outer drop by U_2. Both velocities are assumed parallel and the fluid particles are located eccentrically along the axis of symmetry (see Fig.7.8.1).

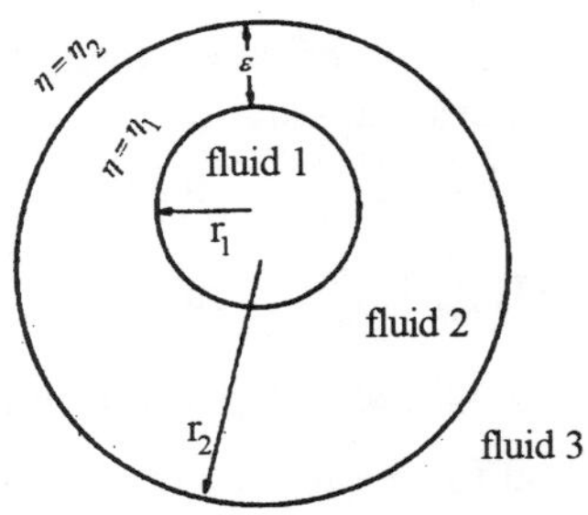

Fig.7.8.1. A sketch of a compound two-phase drop in eccentric configuration.

With r_2 as a typical length, U as a characteristic velocity and $\mu_2 U/r_2$, as a typical pressure, we have the following dimensionless parameters: the Reynolds number $Re = \rho_2 U r_2/\mu_2$, the capillary numbers $Ca_1 = \mu_1 U/\sigma_1$, $Ca_2 = \mu_2 U/\sigma_2$ and the viscosity ratios $\lambda_1 = \mu_1/\mu_2$, $\lambda_2 = \mu_3/\mu_2$, where σ_1 and σ_2 are the surface tensions of the two fluid-fluid interfaces. In the constant velocity formulation of the problem, the characteristic velocity U is one of the velocities U_1 or U_2, while in body force formulation the velocity is taken to be $U = (r_2)^2\rho_2 g/\mu_2$, where g is the acceleration due to gravity. It is also assumed that the fluid motion in each phase is very slow, i.e., Re is almost zero.

In the compound drop there are two interfaces S_1 and S_2, between the phases 1 and 3 (inner interface) and between the phases 3 and 2 (outer interface), respectively. In bispherical coordinate system (η, ξ, φ) the coordinate surface $\eta = \eta_1$=const. < 0 describes the inner interface and $\eta = \eta_2$ =const.< 0, the outer interface ($\eta_1 < \eta_2$).

The velocity conditions are (7.6.1) for the normal velocities components and (7.6.2) for the tangential velocity components, applied on each of them. The tangential stresses on either side of each interface are continuous at constant surface tensions and in the absence of surface viscosities (7.6.4). The normal stresses suffer jumps across the inner and outer interfaces:

$$T_{nn}^3 - T_{nn}^1 = \frac{1}{Ca_1}\left(\frac{1}{R_{11}} + \frac{1}{R_{21}}\right) \text{ on } S_1, \tag{7.8.1}$$

$$T_{nn}^2 - T_{nn}^3 = \frac{1}{Ca_2}\left(\frac{1}{R_{12}} + \frac{1}{R_{22}}\right) \text{ on } S_2, \tag{7.8.2}$$

where R_1 and R_2 are the principle radii of the deformed interfaces in dimensionless form and T_{nn}^i (i = 1, 2, 3) are the dimensionless normal components of the normal stress vectors. Moreover, the velocity inside the inner drop is finite, while that of the continuous phase vanishes at infinity.

The flow field of the problem described above has been investigated in detail by Sadhal and Oguz (1985). The stream function solutions in bispherical coordinates for each phase is given by the expression (7.6.16), with coefficients to be defined from the boundary conditions as stated above.

As only small deformations are considered, then, the shapes of the two interfaces S_1 and S_2 written in bispherical coordinates are:

$$\eta_1^H(\beta) = \eta_1 + H_1(\beta), \qquad \max_\xi |H_1(\beta)| < 1 \text{ on } S_1, \tag{7.8.3}$$

$$\eta_2^H(\beta) = \eta_2 + H_2(\beta), \qquad \max_\xi |H_2(\beta)| < 1 \text{ on } S_2, \tag{7.8.4}$$

where $\beta = \cos\xi$, $H_1(\beta)$ and $H_2(\beta)$ are responsible for the deformation of the two fluid-fluid interfaces. The method for finding the unknown functions $H_1(\beta)$ and $H_2(\beta)$ is similar to that presented for two external drops in section 7.7. The correspondent ordinary differential equations for these functions are similar to (7.7.8) (for more details see Chervenivanova and Zapryanov, 1989) and their solutions are given by formula (7.7.11).

Since the deformations of the two interfaces in a compound drop are due to the normal stress jump $|\Delta T_{\eta\eta}^i|$ (i = 1, 2) across the interfaces, it is convenient to use this quantity as a measure of deformation. The mechanism of the influence of normal stress differences upon the deformation of the fluid interfaces was investigated by Chervenivanova and Zapryanov (1985, 1987, 1988, 1989). In the subsequent typical examples of compound multiphase drops, we shall consider small deformations, which does not necessary mean small capillary numbers. For their part, the deformations depending on the mutual influence of the interfaces may be small at $Ca_i > 1$ and not small at $Ca_i < 0.5$. Since $\max_{\xi} |H_i(\beta)| = H_i(-1)$, for all presented examples (Chervenivanova and Zapryanov, 1989) the capillary numbers are such that $|H_i(-1)| < 0.6$.

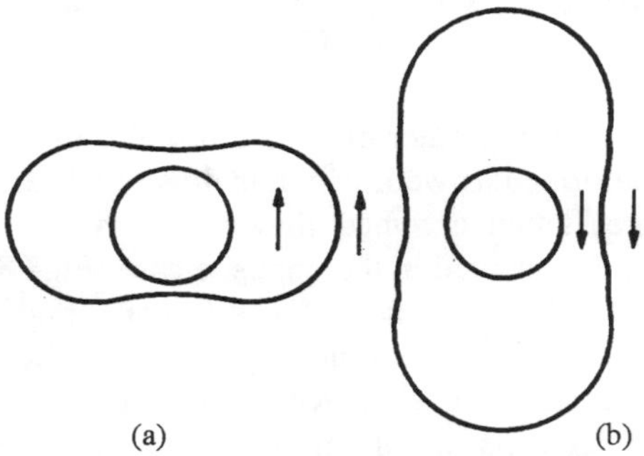

Fig.7.8.2. Shapes of two compound drops with opposite, but equal in magnitude velocities for $\mu_3 = 10\mu_2 = \mu_1$, $r_1 = 0.5$, $\varepsilon = 0.45$, any Ca_1, $Ca_2 = 13.5$ and a) $U_1 = U_2 = 1$; b) $U_1 = U_2 = -1$.

In order to discuss the influence of the double interactions between two fluid-fluid interfaces in a compound drop, we consider the two different classes of flows, as mentioned at the beginning of the present section .

When the droplets inside the compound drop move with constant and parallel velocities, we shall focus on several examples with different relative velocities of the inner drops, different dimensionless minimal distance between the interfaces ε, where $\varepsilon = \dfrac{\varepsilon'}{r_2}$ and different location of the inner drop inside the outer one.

In Fig.7.8.2. and Fig.7.8.3. the shapes of two compound multiphase drops with velocities in opposite directions are shown. Then, the relative velocity of the inner drop is zero. Since the distance between the centres of the drops is very small, the shape of the inner drop is almost spherical. For capillary numbers greater than 10 (Fig.7.8.2.a), the shape of the outer drop is flattened with two "dimples", when the motion is towards the minimum distance ε. For the opposite velocity direction case (Fig.7.8.2.b) the outer drop is elongated by a weak pinching in the middle of the fluid particle.

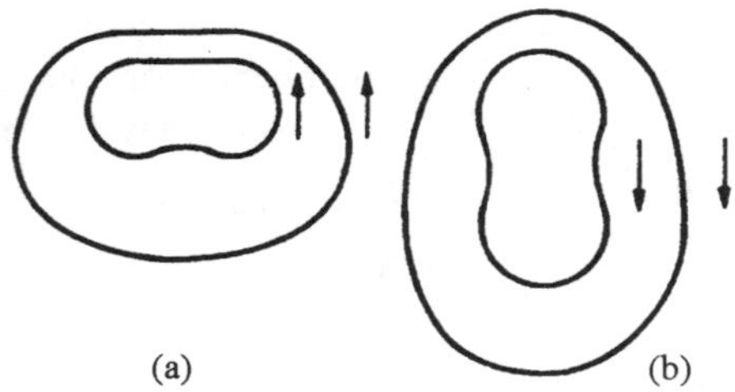

Fig.7.8.3. Shapes of two compound drops with opposite, but equal in magnitude velocities for $\mu_3 = \mu_2 = 10\mu_1$, $r_1 = 0.5$, $\varepsilon = 0.25$, $Ca_1 = 9.6$, $Ca_2 = 1.4$ and a) $U_1 = U_2 = 1$; b) $U_1 = U_2 = -1$.

When the centres of the inner and outer drops are not close sufficiently, i.e., when ε is small, the inner drop is deformed as well, which is illustrated in Fig.7.8.3. The shapes of both drops interfaces are flattened, when they move in the direction of minimum separation (Fig.7.8.3.a) and elongated in the opposite case (Fig7.8.3.b). The deformation of the inner drop appears at $Ca_1 > 15$ and at $Ca_1 < 9$ there is no dimple on the rear part of the inner drop at flattening and no pinching in its middle at elongation. For the presented example, no dimpling is observed on the rear part of the outer drop at flattening, nor any pinching in its middle at elongation. This result indicates that this kind of deformation is a typical consequence of the influence of one drop upon the other at a particular distance ε.

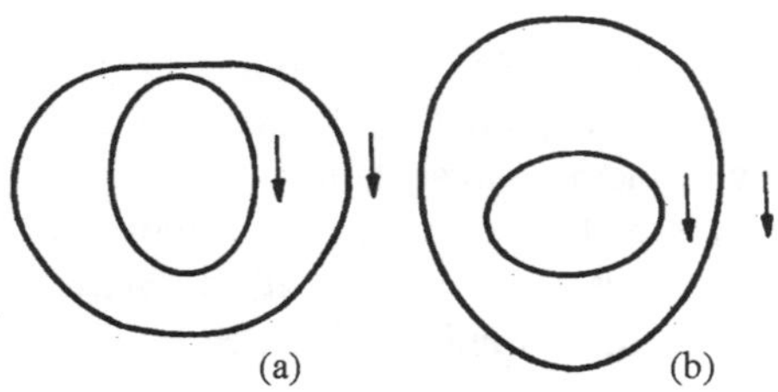

Fig.7.8.4. Shapes of two compound drops with unidirectional, but unequal in magnitude velocities for $\mu_3 = 0.5\mu_2 = 0.5\mu_1$, $r_1 = 0.5$, $\varepsilon = 0.4$, $Ca_1 = 1$, $Ca_2 = 3$, $U_1 = -1.2$, $U_2 = -1$ and a) the inside drop is located in the upper part of the compound drop; b) the inside drop is located in the lower part of the compound drop

Some examples with relative motion between the inner and outer interfaces of a compound drops are shown in Fig.7.8.4. and Fig.7.8.5. When the relative velocity of the inner drop has the same direction as the interfacial velocities, one of the drops is elongated, while the other is flattened. The type of deformation depends on the location of the inner drop inside the outer one. If it is situated in the upper part of the compound drop and its relative motion is downwards directed, then it is elongated and the outer drop is flattened (Fig.7.8.4.a). If the inner drop is placed in the lower part of the compound drop with the same relative motion, then it is flattened, while the outer drop is elongated (Fig.7.8.4.b). In Fig.7.8.5. the relative motion of the inner drop is reversely directed to the motion of the entire compound drop. Comparing the obtained shapes, flattened inner drop and elongated outer drop, with that illustrated in Fig.7.8.4.a, it can be seen that the change of the relative velocity direction leads to a radical change of the shapes of both interfaces.

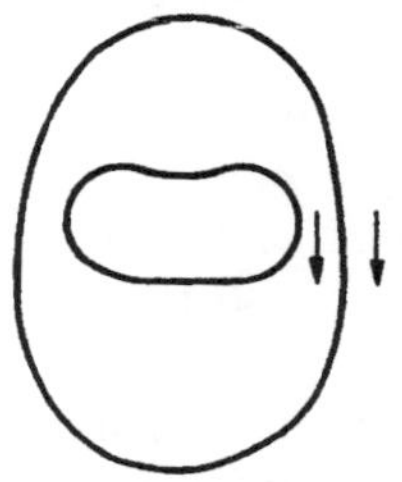

Fig.7.8.5. Shape of a compound drop for $\mu_3 = 0.5\mu_2 = 0.5\mu_1$, $r_1 = 0.5$, $\varepsilon = 0.3$, $Ca_1 = 3.9$, $Ca_2 = 1$, $U_1 = -0.8$, $U_2 = -1$.

Altogether, we can conclude that, when the fluid phases of a compound drop move with constant velocities and there is no relative motion of the inner and outer drops, they both are flattened or elongated, while at any relative motion, one of the drops is flattened and the other one is elongated.

For the second considered class of problems, due to an overall force balance between viscous and buoyant forces on each spherical fluid-fluid interface, the velocities U_1 and U_2 are determined, and the deformations calculated afterwards. It is clear that here the different fluid phases densities ρ_1, ρ_2 and ρ_3 will influence the compound drop deformation by means of the fluid particle velocities U_1 and U_2.

In Fig.7.8.6 - Fig.7.8.8. we show the deformation of compound drops in which the inner droplet is a bubble and phases 2 and 3 are fluids of equal viscosity, but different density, $\rho_3/\rho_2 = 0.5$, 1 and 2, respectively. In both cases the outer drops are elongated, while the inner bubbles are flattened. In the first case (Fig.7.8.6) the bubbles are deformed easier, while the deformation of the outer drops occurs with greater difficulty. Decreasing the minimum separation distance ε the bubble flattens more and at $\varepsilon = 0.3$ a dimple appears in the front part of the particle.

If the densities of the outer drop and continuous phase are equal, i.e., $\rho_3 = \rho_2$, only a smoothing is observed in the front part of the bubble (Fig.7.8.7). For denser outer drop with respect to the continuous phase fluid, i.e., $\rho_3 = 2\rho_2$, both inner and outer droplets are flattened and the dimples can be seen still at $\varepsilon = 0.41$ (Fig.7.8.8). We shall note that the viscosities ratio does not influence essentially the type of deformation, while the density ratio significantly changes the deformation character.

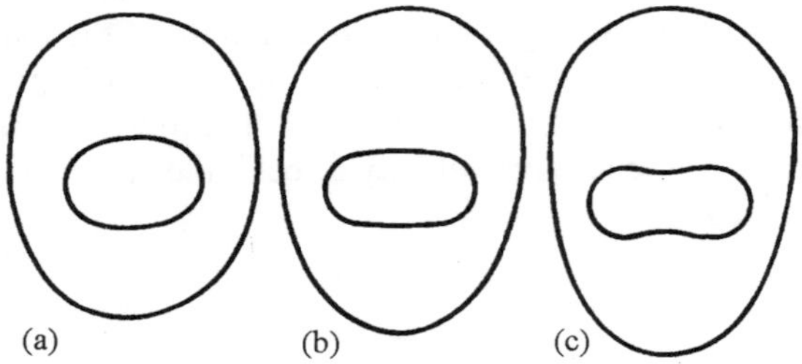

Fig.7.8.6. Shape of a compound drop moving under the action of a constant body force for $\rho_2 = 2\rho_3$, $\rho_1 = 0$, $r_1 = 0.45$, $\lambda_1 = 0$, $\lambda_2 = 1$, $Ca_1 = 23.08$, $Ca_2 = 17.5$ and: a) $\varepsilon = 0.41$, $U_1 = -0.199$, $U_2 = -0.149$; b) $\varepsilon = 0.35$, $U_1 = -0.1968$, $U_2 = -0.1488$; c) $\varepsilon = 0.3$, $U_1 = -0.1926$, $U_2 = -0.1486$.

In the present section it is shown that, unlike the case of a homogeneous drop, a compound multiphase drop in an unbounded Stokes flow will not remain spherical. The complex interaction between the fluid interfaces leads to a radical change of their shapes.

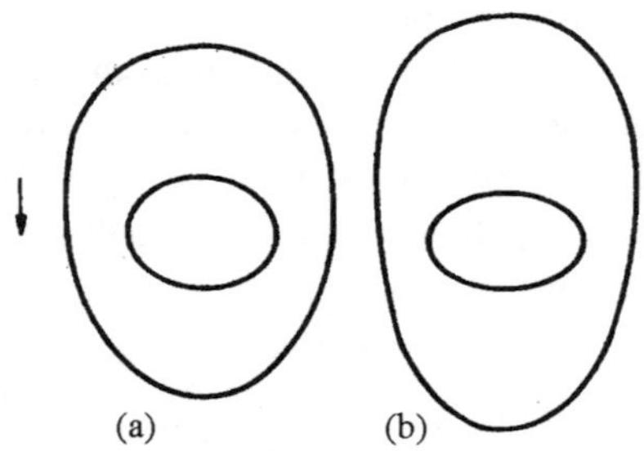

Fig.7.8.7. Shape of a compound drop moving under the action of a constant body force for $\rho_2 = \rho_3$, $\rho_1 = 0$, $r_1 = 0.45$, $\lambda_1 = 0$, $\lambda_2 = 1$, $Ca_1 = 23.08$, $Ca_2 = 17.5$ and: a) $\varepsilon = 0.41$, $U_1 = -0.0662$, $U_2 = -0.0302$; b) $\varepsilon = 0.3$, $U_1 = -0.0648$, $U_2 = -0.0298$.

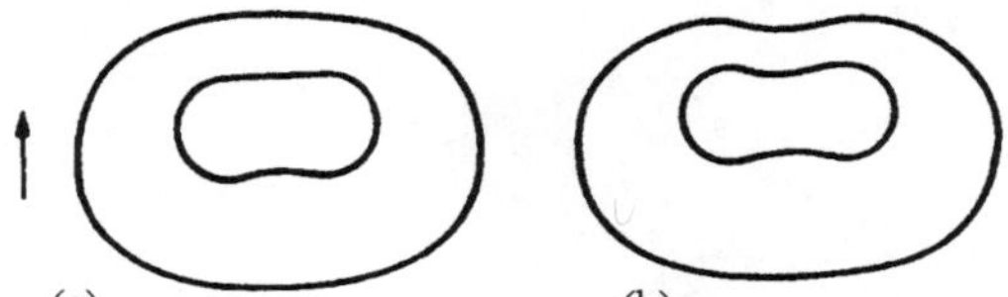

(a) (b)

Fig.7.8.8. Shape of a compound drop moving under the action of a constant body force for $\rho_2 = 2\rho_3$, $\rho_1 = 0$, $r_1 = 0.45$, $\lambda_1 = 0$, $\lambda_2 = 1$, $Ca_1 = 18.2$, $Ca_2 = 13.5$ and: a) $\varepsilon = 0.41$, $U_1 = 0.1989$, $U_2 = 0.2074$; b) $\varepsilon = 0.35$, $U_1 = 0.1994$, $U_2 = 0.2076$.

2 35 – 254

CHAPTER 8.

Boundary Effects on the Motion of a Single Rigid or Fluid Particle

8.1. Introduction

Wall effects on the rate of settling of a rigid or fluid particle are, in many ways, similar to the effects of a second rigid particle, that were considered in chapter 7. By replacing the "wall" with a second particle, the techniques used in the above context are also applied to treat the hydrodynamic interaction of two rigid or fluid particles settling through an otherwise unbounded fluid. If the particle and the wall(s) are sufficiently distant, the hydrodynamic interactions are very weak and the classical solutions for particles in isolation can be used. However, when the particle comes close to the wall, the hydrodynamic interaction is significant.

In Stokes flow the force **F** acting on a rigid or fluid particle moving with velocity **U** through a bounded fluid can be expressed in a general form

$$\mathbf{F} = -6\pi\mu\, a\, \mathbf{K}.\mathbf{U}\,, \tag{8.1.1}$$

where a is a characteristic particle dimension and **K** is a dimensionless symmetric dyadic, which is independent of the fluid properties and of the velocity **U**. However, it depends on the sizes and shapes of particle and boundaries, and on the location of the centre and orientation of the particle relative to the wall. The proof of this and the symmetry of **K** is essentially the same with that given for the case of a particle in an unbounded fluid (Happel and Brenner, 1973). The necessary modifications of the latter proof are trivial since the velocity vanishes on the wall.

As it is common for this book, different methods are applied to a variety of combinations of particle and wall geometry. These methods are based on knowledge of the separate solutions for particles in an infinite medium and for spheres in the proximity of container walls. When it becomes necessary, we resort to a variety of different coordinate systems. These are Cartesian coordinates (x, y, z), spherical coordinates (r, θ, φ) and cylindrical coordinates (ρ, φ, z), each having a common origin at the sphere centre, as we have seen in Chapter 7. Sometimes it is also necessary to utilise bispherical coordinates (ξ, η, φ).

When the particle is far away from the wall, the method of reflections is appropriate and the reflections off the wall are conveniently represented by image singularities. For particles and walls near contact, a combination of lubrication and numerical methods are used. When particles and walls are at moderate separations, numerical techniques are required. For particles and walls of simple shape the boundary-multipole collocation method is recommended. An efficient method for particles and walls of complex shape, as well as for many-body problems, is the boundary integral method.

8.2. Single Rigid or Fluid Particle in the Presence of a Rigid Plane

In this section we consider a moving (or stationary) rigid or fluid particle in viscous fluid, in which there are solid stationary (or moving) walls. Boundary effects on the Stokes flows, when the particle and the walls have particular simple shapes, have been investigated by many authors. Applying the method of reflection, Lorentz (1907) gives an approximate solution of the axisymmetric problem for the constant velocity motion of a rigid sphere in viscous fluid towards a plane, in the case when the distance between the bodies is at least tenfold the particle diameter. Exact solution of this problem in quasisteady Stokes approximation is obtained independently by Brenner (1961) and Maude (1961). For the correction coefficient in the Stokes drag formula (7.2.1) they obtain the expression

$$\Phi = \frac{4}{3}\mathrm{sh}\alpha\sum_{n=1}^{\infty}\frac{n(n+1)}{(2n-1)(2n+3)}\left[\frac{2\mathrm{sh}(2n+1)\alpha+(2n+1)\mathrm{sh}2\alpha}{4\mathrm{sh}^2\left(n+1/2\right)\alpha-(2n+1)^2\mathrm{sh}^2\alpha}-1\right], \qquad (8.2.1)$$

where $\alpha = \mathrm{ch}^{-1}\left(l/a\right)$, a is the particle radius and l is the distance between the particle centre and the wall.

The sphere motion in a viscous fluid parallel to a plane wall is studied by Faxen in his thesis (Faxen, 1921) by the reflection method and the following correction coefficient for the drag is obtained

$$\Phi = \frac{1}{1-\frac{9}{16}\left(\frac{a}{l}\right)+\frac{1}{8}\left(\frac{a}{l}\right)^3-\frac{45}{256}\left(\frac{a}{l}\right)^4-\frac{1}{16}\left(\frac{a}{l}\right)^5}. \qquad (8.2.2)$$

In this expression the effects of the second reflection are taken into account. If the sphere is free to rotate, its angle velocity will be given by

$$\omega = \frac{2U}{32a}\left(\frac{a}{l}\right)^4\left[1-\frac{3}{8}\left(\frac{a}{l}\right)\right]. \qquad (8.2.3)$$

Utilizing the Faxen's method Wakiya (1956, 1957) analyzed two cases of a sphere past by a shear flow between two parallel planes: (i) Poiseuille flow between two parallel stationary planes, induced by a given pressure gradient; (ii) Couette flow in which only one of the planes moves.

Let us denote the distance between the two planes with H and assume that the sphere centre is located at distance l = H/2 from the midplane. Wakiya accepts the origin of the coordinate system z = 0 coinciding with the sphere centre. For the first flow case, he considers a flow between the two planes of the type:

$$u_\infty = U + \frac{2U}{3l} z - \frac{U}{3l^2} z^2, \qquad v_\infty = w_\infty = 0, \tag{8.2.4}$$

where U is the sphere velocity. For the drag and momentum with respect to y axis on a sphere of radius a between two planes, the following formulae are obtained:

$$F_x = \frac{6\pi\mu a U\left[1 - \frac{1}{9}\left(\frac{a}{l}\right)^2\right]}{1 - 0.6526\left(\frac{a}{l}\right) + 0.316\left(\frac{a}{l}\right)^3 - 0.242\left(\frac{a}{l}\right)^4}, \tag{8.2.5}$$

$$T_y = \frac{8}{3}\pi\mu a^2 U\left(\frac{a}{l}\right)\left[1 + 0.0758\left(\frac{a}{l}\right) + 0.049\left(\frac{a}{l}\right)^2\right]. \tag{8.2.6}$$

For the Couette flow case, assuming that the plane nearer to the sphere is stationary and the distant one is moving with velocity U, Wakiya considers the following undisturbed linear flow between the planes:

$$u_\infty = \frac{U}{4} + \frac{U}{4l} z, \qquad v_\infty = w_\infty = 0. \tag{8.2.7}$$

Then, the corresponding drag force is:

$$F_x = \frac{\frac{3}{2}\pi\mu a U}{1 - 0.6526\left(\frac{a}{l}\right) + 0.4003\left(\frac{a}{l}\right)^3 - 0.297\left(\frac{a}{l}\right)^4}, \tag{8.2.8}$$

and the moment is

$$T_y = 4\pi\mu a^2 U\left(\frac{a}{l}\right)\left[1 + 0.0506\left(\frac{a}{l}\right) + 0.033\left(\frac{a}{l}\right)^2\right]. \tag{8.2.9}$$

Dean and O'Neill (1963) and O'Neill (1964) propose a method for solving the 3D problem of rotational and translational motion of rigid sphere parallel to a plane wall.

Using the bispherical coordinate method, Goldman et al. (1967a, b) have found an exact solution for the slow translatory motion of a neutrally suspended rigid spherical particle in Couette flow. This solution is valid for small Reynolds numbers and an arbitrary ratio of the distance between the sphere centre and the plane (excluding the contact case) to the sphere radius.

An exact solution for a viscous flow around a fixed sphere in contact with a fixed plane wall, when the fluid motion in the absence of the sphere is assumed to be a uniform linear shear flow, is given by O'Neill (1968) using tangent-sphere coordinates.

On the basis of the Stimson and Jeffery's approach for solving the Stokes equations in bispherical coordinates, Bart (1968) examines the problem of a fluid drop motion towards a rigid plane or a flat interface separating two immiscible fluids. His solution is incomplete, since he only calculates the drag on the sphere during its motion towards the interface, without calculating the velocity field and pressure.

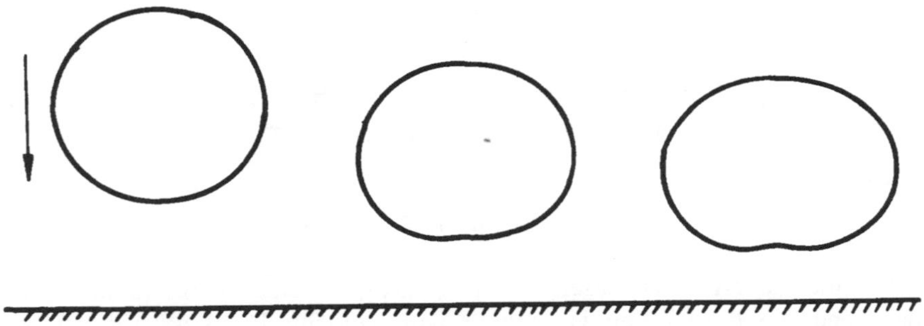

Fig. 8.2.1. Liquid drop approaching a solid plane for $\lambda = 0.5$, $\mathrm{Ca} = 0.1$: a) d=2, l=0.9; b) d=1.5, l=0.8; c) d=1.3, l=0.65.

Zapryanov and Chervenivanova (1983), and Chervenivanova and Zapryanov (1985) have considered some particular cases of a drop receding from or approaching a solid plane. When a single spherical drop moves slowly (at low Reynolds number) in infinite viscous fluid, its shape is preserved. The presence of another fluid particle or of a solid boundary near it might strongly influence its shape. Fig.8.2.1. represents the wall influence over the shape of a drop approaching a solid plane. An important characteristics of the drop deformation is the ratio l_d of the vertical to horizontal deformation. At dimensionless distance from the wall d = 2 this ratio is close to unity, at d = 1.5 a smoothing of the front part is observed (l_d =0.8), and at d = 1.3 an initial stage of "dimple", hollow, on the nearest drop part to the wall (l_d = 0.65) appears.

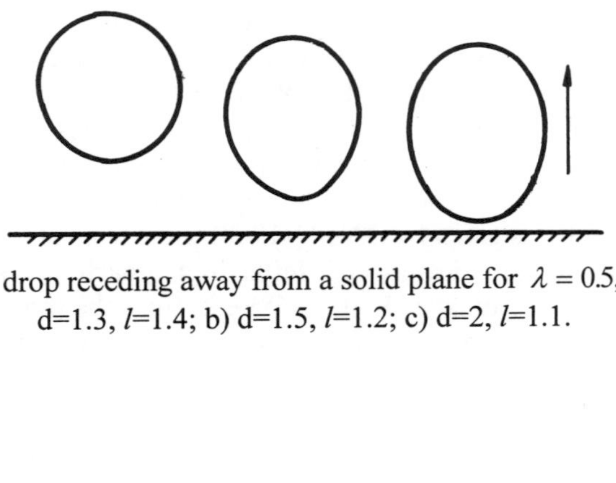

Fig. 8.2.2. Liquid drop receding away from a solid plane for $\lambda = 0.5$, $\mathrm{Ca} = 0.1$ and: a) d=1.3, l=1.4; b) d=1.5, l=1.2; c) d=2, l=1.1.

When the fluid particle recedes from the wall, it elongates and this could be seen in Fig.8.2.2. When the fluid particle moves away from a solid plane and the separation distance between them is small, elongation in vertical direction occurs to yield approximately prolate ellipsoid shape.

The lateral migration due to the drop deformation is mainly studied in the case of small deformations, when the drop is far away from the wall (Leal, 1980). The small deformations allow the solutions to be sought in spherical harmonics form as it is done by Chaffey et al. (1965). A more general theory for the small drop deformations and drop migration is proposed by Chan and Leal (1979) and Shapira and Haber (1990) on the basis of Lorentz reciprocal theorem. Chan and Leal give an approximate formula for the lateral migration velocity in terms of the capillary number Ca, assumed to be

$$U_{lat} = Ca^2\left(\frac{19\lambda+16}{16\lambda+16}\right)\left(\frac{a^2}{l^2}\right)\frac{3(54\lambda^2+97\lambda+54)}{280(1+\lambda)^2}, \tag{8.2.10}$$

where l is distance between the centre of the initially spherical drop and the wall, and λ is the viscosity ratio of the fluid inside and outside the drop. Performing experiments in Couette apparatus, Smart and Leighton (1991) observe larger lateral migration velocities than those predicted in the model of Chan and Leal, which strongly decrease when the drop approaches the wall. Shapira and Haber calculate for small Ca the drop longitudinal velocity and deformation in terms of the dimensionless parameter De = (L-b) / (L+b), where L is the drop length in the direction of the major axis and b, in the minor axis,

$$U_{long} = \frac{Ca^2}{8l^2}\cdot\frac{1+2.5\lambda}{1+\lambda}, \tag{8.2.11}$$

$$De = Ca\frac{19\lambda+16}{16\lambda+16}\left(1+\frac{3a^3}{8l^3}\cdot\frac{1+2.5\lambda}{1+\lambda}\right). \tag{8.2.12}$$

Recently, Uijttewaal et al. (1993) study numerically the homoviscous ($\lambda = 1$) Stokes flow problem of a deformable drop motion in linear shear flow near to a plane wall by means of the boundary integral method. They find that the lateral velocity value is confirmed by the analytical formula (8.2.10) for different wall distances and capillary numbers Ca, except for the case of small distances and large Ca. The latter result is observed, on the other hand, in the experiments of Smart and Leighton, but with overestimated lateral velocity value, probably due to the restrictions of the Couette flow device and quite different viscosity ratio. The longitudinal velocity and deformation obtained by Uijttewaal et al. are in good agreement with (8.2.11) and (8.2.12), respectively, for small Ca and large wall distances. Moreover, the wall effect is expressed, also, in the deformed shape asymmetry as stronger deformation and larger curvature of the drop side, which is nearer to the wall. Such result is not encountered by the authors previously investigating the same problem (Shapira and Haber, 1990).

Interesting theoretical investigations for shear flows are performed by Bretherton (1962a), who applies the method of matched asymptotic expansions for solving the problem of a neutrally suspended cylindrical particle in simple shear flow. Moreover,

Bretherton (1962b) considers also other particles of more complex shape in shear flows and establishes the existence of other motions (modes) in addition to the studied by Jeffery (1922b) and Faxen (1921).

The problems connected with flows around suspended particles near very large collectors can be regarded as stagnation point flows or quadratic shear flows. Goren (1970) studies the case of a small particle touching a plane in an axisymmetric stagnation flow. For spherical collectors:

$$\mathbf{v}_\infty = A\left(z\rho\, \mathbf{e}_\rho - z^2 \mathbf{e}_z\right), \tag{8.2.13}$$

or in a planar stagnation flow for cylindrical collectors,

$$\mathbf{v}_\infty = A\left(2zx\, \mathbf{e}_x - z^2 \mathbf{e}_z\right), \tag{8.2.14}$$

where A is a constant.

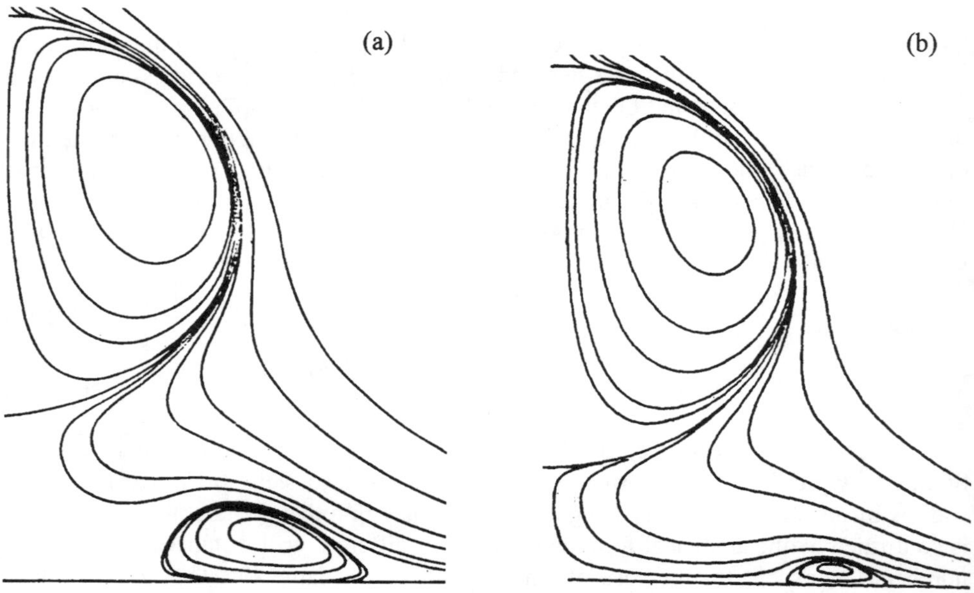

Fig.8.2.3. The streamlines flow patterns around a fluid particle in an axisymmetric stagnation flow near a plane wall with: a) $\lambda = 0.5$, h=1.8a; b) $\lambda = 1.5$, h = 1.6a.

Then the stream function for the axisymmetrical flow at infinity is $\psi_\infty = \frac{1}{2} A z^2 \rho^2$. In the tangential coordinate system (ξ, η, ϕ) connected with the cylindrical coordinate (z, ρ, ϕ) by the relation (7.4.25) the plane is given by $\xi = 0$, while the sphere touching the plane at the origin is expressed by ξ = const. In terms of the tangential coordinates the stream function at infinity is transformed into

$\psi_\infty = \frac{1}{2} A\eta^2\xi^2(\eta^2+\xi^2)^{-4}$. The stream function is obtained by the method of separation of variables of the biharmonic equation in tangential coordinates due to a Fourier analysis.

Then for both cases (axisymmetrical and planar stagnation flow) the hydrodynamic force acting on the sphere of radius a is found to be:

$$F_z = -60.87\mu A a^3 . \quad (8.2.15)$$

A similar axisymmetric problem is considered by Zapryanov (1989a, b) for the case of a fluid spherical particle of radius a near a plane. Its centre is lying on the axis $\rho = 0$ at a distance $h > a$ from the plane $z = 0$. By means of the bispherical coordinate techniques the streamlines flow patterns are illustrated in Fig.8.2.3.a, b. The first one corresponds to the viscosity ratio of the inner to outer fluid viscosity $\lambda = 0.5$; h=1.8a, and two symmetric vortices behind the drop are created. If $h > h_{cr}$, where $6.769a^2 < h^2_{cr} < 7.47a^2$, these vortices disappear. On the second figure, the viscosity ratio is increased, $\lambda = 1.5$, and h = 1.6a, but the vortices still exist, although covering a smaller area. For this case the critical distance is in the limits $7.473a^2 < h^2_{cr} < 8.25a^2$. The correction coefficient to the Stokes drag formula (7.2.1) is calculated for different values of λ. It is shown that when $\lambda \to \infty$, i.e., when the fluid particle becomes rigid, this coefficient, as a function of h, tends to the corresponding coefficient found by Goren and O'Neill (1971) in the case of a rigid sphere near a plane.

A recent review of the main methods used for a low Reynolds number flow, induced by the motion of an arbitrary shaped particle in the presence of a wall or walls, is given by Hsu and Ganatos (1989). There are five different methods: the method of reflections, the boundary collocation and truncated series technique, the finite-element method, the singularity method and the boundary-integral method.

The reflection method has been mentioned before in this section as being suitable for treating such problems. By means of this method Wakiya (1959) solved the problem of the motion of a spheroidal particle parallel to a plane wall with its semi-major axis directed arbitrary with respect to the wall.

Chen and Skalak (1970) applied the boundary collocation, truncated series technique for the problem of axisymmetric flow past a periodic array of spheroidal particles on the centreline of an infinitely long circular cylinder. The case of a torus with axis of symmetry perpendicular to a planar wall has been studied by the same technique by Kucaba-Pietal (1986). For more complicated geometry, such as that analyzed by Hsu and Ganatos (1989), the calculations become difficult and must be performed numerically, which is highly time consuming.

The finite-element method is suitable for arbitrary body shapes, but ineffective for infinite or semi-infinite domains (see section 5.11.). It is applied by Skalak et al. (1972) for the problem of capillary blood flow around biconcave disc-shaped rigid particles located periodically on a circular cylinder axis.

The singularity method substitutes the body with an appropriate set of singularities, satisfying the boundary conditions of the problem. A slender body between walls has been studied by Liron and Mochon (1976a, b) and Liron (1978). The angular velocity of a prolate spheroid adjacent to a plane wall is calculated via this method by Dabros (1985). It is also applied by Yuan and Wu (1987) for an arbitrary prolate body of revolution translating axisymmetrically towards a plane wall.

The boundary integral method is created by Odqvist (1930) and afterwards as implemented by Youngren and Acrivos (1975) to an arbitrary body in a unbounded flow domain by means of body surface discretization into a finite number of surface elements transforms into finite-element method (see section 5.11.). Further on, the flow due the motion of a spherical particle in an infinitely long circular cylinder is obtained by a variation of this method (Lewellen, 1982). The same problem but with a collection of spheroids instead of the spherical particle is treated again by the boundary-integral technique (Tozeren, 1984).

The boundary-integral technique proves to be the most effective one for solving problems of arbitrary shaped bodies oriented arbitrary towards a plane wall. By this method Hsu and Ganatos (1989) calculate the force and torque on an arbitrary body of revolution with axis of symmetry arbitrary oriented to a planar wall at zero Reynolds number. The Stokeslet near a plane wall (3.9.8) instead of the free-space Stokeslet is used in the integral equations (5.11.3), which are transformed into a similar set of linear integral equations for the unknown density functions **f**:

$$\mathbf{v}(\mathbf{r}) = -\frac{3}{2\pi}\overset{PV}{\int\int_{S_p}}\mathbf{K}^W(\mathbf{r},\boldsymbol{\rho})\mathbf{v}(\boldsymbol{\rho})\mathbf{n}(\boldsymbol{\rho})\mathrm{d}\sigma(\boldsymbol{\rho}) + \frac{1}{4\pi\mu}\int\int_{S_p}\mathbf{f}(\boldsymbol{\rho})\mathbf{G}^W(\mathbf{r},\boldsymbol{\rho})\mathrm{d}\sigma(\boldsymbol{\rho}), \quad (8.2.16)$$

where $\mathbf{r} \in S_p$, $\mathbf{G}^W(\mathbf{r},\boldsymbol{\rho})$ and $\mathbf{K}^W(\mathbf{r},\boldsymbol{\rho})$ are defined by (3.9.8) and (3.9.9), respectively, and $\mathbf{v}(\mathbf{r})$ is the velocity on the particle surface. This system is valid for any steady motion of an arbitrary shaped particle near a plane wall at zero Reynolds number and it can be solved only numerically. The unknown density functions are discretized and double numerical integration over the particle surface is performed, because, for bodies of revolution and motions with planar symmetry, they can be represented in advance by Fourier-Legendre series. This procedure decreases significantly the computational costs.

In order to solve (8.2.16) Hsu and Ganatos introduced a residual vector $\mathbf{v}^R(\mathbf{r})$

$$\mathbf{v}^R(\mathbf{r}) = \mathbf{v}(\mathbf{r}) - \mathbf{v}^*(\mathbf{r}), \quad (8.2.17)$$

where $\mathbf{v}(\mathbf{r})$ is the true solution and $\mathbf{v}^*(\mathbf{r})$ is the right-hand side of (8.2.16). If the solution vector **f** is exactly the sought one, then $\mathbf{v}^R(\mathbf{r})$ must be zero. Two different algorithms are constructed on this principle to find an approximate solution of **f**.

The first one is the boundary collocation method, when the residual vector vanishes on discrete points of the body surface. If the density function **f**, represented in double series form, as discussed before, is substituted in the equations (8.2.16), an infinite set of algebraic equations for the series coefficients is obtained. In order to obtain a finite number of equations, the double series of the vector function **f** are truncated to have correspondingly M and N terms in both expansions. Then choosing M x N collocation points on the particle surface to satisfy the no-slip boundary conditions, i.e., applying (8.2.16) on them, yields a finite system of 3 x M x N algebraic equations for 3 x M x N unknown coefficients for **f**. For bodies of revolution sometimes it is possible to calculate analytically the integrals in the symmetry direction (Hsu and Ganatos, 1989).

The weighted residuals method is the second solution method applied for the discussed problem. The residual vector multiplied by the approximated function **f** is integrated over the particle surface and the obtained weighted residual is put equal to zero. The integration is performed as in the collocation method to get a finite linear algebraic equations set for the unknown coefficients of **f**. As shown by Hsu and Ganatos the convergence characteristics of this method are very similar to the boundary-collocation method for the studied cases. However, the computation time is greater than that of the boundary-collocation method and the latter one is chosen by the authors to present their results.

The force and torque coefficients calculated by the boundary-integral method for the case of a spheroid moving in parallel to a wall are compared to those obtained by the reflection technique by Wakiya (1959). At large distance between the spheroid and the wall both results are in good coincidence, while at close spacing there are some discrepancies due to the limitations of the reflection method. For the case of a torus translating in parallel to the wall with is axis of symmetry perpendicular to the wall the force coefficient result is similar to the theoretically found by Kucaba-Pietal (1986) by the multipole technique. In conclusion, boundary-integral method applied to arbitrary shaped particles in the presence of a wall, is also suitable for treating deformable fluid droplets (Rallison and Acrivos, 1978), or deformable interfaces between two immiscible fluids (Leal and Lee, 1982).

The force on a rigid sphere translating parallel to a plane wall in a linear shear flow has a drag and a lift component (Cox and Brenner, 1968). However, the Stokes creeping flow equation, due to its linearity, is not able to predict the lift force component. Therefore, in order to obtain the latter, it is necessary to consider the inertia terms in the full Navier-Stokes equations (5.8.6). At small, but not zero, Reynolds number based on the particle radius, $Re \ll 1$, these equations can be treated asymptotically by successive solution approximations in terms of the Reynolds number as a small parameter. Then, the nonlinear inertia terms, based on the zero order approximation in Re, are taken into account by the first approximation. As a consequence, the force acting on a body near a plane wall has not only a drag component, but a lift component is added.

The method of singular perturbations with matched inner and outer expansions is exploited by Cox and Brenner for the case of a spherical particle near a plane wall, lying in the inner region and when the distance between the particle centre and the wall is large in comparison with the sphere radius. They obtain a general expression for the lift force in terms of a Stokeslet near a wall (see section 3.9) without considering the flow in the outer region, far away from the sphere. The same method was applied by Cox and Hsu (1977) to the flow around a sphere near a wall in a linear shear or quadratic flow.

Their results are generalized by Vasseur and Cox (1976) for the case a sphere between two parallel plane walls.

When the sphere in a shear flow touches the wall, the lift force is shown by Leighton and Acrivos (1985) to be directed away from the wall and is of the order of the fourth power of the sphere radius and the square of the velocity gradient.

The perturbation technique is applied by Drew (1988) to calculate the lift on a sphere translating in a shear flow far away from a plane wall; since the sphere is modelled as a point force. By a similar method Schonberg and Hinch (1989) evaluated the lift force on a sphere in a plane Poiseuille flow for channel Reynolds numbers of the order of unity.

McLaughlin (1993) has studied the inertial lift on a sphere translating in a shear flow bounded by a plane wall and shows that the lift force value tends to that pertaining to Cox and Hsu (1977) as the distance between the sphere and the wall decreases. His analysis is valid when the wall is either very near the particle or far away from it. Cherukat and McLaughlin (1994) treat an analogous problem by an asymptotic analysis, but with the wall lying in the inner region of the disturbance flow caused by the sphere and without any restrictions for the gap and sphere radius ratio. When the sphere is far away from the wall, their results for the lift force coincide with those of Cox and Brenner (1968) and Cox and Hsu (1977) where the sphere is substituted with a point force or a force doublet.

The boundary integral method is applied by Fischer and Rosenberger (1987) to study theoretically and numerically the hydrodynamic interaction of a plane wall and a sphere in uniform and shear flows. In contrast to the works by Cox and Hsu (1977) and Vasseur and Cox (1977), their results cover all values for the distance between the sphere and wall, and are based on the theoretical analysis for the existence and construction of asymptotic expansions for the Navier-Stokes equations at small Re (Fischer, 1983, 1986). The coefficients of the expansions are solutions to different homogeneous and non-homogeneous Stokes problems and can be constructed as single layer potentials. The collocation method is used for the numerical solution of the boundary integrals, similarly to the single body problems (see section 5.11.). Fischer and Rosenberger consider four cases for the velocity at infinity: (i) uniform parallel flow with respect to the wall; (ii) simple shear flow, when the sphere centre is fixed relatively to the wall; (iii) simple shear flow, when the sphere centre is moving with a given slip velocity parallel to the wall, i.e., neutrally buoyant sphere; (iv) two-dimensional Poiseuille flow, when the sphere centre is moving with a given slip velocity parallel to the wall, i.e., neutrally buoyant sphere. They obtain the force and torque in asymptotic expansions with respect to the Reynolds number and numerical values for the first two terms, i.e., with accuracy up to $O(Re^2)$. For both cases (i) and (ii) the drag force comes only from the zero-order term, while the lateral (lift) force comes from the first-order term. For the torque the authors have a non-zero contribution only from the zero-order term. For the flows (iii) and (iv) corresponding to freely suspended sphere, the inverse problem is solved, i.e., the angular and slip velocities are determined via the restriction of zero drag and torque. For the first three cases the lateral force is directed away from the wall, while for the last case there is an equilibrium position. The lateral force strongly increases in value when the sphere is located nearer to the wall, except for the case of uniform parallel flow to the wall, where it oscillates around its limiting value for unbounded flow. The lateral force can be determined also from the reciprocal theorem of Ho and Leal (1974).

Recently Keh and Tseng (1994) have developed a combined analytical-numerical method for steady slow motion of an arbitrary axisymmetric body along its axis of revolution, the latter being normal to a planar surface.

8.3. Single Rigid or Fluid Spherical Particle in the Presence of a Plane Interface

As for the rigid plane wall case, an exact solution of the problem for the hydrodynamic interaction of a particle with a plane free surface, on which the tangential stresses vanish, is obtained by Brenner (1961) and the correction coefficient in the Stokes drag formula (7.2.1) for finite distances between the sphere and the interface is

$$\Phi = \frac{4}{3}\mathrm{sh}\alpha\sum_{n=1}^{\infty}\frac{n(n+1)}{(2n-1)(2n+3)}\left[\frac{4\mathrm{ch}^2(n+1/2)\alpha+(2n+1)^2\mathrm{sh}^2\alpha}{2\mathrm{sh}(2n+1)\alpha-(2n+1)\mathrm{sh}2\alpha}-1\right]. \quad (8.3.1)$$

This result is in good correlation with the drag force coefficient given by Faxen and Dahl (1925) for two spheres moving in parallel to their line of centres in the case of equal spheres' radii and equal approaching velocities. In this case the plane midway between the spheres is a plane of symmetry on which the kinematic and dynamic conditions are as on a free surface.

The investigations of Leal and co-workers, connected with the hydrodynamic interaction of rigid and fluid particles with undeformed or deformed interface in Stokes approximation are extremely diverse and profound. Generalising the Lorentz method (Lorentz, 1907), Lee et al. (1979) obtain an approximate solution (see section 3.9.) of the problem of a sphere approaching a horizontal plane interface, solved by Brenner (1961) and also by Bart (1968) as previously shown.

On the basis of Jeffery's solution (Jeffery, 1912) and Dean and O'Neill's solution (Dean and O'Neill, 1963), Lee and Leal (1980) derive a general exact solution in bispherical coordinates of the Stokes equations for the problems of translation or rotation of a rigid sphere perpendicularly or in parallel direction to an undeformable planar interface. Moreover, a solution of the problems of a rigid sphere rotation about an axis perpendicular or parallel to the horizontal interface, dividing two immiscible fluids, is obtained, too. Since the problems are solved in linear approximation, from these solutions one can get solutions of the problems for a more general spherical particle motion with respect to the undeformable interface.

Following Lee and Leal (1980), we shall determine the fundamental solution of Stokes equations (1.3.6), written in bispherical coordinates in terms of series expansions of eigensolutions for bispherical coordinates. We utilise simultaneously cylindrical (ρ, z, φ) and bispherical coordinates (ξ, η, φ), connected by the relation (7.1.1). In bispherical coordinates the sphere and planar interface equations are given by the expressions, respectively: $\eta = \eta_1 = -\mathrm{ch}^{-1}(l)$ and $\eta = 0$, where l is the distance between the sphere centre and the interface, dimensionalized by the sphere radius a.

Since the pressure p in the Stokes approximation is a harmonic function (1.3.8), i.e.,

$$\Delta \mathrm{p} = 0, \tag{8.3.2}$$

then based on Jeffery (1912), the fundamental solution of (8.3.2) in bispherical coordinates is:

$$\mathrm{p} = \sum_{m=0}^{\infty} \mathrm{p}_m(\xi,\ \eta)\cos(m\varphi + \alpha_m), \tag{8.3.3}$$

where

$$\mathrm{p}_m(\xi,\ \eta) = \frac{1}{c}(\mathrm{ch}\eta - \beta)^{1/2} \sum_{n=m}^{\infty}\left[A_n^m \mathrm{sh}(n+1/2)\eta + B_n^m \mathrm{ch}(n+1/2)\eta\right] S_n^m(\beta), \tag{8.3.4}$$

$$S_n^m(\beta) = a_n^m P_n^m(\beta) + b_n^m Q_n^m(\beta).$$

Here A_n^m, B_n^m, a_n^m, b_n^m and α_m are constants, while $P_n^m(\beta)$ and $Q_n^m(\beta)$ are the associated Legendre functions of first and second kind, respectively, with argument $\beta = \cos\xi$. Since the second kind Legendre function $Q_n^m(\beta)$ is unbounded at $\xi = 0$, correspondent to $z \to \infty$, then $b_n^m \equiv 0$. As a result (8.3.4) is simplified to

$$\mathrm{p}_m(\xi,\ \eta) = \frac{1}{c}(\mathrm{ch}\eta - \beta)^{1/2} \sum_{n=m}^{\infty}\left[A_n^m \mathrm{sh}(n+1/2)\eta + B_n^m \mathrm{ch}(n+1/2)\eta\right] P_n^m(\beta), \tag{8.3.5}$$

where c is the focal distance (see Fig.7.1.1.).

If the velocity components in cylindrical coordinates (ρ, φ, z) are denoted by **v** = (u, v, w), then the Stokes equations (1.3.6) can be given in the form:

$$\left(\Delta - \frac{1}{\rho^2}\right)u - \frac{2}{\rho^2}\frac{\partial v}{\partial \varphi} = \frac{\partial p}{\partial \rho}, \tag{8.3.6}$$

$$\left(\Delta - \frac{1}{\rho^2}\right)v + \frac{2}{\rho^2}\frac{\partial u}{\partial \varphi} = \frac{1}{\rho}\frac{\partial p}{\partial \varphi}, \tag{8.3.7}$$

$$\Delta w = \frac{\partial p}{\partial z}, \tag{8.3.8}$$

$$\frac{1}{\rho}\frac{\partial}{\partial \rho}(\rho u) + \frac{1}{\rho}\frac{\partial v}{\partial \varphi} + \frac{\partial w}{\partial z} = 0, \tag{8.3.9}$$

where $\Delta \equiv \dfrac{\partial^2}{\partial \rho^2} + \dfrac{1}{\rho}\dfrac{\partial}{\partial \rho} + \dfrac{1}{\rho^2}\dfrac{\partial^2}{\partial \varphi^2} + \dfrac{\partial^2}{\partial z^2}$.

The system (8.3.6)-(8.3.9) has a particular solution, given by $u^p = \frac{1}{2} p\rho$, $v^p = 0$, $w^p = \frac{1}{2} pz$. Then, to obtain the general solution of (8.3.6)-(8.3.9), it is only necessary to solve the homogeneous system

$$\left(\Delta - \frac{1}{\rho^2}\right) u^h - \frac{2}{\rho^2}\frac{\partial v^h}{\partial \varphi} = 0, \tag{8.3.10}$$

$$\left(\Delta - \frac{1}{\rho^2}\right) v^h + \frac{2}{\rho^2}\frac{\partial u^h}{\partial \varphi} = 0, \tag{8.3.11}$$

$$\Delta w^h = 0, \tag{8.3.12}$$

together with (8.3.9). The fundamental solution of the Laplace equation (8.3.12), bounded on z, has the same form as that for the pressure (8.3.3) and (8.3.5), i.e.,

$$w^h = \sum_{m=0}^{\infty} w_m(\xi,\ \eta)\cos(m\varphi + \alpha_m) \tag{8.3.13}$$

$$= (\mathrm{ch}\eta - \beta)^{1/2} \sum_{m=0}^{\infty}\sum_{n=m}^{\infty}\left[C_n^m \mathrm{sh}(n+1/2)\eta + D_n^m \mathrm{ch}(n+1/2)\eta\right] P_n^m(\beta)\cos(m\varphi + \alpha_m)$$

For the other two velocity components we assume the similar dependence on the angle φ

$$u^h = \sum_{m=0}^{\infty} u_m(\xi,\ \eta)\cos(m\varphi + \alpha_m), \tag{8.3.14}$$

$$v^h = \sum_{m=0}^{\infty} v_m(\xi,\ \eta)\sin(m\varphi + \alpha_m). \tag{8.3.15}$$

Substitution in equations (8.3.10) and (8.3.11) yields a system of coupled equations for u_m and v_m:

$$\left(\frac{\partial^2}{\partial \rho^2} + \frac{1}{\rho}\frac{\partial}{\partial \rho} + \frac{\partial^2}{\partial z^2} - \frac{m^2+1}{\rho^2}\right) u_m - \frac{2}{\rho^2} m v_m = 0, \tag{8.3.16}$$

$$\left(\frac{\partial^2}{\partial \rho^2} + \frac{1}{\rho}\frac{\partial}{\partial \rho} + \frac{\partial^2}{\partial z^2} - \frac{m^2+1}{\rho^2}\right) v_m - \frac{2}{\rho^2} m u_m = 0. \tag{8.3.17}$$

At m = 0 these equations have the solution

$$u_0 = (\mathrm{ch}\eta - \beta)^{1/2} \sum_{n=1}^{\infty}\left[E_n^0 \mathrm{sh}(n+1/2)\eta + F_n^0 \mathrm{ch}(n+1/2)\eta\right] P_n^1(\beta), \tag{8.3.18}$$

$$v_0 = (\mathrm{ch}\eta - \beta)^{1/2} \sum_{n=1}^{\infty}\left[G_n^0 \mathrm{sh}(n+1/2)\eta + H_n^0 \mathrm{ch}(n+1/2)\eta\right] P_n^1(\beta). \tag{8.3.19}$$

At $m \geq 1$, after addition and subtraction of the equations (8.3.16) and (8.3.17), and replacement of u_m and v_m by $\gamma_m = u_m + v_m$, $\kappa_m = u_m - v_m$, one gets the uncoupled system

$$\left(\frac{\partial^2}{\partial \rho^2} + \frac{1}{\rho}\frac{\partial}{\partial \rho} + \frac{\partial^2}{\partial z^2} - \frac{(m+1)^2}{\rho^2}\right)\gamma_m = 0, \tag{8.3.20}$$

$$\left(\frac{\partial^2}{\partial\rho^2}+\frac{1}{\rho}\frac{\partial}{\partial\rho}+\frac{\partial^2}{\partial z^2}-\frac{(m-1)^2}{\rho^2}\right)\kappa_m=0. \qquad (8.3.21)$$

Following again Jeffery (1912), Lee and Leal express the solutions of (8.3.20) and (8.3.21) in the form:

$$\gamma_m=(\mathrm{ch}\eta-\beta)^{1/2}\sum_{n=m+1}^{\infty}\left[E_n^m\mathrm{sh}(n+1/2)\eta+F_n^m\mathrm{ch}(n+1/2)\eta\right]P_n^{m+1}(\beta), \qquad (8.3.22)$$

$$\kappa_m=(\mathrm{ch}\eta-\beta)^{1/2}\sum_{n=m-1}^{\infty}\left[G_n^m\mathrm{sh}(n+1/2)\eta+H_n^m\mathrm{ch}(n+1/2)\eta\right]P_n^{m-1}(\beta). \qquad (8.3.23)$$

Finally, the general solution of the non-homogeneous system (8.3.6)-(8.3.8) can be given as

$$u=\frac{\rho p}{2}+u_0\cos\alpha_0+\frac{1}{2}\sum_{m=1}^{\infty}(\gamma_m+\kappa_m)\cos(m\varphi+\alpha_m), \qquad (8.3.24)$$

$$v=v_0\sin\alpha_0+\frac{1}{2}\sum_{m=1}^{\infty}(\gamma_m-\kappa_m)\sin(m\varphi+\alpha_m), \qquad (8.3.25)$$

$$w=\frac{zp}{2}+\sum_{m=0}^{\infty}w_m\cos(m\varphi+\alpha_m), \qquad (8.3.26)$$

where p, u_0, v_0, w_m, γ_m and κ_m are expressed in terms of their eigensolutions by (8.3.4), (8.3.13), (8.3.18), (8.3.19), (8.3.22) and (8.3.23). Substitution of (8.3.24)-(8.3.26) into the continuity equation (8.3.9), yields the following equation for p_0, u_0, w_0:

$$\left(3+\rho\frac{\partial}{\partial\rho}+z\frac{\partial}{\partial z}\right)p_0+2\left(\frac{\partial}{\partial\rho}+\frac{1}{\rho}\right)u_0+2\frac{\partial w_0}{\partial z}=0, \qquad (8.3.27)$$

and for p_m, w_m, γ_m and κ_m at $m \geq 1$:

$$\left(3+\rho\frac{\partial}{\partial\rho}+z\frac{\partial}{\partial z}\right)p_m+\left(\frac{\partial}{\partial\rho}+\frac{m+1}{\rho}\right)\gamma_m+\left(\frac{\partial}{\partial\rho}-\frac{m-1}{\rho}\right)\kappa_m+2\frac{\partial w_m}{\partial z}=0. \qquad (8.3.28)$$

The constants participating in the obtained solution must be determined from the boundary conditions and the derived above equations (8.3.27) and (8.3.28), when solving some specific problems. As a result we have 16 independent linear algebraic equations, from which all the coefficients for the fluids above and below the interface may be calculated.

We shall consider in details the formulated four problems.

i) Translation of a sphere perpendicular to an undeformable horizontal interface.

Since the problem is axisymmetrical, its solution does not depend on the coordinate φ and the only non-zero coefficients in the general solution are those at $m = 0$. Moreover, $\alpha_0 = 0$. The boundary conditions on the sphere surface are:

$$u=v=0, \qquad w=1 \qquad \text{at } \eta=\eta_1. \qquad (8.3.29)$$

On the basis of the obtained solution the force **F** and torque **T**, acting on the sphere can be determined:

$$F_x^{T\perp}=F_y^{T\perp}=0, \qquad \mathbf{T}^{T\perp}=0; \qquad (8.3.30)$$

$$F_z^{T\perp} = -\frac{\sqrt{2}}{3}\,\mathrm{sh}\,\eta_1 \sum_n \left[C_n^0 - (n+1/2)\left(A_n^0 - B_n^0\right)\right].$$

ii) Translation of a sphere parallel to an undeformable horizontal interface in $\mathbf{e}_x$ direction.

This problem is 3D and its boundary conditions on the sphere surface have the form:

$$u = \cos\varphi, \qquad v = -\sin\varphi, \qquad w = 0 \qquad \text{at } \eta = \eta_1, \tag{8.3.31}$$

and from it follows, that all the coefficients of the general solution vanish except those corresponding to m = 1 and $\alpha_1 = 0$.

For this case the drag force and the torque yield (O'Neill, 1964):

$$F_y^{T\parallel} = F_z^{T\parallel} = 0\,,\; F_x^{T\parallel} = \Theta, \tag{8.3.32}$$

$$\text{where } \Theta = -\frac{\sqrt{2}}{6}\,\mathrm{sh}\,\eta_1 \sum_n \left[G_n^1 - H_n^1 + n(n+1)\left(A_n^1 - B_n^1\right)\right] \tag{8.3.32a}$$

and

$$T_x^{T\parallel} = T_z^{T\parallel} = 0\,,\; T_y^{T\parallel} = \Sigma,$$

$$\text{where } \Sigma = \frac{\mathrm{sh}^2\eta_1}{12\sqrt{2}} \sum_n \Big[\left(2 + \exp(2n+1)\eta_1\right)\left\{n(n+1)\left(-2C_n^1 - A_n^1 \mathrm{cth}\,\eta_1\right)\right. \tag{8.3.32b}$$

$$\left. - \left(2n+1+\mathrm{cth}\,\eta_1\right)G_n^1\right\} + \left(2 - \exp(2n+1)\eta_1\right)\left\{n(n+1)B_n^1 \mathrm{cth}\,\eta_1 + \left(2n+1+\mathrm{cth}\,\eta_1\right)H_n^1\right\}\Big].$$

iii) Rotation of a sphere perpendicular to an undeformable horizontal interface.

The boundary condition the sphere is:

$$v = \rho, \qquad \text{at } \eta = \eta_1, \tag{8.3.33}$$

and in the general solution only the coefficients with m = 0 and $\alpha_0 = \pi/2$ remain. Then only the second velocity component v is non-zero and for the drag and torque, we have (Jeffery, 1915):

$$\mathbf{F}^{R\perp} = 0, \qquad T_x^{R\perp} = T_y^{R\perp} = 0, \tag{8.3.34}$$

$$T_z^{R\perp} = \frac{\mathrm{sh}^2\eta_1}{\sqrt{2}} \sum_n n(n+1)\left(-G_n^0 + H_n^0\right).$$

iv) Rotation of a sphere parallel to an undeformable horizontal interface, e.g. parallel to the y axis.

The boundary conditions for this case reduce to:

$$u = (z + l)\cos\varphi,\; v = -(z + l)\sin\varphi,\; w = -\rho\cos\varphi \qquad \text{at } \eta = \eta_1, \tag{8.3.35}$$

and the only coefficients in the general solution will be those at m = 1 and $\alpha_1 = 0$.

For the force and torque acting on the sphere we obtain:

$$F_y^{R\parallel} = F_z^{R\parallel} = 0\,,\; F_x^{R\parallel} = \Theta\,, \tag{8.3.36}$$

where Θ is given with the expression (8.3.32a),

$$T_x^{R\parallel} = T_z^{R\parallel} = 0 \text{ and } T_y^{R\parallel} = -\frac{1}{3} - \Sigma\,,$$

where Σ is given by (8.3.32b).

After appropriate numerical computations Lee and Leal (1980) illustrate the obtained flow patterns with respect to the distance l and the viscosity ratio λ between the fluids' viscosities above and below the interface.

A natural extension of these studies is the determination of the horizontal interface deformation, when approached by a rigid sphere. This problem, in the case of small deformations, is solved by Berdan and Leal (1982).

Further on, we shall discuss the problems of the hydrodynamic interaction connected with the motion of slender body particles in the presence of interfaces. This topic has been first studied in the limit of rigid walls by Brenner (1962), Katz et al. (1975) and Lighthill (1975), when the particle is far away or near the wall. For arbitrary particle-wall distances the problem has been treated by de Mestre (1973) and de Mestre and Russel (1975). Their results have been extended later by Fulford and Blake (1983) for the case of a slender body located in the fluid under a flat interface separating two immiscible fluids of different viscosities and densities. Two special slender body orientations are considered: (i) parallel to the interface and moving axially along its symmetry axis, transversely (i.e., normal to its symmetry axis), but still parallel to the interface and normal to the interface; (ii) normal to the interface and moving axially and transversely.

To solve these problems Fulford and Blake employ the perturbation technique with line distributions of Stokeslets (Batchelor, 1970b) and higher-order singularities with a perturbation parameter connected with the body aspect ratio. To overcome the end effects difficulties of the model, a distribution of Stokeslets over the whole body surface is suggested. The interface is assumed planar and undeformable if the analysis of Aderogba and Blake (1978b) is taken into account, which implies that the interface can be regarded as flat, when the interfacial tension or the density difference between the two fluids is very large.

Similarly to the single slender body in a unbounded fluid velocity representation (3.8.4), the velocity field in the presence of an interface is approximated as

$$\mathrm{v}_i(\mathbf{r}) = \int_{-l}^{l} \mathrm{G}_{ij}^{\mathrm{W}}(\mathbf{r},\mathrm{s})\,\mathrm{f}_j(\mathrm{s})\mathrm{d}\mathrm{s}, \tag{8.3.37}$$

where $2l$ is the slender body length, s is centreline coordinate, $\mathbf{f}$ is the integral of the local surface stress vector $\mathbf{t}$ around the closed cross-section perimeter of the particle at fixed s and $\mathbf{G}^{\mathrm{W}}$ is the correspondent Green function. In section 3.9 we have shown the fundamental solution of the Stokes flow for a Stokeslet located in a fluid, denoted as II fluid, separated by another fluid, I fluid, by an interface. There $\mathbf{G}^{\mathrm{W}}$ is given by (3.9.19), since $\mathbf{v}_2 = \boldsymbol{\alpha} \cdot \mathbf{G}^{\mathrm{W}}$.

The unknown force distribution $\mathbf{f}$ in (8.3.37) can be determined by taking into account the no-slip boundary conditions on the slender particle surface. As a result, a system of Fredholm integral equations of first kind for the components of $\mathbf{f}$ is obtained.

When calculating the drag on the slender body, Fulford and Blake find that, if the body is located sufficiently far away from the interface, the particle orientation and shape has no effect and their result is the same as that of Lee et al. (1979) for the sphere case. For small distances between the body and the interface they obtain expressions for the drag which are in good agreement with the results of Blake (1974) for the limiting case of a rigid planar wall. Using the analysis of Fulford and Blake the torque on the particle can also be calculated, but it cannot be used to determine correctly the angular velocity of the particle without studying the particle rotation in the a quiescent fluid. However, an arbitrary oriented slender body motion cannot be predicted from their analysis.

A similar, but free from the mentioned shortcomings, investigation is performed by Yang and Leal (1983) as an extension of a series of works belonging to Leal and collaborators, i.e., Lee et al. (1979), Lee and Leal (1980, 1982) and Berdan and Leal (1982). Particle translation and rotation in three mutually orthogonal directions is considered. Thus, six fundamental problems are solved, which give the basis for the construction of the hydrodynamic resistance tensors relating the force and torque on the particle with its translational and angular velocities. The particle trajectories are further calculated for "sedimentation" of a freely rotating particle under the action of normal or parallel to the interface force. The fundamental solution, as stated in section 3.9, of Lee et al. (1979) together with the slender body theory, as presented by, e.g. Batchelor (1970b), Cox (1970), Keller and Rubinow (1976) and Johnson (1980), is applied to examine the hydrodynamic interaction between an arbitrary oriented slender body particle translating and rotating in a quiescent fluid in the presence of a undeformable planar fluid/fluid interface under the conditions of insignificant deformations.

For the translation of the slender body in parallel or perpendicularly to the interface, the induced torque is only due to the presence of the interface and it is evident that the body will rotate if an opposite torque is not applied to it by some external sources. Then, a freely suspended body will rotate with angular velocity depending totally on the viscosity ratio and the distance and orientation of the body with respect to the interface. For the particle rotation about an axis parallel to the interface, a hydrodynamic force is induced perpendicularly to the axis of rotation due to only the presence of the interface. Yang and Leal show that the dimensionless correspondent of this force necessary to sustain the considered rotation without translation is identical with the dimensionless torque necessary to sustain the correspondent translation without rotation. This result is due to the linearity of Stokes problem and we can conclude, that in the absence of external force in the second direction, the rotation along the first direction will induce a translation in the second direction. When the particle is oriented normally to the interface and rotates about its axis of symmetry, then the total torque is of the order of the body aspect ratio and the ratio between the body radius and distance to the interface, and the induced force is zero. This result is confirmed by the Batchelor's study (Batchelor, 1970b) of a slender body in an unbounded fluid. For arbitrary oriented body, the presence of the interface creates a torque in a direction perpendicular to the rotation axis and its value is absolutely the same as the additional torque, when the axis of rotation is parallel to the interface. Here again we have interchanged values of the induced force, normal to the axis of rotation, and torque, induced by the translation in the same direction. The direction of this force depends on the body orientation with respect to the interface.

The obtained fundamental results by Yang and Leal can be better understood as the trajectories of a freely rotating slender particle are calculated, when a force parallel or perpendicular to a planar interface acts on the particle. This problem is connected with the processes of particle sedimentation near a planar interface or of particle capture in floatation of drops or bubbles.

The hydrodynamic interaction of a floating bubble with a suspension particle is extremely important if the distance between the particle and bubble surface is less or of the order of the particle radius (Derjaguin and Dukhin, 1960-1961). Dukhin and Rulev (1977) give an exact solution of this problem in Stokes approximation, when the spherical particle radius is small compared to that of the bubble and the latter one is considered locally flat. In this respect, one can formulated the problem of an inertia-free rigid sphere of radius a near a stagnation point of an axisymmetrical velocity field of a viscous flow with dynamic viscosity μ bounded by an infinite planar free surface S, normal to the symmetry axis. The distance between the sphere centre and the surface S is h and the velocity field in the absence of the sphere is a pure straining flow near a gas/liquid interface, i.e., $\mathbf{V} = E(\rho\, \mathbf{e}_\rho - 2z\mathbf{e}_z)$, where $\mathbf{V}$ is the velocity vector written in cylindrical coordinate system (ρ, z, φ), $\mathbf{e}_\rho$ and $\mathbf{e}_z$ are the unit vectors in the corresponding directions and E is a constant. Dukhin and Rulev express the hydrodynamic interaction as a superposition of two forces: the first one is due to the action of the normal component of the bubble buoyancy force, which enhances the bubble - particle approach and the second one reflects the viscous resistance of the liquid layer separating the particle and the bubble. Then the velocity satisfying the Stokes equations together with the continuity equation (1.3.6) is represented as a sum of two parts

$$\mathbf{u} = \mathbf{u}_1 + \mathbf{u}_2, \tag{8.3.38}$$

where $\mathbf{u}_1$ is the disturbance velocity, i.e., the flow velocity when a sphere moves in the vicinity of a planar free surface in quiescent fluid with velocity $\mathbf{v}$, which is determined from the induced force due to the second velocity field $\mathbf{u}_2$. The exact solution of the first velocity field is the Brenner's solution (Brenner, 1961) , i.e.,

$$\mathbf{F} = 6\pi\, \mu\, a\mathbf{v}\Phi_1 \tag{8.3.39}$$

with correction coefficient for the drag Φ_1, given at the beginning of the present section (8.3.1).

The second velocity field is a solution of the problem of a pure straining flow around a stationary sphere and interface and is expressed by the Stimson and Jeffery's fundamental solution in bispherical coordinates (Stimson and Jeffery, 1926). For the pure straining flow at infinity, Dukhin and Rulev show that the Stokes drag formula has the form:

$$\mathbf{F} = -12\pi\, \mu\, ahE\Phi_2\mathbf{e}_z\,, \tag{8.3.40}$$

where

$$\Phi_2 = \frac{2sh^2\alpha}{3ch\alpha}\sum_{n=1}^{\infty}\frac{n(n+1)}{\exp(n+1/2)\alpha}\{[2ch\alpha-(2n+1)sh\alpha][ch(n-1/2)\alpha - sh(n+3/2)\alpha]$$
$$+sh\alpha[(2n+3)ch(n+3/2)\alpha-(2n-1)ch(n-1/2)\alpha]\}$$
$$/[(2n+3)sh(2n-1)\alpha ch(n+3/2)\alpha-(2n-1)ch(2n-1)\alpha sh(n+3/2)\alpha]$$

The sphere velocity **v** is simply obtained from (8.3.39) and (8.3.40) to be:

$$\mathbf{v} = -\Phi_2/\Phi_1 2Eh\mathbf{e}_z. \qquad (8.3.41)$$

The authors calculate the ratio Φ_2/Φ_1 for different dimensionless distances h/a and they conclude that the sphere velocity exactly coincides with the normal velocity component of the undisturbed flow, if the particle is located out of the region $0< h/a <3$. In this region the hydrodynamic interaction forces start to act. They are represented by the ratio Φ_2/Φ_1 and hinder the particle approach towards the bubble during finite time.

A similar problem for a sphere or a slender body and a flat fluid/fluid interface is treated by Yang and Leal (1984) by use of spatial distributions of basic singularities. From the obtained solutions the hydrodynamic "resistance" tensor components are afterwards constructed for different undisturbed flows $\mathbf{U}_i$, of fluid i (i = 1, 2): pure straining $\mathbf{U}_i = \mathbf{E.r}$ [$E_{11} = E_{22} = E$, $E_{33} = -2E$, $E_{ij} = 0$ (i≠j)] or simple shear flow $\mathbf{U}_1 = \lambda^{-1}\mathbf{\Gamma.r}$ ($\lambda = \mu_1/\mu_2$ is the viscosity ratio) and $\mathbf{U}_2 = \mathbf{\Gamma.r}$ ($\Gamma_{13}, \Gamma_{23} \neq 0$, the remaining $\Gamma_{ij} = 0$) and different particle velocities, translational or angular velocity. This work is an extension of the previous work of the same authors (Yang and Leal, 1983) as discussed before in the present section.

When the spherical particle is situated in fluid 2 and in the limiting case of a small sphere radius compared with the distance to the interface, a<< h, the disturbance flow velocities are obtained by the method of reflections (see section 3.9). Only the velocity $\mathbf{v}_2$ has a more complicated form in order to satisfy approximately, up to $O(\delta^4)$ with δ=h/a, the no-slip boundary conditions on the sphere surface.

To fulfil this requirement, when the sphere is located in a pure straining flow, several singularities are placed in the sphere centre: a Stokeslet, a potential dipole, a stresslet and a potential quadruple (see section 3.4), which are expanded in power series of the small parameter δ. The correction coefficient Φ_2 in the drag force expression (8.3.40) is given in an expansion in δ:

$$\Phi_2 = 1 + \sum_{n=1}^{3}\left(\frac{3}{8}\delta\frac{2+3\lambda}{1+\lambda}\right)^n - \frac{1+4\lambda}{8(1+\lambda)}\delta^3 - \frac{5}{12h}\sum_{n=1}^{2}\delta\left(\frac{3}{8}\delta\frac{2+3\lambda}{1+\lambda}\right)^n + O(\delta^4). \quad (8.3.42)$$

The upper formula shows a very good agreement with the exact solution of Dukhin and Rulev at $\lambda = 0$, except for the region very near to the interface, h = a, due to the poor convergence of the series (8.3.42) there. However, the maximum error is sufficiently small: 2.72% for h =1.001a, the smallest value considered by Dukhin and Rulev, and at h>1.5a, the error is less than 0.98%.

If the sphere is passed by a simple shear flow, the singularities placed in the sphere centre, satisfying up to $O(\delta^4)$ the no-slip boundary conditions on the sphere surface, are the same as for the pure straining flow plus a rotlet. Then the asymptotic expansion of the correction coefficient Φ_3 in the drag force expression $\mathbf{F} = -6\pi\mu_2\Gamma_{13}ha\Phi_3\mathbf{e}_1$ is:

$$\Phi_3 = 1 + \sum_{n=1}^{3}(-1)^n\left(\frac{3}{16}\delta\frac{2-3\lambda}{1+\lambda}\right)^n - \frac{1+2\lambda}{16(1+\lambda)}\delta^3 + \frac{1}{16h}\delta^2\frac{2-5\lambda}{1+\lambda}\left[1-\frac{3}{16}\delta\frac{2-3\lambda}{1+\lambda}\right] + O(\delta^4), \tag{8.3.43}$$

where for simplicity, but without loss of generality due to the sphere-interface symmetry, the only non-zero component of the simple shear tensor is Γ_{13}. At $\lambda \to \infty$ this result coincides with the exact solution of Goren and O'Neill (1971) for a rigid sphere near a plane wall and the relative error for $h > 1.5a$ is about 2.6%.

The problem for a slender body in a pure straining or simple shear flow is treated by Yang and Leal in an analogous manner as that for the sphere. On the basis of the obtained results for the force and torque, the general trajectory equations for an arbitrary viscosity ratio are constructed and they are well compared with the exact solution results, except for $h \approx a$, e.g. the results of Goldman et al. (1967a, b) for the translational and angular velocities of neutrally buoyant sphere moving in a shear flow near a plane wall at $\lambda \to \infty$. There is a good coincidence with the experimental data of Darabaner and Mason (1967) for the angular velocity of a neutrally buoyant sphere in a Couette viscometer.

A similar study of the interaction between a small drop and a large spherical drop, when both drops translate in a quiescent fluid, is performed by Stoos et al. (1992). The matched asymptotic expansions method is applied as the surface of the large sphere locally regarded planar and the ratio between the small and large spherical drops radii is taken as a small parameter for the expansion. The fluid collector problem is reduced to a series of component problems, each of them treated by a particular version of the reflection method outlined in Lee et al. (1979) and Yang and Leal (1990) (for details see section 3.9.). As a result, the small droplet trajectory is found, which provides the droplet to be attracted closer by the fluid collector than by a solid collector. The repulsion force between the two drops becomes weaker with the reduction of the droplet viscosity and depends strongly on the separation distance from the larger drop surface.

Recently Danov et al. (1995a, b) have investigated the hydrodynamic interaction of a fluid particle with a fluid interface in surfactant presence. These authors have shown that there is a strong influence of both shear and dilational interfacial viscosities on the fluid particle motion when the particle-interface distance is approximately equal to or smaller than the particle radius.

8.4. Small Deformations of One or Two Drops in the Presence of a Deformable Interface

As mentioned in section 8.2., Bart (1968) examines the problem of a fluid drop motion towards a rigid plane or plane interface separating two immiscible fluids by means of the bispherical coordinates system analysis. The influence of a flat fluid/fluid interface on the thermocapillary migration of a spherical bubble moving perpendicularly to the interface in a space laboratory is investigated by Meyyappan et al. (1981a, b).

The slow translation of a rigid sphere normally to an initially flat but deformable interface has been studied approximately by O'Neill and Ranger (1983) and analytically and numerically by Berdan and Leal (1982), Lee and Leal (1982) and Geller et al. (1986). In the last two works the boundary integral technique is applied without any restrictions on the magnitude of the interface deformation.

As stated by Lee et al. (1979), when a sphere of radius a is moving slowly perpendicularly to an interface at a distance $l \geq O(a)$, the small interface deformation restriction means that the capillary number is small, i.e., $Ca = \mu^{(2)}U/\sigma \ll 1$ or its ratio with the Bond number is also small, i.e., $Ca/Bo = \mu^{(2)}U/\rho\, ga^2 \ll 1$, where the superscript (1) refers to the fluid above the interface and (2) refers to the fluid under it, ρ is the density difference between the densities of both fluids, σ is the interfacial tension and U is the sphere velocity. At small but finite Ca or Ca/Bo, Berdan and Leal (1982) consider the effects of the interface deformation, which are quasi-linear and are presented as an asymptotic expansion on a small parameter $\varepsilon = Ca/(1 + Bo) \ll 1$. They study only the case of $l \gg a$ and the interface shape, presented in cylindrical coordinates (ρ, φ, z) is sought in the form of the asymptotic series:

$$F \equiv z - f(\rho, \varphi) = 0, \tag{8.4.1}$$

where $f = \varepsilon f_1 + \varepsilon^2 f_2 + \ldots$.

Thus, the solutions of Lee et al. (1979) and Lee and Leal (1980) represent the zero-order term in this asymptotic expansion, i.e., a flat interface z = 0, when omitting the normal viscous stress balance on the interface. The higher-order corrections f_1, f_2, etc. to the interface shape can be calculated via the condition:

$$\mathbf{n}.\left[\lambda\, \mathbf{n}.\mathbf{T}^{(1)} - \mathbf{n}.\mathbf{T}^{(2)}\right]_{z=0} = \frac{Bo}{Ca} f + \frac{1}{Ca}\left(\frac{1}{R_1} + \frac{1}{R_2}\right), \tag{8.4.2}$$

where R_1 and R_2 are the principal radii of the interface curvature.

For the velocity, pressure and viscous stress tensor of either fluids, analogous asymptotic expansions are assumed:

$$\mathbf{v}^{(i)} = \mathbf{v}_0^{(i)} + \varepsilon \mathbf{v}_1^{(i)} + \varepsilon^2 \mathbf{v}_2^{(i)} + \ldots, \tag{8.4.3}$$

$$p^{(i)} = p_0^{(i)} + \varepsilon p_1^{(i)} + \varepsilon^2 p_2^{(i)} + \ldots,$$

$$\mathbf{T}^{(i)} = \mathbf{T}_0^{(i)} + \varepsilon \mathbf{T}_1^{(i)} + \varepsilon^2 \mathbf{T}_2^{(i)} + \ldots,$$

where i =1, 2. The first order corrections to the velocity, pressure and viscous stress field are calculated after defining the interface deformation f_1 from the normal viscous stress balance (8.4.2). Since $\mathbf{n} = \mathbf{e}_z$ for the zero-order interface shape z = 0, then **n.n.T** = T_{zz} and the zero-order normal viscous stress jump is given by $-\Delta T_{zz} = \left[\lambda\, T_{zz}^{(1)} - T_{zz}^{(2)}\right]_{z=0}$. The values of this jump for four fundamental cases of sphere translation and rotation in parallel and perpendicularly to the interface can be found in (Lee et al., 1979):

$$-\Delta T_{zz} = \frac{9l^3}{R_0^5}\left(1 + \frac{3(2+3\lambda)}{8(1+\lambda)l}\right) + O\left(l^{-4}\right) \quad \text{for normal translation,} \tag{8.4.4}$$

$$-\Delta T_{zz} = \frac{9l^3 \rho\cos\varphi}{R_0^5}\left(1 - \frac{3(2-3\lambda)}{16(1+\lambda)l}\right) + O\left(l^{-5}\right) \quad \text{for parallel translation,} \tag{8.4.5}$$

$$-\Delta T_{zz} = 0 \qquad \text{for normal rotation,} \tag{8.4.6}$$

$$-\Delta T_{zz} = \frac{12\rho\sin\varphi}{R_0^5}\left(1 + \frac{3}{16(1+\lambda)l}\right) + O\left(l^{-7}\right) \quad \text{for parallel rotation,} \tag{8.4.7}$$

where $R_0 = \left(\rho^2 + l^2\right)^{1/2}$ and ρ and l are dimensionalized by the sphere radius a.

From (8.4.6) it is clear that the interface shape is not deformed when the sphere rotates normally with respect to the interface. For the other three cases, the solution for f_1 is subjected to the equation:

$$-\Delta T_{zz} = \frac{Bo}{Ca}\varepsilon f_1 - \frac{\varepsilon}{Ca}\left(\frac{\partial^2 f_1}{\partial\rho^2} + \frac{1}{\rho}\frac{\partial f_1}{\partial\rho} + \frac{1}{\rho^2}\frac{\partial^2 f_1}{\partial\varphi^2}\right) \tag{8.4.8}$$

and to the zero boundary conditions as $\rho \to \infty$. The first-order corrections to the velocity and pressure are afterwards used to calculate the force and torque corrections, too.

Another way to calculate the velocity and pressure corrections is to approximate the boundary conditions on the deformed boundary by corresponding conditions applied on the undeformed boundary as shown by Chan and Leal (1979). Using Lorentz reciprocal theorem Berdan and Leal indicate that the force and torque corrections depend only on the interface shape correction f_1 and on the velocity and pressure belonging to the flat interface shape, i.e., $\mathbf{v}_0^{(i)}$ and $p_0^{(i)}$.

For all cases the results show that the interface deformation is highly sensitive to the value of Bond number Bo. When Bo >> 1 the interface deformation f_1 during the sphere translational motion perpendicular or parallel to the interface is exactly the right-hand side of (8.4.4) or (8.4.5). At the other limiting case when Bo << 1, the deformation is mainly due to the surface tension action and it vanishes at the normal translation, while at the parallel translation f_1 is given by:

$$f_1(\rho,\varphi) = \frac{3l\cos\varphi}{\rho}\left(1 - \frac{1}{R_0}\right)\left(1 - \frac{3(2-3\lambda)}{16(1+\lambda)l}\right). \tag{8.4.9}$$

For both cases of translation the viscosity ratio across the interface has a very small effect on the interfacial deformation. The interfacial deformation due to parallel rotation is similar to that of the parallel translation. If the force and torque are expressed by means of the resistance tensors **A**, **B**, **C** (see section 5.10.), then they are developed in asymptotic series in ε. The zero-order tensor **A** at parallel translation has only non-zero diagonal elements, while its first correction has non-zero elements not only on the diagonal, $A_{1,zx}$ and $A_{1,zy}$, which indicates the existence of a lift force normal to the direction of particle motion. This form of lateral migration resembles the one of a drop in shear flow (Chaffey et al., 1965) as it will be discussed in section 8.5. It is also analogous to the lateral migration observed by Chan and Leal (1979) of a settling drop away from a vertical plane wall due to the drop deformation. Another interesting result obtained by Berdan and Leal is the coupling between the particle parallel rotation and the force acting on the particle, leading to the particle migration away from the interface. Moreover, there are no corrections to the torque of order $O(\varepsilon)$. Finally, we can resume, that the force corrections reduce the drag for the normal translation towards the interface and increase it, when the motion is away from the interface. The parallel translation and rotation lead to a lift force away from the interface.

Chervenivanova and Zapryanov (1988) solve the quasisteady Stokes flow problem of one or two droplets slowly moving perpendicularly to a deformable fluid/fluid interface. Here, we shall briefly present the formulation, the method and the main results of their problem.

Two immiscible fluids (3) and (4) and a droplet of phase (1) rising (falling) through the fluid (3) and/or a droplet of phase (2) falling (rising) through the fluid (4) are considered, as shown in Fig.8.4.1.a) and b). The droplets motion with settling velocities $\mathbf{U}_1$ and $\mathbf{U}_2$ is along their line of centres perpendicularly to the initial position of the fluid/fluid interface. The droplet radii before deformation are respectively r_1 and r_2.

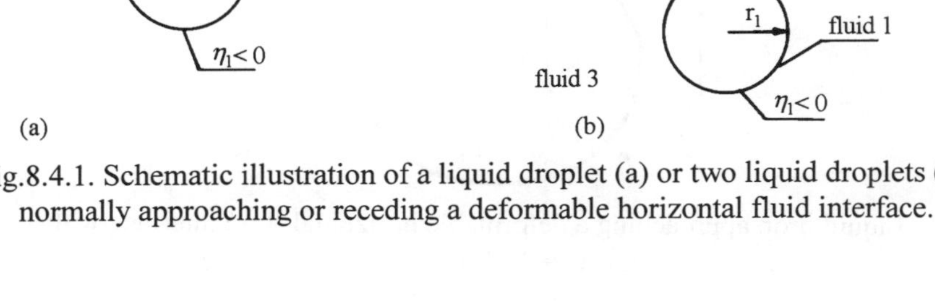

Fig.8.4.1. Schematic illustration of a liquid droplet (a) or two liquid droplets (b), normally approaching or receding a deformable horizontal fluid interface.

The problem parameters are introduced as follows: Reynolds number $Re = \rho_3 U_1 r_1/\mu_3$, the capillary numbers for the interfaces $Ca_1 = \mu_3 U_1/\sigma_1$, $Ca_2 = \mu_4 U_1/\sigma_2$, $Ca_3 = \mu_3 U_1/\sigma_3$, and the viscosity ratios $\lambda_1 = \mu_1/\mu_3$, $\lambda_2 = \mu_2/\mu_4$, $\lambda_3 = \mu_4/\mu_3$. Here μ_i (i =1, 2, 3, 4) are the dynamic viscosities of the fluids and σ_i (i =1, 2, 3) are the surface tensions of the three interfaces. The interface and droplets shapes are assumed steady (see Lee and Leal, 1982) and the Reynolds number is supposed sufficiently small, i.e., $Re < \varepsilon$ where $\varepsilon = \varepsilon'/r_1$ is the dimensionless minimum distance between the droplet and the flat interface.

The velocity and pressure in the four phases satisfy the Stokes equations with the corresponding boundary conditions on each of the threes interfaces: continuous tangential velocities and stresses, as in section 7.6., formulae (7.6.2) and (7.6.4); normal velocities equal to the interface normal velocity (for interfaces (1) and (2) the interface velocities are U_{1n} and U_{2n}, while for (3) the interface velocity is zero). The normal stress balance on every interface S_j is given by

$$T_{nn}^{(e)} - T_{nn}^{(i)} = \frac{1}{Ca_j}\left(\frac{1}{R_{1j}} + \frac{1}{R_{2j}}\right), \qquad (j=1, 2, 3) \qquad (8.4.10)$$

where Ca_j is the interface capillary number, R_{1j} and R_{2j} are the principal radii of curvature of the deformed interface S_j, and $T_{nn}^{(e)}$, $T_{nn}^{(i)}$ are the dimensionless normal stress vectors from both sides of the interface.

In the studied problem the authors restrict only to small interface deformations, as the solutions search starts with spherical droplets shapes and fluid/fluid interface planar shape. The bispherical coordinates method is applied as in section 7.7. for two droplets deformation. The droplets interfacial deformation is again sought in the form (7.7.7), while for the fluid/fluid interface:

$$\eta_3^H(\beta) = H_3(\beta), \qquad (8.4.11)$$

with assumed $\max_{\xi} |H(\beta)| < 1$ irrespective of the values of λ_3 and Ca_3.

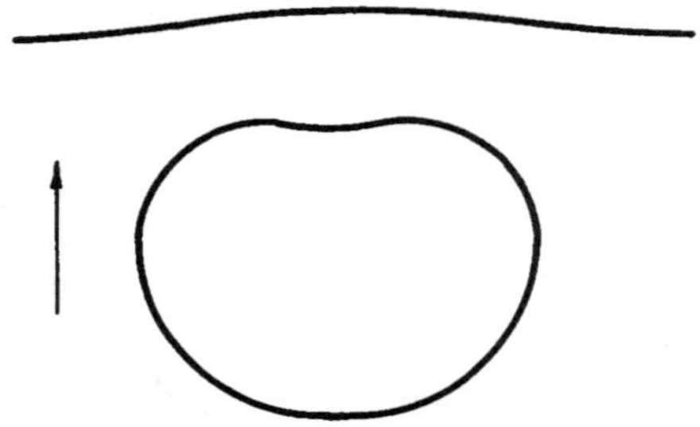

Fig.8.4.2. Liquid drop approaching a deformable horizontal fluid interface with $\varepsilon = 0.2$, $\lambda_1 = \lambda_3 = 0.5$, $Ca_1 = 0.13$, $Ca_3 = 0.85$, $Bo^{\Gamma} = 1.5$.

It is interesting to note that for this interface (j = 3) the equation (8.4.10) has no bounded solution in the absence of gravity (see Aderogba and Blake, 1978a; Berdan and Leal, 1982). To overcome this singularity, the hydrostatic pressure difference on the flat interface is introduced with an interfacial layer density assumed different from the bulk densities. If the excess density of the interface is Γ, the interfacial Bond number is given by $Bo^{\Gamma} = g\Gamma r_1^2/\sigma_3$. The Fourier analysis performed for each interface yields an expansion of the deformation function $H(\beta)$ in Legendre polynomials. For its coefficients recurrence equations are obtained, which are analogous to (7.7.14)-(7.7.17) for the droplets interfaces, while for the fluid/fluid interface their structure is more complicated and it is analysed by Chervenivanova and Zapryanov (1988).

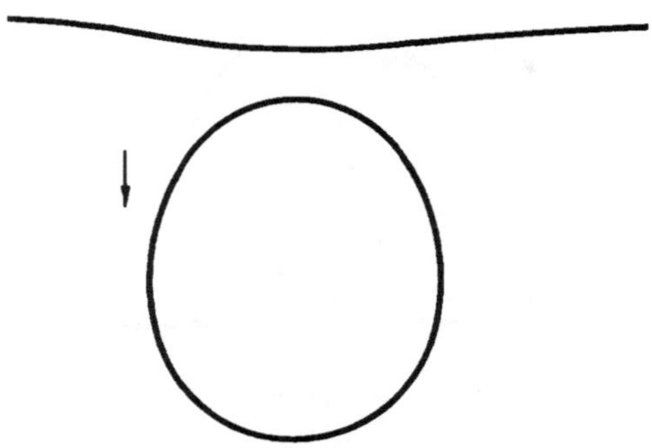

Fig.8.4.3. Gas bubble receding from a deformable horizontal fluid interface with $\varepsilon = 0.7$, $\lambda_1 = 0$, $\lambda_3 = 0.5$, $Ca_1 = 0.36$, $Ca_3 = 0.2$, $Bo^{\Gamma} = 1.5$.

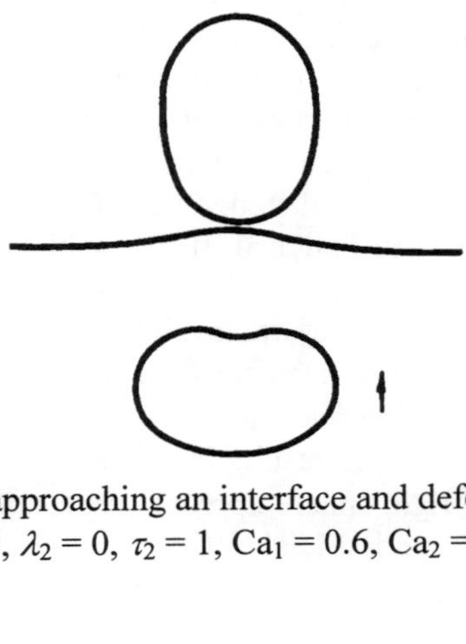

Fig.8.4.4. Liquid droplet approaching an interface and deformable stationary bubble: $\varepsilon_1 = \varepsilon_2 = 0.6$, $\lambda_1 = \lambda_3 = 0.5$, $\lambda_2 = 0$, $\tau_2 = 1$, $Ca_1 = 0.6$, $Ca_2 = 0.9$, $Ca_3 = 0.1$, $Bo^{\Gamma} = 0.5$.

For a constant droplets velocity normal to the interface and initially motionless fluids bounded by the flat interface, the following results are obtained. If the distance between the deformable droplet and the initially flat interface is very large, the interface is not deformed. As the separation distance decreases, the deformable flat interface gradually modifies the initially spherical drop shape, which, on its part, also influences the initially flat interface shape. At small separation distance, as shown in Fig.8.4.2., the droplet is flattened, while the interface is inflated in the direction of the droplet motion (l_1= 0.75). In Fig.8.4.3. a deformable bubble with an approximate shape of a prolate ellipsoid (l_1 = 1.25) receding from an infinite interface is illustrated. As a result the interface sinks into the fluid which contains the bubble.

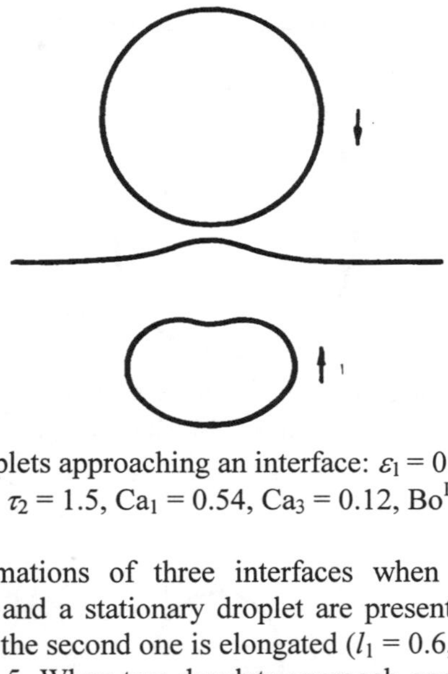

Fig.8.4.5. Two liquid droplets approaching an interface: $\varepsilon_1 = 0.5$, $\varepsilon_2 = 0.4$, $\lambda_1 = \lambda_3 = 0.5$, $\lambda_2 = 2$, $\tau_2 = 1.5$, $\mathrm{Ca}_1 = 0.54$, $\mathrm{Ca}_3 = 0.12$, $\mathrm{Bo}^\Gamma = 0.5$.

The shape deformations of three interfaces when a droplet approaches a deformable flat interface and a stationary droplet are presented in Fig.8.4.4. The first droplet is flattened, while the second one is elongated ($l_1 = 0.6$, $l_2 = 1.37$). An interesting result is shown in Fig.8.4.5. When two droplets approach one another it is possible for the second droplet to preserve its spherical shape for some special values of the problem parameters, while the first droplet is flattened ($l_1 = 0.62$, $l_2 = 1$); the interface towards the fluid of lower viscosity is also flattened. If two bubbles move in the same direction with the same velocity, as given in Fig.8.4.6., then the front bubble is elongated ($l_1 = 1.4$) due to the "suction" of the deformable interface and the rear bubble is flattened ($l_2 = 0.7$), while the interface is inflated towards the elongated bubble. Both droplets flatten ($l_1 = l_2 = 0.62$) and the interface sinks towards the fluid of lower viscosity, as presented in Fig.8.4.7. The obtained results by Chervenivanova and Zapryanov (1988) are in good agreement for a rigid sphere approaching a deformable interface ($\lambda_1 \to \infty$) with the numerical results of Lee and Leal (1982).

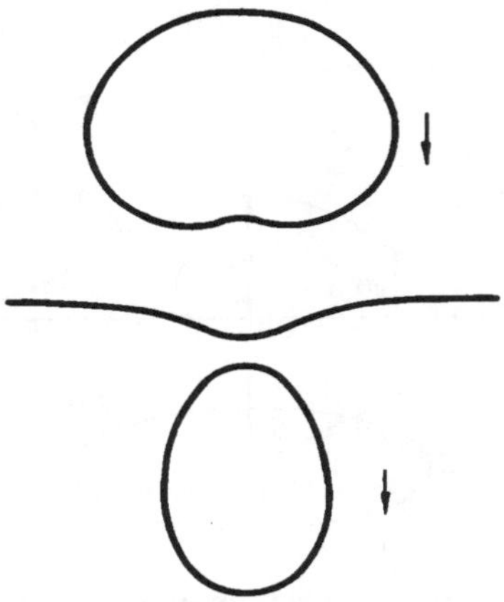

Fig.8.4.6. Two bubbles moving in the same direction with zero relative velocity: $\varepsilon_1 = 0.8$, $\varepsilon_2 = 0.5$, $\lambda_1 = \lambda_3 = 0$, $\lambda_2 = 0.5$, $\tau_2 = 1.5$, $Ca_1 = 0.4$, $Ca_2 = 0.17$, $Ca_3 = 0.08$, $Bo^{\Gamma} = 0.5$.

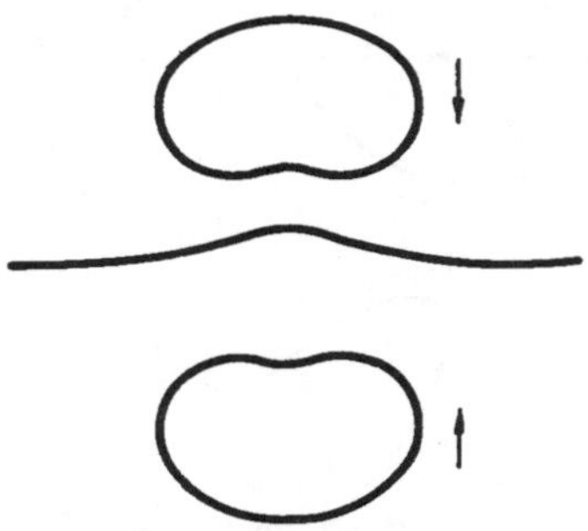

Fig.8.4.7. Two liquid droplets approaching an interface: $\varepsilon_1 = \varepsilon_2 = 0.5$, $\lambda_1 = \lambda_3 = 0.5$, $\lambda_2 = 2$, $\tau_2 = 1$, $Ca_1 = 0.7$, $Ca_2 = 0.46$, $Ca_3 = 0.4$, $Bo^{\Gamma} = 0.5$.

Using the boundary integral technique, Chi and Leal (1989) study the translation of a buoyant deformable drop normal to a deformable interface, where the deformation magnitude is not initially prescribed. However, from the numerical solutions it is quite difficult to evaluate the effect of the different problem parameters on the deformations of either interfaces. Further on, we shall present some of the analytic results obtained by Yang and Leal (1990), when solving the general three fluids problem for normal and parallel translation of a deformable drop with respect to a deformable interface, in the limit of small deformation.

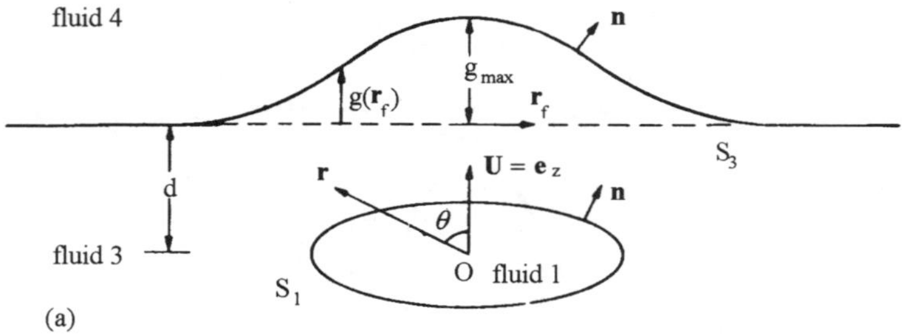

Fig.8.4.8.a) A sketch of the normal translation of a deformable drop with respect to a deformable interface.

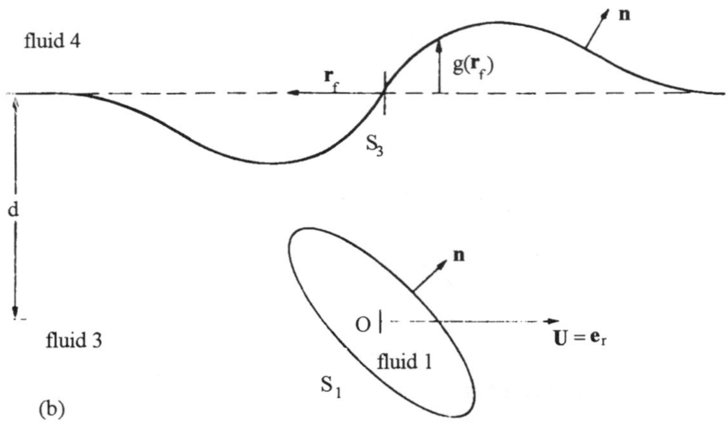

Fig.8.4.8.b) A sketch of the parallel translation of a deformable drop with respect to a deformable interface.

A schematic geometry of the two considered cases is shown in Fig.8.4.8.a) and b). The small deformations of either interface implies, like in (Berdan and Leal, 1982), small but finite Ca_i (i=1, 3) or Ca_i/Bo_i and large dimensionless distance between the drop and the interface $\varepsilon = 1/d \ll 1$. Then the drop shape and the fluid/fluid interface shape, presented in cylindrical coordinates (ρ, φ, z) are sought in the form of the asymptotic series:

$$r-1-f(\rho,\ \varphi)=0, \qquad \text{for the drop interface } S_1, \qquad (8.4.12)$$

$$z-d-g(\mathbf{r}_f)=0, \qquad \text{for the fluid/fluid interface } S_3, \qquad (8.4.13)$$

where $f=\varepsilon^2 f_1+\varepsilon^3 f_2+\ldots$, $g=\varepsilon^2 g_1+\varepsilon^3 g_2+\ldots$ and $\mathbf{r}_f$ is the position vector of points lying in the undeformed interface plane. The spherical drop and the flat interface are the zero-order approximations of (8.4.12) and (8.4.13). Here, again as in the previous discussed works, the normal stress balance condition (8.4.10) can be used to determine the higher-order approximations to the deformations of the two interfaces.

For zero-order shapes of the interfaces, the problem of a parallel or perpendicular translation of the drop to the fluid/fluid interface is solved by the method of reflections of singularities located at the drop centre. Further on the zero-order solution for both translations is used to calculate the first order approximation of the droplet and fluid/fluid interface shapes. Due to the Lorentz reciprocal theorem the lateral migration velocity can be determined only from the zero-order solution, without specifying the first-order corrections to velocity and pressure. For this reason, a complementary Stokes problem is solved for the droplet and interface by the method of domain perturbations, replacing the boundary conditions on the deformed interfaces with modified boundary conditions on the initially undeformed surfaces, i.e., spherical drop and flat interface, as done by Chan and Leal (1979) and Berdan and Leal (1982). This can be done successfully, since the effects of small drop and interface deformations on the migration velocity are additive to each other.

In the case of a drop perpendicular motion to the fluid/fluid interface, the approach of the complementary problem requires at the spherical drop centre a Stokeslet and a potential dipole to be situated. However, these two singularities account only for the drop in an unbounded fluid, while the presence of the interface in the flow needs a correction. In the limit of $\varepsilon \ll 1$, it is expressed as a steady streaming in direction opposite to that of the drop, an axisymmetric uniaxial extensional flow with a stagnation point at the drop centre and the following additional singularities at the drop centre: a Stokeslet, a potential dipole, a stresslet and a potential quadrupole. The drag obtained by Yang and Leal is represented in an asymptotic expansion in the small parameter ε and is in very good agreement with the one found by Bart (1968), except for small distances between the drop and the interface. The drop deformation is given by:

$$De=\frac{3Ca_1G_{ex}\dfrac{16+19\lambda_1}{16+16\lambda_1}}{1-Ca_1G_{ex}\dfrac{16+19\lambda_1}{16+16\lambda_1}}\approx 3Ca_1G_{ex}\frac{16+19\lambda_1}{16+16\lambda_1} \quad \text{as } Ca_1\varepsilon^2\to 0, \qquad (8.4.14)$$

where $G_{ex}=\varepsilon^2\dfrac{9}{32}A(\lambda_1)A(\lambda_3)$ and $A(\lambda_i)=\dfrac{2+3\lambda_i}{3(1+\lambda_i)}$.

The expression of the interface deformation is in a more complicated form (eq.[36], Yang and Leal., 1990) and up to $O(\varepsilon^2)$ it is independent of the viscosity ratio λ_3 and its magnitude is of order $O(\varepsilon^2 Ca_3)$ if $Bo_3 > 0$. However, when $Bo_3 \to 0$, its magnitude is of $O(Ca_3 \ln Bo_3)$ for fixed ε, it has a logarithmic singularity at $Ca_3=O(1)$. In the gravity case when $Bo_3 \to \infty$, the interface deformation is presented by the formula:

$$g(\mathbf{r}_f) \approx \varepsilon^2 \frac{9A(\lambda_1)Ca_3 d^5}{\left[\rho^2 + d^2\right]^{5/2} Bo_3}. \tag{8.4.15}$$

For the drop translation parallel to the interface the singularities located at the drop centre are similar to those for the normal translation, but the presence of the interface is accounted by a steady streaming flow parallel to the interface and a linear shear flow either normal or parallel to the interface. The drag calculated as an asymptotic series on ε for the limiting case of a rigid sphere, i.e., $\lambda_1 \to \infty$, is identical with the drag results of Lee et al. (1979) and Yang and Leal (1984) up to $O(\varepsilon^2)$. If $\lambda_3 > 2/3$ the drag force on the droplet in the presence of the interface is larger than that on a droplet in an unbounded fluid, while at $\lambda_3 < 2/3$ it is less. The drop deformation in the translation parallel to the interface is asymmetrical, in contrast to the normal translation axisymmetric deformation. Then the deformation parameter becomes:

$$De = \varepsilon^2 Ca_1 A(\lambda_1) \frac{9}{16} \cdot \frac{16+19\lambda_1}{16+16\lambda_1} \cdot \frac{\lambda_3}{1+\lambda_3} \quad \text{as } Ca_1 \varepsilon^2 \to 0. \tag{8.4.16}$$

The interface deformation is like the one sketched in Fig.8.4.8.b) and although at first sight it is very different from the interfacial deformation of the drop normal translation, there are some similarities: the deformation increases at smaller distance between the droplet and the interface; the viscosity ratio λ_3 does not influence on the degree of deformation; the density difference has a stronger effect on the deformations than the interfacial tension. However, in this case, the shape solution has no singularities at $Bo_3 \to 0$

$$g(\mathbf{r}_f) \approx \frac{3Ca_3 A(\lambda_1) r_f d}{\rho^2} \left[1 - \frac{d}{\left(\rho^2 + d^2\right)^{1/2}} \right], \tag{8.4.17}$$

and at gravity action $Bo_3 \to \infty$, it becomes

$$g(\mathbf{r}_f) \approx \varepsilon^2 \frac{9A(\lambda_1)Ca_3 r_f d^5}{\left[\rho^2 + d^2\right]^{5/2} Bo_3}. \tag{8.4.18}$$

The performed analysis shows a migration away from the interface with velocity of order $O(\varepsilon^6)$. Yang and Leal study also the effects of viscosity ratios, the capillary numbers and the Bond numbers on the two interfaces deformations, as well as on the drop migration. In addition, it can be noted that if higher-order approximations are necessary, then more complicated singularities must be located at the drop centre, such as Stokes quadrupoles and potential octupoles.

8.5. Rigid or Fluid Particles in Tubes and Channels

The behaviour of rigid or fluid particles immersed in a fluid bounded by a circular cylindrical tube has certain particular features. A complete review of the works till 1972 on the hydrodynamic interaction between rigid or fluid particles in circular tubes filled with viscous fluid is given by Brenner (1966, 1971) and Cox and Mason (1971). A review of more recent theoretical investigations can be found in (Clift et al., 1978) and (Grace, 1983). The effects of inertia and deformation on the behaviour of single particles or particles near walls are studied by Leal (1980).

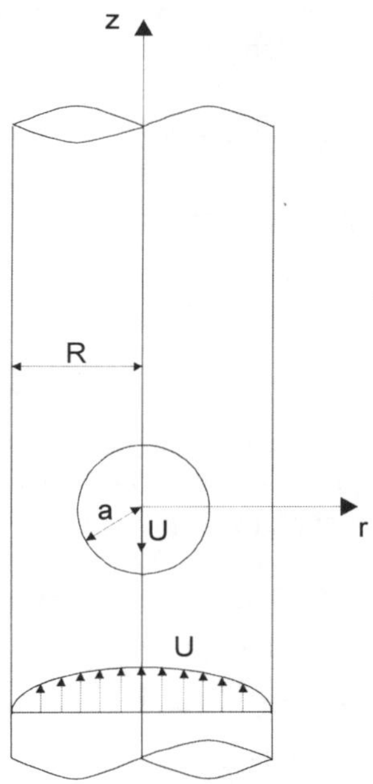

Fig.8.5.1. A sketch of a sphere falling in a tube.

Rigid particles studies. The case of greatest applied interest is when the particle size is small compared with the cylinder radius (see Fig.8.5.1.). This problem can be viewed as one of the remarkable examples for the importance of the wall effects. The magnitude of hydrodynamic interaction effect depends essentially upon the size and lateral position of the particle with respect to the tube wall. Landenburg (1907) examines the problem of the motion of a rigid sphere with constant velocity U in the axis direction of a circular cylindrical tube. Using an approximate method, he obtains the following formula for the drag experienced by the sphere during its motion in the tube:

$$F = 6\pi \mu aU(1 + 2.4\beta), \tag{8.5.1}$$

where r_0 is the tube radius, a is the particle radius and $\beta = a/r_0$. The same problem has been solved by Faxen (1921) on the basis of his own method and he also finds that, due to the wall effect, the drag on the particle in the tube is larger than that at its motion in a unbounded fluid. With this he confirms the conclusion, made by Landenburg, about the retaining effect of the cylindrical wall upon the particle motion. Applying the Faxen's method Wakiya (1953) solves the same problem, when the spherical particle is moving along the axis of a circular tube, in which there is a developed Poiseuille flow. For the drag on the sphere he obtains

$$F = 6\pi \mu aU \frac{1 - 2/3\,\beta}{1 - 2.104\beta + 2.09\beta^2 - 1.11\beta^3}. \tag{8.5.2}$$

To compare with the Landenburg's results, Wakiya analyses the case when the sphere moves with a constant velocity U along the tube axis and reaches to a formula more accurate than (8.5.1) for the drag:

$$F = 6\pi \mu aU(1 + 2.1\beta). \tag{8.5.3}$$

By the reflection method Brenner (1966), as well as Greenstein and Happel (1968), define the force and torque acting on the sphere in the following form:

$$F_z = 6\pi\mu aU\left[\frac{U - U_m(1-\alpha^2) + 2/3\,U_m\beta^2}{1 - \beta f(\alpha)} + U_m O(\beta^3)\right], \tag{8.5.4}$$

$$T_x = 8\pi\mu a^2\left\{(a\Omega - U_m\alpha\beta) + \left[U - U_m(1-\alpha^2)\right]g(\alpha)\left[1 + \beta f(\alpha)\right]\alpha^2 + U_m O(\beta^4)\right\}$$

where $U\mathbf{e}_z$ and $\Omega\mathbf{e}_x$ are the translational and angular velocity of the sphere, $\alpha = 1 - h/r_0$ and the requirement for the sphere to be located away from the walls is given by $\beta \ll 1 - \alpha$. The functions $f(\alpha)$ and $g(\alpha)$ represent the wall effect and they have been first determined by Brenner and Happel (1958) and tabulated by Famularo and Happel (1965) and Greenstein and Happel (1968). The main properties of these functions are:

$$f(\alpha) = 2.104 - 0.6977\alpha^2 + O(\alpha^4), \quad \text{at } \alpha \to 0 \tag{8.5.5}$$

$$g(\alpha) = 1.296\alpha + O(\alpha^3) \qquad \text{at } \alpha \to 0,$$

$$f(\alpha) \approx 9/16(1-\alpha)^{-1} \qquad \text{at } \alpha \to 1,$$

$$g(\alpha) = O(1-\alpha)^{-1} \qquad \text{at } \alpha \to 1.$$

If the particle is located at distance h of O(a) from the tube wall, then the above expressions (8.5.4) and (8.5.5) are no longer valid. In this case, however, the tube wall may be regarded flat, i.e., the results of Dean and O'Neill (1963) and those of Goldman et al (1967a, b) for the motion of a sphere near a planar wall may be utilised.

Since there does not exist a special coordinate system to describe the tube walls and the sphere as coordinate surfaces, the method of reflections proves to be a powerful tool for solving creeping flow problems usually in asymptotic series in the parameter a/h, where h is the distance from the particle centre to the nearest tube wall. If $a/h < 0.2$ the solution converges quite rapidly, while for $a/h > 0.8$ converges poorly or diverges totally. This method is preferred by many authors, e.g., Hetsroni et al. (1970a, b) use it to solve the problem of a droplet moving in Poiseuille flow. Ho and Leal (1974) study the weak inertia effects for a neutrally buoyant, freely rotating sphere in Couette and plane Poiseuille flow by a perturbation method based again on the reflection method. They find an expression for the lift force on the sphere, which is valid when the sphere is not close to the walls. The more general problem of a sphere moving between nonparallel walls has been solved by Sano and Hasimoto (1978) via the reflection method.

The collocation method presented in section 8.2. has been successfully exploited by some authors for solving the problems discussed in this section. Ganatos et al. (1980a,b) study by this method the motion of a sphere in a general direction between two parallel walls. The collocation method can be used for all range of the ratio a/h and the results obtained by it are in good agreement with those found by the reflection method for $a/h < 0.2$. The method of matched asymptotic expansion is applied to the problem of a sphere translated in a shear flow between parallel planar walls for small but finite Reynolds numbers by Cherukat et al. (1994) and it is an extension of a similar work by McLaughlin (1993) for the inertial lift on a sphere translated in a shear flow bounded by a single plane wall.

Cox and Mason (1971) prove that at zero Reynolds numbers there is no radial migration of the moving single spherical particle along the axis of the tube. It can be shown that this result is valid also for neutrally buoyant particle and for cylindrical tubes with arbitrary cross section. This is confirmed experimentally by Goldsmith and Mason (1962) for Reynolds numbers of the order 0.001. Tachibana (1973) observes experimentally, that a rigid sphere in a tube moves under the action of gravity almost parallel to the walls at small, but finite Reynolds numbers, i.e., at Re = 5 to 8. To the same results Ho and Leal reach in their theory (Ho and Leal, 1974). Miyamura et al. (1980) perform similar to Tachibana's experiments for rigid spheres translated between two parallel infinite walls and in conduits of square and triangular cross section.

When performing experiments for the slow motion of a rigid spherical particle neutrally suspended in Poiseuille flow in a circular cylindrical tube, Segre and Silberberg (1961, 1962a) observe a lateral migration of the particle, whose centre is established in equilibrium position at a distance approximately 0.6 of its radius from the cylinder axis, independently of the initial particle position. If the particle is suspended near to the wall, it migrates inwards, while if it is suspended near to the axis, it migrates towards the wall. This effect of lateral migration is investigated by other authors [Karnis et al. (1963), Goldsmith and Mason (1964), Jefferey (1965)], who establish not very high sensibility of the Reynolds number value (calculated with the tube diameter) and the ratio of the tube diameter $2r_0$ and particle diameter 2a. Leal (1980) states that if only a small degree of inertia is present, then a lateral migration of settling rigid spheres to a fixed position midway between the walls of the tube and between the walls and the centreline in a simple shear flow is observed. For the motion of neutrally floating spherical particles of a dilute suspension in a circular cylindrical tube Serge and Silberberg note the presence of transverse forces tending to shift the particles, which are near to the axis, towards the wall and vice versa: the particles near to the wall are shifted towards the tube axis. As a result of these forces, independently of their initial position, the particles are concentrated in an annular band situated almost in the middle of the tube half space (between the tube axis and wall). This effect is called "tubular - pinch" effect of Serge and Silberberg for dilute suspensions.

Following Leal (1980) the lateral migration can be divided into two classes: (a) in undisturbed shear flow and (b) in sedimentation through a stationary fluid. Moreover, depending on the particle position with respect to the tube's walls, these classes split further into two subclasses: when the particle is very close to the walls with viscous forces dominant, and when the particle is away from the walls with inertia forces dominant. For the first subclass Cox and Brenner (1968) give the general form of the lateral velocity, but without explicitly evaluating its coefficients. In the case of settling spheres this is done by Cox and Hsu (1977) and Vasseur and Cox (1977). The shear flow passing a sphere between two parallel plane walls is studied in (Vasseur and Cox, 1976) by means of the integral-transform technique modelling the sphere as a point force (see section 3.9.). As a result, it is shown that the particle always migrates away from the walls towards the midpoint between the plane walls, which is in qualitative agreement with the experiments of Karnis et al. (1966a, b), although the theory is for the 2D case. For a sphere settling in a quiescent fluid between two plane walls, a similar result is obtained by Vasseur and Cox (1977) by the method of matched asymptotic expansions when the particle is sufficiently close to one of the walls to lie in the viscous region of expansion. The authors propose a qualitative experimental verification of this statement. The lateral migration of a sphere in quadratic flow is studied by Ho and Leal (1974), when the sphere is not close to the parallel walls. Their result is quite similar to the one of Vasseur and Cox (1976), except in the region near the walls. The resulting equilibrium point for the quadratic flow is located as in the plane Poiseuille flow at about 0.6 of the half distance between the walls away from their midplane. Again this prediction is qualitatively confirmed by the measurements of Segre and Silberberg (1962a, b) for 3D flow in a tube. However, the equilibrium points position and the trajectories found by Ho and Leal are in excellent agreement with the experiments of Halow and Wills (1970) performed in a Couette flow device. Finally, Leal (1980) concludes in his review on the wall effect of shear flows, that there are no investigations for non-spherical particles or some wall geometry different from infinite parallel plane walls, or for oscillatory or pulsating flows.

The experiments of Segre and Silberberg are continued by Eichhorn and Small (1964), Day and Genetti (1964), Denson et al. (1966), etc., for rigid particles, which are not neutrally suspended in Poiseuille flow in circular tube, but have greater or smaller density than the surrounding fluid. From the investigations of these authors it follows, that if the density of the spherical particle (in Poiseuille flow directed downwards in a vertical circular tube) is larger than that of the surrounding fluid, the sphere velocity due to gravity is added to the flow velocity and the particle migration is towards the tube wall. If its density is smaller than the fluid density, the two velocities have opposite directions and the particle migration is towards the tube axis. The theoretical explanation of these experimental results meats significant difficulties. On the basis of the matched asymptotic expansions method Rubinow and Keller (1961) show that a translating and simultaneously rotating sphere in a stationary viscous fluid experiences a force perpendicular to its direction of motion, i.e., a lift force $\mathbf{F}_L$:

$$\mathbf{F}_L = \pi a^3 \rho \mathbf{\Omega} \times \mathbf{U}[1 + O(\mathrm{Re})]. \tag{8.5.6}$$

Here $\mathbf{U}$ is the translationary motion velocity, $\mathbf{\Omega}$ is angular velocity, ρ is fluid density and $\mathrm{Re} = \rho\, \mathrm{U}a/\mu$ is Reynolds number. Using this result, Rubinow and Keller try to explain theoretically the experiments of Segre and Silberberg, but reach to the conclusion that the lateral force is directed in such way, that the neutrally suspended in Poiseuille flow spherical particle always migrates towards the tube axis. In this way, the account of the inertia terms in the equation of motion still cannot give a satisfactory theoretical explanation of the indicated effect. Bretherton (1962b) investigates theoretically the migration of differently shaped rigid particles in Poiseuille flow and supposes that this phenomenon is connected with the inertial terms in the Navier-Stokes equations. In the experiments of Segre and Silberberg (1961, 1962a, b) the Reynolds number, calculated with the tube diameter $2r_0$ as a characteristic length, is around 30, although the Reynolds number (with the spherical particle diameter) is less than one. Goldsmith and Mason (1962) analyse neutrally suspended spherical particles in Poiseuille flow, when the Reynolds number is less than one, i.e., $\mathrm{Re} = 2r_0\rho\, \mathrm{U}/\mu << 1$, and establish experimentally that at such Reynolds numbers there is no particle migration. They do not also observe migration for neutrally suspended rigid particles with shapes of disks and rods with circular cross section. However, for some particles with more particular shape Bretherton finds that still at $\mathrm{Re} << 1$ a lateral migration exists. The inertia influence on the non-spherical particles motion in tubes has been studied experimentally by Karnis et al. (1963) and they observe a radial migration similar to the spherical particles case. If the fluid in the tube flows with sufficiently high velocity, then the particle starts to rotate while simultaneously oscillating in radial direction (Cox and Mason, 1971).

The axisymmetric problems of a sphere settling in a tube filled with quiescent viscous fluid and in a developed Poiseuille flow for moderate Reynolds numbers has been recently studied by Tabakova and Kolemanov (1998) via the finite element method exploiting the Polyflow software package (Polyflow User's Manual, 1989-1997). Their results show that the Reynolds number corresponding to the onset of flow separation at the sphere rear is the same as the experimentally predicted one by Coutanceau (1972).

Fluid particles studies. A detailed study of the motion of a rigid and fluid spherical particles in cylindrical tube is performed by Haberman and Sayre (1958). If $\beta < 0.3$ deformation due the container walls is negligible and the drag correction coefficient for rigid spheres may be used. For fully mobile droplets settling in a circular tube, Haberman and Sayre obtain approximate solutions for the drag correction coefficient Φ_d, which is smaller than the corresponding one for rigid spheres:

$$F_D = 2\pi\mu\, r_0 U\Phi_d \frac{(2+3\lambda)}{(1+\lambda)}, \tag{8.5.7}$$

where

$$\Phi_d = \frac{1+2.2757\beta^2\left(\dfrac{1-\lambda}{2+3\lambda}\right)}{1-0.7017\beta\left(\dfrac{2+3\lambda}{1+\lambda}\right)+2.0865\beta^3\left(\dfrac{\lambda}{1+\lambda}\right)+0.5689\beta^5\left(\dfrac{2-3\lambda}{1+\lambda}\right)-0.72603\beta^6\left(\dfrac{1-\lambda}{1+\lambda}\right)}$$

When β is increased, the presence of the walls provokes droplet deformation; its shape changes from spherical to prolate ellipsoidal shape in vertical direction along the tube axis. The value of Φ_d from (8.5.7) is still applicable for β up to 0.5, although the droplet shape is seriously deformed.

Brenner (1971) shows that if the size ratio β is very small and $\lambda \leq 0.48$, then the drop will have the same velocity as the continuous phase, but with a smaller pressure gradient than is required, corresponding to the case when there is no drops in the suspending fluid.

The arbitrary motion of a small particle at an arbitrary position in a tube is studied by Hasimoto (1976) and Liron and Shahar (1978) whose results are represented graphically, while Hirschfeld et al. (1984) give numerical values for the solution of the same problem and Falade and Brenner (1985) present analytical expressions for particle near the tube walls. The asymmetric problem has been also treated by Tozeren (1982, 1983) and Cox (1974) in the small particle restriction; it stops to be valid for sphere sizes comparable with the tube inside radius.

The finite size spherical particles arbitrary motion at arbitrary positions in a cylindrical tube is numerically investigated by Higdon and Muldowney (1995). The particle motion is translational along each of the coordinate axes and is also due to pressure gradient. Rigid particles as well as droplets are considered at λ =0 or 1, dimensions ratios $0 < \beta \leq 0.9$ and positions ratios $0 < \alpha < 1$. For the near contact case the lubrication theory is used. The boundary integral method, as described in section 5.11. and section 8.2., is applied when solving problems of both rigid and fluid particles motion. Following these authors, the resulting force **F** is expressed by means of the dimensionless resistance coefficients **A** (see section 5.10.), which are functions of β, α and λ:

$$\begin{pmatrix} F_x \\ F_y \\ F_z \end{pmatrix} = C\mu a \begin{pmatrix} A_x & 0 & 0 \\ 0 & A_y & 0 \\ 0 & 0 & A_z \end{pmatrix} \cdot \begin{pmatrix} U_x \\ U_y \\ U_z \end{pmatrix} + C\mu a A_p \begin{pmatrix} 0 \\ 0 \\ U_p \end{pmatrix}, \tag{8.5.8}$$

where C is a constant different for the different cases, e.g., $C = 6\pi$ for rigid particles and $C = 2\pi\,(2+3\lambda)/(1+\lambda)$ for droplets. The numerical results for the resistance coefficients are approximated to give algebraic expressions for three special regimes, which are compared afterwards with known analytical formulae. For the first regime corresponding to particles located near the tube centre, $\alpha \ll 1$, the found algebraic expressions show a good agreement with the approximate solutions of the reflection method and the asymptotic analysis (Happel and Brenner, 1973) represented by even functions of β up to $\beta = 0.5$. The larger particles solutions conform well with the results obtained by Coutanceau and Thizon (1981). The second regime refers to small particles near the wall, $\alpha = O(1)$, $\beta / (1 - \alpha) \ll 1$ and the resistance coefficients agree with those of Hasimoto (1976) and Hetsroni et al. (1970a, b) in 1% error. The third regime corresponds to $h/a \ll 1$, that is, to a narrow gap between particle and tube, and the lubrication theory is applied. Extending the analysis of Cox (1974) from rigid particles near rigid boundaries to droplets near rigid wall, the correspondence for the two cases with the extrapolated numerical results for $h \to 0$ is within 1%. For large particle sizes $\beta \to 1$ there is an agreement again within 1% error with the asymptotic theory of Bungay and Brenner (1973).

For neutrally buoyant and concentrically located particles of a size comparable to that of the tube Bungay and Brenner (1973) analyse the rigid spheres case, while Hyman and Skalak (1972a, b) calculate the pressure and velocity of deformable and non-deformable drops moving in the tube. Goldsmith and Mason (1963) also measured the velocity and the minimum gap width between the drop and the tube wall for very large drops in Stokes flow. Ho and Leal (1975) consider the case of liquid drop of a diameter comparable to that of the tube, moving in a straight walled tube filled with Newtonian or viscoelastic fluid. Sigli and Coutanceau (1977) measure the drag on rigid spheres, comparable in size with the tube, falling through a cylindrical tube filled with viscoelastic fluid. Olbricht and Leal (1982) perform experiments of non-neutrally buoyant drops in horizontal tube filled with either Newtonian and viscoelastic fluid. Due to the difference in density inside and outside the drop, the latter takes up an eccentric lateral position with respect to the tube axis. The measured additional pressure difference, the drop velocity and shape and the minimum gap width between the drop and the tube are found as functions of the material and flow parameters: viscosity, density and size of the drop, flow rate and fluid rheology. These measured quantities are correlated well with the predicted ones from a simple lubrication approach. When compared with the results of Ho and Leal for concentric neutrally buoyant drops, it can be concluded that even a very small density difference leads to serious eccentricity in the non-neutrally buoyant drop position. For Newtonian suspending fluid the obtained data are related to some theoretical predictions [e.g. Taylor (1932), Fitz-Gerald (1969) and Hyman and Skalak (1972a)] accounting the drop eccentric position.

The studies of the droplet deformation in non-uniform flows start with the pioneer experimental work of Taylor (1934) for Couette and hyperbolic flows that observes an ellipsoidal shape of the droplet, when the ratio between the shear forces and the surface tension is less than 0.2. When the distortion parameter is small, $Ca \ll 1$, the droplet spherical shape can be used as a zero-order approximation (see section 6.2.).

The result for migration similar to that for rigid spheres on the conditions of slow viscous motion of deformable droplets is registered experimentally by Goldsmith and Mason (1962) for Poiseuille flow and by Karnis and Mason (1967) for Couette flow. These observations are done at sufficiently small Reynolds numbers, when no inertia is present, thus the deformed droplets lateral migration is not of the inertia type of Segre and Silberberg. The registered radial translation is increased with the distortion parameter, which is in good agreement with the Taylor's solutions (1932). Brenner (1971) shows that, when a droplet is deformed, it is asymmetrically oriented with respect to the tube walls and this anisotropy causes the lateral migration of a neutrally buoyant droplet towards the tube axis. The only possible stable equilibrium position of a deformed droplet is on the tube axis, since there it is symmetrically oriented. When critically analysing the discussed phenomenon, Saffman (1965) supposes that at some circumstances not only the inertial effects, but also the non-Newtonian effect may be extremely significant. In conclusion, it can be stated that the transverse motion in shear flows is characteristic for all deformable and flexible particles, such as fibre threads (Goldsmith and Mason, 1962) and red blood cells (Goldsmith, 1968, 1971), and for steady or time-dependent flows (Goldsmith and Mason, 1967).

However, as pointed by Leal (1980), the mechanisms of migration are more complicated. For neutrally buoyant drops, there are two reasons for lateral migration: the hydrodynamic interaction between the drop and the tube walls and the interaction between the drop deformation and the shear rate gradients in the undisturbed flow. When the drop is not neutrally buoyant, there is an additional lateral velocity component due to sedimentation. The motion of a deformable neutrally buoyant or non-neutrally buoyant drops in shear gradient flows in tubes is studied first by Haber and Hetsroni (1971) and Hetsroni et al (1970a, b). The plane and axisymmetric Poiseuille flow cases with neutrally buoyant drops are analysed by Wohl (1976) and Wohl and Rubinow (1974), who investigate a more general problem, from which the Haber and Hetsroni's problem is obtained as a special case. In all these treatments the influence of the hydrodynamic interaction between the drop and the tube walls on the migration is neglected and a small drop deformation up to first order in the deformation parameter is included. The transverse velocity direction for neutrally buoyant drops is predicted to be towards the centreline of the tube; only in the work of Haber and Hetsroni it is directed outwards in the Poiseuille flow, which may be due to some calculation errors (see Wohl and Rubinow, 1974; Chan and Leal, 1979). For non-neutrally buoyant drop, however, the migration direction can be either way: towards the centreline or towards the walls, depending on the drop density. For example, in a vertical tube with Poiseuille flow directed upwards, the drop lateral migration is towards the tube axis, if the drop is less dense than the continuous fluid phase, and towards the walls for denser drop. If the Poiseuille flow is directed vertically downwards, then the situation is reversed. Here, the sedimentation velocity plays an important role, since it has a transversal component, when the particle is non-spherical (Brenner, 1963, 1964a) or deformed. Although Wohl states that the lateral migration problem for deformable drops is fully examined, there are still open problems (Leal, 1980) connected with computation inaccuracy, experimental trajectories confirmation, etc.

The drop migration in an unbounded quadratic flow and in bounded linear shear flow between two planar walls is investigated by Chan and Leal (1979). The authors show that the migration in the unbounded flow occurs only when the velocity profile of the undisturbed flow is non-linear, as the wall effect is of higher order of the length ratio β. The rate of migration depends mainly on deformation, viscosity ratio λ and the parameter β. On the other hand, the migration direction changes with the viscosity ratio, i.e., if $\lambda < 0.5$ or $\lambda > 10$, then the migration is towards the tube centreline, if $0.5 < \lambda < 10$, it is towards the nearest wall. The last result is not experimentally verified, while the first one is consistent with the existing measurements. In the linear shear flow case, however, the migration is only due to the hydrodynamic interaction between the deformed drop and the wall, and is directed again towards the centreline for all values of λ. The equilibrium position is near the centreline, which agrees with the experimental observations of Karnis and Mason (1967). The magnitude of this migration is smaller, (e.g., by a factor of order β) than the migration magnitude correspondent to the quadratic flow. The flow trajectories found by Chan and Leal (1979) coincide well with the detected trajectories in Poiseuille flow (Goldsmith and Mason, 1962) and in Couette flow (Karnis and Mason, 1967).

In the case when the drop is suspended in a Poiseuille flow near a flat wall, Chaffey et al. (1965) study the hydrodynamic interaction between the drop, the wall and the Poiseuille flow. This problem approaches the more general problem of a Poiseuille flow in a tube of large radius, when the suspended drop is very near to the tube walls. The obtained results for the transverse velocity are wrong, but afterwards corrected by the authors (Chaffey et al., 1967) and confirmed by the experiments of Karnis et al. (1963). Chaffey and Brenner (1967) have developed a method to analyse the deformations of a drop immersed in a Couette flow up to second-order accuracy. An interesting result is found: the drop's shape is no longer elliptical, when the second approximation of the drop deformation is taken into account. The deviation from spherical droplet shape increases with the distortion parameter as shown by Ho and Leal (1975) in their work of droplet translation in a tube under gravity, which confirms the results of Hetsroni et al. (1970a, b).

Coutanceau and Thizon (1981) have investigated experimentally and theoretically the bubble rise in a tube. For small values of the Reynolds number and at $a/h < 0.2$ (a is the bubble radius, h is the minimum distance from the particle centre to the tube wall), they have established that the bubble becomes almost spherical. At $a/h > 0.6$ the bubble has a cylindrical shape finishing at the both ends with an "umbrella" shape. At $0.2 < a/h < 0.6$ the bubble is an ellipsoid.

Shapira and Haber (1988) have analysed the droplet deformation during the slow motion of a spherical droplet between two parallel plates by the reflection method given in (Hetsroni and Haber, 1970) for small deviations from sphericity. They show that the wall effect on the drag additive factor is symmetrical with respect to $\alpha = 0.5$ and hinders the parallel droplet motion with respect to the plates. The drag is an increasing function of α, with its minimum at $\alpha = 0.5$ and singularity at $\alpha = 1$. The results are sensible for $\alpha = 0.2$ to 0.8. For droplet of positions close to the plates, the method of reflections is poorly converging.

A numerical study of the axisymmetric creeping motion of a neutrally buoyant deformable drop in a circular tube filled with Newtonian viscous fluid is performed by Martinez and Udell (1990). There is no limitations on the viscosity ratio λ of the inside and outside fluid and on the ratio β of the drop size to tube size. However, the most interesting cases are those for drop radii comparable to the tube radius, i.e., $0.5 < \beta < 1.15$. The boundary integral technique applied on discrete boundary elements, as presented in section 5.11., is used for the velocity and surface stress force in a domain including the unknown liquid/liquid interface, which is obtained iteratively from the interface kinematic condition. The sought quantities: drop speed, deformation and additional pressure loss, are tabulated as functions of viscosity ratio λ, drop size ratio β and capillary number Ca. Martinez and Udell show that at ratios $\beta < 0.7$, the capillary number and viscosity ratio have no effect on the additional pressure and drop speed, except for Ca > 0.25, which coincides with the results due to the small deformation approaches [Hetsroni et al. (1970a, b) and Hyman and Skalak (1972a)]. The deformation increases with Ca and β, but the deformation dependence on Ca is stronger, and when $\beta \approx 0.726$ a negative curvature at the drop rear develops for Ca $\approx$ 0.75. The drop speed and shape are practically independent on the size ratio if it is $\beta \geq 1.1$. The computed results for drops comparable in size with the tube radius are in good agreement with the experimental observations of Ho and Leal (1975).

The studies of drops in general quadratic flows can enlighten the not very well understood problems of cross-stream migration of suspended particles. The corresponding Stokes flow problem written in its invariant vectorial form is solved analytically by Nadim and Stone (1991). They use the singularity method of Hinch (1988) based on the Lamb's general solution in a vectorial form in order to obtain in a simple way the velocity and pressure fields and the drop deformation for small Ca, as well. The undisturbed velocity near the particle is represented in Taylor series

$$\mathbf{u}^{\infty}(\mathbf{r}) = \mathbf{U}^{\infty} + \boldsymbol{\Omega}^{\infty} \times \mathbf{r} + \mathbf{r}.\mathbf{E} + \mathbf{rr}{:}\mathbf{K} + \ldots, \tag{8.5.9}$$

where $\mathbf{U}^{\infty}$ and $\boldsymbol{\Omega}^{\infty}$ are the translational and angular velocities, while $\mathbf{E}$ is the rate of the strain tensor, all evaluated at the particle centre. The quadratic flow tensor $\mathbf{K}$ can be given as $\frac{1}{2}\nabla\nabla\mathbf{u}^{\infty}$ and thus it is symmetric and traceless with respect to its last two indices. The third rank tensor $\mathbf{K}$ is decomposed into three irreducible tensor components of decreasing orders $\boldsymbol{\gamma}$, $\boldsymbol{\theta}$, $\boldsymbol{\tau}$ [see (2.14a) of Nadim and Stone, 1991]. The spherical harmonics or the potential singularities, point source, potential dipole, potential quadrupole (see section 3.4.), are applied to obtain the translation velocity of a force free drop

$$\mathbf{U} = \mathbf{U}^{\infty} + \frac{\lambda}{2+3\lambda}\mathbf{I}{:}\mathbf{K}, \tag{8.5.10}$$

and its shape deformation only due to the quadratic flow

$$\mathrm{r} = 1 + \mathrm{Ca}\frac{10+11\lambda}{8(1+\lambda)}\mathbf{eee}(.)^{3}\boldsymbol{\gamma}, \tag{8.5.11}$$

where $\mathbf{r} = \mathbf{e} = \mathbf{n}$ on $r = 1$ and γ is the third rank irreducible decomposition tensor of **K**. It is clear, that only a bubble with $\lambda = 0$ will be translated with the velocity of the undisturbed flow. There is an additional droplet surface deformation in (8.5.11) due to the straining given by $(16 + 19\lambda)$ $\mathbf{ee{:}E}$ / $[8(1 + \lambda)]$. The flow rotation has no contribution to the dynamic effects on the droplet.

As a first example of a quadratic flow, Nadim and Stone consider the unidirectional Poiseuille flow in a cylindrical tube of radius R:

$$\mathbf{u}^{\infty} = U^{*}\left[1-\left(\frac{\rho}{R}\right)^{2}\right]\mathbf{k}, \tag{8.5.12}$$

where **k** and U^{*} are the unit vector and the maximum velocity along the axis, respectively and ρ is the distance from the axis. Thus the quadratic flow tensor **K** is simply

$$\mathbf{K} = -\frac{U^{*}}{R}(\mathbf{I}-\mathbf{kk})\mathbf{k}. \tag{8.5.13}$$

Then, the dimensionless axial migration velocity of a droplet of radius a and centre at a distance ρ from the tube centreline becomes

$$\mathbf{U} = \left[1-\left(\frac{\rho}{R}\right)^{2}\right]\mathbf{k} - \frac{2\lambda}{2+3\lambda}\left(\frac{a}{R}\right)^{2}\mathbf{k}, \tag{8.5.14}$$

while its deformed shape is shown to be

$$r = 1 + Ca\left(\frac{a}{R}\right)^{2}\frac{10+11\lambda}{8(1+\lambda)}\left(\cos^{3}\theta - \frac{3}{5}\cos\theta\right), \tag{8.5.15}$$

with $\cos\theta = \mathbf{e.k}$.

The second case treated by Nadim and Stone concerns the Poiseuille flow between infinite parallel plates at distance 2h. The undisturbed flow velocity is then given by

$$\mathbf{u}^{\infty} = U^{*}\left[1-\left(\frac{x}{h}\right)^{2}\right]\mathbf{k}, \tag{8.5.16}$$

where **k** is the unit vector parallel to the walls and x is the lateral coordinate. The third order dimensionless tensor of the quadratic flow transforms into

$$\mathbf{K} = -\left(\frac{a}{h}\right)^{2}\mathbf{iik}, \tag{8.5.17}$$

where **i** is the lateral unit vector. The dimensionless translational velocity for a droplet, whose centre is at a distance x from the midplane between the walls yields

$$\mathbf{U} = \left[1-\left(\frac{x}{h}\right)^{2}\right]\mathbf{k} - \frac{2\lambda}{2+3\lambda}\left(\frac{a}{h}\right)^{2}\mathbf{k}, \tag{8.5.18}$$

and the formula for the deformed droplet shape becomes:

$$r = 1 + Ca\left(\frac{a}{h}\right)^2 \frac{10+11\lambda}{8(1+\lambda)} f(\theta,\phi), \tag{8.5.19}$$

where the shape correction function is given by

$$f(\theta,\phi) = \frac{1}{5}\cos\theta\left(\cos^2\theta + \sin^2\theta\sin^2\phi - 4\sin^2\theta\cos^2\phi\right)$$

and $\sin\theta\cos\phi$ =**e.i**. In both examples Nadim and Stone do not account for the droplet interaction with the walls and the droplet lateral migration velocity is not determined.

Bozzi et al. (1997) consider the steady axisymmetric flow around a deformable drop falling or rising under the gravity force in a vertical tube at small and intermediate Reynolds numbers Re. A numerical analysis of this free boundary problem is performed, based on the finite element method for in a wide range of the parameters Re, and Ca and at $\beta = 0.5$, $\lambda = 1$ (homoviscous case) and density ratio $\rho_{drop}/\rho_{tube} = 6/5$ and 5/6. On the base of the computation results the authors conclude that, at Ca $\leq$ 1, the drop deformation is much stronger and non-convex than found in previous works (Dandy and Leal, 1989), (Haywood et al., 1994a, b) at non-zero Reynolds number. Bozzi et al. show that there exists a single disjoint recirculation zone, similar to that observed by Dandy and Leal, attached to the drop with the increase of Re, further followed by the drop division into two adjacent recirculation regions. The calculated drag correction coefficient Φ_d for Re = 0 is in good agreement with the approximate value of Haberman and Sayre (1958), given by (8.5.7) for the considered case, and confirms excellently the calculations of Coutanceau and Thizon (1981).

Greenstein (1972) considers the problem of the slow motion of two spherical drops symmetrically situated with respect to the axis of a circular tube. The direction of the particles motion is perpendicular to their line of centres, which complicates extremely the solution of the problem.

The well-known Hele-Shaw device, called "Hele-Shaw cell" (Hele-Shaw, 1898) produces a 2D Poiseuille flow between two vertical parallel plates very near to one another, which is driven by the pressure gradient applied to the upper and lower end of the cell. Since the distance between the plates is very small, the velocity field in a plane parallel to the walls is potential. Then the no-slip condition on the rigid boundaries, i.e., the outside ones and on the cylindrical obstacle, placed between the plates, is not fulfilled. This phenomenon in the Hele-Shaw cell has its application to visualise the streamlines pattern around different 2D obstacles by the ideal potential flow. In fact, the Stokes equations with normal fixed to the plates coordinate lead to the potential flow equations for a plane parallel to the plates. If the Reynolds number is increased, i.e., Re > 1, then the potential flow streamline picture is destroyed, and the inertia terms must be added to the Stokes equations as a higher order approximation. Exact solutions for the steady rise of bubbles in Hele-Shaw cell filled with viscous fluid are derived by Taylor and Saffman (1981). They prove that from these solutions the stable one is that correspondent to a bubble whose width is two times smaller than the walls distance. Interesting results connected with flows in the Hele-Shaw cell are presented in the papers of Bretherton (1961), Park and Homsy (1984), Schwartz et al. (1986) and Reinelt (1987).

Bretherton (1961) investigates either theoretically and experimentally two problems describing the motion of a long bubble in a tube at small Reynolds numbers with contact angle at the wall equal to zero. The first case considered is for a capillary tube, while the second one is for a vertical tube sealed at one end and for this case the bubble rises due to gravity if the tube Bond number $Bo = \rho g r_0^2/\sigma < 0.842$, where ρ is the density difference between the density of the fluids inside the bubble and outside it and σ is the surface tension.

An extension of this work is made by Schwartz et al. (1986), who study experimentally the average thickness of the wetting film obtained by the passage of an air bubble in a capillary tube filled with water. It occurs that for bubbles many times longer than the tube radius the new measurements show that the asymptotic result of Bretherton underpredicts seriously the film thickness. However, for bubbles shorter at least 20 times the tube radius the Bretherton theory is confirmed.

Reinelt (1987) approximates the rise of a long bubble in a vertical tube by the fluid drainage out of the unsealed end, when a finger of air rises axisymmetrically under the gravity with a constant velocity U. The interface shape and the rate of rising are calculated numerically for different values of the Bond number and capillary number $Ca = \mu U/\sigma$. The obtained results are in good agreement with the asymptotic predictions of Bretherton for small Ca and they extend the region of applicability of Bo and Ca, where the asymptotic analysis is not valid.

Recently Pozrikidis (1992b) has studied the hydrodynamic interactions between deformable drops moving train-like in a cylindrical tube.

CHAPTER 9

Many-Particles Hydrodynamic Interactions. Sedimentation.

9.1. Introduction

A knowledge of the hydrodynamic action on an assemblage of small rigid or fluid particles suspended or dispersed in a viscous fluid in motion is of fundamental importance in a variety of engineering applications, including advanced material processing, enhanced oil recovery, waste treatment, food processing, pharmaceutical manufacturing and so on. Any theory that attempts to describe the dynamics of a system of rigid or fluid particles suspended (or dispersed) in a moving fluid must address the issue of hydrodynamic interactions among particles. However, the only exact general solutions for hydrodynamic interactions known to date are those for two rigid or fluid particles systems (see Chapter 7).

The many body hydrodynamic interactions are traditionally studied approximately via the method of reflections, inaugurated by Smoluchowski (1911) for those situations in which the fluid can adequately be described by the Stokes equations.

Using a real-space multiple expansion method Kim (1987) solved exactly the particular problem of three identical spherical rigid particles located at the corners of an equilateral triangle falling perpendicularly to the plane of the triangle. On the basis of the reflection method Kynch (1959) showed that the third and fourth body hydrodynamic effects do not appear before $O(R^{-4})$ and $O(R^{-7})$, respectively, where R is a characteristic particle spacing.

In this chapter we shall focus on many-body hydrodynamic interactions between rigid or fluid particles which are sufficiently small to be affected by Brownian motion, but still sufficiently large to let the fluid be treated as a continuum.

Non-equilibrium processes of multiphase systems where continuum phase is a liquid and the discontinuous phase is in the approximate range of 0.1 to 10 μm have been termed microhydrodynamics (Batchelor, 1976b). Microhydrodynamics, then refers to the description of particle motions where inertial forces can be neglected, but the effects of Brownian motion usually cannot.

The relative motion of the fluid and particles (rigid or fluid) and the velocity of fluid flow with respect to various boundaries is of special importance when the particles settle due to gravity, without net fluid motion. Then the particles movement is designated as sedimentation.

Due to the reversibility of Stokes flows, the flow past two equal spherical particles must be symmetric and there is no tendency for the particles to move toward, or away from, each other. Interaction, however, causes the particles to rotate as they fall. Strong interference effects occur when the particle is in a closed vessel or near a wall. For example, when a rigid spherical particle is moving in direction parallel to a rigid wall there is no tendency for the particle to move toward or away from the wall, but it does experience a torque, which induces rotation.

Matrix relations between resistance and velocity for the pure translational and rotational motions of the members of a general multiparticle system involved in a linear shear flow are given by Brenner and O'Neill (1972).

Swarms of rigid particles usually experience an increase in drag. While a pair of particles in an infinite fluid may fall faster than a single particle, an assemblage of particles falls slower. This effect is called "hindered settling". Particles in the flow settle more slowly than isolated particles, so that any particle that is left behind has the ample opportunity to catch up.

In order to characterise a fluid-particle system the following factors must be considered:

(i) The fluid nature, which may be often specified by information about its viscosity and density;

(ii) The nature of the rigid particles, which are characterised by their size, shape and density;

(iii) The proportion of solid to fluid, which is usually specified in terms of the fractional void volume;

(iv) The motion of fluid and solids (relatively to one another) with respect to the container vessel walls.

The general motion of a settling particle can be quite complicated. The velocity and direction of each particle motion in a settling assemblage of particles is constantly changing because the fluid flow approaches this particle on different conditions. Particles with somewhat regular but asymmetric shapes can make a graceful spiral as they settle through the fluid.

It will be useful at the outset of the chapter to say what we mean by the terms dispersions, suspensions and emulsions and to specify the nature of the problems to be discussed. A colloidal system represents a multiphase system, in which at least one of the phases exists in the form of very small particles. The term dispersion (disperse system) is more general because it also includes bicontinuous systems (in which none of the phases is split into separate particles) and systems containing larger, non-Brownian, particles.

A suspension is a dispersion of small rigid particles immersed in a continuous liquid, while an emulsion contains two liquid immiscible phases, one of which is dispersed in the other in the form of droplets. Emulsions could be oil in water (o/w) type or water in oil (w/o) type. In the former case oil is the dispersed phase, while in the latter case, water is the dispersed phase. Since in studying the dynamics of these dispersions, there are many common problems, the common denomination "suspension" is often used.

Foams are widely familiar gas/liquid disperse systems. Like other disperse systems, such as suspensions and emulsions, the foams are characterised by highly developed interface determining its properties. The contact between the gas bubbles, emulsion drops or suspended particles in the liquid medium occurs through various thick and thin liquid layers (films). [For the experiments and theory in the area of films see Reynolds (1886), Mysels et al. (1959), Ivanov (Ed.) (1988), Exerova and Kruglyakov (1998), Tabakova and Carotenuto (1994) and Tabakova (1998)].

It is well known that in addition to the hydrodynamic forces, non-hydrodynamic forces also act on colloidal particles (of diameter less than a micron) in suspensions. These forces are described in the Derjaguin-Landau-Verwey-Oberbeek (DLVO) theory of colloid stability (Derjaguin and Landau, 1941; Verwey and Oberbeek, 1948) and consist of electrical forces arising from the particles charges and van-der-Waals forces. As it is noted in the Preface, we consider only non-colloidal particles in the present book. So, the effects of electrical and van-der-Waals forces between particles will not be included in our analysis.

An important parameter in sedimentation and suspension rheology is the volume fraction of the particles, defined by

$$\Phi = \frac{\text{Volume occupied by particles}}{\text{Total volume of suspension}}.$$

The particles density number in a suspension N, is the number of particles in a unit volume. The dimensionless number that measures the relative importance of the flow and the diffusion is the Peclet number, determined as

$$\mathrm{Pe} = \frac{\mathrm{Ua}}{\mathrm{D_0}} = \frac{4\pi\,\mathrm{ga}^4\Delta\rho}{3\mathrm{k_B T}},$$

where U is the flow velocity, a is the characteristic particle length, D_0 is the diffusion coefficient at infinite dilution, $\Delta\rho$ is the density difference between solvent and particle, k_B is the Boltzmann constant and g is the acceleration due to gravity.

With respect to the assumption of the geometrical distribution of the particles in the fluid and the character of their interaction, the theoretical studies carried out up to now are mainly based on four approaches.

The first approach is known as the "cell model" approach, proposed at first by Simha (1952) and later modified by Happel (1957, 1958), Kuwabara (1959) and others (see section 9.2).

In the second approach, the particles (spheres) centres are assumed to be periodically located, e.g., in a cubic array with characteristic length $a\Phi^{-1/3}$, where a is the particle radius and Φ is the particles volumic concentration. Hasimoto (1959) is a notable representative of this trend (see 9.5).

In the third approach it is accepted that the influence of all particles on a given particle could be modelled by its motion in a porous medium with phenomenological coefficients depending on the particles volumic concentration Φ. By means of this approximation Brinkman (1949) obtained the formula for the mean interaction experienced by a given particle at its passage through a layer of many stationary particles. The ideas of this method reach a further development in the works of Tam (1969), Lundgren (1972), Childress (1972), Howells (1974) and in the works of Buyevich and collaborators (Buyevich, 1971a, b, c, 1972), (Buyevich and Markov, 1972, 1973a, b). The main result of these works is the sedimentation velocity

$$\frac{\Delta U}{U_0} = \frac{3\sqrt{2}}{2}\Phi^{1/2}, \tag{9.1.1}$$

where $\Delta U = U_0 - U$, with U and U_0 - respectively the mean sedimentation velocity of the particles and the sedimentation velocity of an isolated particle in a viscous fluid.

The latter approach deals with statistical methods, assuming that the suspension is a statistically homogeneous medium, i.e., that the particles move freely and their centres could occupy a statistically arbitrary place in the continuous phase (see 9.3).

9.2. Stokes Flow through Assemblages of Rigid or Fluid Particles. Cell Models

The determination of the hydrodynamic interaction between the particles is the fundamental problem of the mathematical modelling of concentrated suspensions. One of the first methods to calculate the hydrodynamic interaction of any particle of the dispersed phase with the remaining particles is the so-called “single cell method”. It is based on the idea, that the dispersed system could be divided into many identical cells in such a way that each of them has only one particle. This description is adequate when the particles (dispersed phase) are periodically located in the liquid (continuous phase), but it may be also used in some statistical sense for a chaotic particles aggregate. The fluid motion inside the cell fulfils the no-slip condition on the rigid particle surface and a suitably chosen condition for an artificially imposed outer boundary of the liquid layer, as well. When the outer boundary is spherical its radius is taken equal to $a\Phi^{-1/3}$, with a being the particle radius.

The main assumption, on which the “cell models” are based, is that the hydrodynamic interaction of all particles with a given (fixed) particle could be approximated by the effect of the outer cell boundary on the particle itself. In this way, the boundary-value problem of the considered model is reduced to the problem of a single particle (usually spherical) surrounded by a spherical liquid layer. On the outer cell boundary different boundary conditions could be imposed.

The “cell model” is first proposed by Simha (1952). The cell consists of a nucleus, in which a spherical particle of radius a, a liquid layer around it and a concentric rigid envelope of radius b are situated. It is assumed that the disturbance flow caused by the particles outside the cell does not affect the dilatational motion inside it. The hydrodynamic interaction between a given particle and the neighbouring particles is accounted above all.

In order to improve the Simha's model, Happel (1958) relaxed the boundary conditions on the envelope surface as leaving only the normal velocity component and shear stress equal to zero. Happel chose the radius of the cell free surface (envelope) such that the cell encircles a liquid volume fraction equivalent to the liquid volume fraction of the suspension. This model of Happel obtains the name "free surface model". Since it is used by a lot of investigators, we shall study it here in more details.

The cell model has a wide application at the analyses connected with the so-called "hindered settling". Let us consider a system of spherical droplets of radius a homogeneously dispersed in a liquid and settling under the action of gravity. In order to determine the velocity of the fluid inside and outside a given droplet, a spherical coordinate system (r, θ, φ) is chosen in such a way that its origin coincides with the particle centre and the direction of the polar axis coincides with the direction of the gravity acceleration g. According to the cell model, in a Stokes approximation, the stream functions inside and outside the drop $\hat{\Psi}$ and Ψ must satisfy the equations (1.3.21), the boundary conditions (1.4.14)-(1.4.18) on the droplet surface r = a and the free surface conditions at $r = a\Phi^{-1/3}$. In the limits of the considered cell, the velocity distribution inside and outside the droplet is given via (5.2.22), (1.3.14) and (1.3.17) by the following formulae:

(i) inside the drop $r \in [0, a]$

$$\hat{v}_r = U_0 \frac{1}{2(\lambda+1)}\left(1-\Phi^{1/3}\right)\left(1-\frac{r^2}{a^2}\right)\cos\theta, \quad \hat{v}_\theta = U_0 \frac{1}{2(\lambda+1)}\left(1-\Phi^{1/3}\right)\left(2\frac{r^2}{a^2}-1\right)\sin\theta; \tag{9.2.1}$$

(ii) outside the drop $r \in [a, a\Phi^{-1/3}]$

$$v_r = \frac{U_0\lambda}{2(1+\lambda)}\left[\frac{r^2}{a^2}\Phi^{5/3} - \frac{2(1+\lambda)}{\lambda} + \frac{2+3\lambda}{\lambda}\frac{a}{r} - \frac{a^3}{r^3}\right]\cos\theta,$$
$$v_\theta = \frac{U_0\lambda}{2(1+\lambda)}\left[-\frac{2r^2}{a^2}\Phi^{5/3} + \frac{2(1+\lambda)}{\lambda} - \frac{2+3\lambda}{2\lambda}\frac{a}{r} - \frac{a^3}{2r^3}\right]\sin\theta \tag{9.2.2}$$

In these expressions U_0 is the sedimentation velocity of an isolated drop, i.e., the velocity obtained through the Hadamard-Rybczynski theory (3.3.1) $U_0 = \frac{2ga^2(\hat{\rho}-\rho)}{3\mu}\frac{(1+\lambda)}{(2+3\lambda)}$. From the first formula of (9.2.2) Gal-Or and Waslo (1968) find that the sedimentation velocity of the drop in a stationary fluid $\langle U \rangle$ on the conditions of "hindered settling" is reduced to the formula

$$\langle U \rangle = U_0\left(1 - \frac{2+3\lambda}{2(1+\lambda)}\Phi^{1/3}\right), \tag{9.2.3}$$

which is known as the formula of Gal-Or and Waslo. Here the symbol $\langle U \rangle$ stands for the average velocity. At $\lambda \to 0$ the above formula becomes

$$\langle U \rangle = U_0\left(1-\Phi^{1/3}\right), \tag{9.2.4}$$

which gives the bubble velocity on the conditions of "hindered settling". At $\lambda \to \propto$, the rigid spherical particle velocity is obtained

$$< U >= U_0\left(1 - \frac{3}{2}\Phi^{1/3}\right), \tag{9.2.5}$$

in the "hindered settling" flow.

Far-field hydrodynamic interactions for three or more spherical particles was also studied analytically by Mazur and Van Saarloos (1982). These authors developed a Fourier-space multipole expansion method for calculating the sphere mobility functions for a finite system of spheres as a power series in inverse spacing up to order R^{-7}. Their expressions are in agreement with those previously obtained by Kynch (1959). Since Mazur and Van Saarloos' analysis is quite general, it could in theory be extended to include higher-order many-body contributions. These calculations are by no means exact for all particle-particle separations, but they do give some indication on the importance of three-body and higher order effects (Beenakker, 1984) (see Chapter 10).

Another method for calculating the many-body hydrodynamic interactions when the particles are far apart is the point-force approximation technique. This method developed first of all by Burgers (1938a,b, 1941, 1942) required the disturbance produced by a submerged object to be replaced by one or more point forces located at the object foci. In the papers of Burgers (1938a,b, 1941, 1942), McNown and Lin (1952), Tchen (1954), Broersma (1960) and Tam (1969) this technique approximates the exact viscous no-slip boundary condition by requiring the velocity over the rigid spherical particle surface to vanish in some average sense. It is worth noting that this approximate technique has also been used in conjunction with the method of reflections to describe hydrodynamic interaction problems by Burgers (1941, 1942) and Kynch (1959). It is regrettable, however, that the accuracy of the point-force representation quickly diminishes, as the spherical particles approach one another.

The contemporary general methods for treating many-body hydrodynamic interactions can be divided into three classes at least: (a) the multipole collocation technique; (b) the boundary integral technique; (c) the multipole-moment method (Weinbaum et al, 1990). Here we shall discuss briefly each of them.

Gluckman et al. (1971) developed a new technique, called the multipole collocation method, for handling concentrated axisymmetric systems of particles. With this method the no-slip boundary conditions are more accurately satisfied than with the point-force approximation and it also converges more rapidly than the reflection method. Comparison between the solutions obtained via this method and the exact solution for the two sphere problem demonstrates the convergence fastness of the multipole technique even when the spheres are touching each other. Ganatos et al. (1978) showed how the collocation technique, developed by Gluckman et al. for the multiparticle axisymmetric Stokes flows, can be extended to operate with a wide variety of non-axisymmetric creeping motion problems with planar symmetry. Basically, this method is best suited for problems involving a finite number of identical particles positioned in a very symmetric arrangement. Recently Hassonjee et al. (1992) generated the multipole collocation technique for an arbitrary motion of spherical particles 3D clusters. Surface averaged properties such as drag can be calculated accurately, however as the particle separation decreases, many terms must be retained in Lamb's solution, on which the method is based. Weinbaum et al. (1990) showed that the used "multiglobular disturbances" correspond to moments of the stress density on the particle surface, i.e., to the Stokeslet, rotlet, stresslet, quadrupole and octupole. This suggests an inherent deficiency of the method for non-spherical particles, since even for an isolated ellipsoid it creates a disturbance which can be correctly reproduced only by an infinite series of Stokes singularities at its geometrical centre.

A new method has been developed by Clayes and Brady (1993a) for solving the zero Reynolds number flow problems involving elongated particles (prolate spheroids) in an unbounded fluid. The proposed technique extends the multipole moment expansion method to ellipsoidal shape. It is applied to problems of sedimenting spheroids and spheroidal particles in a simple shear flow. The simulation method for prolate spheroids in Stokes flow is extended by Clayes and Brady (1993b, c) to handle statistically homogeneous unbounded suspensions.

Dabros (1985) offered a singularity method for calculating hydrodynamic forces and particle velocities at zero Reynolds number flows. The basic functions utilised are a point force and a point source. The least-square approach is used in order to find the intensities of these singularities. The method of Dabros is actually a variant of the general method explained by Mathon and Johnston (1977), but he does not refer to this earlier work.

The application of the boundary element method for solving the problems of many-body hydrodynamic interactions is represented by Ingber et al. (1989), Tran-Cong and Phan-Thien (1989), Tran-Cong et al. (1990) and Phan-Thien et al. (1991).

9.3. Sedimentation of a Dilute Suspension of Spherical Particles. Statistical Approach.

In the mechanics of heterogeneous media the macroscopic properties of suspensions, two-phase flows and composite materials are investigated by means of some given properties of their components. The difficulties arising at the suspensions mathematical modelling result from the fact, that their macroscopic properties depend not only on their components microscopic structure, but also on their hydrodynamic interaction and on the processes passing in the environment around the separate inclusions (particles). Since the relative position and orientation of the separate particles in the different parts of the disperse systems change arbitrary, the most suitable method to describe their macroscopic properties is the statistical one. The wide usage of the probability theory methods when studying the various kinds of suspensions is a strong contemporary tendency.

In fluid mechanics several divers ways are practised for averaging the governing equations in order to get equations, which do not contain any details of the observed flows, but only account for their fundamental laws. The averaging procedure of the hydrodynamic and gasdynamic values and the corresponding equations is performed in the turbulence theory, kinetic theory of gases, dynamics of multiphase systems, etc. The following types of averaging are well known: averaging with respect to time, to some volume of the considered medium, to an "ensemble of realizations", asymptotic averaging and others.

As an "ensemble" it is understood a set of a large number of systems, which are different regarding some of their microscopic details, but are identical in macroscopic sense. In the mechanics of heterogeneous media the term "realization" of a given suspension containing N spherical particles means the moment configuration of the system of the N particles determined by the location of the radius vectors of their centres. The aggregate of all physically admissible configurations, assumed to be of equal probability, is called an "ensemble of realizations". When averaging about the "ensemble of realizations" the radius vectors of the particles centres are looked as independent.

The concepts "ensemble" and averaging about "ensemble" are introduced for the first time by Gibbs (McQuarrie, 1976) in thermodynamics. The expedience of the averaging about an "ensemble of realizations" at the mathematical modelling of the heterogeneous media is underlined by a lot of scientists (Van De Ven, 1989), (Russel and Gast, 1986).

When one considers a homogeneous suspension without agglomeration, then the fluid flow inside the volume V is determined by the momentary positions and velocities of suspension particles and by the forces acting on the volume boundaries at the absence of inertia. If the volume V contains a large number of spherical particles and the conditions on the volume boundary are compatible with the assumption of zero mean fluid velocity inside the volume, then the velocity at each point of the volume V does not depend on the size and shape of the chosen domain and on the boundary conditions on the vessel walls. On this account, we shall suppose that the velocity and the other quantities at any point r for a given distribution (realization) of the suspension particles are defined only by the momentary positions of the centres of the N spheres in the volume V containing the point r. From this assumption it follows that the average (about an ensemble) value of any suspension parameter at a given point r is determined from the average (about an ensemble) spheres position with respect to the point r.

It is known that a rigid spherical particle of radius a and density ρ_p falls under the gravity action in an unbounded fluid with velocity

$$\mathbf{U}_0 = \frac{2}{9}\frac{a^2(\rho_p - \rho)}{\mu}\mathbf{g}, \tag{9.3.1}$$

where ρ is the fluid density, μ is its viscosity, $\mathbf{g}$ is gravity acceleration. We denote the velocity of an arbitrary suspension particle by U, which is obtained as a result of the gravity force action and the hydrodynamic interaction with other particles. Since U is a random quantity, the difference $U - U_0$ is also a random quantity. The mean value of this difference must be found in dependence on the particles volume fraction of the Φ.

Burgers (1942) showed that for a dilute suspension of particles, which are not grossly aggregated, the sedimentation velocity, U, is given by an equation of the form

$$\frac{U}{U_0} = \frac{1}{1 + k\Phi} \approx 1 - k\Phi + \ldots, \tag{9.3.2}$$

where k is a constant. In an experimental study of monodisperse polystyrene lattice Cheng and Schachman (1955) found

$$\frac{U}{U_0} = 1 - 5.1\Phi. \tag{9.3.3}$$

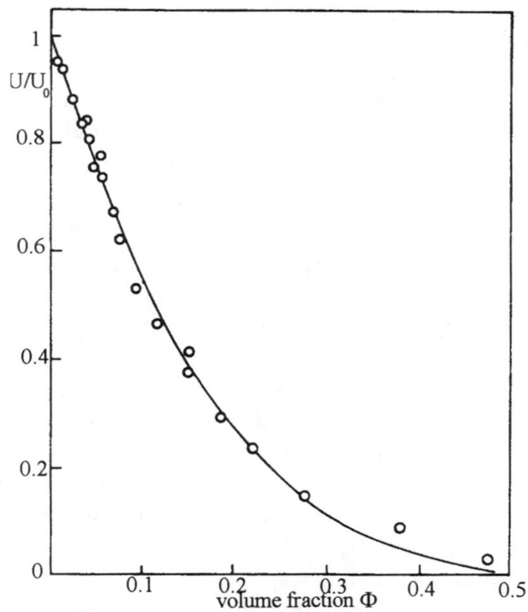

Fig.9.3.1. The relative settling velocity U/U_0 against Φ: (——) calculated by Eq. (9.3.4) with p = 0.58 and k = 5.4; (O) experiments with latex volume fraction in 10^{-3} mole dm^{-3} sodium chloride solution (from Buscall et al., 1982).

Another experimental investigation was made by Buscall et al (1982), who considered dilute and concentrated suspensions of spherical particles of radii 1.5 μm. They found that the relative velocity of settling U/U_0 for considered systems could be represented by the equation

$$\frac{U}{U_0} = \left(1 - \frac{\Phi}{p}\right)^{kp}, \tag{9.3.4}$$

where p is the latex volume fraction at closed packing, Φ is the volume fraction of the latex, and k is a constant. Buscall et al (1982) compared their experimental results with the results obtained from (9.3.4) with p=0.58 and k=5.4 (see Fig.9.3.1).

It can be stated without exaggeration that the progress in the investigations of suspensions after 1970 is directly connected with the successful application of the so called approach "averaging about an ensemble" or "ensemble averaging" utilised initially in the suspension theory by Hashin (1964) and Batchelor (1970a).

Since the hydrodynamic interactions between particles diminish quite slowly with the distance between them, the long range components of the interactions can lead to divergent integrals when summing their contribution to the velocity at a point in the suspension from the indefinitely large number of particles distributed randomly throughout the suspension. Batchelor (1972) overcame the convergence difficulty and calculated the first effects of hydrodynamic interactions on the sedimenting velocity in a dilute, random, monodisperse suspension of spherical particles. To this end, he constructed a special renormalisation procedure. For a "well stirred" suspension of identical spherical particles Batchelor (1972) derived the formula

$$< U >= U_0(1 - 6.55\Phi) \quad \text{as } \Phi \to 0, \tag{9.3.5}$$

where $<U>$ is the mean settling velocity. It should be noted that the correction -6.55Φ applies only to a random array of particles with uniform probability. So, the formula (9.3.2) can be expected to apply to a suspension of Brownian particles that possess the uniform pair probability stipulated in Batchelor's calculations, but not to a suspension of non-Brownian particles that possess a non-uniform structure.

Batchelor's method for treating the sedimentation of a statistically homogeneous, dilute suspension of monodisperse spherical particles has been modified by Feuillebois (1984) in order to investigate the sedimentation of monodisperse spheres in a dilute suspension that is homogeneous in any horizontal plane, but in vertical direction the concentration is prescribed. Another approach was suggested by Hinch (1977), who constructed a hierarchy of equations describing the behaviour of a two-phase macroscopically homogeneous material and showed how the sedimentation velocity can be computed in a systematic way.

Batchelor (1972) has indicated that the $O(\Phi)$ coefficient is sensitive to the details of the random configuration of the suspension. Hinch (1977) gave a compact expression for $O(\Phi)$ coefficient for which the radial-distribution function characterising the microstructure was involved.

Much of the theoretical research in a quiescent sedimentation of particles has been focused on determining the so called "hindered settling function" $f(\Phi)$, defined such that the statistical average particle settling velocity relative to the bulk suspension velocity, which is zero for batch settling, is given by $U_s = U_0 f(\Phi)$, where U_0 is the settling velocity of an isolated particle and Φ is the local volume fraction of particles in the suspension. We can see that there is a considerable uncertainty on the quantitative functional form of the function f(Φ) (see Fig.9.3.2). Saffman (1973) was the first who tried to explain this point in detail. If the particles are arranged in a regular lattice, then the relative reduction in the velocity $\frac{U_0 - U_s}{U_0} = [1 - f(\Phi)]$ is proportional to $\Phi^{1/3}$, as $\Phi \rightarrow 0$. If the particles are arranged in a random fashion, with their relative positions fixed, then 1 - f(Φ) is proportional to $\Phi^{1/2}$, as $\Phi \rightarrow 0$. Finally, if the particles are arranged in a random fashion, with their relative positions free to change in order to maintain a given net force, then 1 - f(Φ) is proportional to Φ, as $\Phi \rightarrow 0$. So, the nature of the dependence of 1 - f(Φ) on Φ is dictated by the suspension microstructure, i.e., the details of the relative positions of the particles. Undoubtedly, the case of random free suspension is the one most likely to occur in processes of particles interest.

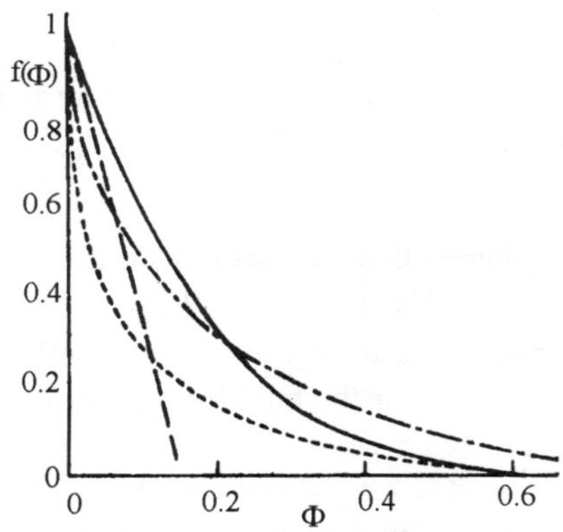

Fig.9.3.2. The hindered settling function f(Φ): (———) empirical correlation by Richardson and Zaki (1954); (- · — · -) empirical correlation by Barnea and Mizrahi (1973); (-------) Batchelor's (1972) theory for dilute suspensions; (···········) exact theoretical results of Sangani and Acrivos (1982) for simple cubic arrays (from Davis and Acrivos, 1985).

Another theory for sedimentation at finite concentration has been developed by Reed and Anderson (1980). In accounting for multiparticle hydrodynamics, they removed the test particle from the suspension and pre-average the neighbour-neighbour interactions. In doing so, the authors of this theory took the effective force on each neighbour to be $F = 6\pi \mu a \langle U \rangle$. The test particle was then placed in the suspension and the average sedimentation velocity was computed as in the dilute limit, except that the environment velocity affecting the test particle was different.

The settling velocity of a drop in an unbounded uniform fluid flow is given by the well known formula due to Hadamard-Rybczynski theory

$$U_0 = \frac{2ga^2(\hat{\rho}-\rho)}{3\mu}\frac{(1+\lambda)}{(2+3\lambda)}, \tag{9.3.6}$$

where μ is the viscosity of the suspending fluid, ρ and $\hat{\rho}$ are, respectively, the densities of the suspending fluid and the drop, a is the radius of the spherical drop, g is the gravity acceleration and $\lambda = \frac{\hat{\mu}}{\mu}$ is the ratio of the fluids viscosities. If λ tends to infinity, equation (9.3.6) gives the well known Stokes formula (9.3.1) for the settling velocity of a rigid spherical particle. Wacholder (1973) and Haber and Hetsroni (1981) included the particle-particle interactions in an emulsion to determine the $O(\Phi)$ correction to the settling velocity expressed by (9.3.6). Their calculations required the determination of the pairwise particle distribution, which in general was rather complicated and limited to the emulsions with very small volume fractions of the drops. Since in Wacholder's result there is some error, we shall give here only Haber and Hetsroni's formula for the sedimentation velocity

$$< U >= U_0\left[1-\Phi\left(2+\frac{2+3\lambda}{1+\lambda}+\frac{(2+3\lambda)(8+15\lambda)}{24(1+\lambda)^2}-\frac{(17+12\lambda)(2+3\lambda)^2}{384(1+\lambda)^3}\right)\right]+O(\Phi^2) \tag{9.3.7}$$

When $\lambda \to \propto$ (rigid spheres) this formula transforms into

$$< U >= U_0(1-6.59\Phi), \tag{9.3.8}$$

while Batchelor (1972) has obtained a slightly different numerical coefficient (6.55). The discrepancy is less than 0.7%. For the bubbles case as $\lambda \to 0$, the equation (9.3.7) yields

$$< U >= U_0(1-4.49\Phi). \tag{9.3.9}$$

Besides the monodisperse mixtures, the polydisperse mixtures are also used in a lot of chemical technologies. Their study began with the works of Tam (1969), Buyevich and Markov (1972) and Lundgren (1972).

Batchelor (1976a) studied the motion of a polydisperse suspension by taking into account the hydrodynamic interaction of the particles at their Brownian motion. Generalising the method, applied to analyse the sedimentation of monodisperse suspension particles (Batchelor, 1972), Batchelor (1982) calculated the additional velocity of a particle of type i, due to the presence of a particle of type j, by means of the functions characterising their mobility. He found the probability density $P_{ij}(r)$ of the relative position of the i-th and j-th particle by solving a differential equation of the type of Fokker-Plank equation. This equation describes the effects of the particles relative motion due to gravity, their interaction force and Brownian diffusion.

The average velocities of the two kinds of particles, marked by indices i and j, are found from the general formulae

$$< U_i >= U_i^{(0)}\left(1 + S_{ii}\Phi_i + S_{ij}\Phi_j\right),$$
$$< U_j >= U_j^{(0)}\left(1 + S_{ji}\Phi_i + S_{jj}\Phi_j\right). \tag{9.3.10}$$

Here S_{ii} and S_{jj} define the hydrodynamic interaction of the same kind particles and S_{ij} depend on $\alpha = \frac{a_j}{a_i}$ and $\gamma = \frac{\rho_j - \rho}{\rho_i - \rho}$, where ρ is the fluid phase density.

The obtained in (Batchelor, 1982) formulae accounting for the hydrodynamic interaction between two particles in the statistically treated homogeneous diluted polydisperse suspension were analysed numerically by Batchelor and Wen (1982). For a suspension with α =0.25 and γ =1, the average velocities according to Batchelor and Wen are:

$$< U_i >= U_i^{(0)}\left(1 - 6.55\Phi_i - 3.83\Phi_j\right),$$
$$< U_j >= U_j^{(0)}\left(1 - 24.32\Phi_i - 6.55\Phi_j\right). \tag{9.3.11}$$

Utilising the Batchelor approach, Feuillebois (1984) studied a suspension, which is non-homogeneous in vertical direction and reached to a formula for the average velocity of the spherical particles sedimentation. If the suspension is homogeneous, as a particular case the formula (9.3.5) of Batchelor yields. A competent review of the investigations dedicated on the sedimentation of non-colloidal particles in suspensions is done by Davis and Acrivos (1985).

9.4. Resistance Tensors of N Particles

In 5.10. and 7.3. we define the resistance and mobility tensors of one and two particles, respectively. Here we shall generalise these concepts for N particles.

Suppose that N particles of arbitrary shape with surfaces S_i (i=1, 2,...,N) move in a viscous fluid. The translational and rotational velocities at fixed points O_i of the particles are denoted respectively by $\mathbf{U}_i$ and $\boldsymbol{\omega}_i$. A flow is assumed to pass the particles

$$\mathbf{v}^\infty = \mathbf{U}^\infty + \boldsymbol{\Omega}^\infty \times \mathbf{r}_0 + \mathbf{E}^\infty.\mathbf{r}_0, \tag{9.4.1}$$

where O is a fixed fluid point and $\mathbf{r}_0$ is the radius vector of an arbitrary fluid point with respect to the point O.

Taking into account that the Stokes equation and the continuity equation are linear, the velocity $\mathbf{v}$ and the pressure p in the fluid at the particles presence could be written in the form

$$\mathbf{v} = \mathbf{v}^\infty + \sum_{k=1}^{N} \mathbf{v}_k, \qquad p = \sum_{k=1}^{N} p_k, \tag{9.4.2}$$

where $\mathbf{v}^\infty$ is determined by (9.4.1) and $\mathbf{v}_k$ satisfy the boundary conditions

$$\mathbf{v}_i = \mathbf{U}_i - \mathbf{U}^\infty + \left(\boldsymbol{\omega}_i - \boldsymbol{\Omega}^\infty\right) \times \mathbf{r}_i - \mathbf{E}^\infty.\mathbf{r}_i. \tag{9.4.3}$$

Here $\mathbf{r}_i$ is the radius vector of an arbitrary point of the i-th particle to some fixed point O_i of it.

Then similarly to the single particle and two particles cases, we have

$$\mathbf{F}_i = -\mu\left\{\sum_{j=1}^{N}\left[\mathbf{K}_{ij}\cdot\left(\mathbf{U}_j - \mathbf{v}_j^\infty\right)\right] + \widetilde{\mathbf{C}}_{ji}\cdot\left(\boldsymbol{\omega}_j - \boldsymbol{\Omega}^\infty\right) + \boldsymbol{\phi}_i : \mathbf{E}^\infty\right\}, \tag{9.4.4}$$

$$\mathbf{M}_i = \left\{\sum_{j=1}^{N}\left[\mathbf{C}_{ij}\cdot\left(\mathbf{U}_j - \mathbf{v}_j^\infty\right)\right] + \boldsymbol{\Omega}_{ij}\cdot\left(\boldsymbol{\omega}_j - \boldsymbol{\Omega}^\infty\right) + \boldsymbol{\tau}_i : \mathbf{E}^\infty\right\}, \tag{9.4.5}$$

where $\mathbf{K}_{ij}$, $\mathbf{C}_{ij}$ and $\boldsymbol{\Omega}_{ij}$ are dyads, while $\boldsymbol{\phi}_i$ and $\boldsymbol{\tau}_i$ are triads. These dyads and triads characterise only the specific properties of the momentary geometrical configuration of the N particles system. In particular, they depend on the particles dimensions, their shape, on the mutual momentary position of the fixed points O_i in them and on their relative orientation. Here the dyads $\mathbf{K}_{ij}$, C_{ij} and $\boldsymbol{\Omega}_{ij}$ have the form:

$$(\mathbf{K}) = \begin{pmatrix} \mathbf{K}_{11} & \mathbf{K}_{12} & \cdots & \mathbf{K}_{1N} \\ \mathbf{K}_{21} & \mathbf{K}_{22} & \cdots & \mathbf{K}_{2N} \\ \cdots & \cdots & \cdots & \cdots \\ \mathbf{K}_{N1} & \mathbf{K}_{N2} & \cdots & \mathbf{K}_{NN} \end{pmatrix}, \tag{9.4.6}$$

as each $\mathbf{K}_{ij}$ is 3x3 matrix with 9 elements, whilst $\mathbf{F}_i$, $\mathbf{M}_i$, $U_i - \mathbf{v}^\infty$ and $\boldsymbol{\omega}_i - \boldsymbol{\Omega}^\infty$ are 3x1 column matrices. Brenner (1964a), proved that the 9 element matrices for every $\mathbf{K}_{ij}$ and $\boldsymbol{\Omega}_{ij}$ were symmetric, from where it followed that the matrices of the type (9.4.6) were symmetric, too. We shall write down the triads $\boldsymbol{\phi}_i$ and $\boldsymbol{\tau}_i$, which are 6Nx6 matrices, concisely as

$$(\boldsymbol{\phi}) = \left[(\boldsymbol{\phi}_1),(\boldsymbol{\phi}_2),\ldots(\boldsymbol{\phi}_N),(\boldsymbol{\tau}_1),(\boldsymbol{\tau}_2),\ldots(\boldsymbol{\tau}_N)\right]. \tag{9.4.7}$$

After introducing the notations

$$(\mathbf{R}) = \begin{bmatrix} (\mathbf{K}) & (\widetilde{\mathbf{C}}) \\ (\mathbf{C}) & (\boldsymbol{\Omega}) \end{bmatrix}, \quad (\mathbf{F}) = \begin{bmatrix} (\mathbf{F}) \\ (\mathbf{M}) \end{bmatrix}, \quad (\mathbf{U}) = \begin{bmatrix} (\mathbf{U} - \mathbf{v}^\infty) \\ (\boldsymbol{\omega} - \boldsymbol{\Omega}^\infty) \end{bmatrix}, \quad (\mathbf{E}) = \begin{bmatrix} \mathbf{E}_1^\infty \\ \vdots \\ \mathbf{E}_6^\infty \end{bmatrix}$$

We could transform the equations (9.4.4) and (9.4.5) into more compact form

$$(\mathbf{F}) = -\mu\left[(\mathbf{R})(\mathbf{U}) + (\boldsymbol{\phi})(\mathbf{E})\right], \tag{9.4.8}$$

Since the matrix **R** is not singular, from (9.4.7) one finds

$$(\mathbf{U}) = (\mathbf{R})^{-1}(\boldsymbol{\phi})(\mathbf{E}). \tag{9.4.9}$$

9.5. Stokes Flow through Periodic Arrays of Rigid or Fluid Particles

Hasimoto (1959) was the first to derive in a satisfactory manner a correction to the Stokes sedimentation velocity due to the presence of other particles. As we shall see further, he considered a dilute cubic array of spherical particles and an uniform pressure gradient which exactly counterbalanced the weight of the particles. The periodicity of the array and the chosen pressure gradient enabled him to formulate the problem for a single period cell, without introducing divergent sums. Hasimoto solved the cell problem by using expansion on the powers of the ratio of sphere radius to lattice spacing and showed that the first correction to the Stokes sedimentation velocity was proportional to $\Phi^{1/3}$. In this way his results are valid for small Φ (dilute concentration).

Hasimoto's basic cell is defined by three linearly independent vectors $\mathbf{a}_1$, $\mathbf{a}_2$ and $\mathbf{a}_3$. With the origin at a corner of a basic cell, the centre of any particle will be given by the vector

$$\mathbf{r}_N = n_1\mathbf{a}_1 + n_2\mathbf{a}_2 + n_3\mathbf{a}_3, \qquad (n_1, n_2, n_3 = 0, \pm 1, \pm 2, \dots). \tag{9.5.1}$$

The volume of the basic cell is given by

$$\tau_0 = \mathbf{a}_1.(\mathbf{a}_2 \times \mathbf{a}_3). \tag{9.5.2}$$

Further, Hasimoto restricted his treatment to dilute suspensions by replacing each particle by a point force retarding the motion of fluid. In this way, Hasimoto's periodic solution is obtained by solving the following Stokes equation and continuity equation:

$$\mu\,\Delta\mathbf{v} = \nabla p + \mathbf{F}\sum_{n_1}\sum_{n_2}\sum_{n_3}\delta(\mathbf{r} - \mathbf{r}_N), \tag{9.5.3}$$

$$\nabla.\mathbf{v} = 0, \tag{9.5.4}$$

where $\Delta \equiv \frac{\partial^2}{\partial x_1^2} + \frac{\partial^2}{\partial x_2^2} + \frac{\partial^2}{\partial x_3^2}$ is the correspondent Laplace operator, $\mathbf{v}$ is the velocity, p - the pressure, μ - the dynamic viscosity, $\mathbf{F}$ - the force acting on one of the particles, (x_1, x_2, x_3) are the Cartesian coordinates of the position vector $\mathbf{r}$ and $\delta(\mathbf{r} - \mathbf{r}_N)$ is the Dirac delta function.

Since velocity $\mathbf{v}$ and $-\nabla p$ are periodic functions, they can be expanded in Fourier series

$$\mathbf{v} = \sum_k \hat{\mathbf{v}}_k \exp[-2\pi i(\mathbf{k}.\mathbf{r})], \tag{9.5.5}$$

$$-\nabla p = \sum_k \hat{p}_k \exp[-2\pi i(\mathbf{r}.\mathbf{k})], \tag{9.5.6}$$

where

$$\mathbf{k} = n_1\mathbf{b}_1 + n_2\mathbf{b}_2 + n_3\mathbf{b}_3 \qquad (n_1, n_2, n_3 = 0, \pm 1, \pm 2, \dots) \tag{9.5.7}$$

are vectors in the reciprocal lattice. They satisfy the equations $\mathbf{k}.\mathbf{a}_j = n_j$, where j =1, 2, 3. The basis vectors $\mathbf{b}_1$, $\mathbf{b}_2$, $\mathbf{b}_3$ in the reciprocal lattice are given by

$$\mathbf{b}_1 = \frac{\mathbf{a}_2 \times \mathbf{a}_3}{\tau_0}, \; \mathbf{b}_2 = \frac{\mathbf{a}_3 \times \mathbf{a}_1}{\tau_0}, \; \mathbf{b}_3 = \frac{\mathbf{a}_1 \times \mathbf{a}_2}{\tau_0}. \tag{9.5.8}$$

Multiplying (9.5.3) and (9.5.4) by $\exp[2\pi i(\mathbf{k}.\mathbf{r})]$ and integrating over the unit cell in physical space, we have

$$-4\pi^2\mu|\mathbf{k}|^2\hat{\mathbf{v}}_k = -\hat{\mathbf{p}}_k + \frac{\mathbf{F}}{\tau_0}, \; (|\mathbf{k}|^2 = \mathbf{k}.\mathbf{k}), \tag{9.5.9}$$

$$\mathbf{k}.\hat{\mathbf{v}}_k = 0. \tag{9.5.10}$$

Taking curl of (9.5.6), we obtain

$$\hat{\mathbf{p}}_k \times \mathbf{k} = 0. \tag{9.5.11}$$

Then from (9.5.9), (9.5.10) and (9.5.11) for $k^2 = |\mathbf{k}|^2 \neq 0$, it follows that

$$\mathbf{k}.\hat{\mathbf{p}}_k = \frac{1}{\tau_0}(\mathbf{k}.\mathbf{F}), \tag{9.5.12}$$

$$\hat{\mathbf{p}}_k = \frac{(\mathbf{k}.\mathbf{F})\mathbf{k}}{\tau_0 k^2}. \tag{9.5.13}$$

Substituting (9.5.13) into (9.5.9) we reach to the formula

$$\hat{\mathbf{v}}_k = \frac{1}{4\pi^2 \mu \tau_0}\left[\frac{(\mathbf{k}.\mathbf{F})\mathbf{k}}{k^4} - \frac{\mathbf{F}}{k^2}\right], \ (k \neq 0) \tag{9.5.14}$$

In this way, the periodic fundamental solutions of the Stokes equations for the flow past a periodic array of spherical particles are given by equations (9.5.5) and (9.5.6) with (9.5.13) and (9.5.14). The components of the velocity **v** and the pressure gradient ($-\nabla p$) in Cartesian coordinates are calculated by the formulae

$$v_j = v_{0j} - \frac{1}{4\pi\mu}\left(F_j S_1 - \sum_{l=1}^{3} F_l \frac{\partial^2 S_2}{\partial x_l \partial x_j}\right), \tag{9.5.15}$$

$$(-\nabla p)_j = \frac{F_j}{\tau_0} - \frac{1}{4\pi}\sum_{l=1}^{3} F_l \frac{\partial^2 S_1}{\partial x_l \partial x_j}, \tag{9.5.16}$$

where $S_1 = \frac{1}{\pi\tau_0}\sum_{k\neq 0}{}' \frac{\exp[-2\pi i(\mathbf{k}.\mathbf{r})]}{k^2}$, $S_2 = \frac{1}{4\pi^3\tau_0}\sum_{k\neq 0}{}' \frac{\exp[-2\pi i(\mathbf{k}.\mathbf{r})]}{k^4}$.

It should be noted that the k = 0 terms are omitted from the summations, as it is assumed that there is no net (k = 0) force on the system.

Then, Hasimoto used a procedure suggested by Burgers, requiring that the mean velocity over the surface of each spherical particle vanish, i.e.,

$$<\mathbf{v}> = \frac{1}{4\pi a^2}\iint_{\text{sphere}} \mathbf{v}\, d\sigma. \tag{9.5.17}$$

In this manner, Hasimoto obtains the following expression for the drag force of a flow through a simple cubic array of spherical particles of radii a:

$$\mathbf{F} = \frac{6\pi\mu a \mathbf{U}_s}{1 - 1.760\Phi^{1/3}}, \tag{9.5.18}$$

where Φ is the volume fraction. Since for $\Phi \ll 1$ we have

$\frac{1}{1-1.760\Phi^{1/3}} = 1 + 1.7601\Phi^{1/3} + O(\Phi^{2/3})$, the equation (9.5.18) takes the form

$$\mathbf{F} = 6\pi\mu a \mathbf{U}_s(1 + 1.7601\Phi^{1/3}) + O(\Phi^{2/3}). \tag{9.5.19}$$

Therefore, the settling velocity of a cubic dilute suspension is given by

$$\mathbf{U}_s = \frac{\mathbf{U}_0}{1 + 1.7601\Phi^{1/3}}. \tag{9.5.20}$$

There exist divergent types of periodic lattices. There are, for example, three types of cubic lattices:

Simple cubic lattice

$$\mathbf{a}_1 = l(1, 0, 0);\ \mathbf{a}_2 = l(0, 1, 0);\ \mathbf{a}_3 = l(0, 0, 1);$$

Body-centred cubic lattice

$$\mathbf{a}_1 = l(1, 0, 0);\ \mathbf{a}_2 = l(0, 1, 0);\ \mathbf{a}_3 = l(1/2, 1/2, 1/2);$$

Face-centred cubic lattice

$$\mathbf{a}_1 = l(1/2, 1/2, 0);\ \mathbf{a}_2 = l(-1/2, 1/2, 0);\ \mathbf{a}_3 = l(1/2, 0, 1/2),$$

where $l = \sqrt[3]{\mathbf{a}_1.(\mathbf{a}_2 \times \mathbf{a}_3)}$.

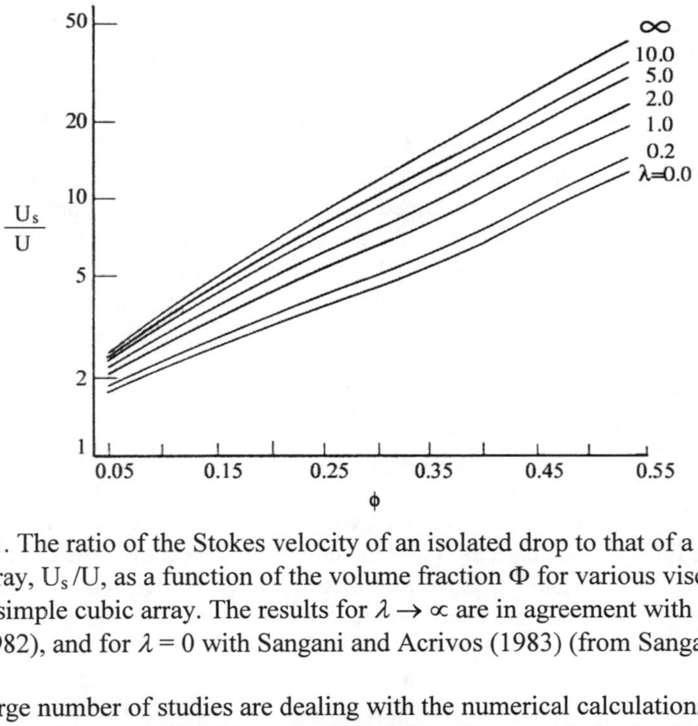

Fig.9.5.1. The ratio of the Stokes velocity of an isolated drop to that of a drop in a periodic array, U_s/U, as a function of the volume fraction Φ for various viscosity ratios λ and a simple cubic array. The results for $\lambda \to \infty$ are in agreement with Zick and Homsy (1982), and for $\lambda = 0$ with Sangani and Acrivos (1983) (from Sangani, 1987a).

A large number of studies are dealing with the numerical calculations of the flow through a periodic array of rigid or fluid particles. These include the calculations of settling velocities or, equivalently, the drag force on a fixed array of rigid or fluid particles (Zick and Homsy, 1982; Sangani and Acrivos, 1982, 1983). These authors extended Hasimoto's results for all concentrations, up to the close packing value, by solving the single-cell problem numerically. Zick and Homsy (1982) employed a method of integral equations based on a periodic fundamental solution of Stokes equations of motion. Sangani and Acrivos (1982, 1983), on the other hand, utilised a method of singularity distribution and determined the settling velocities of rigid particles as well as spherical bubbles. Recently Sangani (1987) extended the method of Sangani and Acrivos (1983) to treat the problems involving drops with finite viscosity. The calculated results agree with the previous analyses for the two limiting cases of the rigid particles ($\lambda \to \infty$) and spherical bubbles ($\lambda \to 0$) (see Fig.9.5.1). Phan-Thien et al. (1991) used boundary elements method to solve Stokes equations for periodic arrays of force-free and torque-free rigid particles. Simple cubic array of spheres, spheroids, cubes and clusters of spheres are subjected to a bulk simple shearing flow. It should be noted that the methods of Hasimoto (1959), Zick and Homsy (1982), of Sangani and Acrivos (1982) and Phan-Thien et al. (1991) cannot be used directly in random configurations of particles. For periodic configurations the effect of the particles on the flow field has usually been represented as a sum of flows due to individual particles. Unfortunately, this sum is not absolutely convergent when there is an infinite number of particles distributed over an infinite domain.

Ladd (1988, 1989, 1990) developed a computational method for estimating the hydrodynamic interactions in a suspension of spherical particles. The method is based on the moment expansion of the induced force density devised by Mazur and van Saarloos (1982), but modified for periodic boundary conditions. Ladd demonstrated that many force moments need to be retained in the description of the particle dynamics in order to achieve quantitative accuracy for the sedimentation rate, even if the lubrication interactions were explicitly included. Recently, Revay and Higdon (1992) have shown that truncation at the dipole level might give accurate sedimentation velocities, if sufficient statistical data are obtained in order to properly extrapolate to infinite number of particles N.

Two methods, based on the idea of multiple scattering, were used by Rubenstein and Keller (1989) to calculate the sedimentation velocity of a dilute and semi-dilute suspension of identical rigid spherical particles. Both methods begin with a finite number of particles N in a finite region with a rigid boundary. The authors split the total flow field into a part scattered by the particles and a part reflected from domain walls. Although the limit of each part, as N and the domain size become infinite, may contain a divergent integral, their sum is a well-defined quantity. Rubenstein and Keller (1989) first solved the equations and then averaged the results, whereas Hinch (1977) first averaged the equations and afterwards solved them. For dilute simple cubic arrays of particles the first method (which is valid when the correlation length l of the particle distribution is large compared to the sphere radius a) yields Hasimoto's (1959) result, while their second method [valid for l = O(a)] when applied to a dilute uniform homogeneous distribution yields Batchelor's (1972) result. It must be noted that both methods are applicable to other distributions and to other flows of dilute suspensions, too.

The known results for the force on a periodic array due to flow at zero Reynolds number has been generalised by Cheng and Papanicolaou (1997) for the case of small but finite Reynolds numbers. These authors used a generalisation of Hasimoto's (1959) approach that is based on an analysis of periodic Green's functions. They compared their results to the phenomenological ones of Kaneda (1986) for viscous flow past a random array of spherical rigid particles.

9.6. Dynamic Simulation of Suspension Flows. Diffusion.

Dynamic simulation is a powerful tool in the investigation of suspensions and emulsions of particles interacting hydrodynamically in Stokes flow regime. An impressive and very efficient method for treating many-body interactions of closely spaced particles is the multipole-moment technique that has been developed and named by Brady and co-workers "Stokesian dynamics" (Bossis and Brady, 1984; Brady and Bossis, 1985; Durlofsky et al., 1987; Brady et al., 1988). Excellent reviews of the applications of this method are given by Brady and Bossis (1988) and Weinbaum et al. (1990).

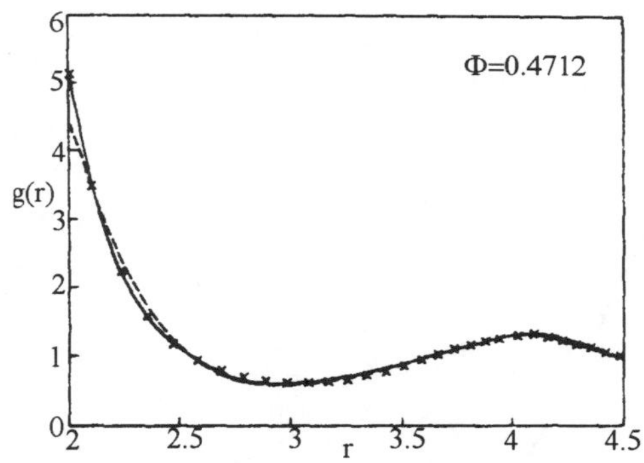

Fig.9.6.1. The hard-sphere radial distribution function g(r) at $\Phi = 0.4712$ computed by: (×) Monte Carlo simulation (Barker and Henderson, 1971); (··········) the Perkus-Yevick equation (Throop and Bearman, 1965); (——) adjusted Perkus-Yevick equation (from Brady and Durlofsky, 1989)

The essential value of this method is that it provides the means for solution of the interaction of many particles and for determination of the suspension microstructure evolution. Stokesian dynamics uses a molecular dynamics-like approach to follow the time evolution of the particles positions in the suspension. For computing the interactions between particles one can use a pairwise additivity of forces or of velocities. The pairwise additivity of forces is the preferred method as it preserves the hydrodynamic lubrication forces which prevent particles from overlapping. Near-field lubrication effects are introduced into the problem solution in a pairwise manner via two-body results. In the algorithm of Stokesian dynamics, the impact of the velocity disturbance on the motion of all particles is determined by applying the generalised Faxén laws. These laws express the motion of a freely suspended particle as a function of the unperturbed fluid velocity field, in which it is immersed. The Stokesian dynamics method takes into account not only the hydrodynamic interaction forces, but also the Brownian forces, the inter-particle attractive and repulsive forces and the external forces. The key feature of the method is, however, the hydrodynamic interaction. It represents an infinite suspension as a spatially periodic array of identical cells in order to compute the many-body interactions by numerical simulation. The many-body mobility matrix is employed to relate the particle velocities to the forces and torques that they exert on the fluid. After solving the equations including the mobility matrix, one obtains the motion of the particles in a given configuration at a given time. Then using a time stepping procedure the particle locations and orientations are determined at the next step. More details of the Stokesian dynamics method can be found in Bossis and Brady (1984) and Brady et al. (1988).

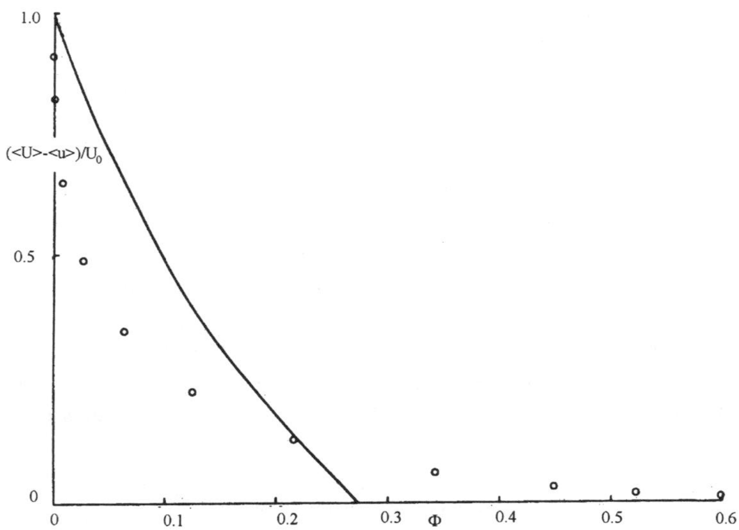

Fig.9.6.2. Reduced sedimentation velocity $\frac{\langle U\rangle - \langle u\rangle}{U_0}$ as function of Φ, where <u> is the mean fluid velocity, calculated by: (——) Glendinning and Russel (1982); (O) Zick and Homsy (1982) (from Glendinning and Russel, 1982).

Since moderate to high particles volume fractions are frequently encountered in real systems, many researchers began extending the existing theories for dilute and semi-dilute to higher concentrations in a way, which recognises the critical role of the suspension microstructure. It is well known that for pairwise additive interactions, the equilibrium radial distribution function g(r) (central quantity for the dense suspensions theory) fully characterises the suspension microstructure. That is why Glendinning and Russel (1982) confined their attention to hard spheres for which accurate analytical approximations for g(r) are available (see McQuarrie, 1976 and Fig.9.6.1). Using a pairwise additive description of hydrodynamic interactions and an equilibrium radial distribution function characterising the hard spheres microstructure Glendinning and Russel (1982) developed a theory for the sedimentation velocity and the mutual and self-diffusion coefficients in concentrated suspensions of monodisperse spherical particles. The results for the sedimentation velocity are shown on Fig.9.6.2., where the exact calculations of Zick and Homsy (1982) are given for comparison. The resulting predictions are reduced to known dilute limits and deviate in a realistic fashion at slightly higher concentrations. Unfortunately, the suggested approximation fails completely; it predicts physically impossible negative sedimentation velocity above a volume fraction $\Phi \approx 0.27$. This failure was attributed by Glendinning and Russel to the lack of many-body hydrodynamics in the particles interactions approximation.

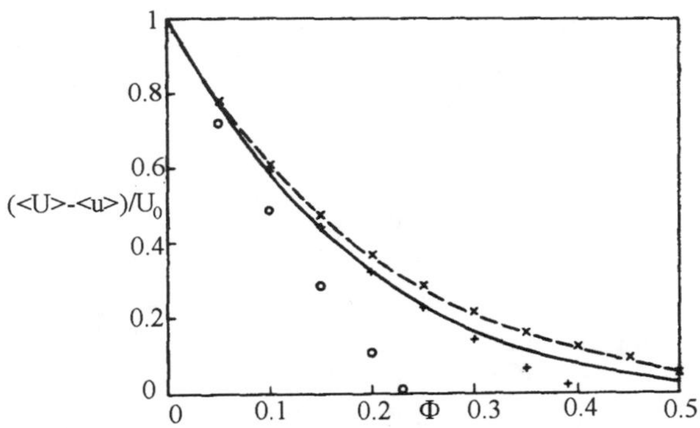

Fig.9.6.3. Reduced sedimentation velocity of a disordered system of monodisperse spherical particles as a function of Φ according to: (———) the empirical results of Garside and Al-Dibouni, (1977); (········) the equation (9.6.1); (×) the theoretical results of Brady and Durlofsky (1988) for the case of sphere-sphere interactions approximation via the Rotne-Prager approximation; (+) the stresslet-free approximation; (O) the strict pairwise additive approximation for sphere-sphere approximation (from Brady and Durlofsky, 1988).

Brady and Durlofsky (1988) derived an explicit expression for the sedimentation velocity in concentrated suspension:

$$< U > - < u >= 1 + \Phi - \frac{1}{5}\Phi^2 - \frac{6}{5}\Phi\left(\frac{5 - \Phi + \frac{1}{2}\Phi^2}{1 + 2\Phi}\right), \tag{9.6.1}$$

where <u> is the mean fluid velocity. The derivation follows the method of O'Brein (1979) for constructing convergent expressions of the particles interactions in suspensions. Brady and Durlofsky (1988) showed that the Rotne-Prager (Rotne and Prager, 1969) approximation gives a very accurate prediction for the sedimentation velocity of random suspensions from the dilute limit all the way up to close packing. It means that Rotne-Prager approximation with the Percus-Yevick hard-sphere distribution actually captures the correct features of the many-body interactions in sedimentation (see Fig.9.6.3.). On the same figure it is also observed that negative sedimentation velocities are predicted by the strict pairwise additive approximation for $\Phi > 0.23$ and by the stresslet-free approximation for $\Phi > 0.39$.

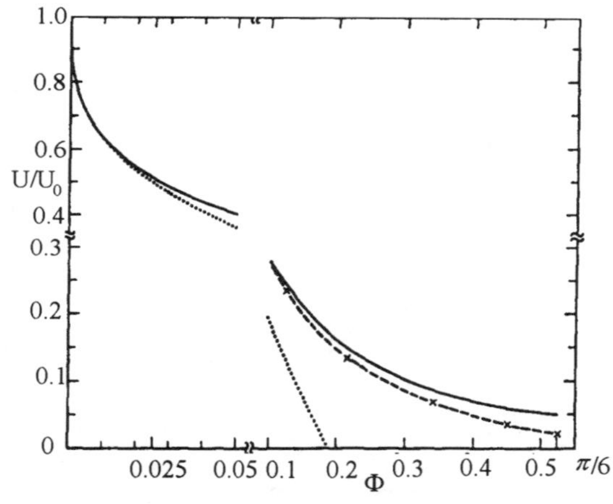

Fig.9.6.4. Non-dimensional sedimentation velocity of a simple cubic array of spheres as a function of Φ according to: (——) the Stokesian-dynamics method (Brady et al., 1988); (----x--) the exact results of Zick and Homsy (1982); (……) the point-force solution of Saffman (1973) (from Brady et al., 1988).

Brady et al (1988) presented a general and rigorous method for calculating hydrodynamic interactions in infinite suspensions. The method allows to study the problems connected with sedimentation, self-diffusion and rheology. Fig.9.6.4. shows some sedimentation results obtained by using this method. Durlofsky et al. (1987)

improved the approximation of Glendinning and Russel (1982) and Brady and Bossis (1985) by proposing a method that includes many-body, far-field interactions among finite numbers of spherical particles and using pairwise additivity only to account for near-field interactions, i.e., for lubrication interactions. Exploiting periodic boundary conditions Brady et al (1988) extended this method for simulating infinite suspensions of spherical particles. It is shown that this method (Stokesian dynamics method) produces results that are in good agreement with the known exact solutions for shear and sedimentation velocity (Zick and Homsy, 1982; Zuzovsky et al., 1983; Nunan and Keller, 1984).

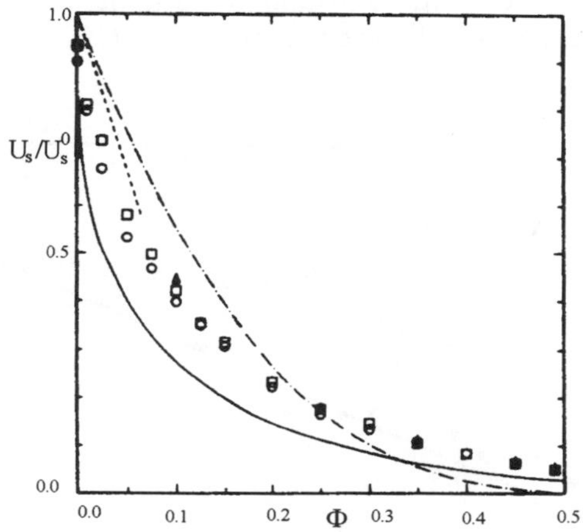

Fig.9.6.5. The sedimentation velocity, non-dimensionalized by the infinite dilution value U_s^0, as a function of Φ according to: (——) the exact result for simple cubic lattice (Zick and Homsy, 1982); (-----) the experiments of Buscall et al. (1982); (........) the low Φ asymptotic results of Batchelor (1972); Stokesian dynamics results for 14 (O), 27 (□) and 64 (Δ) spheres in the unit cell (from Phillips et al., 1988a).

All hydrodynamic properties (sedimentation, diffusion, permeability, rheology) can be determined via the Stokesian dynamics method for any microstructural arrangement of spherical particles. Phillips et al. (1988a) presented Stokesian dynamics results on the transport properties of suspensions of freely mobile particles. Particle distributions are generated by a Monte-Carlo technique and both the far- and near-field hydrodynamic interactions are calculated by Stokesian dynamics simulation. The effects of changing the particles number in the simulation cell are investigated and comparisons of the simulation results with experiment and theory are performed. It should be noted that inserting periodicity into a disordered medium model is a severe restriction in the sedimentation problem. The results of simulations with N values of 14, 27 and 64 particles are plotted on Fig.9.6.5. together with the results of Zick and Homsy (1982) and Batchelor's low volume fraction results. A correlation of experimental data reported by Buscall et al. (1982) for lattice is also included in Fig.9.6.5.

Settlers are large pieces of equipment that typically occupy a considerably large area. That is why it is desirable to scale down the size of settling tanks without reducing their operational capacity. The utilisation of close-space inclined parallel plates greatly enhance the sedimentation process. The steady sedimentation of a suspension over an inclined surface has been analysed by Nir and Acrivos (1990). They showed that for a given value of the particle volume fraction in the unsettled suspension, a steady flow of the sediment can be maintained only if the angle of inclination exceeds a minimum value. High particle volume fractions within the sediment are predicted when the unsettled suspension is either very dilute or very concentrated.

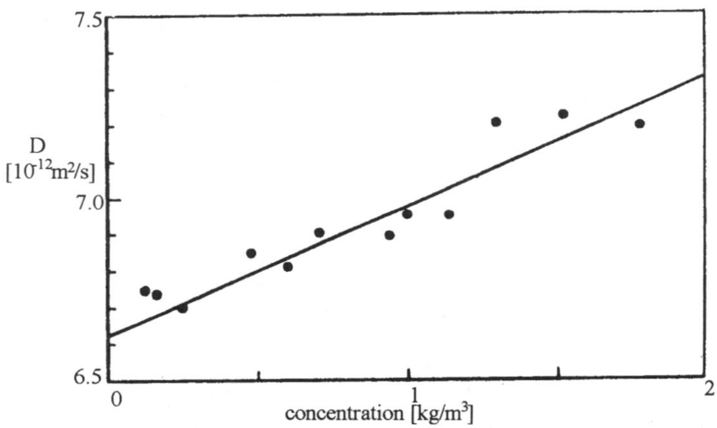

Fig.9.6.6. Concentration dependence of the gradient-diffusion coefficient D for fd bacteriophage DNA in water: (●) light scattering data of Newman et al. (1974); (———) according to Eq.(9.6.1) (from Russel, 1981).

Another area of considerable research activity during the past few decades has involved the diffusion problems at sedimentation. Batchelor (1976a) analysed the sedimentation-diffusion equilibrium by equating the external force acting on a particle to an apparent thermodynamic force equal to the chemical potential gradient. He indicates that multi-particle displacements due to Brownian motion have statistics of Gaussian form and that the joint probability density functions describing particle displacements satisfy the generalized Stokes - Einstein equation. The results of Batchelor elegantly separate the thermodynamic and hydrodynamic effects without restriction on Φ. When $\Phi \ll 1$, Batchelor derived the following formula for the diffusivity tensor **D**:

$$\mathbf{D} = \mathbf{D}_0(1 + 1.45\Phi), \tag{9.6.2}$$

where $\mathbf{D}_0 = \frac{kT}{6\pi\mu a}\mathbf{I}$ is the classical expression for the diffusivity of a very dilute dispersion of independent spheres of radius a in a fluid of viscosity μ (here k is Boltzmann's constant and T is the absolute temperature) derived by Einstein (1905).

Felderhof (1978) and Wills (1979) independently have confirmed the Batchelor's results for hard spheres (i.e., for spheres which are rigid and exert no force on each other when they are not touching) in the dilute limit but have not extended their conclusions to higher concentrations. Attempts to verify these results have been performed by many researchers. Concentration dependence on the gradient-diffusion coefficient for fd bacteriophage DNA in water was presented by Newman et al. (1974) (see Fig.9.6.6).

Felderhof (1977a, 1979) suggested another method in which he used the modified Smoluchowski equation for the particle, and two particle distribution functions in order to represent the diffusion tensor as a sum of contributions due to isolated particles and pairs of particles. In spite of the fact that he applied the point-particle approximation, his results are extremely close to those of Batchelor (1976a).

The self-diffusion coefficient of spherical particles in suspension is also concentration dependent, due to a direct hard-sphere interactions between particles and due to coupling of their motion via the fluid. Beenakker and Mazur (1982, 1983) calculated the concentration dependence of the self-diffusion coefficient for a suspension, including second order terms in the particle volume concentration Φ

$$\mathbf{D} = \frac{kT}{6\pi\mu a}\mathbf{I}\left(1 - 1.73\Phi + 0.88\Phi^2\right). \qquad (9.6.3)$$

This formula contains two- and three-sphere hydrodynamic interaction effects that are of comparable size. For low values of the volume fraction Φ the virial expansion is appropriate. Only two-body hydrodynamic interactions contribute to the well-known term of order Φ (Batchelor, 1976a; Felderhof, 1978), which dominates if the suspension is very dilute. At higher volume concentrations, however, the many-body hydrodynamic interactions cannot be ignored. Beenakker and Mazur showed that a neglect of three-body contributions would give a value of $-0.93\Phi^2$ instead of $+0.88\Phi^2$, the last term in (9.6.3). It should be noted that a virial expansion is not appropriate at high particle volume concentrations. In a concentrated suspension it is therefore essential to fully take into account the many-body hydrodynamic interactions of an arbitrary number of particles.

It is well known that the phenomenon of hydrodynamic diffusion of a particle in a suspension arises from hydrodynamic interactions with other particles and is not related to Brownian diffusion due to the thermal motion of the fluid molecules surrounding each particle. Hydrodynamic diffusion during batch sedimentation of nearly monodisperse suspensions has been observed experimentally by Davis and Hanssen (1988) and Ham and Homsy (1988). Whereas Davis and Hanssen have measured the rate of the interface spreading at the top of suspension, Ham and Homsy have registered the time variance for a marked sphere to fall at a given distance in the suspension interior.

The hydrodynamic interactions of individual particles may cause a net particle migration from regions of high concentration to regions of low concentration, and from high shear regions to low shear regions (Leighton and Acrivos, 1987a). The first experimental study of shear induced diffusion was presented by Eckstein et al. (1977), who observed a random lateral motion, i.e., normal to the fluid velocity, of a tagged spherical particle in a suspension undergoing shear in a Couette device and thereby computed the lateral self-diffusion coefficient. Leighton and Acrivos (1987a) improved their technique and were able to infer values for the diffusion coefficients (normal and tangent to the plane of shear) for the case of particle migration due to gradients in shear and gradients in volume fraction, as well. The shear-induced particle migration mechanism described by Leighton and Acrivos (1987a) might be responsible for the classic findings of Karnis et al (1966a), for the velocity profile blunting (Fig.9.6.7) in the motion of a concentrated suspension in a tube. The particles migrate away from the walls, where the shear rate is the highest and this flattens the velocity profile.

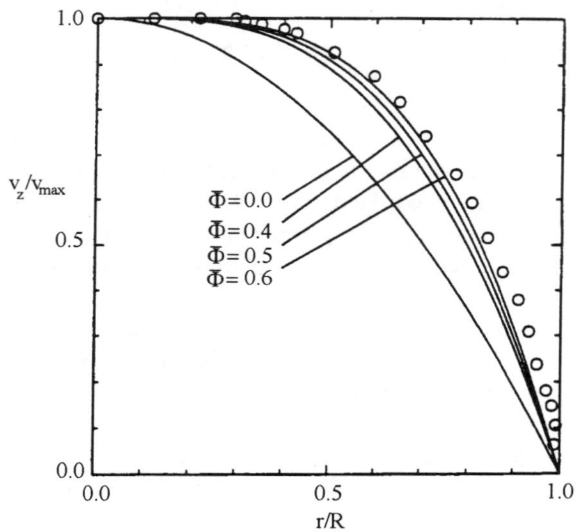

Fig.9.6.7. Predicted velocity profiles from: (——) Phillips et al. (1992) for pure fluid ($\Phi = 0$) and three average volume fractions for Poiseuille flow in a tube; (O) experiments of Karnis et al. (1966b) for an average volume fraction $\overline{\Phi} = 0.38$ (from Davis, 1993).

A numerical evaluation of the self-diffusion coefficient is given by Glendinning and Russel (1982). The self-diffusion coefficient curve, shown on Fig.9.6.8, lies below the dilute limit result of Batchelor (1976a) expressing the effect of the distribution function crowding particles into regions of increased hindrance due to near-field hydrodynamics.

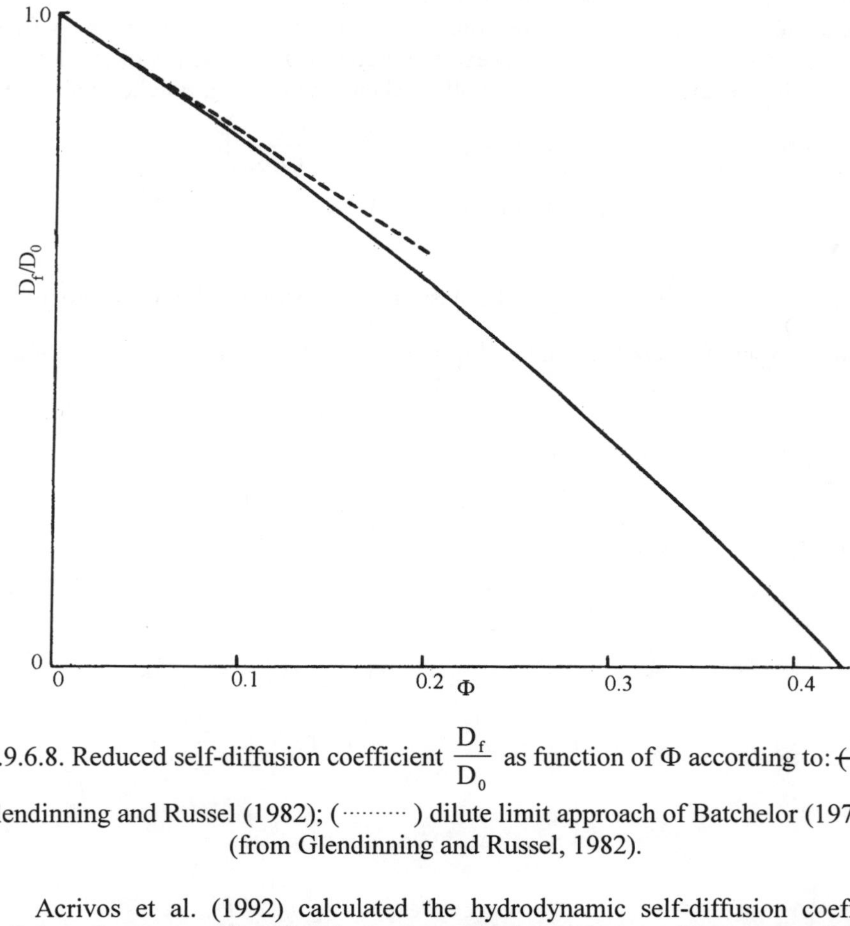

Fig.9.6.8. Reduced self-diffusion coefficient $\frac{D_f}{D_0}$ as function of Φ according to: (——) Glendinning and Russel (1982); (·········) dilute limit approach of Batchelor (1976a); (from Glendinning and Russel, 1982).

Acrivos et al. (1992) calculated the hydrodynamic self-diffusion coefficient (parallel to the stream-wise direction) of a tagged particle in a dilute mono-dispersed suspension of small neutrally buoyant spheres undergoing a steady simple shearing motion. The longitudinal shear inducing self-diffusion coefficient was found to be $0.267a^2 G\ [\Phi \ln \Phi^{-1} + O(\Phi)]$, where G denotes the applied shear rate, $\mathbf{v} = G(x_2, 0, 0)$, a is the spherical particle radius and Φ is the particles volume concentration. While longitudinal diffusion coefficient is determined by pair-wise particle interactions, which are tractable, lateral diffusivity coefficient determination requires the solution of a three-sphere hydrodynamic problem.

Gradient diffusion of a dilute polydisperse system of spherical particles was investigated by Batchelor (1983). When m different species of small particles are dispersed in fluid the existence of a small spatial gradient of concentration of particles of type j is accompanied by a flux of particles of type i (due to the particles Brownian motion). If the total particles volume fraction is small, the diffusivity tensor D_{ij} is approximately a linear function of the volume fractions Φ_1, Φ_2, ... Φ_m, with coefficients which depend on the interactions between the particles. For a system of rigid spherical particles, which exert no force on one another when not touching, Batchelor obtained

$$D_{ij} = D_i^0\left(1 + 1.45\Phi_i - \sum_{k\neq i}\frac{\Phi_k}{1+0.6\alpha_{ik}}\right)$$
$$\underset{(i\neq j)}{D_{ij}} = D_i^0\Phi_i\left(\alpha^3 + 2\alpha^2\right), \tag{9.6.4}$$

where $\alpha = \dfrac{a_j}{a_i}$, $\alpha_{ik} = \dfrac{a_k}{a_i}$ and D_i^0 is the classical expression for the diffusivity of a very dilute dispersion of independent spheres of radius a_i first derived by Einstein (1906).

CHAPTER 10

Hydrodynamic Interaction between Particles and Effective Viscosity of Suspensions and Emulsions

10.1. Introduction

In a two component disperse system consisting of a fluid and suspended in it particles (rigid or fluid), the average properties of the particles determine the corresponding bulk or collective properties of the disperse system. For some purposes it is useful to think of the suspension or emulsion as an equivalent to a continuous medium with certain effective properties. In order to determine each of these collective properties of the dispersion, we have to make the next two steps: (i) to calculate the relevant property of a particle of a given size, orientation, neighbour-configuration, etc. and (ii) to take an average of its property over all possible values of the size, etc., with appropriate probability weighted functions.

The problems, connected with the suspensions viscosity determination, are important for the macroscopic particles (e.g., in separation and other technological processes) as well as for very small (colloidal) particles, the size of which is closer to the continuous phase molecular sizes. The colloidal solutions viscosity determination is a task of the colloidal chemistry. The solution viscosity depends on the type of the continuous phase (the pure solvent), the type and concentration of the suspended particles and the shear motion of the dispersed phase in the continuous phase. The viscosity increase is the least for the dilute solutions, when the particles interaction and the eventual collisions between them can be neglected.

There exist two ways for the theoretical description of a given material according to the linear scale, in which the investigations are performed:

(i) macroscopic, when the material is considered as a complex, but homogeneous continuous phase;

(ii) microscopic, when the material is assumed composed of discrete particles, which translate, rotate, deform and interact with one another.

In this chapter we shall consider suspensions and emulsions of non-colloidal particles, i.e., sufficiently large particles for which the colloidal forces such as electro-viscous effects and van-der-Waals forces are unimportant with regard to the viscous forces on a length scale comparable to the particle diameter. The central topic will be the prediction of the rheological behaviour of a dispersion containing rigid particles or gas bubbles, liquid drops and biological cells. Several approaches to the study of dispersions of rigid or fluid particles are possible: physical experiment, analytical theory and numerical simulation.

Rheological properties of the suspensions and emulsions have been the subject of numerous studies. Many of the results and interpretations ensuing therefrom are summarised in the following books and review papers: Happel and Brenner (1973), Batchelor (1974, 1976b), Brenner (1972a, 1974), Jeffrey and Acrivos (1976), Buyevich and Shchelchkova (1978), Herczynski and Pienkowska (1980), Brady and Bossis (1985), Ladd (1990), Davis (1993) and Zapryanov (1989a, 1992). Essentially four types of models exist: (i) statistical models (Batchelor, 1974); (ii) cell models (Simha, 1952; Happel, 1957); (iii) spatially periodic models (Hasimoto, 1959); (iv) numerical models (Loewenberg and Hinch, 1996, 1997). Each of these analyses suffers from its own special limitations. Moreover, because of some of their ad hoc assumptions, they cannot be rationally improved.

It is well known that the relation between the applied stress, τ, and the resulting shear rate ($G = \dot{\gamma}$) in a simple shear (Couette) flow can be expressed in the form

$$\tau = \mu\dot{\gamma} \,. \tag{10.1.1}$$

For example, when a liquid is sheared between two plates parallel to the xy plane, then $\dot{\gamma} = \dfrac{\partial \mathrm{v}_x}{\partial \mathrm{z}}$. Fig.10.1.1.a) shows a typical plot of $\dot{\gamma}$ vs. τ.

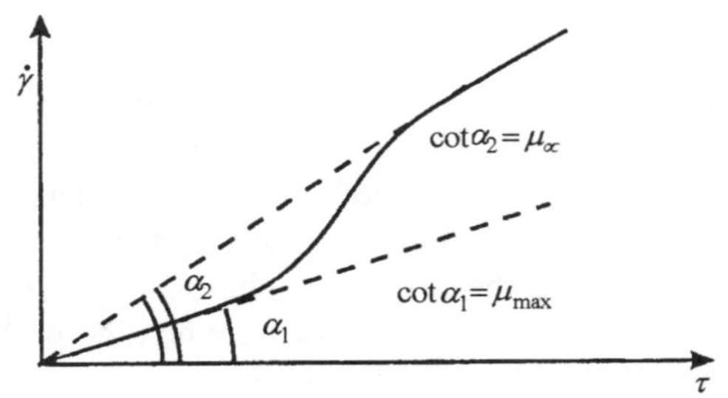

Fig.10.1.1. Rheological properties of suspensions: a) rate of strain, $\dot{\gamma}$, as a function of the applied stress, τ .

It is worth noting that for low and high shear rates and high shear rates, one observes Newtonian behaviour (μ = const.), whereas in the intermediate region, a transition from the lower shear rate viscosity μ_{max} to the higher shear rate viscosity μ_∞ takes place. Fig.10.1.1.b) shows the dependence of the viscosity of a dispersion on the shear rate $\dot{\gamma}$. Note that in the intermediate zone μ has a minimum (Barnes et al., 1989).

Most of the experimental studies performed in this field are of limited interest due to the poorly defined suspension characterisation. Happily, there are exceptions: investigations of Krieger (1972), Krieger and Eguiluz (1976), Hoffman (1972, 1974), Leighton and Acrivos (1987a), Pal and Rhodes (1985, 1989), de Kruif et al. (1985), etc.

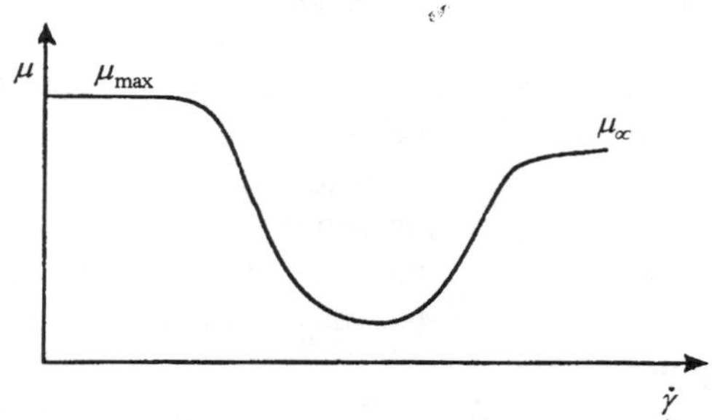

Fig.10.1.1.b) average viscosity of a suspension μ as a function of rate of strain $\dot{\gamma}$.

Suspension rheology is a very developed discipline in terms of the quantity and quality of knowledge currently available in the field. The impetus for the rapid development of the subject over the past fifty years has come largely from the needs of polymer science. Other applications of interest exist in the areas of slurry transport in pipelines, emulsion rheology, manufacture of paper from pulp, paint technology, blood flow, ferro-fluid rheology and many other fields involving Newtonian and non-Newtonian technologies. In recent years there have been many scientific achievements in the domain of dilute and semi-dilute suspensions, but the behaviour of the concentrated and highly concentrated suspensions remains a challenge for rheologists [see Goldsmith and Mason (1967), Russel et al. (1989) and Van de Ven (1989)].

It is clear now, that many of the complex phenomena associated with a flowing suspension cannot be explained by using a classical Newtonian description of a fluid with an effective viscosity depending solely on the volume fraction of the particles. Thus, suspensions have to be treated as non-Newtonian fluids, whose rheological properties are influenced by a large number of variables. It should be noted that older models used for the calculation of suspension properties are at present being displaced by more exact methods which take into account the many factors determining the suspension rheology. The presence or absence of non-Newtonian effects depends primary on the rate of diffusion compared to the relative convection of particles by an imposed flow. The ratio of the rate of diffusion to the relative convection comprises the Peclet number, which must be O(1) in order for the flow to be able to disrupt significantly the suspension microstructure and to produce a non-linear time-dependent rheology. Further progress of suspension rheology requires the development of more powerful theoretical tools for describing and predicting the flows in suspensions, especially highly concentrated ones in which the non-Newtonian effects are most pronounced.

The present chapter may be considered, in essence, as a review, when studying the problems related to the effective viscosity of suspensions and emulsions.

10.2. Effective Viscosity of Dilute and Semi-dilute Suspensions and Emulsions.

Theoretical investigations of suspension and emulsion behaviour have been limited, for the most part, to dilute and semi-dilute systems at zero particle Reynolds number where single or two-particle interactions dominate. They have sought to determine the macroscopic properties as an expansion in volume fraction or number density. Although being limited to low concentrations, these approaches have identified many of the fundamental mechanisms operating in suspensions and provide an important foundation upon which the further studies to be based.

The first theoretical research of the suspensions has been done by Einstein (1906, 1911). He considers limiting case of equal spherical particles which are so distant from one another that the motion of each particle can be assumed as a motion of a single particle in an infinite fluid. For very dilute suspensions Einstein obtains the formula

$$\mu^* = \mu\left(1 + \frac{5}{2}\Phi\right), \tag{10.2.1}$$

where μ is the solvent viscosity, Φ is the ratio of the dispersed phase volume to the total suspension volume and μ^* is the effective viscosity. Einstein's analysis in derivation of (10.2.1) is based upon a computation of the additional rate of mechanical energy dissipation by the introduction of a single spherical particle into a homogeneous shear flow. More recent theories approach the subject via dynamic, rather than energetic, methods. We shall note that Einstein's celebrated formula (10.2.1), which applies to dilute ($\Phi \le 0.03$) suspension of rigid particles, gives the first effect of the particles presence on the suspension viscosity. The wonderful in this formula is that the effective viscosity μ^* does not depend on the suspension dispersity. The experimental measurements of the viscosity of different dilute suspensions performed with ordinary viscometers (Rutgers, 1962b; Thomas, 1965) show that the coefficient before Φ in (10.2.1) may take values between 1.5 and 5. Although the small particles remain suspended in the ambient fluid and move together with it, a lot of interesting questions arise, as for example, about the motion (rotation) of non-spherical particles and their diversion from the fluid translational motion, the changes in the shape of the fluid particles (drops or bubbles), the motion of the elongated macro-molecules, the hydrodynamic and other interactions between the neighbour particles, etc.

The effective viscosity of very dilute suspensions (emulsions) of spherical fluid drops is calculated by Taylor (1932, 1934) by means of the Einstein's method. Taylor assumes that the surface tension is high (then the fluid particles have spherical shape) and no adsorbed layer of surfactant forms around the particles to diminish the motion transfer by the tangential stress inside them. The shear flow passing the fluid particles induces a circulating fluid flow inside the particles and the suspension effective viscosity μ^* becomes

$$\frac{\mu^*}{\mu_0} = 1 + \frac{\frac{5}{2}\hat{\mu} + \mu_0}{\hat{\mu} + \mu_0}\Phi. \tag{10.2.2}$$

Here μ_0 and $\hat{\mu}$ are the viscosity of the continuous and dispersed phase, respectively. If $\frac{\hat{\mu}}{\mu_0} \to \infty$, i.e., for suspension of rigid particles. The Einstein's formula (10.2.1) is obtained from the Taylor's formula (10.2.2) as a special case. When $\frac{\hat{\mu}}{\mu_0} \to 0$, i.e., for suspension of gas bubbles, from (10.2.2) one gets

$$\mu^* = \mu_0(1 + \Phi). \tag{10.2.3}$$

Not surprisingly, this viscosity is considerably lower than the corresponding for rigid spherical particles suspension. Many attempts have since been made to extend Einstein's analysis to higher concentrations. Most authors propose a power series of the general form:

$$\mu^* = \mu_0\left(1 + \frac{5}{2}\Phi + k_1\Phi^2 + k_2\Phi^3 + \ldots\right), \tag{10.2.4}$$

where k_1, k_2, etc., are constants.

Using low and high shear limiting viscosities, μ_∞ and μ_{max}, de Kruif et al. (1985) suggested the empirical expansions:

$$\frac{\mu_{max}}{\mu_0} = 1 + 2.5\Phi + (4 \pm 2)\Phi^2 + (42 \pm 10)\Phi^3 + \ldots,$$

$$\frac{\mu_\infty}{\mu_0} = 1 + 2.5\Phi + (4 \pm 2)\Phi^2 + (25 \pm 7)\Phi^3 + \ldots,$$

where $\mu \to \mu_{max}$ as shear rate $\dot{\gamma} \to 0$ and $\mu \to \mu_\infty$ as shear rate $\dot{\gamma} \to \infty$ (see Fig.10.1.1.b). For the whole range of values of Φ de Kruif et al. (1985) proposed two other empirical expressions:

$$\frac{\mu_{max}}{\mu_0} = \left(1 - \frac{\Phi}{0.63}\right)^{-2}, \qquad \frac{\mu_\infty}{\mu_0} = \left(1 - \frac{\Phi}{0.71}\right)^{-2}.$$

As mentioned before, the formula (10.2.1) was derived by calculating the dissipation of kinetic energy in a homogeneous suspension of non-interacting spherical rigid particles. A new method of deriving the classical Einstein's formula has been suggested by Landau and Lifshitz in their book on Fluid Mechanics (1959). They employed volume averaging of the stress tensor in the suspension and showed that this can be done by calculating the average stress in a large spherical volume of fluid encompassing the test particles, without knowledge on the stress inside the rigid particles. Landau and Lifshitz (1959) obtained the constitutive equation of the suspension by relating the averaged stress tensor to the averaged rate of strain tensor. We shall note that in their analysis the problem of non-absolutely converging integrals has not been overcome.

Peterson and Fixman (1963) used the Landau and Lifshitz's method as a starting point for calculating the Huggins coefficient k_H appearing in the following expression of the effective viscosity

$$\mu^* = \mu_0\left(1+\frac{5}{2}\Phi + k_H\left(\frac{5}{2}\right)^2 \Phi^2 + \ldots\right). \tag{10.2.5}$$

The importance of their paper lies not so much in determining k_H, as in applying concepts of response functions and multipole expansions and in mentioning the possible use of a grand canonical ensemble.

Batchelor (1970a) and Batchelor and Green (1972b) realised that Landau and Lifshitz's approach, if refined, could be a powerful tool in the theory of suspensions. These refinements were connected with: (i) a detailed analysis of the pair interaction of spherical particles; (ii) the relation of the probability density function of particles separation to the flow field, and (iii) the introduction of a certain method to bypass the non-absolute convergence of the integrals appearing in the theory.

It is possible to define the average stress by either ensemble averaging or volume averaging, but for the suspensions that are statistically homogeneous the definitions are equivalent and, therefore, the simpler volume averaging could be used. In a systematic manner Batchelor and Green (1972b) furnished a rigorous two-body treatment, by incorporating the probability density for the relative distance between the centres of two sphere in semi-dilute suspension and reached to the formula

$$\mu^* = \mu_0\left(1+\frac{5}{2}\Phi + 7.6\Phi^2\right) \tag{10.2.6}$$

for pure straining motion and $\mathrm{Pe} = \dfrac{\mathrm{Ga}^3\mu_0}{\mathrm{kT}} \to \infty$.

It should be noted that an analysis based on pairwise particle interactions is insufficient for simple shear flow, because of the existence of closed orbits. In the presence of a strong translational Brownian motion ($\mathrm{Pe} = \dfrac{\mathrm{Ga}^3\mu_0}{\mathrm{kT}} \to 0$) Batchelor (1977) found the expression

$$\mu^* = \mu_0\left(1+\frac{5}{2}\Phi + 6.2\Phi^2\right) \tag{10.2.7}$$

regardless of the specific nature of the bulk flow.

The theoretical studies of dilute and semi-dilute suspensions have elucidated the fundamental mechanisms operating in such systems, and provide an important foundation upon which to base further research. However, we should know that the relative viscosity predicted by these theories agree with experimental data only up to volume fraction of the order of 15-20% (Brady and Bossis, 1985).

For rigid spherical particles several accurate analytical approximations are available. They deduce the bulk properties of semi-dilute suspensions from the local interactions and the microstructure. Among these are the group expansion method of Jeffrey (1974) and the method of Saffman, who uses Fourier integrals and generalised functions. Hinch (1977) found that the divergence difficulties could be overcome by first averaging the fluid equations with respect to the probability distribution of particle positions, with either no particle fixed or one particle fixed. In this way he constructed a hierarchy of equations describing the behaviour of a two-phase macroscopically homogeneous material. By solving the hierarchy of equations using a low concentration expansion, Hinch obtained convergent expressions. An alternative, and equally rigorous, method of O'Brien (1979) employs an integral representation of the solution to the governing equations. O'Brien showed that the renormalisation of the conditionally convergent integral could be effected automatically by applying Green's theorem to a bounded domain. He combined the surface integral over the boundary of the domain with the sum over the particles within the domain and found a convergent expression.

During the 70's and 80's there was considerable research activity on the rheology of dilute suspensions (Jeffrey and Acrivos, 1976; Herczynski and Pienkowska, 1980). Part of this research was on the interplay between the orienting effects of the flow and the disordering effects of Brownian motion, with considerable progress made on predicting the bulk rheological properties of dilute suspensions of spherical particles from the resulting distributions of particle orientations (Leal and Hinch, 1971).

Felderhof (1976a,b, 1978) applied the theory of average local velocity to suspensions of spherically symmetric polymers and analysed the concentration dependence of the translational and rotational friction coefficient of a polymer, and the concentration dependence of the solution viscosity. He represented the diffusion tensor as a sum of contributions due to isolated particles and pairs of particles. Using the results of his previous work concerning hydrodynamic interaction of two spheres in the point-particle approximation Felderhof (1977a, b) derived the diffusion tensor as a function of volume fraction up to the order Φ. In spite of the use of the point-particle approximation, his results are extremely close to those of Batchelor (1976a). Schmitz and Felderhof (1982a, b) solved the creeping flow equations of an incompressible fluid in the presence of two spherical particles as a problem in the hydrodynamic scattering theory. The flow pattern outside the particles was analysed in a two-centre expansion of spherical waves generated by each particle.

Felderhof (1989) studied also the hydrodynamic interactions in suspensions with periodic boundary conditions with the purpose of performing computer simulations with many particles per unit cell. The use of periodic boundary conditions in suspensions minimises surface effects but gives rise to conceptual difficulties due to the long range of the hydrodynamic interaction. The Oseen-Burgers tensor, which describes the interactions, decays as the first power of the inverse distance between the particles.

Moreover, Felderhof showed how macroscopic considerations may be applied to arrive at a periodic solution of the Stokes equations. In terms of Ewald sums, Felderhof (1989) derived expressions for the mobility tensors of freely moving particles, as well as for the resistance tensor of rigid array. He discussed the high frequency effective viscosity of suspensions with periodic boundary conditions

The rheology of a suspension of ellipsoids of revolution dispersed in a simple shearing flow was investigated by Jeffery (1922a, b) for the case of negligible effect of Brownian motion. In the absence of particle and fluid inertia, he showed that each spheroid undergoes a periodic rotation in one of the families of closed orbits around the vorticity axis. Jeffery (1922a, b) calculated the time average viscosity of the suspension by utilising additional energy dissipation arguments and by integrating this instantaneous, orientation-dependent quantity over one period. Unfortunately, this "viscosity" is a function of the initial distribution of particle orientations, and cannot be regarded as an intrinsic property of the fluid-particle suspension. (Goldsmith and Mason, 1967; Brenner, 1974). Including the effects of rotary Brownian motion, Saito (1952) was the first to determine correctly the viscosity of a suspension of spheroids in a simple shear flow. This is possible, because in the presence of rotary Brownian motion the initial distribution of particle orientations at zero time is unimportant (see Fig.10.2.1).

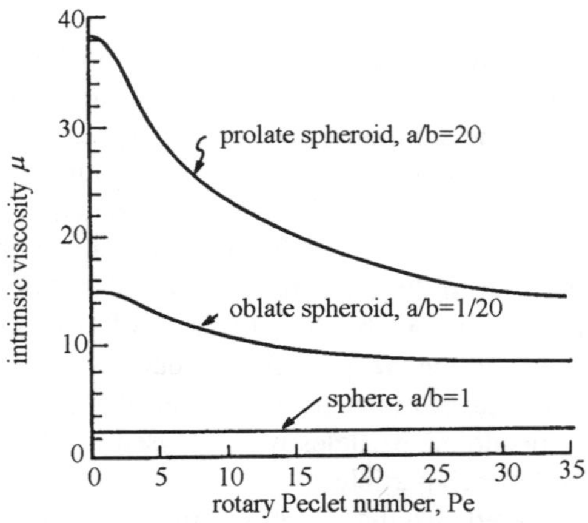

Fig.10.2.1. Intrinsic viscosity as a function of rotary Peclet number for prolate and oblate spheroids in simple shear flow (from Brenner, 1972a).

Another theory of the effective viscosity of a dilute suspension of ellipsoidal particles in extensional flow, incorporating the rotary Brownian motion, was developed by Pokrovskii (1967, 1968). Unfortunately, his theory contains a serious mistake, since in the particular case of suspended spherical particles it yields $\mu^* = \mu_0\left(1 + \frac{3}{2}\Phi\right)$, rather than the correct Einstein's formula (10.2.1).

Suspensions of rods are essential media in many production processes, ranging from the fabrication of fibre-reinforced composites to the manufacture of paper from pulp. It is also well known that rod suspensions can display radically different transport properties than the solvent, even at such low concentrations of solid matter that most other characteristics of the sample remain unaltered. The viscosity of suspensions of rod-shaped particles increases very rapidly with the solid phase concentration. At very low concentrations the differences between the viscosity of suspensions of variously shaped particles are small, while the frequency of contacts between adjacent particles increases with concentration and affects the viscosity value.

First analysis of the rheology of slender rods suspensions (Burgers, 1938a; Simha, 1949) focus on infinite dilution. However, Batchelor (1971) was the first to recognise that the extensional viscosity of a dilute suspension of fibres is increased by $O(nl^3/\ln r)$ over its pure fluid value, where n is the fibre number density, l is the fibre half-length, 2b is the characteristic fibre thickness, $r = 2l/b$ is the aspect ratio and the term dilute here means that $nl^3 \ll 1$. Batchelor (1971) developed a cell model, for which he concluded that the extensional viscosity of a semi-dilute suspension of aligned fibres should be scaled with $nl^3/\ln(1/\Phi)$. Here $nl^3 \gg 1$ and $\Phi \ll 1$, with Φ being the volume fraction of the included fibrous material. Experiments have verified the qualitative form of this scaling (Weinberger and Goddard, 1974; Mewis and Metzner, 1974). Some other authors have analysed the rheology of slender rods and considered dilute concentrations (Hinch and Leal, 1972; Brenner, 1974) or semi-dilute concentrations (Doi and Edwards, 1978a, b; Jain and Cohen, 1981). The former assume $nl^3 \ll 1$, while the latter require $1 \ll nl^3 \ll \left(\frac{l}{a}\right)^2$ and account for interactions via physically compelling but nonetheless, ad hoc added representation of the hindered diffusion process. The effects of far-field hydrodynamic interactions between rods on the orientation distribution and bulk stress for a general steady shear flow have been calculated by Berry and Russel (1987). Their solutions have been obtained up to the order of $O(Pe^3)$, as a regular perturbation expansion. The theoretical predictions for the Huggins coefficient agree qualitatively with the data for a semi-rigid biopolymer (see Chauveteau, 1982).

Shaqfeh and Fredrickson (1990) presented a theory for the momentum transport properties of suspensions containing randomly placed slender fibres. The theory is based on a diagrammatic representation of the multiple scattering expansion for the averaged Green's function. For dilute systems, $nl^3 \ll 1$, solutions for the wavenumber dependent viscosity and the long wavelength (or simply effective) viscosity were derived via a perturbation expansion on the powers of nl^3. Since the complete hydrodynamic interaction between high aspect ratio fibres were employed at every step of calculation, Shaqfeh and Fredrickson (1990) were able to correct the results of Berry and Russel

(1987), who used only the far-field interactions. Since Shaqfeh and Fredrickson (1990) included multiparticle interactions at every level of approximation, their theory is most useful in describing non-dilute systems and, in particular, semi-dilute systems.

A systematic analysis of the effect of hydrodynamic interactions on the orientation of large aspect ratio fibres undergoing simple shear flow has been presented recently by Rahnama et al. (1995). Using the hydrodynamic diffusitives they derive the distribution function by solving Burgers diffusion equation. The theoretically predicted orientation distributions for semi-dilute suspensions were in good agreement with the experimentally observed results by Stover et al. (1992).

Another branch of the disperse systems rheology is the research connected with the non-Newtonian properties of the flowing suspensions, e.g., the viscosity dependence on the rate of strain tensor, the relaxation effects, etc., which arise already at small values of the volume concentration. These suspensions properties are studied by Jeffery (1922b) for ellipsoids, Oldroyd (1953, 1958) for fluid drops, Goddard and Miller (1967) and Roscoe (1967) for elastic spheres. At these investigations, the linear scale of suspension motion is much greater than the particles size. That is why the suspension could be considered as a continuum for which the macroscopic properties are obtained via an ensemble averaging of the corresponding microscopic quantities. After having the detailed field around each particle, it is possible to get the rheological constitutive equation, i.e., the functional relation between the stress and the respective physical quantities. This approach creates significant difficulties, since founding the flow picture around every particle is almost impossible. Therefore, it is preferable to obtain a more general relationship between the stress and strain, for example, the equation of Rivlin-Eriksen (Rivlin and Eriksen, 1955), the Oldroyd equation (Oldroyd, 1950), the equation of Hand (Hand, 1961) and others.

Although the interaction of the suspension rheology with the phenomenological theories is quite feeble, we have to mention the works of Gordon and Schowalter (1972) and Hand (1962) concerning this topic. Comparing some phenomenological equations with the relations between stress and strain, Hand shows that the average stress of dilute suspensions of rigid ellipsoids, elastic spheres and fluid drops satisfies the phenomenological constitutive equation proposed in his work.

The studies of Taylor (1932, 1934) modelling suspensions (emulsions) of fluid drops are prolonged by Schowalter et al. (1968) and Frankel and Acrivos (1970). Using the method of règular expansions, Frankel and Acrivos solve the hydrodynamic problem for a deformable fluid drop in a shear flow and using the approach of Batchelor (1970) calculate the average stress of emulsion and its viscosity, as well.

The results of Frankel and Acrivos (1970) are generalised by Barthes-Biesel and Acrivos (1973a) to account for a higher-order deformation of drops. Their analysis for the rheology of dilute emulsions of deformed fluid drops allows to obtain two different sets of equations: an equation connecting the stress with the rate of strain via an expression describing the anisotropy value, and a set of differential equations modelling the anisotropy change as a function of time and rate of strain. We have to note that all terms participating into the phenomenological equation of Hand (1962) arise naturally in the problem treated by Frankel and Acrivos (1970). Moreover, the dilute suspensions must be regarded as a "pseudo anisotropic" fluid, since their anisotropy is due to the motion. From this it follows that their constitutive equation has to be reduced to the non-Newtonian "simple fluid".

The situation with the dilute suspensions of elastic spheres, studied by Goddard and Miller (1967), is similar, i.e., the suspensions are also "pseudo anisotropic", as their anisotropy manifests itself only at their motion.

The equation obtained by Hand (1961) for suspensions of ellipsoidal rigid particles, is later found also by Batchelor (1970a) via a more general and more effective method. By use of this method the average (macroscopic) stress is calculated as "averaged by the ensemble of configurations" microscopic stress over a representative volume V containing N particles:

$$T_{ij} = -p\delta_{ij} + 2\mu E_{ij} + \frac{4\pi\mu_0}{V}\sum S_{ij}, \qquad (10.2.8)$$

where all the particles are included in the summation and S_{ij} is the stress caused by their presence in the suspension.

In most of the published till now studies of suspensions the inertial forces are neglected, i.e., the equations of Stokes are utilised. Through the method of matched asymptotic expansions applied for the full Navier-Stokes equations at small Reynolds numbers, Lin et al. (1970a) solved approximately the problem of a suspended in a viscous fluid sphere and found that the flow pattern was asymmetric with respect to the particle, due to the inertial forces. This leads to anisotropy in a flowing dilute suspension of rigid spherical particles at non-zero Reynolds number. Therefore when modelling such type of flowing suspensions, it is necessary to employ the generalised equation of Hand, which involves a tensor of a higher than second rank. The performed analysis and velocity distribution obtained by Lin et al. (1970a) allowed the authors to calculate the inertial forces influence on the rheology of a suspension of this type. For the effective viscosity they got the formula:

$$\frac{\mu^*}{\mu_0} = 1 + \frac{5}{2}\Phi + 1.34\,\mathrm{Re}^{3/2}\,\Phi, \qquad (10.2.9)$$

where Re is the Reynolds number.

The interest to the investigations of the hydrodynamic interaction of Brownian particles has recently grown very much. This is due to its great effect on solutions with big macromolecules, where its account leads to significant divergence with respect to the treatment of each Brownian particle as a single one. From hydrodynamic point of view, the Brownian motion influence is of major importance for the averaged characteristics of the suspensions: stress, viscosity, heat-transfer, etc.

It is known, that the orientation of the non-spherical particles (e.g., ellipsoidal particles) in a dilute suspension depends on the type of the flow, in which it takes part. Apart of the fluid flow, which "arranges" in some definite order the suspension particles, the Brownian motion leads to their random disposition. The interaction of these two effects is described by the equation of Fokker-Plank. In particular, the coefficient k_1 in the formula $\mu^* = \mu_0(1 + k_1\Phi)$ for the effective viscosity depends on the particles shape and on the type of the flow, in which they participate.

The first systematic study of the rheology of dilute suspensions of equal, sufficiently small particles with arbitrary shape, for which the Brownian pairwise forces are important, is performed by Giesenkus (1962). The Brownian motion influence on the rheological properties of suspensions of non-spherical particles is also analysed by Hinch and Leal (1972) and Leal and Hinch (1973).The effect of the Brownian rotation of equal arbitrary-shape particles in a dilute suspension on its rheology is considered by Rallison (1978a). On the basis of Batchelor's approach, Rallison deduced the constitutive equations describing the unsteady flow generated by the Brownian pairwise forces, acting on the dilute suspension of equal arbitrary-shape particles. A full survey of the investigations, connected with the particles interaction at their Brownian motion in suspensions, is presented by Russel (1981). The suspensions stability is of great importance for a lot of technological processes. The ability, to preserve any system status (including suspensions) unchanged in the whole volume, is usually termed as system stability. The instability of the roughly dispersed systems is mainly due to the considerable sedimentation particles velocity under the gravity action. The process of direct particles adhering at the collisions is substantial and is called a coagulation or flocculation.

In parallel to the other questions, the problem of utilising the principles of variations for searching the upper and lower limit of the scalar parameters (effective viscosity and effective heat-transfer coefficient) gradients, characterising the suspensions transfer properties, is considered in the reviews of Jeffrey and Acrivos (1976), and Herczynski and Pienkowska (1980). It is worth noting the work of Hashin (1964) for the possible mathematical limits of the effective viscosity of an isotropic suspension of arbitrary-shape particles. If the viscosity of the medium in the particles then is $\hat{\mu}$, and the interfacial tensions are zero, for the lower and upper limit Hashin gives the following respective expressions:

$$\mu + \frac{\Phi\mu_0}{\frac{2}{5}(1-\Phi) + \mu_0(\hat{\mu} - \mu_0)} \quad \text{and} \quad \hat{\mu} + \frac{(1-\Phi)\hat{\mu}}{\frac{2}{5}\Phi + \hat{\mu}(\mu_0 - \hat{\mu})}. \tag{10.2.10}$$

In Walpole (1971) it is suggested that the particles spherical shape in an isotropic suspension is an extreme case in a way, that if one of the Hashin's limits is accomplished, this definitely means that the suspension consists of spherical particles. The limits in which the effective viscosity of isotropic suspension (not necessary dilute) of spherical particles with viscosity $\hat{\mu}$ inside can be changed are calculated by Keller et al. (1967).

It is interesting to note, that the effective viscosity lower limit at small values of the volume concentration of spherical particles suspension coincides with the formula of Einstein (10.2.1). A discussion about the lower and upper limit of the suspension effective viscosity can be found in the survey of Hashin (1964) and the monograph of Beran (1968).

Using the known analogy between the solid deformable body mechanics and the fluid interfaces mechanics of Hashin and Shtrikman (1963) for elastic modules of composite materials, the effective viscosity limits for Newtonian suspensions could be obtained. Keller et al. (1967) found these limits for suspensions of ordered particles. The general case for the effective viscosity limits of suspensions with chaotic particles disposition is still not solved in the scientific literature.

Wiener (1938, 1958) first defined a polynomial functional of a Gaussian process, called a homogeneous chaos, by a kind of multiple stochastic integral with respect to the Brownian motion process. Making use of the concept of multivariable orthogonal polynomials Ogura (1972) presented a theory of the orthogonal functionals of the Poisson process. Following the ideas of these papers Christov and Markov (1985), and Markov and Christov (1992) developed a method taking into account the inter-particle interactions by using the Voltera-Wiener series. On the basis of this method explicit results were obtained analytically to the order $O(\Phi^2)$ (Markov, 1989) for the effective conductivity of a dispersion and they are identical to those of Jeffrey (1973). It should be noted that the used functional series represent a rich and convenient source of trial fields when combined with an appropriate variational principle in order to produce variational estimates for effective properties of dispersions (see Markov, 1987a, b).

10.3. Effective Viscosity of Concentrated Suspensions and Emulsions

Numerous attempts to make a similar theory as that of Einstein (1906) and Batchelor and Green (1972b) (for dilute and semi-dilute suspensions) for suspensions of high or moderate concentration have not achieved, on the whole, an appreciable success up to the present. Effective parameters of such suspensions are obtained, as a rule, on the basis of the assumption that the particles are located in some regular (spatially periodic) manner or on the basis of various semi-empirical models, among which the first place belongs to the "cell" model (see sections 9.2 and 10.4). In particular, a large number of equations have been proposed for the relative viscosity of a suspension undergoing shear as a function of the volumetric particle concentration Φ.

For concentration $0.15 \le \Phi \le 0.6$ Thomas (1965) proposed a non-linear relationship:

$$\mu_r = \frac{\mu^*}{\mu_0} = 1 + \frac{54}{4a^3} \frac{\Phi}{\left(1 - \Phi/\Phi_{max}\right)^3}, \tag{10.3.1}$$

where a depends on Φ and $\Phi_{max} = 0.625$ for spherical particles. The values of a and μ_r as a function of Φ were given in the following Thomas' table:

TABLE 10.3.1.

Φ	0.2	0.3	0.4	0.5	0.6
a	1.136	1.49	1.953	2.903	9.583
μ_r	1.95	3.09	5.97	14.08	70.0

For $0 \le \Phi \le \Phi_{max}$ Thomas (1965) offered another formula which contained an exponential term:

$$\mu_r = 1 + 2.5\Phi + 10.05\Phi^2 + 2.73 \times 10^{-3} \exp(16.6\Phi). \tag{10.3.2}$$

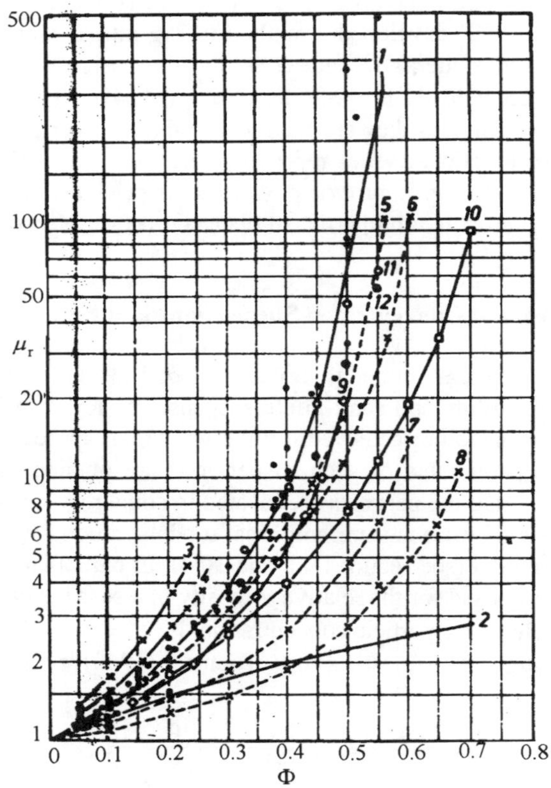

Fig.10.3.1.a. Relative viscosity μ_r as function of volume fraction Φ: (1) is according to the Rutgers' analysis. The remaining curves are given in the paper of Rutgers (1962a).

The spread of experimental values of the relative viscosity as a function of suspension concentration is given in Fig.10.3.1.a (see also Fig.1. in Thomas, 1965). In this Figure the experimental results of 16 investigators were represented by Rutgers (1962a, b). The scatter of the data is the most striking feature of this plot, especially at the higher values of Φ.

After discussing Vand's (1948a, b) results Ford (1960) has presented the following expression:

$$\frac{1}{\mu_r} = 1 - 2.5\Phi + 11\Phi^5 - 11.5\Phi^7, \qquad (10.3.3)$$

in which the term $\frac{1}{\mu_r}$ expresses the "fluidity" of the suspension.

At concentrations lying beyond the two-body range, the power series (10.3.3) fails to accord with experimental data. That is why Mooney (1951) suggested the equation

$$\mu_r = \exp\left[\frac{2.5\Phi}{1 - k\Phi}\right], \qquad (10.3.4)$$

wherein the coefficient 2.5 is simply established by comparison with Einstein's formula (10.2.1) in the limiting case $\Phi \to 0$. The remaining coefficient k is determined experimentally and usually is in the range $1.35 < k < 1.91$. Using a concentric-sphere cell model Simha (1952) obtained the approximate formula

$$\lim_{\Phi \to \Phi_{max}} \frac{\mu^*}{\mu_0} = 1 + \frac{54}{3f^3}\left[\frac{\Phi^2}{\left(1 - \Phi/\Phi_{max}\right)^3}\right], \qquad (10.3.5)$$

where f is a semi-empirical quantity that increases slowly with concentration. Other cell models have been applied by Happel (1957) and Kuwabara (1959). Based on some experimental data given by Rutgers (1962a, b) one can state that for $\Phi \to 0.6$, $\mu_r \to 500$. Because of the large differences in the values of the suspension viscosities at large Φ (see Table 10.3.1. and Fig.10.3.1.a) it is difficult to select the relationship which best describes the phenomenon. Nevertheless, many authors have continued to suggest new relationships.

For the viscosity of hard-sphere dispersions Krieger and Dougherty (1959) suggested the following expansion

$$\frac{\mu}{\mu_0} = \left(1 - \frac{\Phi}{\Phi_{max}}\right)^{-[\mu]\Phi_{max}},$$

where $[\mu]$ is the dimensionless intrinsic viscosity, which has a theoretical value of 2.5 for monodisperse rigid spheres, and Φ_{max} is the maximum packing volume fraction for which the viscosity μ diverges. The maximum value Φ_{max}, depends on the type of packing of the particles. The Table 10.3.2. presents the values of Φ_{max} for various arrangements of monodisperse spheres.

Application of the "lubrication theory" technique to the rheology of concentrated suspensions was pioneered by Frankel and Acrivos (1967). They have determined theoretically, for concentrated suspensions of rigid spherical particles, the following formula

$$\mu_r = 1 + \frac{9}{8}\frac{\left(\Phi/\Phi_{max}\right)^{1/3}}{1-\left(\Phi/\Phi_{max}\right)^{1/3}}. \tag{10.3.6}$$

This formula as well as formulae (10.3.1) and (10.3.5) diverge when the volume fraction Φ approaches the maximum Φ_{max} for a particular lattice. Φ_{max} can be determined experimentally by plotting $\frac{1}{\mu_r}$ versus Φ and extrapolating the data to the point where $\frac{1}{\mu_r}$ is zero. This Φ_{max} will depend on the particle size and the size distribution of the dispersed phase.

TABLE 10.3.2.

Arrangements	Φ
1. Simple cubic	0.52
2. Minimum thermodynamically stable configurations.	0.5448
3. Hexagonally packed sheets just touching	0.605
4. Random close packing	0.637
5. Body-centred cubic packing	0.68
6. Face-centred cubic/hexagonal cubic packing	0.74

Noting that particle interactions are effective primarily through lubrication forces, and integrating the dominant component of the particle stress tensor, Frankel and Acrivos (1967) have demonstrated that the asymptotic behaviour in the limit of maximum packing is correctly captured by their model. Frankel and Acrivos' model has been criticised for steady force (Batchelor, 1974) but should remain valid for small amplitude oscillations (Goddard, 1977). Fig.10.3.1.b compares the theoretical results of Batchelor (1977) and, Frankel and Acrivos (1967) with experimental data of Saunders (1961) and Krieger (1972). This figure gives us the shear viscosity dependence on the concentration for $0.01 \leq \Phi \leq 0.50$. It is seen that the $O(\Phi^2)$ Batchelor's theory remains reasonable up to $\Phi \sim 0.20$ - 0.25. Since Frankel and Acrivos (1967) have treated the suspension microstructure as a regular array and neglected the Brownian effects there, it is clear that there should exist some difference between the theoretical and experimental results.

The concentration affects not only the Newtonian viscosity, but also causes non-Newtonian behaviour. At small concentration values the viscosity μ^* generally increases linearly with the concentration Φ. However, after a certain concentration value has been exceeded, the viscosity increases significantly and the suspension behaviour is non-

Newtonian. According to Vand (1945) the Newtonian concept cannot be applied to suspensions having a volume concentration in excess of 37%. It has been experimentally determined (Vand, 1945; Karnis et al., 1966a, b) that certain suspensions of rigid particles in a Newtonian liquid behave as a viscoplastic material even for $\Phi \sim 0.12 - 0.20$. Oldroyd (1953) has established that the emulsions and suspensions of elastic spheres dispersed in a Newtonian fluid, behave rheologically as a viscoelastic material. Pokrovskii (1968) has discovered on theoretical grounds that a suspension of ellipsoidal particles in a Newtonian liquid is viscoelastic.

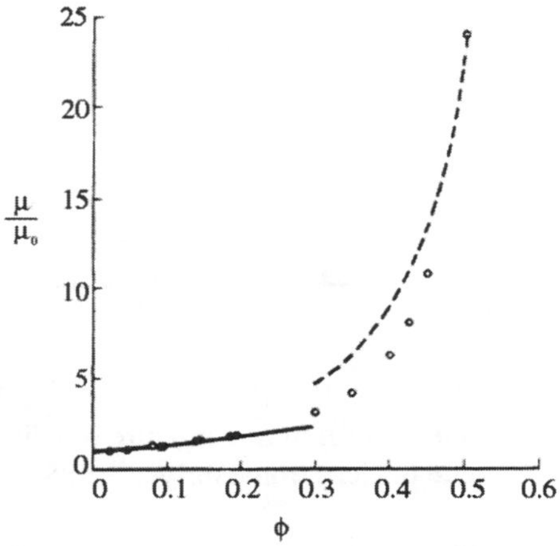

Fig.10.3.1.b. Shear viscosity as function of volume fraction Φ from: (———) pair interaction theory (Batchelor, 1977); (------) lubrication theory (Frankel and Acrivos, 1967); experiments for polystyrene lattices (O) from Krieger (1972) and (●) Saunders (1961).

Theoretical approaches to the study of suspensions and emulsions with higher concentrations pose at least two problems. The first is the determination of the many-body interactions between rigid or fluid particles, while the second problem is the determination of the spatial and temporal distribution of the particles, i.e., the suspension (emulsion) microstructure. These are extremely complex problems. Only recently the progress in analytically computing many-body hydrodynamic interactions has been made (Mazur and van Saarloos, 1982; van Saarloos and Mazur, 1983). The calculations of the cited authors are by no means exact for all particle separations, but they do give some indication of the importance of three-body and higher order effects (Beenakker and Mazur, 1983, 1984; Beenakker, 1984a, b). In addition to many-body interactions, lubrication forces are important. They result from the thin layer of viscous fluid that separates the surfaces of nearly touching particles. So, the effect of these forces need to be addressed, as well.

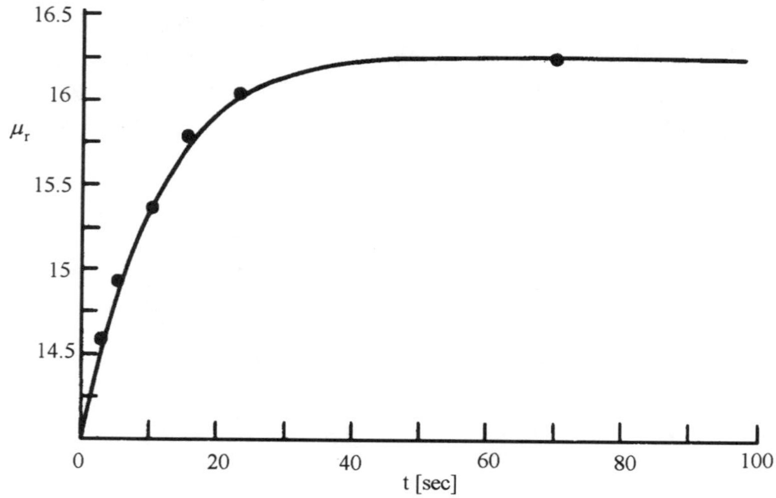

Fig.10.3.2. Short-time rise in viscosity μ at $\dot{\gamma}$ = 2.4 s^{-1} and Φ = 0.45 of a suspension of polystyrene spheres of diameter 40-50μm in a mixture of silicone oil, sheared in the Couette device (from Acrivos, 1985).

Little advance (beyond the two-particle limit) has been made up to present concerning the determination of the spatial and temporal distribution of the particles. This is due, to a large part, to the fact that the particles distribution cannot be specified a priori but rather must be found as a part of the complete problem. This aspect fixes the sedimentation and effective viscosity because at permeability and conductivity process the particles distribution may be supposed known or given at the onset. The only analyses beyond the two-body limit and treating non-dilute suspensions are those for spatially periodic, lattice models.

After performing series of measurements for concentrated suspensions of non-colloidal spherical particles Gadala-Maria and Acrivos (1980) and Leighton and Acrivos (1987a) reported the puzzling phenomena of a short-time rise in the apparent viscosity (see Fig.10.3.2) followed by a long-time decrease in the apparent viscosity of the suspension (see Fig.10.3.3). In the theoretical part of their studies they demonstrated that the short-term increase in the viscosity of concentrated suspensions can be attributed to the diffusive migration of particles due to the irreversible inter-particle interactions that occur in such suspensions. They demonstrated also that the decrease in the concentrated suspension viscosity after long periods of shearing is due to the shear-induced migration of particles. It should be noted that the observed viscosity increase is over very short time scales, much shorter than those corresponding to the long-term viscosity decrease (compare the time scales in Fig.10.3.2 and Fig.10.3.3).

The transient viscosity increase and its subsequent long-term decrease enabled the authors who discovered these phenomena to infer the values of the shear-induced diffusion coefficients which are important in determining the particle concentration profile established in shear flows. It must be emphasised that the shear-induced diffusive drift, both normal to the plane of shear and parallel to the fluid velocity gradients within the plane of shear, represent a new explanation of the effect of blunting of the velocity profiles in the suspensions flow through tubes (see Fig.9.6.8). The phenomena, we are discussing, persist at zero Reynolds number, unlike the well known sources of particle migrations such as Segre-Silberberg's effect for flow through tubes which are caused by small inertial effects (see section 9.6). Thus at zero Reynolds number flows of concentrated suspensions, they are likely to represent the dominant source of drift across streamlines.

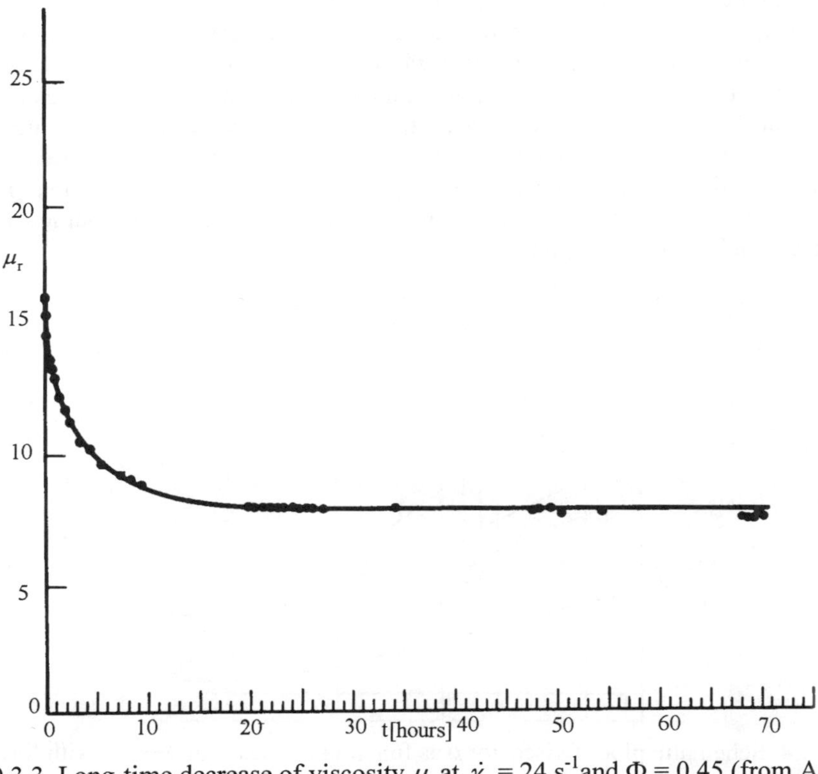

Fig.10.3.3. Long-time decrease of viscosity μ_r at $\dot{\gamma}$ = 24 $s^{-1}$and Φ = 0.45 (from Acrivos, 1985).

Surface roughness has also been proposed as a mechanism leading to a variety of phenomena observed in concentrated suspensions (Leighton and Acrivos, 1987a). Arp and Mason (1977) found that the small degree of surface roughness present on particles used in their experimentation was sufficient to eliminate the closed orbits of interacting spheres that were predicted for smooth particles. Smart and Leighton (1989) showed that a small amount of particle surface roughness would prevent the approach of the nominal surface of two particles closer than the roughness. However, it would not prevent them from receding from each other, once the pair reaches an orientation where hydrodynamic forces pull them apart rather than pushing them together. Parsi and Gadala-Maria (1987) demonstrated that the near-contact interactions of particles would lead to anisotropic suspension microstructure since the surface roughness would cause higher values for the pair distribution function on the approaching side of the reference sphere than on the receding side. Leighton and Acrivos (1987b) showed that the shear-induced coefficient of self-diffusion in the direction normal to the plane of shear is of the form $D = \dot{\gamma} a^2 \hat{D}(\Phi)$, where $\dot{\gamma}$ is the shear rate, a is the sphere radius and $\hat{D}(\Phi)$ is the dimensionless diffusivity. The short-time viscosity increase was modelled by Leighton and Acrivos (1987b) as due to an one-dimensional diffusion process in direction parallel to the viscosity gradient within the plane of shear, whereas the long-time viscosity decrease, as due to an one-dimensional diffusion perpendicular to the time of shear. Since its values for concentrated suspensions were found to be nearly an order of magnitude higher than the corresponding coefficient of self-diffusion, the authors attributed this difference to the presence of a concentration gradient that induces a migration of particles from region of high to low concentration in addition to that provided by random self-diffusion.

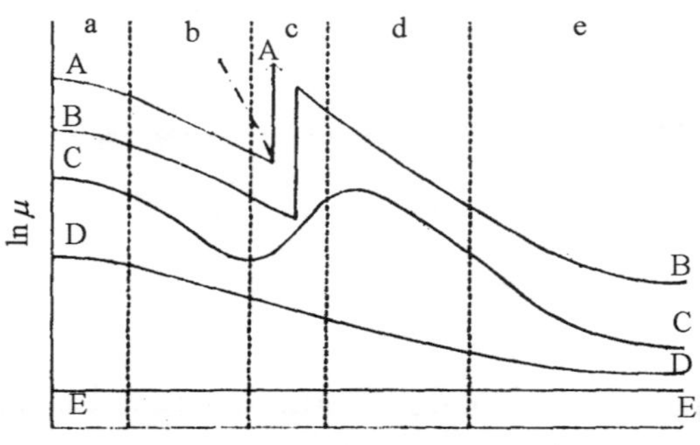

Fig.10.3.4. Schematic plot of viscosity μ as function of shear rate (——) with lines of constant stress (·········). The regions a, b, c, d, e subdivide the range of shear rates (from Chaffey, 1977).

Before the paper of Acrivos et al. (1992) there was no theoretical calculation of any of the above coefficients from the basic principles, mainly owing to the difficulty of analysing the hydrodynamic interaction of more than two particles. Acrivos et al. (1992) derived an expression for the coefficient of shear-induced diffusion in direction of the fluid velocity in a simple shear flow for low values of the concentration Φ, since in contrast to the transverse diffusivity longitudinal diffusivity is determined by pairwise particle interactions, which are tractable. The longitudinal shear-induced self-diffusion coefficient is found to be $0.267a^2\dot{\gamma}\left[\Phi \ln \Phi^{-1} + O(\Phi)\right]$.

Some recent studies of non-colloidal and colloidal suspensions have shown a shear thinning behaviour for small Peclet numbers to the relative decrease in the Brownian contribution to the stress with increasing shear rate. The various stresses in suspensions of rigid nonspherical particles tend to align the particles in such a manner as to alleviate these stresses and hence to reduce the viscosity. Although that the different curves on Fig.10.3.4. are not all seen in anyone system or family of systems, increasing the volume fraction Φ of particles tends to transform the viscosity curve from shear-thinning (curve D) to one showing gradual (curve C, regions b, c) or abrupt (curve A) shear-thickening due to hydrodynamic interactions and cluster formation at high Peclet numbers.

The simplest system of hard spheres is affected by viscous forces, Brownian motion, and excluded volume of the particles. Krieger's (1972) measurements of the relative viscosity of many suspensions, all having the same volume fraction Φ, are shown on Fig.10.3.5. plotted against Peclet number Pe. The data for shear viscosity in Fig.10.3.5 demonstrate the shear thinning generally observed between the well-defined low and high shear limiting viscosities, denoted by μ_{max} and μ_{∞}, respectively. Graphs of $\mu_{max}(\Phi)$ and $\mu_{\infty}(\Phi)$ can thus be drawn, and these are presented in Fig.10.3.6.

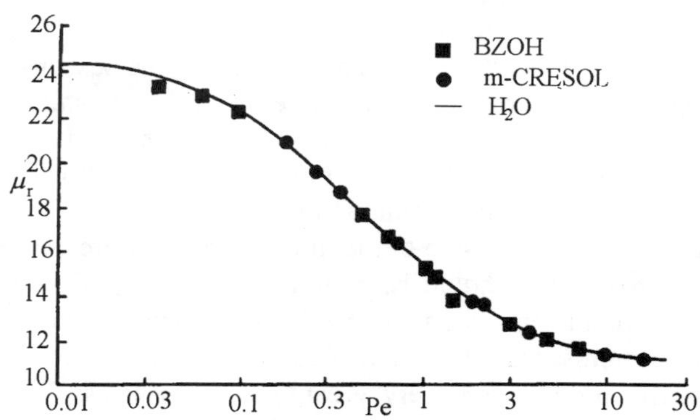

Fig.10.3.5. Relative viscosity μ_r as a function of Peclet number Pe for $\Phi = 0.5$: (●) and (■) for 0.1-0.5 μm latex particles dispersed in two solvents; (——) for the same sized particles in water (from Barnes et al., 1987).

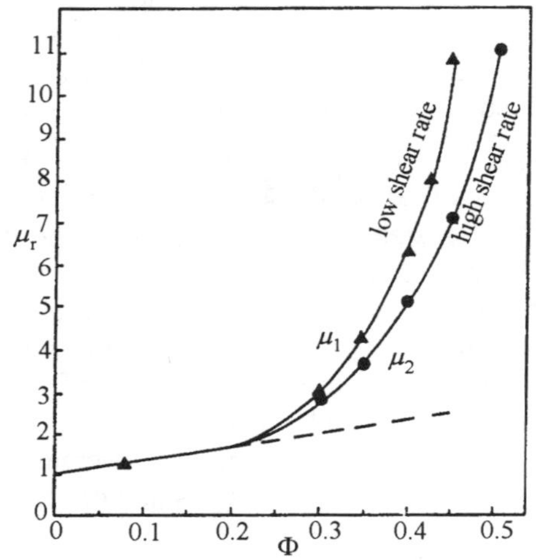

Fig.10.3.6. Relative viscosity μ_r as a function of volume fraction Φ for high and low shear asymptotic values (from Barnes et al., 1987).

A theory for the non-equilibrium structure and stresses in a sheared suspension with a fluid rest state has been formulated by Russel and Gast (1986). Their theory included the many-body interactions in the thermodynamic sense but truncated at the pair level for the hydrodynamics. The calculations on the basis of this theory for hard spheres in weak flows demonstrated the importance of stresses arising from the non-equilibrium structure and explained the shear rate dependence observed at volume fractions greater than 0.25 - 0.30 (see Fig.10.3.7). Felderhof (1987) and Felderhof and Jones (1987a,b) formally defined the equations governing the suspension microstructure and bulk rheology on the Smoluchowski time scale. They studied the frequency-dependent viscosity of a suspension of spherical particles on the basis of the linear response theory applied to the generalised Smoluchowski equation. The hydrodynamic interactions were fully taken into account and expressions for the frequency-dependent viscosity based on a cluster expansion of the linear response were derived.

Another approach that should be mentioned is presented by Buyevich and co-workers. In order to find a firm basis for the theory of dense suspensions he considers each phase of suspension separately. The difficulties in solving the equations derived by him for any specific problem are very great, but nevertheless his results are quite promising. A complete review of these results can be found in Buyevich and Shchelchkova (1978).

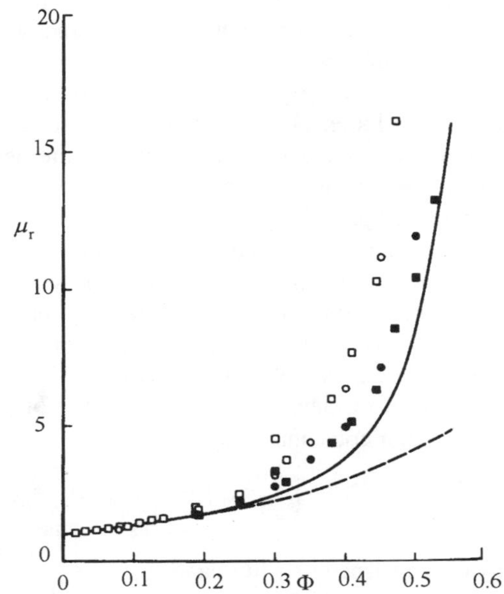

Fig.10.3.7. Relative viscosity μ_r as a function of volume fraction Φ for hard spheres for: (——) the low shear limiting viscosity; (·········) viscous contribution alone; and compared with data for lattices in water (○) Pe → 0, (●) Pe →∝ (Krieger, 1972) and silica spheres in cyclohexane (□) Pe → 0, (■) Pe →∝ (de Kruif et al., 1985) (from Russel and Gast, 1986)

10.4. Effective Viscosity of Periodic Suspensions and Emulsions. Numerical Simulations

Lattices of rigid or fluid particles provide us with useful models for studying the flow through different aggregates of particles and the rheology of dilute and non-dilute suspensions and emulsions. There have been numerous experimental and theoretical investigations for the macroscopic behaviour of a flowing dispersion of rigid or fluid particles in a fluid, in which the motion is a result either of an applied flow or an applied external force. The macroscopic properties of such dispersions are related to their microstructure which may be changed by particle orientation or by relative positions of particles. In flowing dispersions the (rigid or fluid) particles concentration varies with position. In a shear flow, fluxes of particles across the shear flow may occur in a diffusion-like manner due to particle-particle interactions. The solution of many-body problem and the determination of the microstructure is complicated but essential for modelling dispersion rheology.

The numerical approximation is complementary to the theoretical and experimental approaches. With such approach, the simultaneous investigation of the deformation processes and the bulk properties is possible.

A real dispersion of particles (rigid or fluid) contains, of course, a very large number of particles per unit volume. Direct simulation of such a system is presently impossible due to the immense amount of computational time needed. To circumvent this problem, one models a dispersion as periodic, i.e., instead of taking one box filled with millions of randomly placed particles one considers a finite number of particles in a smaller domain and copies this cluster in the direction of the lattice vectors. In this way, one can both model the macroscopic domain of a real dispersion and take into account all particles interactions, as everything done in a reasonable computational time.

Nunan and Keller (1984) extended the results of dilute suspensions for suspensions with moderate and large volume fractions and showed that the instantaneous effective stress tensor may be expressed in terms of a fourth rank viscosity tensor, which involves two scalar constants α and β:

$$\mu^{*}_{ijkl} = \frac{1}{2}\mu_0(1+\beta)\left(\delta_{ik}\delta_{jl} + \delta_{il}\delta_{jk} - \frac{2}{3}\delta_{ij}\delta_{kl}\right) + \mu_0(\alpha-\beta)\left(\delta_{ijkl} - \frac{1}{3}\delta_{ij}\delta_{kl}\right), \quad (10.4.1)$$

where μ_0 is the viscosity of the suspending fluid and $\delta_{ijkl} = \begin{cases} 1 & \text{if } i=j=k=l \\ 0 & \text{otherwise} \end{cases}$.

For low concentration suspension Zuzovsky (1976) obtained the following results, which were identified by Nunan and Keller as viscosity tensor

$$\begin{aligned} \alpha &= \frac{5}{2}\Phi\left[1-(1-60b)\Phi + 12a\Phi^{5/3} + O\left(\Phi^{7/3}\right)\right]^{-1} \\ \beta &= \frac{5}{2}\Phi\left[1-(1+40b)\Phi - 8a\Phi^{5/3} + O\left(\Phi^{7/3}\right)\right]^{-1} \end{aligned} \quad (10.4.2)$$

where a and b depend on the lattice geometry and are given on Table 10.4.1.

TABLE 10.4.1.

	Simple cubic	Body-centred	Face-centred
a	0.2857	-0.0897	-0.0685
b	-0.04655	0.01432	0.01271

For high concentration suspension Nunan and Keller (1984) gave the asymptotic formulae:

(i) simple cubic lattices

$$\alpha = \frac{3}{16}\pi\varepsilon^{-1} + \frac{27}{80}\ln\varepsilon^{-1} + 3.1 + 0.25\varepsilon\ln\varepsilon^{-1} + O(\varepsilon)$$
$$\beta = \frac{1}{4}\pi\ln\varepsilon^{-1} + 0.63 + 0.0\varepsilon\ln\varepsilon^{-1} + O(\varepsilon) \tag{10.4.3}$$

(ii) Body-centred cubic lattices

$$\alpha = \frac{1}{4}\sqrt{3}\pi\ln\varepsilon^{-1} - 1.73 + 12.3\varepsilon\ln\varepsilon^{-1} + O(\varepsilon)$$
$$\beta = \frac{1}{8}\sqrt{3}\pi\varepsilon^{-1} + \frac{37}{120}\sqrt{3}\pi\ln\varepsilon^{-1} + 12.8 - 35\varepsilon\ln\varepsilon^{-1} + O(\varepsilon) \tag{10.4.4}$$

(iii) Face-centred cubic lattices

$$\alpha = \frac{3}{32}\sqrt{2}\pi\varepsilon^{-1} + \frac{87}{160}\sqrt{2}\pi\ln\varepsilon^{-1} + 9.7 - 15.5\varepsilon\ln\varepsilon^{-1} + O(\varepsilon)$$
$$\beta = \frac{3}{16}\sqrt{2}\pi\varepsilon^{-1} + \frac{47}{80}\sqrt{2}\pi\ln\varepsilon^{-1} + O(1), \tag{10.4.5}$$

where $\varepsilon = 1 - \left(\Phi/\Phi_{max}\right)^{1/3}$ and $\Phi_{max} = \frac{\pi}{6}$ for a simple cubic lattice, $\frac{1}{8}\sqrt{3}\pi$ for a body-centred cubic lattice and $\frac{1}{6}\sqrt{2}\pi$ for a face-centred cubic lattice. Nunan and Keller (1984) computed numerically the two parameters α and β for simple, body-centred and face-centred cubic lattice of spherical particles with solute concentrations up to 90% of the close-packing concentration.

Using the method of successive approximations (see Sangani and Acrivos, 1982) Sangani and Lu (1987) calculated the effective viscosity of an emulsion of small neutrally buoyant spherical drops in the case of simple cubic lattices. They computed the instantaneous effective viscosity tensor from dilute limit to maximum packing. For the scalars α and β in the effective viscosity tensor μ^*_{ijkl} in the case of dilute array they got the expressions:

$$\alpha = \frac{5}{2}\Phi\frac{\lambda + 2/5}{\lambda + 1}\left[1 + \frac{\lambda + 2/5}{\lambda + 1}\lambda_1\Phi + O\left(\Phi^{5/3}\right)\right]^{-1}$$
$$\beta = \frac{5}{2}\Phi\frac{\lambda + 2/5}{\lambda + 1}\left[1 + \frac{\lambda + 2/5}{\lambda + 1}\lambda_2\Phi + O\left(\Phi^{5/3}\right)\right]^{-1} \tag{10.4.6}$$

The numerical values of λ_1 and λ_2 for the three cubic arrays are represented in Table10.4.2.

TABLE 10.4.2.

	Simple cubic	Body-centred	Face-centred
λ_1	-3.793	-0.1408	-0.2374
λ_2	0.8138	-1.5084	-1.5728

The results from Sangani and Lu's numerical solutions, given for simple, body-centred and face-centred cubic arrays, are in good agreement with those of Nunan and Keller (1984) for rigid particles ($\lambda \to \infty$), except for β for the face-centred cubic array. The accuracy of these numerical results is also adequate in most cases to yield the formulae for concentrated arrays of very viscous drops ($\lambda >> 1$).

Tran-Cong et al. (1990) have developed a boundary element method to compute the volume-average rheological properties of layered suspensions in a mean simple shearing flow up to a volume concentration of 0.5. Although the BEM is of general applicability, they have considered only the simple shear flow past a periodic array (in x and y directions) of force-free and torque-free rigid spherical particles (layered suspensions). Instantaneous rheological properties are computed by relevant volume averages. Nevertheless, the layered structure is an idealised model for suspension, it is a realizable suspension flow. For such systems, experiments (Bouillot et al., 1982) and other numerical simulations (Brady and Bossis, 1985) have been performed.

The instantaneous behaviour of suspensions of force-free, torque-free and spinning spherical rigid particles arranged on cubic lattices was studied first by Kapral and Bedeaux (1978) and later in more details in a sequence of papers by Zuzovsky et al. (1983), Adler (1984), Adler and Brenner (1985) and Adler et al.(1985).

It is well established (see Adler et al., 1985) that transient effects are important and the time-averaged behaviour will be substantially different from the instantaneous behaviour pertaining to a particular type of lattice at a certain instant of time. So, on a macroscopic level, the motion of a suspension is properly described in terms of the time-average behaviour over a period of time which is long, compared to the time scale of the microstructure evolution, but sufficiently short to allow the meaningful computation of time averages.

Pozrikidis (1993) constructed numerical simulations using boundary integral calculations which describe the transient motion of ordered emulsions. He considered a linear flow $\mathbf{v} = \mathbf{A.r}$ past a 3D lattice of force-free and torque-free spherical drops of radius a and viscosity $\lambda\mu$ suspended in a fluid of viscosity μ. Pozrikidis compared his numerical and asymptotic results with those derived by previous numerical and asymptotic analysis. He found that the particle stress tensor is given by $2\alpha\,\mathbf{E}$ and $2\beta\,\mathbf{E}$, for the simple shear and purely straining flow, respectively, where $\mathbf{E}$ is the symmetric component of $\mathbf{A}$. The coefficients α and β are given as asymptotic expansions:

$$\begin{aligned} \alpha &= \frac{5}{2}\Phi\frac{\lambda+2/5}{\lambda+1}\left[1+\frac{\lambda+2/5}{\lambda+1}\alpha_1\Phi-\frac{\lambda}{\lambda+1}\alpha_2\Phi^{5/3}+\cdots\right]^{-1} \\ \beta &= \frac{5}{2}\Phi\frac{\lambda+2/5}{\lambda+1}\left[1+\frac{\lambda+2/5}{\lambda+1}\beta_1\Phi-\frac{\lambda}{\lambda+1}\beta_2\Phi^{5/3}+\cdots\right]^{-1}, \end{aligned} \tag{10.4.7}$$

where the coefficients α_1, α_2, β_1 and β_2 are shown on Table10.4.3.

Most of Pozrikidis' results are in perfect agreement with those given by Zuzovsky et al. (1983) and Sangani and Lu (1987). Pozrikidis (1990a) investigated the effect of volume fraction and drop deformation, as well. He found that the drops were deformed in an oscillatory manner about a mean value, causing a corresponding oscillation of the particle stress tensor.

TABLE 10.4.3.

	Simple cubic	Body-centred	Face-centred
α_1	-3.787	-0.1408	-0.2361
α_2	-3.4267	1.0788	0.8232
β_1	0.858	-1.5728	-1.5093
β_2	-1.5728	-0.7192	-0.5488

It is worth noting that studies of ordered suspensions are not expected to provide direct information on the rheology of general dispersed systems. They could only extract information on some specific aspects of suspension rheology like simultaneous effect of particle shapes, particle deformations, and particle interactions.

Stokesian dynamics may be successfully used to predict rheological properties. To this end, the complete instantaneous and time-averaged bulk stress should be determined. On these lines Brady et al (1988) based their analysis on a method devised by O'Brien (1979) to deal with the problem of long range interactions. In the beginning the authors consider solutions of the flow equations in infinite periodic systems. Further on they handle convergence difficulties by considering the difference of the flow for a particular realisation and the flow averaged over realisations.

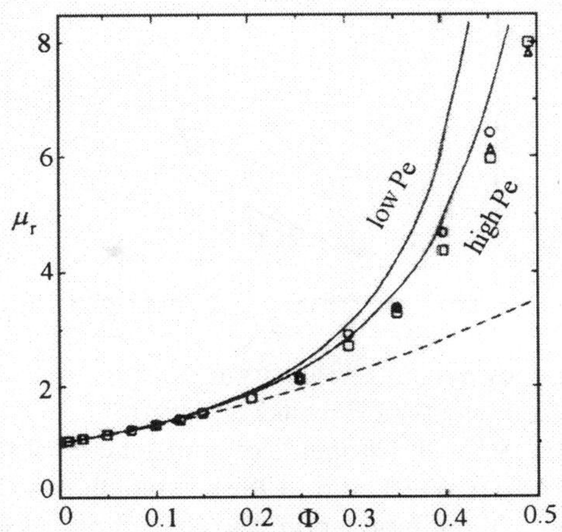

Fig.10.4.1. Relative viscosity μ_r as function of volume fraction Φ from: (——) the experiments of Krieger (1972) for low and high Pe limit, respectively; (----) the theoretical results of Batchelor and Green (1972); Stokesian dynamics results for 14 (O), 27 (□), and 64(Δ) spheres (from Phillips et al., 1988a).

On the basis of the Stokes dynamics method Phillips et al. (1988a, b) calculated the transport properties of hard-sphere dispersions for volume fractions Φ spanning the dilute limit up to the fluid-solid transition at Φ=0.49. Particle distributions were generated by a Monte Carlo technique and the number of particles in the simulation cell was investigated. Comparison of the simulation results is made with both experiments and theory (Figs.10.4.1. and 10.4.2.). It is seen that the obtained suspension viscosity results are in excellent agreement with experiment.

If suspension particles interact through purely hydrodynamic forces then the effective viscosity can be expressed solely as a function of the volume fraction. If in addition to the hydrodynamic shear forces the particles interact through repulsive DLVO-type forces, the effective viscosity depends on three parameters: (i) the volume fraction; (ii) the shear rate; (iii) a parameter characterising the range of the colloidal forces. Molecular dynamics has proved an effective means for determining the configuration of particles, in particular, the pair distribution function g(**r**), from which most properties of dense suspensions can be calculated.

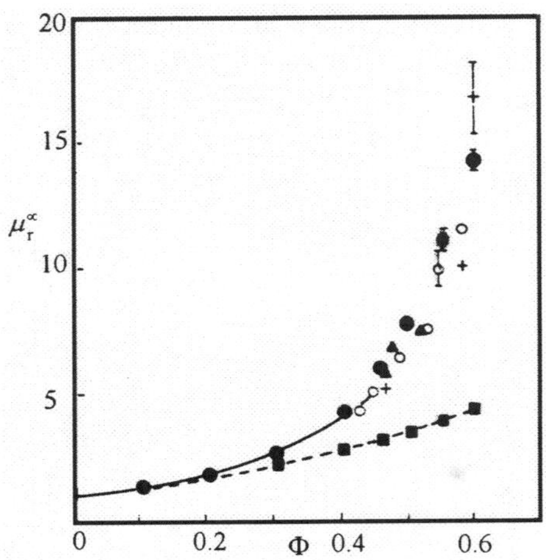

Fig.10.4.2. High-frequency dynamic viscosity μ_r^∞ as function of volume fraction Φ from: (——) the theoretical results of Beenakker (1984b); (-----) the theoretical results of Batchelor and Green (1972); experiments of hard spheres of radii 28nm (+), 46nm (O), and 76nm (▲) (Van der Werff et al., 1989); simulation with 27 and 32 particles (●); simulation without lubrication (■) (from Phillips et al., 1988a).

Recently Stokesian dynamics has been used by Chang and Powell (1993) to simulate the dynamics of a monolayer of a suspension of bimodally distributed spherical particles subjected to simple shearing (see Fig.10.4.3.) The near-field effects are calculated from the exact equations for the interaction between two unequally sized spheres, whereas the many-body far-field effects are calculated using the inverse of the grand mobility matrix. These authors simulated the flow of an "infinite" suspension by considering 25, 49, 64 and 100 particles to be "one" cell of an infinite periodic array. They found that for $\Phi_a > 0.4$, and for a fixed fraction Φ_b of small spheres, the bimodal suspensions generally had lower viscosities than mono-dispersed suspensions, with the size of this effect increasing with Φ_a. Chang and Powell (1993) showed that at the microstructural level, viscosity reduction is related to the influence of particle size on the average number of particles in clusters. Dependence of relative viscosity on volume fraction and particle size ratio, as the fraction of the small spheres is fixed, is given in Fig.10.4.4.

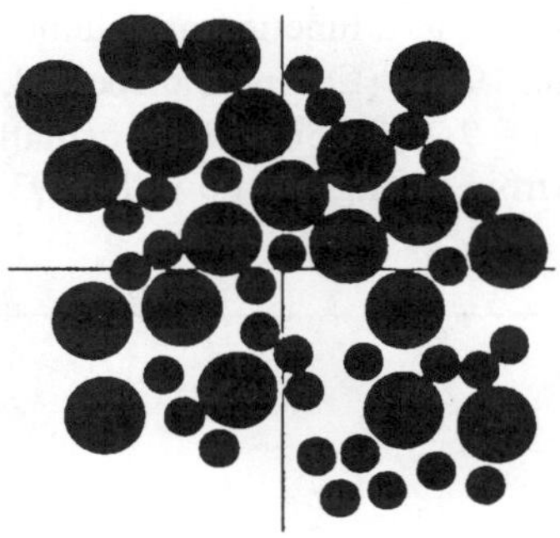

Fig.10.4.3. A configuration for a randomly distributed bimodal suspension. (from Chang and Powell, 1994a).

The high-frequency dynamic viscosity of a bimodal suspension of unequally sized hard spheres was studied in other papers of Chang and Powell (1994a). Particle distributions of a suspension monolayer were generated by a Monte Carlo technique and applied periodic boundary conditions to represent an infinite suspension. Some of the obtained results are presented in Fig.10.4.5. and a good agreement between the results of 2D simulations, experiments, and 3D simulations of monodispersed suspensions can be observed.

Loewenberg and Hinch (1996, 1997) have developed a 3D numerical method for computer simulation of a concentrated emulsion in shear flow at low Reynolds number and finite capillary number conditions. They have presented numerical results for transient and steady-state rheology of concentrated emulsions with volume fractions up to 30% and dispersed phase viscosity ratios in the range $0 \le \lambda \le 5$. Their results have been obtained by using an efficient boundary integral formulation with periodic boundary conditions and up to twelve drops in each periodically repeated unit cell. Loewenberg and Hinch have found that the viscosity of an emulsion is a moderate, approximately linear, function of the dispersed phase volume fraction, at least up to

Φ = 30%, unlike the sharply increasing viscosity of suspensions of rigid particles or undeformed drops.

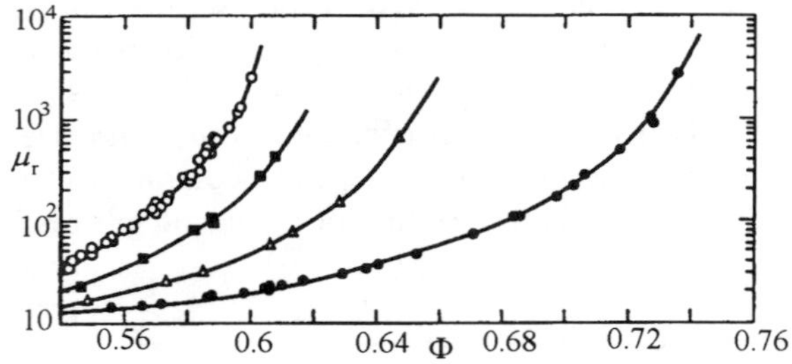

Fig.10.4.4. Relative viscosity μ_r as a function of volume fraction Φ and particle size ratio λ (from Chong et al., 1971): (○), monodispersed, $\lambda = 1$; (■), $\lambda = 2.1$; (Δ), $\lambda = 3.19$; (●), $\lambda = 7.25$. The fraction of small spheres is 0.25 (from Chang and Powell, 1993).

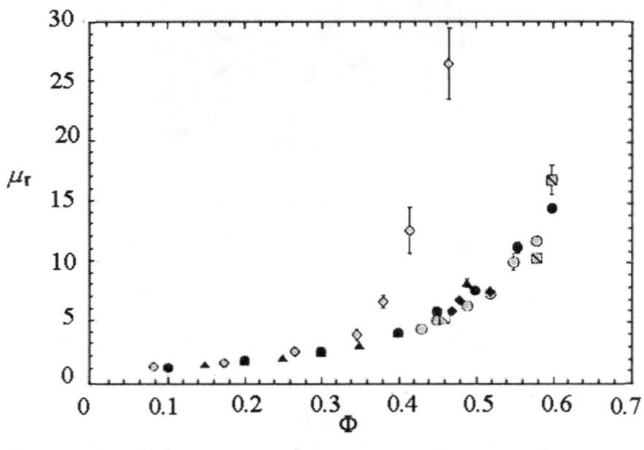

Fig.10.4.5. Relative viscosity μ_r as a function of area fraction $\Phi_a = 2/3\Phi$ from: (▲) Monte Carlo simulations of Chang and Powell (1994a); experiments of hard spheres of radii 28nm (□), 46nm (○), and 76nm (◆) (Van der Werff et al., 1989); (●) 3D simulation results (Phillips et al., 1988a) for monodispersed suspensions; (✧) dynamic simulation (Chang and Powell, 1993), (from Chang and Powell, 1994a).

The known results for the force on a periodic array due to flow at zero Reynolds number has been generalised by Cheng and Papanicolau (1997) for the case of small but finite Reynolds numbers. These authors use a generalisation of Hasimoto's (1959) approach based on an analysis of periodic Green's functions. They compare their results with the phenomenological ones of Kaneda (1986) for viscous flow past a random array of spherical particles.

PART IV.

UNSTEADY FLOWS. HYDRODYNAMIC INTERACTIONS BETWEEN DROPS, BUBBLES AND RIGID PARTICLES

CHAPTER 11.

Unsteady Motion of Rigid or Fluid Particles in Stokes Approximation

11.1. Introduction

In the book, up to this point, we have restricted our attention to steady or quasi-steady flows. However, when a rigid (or fluid particle) is transferred by a flow, it becomes subject to unsteady hydrodynamic forces because of its continuously changing environment. When the rate of change of the flow is low, we may assume that the velocity of the particle is determined exclusively by the instantaneous flow conditions and boundary configuration, i.e., we may assume that the particle is in a state of quasi-steady motion. In contrast, when the flow changes at an appreciable rate, unsteady forces become important and the full history of the motion must be taken into consideration.

The existing investigations show that the unsteady hydrodynamic problems solutions are considerably more complicated than those of the steady ones. Contrary to the steady Stokes equations for creeping motion, the time-dependent linearized Navier-Stokes equations have only been solved for the spherical rigid (or fluid) particle case, which is the sole analytical solution for an isolated finite body. In fact, the unsteady Stokes equations were first solved by Stokes (1851) for the oscillatory motions of a sphere along a diameter, a cylinder along a diameter, and a flat wall in its own plane. For the hydrodynamic force on an oscillating spherical particle in a viscous fluid Stokes found the following formula

$$\mathbf{F} = \mathrm{Real}\left\{ -6\pi\mu a\mathbf{U}\left(1 + \mathrm{M} + \frac{1}{9}\mathrm{M}^2\right)\exp(-i\omega t)\right\}, \qquad (11.1.1)$$

in which ω is the oscillation frequency, $M^2 = -\frac{i\omega a^2}{\nu}$ where Real{M} > 0 is the so called frequency parameter, ν and μ are the kinematic and dynamic viscosities of the fluid and **U** and a are the peak velocity and the radius of the particle. Since

$$M = \left(\frac{\sqrt{2}}{2} - i\frac{\sqrt{2}}{2}\right)\sqrt{\frac{\omega a^2}{\nu}} = \sqrt{\frac{\omega a^2}{\nu}}\exp\left(\frac{-i\pi}{4}\right) \text{ and } M^2 = -\frac{i\omega a^2}{\nu} = \frac{\omega a^2}{\nu}\exp\left(\frac{-i\pi}{2}\right),$$

the expression (11.1.1) can be written in the form

$$\mathbf{F} = \mathrm{Real}\left\{-6\pi\mu a\mathbf{U}\left(1 + a\sqrt{\frac{\omega}{\nu}}\exp\left(\frac{-i\pi}{4}\right) + \frac{1}{9}\frac{\omega a^2}{\nu}\exp\left(\frac{-i\pi}{2}\right)\right)\exp(-i\omega t)\right\} \quad (11.1.2)$$

In the above formula the first term is in phase with **U**, the middle term has a phase lag of $\pi/4$, and the last term is $\pi/2$ out of phase with respect to **U**. Very few new works have appeared for the unsteady linearized equations since these celebrated Stokes solutions were published.

A significant generalisation of the Stokes problem for the uniform motion of a spherical rigid particle in a viscous fluid was made by Basset (1888) and Boussinesq (1903), who considered the case of a spherical particle velocity dependent on time and transient flow. In order to obtain the force on a spherical particle in an arbitrary time-dependent motion with velocity **U**(t), Basset (1888) and Boussinesq (1903) integrated (11.1.1) for all frequencies

$$\mathbf{F} = -6\pi\mu a\mathbf{U}(t) - 6\pi\mu a^2\frac{1}{\sqrt{\pi\nu}}\int_0^t \frac{d\mathbf{U}}{d\tau}(t-\tau)^{-\frac{1}{2}}d\tau - \frac{2}{3}\pi\rho a^3\frac{d\mathbf{U}}{dt}. \quad (11.1.3)$$

Here ρ is the fluid density. The second term on the right-hand side of (11.1.3) is the so-called Basset force, which describes the history of the particle motion, while the third term describes the non-dissipative added-mass force for the spherical particle.

A more detailed investigation of this problem was performed by Villat (1943) and Ockendon (1968). In general, previous investigations of unsteady flows may be classified in two categories that address external and internal flows, respectively. Typical ones in the first category are the studies of flow past isolated rigid and fluid particles or collections of them. (Riley, 1967; Pienkowska, 1984).

The solution of time-dependent linearized flow problems may be facilitated by the application of Laplace or Fourier transforms to the governing unsteady Stokes equations. In this way one eliminates the time as an explicit variable and converts the problem into one of oscillatory motion. Such approach was used by Buchanan (1891), Kanwal (1955), Lai and Mockros (1972), and Lawrence and Weinbaum (1986, 1988) for particles of spheroidal shape. Kanwal (1964), Williams (1966) and Batchelor (1967) obtained some asymptotic results for particles of arbitrary shape in low and high-frequency oscillatory motion.

Sy et al. (1970) solved the problem for the unsteady motion of a bubble in a viscous fluid. The same problem was generalised by Sy and Lightfoot (1971) for the fluid drop case, but the obtained solution is wrong due to the incorrect problem postulation. The correct solution of the unsteady motion of a drop in a viscous fluid is given later by Chisnell (1987).

Equation (11.1.3) may be used to derive an integro-differential equation describing the gravitational settling of a spherical particle that has been released from rest at t = 0 (see Yih, 1979, and Clift et al., 1978). Balancing the drag, the weight and the buoyancy force of a rigid spherical particle with the rate of change of the momentum of the particle we obtain the equation

$$\mathrm{V_s}\frac{d\mathbf{U}}{dt} = (\rho_s - \rho)\mathrm{V_s}\mathbf{g} - 6\pi\mu a\mathbf{U}(t) - 6a^2\sqrt{\pi\rho a}\int_0^t \left(\frac{d\mathbf{U}}{dt}\right)_{t=s} \frac{1}{\sqrt{t-\tau}}d\tau, \quad (11.1.4)$$

where ρ_s is the density of the particle and V_s is the volume. It is worth noting that the equation (11.1.4.) can be recast into the equivalent form of a second-order ordinary differential equation in time for **U**(t). Excellent reviews of the problem that we consider in this chapter are given by Vojir and Michaelidis (1994) and Pozrikidis (1995).

On the basis of the unsteady Stokes equations Zapryanov and Chervenivanova (1982) studied the interaction of a rotationally oscillating circular disk and an interface, which is covered by a surfactant monolayer. By use of the proposed by Williams (1962) method, the obtained first kind integral equation of Fredholm is reduced to an second kind integral equation of Fredholm, which is solved asymptotically and numerically. The problem solved by Zapryanov and Chervenivanova (1982) is analysed independently by Chakrabarti et al. (1982).

Nyborg (1953) and Elder (1959) investigated experimentally the stationary (acoustic) streaming created around an oscillating bubble. A theoretical study of the steady streaming appearance and structure induced by an oscillating bubble was done by Davidson and Riley (1971). Zapryanov and Stoyanova (Tabakova) (1978a) gave an exact solution of the steady streaming in Stokes approximation for the problem of the translational harmonic oscillations performed by a spherical drop immersed in another viscous fluid. Applying the method of matched asymptotic expansions for the full Navier-Stokes equations, they obtained the vorticity structure of the induced steady streaming and the drag suffered by the fluid particle, as well. Their drag formula was a generalisation of the Taylor and Acrivos' drag formula for the steady drop motion at small Re.

The observation of the growth and subsequent collapse of a cavitational bubbles dates about a century back, when Reynolds (1894) discovered their formation in water flowing through constricted tubes. The bubbles form because of a local lowering of the dynamic pressure below the saturated vapour pressure. Lord Rayleigh (1917) studied growth and collapse of a spherical bubble in an infinite fluid and showed that tremendous dynamic pressures originate during bubble breakup.

Bubbles play a specific role in some areas of technology as propeller-induced cavitation in ships, cavitation in fluid machinery, nucleate boiling in reactors and in many devices as centrifuges and mixers used in chemical process industry. A wide variety of research, both fundamental and of more applied nature has been going on for many years. Excellent reviews in this area are given by Plesset and Prosperetti (1977) and Feng and Leal (1997). Another related review, written by Blake and Gibson (1987), considered the problem of a bubble in a quiescent fluid near a plane wall.

11.2. Unsteady Motion of a Spherical Particle in a Viscous Fluid

Basset (1888) generalised the Stokes problem for the steady flow past a sphere of radius a by a uniform viscous flow for the case, when the sphere moves with unstationary velocity **U**(t) in the fluid. The problem is to solve the unstationary Stokes equation (1.3.9)

$$\frac{\partial \mathbf{v}}{\partial t} = -\frac{1}{\rho}\,\mathrm{grad}p + \nu \Delta \mathbf{v} \qquad (11.2.1)$$

and the continuity equation

$$\nabla . \mathbf{v} = 0 . \qquad (11.2.2)$$

If a spherical coordinate system (r, φ, θ)is introduced with origin coinciding with the sphere centre, the boundary condition at infinity is

$$\mathbf{v} = 0, \qquad \text{at } r \to \infty, \qquad (11.2.3)$$

and on the particle surface reads

$$\mathbf{v} = \mathbf{U}, \qquad \text{at } r = a. \qquad (11.2.4)$$

If the flow is not periodic in time, then some initial condition has to be imposed, e.g.

$$\mathbf{v} = 0, \qquad \text{at } t = 0. \qquad (11.2.5)$$

The original problem treated by Basset was for a spherical particle oscillating with velocity which is a sufficiently smooth function of time. Let us assume an exponentially decaying function of time, and the boundary condition (11.2.4) becomes

$$\mathbf{U}(t) = \mathbf{U}_0 \exp(-i\omega t) \qquad \text{at } r = a, \qquad (11.2.6)$$

where ω is the oscillation frequency and $\mathbf{U}_0$ is constant vector.

In order to eliminate the pressure, the operator rot is applied on both sides of (11.2.1)

$$\frac{\partial}{\partial t}(\mathrm{rot}\,\mathbf{v}) = \nu \Delta(\mathrm{rot}\,\mathbf{v}) \qquad (11.2.7)$$

and the solution is sought in the form

$$\mathbf{v} = \exp(-i\omega t)\ \mathrm{rot\,rot}\left(f\mathbf{U}_0\right), \qquad (11.2.8)$$

where f = f(r).

The function f will be defined from equation (11.2.7). Taking into account (11.2.8) we have

$$\mathrm{rot}\,\mathbf{v} = \exp(-i\omega t)\ \mathrm{rot\,rot\,rot}\left(f\mathbf{U}_0\right) \qquad (11.2.9)$$

$$= \exp(-i\omega t)\left(\mathrm{grad\,div} - \Delta\right)\mathrm{rot}\left(f\mathbf{U}_0\right) = -\exp(-i\omega t)\Delta\mathrm{rot}\left(f\mathbf{U}_0\right)$$

$$= -\exp(-i\omega t)\Delta\left(\nabla f \times \mathbf{U}_0\right).$$

Due to the upper expression and after integrating (11.2.7), it transforms into the necessary equation for the function f

$$\Delta^2 f + \frac{i\omega}{\nu}\Delta f = 0 . \qquad (11.2.10)$$

Then the solution of Δf is given by

$$\Delta f = \frac{C}{r}\exp(ikr) , \qquad (11.2.11)$$

where C is an arbitrary constant and $k = \sqrt{\frac{\omega}{\nu}}$.

Therefore

$$\frac{df}{dr} = \frac{C_1}{r}\exp(ikr)\left(r - \frac{1}{ik}\right) + \frac{C_2}{r^2} , \qquad (11.2.12)$$

where C_1 and C_2 are determined by the boundary condition (11.2.6):

$$C_1 = -\frac{3a}{2ik}\exp(-ika) , \qquad C_2 = -\frac{a^3}{2}\left(1 - \frac{3}{ika} - \frac{2}{k^2a^2}\right) . \qquad (11.2.13)$$

From (11.2.13) it follows that for high frequencies ($\omega \to \infty$) $C_1 \to 0$, $C_2 \to -\frac{a^3}{2}$ and

$$\frac{df}{dr} = -\frac{a^3}{2r^2} . \qquad (11.2.14)$$

This flow corresponds to the potential motion of a sphere in an ideal fluid.

The drag force exerted by a translationally oscillating spherical particle in a viscous fluid is defined by the formulae (11.2.8) and (11.2.12). The final result is

$$\mathbf{F} = 6\pi\mu a\left(1 + \frac{\sqrt{2}ak}{2}\right)\mathbf{U} + 3\pi a^2\sqrt{\frac{2\mu\rho}{\omega}}\left(1 + \frac{\sqrt{2}ak}{9}\right)\frac{d\mathbf{U}}{dt} . \qquad (11.2.15)$$

The Stokes formula (11.1.1) follows from (11.2.15) at $\omega = 0$, while for high frequencies ($\omega \to \infty$) we have

$$\mathbf{F} = \frac{2\pi}{3}\rho a^3\frac{d\mathbf{U}}{dt} + 3\pi a^2\sqrt{2\mu\rho\omega}\,\mathbf{U} . \qquad (11.2.16)$$

The drag force formula for the case when the spherical particle velocity dependence on time is arbitrary, i.e., $\mathbf{U} = \mathbf{U}(t)$, will be derived by use of the Fourier transform applied to (11.2.15). The Fourier transform of $\mathbf{U}(t)$ is $\mathbf{U}^*(\omega)$:

$$\mathbf{U}(t) = \int_{-\infty}^{\infty}\mathbf{U}^*(\omega)\,\exp(-i\omega t)\,d\omega , \qquad (11.2.17)$$

where

$$\mathbf{U}^*(\omega) = \frac{1}{2\pi}\int_{-\infty}^{\infty}\mathbf{U}(t)\,\exp(i\omega t)\,dt . \qquad (11.2.18)$$

The linearity of the equations describing the flow induced by the motion of the spherical particle in the viscous fluid allows us to express the full drag force as an integral of the drag forces of a motion of velocities equal to the Fourier components $\mathbf{U}^*(\omega)\exp(-i\omega t)$. These forces are given by formula (11.2.15) for the considered in the present chapter problem of a spherical particle translationally oscillating in a viscous fluid. Then the result is

$$\pi\rho a^3\,\mathbf{U}^*\exp(-i\omega t)\left[\frac{6\nu}{a^2}-\frac{2i\omega}{3}+\frac{3\sqrt{2\nu}}{a}(1-i)\sqrt{\omega}\right]. \tag{11.2.19}$$

Taking into account that

$$\left(\frac{d\mathbf{U}}{dt}\right)^* = -i\omega\,\mathbf{U}^*, \tag{11.2.20}$$

the expression (11.2.19) can be written in the form

$$\pi\rho a^3\exp(-i\omega t)\left[\frac{6\nu}{a^2}\mathbf{U}^*+\frac{2}{3}\left(\frac{d\mathbf{U}}{dt}\right)^*+\frac{3\sqrt{2\nu}}{a}\left(\frac{d\mathbf{U}}{dt}\right)^*\frac{(1+i)}{\sqrt{\omega}}\right]. \tag{11.2.21}$$

From (11.2.17) and (11.2.18) it follows, that after integrating the expression (11.2.21) on $d\omega$, we shall obtain from its first and second terms expressions which are proportional to $\mathbf{U}(t)$ and $\frac{d\mathbf{U}}{dt}$, respectively. In order to integrate the third term of (11.2.21), containing $\sqrt{\omega}$ in its denominator, we must to take into account that for negative values of ω, we have $\frac{1+i}{\sqrt{\omega}}=\frac{1-i}{\sqrt{|\omega|}}$. Then the integration on $d\omega$ within limits from $-\infty$ to ∞ can be substituted by the doubling of the real of the integral with limits from 0 to ∞. In order to reform the expression

$$J = 2\,\mathrm{Real}\left[(1+i)\int_0^\infty\left(\frac{d\mathbf{U}}{dt}\right)^*\exp(-i\omega t)\frac{d\omega}{\sqrt{\omega}}\right],$$

we substitute $\left(\frac{d\mathbf{U}}{dt}\right)^* = \frac{1}{2\pi}\int_{-\infty}^{\infty}\frac{d\mathbf{U}(\tau)}{d\tau}\exp(-i\omega\,\tau)\,d\tau$ and reach

$$J = \frac{1}{\pi}\,\mathrm{Real}\left[(1+i)\int_{-\infty}^{\infty}\int_0^{\infty}\frac{d\mathbf{U}(\tau)}{d\tau}\exp[i\omega(\tau-t)]\frac{d\omega}{\sqrt{\omega}}\,d\tau\right]$$

$$=\frac{1}{\pi}\,\mathrm{Real}\left[(1+i)\int_{-\infty}^{t}\int_0^{\infty}\frac{d\mathbf{U}(\tau)}{d\tau}\exp[i\omega(\tau-t)]\frac{d\omega}{\sqrt{\omega}}\,d\tau+(1+i)\int_t^{\infty}\int_0^{\infty}\frac{d\mathbf{U}(\tau)}{d\tau}\exp[i\omega(\tau-t)]\frac{d\omega}{\sqrt{\omega}}\,d\tau\right]$$

$$=\sqrt{\frac{2}{\pi}}\,\mathrm{Real}\left[\int_{-\infty}^{t}\frac{d\mathbf{U}(\tau)}{d\tau}\frac{d\tau}{\sqrt{t-\tau}}+i\int_t^{\infty}\frac{d\mathbf{U}(\tau)}{d\tau}\frac{d\tau}{\sqrt{\tau-t}}\right]=\sqrt{\frac{2}{\pi}}\int_{-\infty}^{t}\frac{d\mathbf{U}(\tau)}{d\tau}\frac{d\tau}{\sqrt{t-\tau}}.$$

To derive the above expression, we have used the formula $\int_0^{\infty} \frac{\exp(-cx)}{\sqrt{x}} dx = \sqrt{\frac{\pi}{c}}$, where c is some complex number.

Therefore, the drag force suffered by the particle during its unsteady motion in a viscous fluid is given by

$$\mathbf{F} = 6\pi\mu a \left[\mathbf{U}(t) + \frac{1}{9}\frac{a^2}{\nu}\frac{d\mathbf{U}}{dt} + \frac{a}{\sqrt{\pi\nu}} \int_{-\infty}^{t} \frac{d\mathbf{U}(\tau)}{d\tau} \frac{d\tau}{\sqrt{t-\tau}} \right], \tag{11.2.22}$$

which is the drag formula of Basset (11.1.3). The first term on the right-hand side of (11.2.22) characterises the Stokes friction force, the second term is the force necessary to accelerate the added-mass and the third term is the Basset force accounts for the unsteady character of the particle motion, i.e., the particle motion history. The Basset force increases the momentary drag force. In the moment $\tau = t$ the integrand in (11.2.22) tends to infinity as $\lim_{s \to 0} s^{-0.5}$. When the particle accelerates rapidly under the action of a big external force, the Basset force can become significant and can exceed several times the drag force, correspondent to the steady motion of the particle.

It is easy to show that the Stokes drag formula can be obtained from the Basset formula (11.2.22) for $\mathbf{U}(t)$ = const. The drag formula (11.2.15) for the oscillation of a particle in a viscous fluid can also be obtained from (11.2.22) as a special case.

It is worth to define from (11.2.22) the drag for a spherical particle, starting uniformly to accelerate from rest (at the moment t = 0) by the law $\mathbf{U}(t) = \boldsymbol{\alpha} t$, ($\boldsymbol{\alpha}$ = const.). If in (11.2.22) we substitute $\mathbf{U}(t) = 0$ at $t < 0$ and $\mathbf{U}(t) = \boldsymbol{\alpha} t$ at $t > 0$, then

$$\mathbf{F} = 6\pi\mu\boldsymbol{\alpha} \left[t + \frac{1}{9}\frac{a^2}{\nu} + 2a\sqrt{\frac{t}{\pi\nu}} \right]. \tag{11.2.23}$$

As another special case we shall present the drag of a spherical particle, starting impulsively to move from rest with uniform velocity $\mathbf{U}(t) = \mathbf{U}_0$ = const. in a viscous fluid. It is assumed that $\mathbf{U}(t) = 0$ at $t < 0$ and $\mathbf{U} = \mathbf{U}_0$ = const. at $t > 0$. Then the derivative of $\mathbf{U}(t)$ is everywhere zero except at t = 0, where it is infinity. This is the reason to use the Dirac delta function δ (t) instead of the velocity derivative, i.e., $\frac{d\mathbf{U}}{dt} = \mathbf{U}_0 \delta(t)$. From (11.2.22) for the drag force we get

$$\mathbf{F} = 6\pi\mu a \mathbf{U}_0 \left[1 + \frac{1}{9}\frac{a^2}{\nu}\delta(t) + a\frac{1}{\sqrt{\pi\nu t}} \right]. \tag{11.2.24}$$

11.3. Unsteady Motion of a Spherical Drop in a Viscous Fluid

Suppose that a spherical drop of radius a is moving under the action of the gravity force with velocity V(t). With respect to the coordinate system connected to the drop, the unsteady Stokes equation is written in the form

$$\frac{\partial \mathbf{v}}{\partial t} = -\frac{1}{\rho}\operatorname{grad} p + \nu \Delta \mathbf{v} - g\mathbf{k} + \frac{dV}{dt}\mathbf{k}. \qquad (11.3.1)$$

Here **k** is a unit vector directed upwards, g is the earth acceleration, (ν, ρ) and $(\hat{\nu}, \hat{\rho})$ are the respective kinematic viscosity and density of the fluid outside and inside the drop. Because of the problem axisymmetry, a stream function Ψ is introduced, which is connected with the velocity components in spherical coordinate system (r, θ, φ) by the relations (1.3.17) and (1.3.14), identically satisfying the continuity equation $\nabla.\mathbf{v} = 0$. Then the equation (11.3.1) changes to the "unsteady variant" of the biharmonic equation [see (5.2.1)] for Ψ

$$\frac{\partial}{\partial t}\left(E^2\Psi\right) = \nu E^4\Psi, \qquad (11.3.2)$$

where $E^2\Psi$ is given by (5.2.2).

According to the uniform flow $\mathbf{v} \to V(t)\mathbf{k}$ at infinity; it is evident that $\Psi \to \frac{1}{2}V(t)r^2\sin^2\theta$ and the solution of (11.3.2) is sought in the form

$$\Psi = f(r,t)\sin^2\theta. \qquad (11.3.3)$$

Function f(r, t) has to satisfy the equation

$$L_1L_2f = 0, \qquad (11.3.4)$$

where L_1 and L_2 are the following operators:

$$L_1 \equiv \frac{\partial^2}{\partial r^2} - \frac{2}{r^2}, \qquad L_2 \equiv \frac{\partial^2}{\partial r^2} - \frac{2}{r^2} - \frac{1}{\nu}\frac{\partial}{\partial t}.$$

For the stream function inside the drop we suppose an analogous form to (11.3.3), i.e., $\hat{\Psi} = \hat{f}(r,t)\sin^2\theta$, where $\hat{f}$ must satisfy an equation of the type (11.3.4).

The boundary conditions for the fluid flow outside and inside the drop (1.4.14)-(1.4.17) written in terms of the function f(r, t) will give

$$f \to \frac{1}{2}V(t)r^2 \quad \text{as } r \to \infty, \qquad (11.3.5)$$

$$\left|\frac{f(r,t)}{r^2}\right| < \infty \qquad \text{at } r = 0, \qquad (11.3.6)$$

$$f(r,t) = 0, \qquad \hat{f}(r,t) = 0 \qquad \text{at } r = a, \qquad (11.3.7)$$

$$\frac{\partial f}{\partial r} = \frac{\partial \hat{f}}{\partial r} \qquad \text{at } r = a, \qquad (11.3.8)$$

$$\mu\left[\frac{\partial^2 f}{\partial r^2}-\frac{2}{r^2}\frac{\partial f}{\partial r}\right]=\hat{\mu}\left[\frac{\partial^2 \hat{f}}{\partial r^2}-\frac{2}{r^2}\frac{\partial \hat{f}}{\partial r}\right] \quad \text{at } r=a. \tag{11.3.9}$$

When the observed motion starts from rest, the following initial conditions must also hold

$$f(r,t)=0\,,\qquad \hat{f}(r,t)=0 \qquad \text{at } t=0. \tag{11.3.10}$$

If we suppose that at $t<0$ till the moment $t=0$ the drop has been moving with velocity V_0 different from the settling (terminal) velocity

$$V_1=\frac{2}{3}\frac{(1+k_1)(\gamma-1)a^2 g}{(3+2k_1)\nu}, \tag{11.3.11}$$

then the initial condition will be the well known steady solution (Hadamard, 1911) and (Rybczynski, 1911):

$$\frac{f(r,0)}{V_0 a^2}=\frac{1}{2}\left(\frac{r^2}{a^2}-\frac{a}{r}\right)+\frac{(3+2k_1)}{4(1+k_1)}\left(\frac{a}{r}-\frac{r}{a}\right), \quad \text{at } r>a \tag{11.3.12}$$

$$\frac{\hat{f}(r,0)}{V_0 a^2}=\frac{k_1 r^2(r^2-a^2)}{4(1+k_1)a^4}, \quad \text{at } r<a. \tag{11.3.13}$$

In formulae (11.3.11) - (11.3.13) $k_1=\dfrac{\mu}{\hat{\mu}}$ and $\gamma=\dfrac{\hat{\rho}}{\rho}$.

In order to solve the initial-boundary value problem (11.3.4) - (11.3.13), Chisnell (1987) makes use of the Laplace transformation

$$f^*(r,s)=\int_0^\infty \exp(-st)\, f(r,t)\, dt\,, \quad (\text{s is a complex number}), \tag{11.3.14}$$

and expresses the function f(r, t) as a sum of two functions $f=f_1+f_2$, where

$$L_1 f_1=0\,,\qquad L_2 f_2=0. \tag{11.3.15}$$

It is assumed that the functions $f_1(r,t)$ and $\hat{f}_1(r,t)$ satisfy the initial condition

$$f_1(r,0)=\frac{1}{2}V_0 a^2\left(\frac{r^2}{a^2}-\frac{a}{r}\right) \quad \text{and} \quad \hat{f}_1(r,0)=0\,, \tag{11.3.16}$$

while the functions $f_2(r,0)$ and $\hat{f}_2(r,0)$ contain the remaining terms of (11.3.12) and (11.3.13). After applying the transformation (11.3.14) to the functions $f_2(r,t)$ and $\hat{f}_2(r,t)$, the following equations are obtained

$$\left(\frac{\partial^2}{\partial r^2}-\frac{2}{r^2}-\frac{s}{\nu}\right) f_2^*(r,s)+\frac{1}{\nu}f_2(r,0)=0\,, \tag{11.3.17}$$

$$\left(\frac{\partial^2}{\partial r^2}-\frac{2}{r^2}-\frac{s}{\nu}\right) \hat{f}_2^*(r,s)+\frac{1}{\nu}\hat{f}_2(r,0)=0\,, \tag{11.3.18}$$

which are ordinary differential equations with corresponding solutions:

$$f_2^* = \frac{V_0 a^2 (3+2k_1)}{4s(1+k_1)}\left(\frac{a}{r} - \frac{r}{a} + \frac{2\nu}{asr}\right) + A\exp\left[-s_1\left(\frac{r}{a}-1\right)\right]\left(\frac{a}{r}+s_1\right), \qquad (11.3.19)$$

$$\hat{f}_2^* = \frac{V_0 k_1}{4a^2 s(1+k_1)}\left(r^4 - a^2 r^2 + \frac{10\nu r^2}{s}\right) + B\left[\frac{a}{r}\sinh\left(\frac{\hat{s}_1 r}{a}\right) - s_1\cosh\left(\frac{\hat{s}_1 r}{a}\right)\right].$$

Here the parameters $s_1 = a\sqrt{\frac{s}{\nu}}$ and $\hat{s}_1 = a\sqrt{\frac{s}{\hat{\nu}}}$ have positive real parts.

Since the equations

$$L_1 f_1 = 0 \qquad \text{and} \qquad L_1 \hat{f}_1 = 0 \qquad (11.3.20)$$

do not contain any time derivatives, they do not change after the transformation (11.3.14). The solutions of equations (11.3.20), which have no singularities for small and large r and together with f_2^* and $\hat{f}_2^*$ satisfy the conditions (11.3.7), are written in the form

$$f_1^* = \frac{1}{2}V^*(s)\left(r^2 - \frac{a^3}{r}\right) - \frac{V_0(3+2k_1)\nu}{2s^2(1+k_1)}\frac{a}{r} - A\frac{a}{r}(1+s_1), \qquad (11.3.21)$$

$$\hat{f}_1^* = \frac{5V_0 k_1 \hat{\nu}}{2s^2(1+k_1)}\frac{r^2}{a^2} - B\frac{r^2}{a^2}\left[\sinh(\hat{s}_1) - \hat{s}_1\cosh(\hat{s}_1)\right].$$

The constants A and B are determined from the boundary conditions (11.3.8) and (11.3.9)

$$A = \frac{3\nu(T_2 + 2k_1 T_1)}{2s\left[T_2 + k_1(3+s_1)T_1\right]}\left(V^* - \frac{V_0}{s}\right), \qquad (11.3.22)$$

$$B = \frac{3k_1 a^2(1+s_1)}{2\left[T_2 + k_1(3+s_1)T_1\right]}\left(V^* - \frac{V_0}{s}\right),$$

where

$$T_1 = 3\hat{s}_1\cosh(\hat{s}_1) - (3+\hat{s}_1^2)\sinh(\hat{s}_1) = -\sum_{n=0}^{\infty}\frac{\hat{s}_1^{5+2n}}{(5+2n)(3+2n)(2n+1)!}, \qquad (11.3.23)$$

$$T_2 = (6+3\hat{s}_1)\sinh(\hat{s}_1) - (6+\hat{s}_1^2)\cosh(\hat{s}_1) = -\sum_{n=0}^{\infty}\frac{\hat{s}_1^{5+2n}}{(5+2n)(2n+1)!}.$$

The functions f^* and $\hat{f}^*$ are expressed by means of the function V^*, which can be determined from the equation of the drop motion

$$M\frac{dV}{dt} = gM - F, \qquad (11.3.24)$$

where $M = \frac{4}{3}\pi\hat{\rho}a^3$ and F is the force exerted on the drop at its motion, i.e.,

$$F = 2\pi a^2 \int_0^{\pi} \left(T_{rr}\big|_{r=a} \cos\theta - T_{r\theta}\big|_{r=a} \sin\theta \right) \sin\theta \, d\theta . \tag{11.3.25}$$

The stress components in (11.3.25) refer to the fluid flow outside the drop. In order to calculate the pressure in the expression of the normal stress component, the *θ* component of the vector equation (11.3.1) is used. Replacing

$$T_{rr} = -p_1 + T'_{rr} \cos\theta, \; T_{r\theta} = T'_{r\theta} \sin\theta,$$

where p_1 is a constant, we obtain for T'_{rr} and $T'_{r\theta}$

$$T'_{rr} = \rho \frac{\partial^2 f}{\partial r \partial t} - \frac{\mu}{a^3}\left(a^3 \frac{\partial^3 f}{\partial r^3} - 6a \frac{\partial f}{\partial r} + 12 f \right) + \rho a \left(g - \frac{dV}{dt} \right), \tag{11.3.26}$$

$$T'_{r\theta} = -\frac{\mu}{a^3}\left(a^2 \frac{\partial^2 f}{\partial r^2} - 2a \frac{\partial f}{\partial r} + 2f \right),$$

$$T'_{rr} - 2T'_{r\theta} = \frac{\hat{\rho} a}{M} F .$$

If the function f^*, expressed by V^* with the formulae (11.3.19) and (11.3.21), is substituted into equations (11.3.24) and (11.3.26), the following formula for the drop motion velocity is got

$$V^* - \frac{V_0}{s} = \frac{g(\gamma - 1) - \dfrac{3\nu(3 + 2k_1)}{2a^2(1 + k_1)}}{s\left[(\gamma + 0.5)s + \dfrac{9\nu(1 + s_1)(T_2 + 2k_1 T_1)}{2a^2\left[T_2 + k_1(3 + s_1)T_1\right]} \right]} . \tag{11.3.27}$$

Knowing V^*, it can be proceeded further with the inverse of the Laplace transformation (11.3.14) in order to find the originals.

The force $F = gM - M\dfrac{dV}{dt}$, acting on the outer drop surface, includes the Archimedes' force $\dfrac{4}{3}\pi a^3 \rho g$ and the drag force

$$D = \frac{4}{3}\pi a^3 \rho g \left[g(\gamma - 1) - \gamma \frac{dV}{dt} \right]. \tag{11.3.28}$$

From (11.3.1) and (11.3.28) the drag coefficient (dimensionalized with the terminal falling velocity of the drop in the gravity field) is found to be

$$c_D = \frac{D}{\frac{1}{2}\rho V_1^2 \pi a^2} = \frac{8(3 + 2k_1)}{Re(1 + k_1)}\left[1 - \frac{\gamma}{g(\gamma - 1)} \frac{dV}{dt} \right], \tag{11.3.29}$$

where $Re = \frac{2aV_1}{\nu}$ is the Reynolds number. We shall note that when the motion is uniform ($\frac{dV}{dt} = 0$), the drag coefficient formula (3.3.1) for the drop steady motion is obtained from (11.3.29).

In Chisnell (1987) it is proved that at unsteady motion in a viscous fluid at zero Reynolds numbers the shape of the spherical drop is preserved. At the assumption that not only the Reynolds number, but also the Weber number is small, the spherical shape preservation at the unsteady drop motion is established in Sy and Lightfoot (1971) and Clift et al. (1978).

After a thorough study Chisnell obtains the following results for the unknown functions V(t), $f(r,t)$ and $\hat{f}(r,t)$:

$$s = -\nu\frac{x^2}{a^2}, \quad \frac{V_1 - V(t)}{V_1 - V_0} = \frac{2(3+2k_1)}{3(1+k_1)\pi}\int_0^\infty \frac{\exp(st)}{D_1^2 + x^2 D_2^2}dx, \tag{11.3.30}$$

$$\frac{f(r,t) - f(r,\infty)}{(V_1 - V_0)a^2} = -\frac{2k_1(3+2k_1)(\gamma+0.5)}{9(1+k_1)\pi}\int_0^\infty \frac{N_1 x^2 \exp(st)}{x^3(D_1^2 + x^2 D_2^2)}dx,$$

$$\frac{\hat{f}(r,t) - \hat{f}(r,\infty)}{(V_1 - V_0)a^2} = \frac{1}{2}\left(\frac{r^2}{a^2} - \frac{a}{r}\right)\frac{(V(t) - V_1)}{(V_1 - V_0)} - \frac{(3+2k_1)}{(1+k_1)\pi}\int_0^\infty \frac{N_2 x^2 \exp(st)}{x^3(D_1^2 + x^2 D_2^2)}dx$$

where

$$D_1 = 1 - \frac{2}{9}x^2(\gamma+0.5)\frac{(T_2 + 3k_1T_1)}{[T_2 + 2k_1T_1]}, \quad D_2 = 1 - \frac{2}{9}x^2(\gamma+0.5)\frac{k_1T_1}{[T_2 + 2k_1T_1]},$$

$$\frac{T_2}{T_1} = \frac{(6-3y^2)\sin y - y(6-y^2)\cos y}{3y\cos y - (3-y^2)\sin y}, \qquad y = x\sqrt{\frac{\nu}{\hat{\nu}}}.$$

The functions N_1 and N_2 participating in (11.3.30) are equal, respectively, to:

$$N_1 = \frac{\frac{a}{r}\sin\left(\frac{r}{a}y\right) - y\cos\left(\frac{r}{a}y\right) - \frac{r^2}{a^2}(\sin y - y\cos y)}{(6-3y^2)\sin y - y(6-y^2)\cos y + 2k_1[3y\cos y - (3-y^2)\sin y]},$$

$$N_2 = \left[x\cos\left(\frac{r}{a}-1\right)x - \frac{xa}{r} - \frac{a}{r}\sin\left(\frac{r}{a}-1\right)x\right]D_1 - x\left[\frac{a}{r}\cos\left(\frac{r}{a}-1\right)x - \frac{a}{r} + \sin\left(\frac{r}{a}-1\right)x\right]D_2$$

The formulae (11.3.30) give the opportunity to study numerically the streamlines inside and outside the drop at different values of the parameters k_1, γ and the time t. In Chisnell (1987) it is also established that at $t \to \propto$ the unsteady velocity tends to the steady terminal velocity (11.3.11).

11.4. Application of the Induced Force Method on the Stokes Problem

Let a rigid spherical particle of radius a moves in a viscous incompressible fluid. The motion of the particle is described by the equation

$$m\frac{d\mathbf{U}(t)}{dt} = \int_{S(t)} \mathbf{T}(\mathbf{r},t).\mathbf{n}dS + \mathbf{\Phi}_{ext}, \tag{11.4.1}$$

where m is the particle mass, $\mathbf{U}(t)$ is the velocity of the particle, i.e., of its centre with radius vector $\mathbf{r}_0$, $\mathbf{\Phi}_{ext}$ is the outer force acting on the particle, $\mathbf{T}(\mathbf{r},t)$ is fluid stress tensor (1.2.10), S(t) is the spherical particle surface at the moment t and $\mathbf{n}$ is the outer unit normal vector acting on the spherical surface S(t).

The fluid flow past the particle can be found from the unsteady Stokes equation

$$\rho\frac{\partial \mathbf{v}(\mathbf{r},t)}{\partial t} = \mathrm{div}\mathbf{T}(\mathbf{r},t) + \mathbf{F}_{ext}, \tag{11.4.2}$$

where $\mathbf{F}_{ext}$ is the outer force density, and the continuity equation (1.3.2) and the boundary conditions

$$\mathbf{v}(\mathbf{r},t) = \mathbf{U}(t), \text{ at } |\mathbf{r}-\mathbf{r}_0(t)| = a, \tag{11.4.3}$$

$$\mathbf{v}(\mathbf{r},t) = 0, \qquad \text{as } \mathbf{r}\to\infty. \tag{11.4.4}$$

The radius vector of the particle centre $\mathbf{r}_0$ participates in the condition (11.4.3); it is in general a non-linear function of the time t. The surface S(t) from equation (11.4.1) is also a non-linear function of the time t. Therefore, the treated problem can be strictly linear, if $\mathbf{r}_0$ and S do not depend on time t. This condition is fulfilled when the coordinate system origin coincides with the particle centre.

When solving the formulated boundary value problem it is convenient to use the same equation holding for the whole space, outside and inside the particle. This is achieved when applying the idea of the particle substitution with a fictitious force on the right-hand side of the equation (11.4.2), which is called induced force and is denoted by $\mathbf{F}_{ind}$ (see Faxen, 1924; Mazur and Bedeaux, 1974). A similar method is often used in the electromagnetic field theory. This alternative problem postulation is quite suitable when solving the posed problem.

The flow equation (11.4.2) in the new formulation takes the form

$$\rho\frac{\partial \mathbf{v}(\mathbf{r},t)}{\partial t} = \mathrm{div}\mathbf{T}(\mathbf{r},t) + \mathbf{F}_{ind} + \mathbf{F}_{ext}. \tag{11.4.5}$$

Here, in contrast to the previous formulation, the fluid velocity and pressure are defined not only outside the sphere and its surface, but also inside it. The induced force $\mathbf{F}_{ind}(\mathbf{r},t)$ is chosen in such a way that outside the particle it is:

$$\mathbf{F}_{ind}(\mathbf{r},t) = 0, \text{ for } r > a, \tag{11.4.6}$$

and inside it the velocity and pressure to satisfy the conditions

$$\mathbf{v}(\mathbf{r},t) = \mathbf{U}(t), \qquad \text{for } r \le a, \tag{11.4.7}$$

$$p(\mathbf{r},t)=0, \qquad \text{for } r<a. \tag{11.4.8}$$

We have to point out that the boundary condition (11.4.3) defines uniquely the hydrostatic pressure on the particle surface. That is the reason (11.4.8) to be satisfied only for r < a.

In this way the solution of the system (11.4.1)-(11.4.4), together with the continuity equation (1.3.2), is also a solution of the system (1.3.2), (11.4.1), (11.4.4)-(11.4.8). Moreover, this solution is unique.

The fact that it is always possible to construct a function $\mathbf{F}_{ind}(\mathbf{r},t)$ with the necessary properties, can be obtained from the following considerations. If we spread the obtained velocity and pressure fields of (11.4.1) - (11.4.4) over the particle inside, defining them there in correspondence with the conditions (11.4.7) and (11.4.8), and substitute them in (11.4.5), then we reach the following expression for $\mathbf{F}_{ind}(\mathbf{r},t)$:

$$\mathbf{F}_{ind}(\mathbf{r},t)=\rho\,\frac{d\mathbf{U}(t)}{dt}, \quad \text{for } r<a. \tag{11.4.9}$$

From the uniqueness of the primary problem solution it follows the uniqueness of the induced force density expression (11.4.9).

We have to note that the elimination of the sphere from the flow and substitution of its effect with an induced force addition in the flow equation (11.4.2) leads to the problem simplification. The particle drag force expression can be obtained if the particle velocity $\mathbf{U}(t)$ and the fluid velocity in the absence of the sphere are known. The explicit forms of $\mathbf{F}_{ind}$ and the true velocity $\mathbf{v}(\mathbf{r},t)$ are not necessary. This simplification leads to the use of generalized functions, as for example $\mathbf{F}_{ind}$ is singular at least at r = a. For this purpose a Fourier transform is applied to the velocity, pressure, $\mathbf{F}_{ind}$ and $\mathbf{F}_{ext}$

$$\mathbf{v}^*(\mathbf{r},\omega)=\int_{-\infty}^{\infty}\mathbf{v}(\mathbf{r},t)\exp(i\omega t)\,dt, \quad \mathbf{F}_{ind}{}^*(\mathbf{r},\omega)=\int_{-\infty}^{\infty}\mathbf{F}_{ind}(\mathbf{r},t)\exp(i\omega t)\,dt, \tag{11.4.10}$$

$$p^*(\mathbf{r},\omega)=\int_{-\infty}^{\infty}p(\mathbf{r},t)\exp(i\omega t)\,dt, \quad \mathbf{F}_{ext}{}^*(\mathbf{r},\omega)=\int_{-\infty}^{\infty}\mathbf{F}_{ext}(\mathbf{r},t)\exp(i\omega t)\,dt.$$

In a similar manner the Fourier transformation is applied to the equations (1.3.2) and (11.4.5). After lengthy calculations (see Protodiakonov et al., 1987), which are dropped here, the pressure and velocity are found to be

$$p^*(\mathbf{r},\omega)=p^*_\infty(\mathbf{r},\omega)-\mu\int G(\mathbf{r}-\mathbf{r}',0)\,\nabla'.\mathbf{F}^*_{ind}(\mathbf{r}',\omega)\,d\mathbf{r}', \tag{11.4.11}$$

$$\mathbf{v}^*(\mathbf{r},\omega)=\mathbf{v}^*_\infty(\mathbf{r},\omega)+\int\left\{G(\mathbf{r}-\mathbf{r}',\omega)+\alpha^{-2}\nabla'\nabla'\left[G(\mathbf{r}-\mathbf{r}',0)-G(\mathbf{r}-\mathbf{r}',\omega)\right]\right\}\mathbf{F}^*_{ind}(\mathbf{r}',\omega)\,d\mathbf{r}' \tag{11.4.12}$$

where $\mathbf{v}^*_\infty(\mathbf{r},\omega)$ and $p^*_\infty(\mathbf{r},\omega)$ are the generalized solutions at the absence of induced force, μ is the fluid viscosity, $G(\mathbf{r},\omega)=(4\pi\mu r)^{-1}\exp(-\alpha r)$ is the Green function at $\alpha=(-i\rho\omega/\mu)^{1/2}$ and Real(α) > 0, and $\nabla'\equiv\dfrac{\partial}{\partial\mathbf{r}'}$.

Now, let us express the force $\mathbf{\Phi}(t)$, with which the fluid acts on the spherical particle, by means of the induced force density $\mathbf{F}_{ind}$. From (11.4.1) and the divergence theorem it follows that

$$\mathbf{\Phi}(t) = \int_S \mathbf{T}(\mathbf{r},t).\mathbf{n}\,dS = \int_{r\le a} \mathrm{div}\mathbf{T}(\mathbf{r},t)\,d\mathbf{r}. \tag{11.4.13}$$

Since inside the sphere $\mathbf{F}_{ext} = 0$, from (11.4.5) together with (11.4.13), it follows that

$$\mathbf{\Phi}(t) = -\int_{r\le a}\left[-\rho\frac{\partial \mathbf{v}(\mathbf{r},t)}{\partial t} + \mathbf{F}_{ind}(\mathbf{r},t)\right]d\mathbf{r}. \tag{11.4.14}$$

Applying on both sides of (11.4.14) the Fourier transformation and taking into account (11.4.7), we reach

$$\mathbf{\Phi}(\omega) = -i\omega\rho\frac{4\pi a^3}{3}\mathbf{U}(\omega) - \int_{r\le a}\mathbf{F}_{ind}(\mathbf{r},\omega)\,d\mathbf{r}. \tag{11.4.15}$$

Now, we shall apply the idea of induced force when solving the problem for the flow past a spherical particle at rest by a slow steady uniform flow, i.e., $\mathbf{v}_\infty = \mathrm{const.}$, $p_\infty = \mathrm{const.}$, $\dfrac{d\mathbf{U}(t)}{dt} = 0$. Then the induced force density is different from zero only on the particle surface r = a. If we assume that the induced force density linearly depends on the boundary conditions at infinity and owing to the problem symmetry, we have

$$\mathbf{F}_{ind}(\mathbf{r}) = \left(C_1\mathbf{v}_\infty + C_2 p_\infty\mathbf{n}\right)\delta(r-a), \tag{11.4.16}$$

where C_1 and C_2 are constants and $\delta(r)$ is the Dirac delta function.

The constants can be determined from the conditions $\mathbf{v}(\mathbf{r}) = 0$ at $r \le a$ and $p(\mathbf{r}) = 0$ at $r < a$, i.e., $C_1 = -\dfrac{3\mu}{2a}$ and $C_2 = 1$. Next we shall show that the induced force density defined as (11.4.16) leads to velocity and pressure expressions which are solutions of the Stokes steady problem. For the sake of brevity the long calculations are omitted again (see Protodiakonov et al., 1987), and only the final results are given:

$$\mathbf{v}(\mathbf{r}) - \mathbf{v}_\infty = C_1\begin{cases}\dfrac{2a}{3\mu}\mathbf{v}_\infty, & r<a\\[2ex] \left\{\dfrac{a^2}{2r\mu}\left(\mathbf{1}+\dfrac{\mathbf{rr}}{r^2}\right) + \dfrac{a^4}{6r^3\mu}\left(\mathbf{1}-3\dfrac{\mathbf{rr}}{r^2}\right)\right\}\mathbf{v}_\infty, & r>a\end{cases} \tag{11.4.17}$$

$$p(\mathbf{r}) - p_\infty = \begin{cases}-p_\infty, & r<a\\ C_1 a^2 r^{-3}(\mathbf{r}.\mathbf{v}_\infty), & r>a\end{cases}. \tag{11.4.18}$$

The formulae (11.4.17) and (11.4.18) represent the solution of the Stokes steady problem (Landau and Lifshitz, 1959) in Cartesian coordinates. The force acting on the spherical particle from the fluid side, can be easily calculated when $\omega = 0$ is put in (11.4.15) and using (11.4.16) - (11.4.18):

$$\boldsymbol{\Phi} = -\int_S \left(-\frac{3\mu}{2a} \mathbf{v}_\infty + p_\infty \mathbf{n} \right) dS = 6\pi\, \mu\, a \mathbf{v}_\infty \,. \tag{11.4.19}$$

The obtained result coincides with the Stokes drag formula (3.2.1).

11.5. Unsteady Motion of an Axisymmetric Body in a Viscous Fluid

First, in this section we shall give an analogous to Payne and Pell's (1960) result for the force on an arbitrary axisymmetric body in oscillating flow. Further on, we shall use this result to calculate the generalised Basset force, whose first component corresponds to the well known Basset force for the sphere problem, while its second component has a new memory integral function dependent on geometry for arbitrary velocity U(t) and it is different from $t^{-1/2}$.

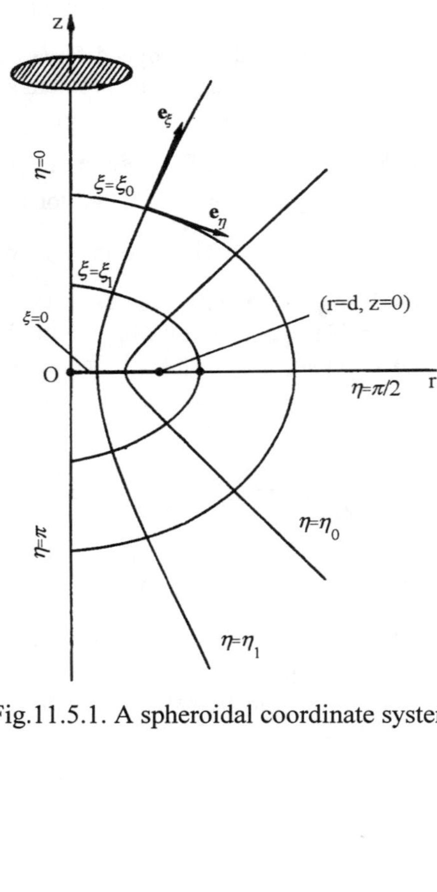

Fig.11.5.1. A spheroidal coordinate system.

Following the works of Lawrence and Weinbaum (1986, 1988) we shall analyse here as axisymmetric bodies spheroids with different aspect ratios and the nearly spherical body problem will be treated as a special case. As shown in Fig.11.5.1, a spheroidal coordinate system (ξ, η, φ) connected with the body is exploited. It is related to a cylindrical coordinate system (r, z, φ) with axis of symmetry Oz through the equations

$$r = d\cosh\xi\,\sin\eta\,, \qquad z = d\sinh\xi\cos\eta, \tag{11.5.1}$$

where d is the focal radius. The results presented here will concern the oblate spheroid case with semi-axes a > b, where $a = d\cosh\xi_0$, $b = d\sinh\xi_0$. Then the spheroid surface is the surface coordinate $\xi = \xi_0$ or $\alpha = \alpha_0$, where $\alpha = \sinh\xi$ and $\alpha_0 = \left[\left(\frac{a}{b}\right)^2 - 1\right]^{-1}$.

The flow domain is the spheroid exterior, i.e., $\xi_0 < \xi < \infty$, $0 \le \eta \le \pi$, $0 \le \varphi < 2\pi$. Due to the axial symmetry only the coordinates ξ and η will be taken into account.

Here we shall consider the case when the spheroid performs translational oscillations along its axis of symmetry with velocity $U\cos\omega t$, while the fluid is stationary at infinity. The conventional stream function is connected to the velocity $\mathbf{v} = \left(\hat{v}_\xi \exp(-i\omega t), \hat{v}_\eta \exp(-i\omega t)\right)$ by the relation (1.3.17), which in the spheroidal coordinates has the form

$$\hat{v}_\xi = -\frac{h}{r}\frac{\partial\Psi}{\partial\eta}, \qquad \hat{v}_\eta = \frac{h}{r}\frac{\partial\Psi}{\partial\xi}, \tag{11.5.2}$$

where $h = \dfrac{1}{d\left(\cosh^2\xi - \sin^2\eta\right)^{1/2}}$.

The dimensionless governing equation for the stream function (11.3.2) is

$$E^4\Psi - c^2E^2\Psi = 0, \tag{11.5.3}$$

where $c^2 = -\dfrac{i\omega d^2}{\nu}$, d is used as a length scale and U, as velocity scale. The boundary conditions for the stream function on the body surface are

$$\Psi = -\frac{1}{2}\left(1 + \alpha_0^2\right)\sin^2\eta, \qquad \Psi_\alpha = -\alpha_0\sin^2\eta \qquad \text{on } \alpha = \alpha_0, \tag{11.5.4}$$

while at infinity $\Psi \to 0$. The symmetry condition for Ψ implies that $\Psi = 0$ and Ψ_β is finite at $\eta = 0$ and $\eta = \pi$ with $\beta = \cos\eta$. Here the symbols Ψ_α and Ψ_β denote the derivatives of Ψ with respect to α and β.

The solution of (11.5.3) is sought as a superposition of two stream functions

$$\Psi = \Psi^P + \Psi^D, \tag{11.5.5}$$

the first of which Ψ^P is a potential function, i.e., $E^2\Psi^P = 0$ and the second Ψ^D is a diffusive function subjected to the equation $\left(E^2 - c^2\right)\Psi^D = 0$. Both terms Ψ^P and Ψ^D

are assumed to satisfy the respective boundary conditions. Then their solutions are given by

$$\Psi^{P} = \sqrt{(1-\beta^{2})(1+\alpha^{2})}\sum_{n=1}^{\infty} A_{n} P_{n}^{1}(\beta) Q_{n}^{1}(i\alpha), \tag{11.5.6}$$

$$\Psi^{D} = \sqrt{(1-\beta^{2})(1+\alpha^{2})}\sum_{n=1}^{\infty} B_{n} S_{1n}(c,\beta) R_{1n}^{(3)}(c,i\alpha), \tag{11.5.7}$$

where A_n and B_n are zero for even values of n; P_n^1 and Q_n^1 are respectively the first and second kind associated Legendre functions first order with P_n^1 given by (5.7.3). The functions S_{1n} are the first kind prolate angle spheroidal wave functions of order one, which are connected with the associated Legendre functions by the relation (Abramowitz and Stegun, 1965)

$$S_{1n}(c,\beta) = \sum_{r=0}^{\infty}{}' d_r^{1n}(c) P_{1+r}^{1}(\beta), \tag{11.5.8}$$

where the prime of summation stands for odd or even values of r with respect of the parity of the (n-1). The functions $R_{1n}^{(3)}$ are the third kind radial spheroidal wave functions of order one, which are connected with the Bessel functions and their explicit definition can be found again in the book of Abramowitz and Stegun (1965).

The no-slip boundary condition on the body surface (11.5.4) when applied to (11.5.5), with (11.5.6) and (11.5.7) taken into account, and after some rearranging leads to a coupled infinite system of linear equations for the unknown coefficients A_n and B_n

$$A_m + \sum_{n=1}^{\infty}{}' \overline{B}_n d_{m-1}^{1n}(c) / Q_m^1(i\alpha_0) = \delta_{1m}\left[\frac{1}{2}\sqrt{(1+\alpha_0^2)} / Q_m^1(i\alpha_0)\right], \tag{11.5.9}$$

$$A_m + \sum_{n=1}^{\infty}{}' \overline{B}_n d_{m-1}^{1n}(c) \frac{DR_{1n}^{(3)}(c,i\alpha_0)}{R_{1n}^{(3)}(c,i\alpha_0) DQ_m^1(i\alpha_0)} = \delta_{1m}\left[\alpha_0 / DQ_m^1(i\alpha_0)\right], \tag{11.5.10}$$

where $\overline{B}_n = B_n R_{1n}^{(3)}(c,i\alpha_0)$, δ_{1m} is the Kronecker symbol and D is the differential operator

$$Df(x) = \frac{d}{dx}\left[f(x)\sqrt{x^2-1}\right], \quad Df(ix) = \frac{d}{dx}\left[f(ix)\sqrt{x^2+1}\right] \tag{11.5.11}$$

with x being a real variable. The above system (11.5.9) - (11.5.10) can be solved simultaneously by truncation and matrix inversion (Lawrence, 1986).

Similarly to the velocity, the force on the spheroid is supposed to have the form $F = \hat{F}\exp(-i\omega t)$, where the amplitude $\hat{F}$ given by (Lawrence and Weinbaum, 1986) is

$$\hat{F} = c^2 V + 4\pi c^2 \lim_{R\to\infty} \frac{R^3}{r^2}\Psi, \tag{11.5.12}$$

with $V = \frac{4}{3}\pi\alpha_0(1+\alpha_0^2)$ the dimensionless spheroid volume and $R = \sqrt{r^2+z^2}$ the distance from the coordinate system origin. The expression (11.5.12) is analogous to the result of Payne and Pell (1960) for steady Stokes flow, i.e., $F = 8\pi \lim_{R\to\infty} \frac{R}{r^2}\Psi$. However, the limit $c \to 0$ is singular and their formula cannot be obtained. The physical reason for this is that even for very slow oscillations the far field from the body is inviscid and the Oseen's flow has to be considered, too.

It occurs that only the potential part of the stream function Ψ^P gives any contribution in the force expression (11.5.12), while Ψ^D decays exponentially at infinity (Lawrence and Weinbaum, 1988). This contribution is connected only with the first coefficient A_1 of the solution series, i.e.,

$$\lim_{R\to\infty} \frac{R^3}{r^2}\Psi = \lim_{\alpha\to\infty} \frac{\alpha^3}{\alpha\left(1-\beta^2\right)}\Psi = -\frac{2}{3}A_1, \tag{11.5.13}$$

which together with (11.5.12) leads to the final form of the drag force

$$\hat{F} = \frac{4}{3}\pi c^2\left[\alpha_0\left(1+\alpha_0^2\right) - 2A_1\right]. \tag{11.5.14}$$

Further, we shall make a comparison between the drag force for the spheroid with the one for the nearly spherical body, discussed in details in (Lawrence and Weinbaum, 1986). The results presented here are dimensionalized with d (focal distance) as length scale. Instead of it, the major semi-axis a will be chosen as a more appropriate length scale, corresponding to a sphere radius. Since the drag force is referred to the steady Stokes drag on a sphere, namely $6\pi\mu Ua$. Then instead of c the convenient frequency parameter $M = c\frac{a}{d}$ will be used.

For small and large M accurate numerical calculations give the following asymptotic approximation of the force

$$-\hat{F} = F_s + MF_1 + O\left(M^2\right) \qquad \text{for } |M| \ll 1, \tag{11.5.15}$$

$$-\hat{F} = M^2 m_a + MB + O(1) \qquad \text{for } |M| \gg 1, \tag{11.5.16}$$

where

$$F_s = \frac{4}{3}\left[\alpha_0 - \left(\alpha_0^2 - 1\right)\cot^{-1}\alpha_0\right]^{-1},\ F_1 = F_s^2,\ m_a = \frac{2}{9}\left(1+\alpha_0^2\right)^{\frac{1}{2}} \frac{1-\alpha_0\cot^{-1}\alpha_0}{\alpha_0 - \left(\alpha_0^2+1\right)\cot^{-1}\alpha_0}$$

$$B = \frac{1}{3}\left[\frac{\alpha_0}{\left(1+\alpha_0^2\right)^{\frac{1}{2}}} - \frac{9}{2}m_a\right]^2 \left(1+\alpha_0^2\right)\left[1-\left(2+\alpha_0^2\right)\left(1+\alpha_0^2\right)^{-\frac{1}{2}}\sinh^{-1}\frac{1}{\alpha_0}\right]. \tag{11.5.17}$$

The term F_s corresponds to the Stokes steady drag, the term m_a - to the added-mass force, while F_1 and B are the first order corrections for low and high frequencies. The low frequency correction F_1 is quadratic to the zero order term. The Basset force is presenting in the high frequency expression, named as B and for the general case is different from F_1. The two different force functions given by (11.5.15) and (11.5.16) have the same slope and curvature, i.e., $B \equiv F_1$ at b = a and are close to each other in the limits 0.1 < b/a <10 (Lawrence and Weinbaum, 1988).

Here it is interesting to note, that for the nearly spherical body with semi-axes $a = b(1+\varepsilon)$, where ε is a small parameter, the dimensionless force is obtained (Lawrence and Weinbaum, 1986) in the form

$$-\hat{F} = F_s + MB + M^2 m_a + \phi(M,\varepsilon), \tag{11.5.18}$$

where $F_s = 1 - \frac{1}{5}\varepsilon + \frac{37}{175}\varepsilon^2 + O(\varepsilon^3)$, $\quad B = 1 - \frac{2}{5}\varepsilon + \frac{81}{175}\varepsilon^2 + O(\varepsilon^3)$,

$$m_a = \frac{1}{9}\left(1 + \frac{1}{5}\varepsilon - \frac{26}{175}\varepsilon^2\right) + O(\varepsilon^3), \qquad \phi(M,\varepsilon) = \frac{8\varepsilon^2}{175}\left(\frac{M^2}{3+3M+M^2}\right) + O(\varepsilon^3).$$

The coefficient F_s is the same as the exact solution for the steady force on a slightly oblate spheroid obtained by Oberbeek (Happel and Brenner, 1973). The added-mass term m_a is also the same as the Green's solution (Green, 1833).

Following the force presentation (11.5.18), it is suitable to express the force on a spheroid with semi-axes ratio in the range 0.1 < b/a <10 in an analogous manner

$$-\hat{F} = F_s + MB + M^2 m_a + M\Phi(M, b/a), \tag{11.5.19}$$

where $\Phi(M, b/a)$ is a complicated function tending to $(F_1 - B)$ for small M and to $O(M^{-1})$ for high M. When the aspect ratio b/a is close to unity, the difference $(F_1 - B)$ is almost zero and the term $\Phi(M, b/a)$ will have very small effect on the total force value. A detailed numerical analysis and some approximations to the last term can be found in (Lawrence and Weinbaum, 1988). The respective term for the nearly spherical body $\phi(M,\varepsilon)$ is of order $O(M^2)$ for small M and of O(1) for large M and its contribution to the force is small.

The semi-axes aspect ratio has a significant influence on the other terms of the representation (11.5.18). At large b/a the added-mass term is small, while the Stokes drag and the Basset force are increasing functions of the body length. Due to the analysis made in (Lawrence and Weinbaum, 1988), we can conclude that for long bodies the induced force depends significantly on the oscillation frequency. At low frequency the flow is almost steady and the force is mainly from the steady Stokes drag. At high frequencies the flow approaches the potential flow behaviour and the force is due to the added-mass term. In the intermediate frequency range the body ends and the far flow do not contribute to the force and it results from the Basset and phase-changing force terms. Batchelor (1954) and Hasimoto (1955) reached similar conclusions, but for infinite cylinders case.

Further on, we shall give the analogous expression of the force for the nearly spherical body moving with an arbitrary time-dependent velocity **U**(t), supposed to be zero for $t \le 0$. Then the equation (11.5.3) be can regarded as a Laplace transform of the unsteady Stokes equation (11.2.1) with transformation variable $s = M^2$. Exploiting the

Basset technique for the sphere problem cited in section 11.2., we can invert (11.5.18) to obtain the dimensionless unsteady force **F**(t):

$$-\mathbf{F}(t) = F_s \mathbf{U}(t) + \frac{B}{\sqrt{\pi}} \int_0^t \frac{dU}{d\tau} \frac{d\tau}{\sqrt{t-\tau}} + m_a \frac{dU}{dt} + \frac{(F_1 - B)}{\sqrt{\pi}} \int_0^t \frac{dU}{d\tau} G(t-\tau), \quad (11.5.20)$$

where $G(t) = \frac{8\varepsilon^2}{175} \mathrm{Im}\left\{\left(\frac{1}{3}\pi\alpha\right)^{1/2} \exp(\alpha t)\mathrm{erfc}(\alpha t)^{1/2}\right\}$, $\alpha = \frac{3}{2}\left(1 + i\sqrt{3}\right)$.

The first three terms are named as the corresponding ones for the sphere case, i.e., the steady Stokes drag, Basset force and added-mass force and tend exactly to them in the limit $\varepsilon \to 0$. The last term of (11.5.20) is a new memory integral representing the history of body motion, which is absent from the drag force formula for the sphere (11.2.22). Its kernel function G(t) is different from the Basset force kernel $t^{-1/2}$. Using the properties of G(t) (Abramowitz and Stegun, 1965), it is obtained that for small times G(t) is finite, i.e., $G(0) = \frac{1}{2}(3\pi)^{\frac{1}{2}}$ and its slope is infinite, while for large times decays as $\frac{1}{4\sqrt{3t^3}}$. For a spheroid with arbitrary semi-axes aspect ratio, this function is more complicated and only some correlation approximations can be given for it (Lawrence and Weinbaum, 1988).

From the above discussion we can conclude that the sphere is a special case when the force can be spilt into parts depending only on the frequency independently of the shape. However, for more complicated geometry, this procedure is inapplicable and the force contains a term accounting for the coupled effect between time dependence and shape.

11.6. Dynamics of a Spherical Bubble. The Rayleigh-Plesset Equation.

When liquid containing dissolved gas is subjected to sudden decompression, cavitation (gas bubbles formation and growth) is observed. The governing equation for this and other related problems was formulated by Rayleigh (1917) and Plesset (1949). The complete solution to the Rayleigh-Plesset equation was obtained for the inviscid case by Ma and Wang (1962). The cavitation pressure of special gas bubbles was correlated to the fluid viscosity in a theoretical investigation of the viscous Rayleigh-Plesset equation by Chang and Chen (1986).

The object of the present section is connected with the inertia-dominant dynamics of a spherical gas or vapour bubble in an incompressible liquid. One of the purposes of this section is to derive the famous Rayleigh-Plesset equation.

Let us suppose that inside the bubble the gas pressure p_b changes with time according to some known law. Then the bubble radius R is also a function of time. If the velocity of the flow caused by the expanding bubble is not large, the liquid could be treated as incompressible.

Since the expanding of a single bubble in an infinite fluid has a spherical symmetry, it is convenient to choose a spherical coordinate system (r, θ, φ) with an origin in the bubble centre. Then the velocity field of the surrounding liquid will have only a radial component, i.e., $v_r = v_r(r, t)$, $v_\theta = v_\varphi = 0$. Due to the flow symmetry from the Navier-Stokes equations (1.3.1) and continuity equation (1.3.2), we get for the radial velocity component the following two equations:

$$\frac{\partial v_r}{\partial r} + \frac{2\, v_r}{r} = 0, \tag{11.6.1}$$

$$\frac{\partial\, v_r}{\partial\, t} + v_r \frac{\partial v_r}{\partial\, r} = -\frac{1}{\rho}\frac{\partial p}{\partial\, r} + \nu\left(\frac{\partial^2 v_r}{\partial\, r^2} + \frac{2}{r}\frac{\partial v_r}{\partial\, r} - \frac{2 v_r}{r^2}\right). \tag{11.6.2}$$

For the interface motion velocity, we have

$$v_r(r,t) = \frac{dR}{dt}, \text{ at } r = R, \tag{11.6.3}$$

while the pressure at infinity $r \to \propto$ is the atmospheric pressure, $p(\propto, t) = p_{at}$.

Integrating (11.6.1), we get

$$v_r\, r^2 = C_1(t),$$

where $C_1(t)$ is obtained from (11.6.3) to be $C_1(t) = \dot{R}R^2$ (here the upper dot denotes differentiating with respect to time). Therefore the radial velocity becomes

$$v_r(r,t) = \frac{C_1(t)}{r^2} = \frac{\dot{R}R^2}{r^2}.$$

In order to obtain the equation of the bubble radius change we find the derivatives

$$\frac{\partial\, v_r}{\partial\, t} = \frac{\ddot{R}R^2 + 2R\dot{R}^2}{r^2}, \quad \frac{\partial\, v_r}{\partial r} = \frac{-2\dot{R}R^2}{r^3} \quad \frac{\partial^2\, v_r}{\partial\, r^2} = \frac{6\dot{R}R^2}{r^4}$$

and after substitution in (11.6.2) we reach the following equation for the pressure

$$\frac{1}{\rho}\frac{\partial p}{\partial\, r} = -\frac{\ddot{R}R^2 + 2R\dot{R}^2}{r^2} + \frac{2\dot{R}^2R^4}{r^5}.$$

Integrating it with respect to r, we obtain

$$\frac{p}{\rho} = \frac{R\left(\ddot{R}R + 2\dot{R}^2\right)}{r} - \frac{1}{2}\frac{\dot{R}^2R^4}{r^4} + C, \tag{11.6.4}$$

where the value of the constant C follows from the pressure condition at infinity $r \to \propto$, i.e., $C = \dfrac{p_{at}}{\rho}$. Finally (11.6.4) has the form

$$\frac{1}{\rho}\left(p - p_{at}\right) = \frac{R\left(\ddot{R}R + 2\dot{R}^2\right)}{r} - \frac{1}{2}\frac{\dot{R}^2R^4}{r^4}. \tag{11.6.5}$$

The normal stresses boundary condition (1.4.18) is written as

$$T_{rr} = \hat{T}_{rr} + \frac{2\sigma}{R}, \tag{11.6.6}$$

where the stresses inside the bubble are denoted with a cap.

Since the gas viscosity inside the bubble is much smaller than the liquid viscosity and the gas motion inside the bubble can be neglected, the normal stresses are expressed by

$$T_{rr} = -p + 2\mu\frac{\partial v_r}{\partial r}, \qquad \hat{T}_{rr} = -p_b,$$

where p_b is the gas pressure inside the bubble. When the bubble expands, its pressure falls according to an adiabatic law. At the final stage of expansion the pressure p_b becomes much smaller than p_{at}. From (11.6.6) it follows that

$$p + 4\mu\frac{\dot{R}}{R} = p_b - \frac{2\sigma}{R}, \qquad \text{at } r = R,$$

or

$$p(R+0,t) = p_b - \frac{2\sigma + 4\mu\dot{R}}{R}.$$

Substituting r with R + 0 and p with p(R+0, t) in (11.6.5), we obtain the so-called Rayleigh-Plesset equation

$$R\ddot{R} + \frac{3}{2}\dot{R}^2 + \frac{2\sigma}{\rho R} + \frac{4\mu\dot{R}}{\rho R} = \frac{p_b - p_{at}}{\rho}. \tag{11.6.7}$$

In order to solve this differential equation, we need two initial (boundary) conditions. Usually R(0) and $\dot{R}(0)$ are prescribed and the equation (11.6.7) is solved numerically by means of some of the standard procedures at given function $p_b(t)$. Multiplying (11.6.7) by $\dot{R}R^2$ we get the first integral

$$\frac{1}{2}\frac{d}{dt}\left(R^3\dot{R}^2\right) + \frac{2\sigma}{\rho}\dot{R}R + \frac{4\mu}{\rho}R\dot{R}^2 = \frac{p_b - p_{at}}{\rho}R^2\dot{R}. \tag{11.6.8}$$

Neglecting the influence of viscosity, bubble pressure p_b change and surface tension σ, the integral of (11.6.8) gives

$$R^3\dot{R}^2 = \frac{2}{3}\frac{p_b - p_{at}}{\rho}\left(R_m^3 - R^3\right)R, \tag{11.6.9}$$

where R_m is the bubble maximum radius. The equation (11.6.9) describes the bubble behaviour in the critical stage of flow development, when the bubble ceases to expand and begins to shrink and $p_{at} >> p_b$. The relation between the radius of such effectively empty bubble and the time t is determined by the equality

$$\left|t - t_m\right| = \left(\frac{3\rho}{2p_{at}}\right)^{1/2}\int_R^{R_m}\frac{dR}{\left[\left(\frac{R_m}{R}\right)^3 - 1\right]^{1/2}}, \tag{11.6.10}$$

where t_m is the moment in which $R = R_m$. It is interesting to note that the formula (11.6.10) is applicable for the stage of expansion $t < t_m$, as well as for the shrinking stage $t > t_m$. The motion at shrinking is obtained by reversing the direction of motion at expansion.

R_m can be evaluated in the case of an inviscid fluid. Then the bubble is considered as effectively empty ($p_b = 0$) and the energy loss is neglected because of the distance and heat transfer.

Let us apply equation (11.6.7) to study the gas bubble shrinkage. At high pressure gradient regions (caverns) filled with vapour can develop in the liquid phase. If afterwards the fluid pressure starts to increase, the caverns soon shrink and collapse. In the simplest case it can be supposed that the collapsing cavern preserves its spherical shape and the pressure difference $p_{at} - p_b$ remains constant. At these assumptions the cavern radius dependence on time t is given by the equation (11.6.10). After performing numerical integration for the moment t_0 when the radius becomes zero, we find this moment to be

$$t_0 = 0.915\, R_m \left(\frac{\rho}{p_{at} - p_b} \right)^{1/2}.$$

For the relative quantities t/t_0 and R/R_m we get the relation

$$\frac{t}{t_0} = 1.34 \int_{R/R_m}^{1} \frac{dx}{\left(x^{-3} - 1\right)^{1/2}}, \tag{11.6.11}$$

which correlates well with the observations of collapsing caverns at $p_{at} - p_b = \text{const}$. The analysis of the final stage of collapse requires complicating the model by means of including other important physical processes.

We shall note that the Rayleigh-Plesset equation is strictly applicable only to isothermal growth of gas bubbles driven by mechanical pressure differences. For evaporation-driven vapour bubble growth in a superheated liquid, Forster and Zuber (1954) added an integral term to the Rayleigh-Plesset equation, which accounts for the heat transfer effects.

When the bubble radius oscillates close to the stable equilibrium, the bubble dynamics can be approximated by a weakly nonlinear oscillator and it is possible to use perturbation techniques such as the method of multiple time scales. A complete resonance analysis of the Rayleigh-Plesset equation has been given by Francescutto and Nabergoj (1983). It is worth noting that Lauterborn (1991) observed experimentally that the sound produced by the bubble has a component at half of the frequency of the acoustic wave applied to the liquid.

When the bubble deviation from the equilibrium radius is not small or the forcing amplitude is not small, the bubble dynamics can only be studied using numerical integration of the Rayleigh-Plesset equation. Numerical investigation of the bifurcation structure based on this equation was given by Lauterborn (1976). Although an explicit solution for the bubble radius as a function of time can only be obtained numerically, Ma and Wang (1962) found some interesting qualitative results by analytically examining the solution trajectories in the phase space (R, $\dot{R}$). In fact, Ma and Wang have shown that the solution trajectories in the phase space (R, $\dot{R}$) can be derived from a Hamiltonian function in the inviscid limit. The discovery of this Hamilton function tremendously simplifies the analysis for the free oscillations of the gas and vapour bubbles. Using Melnikov (1963) theory to perturb the first integral of the inviscid equation, Chang and Chen (1986) studied the breakup of the homoclinic orbits due to perturbations in the external pressure and confirmed the correlation between the cavitation pressure and the fluid viscosity discovered experimentally by Bull (1956).

If the forcing frequency is near one half of the bubble resonance frequency, the experimental data deviate considerably from the quantitative predictions of the Rayleigh-Plesset equation. The reasons for this discrepancy are mainly connected with the facts that the Rayleigh-Plesset equation assumes the liquid to be incompressible and that the spherical shape of the bubble can be unstable at large oscillation amplitudes. Thermal effects as energy loss due to heating and cooling of the gas can also be a reason for such kind of discrepancy (Flynn, 1975). Models that include liquid compressibility and thermal effects in the spherical bubble dynamics have been recently summarised by Prosperetti (1993).

11.7. Shape Oscillations of a Drop or a Bubble. Coupling between Volume and Shape Oscillations for a Gas or a Vapour Bubble

The first investigation of the problem about the oscillations of a drop in an inviscid fluid under the action of interface tension is done by Rayleigh (1892). Lamb (1945) found the natural frequencies of the oscillations of an inviscid drop in another inviscid fluid. The oscillations are induced by the action of the surface tension of the drop interface. Lamb (1945) considered also the first approximation of the oscillations of a low viscosity drop. For arbitrary values of viscosity Chandrasekhar (1959) has given detailed results for the aperiodic modes of decay for n = 1, 2, 3 and 4, where n is the order of the spherical harmonic deformation considered. A full study of the oscillations of a viscous drop in a viscous fluid is performed by Reid (1960) and Miller and Scriven (1968).

It is well known that the oscillation of a bubble can be treated using the linear theory as a superposition of its normal modes. If n is the order of the Legendre polynomial $P_n(\cos\theta)$ then each normal mode is associated with the corresponding polynomial (Lamb, 1945). The mode corresponding to the lowest order polynomial (n = 0) is related to the oscillation of the bubble volume, while all higher order modes are referred to as shape modes. The volume mode (n = 0) is often called volume mode or "breathing mode". Since each mode oscillates at its own resonance frequency, the last frequency is called natural or fundamental. All higher order modes are also referred as shape modes or asymmetric modes.

The newly created bubbles are normally deformed from a spherical shape during the generation and injection process. This is the principle origin of bubble-related noise in oceans. As a result of the newly created bubbles the surface waves of relatively large amplitude exist on these bubbles.

The breathing frequency ω_0 and shape frequency ω_n (n > 0) can be calculated by the following formula

$$\omega_n = \begin{cases} \dfrac{3\gamma\, p_0}{\rho\, a^2} - \dfrac{2\sigma}{\rho\, a^3}, & n = 0 \\ (n-1)(n+1)(n+2)\dfrac{\sigma}{\rho\, a^3}, & n > 0 \end{cases} \tag{11.7.1}$$

where p_0 is the ambient pressure, γ is the ratio of the specific heats, a is the bubble radius, ρ is the liquid density and σ is the surface tension. The coupling between the volume mode and shape modes reaches maximum at resonance

$$\omega_0^2 = 4\omega_n^2 \quad (n = 2, 3, \ldots). \tag{11.7.2}$$

Longuet-Higgins (1989a) developed a second-order theory of the second-harmonic generation by quadratic coupling. He examined how shape oscillations of a bubble with natural frequency ω_n can excite the breathing mode which radiates sound at $2\omega_n$. Longuet-Higgins (1989b) gave also the transient solution for given initial shape distortion and calculated the damped oscillation of the second-order breathing mode and the radiated sound. He considered various damping mechanisms: acoustic, thermal and viscous, so that the amplitude of the breathing mode is finite even if the resonance condition $\omega = 2\omega_n$ is satisfied exactly, where ω_n is the fundamental (natural) frequency of the n-th distortion mode.

In an experimental research on the behaviour of small gas bubbles in liquids, subject to acoustic standing waves, Gaines (1932) and Kornfeld and Suvorov (1944) observed a curious phenomenon. It was called "dancing bubbles" by Gaines, the bubbles moved in zig-zag paths reminiscent of Brownian motion.

The parametric excitation of shape oscillation in free bubbles pulsating radially in a sound field was proposed by Benjamin and Strasberg (1958), and Strasberg and Benjamin (1958) as a cause of the erratic motions observed. The precise reason why the excited bubbles may be propelled along an erratic path has been exhibited by Benjamin and Ellis (1990). Their analysis is based on an exact integral relation derived first by Benjamin (1987).

Chaotic oscillations of bubbles were first shown in the pioneering works of Lauterborn and collaborators (see Lauterborn and Parlitz (1988) for a review) who studied isotropic modes of a spherical bubble with finite oscillation amplitude. The work of Mei and Zhou (1991) suggests that simple harmonic forcing can also excite chaotic oscillations of the isotropic mode through its interaction with one or more shape modes, and can lead to radiation of random signals to the far field. These authors show that a combination with the theory of Benjamin and Ellis (1990) provides a possible explanation of the erratic dancing of bubbles caused by sound.

It is worth noting that a major shift in our recent understanding of the dynamics of gas or vapour bubble has been the realisation of the importance of coupling between purely volume oscillations of the spherical bubble theories, and one or more of the shape modes. Mei and Zhou (1991) studied also the resonant interaction between a spherical (pure volume) oscillation and one or two distorial modes for a bubble in water when spherical mode is forced by periodic oscillation of an isotropic pressure field. They have found that in the considered case there exist chaotic bubble oscillations but they do not explore the bifurcation sequence that leads to this chaotic motion. Feng and Leal (1994) have discovered that when the pressure forcing is anisotropic the bubble response is quite different. They have found that for sufficiently large amplitudes of oscillation the volume mode may lose stability and the subsequent instability may lead to chaotic oscillations of both the volume and shape modes.

In the so-called normal-mode technique (Chandrasekhar, 1961) the separation of the time variable is conducted. Although perfectly justified from a mathematical view point, this technique has very strong practical limitations in the solution of initial-value problems because of the complexity of the operations required. That is why an alternative technique based on the use of Laplace transform has been developed by Prosperetti (1976, 1980a, b). By use of this technique it was shown that the free oscillations of droplets and bubbles about the spherical shape cannot be represented in terms of a single value of the frequency and the damping parameter. If the interface separating two fluids has the form

$$r(\theta,\varphi) = a + a_n(t) Y_n^m(\theta,\varphi), \tag{11.7.3}$$

then rather a representation in terms of modulated oscillations of the form $a_n(t) \sim \exp[-b(t) \pm \omega(t)]t$ would appear, to represent more closely the process, both mathematically and physically. Explicit expressions for $\omega(t)$ and b(t) at $t \to 0$ are available only for two cases, the free drop and the gas bubble.

A straightforward estimate of the order of magnitude of the convective term in the momentum equation shows that it will be negligible compared with $\frac{\partial u}{\partial t}$ provided that $|a_n(t)| << a$, as it is to be expected. In this case second order effects, such as microstreaming are negligible (see for example Riley, 1967 and chapter 13 of the present book).

Asaki and Marston (1994) developed an extension of the previous theoretical models (Miller and Scriven, 1968; Marston, 1980; Prosperetti, 1980a, b), which describe more accurately the gas bubbles dynamics in a liquid and the liquid drops dynamics in air. In addition to the solutions of the characteristic equation, they presented (in support to the new calculations) an experimental study of the free decay of small-amplitude quadrupole shape oscillations of bubbles in clean water. Asaki and Marston (1995) derived their characteristic equation for the decay of shape oscillations by linearisation of the Navier-Stokes equations, requiring the amplitude of distortions to be small compared to the bubble or drop size. The obtained damping rates give insight into the importance of nonlinear mode coupling rates approximated for bubbles in inviscid fluids (Longuet-Higgins, 1992; Yang et al., 1993).

The dynamics of gas and vapour bubbles is strongly influenced by the mean bubble deformation. From a theoretical point of view, the problem of bubble volume and shape oscillations in the presence of mean deformation is very difficult and as a consequence of this, analytical formulae for the bubble shape are only possible when the mean deformation is asymptotically small. When the mean deformation is not small, in order to determine the bubble shape, numerical calculations must be performed. The dynamics of a deformable bubble in quiescent fluid due to impulsive application of axisymmetric, non-uniform pressure at the bubble surface were first studied by Yang et al. (1993) and Feng and Leal (1993). The more realistic problems that have been studied are the shape oscillations of gas bubbles in a flow. Using a domain perturbation technique Feng (1992) considered the shape oscillations of a gas bubble in an uniform flow. Kang and Leal (1988a, 1989, 1990, 1991) have studied the shape oscillations of a bubble in a uniaxial straining flow.

The numerical simulation of an oscillating bubble is complicated by the fact that the discretization must follow the evolving free surface within the calculation domain. Using the boundary integral method, which is suited very well to this class of problems, McDougald and Leal (1994, 1996) and Chahine (1994) have studied the single bubble dynamics in an infinite fluid. However, the transformation through Green's functions from the equations of motion in differential form to an equivalent form involving only boundary integrals is possible only in the creeping and potential flow limits. On the other hand, many of the most commonly encountered bubble dynamics problems involve air or vapour bubbles in water, where Reynolds number is actually quite large. That is why the restriction of boundary integral method to potential bubble dynamics flow problems precludes an exact accounting of the role of viscous effects in dynamics of oscillating bubbles.

A main difficulty in solving the exact Navier-Stokes equations with appropriate boundary conditions for bubble dynamics problems arises from the changing in time problem geometry as the bubble oscillates in volume and shape. Using boundary-fitted orthogonal coordinates to map the physical domain to a square computational grid, Ryskin and Leal (1984a, b, c) applied the finite difference method for solving steady free-boundary problems. This method of simulation was generalised by Kang and Leal (1987, 1989) to time dependent bubble deformation without volume change. Since this simulation technique was found to exhibit some artificial dissipation that gave rise to a small phase lag in the bubble response to externally forced changes of shape, all comprehensive studies of time-dependent motions of compressible bubbles have so far been studied in the inviscid limit via the boundary integral method.

CHAPTER 12.

Application of the Singularity Method for Unsteady Flows Past Rigid or Fluid Particles

12.1. Introduction

The singularity method for the computation of unsteady viscous flows involving rigid or fluid particles was developed by direct analogy with the application of the same method for the corresponding problems of steady Stokes flows. This has been done by Williams (1966), Kanwal (1971), Kim and Russel (1985), Pozrikidis (1989a) and many others. As in the steady case the basic idea is to represent the flow in terms of discrete or continuous distributions of fundamental solutions to the governing unsteady Stokes equations. Depending on the nature of the flow and the geometry of the particles, the type of fundamental solutions is determined. Approximate solutions for slender particles can be obtained by using a line distribution of fundamental solutions along a properly defined particle axis. In order to account for inter-particle or particle-boundary interactions, it is convenient to combine the singularity method with the method of reflections (Kim and Russel, 1985).

We shall note that for the unsteady case the specific equations for the unsteady singularities including the oscillating Stokeslet and dipole are more involved than the corresponding formulae for the steady case and the computation of singularity representations becomes considerably more cumbersome. That is why exact singularity solutions for unsteady Stokes flows are limited to those for flow induced by the longitudinal or rotational oscillations of a rigid or fluid spherical particle, and for the flow due to a rigid spherical particle embedded in a linear ambient flow (Kim and Karrila, 1991, Chapter 6; Pozrikidis, 1989a). Approximate singularity representations for the flow due to the longitudinal oscillations of a prolate spheroidal particle have been obtained, as well (Pozrikidis, 1989a,b).

Kim (1985b) and Pozrikidis (1989a) have shown that the derivation of the Lorentz reciprocal theorem for unsteady flows is quite similar to that given for steady flows and the procedure of obtaining the integral representation for unsteady disturbance fields is analogous to that used for steady flows. Furthermore, using Kim's findings (Kim, 1985b) for the remarkable connection between the singularity representation and Faxen's laws for steady flow Pozrikidis (1989a) has shown that the force on a particle may be derived from the total strength of the distributed oscillating Stokeslets necessary to represent the flow, i.e., he derived Faxen's laws for rigid and fluid particles in terms of unsteady singularity representations.

The application to unsteady drop deformations of the boundary integral method presented in section 5.11. makes possible the thorough analysis of the complicated phenomena of drop breakup and drop stability. A good review of the most recent advances in this topic is presented by Stone (1994).

12.2. Unsteady Fundamental Solutions

We shall recall the time-dependent Stokes equation (1.3.9). For the oscillating flow it is convenient to put in it

$$\mathbf{v} = \hat{\mathbf{v}} \exp(-\mathrm{i}\omega t) \text{ and } \mathrm{p} = \hat{\mathrm{p}} \exp(-\mathrm{i}\omega t), \tag{12.2.1}$$

where ω is the angular velocity of oscillations, and afterwards to reach to

$$\mu \nabla^2 \hat{\mathbf{v}} - \mu \alpha^2 \hat{\mathbf{v}} = \nabla \hat{\mathrm{p}}, \tag{12.2.2}$$

where $\alpha^2 = -\dfrac{\mathrm{i}\omega}{\nu}$. Using a for a characteristic length scale and $\mathrm{M}^2 = -\dfrac{\mathrm{i}\omega\, \mathrm{a}^2}{\nu}$ as a dimensionless group, we transform equation (12.2.2) in the dimensionless form

$$\left(\nabla'^2 - \mathrm{M}^2\right)\hat{\mathbf{v}}' = \nabla \hat{\mathrm{p}}'. \tag{12.2.3}$$

Further on, for simplicity, we shall drop the primes but from the context it will be clear if we are dealing with dimensional or dimensionless variables.

In order to compute the free-space Green's function corresponding to the unsteady Stokes flow, we shall represent the solutions of the continuity equation and the singularity forced unsteady Stokes equation

$$\left(\nabla^2 - \mathrm{M}^2\right)\hat{\mathbf{v}} = \nabla \hat{\mathrm{p}} - \mathbf{F}^{(c)} \delta\left(\mathbf{r} - \boldsymbol{\rho}^0\right), \tag{12.2.4}$$

where $\mathbf{F}^{(c)}$ is a constant point force, $\hat{\mathbf{r}} = \mathbf{r} - \boldsymbol{\rho}^0$ and $\mathrm{R} = |\hat{\mathbf{r}}|$. Proceeding as in section 2.3., when dealing with Green's function for the steady Stokes flow, we can derive the pressure and velocity expressions by making use of the delta function. Similarly to (2.3.11) and (2.3.28), here we have

$$\nabla^2\left(\frac{1}{\mathrm{R}}\right) = -4\pi\, \delta(\hat{\mathbf{r}}), \tag{12.2.5}$$

and

$$\hat{\mathrm{p}} = \mathbf{F}^{(c)} . \left[-\frac{1}{4\pi} \nabla\left(\frac{1}{\mathrm{R}}\right)\right]. \tag{12.2.6}$$

Substituting the latter in (12.2.4) gives

$$\left(\nabla^2 - \mathrm{M}^2\right)\hat{\mathbf{v}} = -\frac{1}{4\pi} \mathbf{F}^{(c)} . \left(\nabla\nabla - \mathbf{I}\nabla^2\right)\left(\frac{1}{\mathrm{R}}\right). \tag{12.2.7}$$

The solution of this equation we seek in the form

$$\hat{\mathbf{v}} = \mathbf{F}^{(c)} . \left(\nabla\nabla - \mathbf{I}\nabla^2\right)\mathrm{H}(\mathrm{R}), \tag{12.2.8}$$

where H(R) is a unknown scalar function. Substituting (12.2.8) into (12.2.7), the following equation is found

$$\left(\nabla^2 - M^2\right)\left[\mathbf{F}^{(c)}.\left(\nabla\nabla - \mathbf{I}\nabla^2\right)H(R)\right] = -\frac{1}{4\pi}\mathbf{F}^{(c)}.\left(\nabla\nabla - \mathbf{I}\nabla^2\right)\left(\frac{1}{R}\right), \quad (12.2.9)$$

which is satisfied by any solution of the equation

$$\left(\nabla^2 - M^2\right)H(R) = -\frac{1}{4\pi R}. \quad (12.2.10)$$

Taking into account (2.3.10) we find that H(R) is the fundamental solution of the modified biharmonic equation

$$\nabla^2\left(\nabla^2 - M^2\right)H(R) = \delta(\hat{\mathbf{r}}). \quad (12.2.11)$$

Since the solution of this equation is (Pozrikidis, 1992a)

$$H(R) = \frac{1}{4\pi M^2 R}\left[1 - \exp(-MR)\right] = \frac{1}{4\pi Ms}\left[1 - \exp(-s)\right],$$

where s = MR, we have

$$\hat{\mathbf{v}} = \frac{1}{8\pi}\mathbf{F}^{(c)}.\left(\nabla\nabla - \mathbf{I}\nabla^2\right)\left\{\frac{2}{M^2 R}\left[1 - \exp(-MR)\right]\right\}. \quad (12.2.12)$$

Introducing

$$\nabla^2 H = M^2 H - \frac{1}{4\pi R} = -\frac{1}{4\pi R}\exp(-MR)$$

in (12.2.10), simplifies the expression for $\hat{\mathbf{v}}$

$$\hat{\mathbf{v}} = \frac{1}{8\pi}\mathbf{F}^{(c)}.\left\{\left[2\exp(-s)\left(1 + \frac{1}{s} + \frac{1}{s^2}\right) - \frac{2}{s^2}\right]\frac{\mathbf{I}}{R} + \left[-2\exp(-s)\left(1 + \frac{3}{s^2} + \frac{3}{s}\right) + \frac{6}{s^2}\right]\frac{\hat{\mathbf{r}}\hat{\mathbf{r}}}{R^3}\right\}$$

$$= \frac{1}{8\pi}\mathbf{F}^{(c)}.\overline{\mathbf{B}}(s),$$

where the tensor $\overline{\mathbf{B}}$

$$\overline{B}_{ij} = A(s)\frac{\delta_{ij}}{R} + B(s)\frac{\hat{r}_i\hat{r}_j}{R^3}, \quad (12.2.13)$$

$$A(s) = 2\exp(-s)\left(1 + \frac{1}{s} + \frac{1}{s^2}\right) - \frac{2}{s^2}, \quad (12.2.14)$$

$$B(s) = -2\exp(-s)\left(1 + \frac{3}{s^2} + \frac{3}{s}\right) + \frac{6}{s^2}$$

is called the oscillating Stokeslet.

When $s \to 0$ the coefficients A and B tend to 1 and $\overline{\mathbf{B}}$ is reduced to the Oseen-Burgers tensor **B** for steady Stokes flow. In addition to that, at very large s, $\overline{\mathbf{B}}$ behaves like a potential dipole, suggesting that far away from the oscillating Stokeslet the flow tends to be irrotational. The stress tensor associated with the oscillating Stokeslet is denoted by $\overline{\mathbf{K}}$.

Expanding $\overline{\mathbf{B}}$ in an asymptotic series for small s = MR, we obtain the following relationship between the oscillating and steady Stokeslets

$$\overline{\mathbf{B}} = \mathbf{B} - \frac{4s}{3}\mathbf{I} + \frac{1}{4}s^2\left(\frac{3}{R}\mathbf{I} - \frac{\hat{\mathbf{r}}\hat{\mathbf{r}}}{R^3}\right) + \ldots,$$

where **I** is the unit tensor.

The pressure, stress and vorticity fields corresponding to an oscillating Stokeslet are, respectively (Pozrikidis, 1989a; Pozrikidis, 1997):

$$p = \frac{1}{8\pi}P_j F_j^{(c)}, \qquad T_{ik} = \frac{1}{8\pi}\overline{K}_{ikj}F_j^{(c)}, \qquad \omega_i = \frac{1}{8\pi}\Omega_{ij}F_j^{(c)}, \qquad (12..2.15)$$

where

$$P_j = 2\frac{\hat{r}_i}{R^3}, \qquad \Omega_{ij} = 2\varepsilon_{ijk}\frac{\hat{r}_k}{R^3}\exp(-s)(s+1), \qquad (12.2.16)$$

$$\overline{K}_{ikj} = -\frac{2}{R^3}\left(\delta_{ik}\hat{r}_j + \delta_{jk}\hat{r}_i\right)\left[\exp(-s)(s+1) - B(s)\right]$$

$$-\frac{2}{R^3}\delta_{ij}\hat{r}_k\left[1 - B(s)\right] - 2\frac{\hat{r}_i\hat{r}_k\hat{r}_j}{R^5}\left[5B(s) - 2\exp(-s)(s+1)\right].$$

Furthermore, the next fundamental solution of unsteady Stokes equations is the symmetric Stokeslet quadrupole $\overline{\mathbf{B}}^q$ obtained by twice differentiating of $\overline{\mathbf{B}}$ with respect to $\boldsymbol{\rho}^0$

$$\overline{B}_{ij}^q = -\frac{1}{2}\nabla_0^2\overline{B}_{ij} = -C(s)\frac{\delta_{ij}}{R^3} + D(s)\frac{3\hat{r}_i\hat{r}_j}{R^5}, \qquad (12.2.17)$$

where

$$C(s) = \exp(-s)\left(1 + s + s^2\right) \text{ and } D(s) = \exp(-s)\left(1 + s + \frac{s^2}{3}\right).$$

It may be shown that C(0) = D(0) =1. Expanding $\overline{\mathbf{B}}^q$ in an asymptotic series for small s = MR yields the following relationship between the oscillating and steady dipoles

$$\overline{\mathbf{B}}^q = \mathbf{D} - \frac{1}{2}M^2\mathbf{B} + \ldots,$$

where **D** and **B** are the steady potential dipole and Stokeslet, respectively.

Further singularities may be obtained by separating the dipole into symmetric and antisymmetric parts. In this fashion the oscillating stresslet and rotlet could be constructed. The resulting velocity due to an oscillating rotlet of strength $\mathbf{F}^{(c)}$ is expressed by

$$\hat{v}_i = \frac{1}{16\pi}\Omega_{ij}F_j^{(c)}, \qquad (12.2.18)$$

where $\boldsymbol{\Omega}$ was given by (12.2.16).

We shall note that the internal flows must be regular at the origin. In order to compute the oscillating Stokeson, we set the pressure, as a harmonic function, equal to

$p = 10\hat{r}_i F_i^{(c)}$, where $\hat{\mathbf{r}} = \mathbf{r} - \boldsymbol{\rho}^0$ and $\boldsymbol{\rho}^0$ is an arbitrary point in space and find that the velocity may be expressed in terms of the tensor $\overline{\mathbf{E}}(\hat{\mathbf{r}})$ (Pozrikidis, 1992a)

$$\hat{v}_i = \overline{E}_{ij} F_j^{(c)}, \tag{12.2.19}$$

where

$$\overline{E}_{ij} = 2\delta_{ij} R^2 Q(s) - \hat{r}_i \hat{r}_j W(s), \tag{12.2.20}$$

$$W(s) = \frac{15}{s^5}\left[\left(s^2+3\right)\cosh s - 3s\sinh s\right], \quad Q(s) = -\frac{1}{4s^5}\left[20s^3 - 30\left(s^2+1\right)\sinh s + 30s\cosh s\right].$$

It can be shown easily, that Q(0) = W(0) = 1.

12.3. The Reciprocal Theorem for Unsteady Flow. Integral Representations. Faxen Laws

The derivation of Lorentz reciprocal theorem for unsteady Stokes flow is similar to that for steady flow (see section 2.4). Since equation (12.2.3) is linear and elliptic, its solution may be expressed as an integral on the flow boundaries. Using the well established theory (Ladyzhenskaya, 1969), we shall consider two arbitrary non-singular dimensionless solutions $(\hat{\mathbf{v}}, \hat{\mathbf{T}})$ and $(\hat{\mathbf{v}}^*, \hat{\mathbf{T}}^*)$ to the unsteady Stokes equation (12.2.3) and write down the corresponding reciprocal relationship

$$\nabla.\left(\hat{v}_i^* \hat{T}_{ik} - \hat{v}_i \hat{T}_{ik}^*\right) = 0. \tag{12.3.1}$$

In the derivation of the theorem we have assumed that the velocity vanishes at infinity at a sufficiently fast rate. For the further analysis we select a point $\boldsymbol{\rho}^0$ in the flow domain and integrate the above equation in a volume occupying the space between the particle, a small spherical surface centred at the Stokeslet $\boldsymbol{\rho}^0$ and a very large spherical surface extending to infinity

$$\iiint_{V-V_\varepsilon} \nabla.\left(\hat{\mathbf{v}}^*.\hat{\mathbf{T}}\right) d\tau(\mathbf{r}) = \iiint_{V-V_\varepsilon} \nabla.\left(\hat{\mathbf{v}}.\hat{\mathbf{T}}^*\right) d\tau(\mathbf{r}), \tag{12.3.2}$$

where V_ε is the volume of the sphere of radius ε and centred at the singular point $\boldsymbol{\rho}^0$. Using the divergence theorem to convert the volume integral into a surface integral and letting $\varepsilon \to 0$, we obtain an integral expression for the dimensionless velocity at the point $\boldsymbol{\rho}^0$ in terms of single-layer and double-layer distributions (see section 2.5), i.e., for a point inside the control volume we obtain

$$\hat{v}_j\left(\boldsymbol{\rho}^0\right) = -\frac{1}{8\pi} \iint_{S_p} \left(\hat{f}_i(\mathbf{r}) \overline{B}_{ij}(\hat{\mathbf{r}}) - \hat{v}_i(\mathbf{r}) \overline{K}_{ijk}(\hat{\mathbf{r}}) \hat{n}_k\right) d\sigma(\mathbf{r}), \tag{12.3.3}$$

where $\hat{\mathbf{f}} = \hat{\mathbf{T}}.\hat{\mathbf{n}}$ and $\hat{\mathbf{n}}$ is the normal unit vector pointing into the fluid. For the points that are located exactly on the particle boundary S_p we have

$$\hat{v}_j(\boldsymbol{\rho}^0) = -\frac{1}{4\pi}\iint_{S_p}\hat{f}_i(\mathbf{r})\overline{B}_{ij}(\hat{\mathbf{r}})d\sigma(\mathbf{r}) + \frac{1}{4\pi}\overset{PV}{\iint_{S_p}}\hat{v}_i(\mathbf{r})\overline{K}_{ijk}(\hat{\mathbf{r}})\hat{n}_k(\mathbf{r})d\sigma(\mathbf{r}). \quad (12.3.4)$$

In second and third chapter of the present book, we examined some Faxen's laws about the drag force determination at the steady flow past solid or fluid particles. There we have shown (see Kim, 1985b, and Kim and Lu, 1987) the direct relation between the velocity singular representation in the steady flow problem and the drag force, experienced by the particles due to this flow. Here, again, we shall demonstrate this relation for the unsteady case.

Let a stationary particle is embedded in a non-uniform linearly oscillating flow with velocity $\mathbf{v}^\infty(\mathbf{r}, t) = \hat{\mathbf{v}}^\infty(\mathbf{r})\exp(-i\omega t)$. If the velocity of the flow disturbed by the presence of the particle is denoted by $\mathbf{v}^D(\mathbf{r}, t)$ and its steady part, respectively with $\hat{\mathbf{v}}^D(\mathbf{r})$, then $\hat{\mathbf{v}} = \hat{\mathbf{v}}^\infty + \hat{\mathbf{v}}^D$. It is assumed that the disturbance flow velocity $\hat{\mathbf{v}}^D$ can be expressed as a distribution of singularities situated inside the particle. Because of this, the velocity $\hat{\mathbf{v}}^D$ is continuously extended in the particle interior and the boundary integral equation (12.3.3) holds for it, where $\boldsymbol{\rho}^0$ is located inside or outside the particle surface S_p. When $\boldsymbol{\rho}^0$ is lying exactly on S_p, the velocity $\hat{\mathbf{v}}^D$ is expressed by (12.3.4).

We shall consider further another flow $\mathbf{v}^*(\mathbf{r}, t) = \hat{\mathbf{v}}^*(\mathbf{r})\exp(-i\omega t)$, induced by the translational oscillations of some particle with velocity $\hat{\mathbf{U}}\exp(-i\omega t)$ and shall assume that it can be expressed by some Stokeslets distribution of the type $\hat{\mathbf{v}}^* = \hat{\mathbf{U}}.\boldsymbol{L}(\overline{\mathbf{B}})$, where $\boldsymbol{L}$ is a linear tensorial integro-differential operator and $\overline{\mathbf{B}}$ is the oscillating Stokeslet.

For the points $\boldsymbol{\rho}^0$ inside the particle $\hat{\mathbf{v}}(\boldsymbol{\rho}^0) = 0$ and from (12.3.3) for $\hat{\mathbf{v}}^\infty(\boldsymbol{\rho}^0)$, it follows

$$\hat{\mathbf{v}}^\infty(\boldsymbol{\rho}^0) = \frac{1}{8\pi}\iint_{S_p}\left[\overline{\mathbf{B}}(\hat{\mathbf{r}}).\hat{\mathbf{T}}^\infty(\mathbf{r}) - \hat{\mathbf{v}}^\infty(\mathbf{r}).\overline{\mathbf{K}}(\hat{\mathbf{r}})\right].\hat{\mathbf{n}}d\sigma(\mathbf{r}), \quad (12.3.5)$$

where $\hat{\mathbf{n}}$ is a normal unit vector to the particle surface directed towards its exterior. When applying the operator $\hat{\mathbf{U}}.\boldsymbol{L}$ to both sides of the identity (12.3.5), we reach to

$$\hat{\mathbf{U}}.\boldsymbol{L}(\hat{\mathbf{v}}^\infty) = \frac{1}{8\pi}\iint_{S_p}\left\{\hat{\mathbf{U}}.\boldsymbol{L}[\overline{\mathbf{B}}(\hat{\mathbf{r}})].\hat{\mathbf{T}}^\infty(\mathbf{r}) - \hat{\mathbf{v}}^\infty(\mathbf{r}).\hat{\mathbf{U}}.\boldsymbol{L}[\overline{\mathbf{K}}(\hat{\mathbf{r}})]\right\}.\hat{\mathbf{n}}d\sigma(\mathbf{r}), \quad (12.3.6)$$

or taking into account that $\hat{\mathbf{v}}^*(\mathbf{r}) = \hat{\mathbf{U}}.\boldsymbol{L}[\overline{\mathbf{B}}(\hat{\mathbf{r}})]$ and $\hat{\mathbf{T}}^*(\mathbf{r}) = \hat{\mathbf{U}}.\boldsymbol{L}[\overline{\mathbf{K}}(\hat{\mathbf{r}})]$

$$\hat{\mathbf{U}}.\boldsymbol{L}(\hat{\mathbf{v}}^\infty) = \frac{1}{8\pi}\iint_{S_p}\left\{\hat{\mathbf{v}}^*(\mathbf{r}).\hat{\mathbf{T}}^\infty(\mathbf{r}) - \hat{\mathbf{v}}^\infty(\mathbf{r}).\hat{\mathbf{T}}^*(\mathbf{r})\right\}.\hat{\mathbf{n}}d\sigma(\mathbf{r}). \quad (12.3.7)$$

Let us now assume, that the particle is solid and apply the Lorentz reciprocal identity for the two flows $(\hat{\mathbf{v}}^D, \hat{\mathbf{T}}^D)$ and $(\hat{\mathbf{v}}^*, \hat{\mathbf{T}}^*)$, both vanishing at infinity

$$\nabla.\left(\hat{\mathbf{v}}^*.\hat{\mathbf{T}}^D - \hat{\mathbf{v}}^D.\hat{\mathbf{T}}^*\right) = 0. \quad (12.3.8)$$

Proceeding further, we shall take a control volume V of the fluid, bounded by the particle surface and a spherical surface, enveloping the particle and having a very large radius. After integrating (12.3.8) over the control volume V and applying the divergence theorem, we get

$$\iint_{S_p} \hat{\mathbf{U}}.\left(\hat{\mathbf{T}}^{D}.\hat{\mathbf{n}}\right)d\sigma(\mathbf{r}) = -\iint_{S_p} \hat{\mathbf{v}}^{\infty}.\left(\hat{\mathbf{T}}^{*}.\hat{\mathbf{n}}\right)d\sigma(\mathbf{r}), \tag{12.3.9}$$

or

$$\hat{\mathbf{U}}.\hat{\mathbf{F}}^{D} = -\iint_{S_p} \hat{\mathbf{v}}^{\infty}.\left(\hat{\mathbf{T}}^{*}.\hat{\mathbf{n}}\right)d\sigma(\mathbf{r}). \tag{12.3.10}$$

Here it is taken into account that $\hat{\mathbf{v}}^{D} = -\hat{\mathbf{v}}^{\infty}$ on the particle surface and that the integral over the large spherical surface tends to zero, when the large sphere radius tends to infinity. Since $\hat{\mathbf{v}}^{*}\big|_{S_p} = \hat{\mathbf{U}}$ and when substituting (12.3.10) in (12.3.7), we find that

$$\hat{\mathbf{U}}.\boldsymbol{L}\left(\hat{\mathbf{v}}^{\infty}\right) = \frac{1}{8\pi}\hat{\mathbf{U}}.\iint_{S_p} \hat{\mathbf{T}}^{\infty}(\mathbf{r}).\hat{\mathbf{n}}d\sigma(\mathbf{r}) + \frac{1}{8\pi}\hat{\mathbf{U}}.\hat{\mathbf{F}}^{D}. \tag{12.3.11}$$

However, $\hat{\mathbf{F}} = \hat{\mathbf{F}}^{\infty} + \hat{\mathbf{F}}^{D}$ and from (12.3.11) it follows that

$$\hat{\mathbf{U}}.\hat{\mathbf{F}} = 8\pi\,\hat{\mathbf{U}}.\boldsymbol{L}\left(\hat{\mathbf{v}}^{\infty}\right),$$

or

$$\hat{\mathbf{F}} = 8\pi\boldsymbol{L}\left(\hat{\mathbf{v}}^{\infty}\right), \tag{12.3.12}$$

which is the Faxen law for the force acting on a solid particle. In an analogous manner the corresponding Faxen law for the torque, the stresslet and the higher order moments could be derived (Pozrikidis, 1989a). Here, we shall fix our attention only on the torque as it will be necessary for our further analysis. If the particle performs rotational oscillations with angular velocity $\boldsymbol{\omega}$ instead of translational ones, then we can write $\hat{\mathbf{v}} = \boldsymbol{\omega}.\boldsymbol{L}\left(\overline{\mathbf{B}}\right)$ and the Faxen laws for the torque simply is

$$\hat{\mathbf{M}} = 8\pi\boldsymbol{L}\left(\mathbf{v}^{\infty}\right). \tag{12.3.13}$$

The next case we are going to discuss here, is the case of a stationary fluid particle immersed in a non-uniform linearly oscillating flow with velocity $\mathbf{v}^{\infty}(\mathbf{r}, t) = \hat{\mathbf{v}}^{\infty}(\mathbf{r})\exp(-i\omega t)$, as in the solid particle case. Here, again the total velocity can be represented in the form $\hat{\mathbf{v}} = \hat{\mathbf{v}}^{\infty} + \hat{\mathbf{v}}^{D}$, where $\hat{\mathbf{v}}^{D}$ is the steady part of the disturbance flow velocity. On the other hand, another flow induced by the translational oscillations of the fluid particle with velocity $\hat{\mathbf{U}}\exp(-i\omega t)$, where $\hat{\mathbf{U}}$ = const., is considered. Let us assume that the velocity of this flow outside the particle $\hat{\mathbf{v}}^{*}_{ext}$ can be expressed by the unsteady Stokeslets distribution $\hat{\mathbf{v}}^{*}_{ext} = \hat{\mathbf{U}}.\boldsymbol{L}\left(\overline{\mathbf{B}}\right)$, where $\boldsymbol{L}$ is a linear tensorial integro-differential operator. Then, following Pozrikidis (1989a), we shall prove that

$$\hat{\mathbf{F}}_{ext} = 8\pi\boldsymbol{L}\left(\hat{\mathbf{v}}^{\infty}\right). \tag{12.3.14}$$

According to the Lorentz reciprocal theorem, for the two flows $\left(\hat{\mathbf{v}}^{\mathrm{D}}_{\mathrm{ext}},\ \hat{\mathbf{T}}^{\mathrm{D}}_{\mathrm{ext}}\right)$ and $\left(\hat{\mathbf{v}}^{*}_{\mathrm{ext}},\ \hat{\mathbf{T}}^{*}_{\mathrm{ext}}\right)$ vanishing at infinity, we have

$$\iint_{S_p} \hat{\mathbf{v}}^{\mathrm{D}}_{\mathrm{ext}} . \left(\hat{\mathbf{T}}^{*}_{\mathrm{ext}} . \hat{\mathbf{n}}\right) \mathrm{d}\sigma = \iint_{S_p} \hat{\mathbf{v}}^{*}_{\mathrm{ext}} . \left(\hat{\mathbf{T}}^{\mathrm{D}}_{\mathrm{ext}} . \hat{\mathbf{n}}\right) \mathrm{d}\sigma . \tag{12.3.15}$$

Since $\hat{\mathbf{v}}^{\mathrm{D}}_{\mathrm{ext}} = \hat{\mathbf{v}}_{\mathrm{ext}} - \hat{\mathbf{v}}^{\infty}$ and $\hat{\mathbf{U}} = \text{const.}$, the identity (12.3.15) can be rewritten in the form

$$\iint_{S_p} \hat{\mathbf{v}}_{\mathrm{ext}} . \left(\hat{\mathbf{T}}^{*}_{\mathrm{ext}} . \hat{\mathbf{n}}\right) \mathrm{d}\sigma - \iint_{S_p} \hat{\mathbf{v}}^{\infty} . \left(\hat{\mathbf{T}}^{*}_{\mathrm{ext}} . \hat{\mathbf{n}}\right) \mathrm{d}\sigma = \iint_{S_p} \left(\hat{\mathbf{v}}^{*}_{\mathrm{ext}} - \hat{\mathbf{U}}\right) . \left(\hat{\mathbf{T}}^{\mathrm{D}}_{\mathrm{ext}} . \hat{\mathbf{n}}\right) \mathrm{d}\sigma + \hat{\mathbf{U}} . \hat{\mathbf{F}}^{\mathrm{D}}_{\mathrm{ext}} . \tag{12.3.16}$$

From (12.3.7) it follows that

$$\frac{1}{8\pi} \iint_{S_p} \hat{\mathbf{v}}^{\infty} . \left(\hat{\mathbf{T}}^{*}_{\mathrm{ext}} . \hat{\mathbf{n}}\right) \mathrm{d}\sigma = -\hat{\mathbf{U}} . \boldsymbol{L}\left(\hat{\mathbf{v}}^{\infty}\right) + \frac{1}{8\pi} \iint_{S_p} \hat{\mathbf{v}}^{*}_{\mathrm{ext}} . \left(\hat{\mathbf{T}}^{\infty}_{\mathrm{ext}} . \hat{\mathbf{n}}\right) \mathrm{d}\sigma$$

or

$$\frac{1}{8\pi} \iint_{S_p} \hat{\mathbf{v}}^{\infty} . \left(\hat{\mathbf{T}}^{*}_{\mathrm{ext}} . \hat{\mathbf{n}}\right) \mathrm{d}\sigma = -\hat{\mathbf{U}} . \boldsymbol{L}\left(\hat{\mathbf{v}}^{\infty}\right) + \frac{1}{8\pi} \iint_{S_p} \left(\hat{\mathbf{v}}^{*}_{\mathrm{ext}} - \hat{\mathbf{U}}\right) . \left(\hat{\mathbf{T}}^{\infty}_{\mathrm{ext}} . \hat{\mathbf{n}}\right) \mathrm{d}\sigma + \frac{1}{8\pi} \hat{\mathbf{U}} . \hat{\mathbf{F}}^{\infty}_{\mathrm{ext}} . \tag{12.3.17}$$

The last expression together with (12.3.16) gives

$$\iint_{S_p} \left[\hat{\mathbf{v}}_{\mathrm{ext}} . \hat{\mathbf{T}}^{*}_{\mathrm{ext}} - \left(\hat{\mathbf{v}}^{*}_{\mathrm{ext}} - \hat{\mathbf{U}}\right) . \hat{\mathbf{T}}^{\infty}_{\mathrm{ext}}\right] . \hat{\mathbf{n}} \mathrm{d}\sigma + 8\pi \hat{\mathbf{U}} . \boldsymbol{L}\left(\hat{\mathbf{v}}^{\infty}\right) = \hat{\mathbf{U}} . \hat{\mathbf{F}}^{\infty}_{\mathrm{ext}} . \tag{12.3.18}$$

It is clear, that equation (12.3.14) will be proved if we are able to show that

$$\iint_{S_p} \left[\hat{\mathbf{v}}_{\mathrm{ext}} . \hat{\mathbf{T}}^{*}_{\mathrm{ext}} - \left(\hat{\mathbf{v}}^{*}_{\mathrm{ext}} - \hat{\mathbf{U}}\right) . \hat{\mathbf{T}}^{\infty}_{\mathrm{ext}}\right] . \hat{\mathbf{n}} \mathrm{d}\sigma = \iint_{S_p} \mathrm{A} \mathrm{d}\sigma = 0 . \tag{12.3.19}$$

From the Lorentz reciprocal theorem applied for the pair of flows $\left(\hat{\mathbf{v}}_{\mathrm{int}},\ \hat{\mathbf{T}}_{\mathrm{int}}\right)$ and $\left(\hat{\mathbf{v}}^{*}_{\mathrm{int}} - \hat{\mathbf{U}},\ \hat{\mathbf{T}}^{*}_{\mathrm{int}}\right)$ inside the drop we have

$$\iint_{S_p} \hat{\mathbf{v}}_{\mathrm{int}} . \left(\hat{\mathbf{T}}^{*}_{\mathrm{int}} . \hat{\mathbf{n}}\right) \mathrm{d}\sigma = \iint_{S_p} \left(\hat{\mathbf{v}}^{*}_{\mathrm{int}} - \hat{\mathbf{U}}\right) . \left(\hat{\mathbf{T}}_{\mathrm{int}} . \hat{\mathbf{n}}\right) \mathrm{d}\sigma . \tag{12.3.20}$$

On the other hand, according to the boundary conditions on the interface S_p the tangential velocities are equal, while the normal velocities vanish, which gives on S_p

$$\mathrm{A} = \hat{\mathbf{v}}_{\mathrm{ext}} . \left(\hat{\mathbf{T}}^{*}_{\mathrm{ext}} . \hat{\mathbf{n}}\right) - \left(\hat{\mathbf{v}}^{*}_{\mathrm{ext}} - \hat{\mathbf{U}}\right) . \left(\hat{\mathbf{T}}_{\mathrm{ext}} . \hat{\mathbf{n}}\right) = \hat{\mathbf{v}}_{\mathrm{int}} . \left(\hat{\mathbf{T}}^{*}_{\mathrm{ext}} . \hat{\mathbf{n}}\right) - \left(\hat{\mathbf{v}}^{*}_{\mathrm{int}} - \hat{\mathbf{U}}\right) . \left(\hat{\mathbf{T}}_{\mathrm{ext}} . \hat{\mathbf{n}}\right)$$

Moreover, the tangential components of the stress tensor are continuous through the interface S_p : $\left(\hat{\mathbf{T}}_{\mathrm{ext}} . \hat{\mathbf{n}}\right)_t = \left(\hat{\mathbf{T}}_{\mathrm{int}} . \hat{\mathbf{n}}\right)_t$ and $\left(\hat{\mathbf{T}}^{*}_{\mathrm{ext}} . \hat{\mathbf{n}}\right)_t = \left(\hat{\mathbf{T}}^{*}_{\mathrm{int}} . \hat{\mathbf{n}}\right)_t$. Therefore,

$$\mathrm{A} = \hat{\mathbf{v}}_{\mathrm{int}} . \left(\hat{\mathbf{T}}^{*}_{\mathrm{int}} . \hat{\mathbf{n}}\right) - \left(\hat{\mathbf{v}}^{*}_{\mathrm{int}} - \hat{\mathbf{U}}\right) . \left(\hat{\mathbf{T}}_{\mathrm{int}} . \hat{\mathbf{n}}\right) \tag{12.3.21}$$

and together with (12.3.20) equation (12.3.19) is proved. As a result the Faxen law (12.3.14) for the force acting on a fluid particle is obtained, which is identical in a functional form with the corresponding one for a solid particle (12.3.12).

We shall note here that the integral expression for the velocity (12.3.4) is used by Pozrikidis (1989b) to construct a numerical procedure by the boundary integral method to study the unsteady Stokes flow due to the oscillations of spheroids, dumbbells and biconcave disks. He obtains a Fredholm integral equation of the first kind for the surface force, which is identical with that of Youngren and Acrivos (1975) for the steady Stokes flow.

12.4. Translational or Rotational Oscillations of a Spherical Rigid Particle

In chapter 3 we have seen how the flows induced by steady translational or rotational motion of a solid spherical particle in a viscous fluid is modelled by means of singular (in mathematical sense) flows. In the present section we shall model the similar flows, but due to translational or rotational oscillations of a solid spherical particle.

Let a solid spherical particle of radius a oscillates in a viscous fluid with velocity $\hat{\mathbf{U}}\exp(-i\omega t)$. In order to find the induced flow in dimensionless form $\hat{\mathbf{v}}$, it is necessary to solve the unsteady Stokes equation (12.2.3) and the continuity equation with boundary conditions $\hat{\mathbf{v}} = \hat{\mathbf{U}}$ at $R = 1$ and $\hat{\mathbf{v}} \to 0$ as $R \to \infty$.

As in section 3.2, for the modelling of the flow induced by the translational oscillations of a solid spherical particle, we shall use the unsteady Stokeslet and the unsteady symmetric quadrupole both located at the sphere centre ($\boldsymbol{\rho}^0 = \mathbf{0}$), i.e., we shall seek the induced flow velocity in the form

$$\hat{\mathbf{v}}(\mathbf{r}) = \mathbf{p}.\overline{\mathbf{B}}(\mathbf{r}) + \mathbf{q}.\overline{\mathbf{B}}^{q}(\mathbf{r}), \tag{12.4.1}$$

where **p** and **q** are unknown constant vectors.

Since the discussed problem is linear, the vectors **p** and **q** depend linearly on the vector $\hat{\mathbf{U}}$, i.e., $\mathbf{p} = C_0\hat{\mathbf{U}}$ and $\mathbf{q} = C_2\hat{\mathbf{U}}$. After substituting the formulae (12.2.13) and (12.2.17) in (12.4.1), we obtain

$$\hat{\mathbf{v}}(\mathbf{r}) = \mathbf{p}\left[A(s)\frac{\mathbf{I}}{R} + B(s)\frac{\mathbf{rr}}{R^3}\right] + \mathbf{q}\left[-C(s)\frac{\mathbf{I}}{R^3} + D(s)\frac{3\mathbf{rr}}{R^5}\right], \tag{12.4.2}$$

where A(s) and B(s) are defined by (12.2.14), and C(s) and D(s) - by (12.2.17).

If the normal unit vector to the spherical particle surface is denoted by **n**, then **r** = **n**. Applying the boundary condition $\hat{\mathbf{v}} = \hat{\mathbf{U}}$ on the particle surface $R = 1$, then from (12.4.2) and $s = M$, it follows

$$\hat{\mathbf{U}} = \mathbf{p}[A(s)\mathbf{I} + B(s)\mathbf{nn}] + \mathbf{q}[-C(s)\mathbf{I} + 3D(s)\mathbf{nn}]$$

or

$$[-1 + C_0A(s) - C_2C(s)]\hat{\mathbf{U}} + [C_0B(s) + 3C_2D(s)]\hat{\mathbf{U}}.\mathbf{nn} = 0. \tag{12.4.3}$$

Taking into account that the vectors $\hat{\mathbf{U}}$ and **n** are independent, we obtain the system

$$\left|\begin{array}{l} C_0A(s) - C_2C(s) = 1, \\ C_0B(s) + 3C_2D(s) = 0, \end{array}\right.$$

from which the dimensionless coefficients are found

$$C_0 = \frac{1}{4}(3 + 3M + M^2), C_2 = \frac{1}{2M^2}[3 + 3M + M^2 - 3\exp(M)]. \tag{12.4.4}$$

Finally, the velocity is

$$\hat{\mathbf{v}}(\mathbf{r})=\left\{\frac{1}{4}\left(3+3\mathrm{M}+\mathrm{M}^2\right)\overline{\mathbf{B}}(\mathbf{r})+\frac{1}{2\mathrm{M}^2}\left[3+3\mathrm{M}+\mathrm{M}^2-3\exp(\mathrm{M})\right]\overline{\mathbf{B}}^{q}(\mathbf{r})\right\}.\hat{\mathbf{U}}\,. \quad (12.4.5)$$

Then, the drag force on the oscillating solid particle (Pozrikidis, 1989a) is

$$\hat{\mathbf{F}}=\left(-8\pi\,\mathrm{C}_0+\frac{4}{3}\pi\,\mathrm{M}^2\right)\hat{\mathbf{U}}=-6\pi\left(1+\mathrm{M}+\frac{\mathrm{M}^2}{9}\right)\hat{\mathbf{U}}\,, \quad (12.4.6)$$

which is the Stokes famous result (11.1.1).

From (12.3.12) and (12.4.5) the Faxen law for the drag force acting on the solid particle is obtained in terms of the undisturbed flow evaluated at the sphere centre

$$\hat{\mathbf{F}}=2\pi\left\{3+3\mathrm{M}+\mathrm{M}^2-\frac{1}{\mathrm{M}^2}\left[3+3\mathrm{M}+\mathrm{M}^2-3\exp(\mathrm{M})\right]\nabla^2\right\}\hat{\mathbf{v}}^{\infty}(0)\,. \quad (12.4.7)$$

As in the steady case (section 3.2), the oscillatory rotation of a solid spherical particle in a quiescent fluid is replaced by a rotlet $\boldsymbol{\Omega}$, given by (12.2.16) and located exactly at the centre of the sphere. Then the steady part of the flow velocity is expressed by

$$\hat{\mathbf{v}}=\mathrm{C}_1\boldsymbol{\Omega}.\boldsymbol{\omega}\,, \quad (12.4.8)$$

where $\boldsymbol{\omega}$ is the particle angular velocity and $\mathrm{C}_1=\dfrac{\exp(\mathrm{M})}{2(1+\mathrm{M})}$, which is obtained from the boundary condition on the particle. The torque on the oscillatory rotating particle equals (Pozrikidis, 1989a)

$$\hat{\mathbf{M}}=-\frac{8\pi}{3}\left[3+\frac{\mathrm{M}^2}{1+\mathrm{M}}\right].\boldsymbol{\omega}, \quad (12.4.9)$$

which can be found in the classical books, e.g. Lamb's book (Lamb, 1945).

Since $\boldsymbol{\Omega}=\nabla\times\overline{\mathbf{B}}$, the Faxen law for the torque (12.3.13) acting on the solid particle reads

$$\hat{\mathbf{M}}=4\pi\frac{\exp(\mathrm{M})}{(1+\mathrm{M})}\nabla\times\mathbf{v}^{\infty}(0)\,. \quad (12.4.10)$$

12.5. Oscillating Spherical Drop

Let us consider a spherical viscous drop of radius a, viscosity $\mu^{(i)}$ and density $\rho^{(i)}$, translationally oscillating with velocity $\hat{\mathbf{U}}\exp(-i\omega t)$ in a stationary fluid of viscosity $\mu^{(e)}$ and density $\rho^{(e)}$. The drop to ambient fluid viscosity ratio is expressed by $\lambda=\dfrac{\mu^{(e)}}{\mu^{(i)}}$. For either fluids the dimensionless variables and parameters are constructed with respect to the corresponding fluid physical properties, i.e., densities and viscosities. Then we obtain two different dimensionless governing equations for each fluid

$$\left(\nabla^2 - M_{ext}^2\right)\hat{\mathbf{v}}^{(e)} = \nabla\hat{p}^{(e)}, \qquad \left(\nabla^2 - M_{int}^2\right)\hat{\mathbf{v}}^{(i)} = \nabla\hat{p}^{(i)}, \tag{12.5.1}$$

where $M_{ext}^2 = \dfrac{-i\omega a^2}{\nu^{(e)}}$ and $M_{int}^2 = \dfrac{-i\omega a^2}{\nu^{(i)}}$; (e) stands for the external (for the particle) flow; the internal flow is denoted by (i).

Similarly to the steady motion of a drop discussed in section 3.3, the outer and inner flows are expressed by means of the fundamental unsteady singularities given in section 2 of the present chapter. The exterior flow field $\hat{\mathbf{v}}^{(e)}$ is modelled as a superposition of an oscillatory Stokeslet and oscillatory symmetric quadrupole, both located at the sphere centre

$$\hat{\mathbf{v}}^{(e)}(\mathbf{r}) = \left[C_0^{(e)}\overline{\mathbf{B}}(\mathbf{r}, M_{ext}) + C_2^{(e)}\overline{\mathbf{B}}^q(\mathbf{r}, M_{ext})\right].\hat{\mathbf{U}}. \tag{12.5.2}$$

The interior flow is replaced by an oscillatory Stokeson combined with an oscillatory uniform field at the drop centre

$$\hat{\mathbf{v}}^{(i)}(\mathbf{r}) = \left[D_0^{(i)}\mathbf{I} + D_2^{(i)}\overline{\mathbf{E}}(\mathbf{r}, M_{int})\right].\hat{\mathbf{U}}. \tag{12.5.3}$$

Applying convenient dimensionless boundary conditions at the interface of the droplet, we shall determine the unknown coefficients $C_0^{(e)}$, $C_2^{(e)}$, $D_0^{(i)}$ and $D_2^{(i)}$. The normal component of the velocity vectors on either side of the interface must be equal to the normal velocity of the droplet, which is given by equations (1.4.14) and (1.4.15), i.e., $\hat{\mathbf{v}}^{(e)}.\mathbf{n} = \hat{\mathbf{U}}.\mathbf{n}$ and $\hat{\mathbf{v}}^{(i)}.\mathbf{n} = \hat{\mathbf{U}}.\mathbf{n}$ at $r = 1$. Then we get the first two equations for the coefficients:

$$C_0^{(e)}\left[A(M_{ext}) + B(M_{ext})\right] + C_2^{(e)}\left[3D(M_{ext}) - C(M_{ext})\right] = 1, \tag{12.5.4}$$

$$D_0^{(i)} + D_2^{(i)}\left[2Q(M_{int}) - W(M_{int})\right] = 1. \tag{12.5.5}$$

Here A and B are the Stokeslet coefficients given by (12.2.14), while the Stokeson coefficients Q and W are expressed by (12.2.20).

The next boundary condition is the continuity condition of the tangential velocity vectors expressed by equation (1.4.16). In our case it means $\hat{\mathbf{v}}^{(e)}.(\mathbf{I} - \mathbf{nn}) = \hat{\mathbf{v}}^{(i)}.(\mathbf{I} - \mathbf{nn})$ at $r = 1$ and transforms into the third equation for the coefficients

$$C_0^{(e)}A(M_{ext}) - C_2^{(e)}C(M_{ext}) - D_0^{(i)} - 2D_2^{(i)}Q(M_{int}) = 0. \tag{12.5.6}$$

The last boundary condition that we shall impose is the continuity of the tangential components of the stress vectors as prescribed by equation (1.4.17). Here it takes the form $\left(\mathbf{E}^{(e)}.\mathbf{n}\right).(\mathbf{I} - \mathbf{nn}) = \lambda\left(\mathbf{E}^{(i)}.\mathbf{n}\right).(\mathbf{I} - \mathbf{nn})$ at $r = 1$, where $\mathbf{E}^{(e)}$ and $\mathbf{E}^{(i)}$ are deformation tensors for both fluids [see (1.2.11)]. After some algebra, the final form of this condition is

$$C_0^{(e)}K(M_{ext}) + C_2^{(e)}E(M_{ext}) - \lambda D_2^{(i)}G(M_{int}) = 0, \tag{12.5.7}$$

where $K(s) = 2\left[B(s) - \exp(-s)(1+s)\right]$, $E(s) = \exp(-s)\left[6 + 6s + 3s^2 + s^3\right]$ and $G(s) = \dfrac{15}{s^5}\left[s(s^2+6)\cosh s - 3(s^2+2)\sinh s\right]$.

We have to note, that the normal stress jump boundary condition is omitted, as it is supposed that the surface tension is sufficiently high to keep the drop spherical. Some authors give interesting analysis of this condition (Pozrikidis, 1989a).

The equations (12.5.4)-(12.5.7) form a system of equations for the unknown coefficients $C_0^{(e)}$, $C_2^{(e)}$, $D_0^{(i)}$ and $D_2^{(i)}$, but for the sake of brevity its solution will not be presented here. If further interested, the reader can find it in (Pozrikidis, 1989a). It is interesting to note that in the limit $\lambda \to \infty$ these coefficients tend to their correspondent values in the solid particle case, i.e., $C_0^{(e)} \to C_0$, $C_2^{(e)} \to C_2$, $D_0^{(i)} \to 1$ and $D_2^{(i)} \to 0$.

Then, the respective Faxen law reads

$$\hat{\mathbf{F}} = 8\pi\left[C_0^{(e)} - \frac{C_2^{(e)}}{2}\nabla^2\right]\hat{\mathbf{v}}^\infty(0). \tag{12.5.8}$$

12.6. Relaxation and Breakup of an Initially Extended Drop. Instability of Translating Drops

This section is a continuation of sections 6.2 and 6.6. dealing with the steady deformation of a drop. As mentioned in section 6.1., the drop breakup occurs when the critical capillary number Ca_{cr} is exceeded and no steady drop shape exists. The resulting transient elongation with subsequent drop fragmentation will be studied in the present section.

Relaxation and breakup of initially extended drops. The study of the interfacial tension-driven fragmentation of a very long fluid filament in a quiescent fluid is a classical problem of fluid mechanics. The problem dynamics is modelled as completely inviscid or is treated as a viscosity dominated Stokes flow. The inviscid case was first investigated by Rayleigh (1892) in the context of jet breakup. The viscous case was first studied by Tomotika (1935) who assumed that the viscous drop has an elongated shape with a long cylindrical midsections and bulbous ends. In order to apply the linear stability theory Tomotika modelled the drop as an infinite fluid cylinder of initially uniform radius that can break up due to capillary wave instabilities. In this way, it is interesting to analyse the effect of viscosity and disturbance wavelength on the drop size and shape. The phenomenon considered is connected with a variety of topics (emulsification of liquids, for example) and is studied both in quiescent and shear flows.

In particular, the problem was examined by Rumscheidt and Mason (1961) and by Flumerfelt (1980) for the case of quiescent fluids and by Mikami et al. (1975) and Khakhar and Ottino (1987) for droplets in a shearing flow.

Since the real processes almost always involve time-dependent flows, Stone et al. (1986) have studied how transient effects can alter the deformation and breakup process. Specifically, they examine experimentally droplet stretching near the critical capillary number in two-dimensional flows generated in a computer controlled four-roll mill and investigated the relaxation dynamics that occurs if the imposed flow is stopped abruptly with the droplet in a stretched non-equilibrium state. After the flow stoppage the drop relaxes back to spherical shape, but later breaks up into smaller drops. A relatively rapid bulbing of the drop end and subsequent break-off of the bulbous end from the drop central part is observed and this behaviour is known as "end-pinching". The mechanism of breakup is different for the different ranges of λ, and the drop and bulk viscosity ratio. For example, for $\lambda > O(1)$, with the increase of λ, larger elongations are necessary to reach breakup.

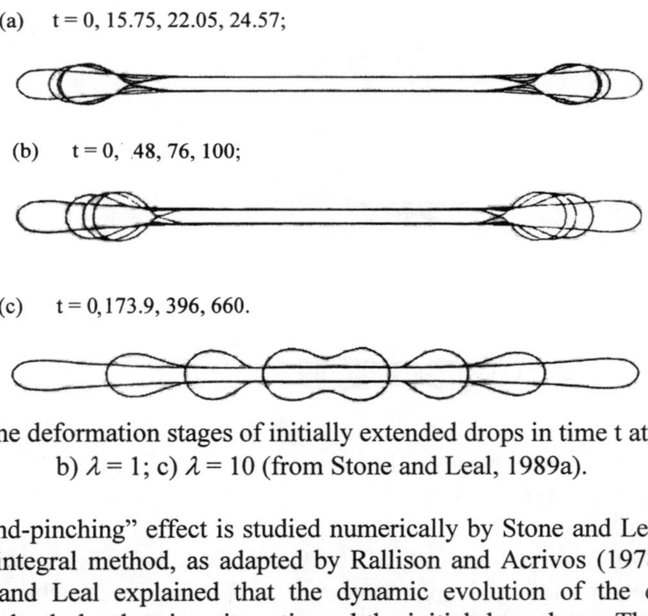

Fig.12.6.1. The deformation stages of initially extended drops in time t at: a) $\lambda = 0.05$; b) $\lambda = 1$; c) $\lambda = 10$ (from Stone and Leal, 1989a).

The 'end-pinching" effect is studied numerically by Stone and Leal (1989a) via the boundary-integral method, as adapted by Rallison and Acrivos (1978) for viscous drops. Stone and Leal explained that the dynamic evolution of the drop shape is governed completely by the viscosity ratio and the initial drop shape. They presented a lot of typical drop shape evolution patterns for different initial drop shapes. In Fig.12.6.1. some of their results for initial shape (taken from experiment at $\lambda = 11.3$) at $L/a = 8.6$, $Ca = 0$ and $\lambda = 0.05$, 1, 10 are shown. It is seen from the figure, that for small λ the pinching starts very quickly after the initial moment (with time scale $R\mu/\sigma$, where R is the initial midsection radius). The central cylindrical thread after the break-off of the bulbous ends will be subjected to similar bulbing and relaxation mechanisms. With the viscosity ratio increase, the relaxation process becomes slower, the drop shortens and a neck in the central part of the initially elongated drop is developed. For high viscosity ratios the drop can be preserved; no breakup is reported (see Fig.12.6.1.c) and the drop returns to its equilibrium spherical shape. There is a very good agreement between the numerical results of Stone and Leal (1989a) with their previous experimental measurement (Stone et al., 1986). On the basis of the observations presented in both works, we shall try to explain the mechanism of end-pinching. Since the pressure gradient in the initially cylindrical droplet shape is almost zero, very slow flow motion occurs. On the other hand, the pressure gradient in the transition zone between the cylindrical and bulbous part is significant to induce a flow towards the end, leading to a neck in the drop shape. The neck is after that transformed into end-pinching.

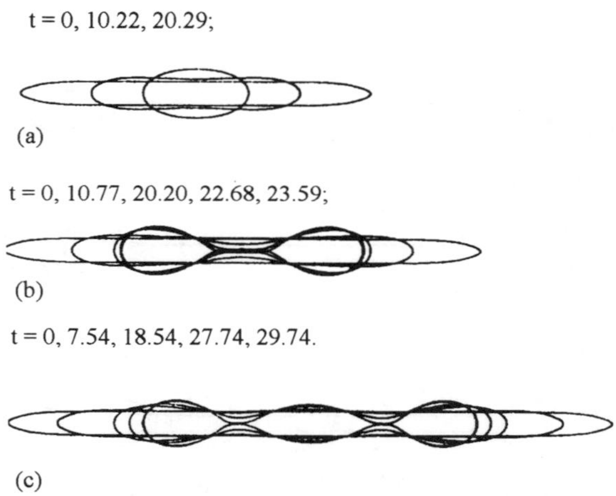

Fig.12.6.2. The deformation stages of low viscosity ratio drops ($\lambda = 0.01$) upon time t at: a) L/a = 5.3; b) L/a = 6.4; c) L/a = 7.5 (from Stone and Leal, 1989a).

For almost inviscid drops at $\lambda < O(0.01)$ the initial drop shapes are long, slender, with nearly pointed ends. The great curvature at the drop ends enhances high local velocities there resulting into quick drop shortening. If the initial elongation is sufficiently strong, the drop may not relax to its equilibrium spherical shape, but deforms at the end region, similarly to the higher viscosity ratio case. These results for $\lambda = 0.01$ and Ca = 0 are shown in Fig.12.6.2. with initial shapes (different in length) taken from experiments at $\lambda = 0.011$. It is interesting to note, as illustrated in Fig.12.6.2.b, that in the thin thread zone a small satellite drop is formed shortly before the two end parts split into fragments. In the last figure, Fig.12.6.2.c, as the initial shape is more elongated, no satellite drops appear, but three almost identical daughter droplets are formed. The experiments also show that the critical initial elongation at $\lambda = 0.01$ necessary to reach drop fragmentation is within the limits $5.4 < L/a < 6.2$.

The formation of satellite drops can be predicted only by means of the non-linear stability theory. The capillary instabilities are responsible for the appearance of satellite and even sub-satellite drops on initially long quiescent viscous thread (Tjahjadi et al., 1992). Both experimentally and numerically via boundary integral method, Tjahjadi et al. studied the satellite creation at three orders of magnitude of λ. For low viscosity ratio $\lambda < O(0.1)$ the breakup mechanism is self-repeating: after neck formation a pinch-off is registered and the process repeats itself.

The viscosity ratio λ determines two major classes of observations of drops number, size and shape, for $0.01 < \lambda < O(1)$ and for $\lambda > O(1)$. For higher λ very long and thin threads are created, which undergo capillary fragmentation into small drops and these drops are not fragmented anymore. However, the lower λ systems have the tendency to repetitive drop fragmentation; the new formed drops continue to elongate and break up.

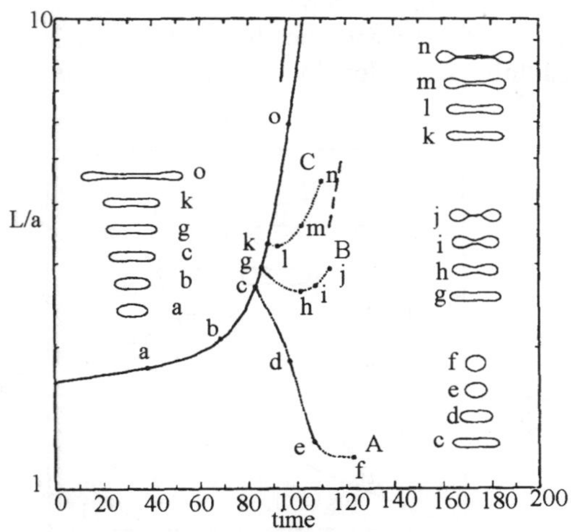

Fig.12.6.3. Results after the numerical simulations of Stone and Leal (1989b) for the deformation upon time after a step change in Ca from Ca_{cr} to $0.5Ca_{cr}$ for three different initial conditions at $\lambda = 1$: (———) $Ca = Ca_{cr}$ (the vertical line is the asymptotic limit of stretching in undisturbed flow); (··········) $Ca = 0.5Ca_{cr}$; (- - - - -) the asymptotic limit of stretching in the new flow conditions.

In a subsequent paper Stone and Leal (1989b) study experimentally and numerically the step change from critical to subcritical flow conditions and its effect on the drop deformation and eventual breakup. The experiments are performed for two-dimensional linear flows created in a four-roll mill, while the numerical calculations refer to a uniaxial extensional flow. The results of Stone and Leal due to the numerical calculations for $\lambda = 1$ are presented in Fig.12.6.3. The subcritical capillary number is changed to $Ca = 0.5Ca_{cr}$ for three different initial conditions. The three modes of behaviour registered also by the experiments are: (i) relaxation to steady spherical shape given by the curve A and schematically the drop shapes in time with (c) - (f); (ii) the breakup mode with shortening, neck formation and fragmentation into a satellite drop and two daughter drops displayed by curve B and the drop shapes , by (g) - (j); (iii) end-pinching of the bulbous ends and simultaneous drop stretching without shortening shown by curve C and the drop shapes, by (k) - (n). For larger viscosity ratios λ the three modes, together with the corresponding drop shapes, are similar, except for the latter stage (j) where the drop tends to elongate faster and tear itself away and for the last mode where the drop stretches without breaking up.

An attempt to apply the drop breakup to real mixing process is made by Tjahjadi and Ottino (1991). They set up an experiment for stretching and breakup of filaments in 2D chaotic flows at low Re created in an eccentric journal bearing apparatus. A number of experiments is performed for the range $0.01 < \lambda < 2.8$ of a viscous drop in a flow with $Ca > Ca_{cr}$. It is observed that the drop stretches quickly into a thin thread, its length increases exponentially with time and the drop fragments into a series of drops of different size dispersed through the flow by the chaotic stirring.

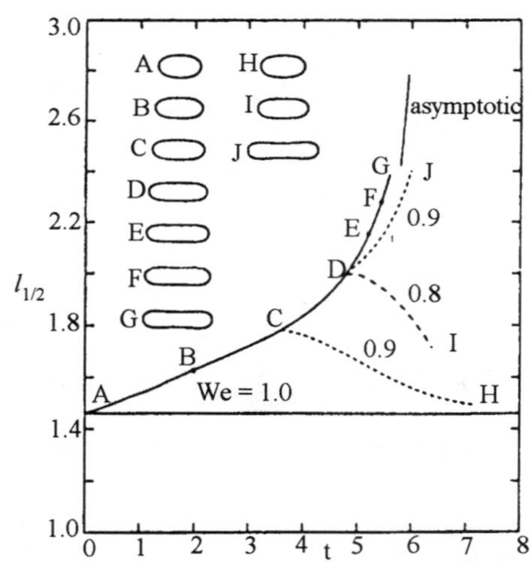

Fig.12.6.4. Bubble deformation upon time at Re = 10 for different We; asymptotic curve is given by $d\,l_{1/2}/dt = l_{1/2}$ (from Kang and Leal, 1987).

We have noted in section 6.3. that Taylor and Acrivos (1964) considered small inertia effects and showed that a viscous drop translating under the action of buoyancy through an unbounded quiescent fluid deforms slightly to an oblate ellipsoid shape for small Reynolds numbers, but the sphere is an exact steady solution for arbitrary capillary number in the limit Re ≡ 0. However, for finite Reynolds numbers the transient approach of the bubble deformations may lead to steady stable or unsteady bubble shape, depending on the initial bubble shape and Weber number. Following the finite-difference numerical method applied on a boundary-fitted orthogonal curvilinear coordinate system as developed by Ryskin and Leal (1984a, b, c), Kang and Leal (1987) solve the unsteady deformation problem of a bubble in a uniaxial straining viscous flow. Starting from deformed bubbles shapes correspondent to slightly subcritical Weber numbers and with slightly supercritical Weber numbers at the initial moment Kang and Leal find different solution branches at the same Re and We. In Fig.12.6.4. the dimensionless length of the semi-axis in x-direction $l_{1/2}$ of the deformed bubble is plotted versus the dimensionless time with respect to G^{-1} for the case Re = 10 and initial We = 1. Separate bubble shapes are shown, some of which are obtained after decreasing Weber number. For example, at $l_{1/2}$ = 1.8, if We is reduced to 0.9, the bubble shape returns to its stable form, as found by Ryskin and Leal (1984c) and shown in Fig.6.8.1. Moreover, after the appearance of waist the bubble elongation increases rapidly, tending asymptotically to a line element of the undisturbed flow. The unsteady deformation of a bubble in start-up from rest (from a spherical shape) is another item studied by Kang and Leal. Although a subcritical initial Weber number, the bubble might not achieve a steady shape, due to the inertial overshoot of shape. It turns out that the minimum value of $l_{1/2}$ necessary for continuous bubble extension for a given We and Re is the same as for the previous case of initially deformed bubbles at their critical steady shape.

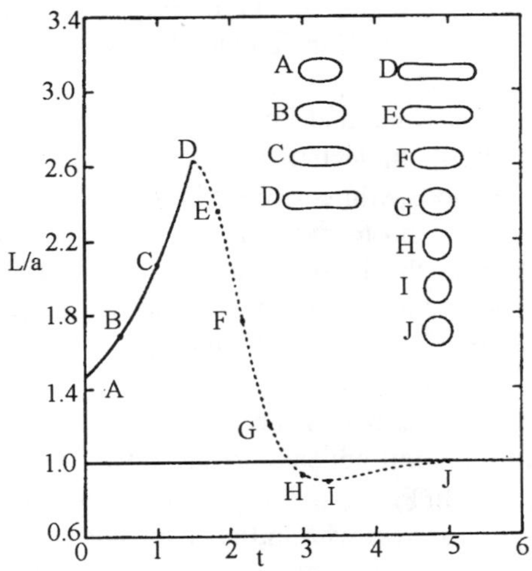

12.6.5. Bubble deformation upon time after a sudden removal of external flow at a very large deformation at: (——) Re = 0.1, We = 10^{-4}; (·········) Re = 0.1, We = 1.5×10^{-4}. (from Kang and Leal, 1987).

The sudden removal of the external flow after large bubble deformation is also studied by Kang and Leal. In Fig.12.6.5. the half-length of the bubble and some of its deformed shapes are shown in the case of starting with the steady shape solution pertaining to Re = 10 and We = 0.9. The transient shape is obtained for Re = 10 and We = 1.5, which is much larger than the critical value We = 1 and ensures faster bubble elongation. The uniaxial flow rate is suddenly reduced 100 times at t = 1.5 and then the new Reynolds and Weber numbers become Re = 0.1 and We = 0.00015, respectively. With these characteristic numbers the steady bubble shape is overshoot even at small Re. The oscillatory motion of a bubble is investigated by Kang and Leal at Re $\rightarrow \propto$, when the flow can be regarded as inviscid and irrotational. They find that with the increase of Weber number the oscillation frequency decreases and becomes zero at the critical Weber number.

Fig.12.6.6. Bubble deformation in biaxial flow at Re = 100 and We = 12 (from Kang and Leal, 1989)

In a subsequent paper Kang and Leal (1989) study the biaxial straining flow at $0 \leq$ Re ≤ 400 and We $\leq O(10)$. In section 6.8. we have presented their results for the steady state bubble deformations, while here we proceed with the unsteady ones for high Re and potential flow. In unsteady biaxial flows the bubble deformation is fundamentally different from the deformations in uniaxial flows, which means that breakup in the two cases occurs in completely different manner. In Fig.12.6.6. the transient bubble deformation at Re = 100 and We = 12, when the initial shape corresponds to the steady shape at Re = 100 and We =10. In the case of potential flow (Re $\rightarrow \propto$) when the Weber number is increased up to some supercritical value, the bubble elongates first in direction opposite to the flow, but later this elongation stops and waist thickening begins. In Fig.12.6.7. the unsteady bubble deformation for t $\in$ [1.7, 4.5] and We = 2.9 at t = 0 is shown with initial shape corresponding to the steady shape for We = 2.7. If the initial Weber number is smaller, for example We = 1, then bubble oscillations are registered having different phases and almost the same frequencies with the similar oscillations pertaining to the uniaxial case.

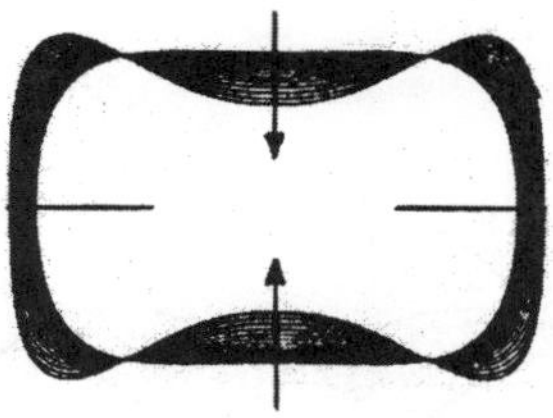

Fig.12.6.7. Bubble deformation in inviscid, biaxial flow at We = 2.9
(from Kang and Leal, 1989)

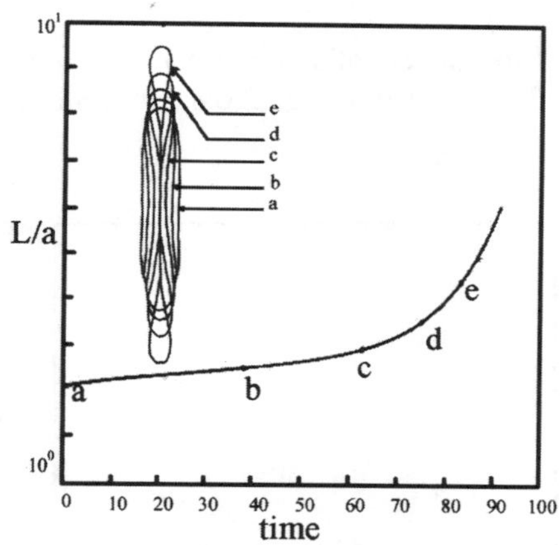

Fig.12.6.8. Drop deformation in biaxial flow as function of time at Ca_{cr} and $\lambda = 1$
(from Stone and Leal, 1989c).

On the basis of the boundary integral method, Stone and Leal (1989c) study the drop deformation and breakup in a time dependent biaxial extensional flow using its steady solution (see section 6.7.). If the drop is initially elongated to the steady shape corresponding to the critical capillary number $Ca_{cr} = 0.41$ at $\lambda = 1$, as shown in Fig.6.7.4., then the drop rapidly stretches and uniformly thins at the middle, while its ends represent large bulbous rings. Some separate stages of drop deformation and the dimensionless drop radius L/a as a function of time are sketched in Fig.12.6.8. If afterwards the flow is suddenly halted, the drop shape is relaxed to its steady stable spherical shape. This process can be observed in Fig.12.6.9. In contrast to the uniaxial flow where breakup is always observed if the drop is extended sufficiently, the biaxial flow does not lead to breakup, although the drop is strongly deformed. Stone and Leal show also that at step changes to subcritical capillary numbers the drop may recover its steady shape (cases A and B in Fig.12.6.10) or may thin very rapidly and breakup subsequently (case C in Fig.12.6.10).

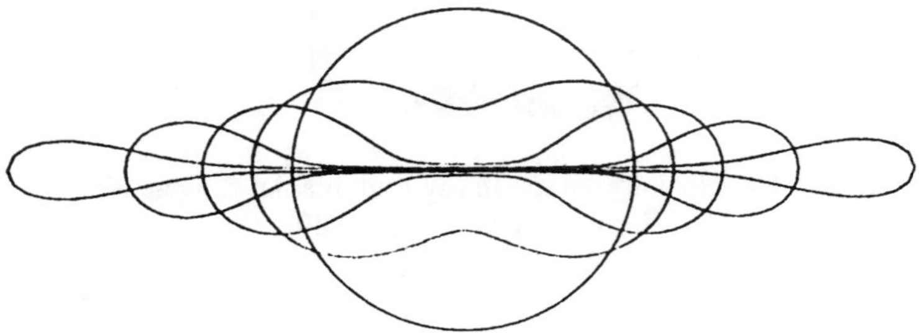

Fig.12.6.9. Relaxation back to a sphere of an initially extended toroidal drop. The initial shape corresponds to the stage (e) of Fig.12.6.8 (from Stone and Leal, 1989c).

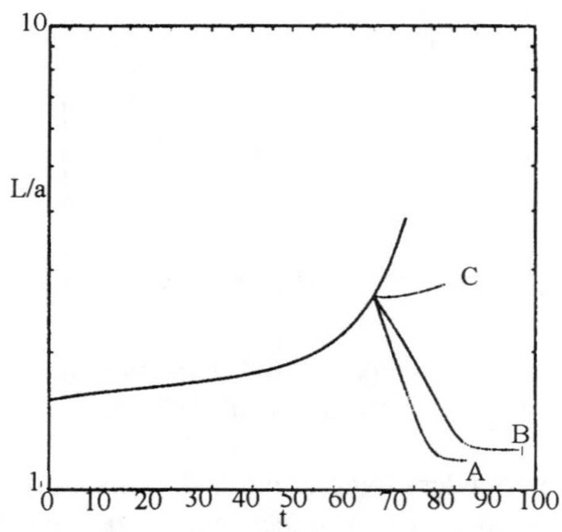

Fig.12.6.10. Deformation of an initially extended toroidal drop at $\lambda = 1$ and step changes from Ca_{cr} to: $0.5Ca_{cr}$ at curve A; 0.65 Ca_{cr} at curve B; $0.8Ca_{cr}$ at curve C (from Stone and Leal, 1989c).

Ascoli et al. (1990) study the buoyancy-driven approach of a deformable drop to a rigid wall by means of the boundary integral method and find the complete time evolution of drop shape from spherical, when the drop is at a distance of 15 drop radii away from the wall, up to highly deformed shape near the wall. The "dimple" generation and its evolution is important for the film drainage theory. The presented results are for

$\lambda = 0.3$, 1 and 3 at Ca = 0.3, 1 and 3 for each λ. The film thickness ε, established via the numerical simulation, is in the range 0.1 - 0.3 radii of the initially spherical drop. This means that the lubrication approximation ($\varepsilon \to 0$), usually applied for thin films analysis, may not give accurate results for the dimple generation. Ascoli et al. show that the dimple can be obtained from several deformation stages, but for each of them the pressure gradient normal to the wall is negative near the point of dimple. However, the lubrication theory neglects this normal pressure gradient and does not predict any local deformation of the flattened drop surface.

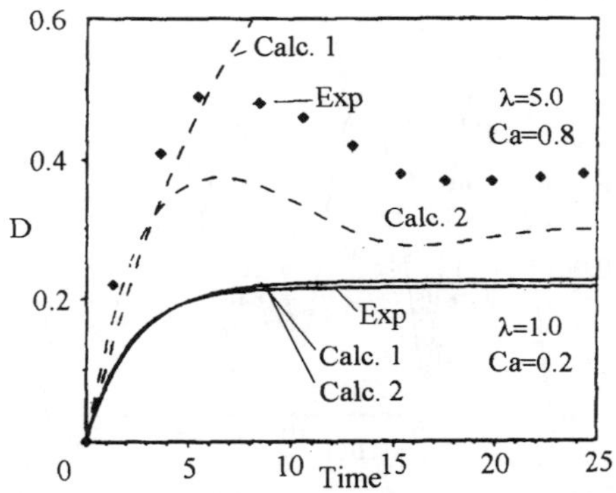

Fig.12.6.11. Drop deformation as function of time from: (Calc.1) numerical results and (Exp.) experiments of de Bruin (1989); (Calc.2) numerical results of Uijttewaal and Nijhof (1995).

As mentioned in sections 6.2 and 8.2 Uijttewaal et al. (1993) presented an analysis of droplet deformations near a plane wall in a shear flow. Owing to the assumption of small capillary number Ca < 0.5 and homoviscosity, $\lambda = 1$, transient effects were not achieved. In a subsequent paper Uijttewaal and Nijhof (1995) extended their numerical model for various λ to account for drop stability and transient motion. They showed that in the absence of a wall the critical capillary number Ca_{cr} increases very rapidly from 3 to 100 when $3 < \lambda < 4$, which is in good agreement with the experiments of Torza et al. (1972) and de Bruin (1989). The droplet deformation evolution in time, when a droplet of initially spherical form is instantaneously subjected to a constant shear flow, is presented in Fig.12.6.11. for two cases: Ca = 0.8 and $\lambda = 5$, and Ca = 0.2 and $\lambda = 1$. The calculations of Uijttewaal and Nijhof are compared with the experimental and numerical results of de Bruin. The time necessary for the drop to reach its steady shape is comparable to the relaxation time. For the first cases of parameters this time is a little overestimated with respect to the experimental data, while the time predicted by de Bruin has no sense for relaxation (it indicates a fast droplet breakup).

The "wobbling" effect indicated still by Cox (1969) is also found by Uijttewaal and Nijhof for more viscous drops. In Fig.12.6.12. their results from numerical simulation for $\lambda = 25$ and Ca = 1.5 are compared with the experimental ones of Torza et al. (1972) and analytical of Cox (1969). The discrepancy with the experimental readings, Uijttewaal and Nijhof explain by the experimental imperfections and the differences in the conditions of experimental performance and numerical model. At a presence of a wall and $\lambda > 2$, Uijttewaal and Nijhof found the droplet to migrate towards the wall. This droplet behaviour is always transient, while at steady motion it is never observed.

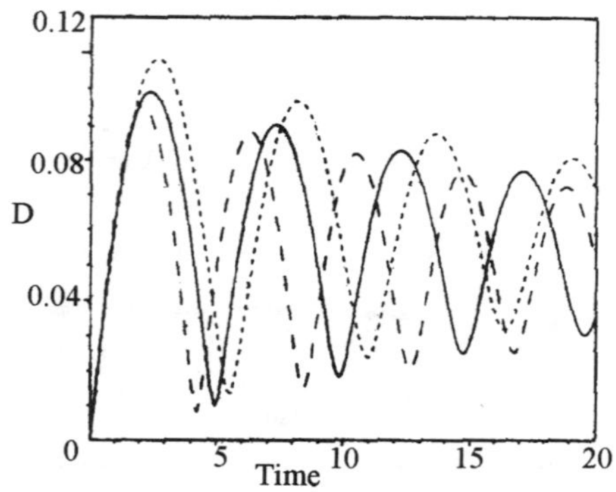

Fig.12.6.12. Drop deformation as a function of time at $\lambda = 25$ and Ca = 1.5 from: (——)numerical results of Uijttewaal and Nijhof (1995); (.........) experiments of Torza et al. (1972); (------) theoretical results of Cox (1969).

Instability of translating drops. Now we shall focus our attention on the instability problems of translating drops, recalling that the initial shape of a non-spherical drop subjected to deformation is regarded as stable if it returns to spherical shape and as unstable if it deforms further as a function of time. The stability of steady, spherical drop shape was first studied experimentally and theoretically by Kojima et al. (1984). These authors observed the behaviour of a heavy drop falling into a pool of a lighter miscible fluid with smaller viscosity than the drop and described several stages of the drop evolution. First, soon after entering the pool the drop develops an elongated tail; second, the tail separates from the main body of the drop and the drop flattens and forms an intrusion at the rear stagnation point; third, the intrusion grows in size, producing an axisymmetric expanding drop ring; and the last, the ring becomes unstable and is dispersed into the ambient fluid. Kojima et al. (1984) undertook a theoretical investigation of the observed drop behaviour in the context of the linear stability analysis and showed that in the absence of interfacial tension the spherical shape is unstable with respect to small disturbances and that at finite capillary numbers the drop will recover a steady spherical shape in case that the initial non-spherical distortion is not very large. Further they showed that the perturbations near the rear stagnation point are amplified and generate a tail for initially prolate distortions and a dimple or cavity for initially oblate distortions.

Moreover, they stated that the spherical drop is unstable to very small disturbances if Ca $\rightarrow\infty$. Unfortunately their results could not explain the observed stages in drop evolution, for example the ring formation from initially prolate drops. This shortcoming of Kojima et al. theory is not overcome even by imposing inertial effects in the model.

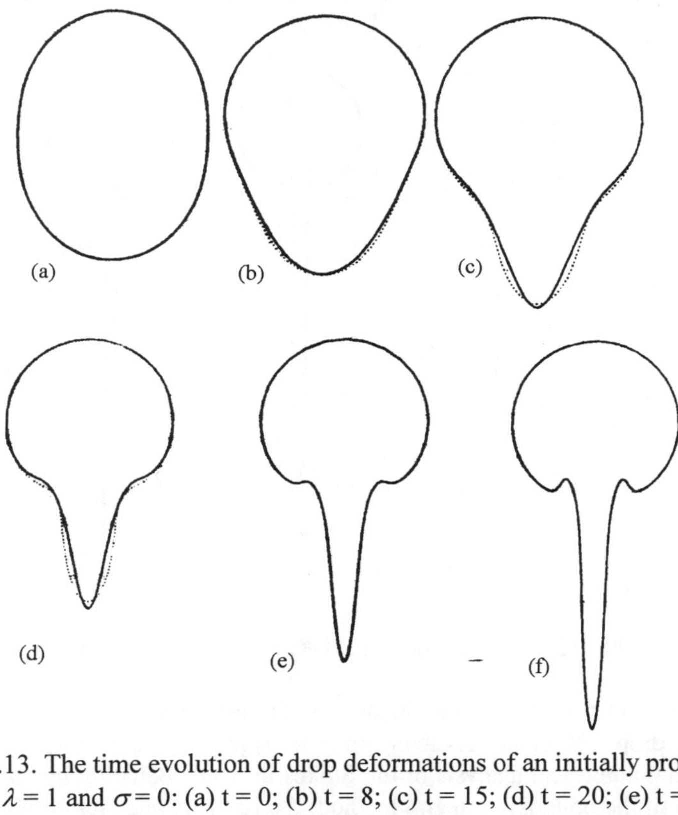

Fig.12.6.13. The time evolution of drop deformations of an initially prolate drop for $\varepsilon = 0.2$, $\lambda = 1$ and $\sigma = 0$: (a) t = 0; (b) t = 8; (c) t = 15; (d) t = 20; (e) t = 25; (f) t = 30; (··········) predictions from the linear theory (from Pozrikidis, 1990a).

The observations of Kojima et al. (1984) discussed here were confirmed by the numerical simulations of Koh and Leal (1989). By means of numerical boundary-integral calculations Koh and Leal studied the evolution of the shape of an initially nonspherical (prolate or oblate) drop in a quiescent unbounded fluid at zero Reynolds numbers and different values of viscosity ratio λ = 0.1, 0.5 and 5. They found that for prolate initially shaped drops, a highly viscous drop is more stable than a less viscous one, while for oblate drops, the less viscous drops are more stable. The critical capillary number Ca_{cr} dependence on λ is different for prolate and oblate drops, as for the latter is stronger. In a subsequent paper Koh and Leal (1990) studied the same problem experimentally and their results are confirmed by their previous numerical calculations. They found that the drops relax to the spherical shape when the initial deformation is sufficiently small. If, however, the initial deformations are significant, the drop deformation will continue. They observe that a prolate rising drop breaks up into many smaller droplets, while an oblate drop becomes a double-emulsion drop.

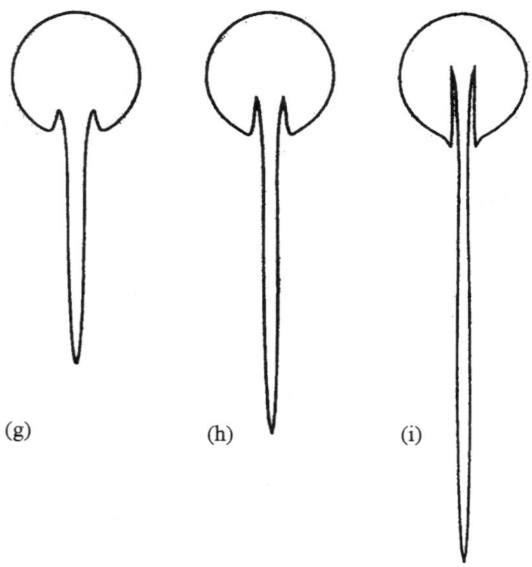

Fig.12.6.13 (continue). (g) t = 35.1; (h) t = 40.7; (i) t = 50.2.

Pozrikidis (1990a) investigated a similar problem for the translation of a spherical drop subjected to axisymmetric (prolate and oblate) perturbations. He performed a numerical analysis of the Stokes unsteady equation by use of two different variations of the boundary integral method: the Rallison and Acrivos (1978) method and the method of interfacial Stokeslets distribution. Although the paper of Pozrikidis (1990a, b) and that of Koh and Leal (1989) are written simultaneously and independently from each other, there is an excellent agreement between the results presented in both works.

The drop evolution is obtained as a function of perturbation amplitude ε , the surface tension σ and the viscosity ratio λ. In Fig.12.6.13. the non-linear drop evolution in time is presented. The drop is initially disturbed by a prolate perturbation of amplitude $\varepsilon = 0.2$ at $\lambda = 1$ and $\sigma = 0$. The first three drop shapes correspond to the first stage of evolution, namely, the undisturbed spherical shape recovering. These shapes are almost similar with the ones predicted by the linear theory. The second stage refers to the next three drop shapes in Fig.12.6.13.d-f illustrating the development of tail. Here, the linear theory gives worse results without being able to predict the tail development. The last stage consisting of the last three drop shapes shows the tail stretching into a thin long continuously elongated thread, and a spike encountered at the drop rear in Fig.12.6.13.i.

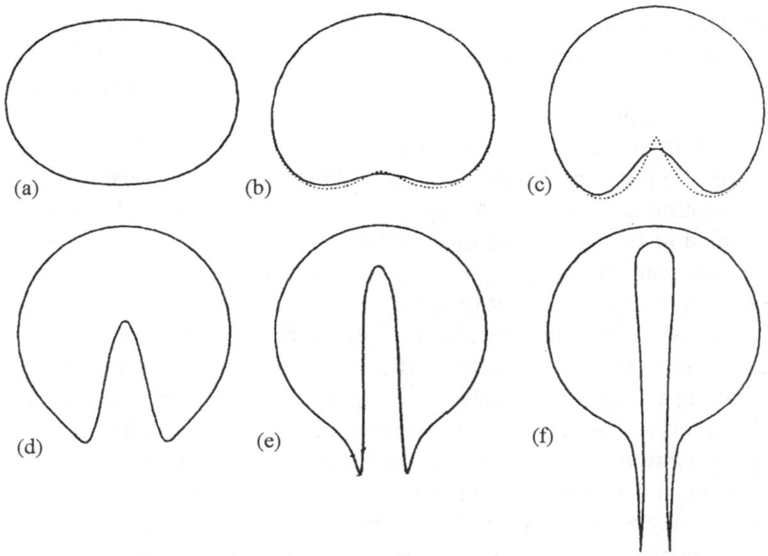

Fig.12.6.14. The time evolution of drop deformations of an initially oblate drop for $\varepsilon = -0.2$, $\lambda = 1$ and $\sigma = 0$: (a) t = 0; (b) t = 8; (c) t = 15; (d) t = 25; (e) t = 35; (f) t = 45; (··········) predictions from the linear theory (from Pozrikidis, 1990a).

The results of Pozrikidis (1990a, b) for the oblate perturbations with $\varepsilon = -0.2$, $\lambda = 1$ and $\sigma = 0$ are sketched in Fig.12.6.14.a-f. Here again three main stages are seen. The first one is connected with a dimple development at the drop rear (see Fig.12.6.14.a-c). The second stage indicates (see Fig.12.6.14.d-f) the ambient fluid penetration inside the dimple till the drop reduction to a ring. The last stage refers to larger times when the drop translation progresses steadily as a symmetric ring. Thus, Pozrikidis performed a parametrical study of the non-linear instability of a drop. Although he was able to reveal the mechanisms of generation of a steady ring and a stretching tail via the non-linear instability analysis, some of the reported by Kojima et al. (1984) observations, such as the drop rear flattening and the expansion of the developed drop ring, remained still unexplained. The account of the inertial effects may help to resolve these problems, but up to our knowledge this is still an open question.

Influence of the surfactants on the drop deformation and breakup. The last topic that we shall discuss in this section is the influence of surfactants on the transient deformation and breakup of drops. The presence of impurities has a noticeable effect on the deformation of viscous drops, because the impurities modify the mean interfacial tension, which alters the normal stress jump across the fluid/fluid interface. Also, surfactant gradients along the drop interface cause tangential stresses that produce fluid motions (Marangoni flows, see section 5.5.). The surfactants change also the interface rheology, its shear and dilational viscosity. The presence of surfactants in the fluid/fluid interfaces complicates the matters, because surfactants distribution is intimately coupled to the drop shape and interface mobility, which in turn affects the time-dependent evolution of the drop shape.

In recent years, there has been a revival of the interest to interfacial flows with surfactants. The paper of Stone and Leal (1990) examined the effects of surfactants on drop deformation and breakup. An analytical result valid for nearly spherical distortions was presented at first and finite drop deformations were studied numerically by use of the boundary integral method in conjunction with the time-dependent convective-diffusion equation for surfactant transport.

In order to quantify the effect of surfactants on free-boundary problems, it is necessary to introduce a convective-diffusion equation which describes the surfactant transport along a fluid/fluid interface. The basic equation for surfactant transport along a deforming interface has been given by Scriven (1960), Aris (1989) and Waxman (1984). A simple derivation of the time-dependent convective-diffusion equation for surfactant transport along a deforming interface was presented by Stone (1990).

Following Stone's paper, let us consider a material surface element S(t) that lies on a two dimensional curved surface which may be deformed. The mass balance for insoluble surfactant in the absence of diffusion, chemical reaction or a flux to the surface from either of the surrounding bulk phases, yields

$$\frac{d}{dt}\iint_{S(t)} \Gamma d\sigma = 0, \qquad (12.6.1)$$

where Γ denotes the mass of surfactant per unit area. Since $\frac{d}{dt}$ is the material derivative and S(t) is an arbitrary surface, (12.6.1) transforms into the differential equation for the surfactant mass on the surface S(t):

$$\frac{\partial \Gamma}{\partial t} + \nabla_s.(\Gamma \mathbf{v}) = 0 . \qquad (12.6.2)$$

If the velocity **v** is decomposed into normal v_n and surface velocity v_s, then (12.6.2) has the form:

$$\frac{\partial \Gamma}{\partial t} + \nabla_s.(\Gamma \mathbf{v}_s) + \Gamma(\nabla_s.\mathbf{n})v_n = 0 . \qquad (12.6.3)$$

If a surfactant diffusion is added to (12.6.3) the final form of the convective - diffusion equation for surfactant transport is:

$$\frac{\partial \Gamma}{\partial t} + \nabla_s.(\Gamma \mathbf{v}_s) + \Gamma(\nabla_s.\mathbf{n})v_n = D_s \nabla_s^2 \Gamma , \qquad (12.6.4)$$

where D_s is the surface diffusivity. This equation is fundamental when studying free surface problems with surfactants. The third term on the left-hand side of (12.6.4) is a source like term caused by the changes in interfacial area. The dimensionless variant of (12.6.4) can be written as

$$\frac{\partial \Gamma / \Gamma_0}{\partial t} + \nabla_s . \left[\frac{\Gamma}{\Gamma_0} \mathbf{v}_s - \frac{1}{\mathrm{Ca}\delta} \nabla_s \frac{\Gamma}{\Gamma_0} \right] + \frac{\Gamma}{\Gamma_0} (\nabla_s.\mathbf{n}) v_n = 0 , \qquad (12.6.5)$$

where $\delta = \sigma_0 a / (\mu D_s)$, σ_0 is the surface tension of a clean interface ($\Gamma = 0$) and $\mathrm{Ca}\delta$ is the surface Peclet number. Drop deformation is coupled with the surfactants flow, while the drop shape depends on the surfactants distribution.

As discussed in section 6.2. for small deformations from the spherical shape the steady state deformation parameter D is given by (6.2.8) (see Stone and Leal, 1990), while for the time-dependent coefficient B a second-order differential equation with constant coefficient is obtained. For moderate and large deformations, the coupled free-boundary and surface transport problem is solved numerically (Stone and Leal, 1990) by the boundary-integral method. The results obtained from the numerical and analytical calculations show that at low Ca the surfactants presence stimulates the drop deformation, while at larger Ca and finite deformations the convective effect of surfactants compete with “dilution” of surfactant due the growth of interface.

Using a numerical method Li and Pozrikidis (1997) studied the effect of an insoluble surfactant on the transient deformation and asymptotic shape of a spherical drop that is subjected to a linear or extensional flow at vanishing Reynolds number. They assumed the viscosity of the drop is equal to that of the ambient fluid, and that the interfacial tension depends linearly on the local surfactant concentration. The 3D drop deformation is affected by non-uniformities in the surface tension due to the surfactant distribution. Li and Pozrikidis’ numerical procedure combines the boundary integral method for computing the velocity field and a finite-difference method for solving the convective-diffusion equation for the surfactant concentration over the evolving interface of the deforming drop.

As reported in section 6.6., de Bruin (1989, 1993) studied the influence of surfactants on the tip streaming. Milliken et al. (1993) addressed the effect of the viscosity ratio λ, time-dependent motion and the significance of a nonlinear equation of state to the surface tension and stated that the tip streaming may be a result of the polymer acting as a surfactant Boulton-Stone (1995) investigated the effect of a surfactant on the behaviour of an axisymmetric bubble bursting at a free surface of a nearly inviscid fluid, also incorporating the effects of sorption kinetics and interfacial viscosity. Milliken and Leal (1994) proceeded to study the effects of surfactant solubility and Pawar and Stebe (1996) investigated the effects of interfacial saturation and intersurfactant cohesion, all for axisymmetric deformations. This substantial research activity reflects a desire to realistically describe the behaviour of physical systems from which organic species and other contaminants have not been removed.

Olbricht and Kung (1992) found that in straight capillary tube small-capillary-number drops will approach a steady state shape, while large-capillary-number drops will largely deform permitting a jet of the suspending fluid to enter into the drop from the trailing interface. If the tube is a constriction tube and the viscosity ratio is small, they observe that the bubble can break into two or more pieces, that is to say the bubble snaps off. Tsai and Miksis (1994) numerically established that there is a finite range of capillary numbers for which the bubble would snap off. In a subsequent paper Tsai and Miksis (1997) studied numerically also the effects of insoluble surfactant on the motion of a drop or bubble as it is driven by a pressure gradient through a capillary tube. For a gas bubble moving through a constricted capillary tube they observed that snapping off can occur and that surfactants enhance the process.

12.7. Finite Deformations of a Drop Moving Through a Fluid Interface. Time-Dependent Interactions Between Two Deformable Drops

The problems of hydrodynamic interactions of two deformable interfaces are fundamental when investigating such processes as coalescence and floatation. The general examples are the interaction between two drops, between a small rigid particle and a bubble or a drop or between a particle or drop and an initially flat fluid interface. The latter case has been thoroughly studied as a limiting model problem for the more general droplet coalescence or particle capture problems and for separation processes in multiphase emulsions or suspensions. In sections 7.5. and 8.4 the steady deformation of two drops or of one or two drops in the presence of a deformable interface has been discussed. Here we shall proceed with the transient motion and drop shape evolution without the restrictions of small deformations.

<u>Finite Deformations of a Drop Moving Through a Fluid Interface.</u> In the papers of Lee et al. (1979), Lee and Leal (1980) and Yang and Leal (1984) the motion of a rigid spherical particle near a plane interface is investigated at $Ca \to 0$ or $Ca/Bo \to 0$. Berdan and Leal (1982) generalise these works for the influence of small deformations of the interface on the force and moment, suffered by the rigid spherical particle at its motion with the prescribed translational or rotational velocity. In the beginning of the 70-s two

modes of the interface deformation due to the rising (falling) motion of a rigid or fluid particle through it were experimentally established (Maru et al., 1971; Shah et al., 1972): (i) between the rigid sphere or drop and the interface a thin liquid layer (film) of the lower (upper) fluid is formed, which thins with time till its "rupture" occurs and the particle passes across the interface; (ii) the layer between the particle and interface is still preserved and the particle passes into the domain of the upper (lower) fluid connected with the lower (upper) fluid of its original domain by a thin "tail" or "column". In the latter case, the tail thinning and rupture is the second mode of deformation in addition to the film drainage. Maru et al. (1971) have observed that the tail forms a node near the sphere rear and the column detaches from the sphere. Then the column becomes unstable and breaks up into small droplets. Afterwards the sphere remains still enveloped by the lower (upper) phase fluid and behaves different from an unenveloped sphere. The study of Maru et al. performed for Re ≈ 0.5 does not give a full explanation of the column formation and whether it is a natural phenomenon for creeping flow regimes. A lot of questions remained open, but the further numerical simulations treating the finite interface deformations gave answers to some of them (Lee and Leal, 1982; Geller et al., 1986).

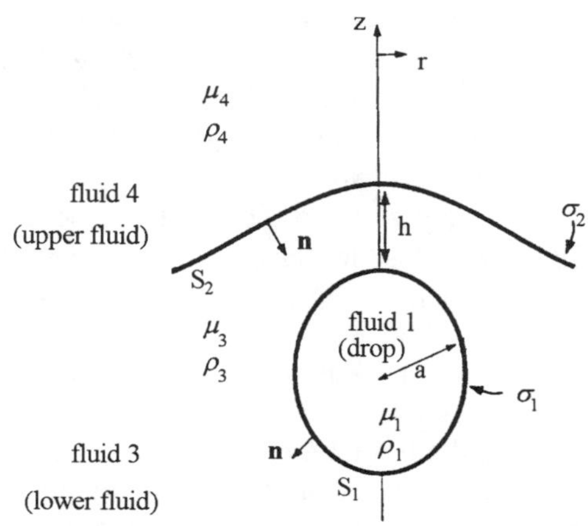

Fig.12.7.1. A schematic sketch of the drop and interface configuration.

Following these last papers in this section we shall consider the translation of a fluid particle perpendicular to an initial flat but deformable interface in the creeping flow limit for a wide range of values of the viscosity ratio, the capillary number and the Bond number that characterise the system (see Fig.12.7.1). The Reynolds number is assumed zero, but no restrictions are placed on the interface deformation magnitude. The full problem of the fluid particle motion is not solely hydrodynamic problem of a deformable interface of constant surface tension, but it has some physico-chemical aspects connected with the presence of soluble or insoluble surfactants, which may lead to surface mobility due to Marangoni convection and to additional interfacial rheological properties, and with the London-van der Waals and electrical double layer forces affecting the thin films behaviour. However, in the presented here analysis we shall ignore all the physico-chemical effects on the particle and interface deformations.

Our major objective will be to delineate the conditions in which the interface shape and the drag force can be approximated for either of the limiting cases of a quasisteady deformed interface with zero normal velocity, and of an instantaneously flat interface with nonzero normal velocity. In these two limiting cases it is assumed that the process of the interface deformation is either very fast, relatively to the time scale of particle motion, $t_p = a/U$, so that the interface shape at any instant t is the steady equilibrium form for a particular value of h(t), or is very slow so that the interface remains all the time almost flat (see Lee and Leal, 1982). It means that in order to distinguish the different possible cases we have to compare the time scale t_p with some "intrinsic" time scale, t_i, which characterises the interface deformation. Although the definition of the latter time scale is arbitrary in detail, one can choose $t_i = f_{max}/v_{max}$, where f_{max} is the maximum deformation, i.e., the deformation at the axis of symmetry of the steady interface shape at each h, v_{max} is the maximum normal velocity to the undeformed interface. Thus the time scale t_i is indicative of the time period that is required to go from a flat, but deforming interface to the steady shape for fixed values of Ca, Ca/Bo, h and λ. In this way the ratio $\tau = t_i/t_p$ will indicate the closeness of the actual unsteady interface shape to one or the other of the two mentioned limiting cases and will indicate also the difference between the exact solution for some particular values of the parameters Ca, Ca/Bo and λ and the two approximate limiting solutions. In general for $\tau \ll 1$, the interface will be very close to the steady shape at any instant, while for $\tau \gg 1$ the interface will remain close to planar at any instant (although $\mathbf{v.n} \neq 0$ as we have noted already). Finally when $\tau = O(1)$ the interface will have a fully unsteady shape.

Leal and Lee (1982) studied the transient evolution of an initially flat interface when a sphere is moving towards it with a constant translational velocity, and the conditions for film drainage and tail formation. Using the boundary integral method the authors find the existence of an elongated tail with a film of the original fluid attached to the sphere enveloped by the same fluid long after the sphere has passed through the plane of the initially undeformed interface (see Fig.12.7.2.). An extension of this work for a similar problem treating a deformable drop motion towards an interface is examined by Chi and Leal (1989) by means of the same numerical techniques. They considered the complete fluid mechanics problem including either the motion inside the drop and in the fluids from both sides of the interface for the special case of homoviscous fluids inside the drop and in upper domain of the interface. As a result three distinct types of film geometry are observed due to the different values of the viscosity ratio parameter λ corresponding to three mechanisms of film drainage. For $\lambda = 0.1$ a rapid drainage occurs, while at $\lambda = 1$ the drainage rate is uniform and a dimpled drainage corresponds to $\lambda = 10$ with minimum film thickness at the rim. These drainage patterns are confirmed by the experimental observations of Hodgson and Woods (1969) and Burrill and Woods (1973). Their results indicate that the dimpled film configuration is inevitable for very viscous drops with respect to the surrounding fluid, while for less viscous drops the drainage mode could be controlled through the introduced surfactant.

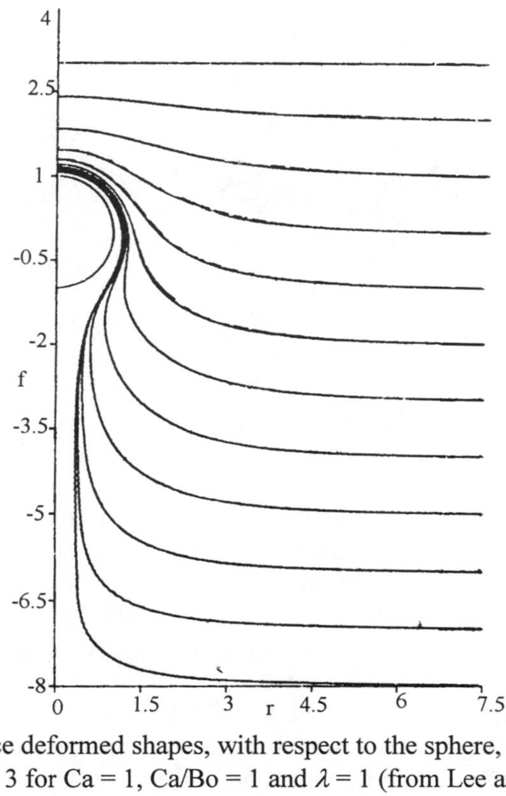

Fig.12.7.2. Interface deformed shapes, with respect to the sphere, starting from a flat interface at h = 3 for Ca = 1, Ca/Bo = 1 and $\lambda = 1$ (from Lee and Leal, 1982).

In a more recent paper Manga and Stone (1995) performed numerical and experimental investigations of the buoyancy-driven translation of a deformable drop towards a deformable interface when the drop is expected to pass through the interface and to generate large interface distortions. If the notation accepted in section 8.4. and

Fig.8.4.1.a) is introduced and if the time scale is $\frac{\mu_3}{(\rho_3-\rho_1)gr_1}$, then the problem parameters are: the Bond numbers $Bo_1 = \frac{(\rho_3-\rho_1)gr_1^2}{\sigma_1}$, $Bo_2 = \frac{(\rho_4-\rho_1)gr_1^2}{\sigma_2}$; the viscosity ratios $\lambda_1 = \mu_1/\mu_3$, $\lambda_2 = \mu_4/\mu_3$; the buoyancy parameter $\beta = \frac{\rho_3-\rho_4}{\rho_3-\rho_1}$. The flow dynamics is characterised by the drop deformation mode, drop rise speed, increase of drop surface area and the drainage rate of fluid enclosed between the drop and interface.

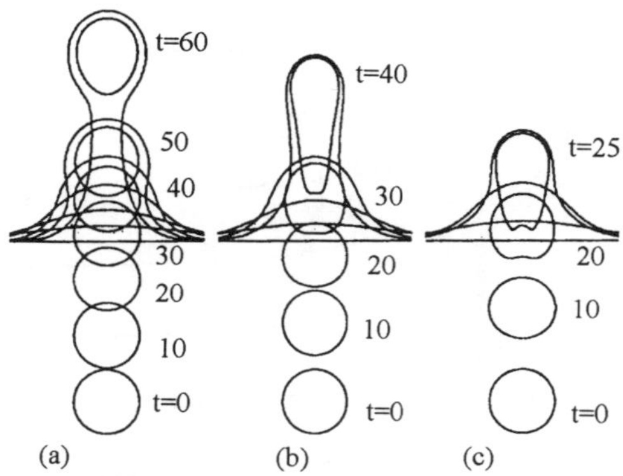

Fig.12.7.3. Drop shapes for different times t at $\lambda_2 = 0.1$, $\beta = 0.2$, $Bo_1 = 20$ and $Bo_2 \to \infty$ for: (a) $\lambda_1 = 10$; (b) $\lambda_1 = 1$; (c) $\lambda_1 = 0.1$ (from Manga and Stone, 1995).

Some of the results from the numerical simulation are shown in Fig.12.7.3. and Fig.12.7.4. In the first figure the effect of viscosity ratio λ_1 on the drop shapes for different times t is presented for three particular cases $\lambda_1 = 10$, $\lambda_1 = 1$ and $\lambda_1 = 0.1$ at $\lambda_2 = 0.1$, $\beta = 0.2$, $Bo_1 = 20$ and $Bo_2 \to \infty$. Decreasing λ_1 the rate of drop passage through the interface increases. At $\lambda_1 = 10$ a thick and almost spherical shell of the lower fluid envelops the drop and connects it with the lower bulk by a cylindrical fluid column, while at $\lambda_1 < O(1)$ only a thin layer of the lower fluid covers the drop. In the latter case the drop is largely extended at the rear end in the final stage and dimple is even formed at $\lambda_1 = 0.1$. In Fig.12.7.4. the cavity formation inside the drop and drop elongation are illustrated for two values of the viscosity ratio of the upper fluid to lower fluid, $\lambda_2 = 1$ and $\lambda_2 = 0.1$ at fixed $\lambda_1 = 1$, $\beta = 0.2$ and $Bo_1 = Bo_2 \to \infty$. From both figures it can be concluded that the drop deformation passing through a deformable interface is similar to that of a single drop. When the drop enters a higher viscosity phase the drop may almost preserve its spherical shape, but if the upper fluid is with lower density, then the drop becomes elongated with a small dimple at the drop rear. If the fluids from both sides of the interface are with equal viscosities, then a large cavity inside the drop is developed with an annular tail. Manga and Leal performed experiments and succeeded to confirm some of the numerical results, for example, that the drops of high viscosity ratio entering a lower viscosity fluid remain enveloped with a thick fluid layer and that a particle passing from a lower to a higher viscosity fluid will entrain a column of the lower fluid across the interface. Moreover, at the latter observation, the drop and the column break up after a long time evolution. The authors carried experiments with multiple particles and arrived at very interesting results, part of which they were able to understand only owing to the numerical simulations.

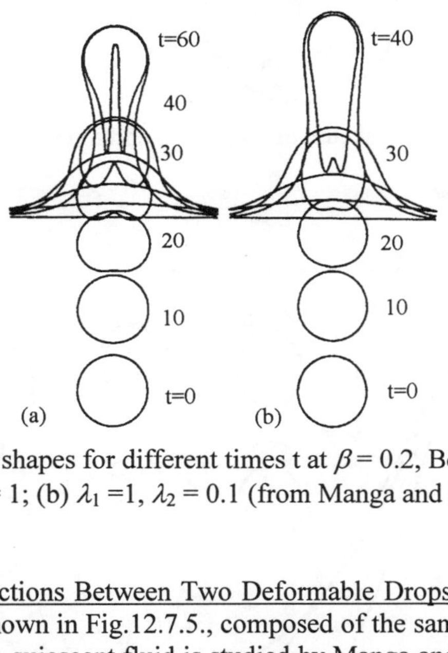

Fig.12.7.4. Drop shapes for different times t at $\beta = 0.2$, $Bo_1 = Bo_2 \to \infty$ for: (a) $\lambda_1 = \lambda_2 = 1$; (b) $\lambda_1 = 1$, $\lambda_2 = 0.1$ (from Manga and Stone, 1995).

Time-Dependent Interactions Between Two Deformable Drops. The interaction of two deformable drops, as shown in Fig.12.7.5., composed of the same fluid translating along their line of centres in a quiescent fluid is studied by Manga and Stone (1993) by means of the boundary integral method. If the trailing drop radius r_1 is chosen as a length scale and $\dfrac{\mu}{(\rho - \rho_1) g r_1}$ as a time scale, the Bond number based on the properties of the trailing drop is $Bo = \dfrac{(\rho - \rho_1) g r_1^2}{\sigma}$, while the second drop Bond number is $Bo(r_2/r_1)^2$. The other problem parameters are the ratio of drop radii r_2/r_1 and the viscosity ratio λ. The rate of film rupture and dimple formation are numerically simulated and some of the obtained results are presented in the next two figures.

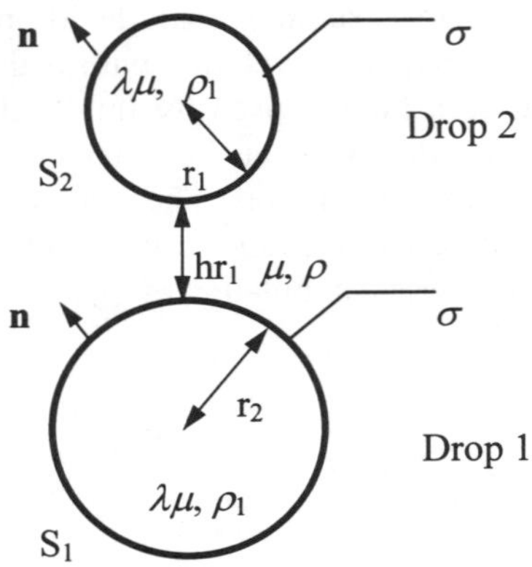

Fig.12.7.5. A schematic sketch of two drops and interface configuration.

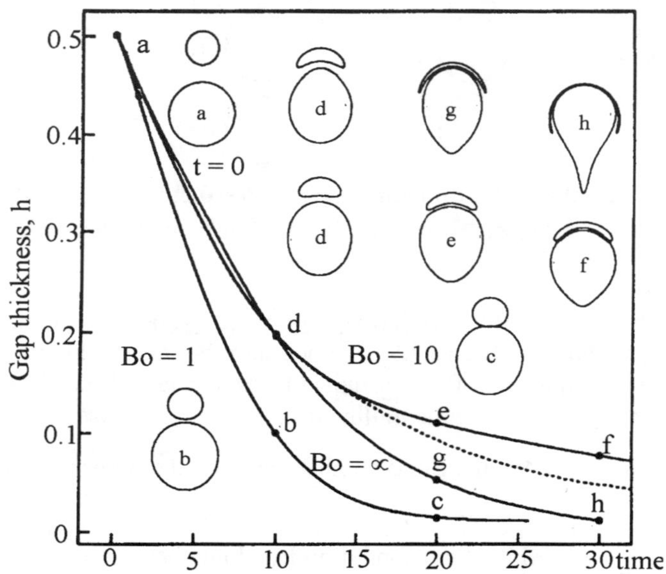

Fig.12.7.6. The rate of film drainage as a function of time at Bo = 1, 10 and ∝, $r_2/r_1 = 0.2$ and $\lambda = 1$ with: (———) film thickness on the axis of symmetry; (··········) minimum film thickness (from Manga and Stone, 1993).

In Fig.12.7.6. the time evolution of the gap thickness along the symmetry axis is plotted with solid line and the minimum gap thickness - with dashed line for three values of Bo = 1, 10 and ∝ at $r_2/r_1 = 0.2$ and $\lambda = 1$. Several drop shapes during the evolution process are given separately. The dimple formation at Bo = 10 corresponds to the branching of the solid and dashed line, when the minimum gap is displaced from the symmetry axis. In other two cases no dimple develops.

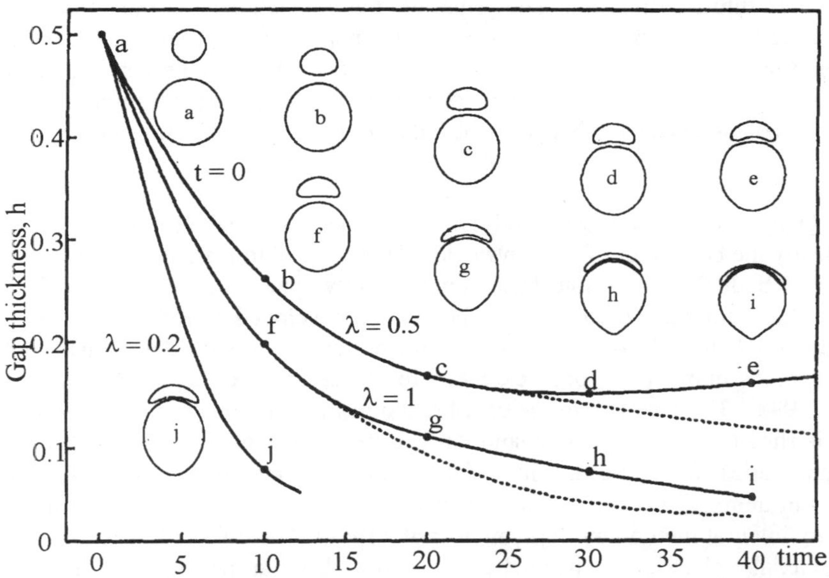

Fig.12.7.7. The rate of film drainage as a function of time at $\lambda = 0.2$, 1 and 5, $r_2/r_1 = 0.2$ and Bo = 10 with: (———) film thickness on the axis of symmetry; (··········) minimum film thickness (from Manga and Stone, 1993).

The gap thickness change in time is shown also in Fig.12.7.7., but for three different values of viscosity ratio $\lambda = 0.2$, 1 and 5, at $r_2/r_1 = 0.2$ and Bo = 10. The dimple formation begins earlier for isoviscous drops. For lower viscosity drops the interfaces are more mobile and the rate of film drainage is faster, while for higher viscosity drops the interfaces become more immobile and the drainage process is slower. Moreover, for very viscous drops, $\lambda = 5$, the gap thickness along the symmetry axis increases in the later evolution stages. On the basis of their laboratory experiments and numerical simulation (see Fig.12.7.7., drop shape h) Manga and Stone conclude that regions of large curvature form at the rear of the trailing drop. The cause of this effect is due to the existence of a locally extensional flow behind the trailing drop and due to tendency of the leading drop to deform the trailing one towards a prolate shape, which is unstable for large Bond numbers (Koh and Leal, 1989; Pozrikidis, 1990a). Another conclusion is that for sufficiently deformable drops the small drop may coat the large drop, whereas for sufficiently large drop viscosities or interfacial tension, the small drop will be rolled over the larger drop.

Up to now the coalescence problems are little investigated except for undeformable spherical drops (Zinchenko, 1984b; Wang et al., 1994). The interaction of two deformable drops in shear flow has been recently studied by Loewenberg and Hinch (1997) in order to obtain the cross-flow self-diffusion coefficients of deformable drops in a sheared dilute suspension. They compute the interactions and trajectories of a pair of deformable real drops (Ca > 0) using the boundary integral method. On the basis of the numerical results they reach to the conclusion that the coalescence of real drops requires the account of van-der-Waals attraction. For small capillary numbers Ca << 1 the drops coalescence is enhanced when the drops are pressed together by the shear flow.

<u>Finite Deformations of Compound Drops.</u> Stone and Leal (1990) have examined analytically the behaviour of concentric double emulsion droplets in linear flows, for the case when both fluid/fluid interfaces remain nearly spherical, and numerically for the effect of finite interface deformation. The method developed by Rallison and Acrivos to solve the problem of the deformation of a single drop due to the exterior shear flow for a zero Reynolds number has been extended to the deformation of a compound drop by Power (1993). The result in terms of a Fredholm integral equation of the second kind, which is shown to have a unique solution is given by means of a uniformly convergent Neumann series. Recently, the finite deformations of non-concentric double emulsion droplets in uniaxial or biaxial extensional flows have been studied numerically via the boundary integral method by Zapryanov et al. (1998). The effect of different flow types (i.e., uniaxial or biaxial extensional flows) shows different behaviour on the inner and outer droplet. Though the outer droplet maintains a steady shape the inner one may breakup.

CHAPTER 13

Hydrodynamic Interactions in Some Unsteady Viscous Flows

13.1. Introduction

As we have seen in the previous chapters, the inclusion of unsteadiness in viscous flow can produce some unusual phenomena, completely at variance with those which arise normally in the steady case. An important class of unsteady fluid flows are those in which a body starts to move from rest either impulsively or with acceleration in a viscous fluid. Another class unsteady fluid flows are the oscillating viscous flows. An increasing interest towards their study is noticed in the recent years. This is connected with the fact, that some chemical technological processes: extraction, absorption, drying, crystallisation, combustion of liquid fuels, etc., run more rapidly in the presence of oscillating flows. In this chapter we shall discuss some problems connected with both classes of unsteady flows.

Unsteady Viscous Flows from Rest. Both the body and the fluid have zero velocities up to a certain instant of time at which the motion begins and the flow relative to the body is at first entirely irrotational. However, since the fluid adjacent to the boundary is immobile with respect to it, a region of very high velocity gradient, in direction normal to the surface, forms near the body and vorticity is present there. As this vorticity is rapidly diffused in the fluid by the molecular action and convected by the flow, a thin vorticity or boundary layer is formed relative to the body. The thickness of this boundary layer increases with time.

On the basis of the assumption that the velocity changes much more rapidly across the boundary layer than along it, Prandtl (1904) reduces the Navier-Stokes equations for viscous flow to a much simpler form. As a result he opened the way to study many local fluid properties such as skin friction and heat transfer between body and fluid, but also to obtain a better understanding of some fundamental phenomena as circulation, lift and drag of bluff bodies.

The problem of the boundary layer growth on a body started from rest in an infinite incompressible viscous fluid has been investigated by many authors. There is no need at present to discuss the general theory of the boundary layers. The reader can find such discussion in the excellent books of Rosenhead (Ed.) (1963) and Schlichting (1964) and the review papers of Stewartson (1957), Riley (1965) and Telionis (1981).

The solutions of the problems involving infinite planes or cylinders are obtained using the full Navier-Stokes equations by means of appropriate similarity variables [see for example: Rayleigh (1911); Howarth (1950); Carslaw and Jaeger (1947); Batchelor (1954); Greenspan and Howard (1963)]. If a finite solid body, immersed in an incompressible fluid at rest, is set in motion, the structure of the flow which comes into being is quite complicated. Then one can use boundary layer equations which are simpler than the Navier-Stokes equations.

The Prandtl equations (Schlichting, 1964) for the boundary layer around a 2D body have the form

$$\frac{\partial u}{\partial t}+u\frac{\partial u}{\partial x}+v\frac{\partial u}{\partial y}=-\frac{1}{\rho}\frac{\partial p}{\partial x}+\nu\frac{\partial^2 u}{\partial y^2}, \tag{13.1.1}$$

$$\frac{\partial p}{\partial y}=0,$$

$$\frac{\partial u}{\partial x}+\frac{\partial v}{\partial y}=0,$$

where x and y are measured in parallel and perpendicular directions to its surface from an origin O fixed to the body forward stagnation point.

If the fluid velocity in the irrotational main stream, just outside the boundary layer, is $U_1(x,t)$, then the first equation in (13.1.1) becomes

$$\frac{\partial u}{\partial t}+u\frac{\partial u}{\partial x}+v\frac{\partial u}{\partial y}=\frac{\partial U_1}{\partial t}+U_1\frac{\partial U_1}{\partial x}+\nu\frac{\partial^2 u}{\partial y^2}. \tag{13.1.2}$$

The boundary conditions to be used with equation (13.1.2) are

$$u=v=0 \text{ at } x\geq 0,\ y=0,\ t\geq 0, \tag{13.1.3}$$

$$u=0 \text{ at } x\geq 0,\ y\geq 0,\ t<0,$$

$$u\to U_1(x,t) \text{ at } x\geq 0,\ y\to\infty,\ t\geq 0.$$

Blasius (1908) was the first who solved the Prandtl equations for the problem of a circular cylinder moving impulsively in a viscous incompressible fluid and reaching its final steady velocity, i.e., $U_1(x,t)=U_1(x)$, very rapidly (short times t). The solution method adopted by Blasius is to expand the dimensionless stream function F in an infinite power series of t:

$$F=F_0(x,\eta)+tF_1(x,\eta)+t^2F_2(x,\eta)+\ldots, \tag{13.1.4}$$

where $F(x,\eta,t)=\dfrac{\Psi}{2\sqrt{\nu t}\,U_1}$, $\eta=\dfrac{y}{2\sqrt{\nu t}}$ and F_n satisfy ordinary differential equations in η. According to Blasius the first two terms in (13.1.4) are expressed as:

$$\frac{\partial F_0}{\partial \eta}=\operatorname{erf}\eta, \qquad F_1(x,\eta)=U_1'(x)g(\eta),$$

where $g(\eta)$ is known. The third term was calculated by Goldstein and Rosenhead (1936). When the undisturbed stream velocity is U_∞, the outer flow velocity is

$U_1(x) = 2U_\infty \sin\frac{x}{a}$, where a is the cylinder radius and x is the arc length. Therefore, the boundary layer separation starts at the point where the surface shear stress $\mu\frac{\partial u}{\partial y}$ first vanishes, i.e., at the point where $U_1'(x) = -2\frac{U_\infty}{a}$, and at time $t_s = \frac{0.351a}{U_\infty}$.

In the same paper Blasius (1908) solved approximately also the problem of the boundary layer growth on a uniformly accelerated circular cylinder. In this case only odd powers of t occur in (13.1.4).

Görtler (1944) extended the theoretical calculation of the boundary layer formation process during acceleration assuming a potential flow of the form $U_1(x,t) = W(x)t^n$ with n = 0, 1, 2, 3, 4.

It is interesting to note that certain classes of accelerating flows can be reduced to a single similarity variable combining both space and time, an example being the Rayleigh variable $\eta = \frac{y}{2\sqrt{\nu t}}$. A large class of similarity solutions in which $U_1(x,t) = At^n$, $A\frac{x}{t}$ and Ae^{nt} are derived and discussed by Watson (1955), Schuh (1953) and Rozin (1957). The formation of the boundary layer on a rotating cylinder started impulsively was investigated by Tollmien (1924).

The process of formation and the initial stage of boundary layer development about an axisymmetric body accelerated impulsively was investigated by Boltze (1908). If the body is a sphere with radius a and the free-stream velocity is U_∞, then $U_1(x) = \frac{3}{2}U_\infty \sin\frac{x}{a}$. Separation begins at the downstream separation point, where $\cos\frac{x}{a} = -1$ and $t_s = 0.424\frac{a}{U_\infty}$. Taking into account the two further terms in the stream function expansion, one can obtain a more accurate instant of separation for a sphere started impulsively, namely $t_s = 0.392\frac{a}{U_\infty}$. The separation point moves from $\theta = \pi$ towards $\theta \approx 110^0$ which is the separation position at the steady flow case.

All the above investigations were based on the Prandtl's boundary layer equations. The Prandtl's boundary layer theory, which is the foundation for the development of the general viscous flow theory, gives us the first order approximation of the Navier-Stokes equations for the limit of large Reynolds numbers. However, many flows that occur in modern technology possess properties that can not be treated within the framework of the Prandtl approximation because it neglects surface curvature and displacement of the external flow by the boundary layer.

On the basis of higher order boundary layer theory short-time solutions for flows past a cylinder starting from rest with several different prescribed motions have been presented by Wang (1967a, b), Zapryanov (1974), Slavchev (1975), Zapryanov and Kalitzova-Kurteva (1976), Simeonov (1977), and Slavchev and Simeonov (1978).

In 1969 Wang (1969) via the method of inner and outer expansions solved the problem of the impulsive start of a sphere in a viscous flow for small times. By use of the so-called "independent principle" (see Rosenhead, 1963) Zapryanov (1975, 1977) applied the higher order approximation of the boundary layer theory to solve the 3D problem of the flow past a yawed cylinder.

The presence of rigid walls, free surfaces and adjacent particles inevitably affects the unsteady motion of rigid particles and/or droplets in viscous fluids. The effect of the boundary walls on the flow past rigid or fluid particles has so far been mostly investigated on the basis of either Stokes or Oseen's linearized equations of motion, to avoid the difficulties arising from the non-linear Navier-Stokes equations (see Zapryanov, 1982).

Using the method of inner and outer expansions, Zapryanov and Lambova (1986) studied the formation of a boundary layer round a fixed circular cylinder started into motion from rest in a semi-infinite fluid. The emphasis of the investigation is on the influence of the plane wall on the boundary layer detachment from the cylinder and the interaction between the cylinder and the flow.

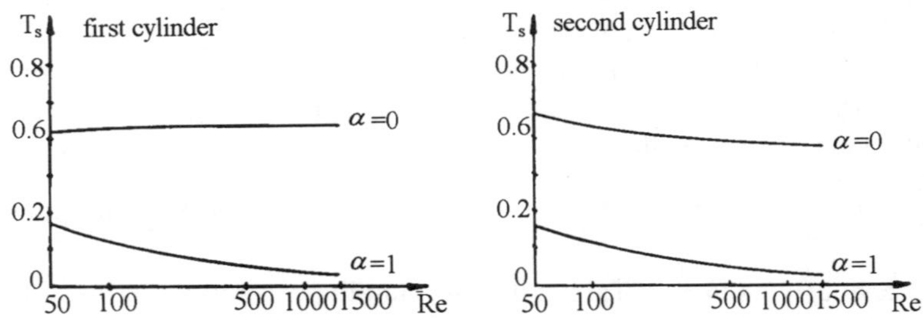

Fig.13.1.1. Dependence of the initial-time boundary layer detachment on Re for impulsive and uniformly accelerated flows around two cylinders.

Kalitzova-Kurteva and Zapryanov (1991) studied the hydrodynamic interaction problem of the unsteady motion of two circular cylinders of radii a and b, and distance d between their closest points in a viscous fluid by the method of matched asymptotic expansions applied for the boundary layer and the outer flow. The cylinders velocity is linear, $V(t) = t^{\alpha}$, in a direction parallel to the plane containing the cylinders axes. The special case of an impulsive start can be obtained at α =0. Due to the solution of Kalitzova-Kurteva and Zapryanov, some dependencies of the first separation time T_{1s} on the parameters Re are shown in Fig.13.1.1. at d/a = 1.6 and b/a = 1 for the cases $\alpha = 0$ and $\alpha = 1$. It is observed that at the impulsive motion the detachment time changes slightly as Re increases, while at the accelerated motion the detachment time strongly decreases. Moreover, for equal cylinders radii, a = b, short times and in the case of unsteady motion started from rest, the drag coefficient for the downstream cylinder is greater than that for the upstream cylinder, in contrast to the steady flow past two circular cylinders (Belov and Kudriavtzev, 1981). As can be expected, when the separation distance d/a tends to infinity, the drag coefficient of the two cylinders tends to that of a single circular cylinder under the same flow conditions. The dependence of the drag coefficients of the two equal cylinders on the time T_1 and the separation distance d/a, for various values of Re and the acceleration parameter α can be found in Fig.7 and Fig.8 of (Kalitzova-Kurteva and Zapryanov, 1991), respectively. It turns out that the drag coefficients of the two cylinders are decreasing functions of Re, and increasing functions of acceleration parameter α.

Oscillating Viscous Flows. The beneficial effects of these flows are mainly due to the appearance of a "steady streaming" along with the unsteady flow. This steady streaming helps for the fully mixing of the fluid particles, as well as for the better contact between them in some heat and mass transfer processes. The experimental verification of enhancement in mass transfer rates due to the existence of steady streaming at solid-liquid systems is given by Nyborg (1965), while effects on mass transfer in liquid-liquid extraction apparatus has been studied by Krasuk and Smith (1963). The latter have shown an analogy between mass and momentum transfer for predicting mass transfer coefficients. Critical reviews of the heat and mass transfer between a solid body and oscillating flow are presented by Richardson (1967) and Al Taweel and Landau (1976), respectively.

The first theoretical investigations of the oscillating viscous flows are connected again with the name of Stokes (1851). The problem, to obtain the hydrodynamic characteristics of the flow induced by the harmonic oscillations of an infinite plate with velocity $U_0 \cos\omega t$ in its own plane (y = 0) in a viscous fluid, was studied first by him and afterwards by Rayleigh (1884). This problem is known as the second problem of Stokes. Due to the simple geometry, a similarity variable $\sqrt{\frac{\omega}{2\nu}}y$ is constructed, by means of which an exact formula for the velocity of the induced oscillating flow is obtained:

$$u = U_0 \exp\left[-\sqrt{\frac{\omega}{2\nu}}y\right]\cos\left(\omega t - \sqrt{\frac{\omega}{2\nu}}y\right) \tag{13.1.5}$$

and the induced oscillating flow around the plate is named "Stokes layer". However, the problem is linear and fully unsteady, i.e. no steady streaming exists.

If the oscillations are perpendicular to the plate plane, the simplifications made are no longer valid and the problem becomes non-linear, which is the cause a steady streaming to be observed in addition to the unsteady one. Faraday (1831) experimentally discovered the presence of a steady streaming around a perpendicularly oscillating plate, while Dvorak (1874) described this phenomenon for oscillating flow in a tube.

Later, in the works of Carriere (1929), Andrade (1931) and Schlichting (1932) the interest in the problem of the steady streaming, induced by the oscillations of a circular cylinder in a viscous fluid, is renewed. The structure of the steady streaming strongly depends on the shape of the oscillating body in the fluid. Due to the unsteady boundary layer theory, Schlichting (1932) has obtained a result that is remarkable with the fact, that in an unsteady flow caused by the harmonic oscillations of the fluid particles a secondary steady streaming appears not only in the boundary layer, but also outside of it. Further, the same problem has been investigated by Andres and Ingard (1953), Holtsmark et al. (1954), Raney et al. (1954), as well.

In connection with the difficulties at the 3D analysis of the steady streaming around a sphere, the study of this problem dates till the last 30 - 40 years. Following the Schlichting's approach, Lane (1955) examined the harmonic oscillations of a sphere in a viscous fluid. Analysing in detail the structure of the steady streaming Stuart (1966) has proved that at given conditions along with the unsteady boundary layer there also exists a second boundary layer, in which the steady streaming decays. This second steady boundary layer depends on the parameter $\mathrm{Re_s} = \dfrac{U_\infty^2}{\omega\nu} = \dfrac{\mathrm{Re}}{\mathrm{St}} = \varepsilon\,\mathrm{Re}$, where $\mathrm{Re} = \dfrac{U_\infty a}{\nu}$ is the Reynolds number, $\mathrm{St} = \dfrac{a\omega}{U_\infty} = \varepsilon^{-1}$ is the Strouhal number and $\varepsilon = \dfrac{U_\infty}{\omega a}$ is the dimensionless amplitude of oscillations. If $\mathrm{Re_s} >> 1$, there exists a steady boundary layer of thickness $O(\mathrm{Re_s}^{-1/2})$, while at $\mathrm{Re_s} << 1$ the steady streaming is of Stokes flow type.

On the basis of the method of matched asymptotic expansions applied for the boundary layer region and outer region, Riley (1965, 1966, 1967) has established a full theory for the harmonic oscillations of a circular cylinder or sphere in a viscous flow. In his work Wang (1968) has analysed the problem of the translational oscillations of a circular cylinder in a viscous fluid, but by means of another approach and has found separate equations and boundary conditions for the steady and unsteady flow.

Zapryanov and Stoyanova (Tabakova) (1977, 1978a) have found exact solutions of the Stokes equations describing the steady flow induced by the presence of a porous rigid sphere and fluid particle, respectively, in an oscillating viscous flow. They have given approximate solutions with second order accuracy of the full Navier-Stokes equations for the same problems. In a similar manner the problem of the harmonic oscillations of an elliptic cylinder in a viscous fluid is solved (Zapryanov and Tabakova, 1978b).

The torsional oscillations of cylinders and spheres have also been treated both theoretically and experimentally. Di Prima and Liron (1976) have pointed out that the torsional oscillations of a sphere in an unbounded viscous fluid induce a secondary steady flow in the planes containing the axis of rotation and have calculated the effect of this flow on the torque acting on the sphere. The problem of the oscillating fluid flow around a slowly rotating sphere is investigated by Bestman (1983).

Using the method of matched asymptotic expansions Tabakova and Zapryanov (1978) and Zapryanov and Tabakova (1979) studied the unsteady and secondary steady streaming between two concentric spheres induced by the torsional oscillations of the inner sphere in the case of high-frequency oscillations. The same problem, but for low-frequency oscillations case, is treated both theoretically and experimentally by Munson and Douglass (1979). The problem, when the inner sphere is fluid, is solved by the matched asymptotic expansions method by Zapryanov and Chervenivanova (1981).

In this chapter we shall present some problems of hydrodynamic interactions between two rigid spheres in fluid flow starting from rest and in oscillating flow, as well. For the second type of flows, several geometrical cases connected with the spheres location to each other will be considered (Tabakova and Zapryanov, 1982a,b, 1987; Kovatcheva et al., 1985, 1988).

13.2. Hydrodynamic Interaction between Two Spherical Particles at Small Times

Here, we shall formulate the mathematical model describing the hydrodynamic interaction between two rigid spheres (Kalitzova-Kurteva, 1987; Kalitzova-Kurteva and Zapryanov, 1990). The gross properties of the separation phenomenon and the resistance forces of the two bodies are investigated. The method of inner and outer expansions makes possible to calculate all basic characteristics of the flow and their time development depending on the Reynolds number, the distance between two spheres, the ratio of the two bodies radii and the acceleration parameter.

Consider two spherical particles of radii a and b immersed in an infinite incompressible viscous fluid of constant density and kinematic viscosity ν. It is further supposed that the fluid flows in a direction parallel to the line connecting the spheres centres with velocity at infinity, which varies with time in a manner $V'(t) = U_0\left(\frac{t'}{T_0}\right)^{\alpha}$.

Then the velocity dimensionless correspondent is $V(t) = \frac{V'(t)}{U_0} = t^{\alpha}$, where $\alpha \geq 0$ is a dimensionless constant, t denotes dimensionless time, T_0 and U_0 are the characteristic time and velocity, respectively. The lengths will be normalised by the first sphere radius a. Bispherical coordinates (ξ, η, φ) are introduced as in section 7.1. by means of the relation (7.1.1) and are shown in Fig.7.1.1. The basic assumptions are that the Reynolds number $Re = \frac{U_0 a}{\nu}$ is large and only short times are considered. It is assumed further that the Strouhal number $St = \frac{a}{U_0 T_0}$ is of order O(Re), so that $\varepsilon = St^{-1} = \gamma Re^{-1}$ is a small parameter and $\gamma = O(1)$.

Since the problem is axisymmetrical, i.e. independent on φ, we can write the Navier-Stokes equations and continuity equation in bispherical coordinates (ξ, η) in the dimensionless version as follows

$$\frac{\partial u}{\partial t}+\varepsilon\left(\frac{u}{h}\frac{\partial u}{\partial \xi}+\frac{v}{h}\frac{\partial u}{\partial \eta}+\frac{uv}{h^2}\frac{\partial h}{\partial \eta}-\frac{v^2}{h^2}\frac{\partial h}{\partial \xi}\right) \tag{13.2.1}$$

$$=-\frac{1}{h}\frac{\partial p}{\partial \xi}-\frac{\varepsilon^2\gamma}{hh_3}\frac{\partial}{\partial \eta}\left\{\frac{h_3}{h^2}\left[\frac{\partial}{\partial \xi}(hv)-\frac{\partial}{\partial \eta}(hu)\right]\right\},$$

$$\frac{\partial v}{\partial t}+\varepsilon\left(\frac{u}{h}\frac{\partial v}{\partial \xi}+\frac{v}{h}\frac{\partial v}{\partial \eta}+\frac{uv}{h^2}\frac{\partial h}{\partial \xi}-\frac{u^2}{h^2}\frac{\partial h}{\partial \eta}\right) \tag{13.2.2}$$

$$=-\frac{1}{h}\frac{\partial p}{\partial \eta}-\frac{\varepsilon^2\gamma}{hh_3}\frac{\partial}{\partial \xi}\left\{\frac{h_3}{h^2}\left[\frac{\partial}{\partial \eta}(hu)-\frac{\partial}{\partial \xi}(hv)\right]\right\},$$

$$\frac{\partial}{\partial \xi}(hh_3u)+\frac{\partial}{\partial \eta}(hh_3v)=0, \tag{13.2.3}$$

where u, v are the velocity components, $h=h_1=h_2=\dfrac{\sinh|\eta_1|}{\cosh\eta-\cos\xi}$, $h_3=\dfrac{\sinh|\eta_1|\sin\xi}{\cosh\eta-\cos\xi}$ and p is normalised by $\rho U_0^2\,St$. The continuity equation will be satisfied if a stream function $\Psi(\xi,\eta,t)$ is introduced as follows

$$u=\frac{1}{hh_3}\frac{\partial \Psi}{\partial \eta},\quad v=-\frac{1}{hh_3}\frac{\partial \Psi}{\partial \xi}. \tag{13.2.4}$$

Since at the initial instant, an inviscid potential flow prevails throughout the entire flow field, the stream function Ψ belonging to the outer region has the form

$$\Psi(\xi,\eta,t)=V(t)Q(\xi,\eta),$$

where $Q(\xi,\eta)$ is the steady potential flow due to a unit velocity at infinity.

Using the expression for $Q(\xi,\eta)$ from (Bentwich and Miloh, 1978) we can write the initial and boundary conditions for the problem under consideration, namely

$$\Psi=\frac{\partial \Psi}{\partial \eta}=0 \quad \text{at } \eta=\eta_1 \text{ and } \eta=\eta_2,\ t>0, \tag{13.2.5}$$

$$\Psi\sim t^{\alpha}\sinh|\eta_1|\frac{\sin^2\xi}{2(\cosh\eta-\cos\xi)^2} \quad \text{as } \xi^2+\eta^2\to 0,\ t>0 \tag{13.2.6}$$

$$\Psi=0 \qquad \text{at } \xi=0,\ \xi=\pi,\ t>0, \tag{13.2.7}$$

$$\Psi=0 \qquad \text{for } t\le 0, \tag{13.2.8}$$

$$\Psi=t^{\alpha}\frac{\sinh^2|\eta_1|}{(\cosh\eta-\cos\xi)^{1/2}}\left\{\frac{\sin^2\xi}{2(\cosh\eta-\cos\xi)^{3/2}}\right. \tag{13.2.9}$$

$$\left.-\sum_{n=1}^{\infty}\left[\hat{a}_n\exp\left[\left(n+\frac{1}{2}\right)\eta\right]+\hat{b}_n\exp\left[-\left(n+\frac{1}{2}\right)\eta\right]\right]V_n(\beta)\right\} \qquad \text{as } t\to 0^+,$$

where

$$\hat{a}_n = \frac{\sqrt{2}}{A} n(n+1) \exp\left[-\left(n+\frac{1}{2}\right)\eta_2\right] \sinh\left[\left(n+\frac{1}{2}\right)\eta_1\right],$$

$$\hat{b}_n = -\frac{\sqrt{2}}{A} n(n+1) \exp\left[\left(n+\frac{1}{2}\right)\eta_1\right] \sinh\left[\left(n+\frac{1}{2}\right)\eta_2\right],$$

$$A = \sinh\left[\left(n+\frac{1}{2}\right)(\eta_2 - \eta_1)\right]$$

and $V_n(\beta)$ are the Gegenbauer polynomials of degree $-\frac{1}{2}$ and order n+1, connected with Legendre polynomials of first kind $P_n(\beta)$ by the relation

$$V_n(\beta) = \frac{P_{n-1}(\beta) - P_{n+1}(\beta)}{2n+1}, \quad (\beta = \cos\xi).$$

The problem (13.2.1) - (13.2.9) is singular and may be treated by means of the method of matched asymptotic expansions. Hence, the solution can be sought in three regions: two boundary layers adjacent to the spheres and an outer flow field.

In the outer region for small ε one can assume that

$$F = F_1 + \varepsilon F_2 + \varepsilon^2 F_3 + \ldots,$$

where F = (Ψ, U, V, P). Since the outer flow is inviscid and irrotational, all Ψ_1, Ψ_2,… are solutions of the equation

$$\frac{\partial}{\partial\xi}\left(\frac{1}{h_3}\frac{\partial\Psi}{\partial\xi}\right) + \frac{\partial}{\partial\eta}\left(\frac{1}{h_3}\frac{\partial\Psi}{\partial\eta}\right) = 0. \tag{13.2.10}$$

For the boundary layer around the sphere $\eta = \eta_1$, we can expand the variable in the form

$$u = u_1 + \varepsilon u_2 + \ldots, \qquad v = \varepsilon v_1 + \varepsilon^2 v_2 + \ldots, \tag{13.2.11}$$

$$p = p_1 + \varepsilon p_2 + \ldots, \qquad \psi = \psi_1 + \varepsilon \psi_2 + \ldots$$

and $\eta = \eta_1 + \varepsilon\zeta$, where ζ is the stretched variable. In the boundary layer around the sphere $\eta = \eta_2$, we use the stretched variable $\eta = \eta_2 - \varepsilon\bar{\zeta}$ and similar expansions for $\bar{\psi}$, $\bar{u}$, $\bar{v}$, $\bar{p}$.

The boundary conditions for the outer and inner solutions are the conditions at infinity and on the two spherical surfaces, respectively, and a matching procedure is applied between the boundary layers solutions and the outer solution.

The expression for the explicit solution and the governing equations for the detachment time can be found in Kalitzova-Kurteva and Zapryanov (1990). On the basis of the formulae obtained there, we shall study the flow properties depending on the acceleration parameter α, the separation between the spheres D = d/a, the radii ratio k = a/b and the Reynolds number Re.

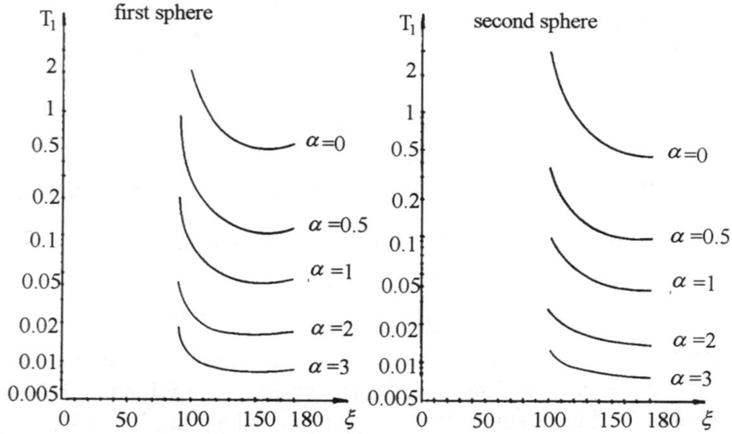

Fig.13.2.1. Movement of the detachment points over the spheres at Re = 1000, D = 1, k= 1 and various α.

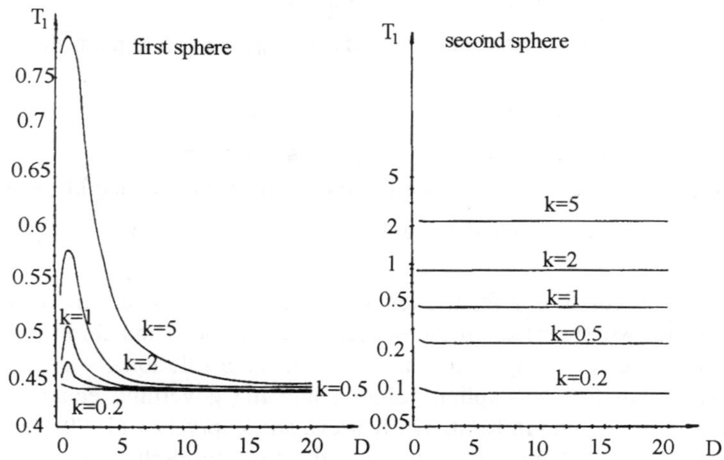

Fig.13.2.2. Initial detachment time as a function of the separation distance D at Re = 1000, $\alpha = 0$ and various k.

Fig.13.2.1. shows the movement of the detachment points from the two spheres at Re = 1000, D = 1, k = 1 and the various values of α. The detachment time is inversely proportional to α. It is interesting to note that the point of the first detachment from the second sphere coincides with the rear stagnation point, while the initial detachment from the first sphere occurs before its rear stagnation point.

The initial detachment times for the two spheres as a function of the separation distance D at Re = 1000, $\alpha = 0$ and various values of k are given in Fig.13.2.2. If the distance D increases above 10, then the two times tend to become equal. The character of changing of the detachment times is different for each sphere. The initial detachment time for the upstream sphere increases for $D \leq 1$ and decreases for other values of D, while the initial detachment time for the downstream sphere does not practically depend on D.

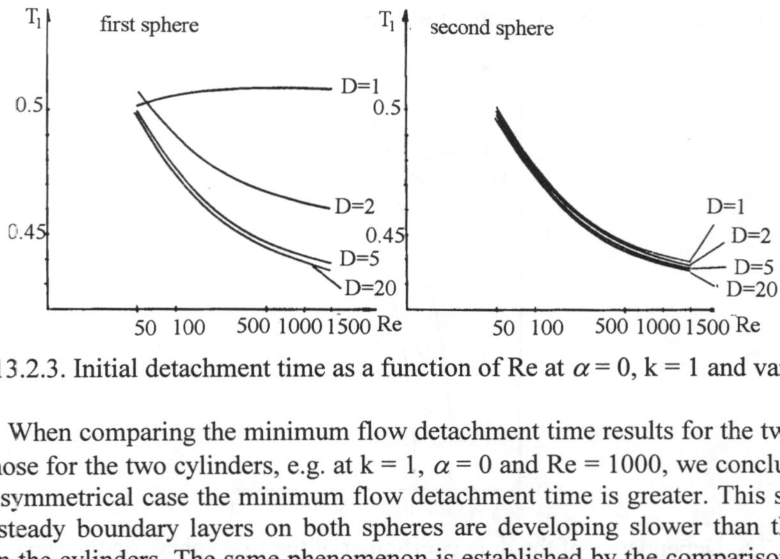

Fig.13.2.3. Initial detachment time as a function of Re at $\alpha = 0$, k = 1 and various D.

When comparing the minimum flow detachment time results for the two spheres with those for the two cylinders, e.g. at k = 1, $\alpha = 0$ and Re = 1000, we conclude that in the axisymmetrical case the minimum flow detachment time is greater. This shows that the unsteady boundary layers on both spheres are developing slower than the similar ones on the cylinders. The same phenomenon is established by the comparison between the single cylinder results and the single sphere results.

Fig.13.2.3 presents the initial time of boundary layer detachment from two equal spheres as a function of Re, at $\alpha = 0$ and various D. It is seen that at D = 1 the initial detachment time is an increasing function of Re for the first sphere and a decreasing function for the second one. For D=20 (practically no interaction between the two bodies), there is a good agreement with Slavchev's results (Slavchev, 1974/1975) for a single sphere in the same flow conditions.

In Fig.13.2.4 the streamline pattern for an impulsive flow past two spheres is illustrated, concerning an early moment before detachment takes place from the spheres, at D = 0.5, k = 2, $T_1 = 0.5$ and Re = 1000.

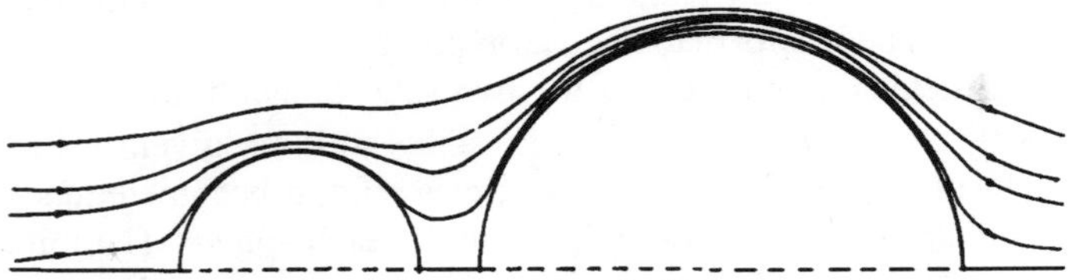

Fig.13.2.4. Streamline pattern for two spheres before detachment at Re = 1000, $T_1 = 0.5$, k =2, $\alpha = 0$ and D = 0.5.

Due to the symmetry of the flow, the integral effect of the stresses acting upon the entire surface of each sphere is only the drag force.

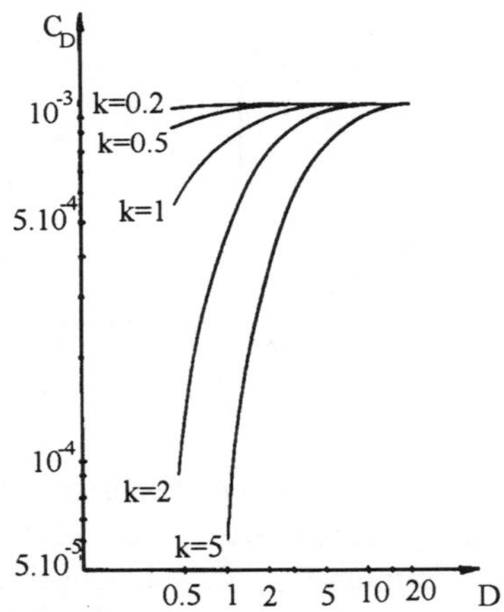

Fig.13.2.5. Dependence of the drag coefficient of the first sphere on the separation distance D at Re = 1000, $T_1 = 0.09$, $\alpha = 0$ and various k.

In order to analyse the hydrodynamic force interaction between the two spheres, we give the drag force coefficients:

$$C_{D_1} = \frac{F_{x_1'}}{\rho U_0^2 a^2 St} = -\pi \sinh|\eta_1| \int_0^{\pi} \left(2\varepsilon\gamma \sinh\eta_1 \left.\frac{\partial u_1}{\partial \zeta}\right|_{\eta=\eta_1} \right. \tag{13.2.12}$$

$$\left. + \sinh|\eta_1| \left.\frac{\partial p_1}{\partial \xi}\right|_{\eta=\eta_1} + \sinh|\eta_1| \varepsilon \left.\frac{\partial p_2}{\partial \xi}\right|_{\eta=\eta_1} \right) \frac{\sin^2 \xi}{(\cosh \eta_1 - \cos \xi)^2} d\xi$$

for the upstream sphere and

$$C_{D_2} = \frac{F_{x_2'}}{\rho U_0^2 a^2 St} = \pi \sinh|\eta_1| \int_0^{\pi} \left(-2\varepsilon\gamma \sinh\eta_2 \left.\frac{\partial \bar{u}_1}{\partial \zeta}\right|_{\eta=\eta_2} \right. \tag{13.2.13}$$

$$\left. + \sinh|\eta_1| \left.\frac{\partial \bar{p}_1}{\partial \xi}\right|_{\eta=\eta_2} + \sinh|\eta_1| \varepsilon \left.\frac{\partial \bar{p}_2}{\partial \xi}\right|_{\eta=\eta_2} \right) \frac{\sin^2 \xi}{\left(\cosh \eta_2 - \cos \xi\right)^2} d\xi$$

for the downstream sphere.

Fig.13.2.5 shows the dependence of the first sphere drag force coefficient on D at Re = 1000, $\alpha = 0$, $T_1 = 0.09$ and various k. One can observe that when the separation distance increases, the first sphere drag coefficient increases and tends to the value of the single sphere drag coefficient at the same flow conditions. The first sphere drag force coefficient is a decreasing function of the second sphere radius b.

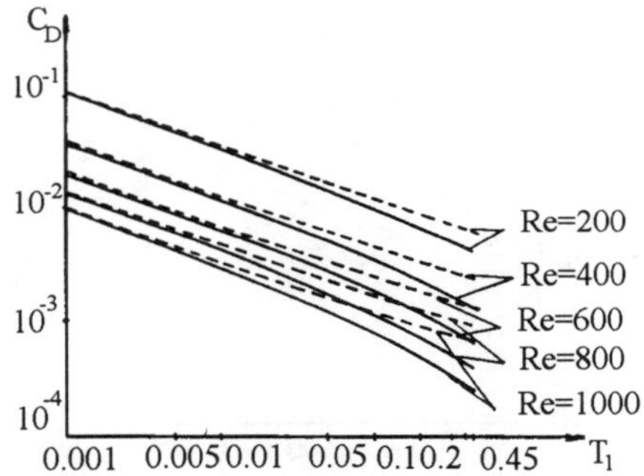

Fig.13.2.6. Dependence of the drag coefficient of the two spheres on the time T_1 at $\alpha = 0$, k = 1, D = 1 and various Re. The curve (———) corresponds to the first sphere, (…………) to the second sphere.

In Fig.13.2.6 the drag coefficients of the two spheres are plotted against the time T_1, at $\alpha = 0$, k = 1, D = 1 and different values of Re. It is seen that the drag coefficients of the two bodies are decreasing functions of T_1 and, in contrast to the steady flow past two spheres (Zapryanov and Toshev, 1985), the drag coefficient of the second sphere for small times is greater.

The drag coefficients of the two spheres as functions of Re at D = 0.5, $T_1 = 0.01$, $\alpha = 0$ and various k are given in Fig.13.2.7 and these functions are decreasing.

There are no results in the literature treating the unsteady interactions between two spheres at small times. For this reason we have compared our results in the limiting case of large distance between the two bodies with the result of Wang (1969) for one sphere. It is seen from Fig.13.2.8 that when there are no interactions between the bodies (D = 30) we have a very good agreement of the two results at large Re (Re = 100).

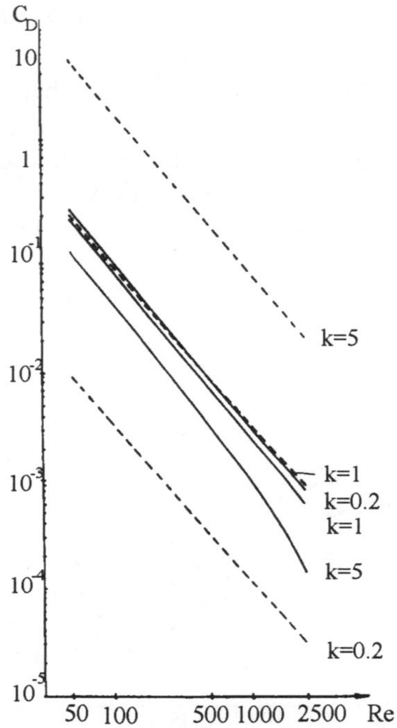

Fig.13.2.7. Drag coefficients of the two spheres as a function of Re at $T_1 = 0.01$, $\alpha = 0$, D = 0.5 and various k. The curve (———) corresponds to the first sphere, (··········) to the second sphere.

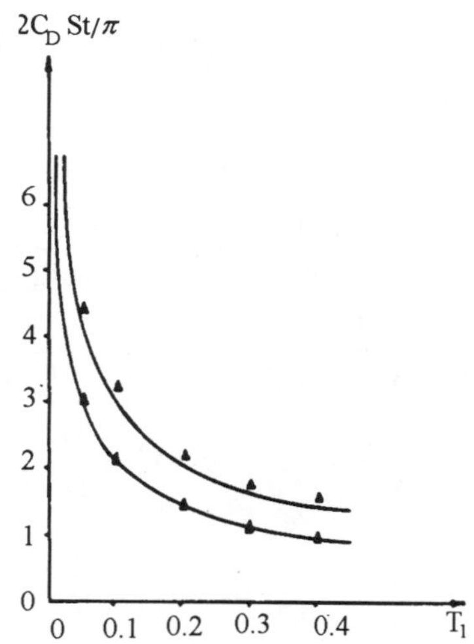

Fig.13.2.8. Dependence of the drag coefficient on the time T_1 at Re = 50 and 100 from: (——) the two spheres results of Kalitzova-Kurteva and Zapryanov (1990) at $\alpha = 0$, k = 1, and D = 30; (▲) the single sphere results of Wang (1969).

13.3. Hydrodynamic Interaction between Two Translatory Oscillating Spherical Particles

In this section we are going to consider two outer problems of oscillatory particles motion:

1) the particles are oscillating in direction parallel to their line of centres and the problem is axisymmetrical;

2) the particles are oscillating perpendicularly to their line of centres and the problem is three dimensional.

Axisymmetric problem. The two spherical particles of radii a and b are supposed to be immobile, while the flow at infinity oscillates with a velocity $U_0 \cos\omega t$ in a direction parallel to their line of centres. A bispherical coordinate system (η, ξ, φ) is connected with the spheres as it has been introduced in section 7.1. Then the equations of the spheres' surfaces have the form: $\eta = \alpha_1 > 0$ and $\eta = \alpha_2 < 0$. Due to the axisymmetry of the problem the third velocity component v_φ is zero. If U_0 is a characteristic velocity, ω^{-1} is characteristic time and the focal distance c (see Fig.7.7.1) is characteristic length, then the dimensionless Navier - Stokes equation for the stream function Ψ has the form:

$$\frac{\partial\left(E^2\Psi\right)}{\partial\tau}-\frac{\varepsilon h_3}{h^2}\frac{\partial\left(\Psi,\frac{E^2\Psi}{h_3^2}\right)}{\partial(\eta,\xi)}=M^{-2}E^4\Psi, \tag{13.3.1}$$

where

$$E^2\equiv\frac{h_3}{h^2}\left[\frac{\partial}{\partial\eta}\left(\frac{1}{h_3}\frac{\partial}{\partial\eta}\right)+\frac{\partial}{\partial\xi}\left(\frac{1}{h_3}\frac{\partial}{\partial\xi}\right)\right], \tag{13.3.1'}$$

$$h=h_1=h_2=\frac{1}{(\cosh\eta-\cos\xi)},\qquad h_3=\frac{\sin\xi}{(\cosh\eta-\cos\xi)} \tag{13.3.1''}$$

and $\tau=\omega t$, $M^2=\frac{\omega c^2}{\nu}$, $\varepsilon=\frac{U_0}{\omega c}$, $Re_s\equiv\varepsilon^2M^2=\frac{U_0^2}{\omega\nu}$.

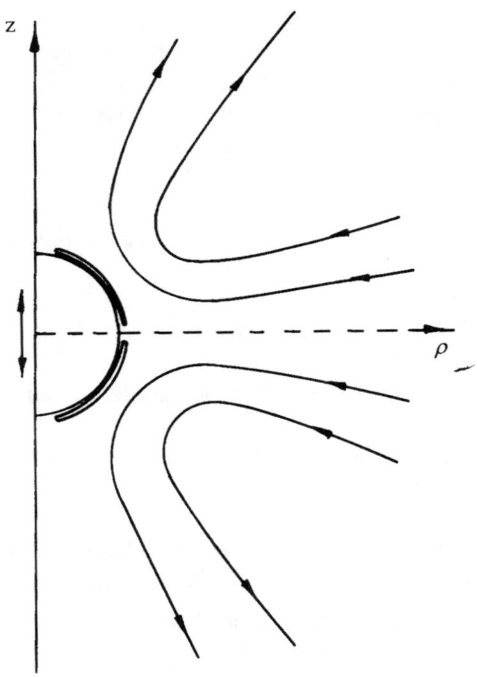

Fig.13.3.1. Sketch of the steady flow structure in the case of a single sphere at $\varepsilon \ll 1$, $M \gg 1$, $Re_s \ll 1$.

The boundary conditions are:

$$\Psi = \frac{\sin^2 \xi}{2(\cosh\eta - \cos\xi)^2}\exp(i\tau) \qquad \text{as} \qquad \eta^2 + \xi^2 \to 0, \tag{13.3.2}$$

$$\Psi = \frac{\partial \Psi}{\partial \eta} = 0 \qquad \text{at} \qquad \eta = \alpha_1 \,, \ \ \eta = \alpha_2 \,. \tag{13.3.3}$$

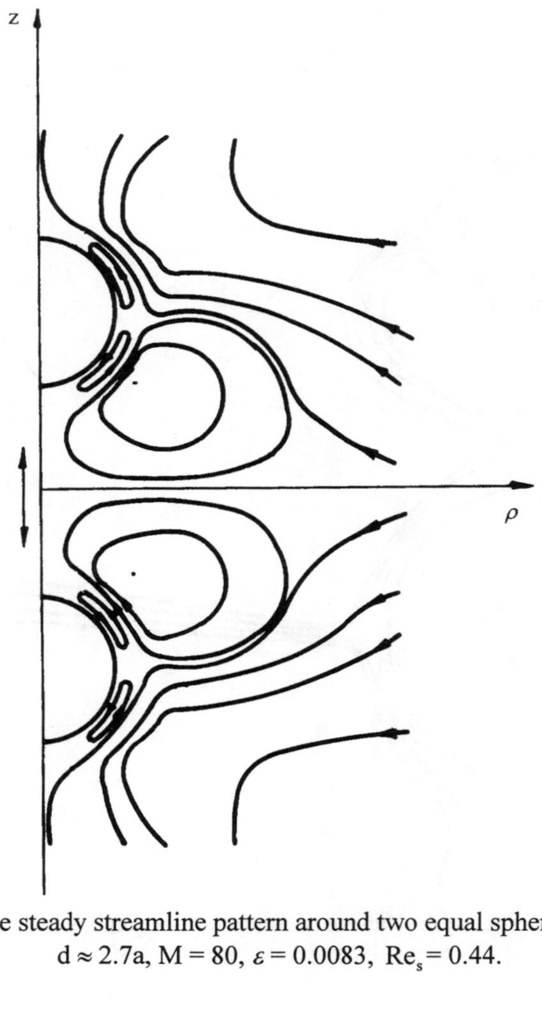

Fig.13.3.2. The steady streamline pattern around two equal spheres at a=b=0.47, $d \approx 2.7a$, $M = 80$, $\varepsilon = 0.0083$, $Re_s = 0.44$.

The formulated problem (13.3.1)-(13.3.3) will be solved by the method of matched asymptotic expansions.

It is interesting to analyse the vortex structure of the flow and the hydrodynamical interaction of the two spheres as a function of the distance between them, the frequency parameter M and the radii ratio. The vortex pattern of the steady streaming induced by the harmonic oscillations of a single sphere in a viscous fluid at $\varepsilon << 1$, $M >> 1$, $Re_s << 1$ (Riley, 1966) is given in Fig.13.3.1.

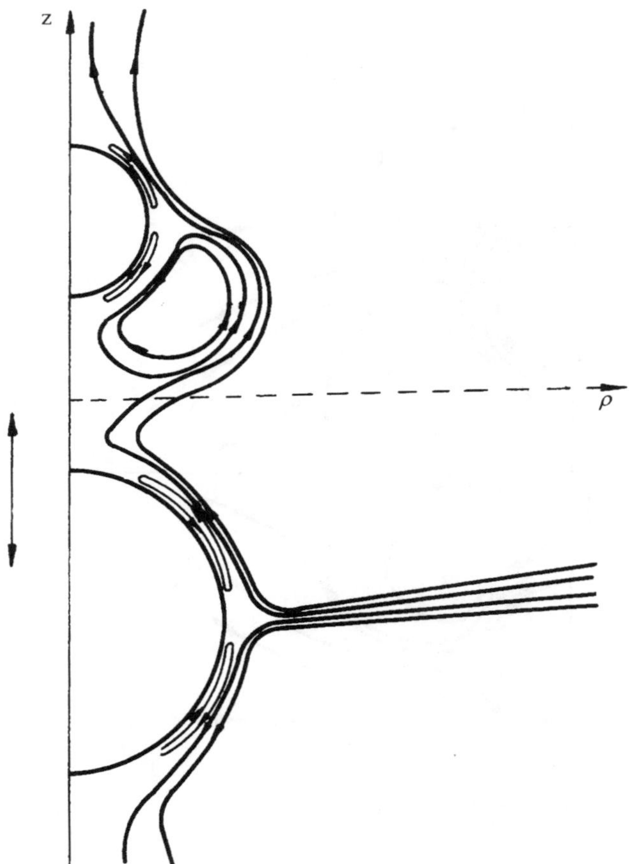

Fig.13.3.3. The steady streamline pattern around two unequal spheres at a = b/2 = 0.47, $d \approx 2.2a$, M = 80, $\varepsilon = 0.0083$, $Re_s = 0.44$.

The obtained results for the vortex structure of the secondary steady streaming during the harmonic oscillations of two spheres parallel to their central line are illustrated in Fig.13.3.2. and Fig.13.3.3. The first figure represents the steady flow in the case of two equal spheres and it could be seen that, besides the Schlichting's vortices in the Stokes boundary layers around the spheres, in the external flow there exist four vortices with finite centres and four vortices, which diffuse at infinity. When the distance between the two spheres becomes very large, the interaction between the particles vanishes and the streamline pattern around each of the spheres gets the structure obtained by Riley (1966) and shown in Fig.13.3.1.

When a = b the steady streaming is directed symmetrically toward both spheres, while at a ≠ b it is directed towards the larger sphere and the flow pattern is changed significantly (Fig.13.3.3.). Due to the symmetry of the obtained solution, a drag force acting only on the spheres exists. Its' coefficient is given by the formula:

$$F_D = \frac{F_z}{c^3 \bar{\rho} \omega U_0} = \mp \frac{\pi}{M^2} \int_{-1}^{1} \frac{(1-\beta^2)}{(\cosh\eta - \beta)^2} \frac{\partial}{\partial \eta} \left[\frac{(\cosh\eta - \beta)^2 E^2 \Psi}{(1-\beta^2)} \right]_{\eta=\alpha} d\beta ,$$

where the sign (-) corresponds to $\alpha > 0$ and (+) corresponds to $\alpha < 0$; $\bar{\rho}$ is the fluid density and $\beta = \cos\xi$. Making use of the stream function asymptotic expansion in M^{-1} and ε, the drag force coefficient could be expressed as:

$$F_D^k(\alpha_k) = -\frac{\pi}{2} \int_{-1}^{1} \left\{ \hat{\gamma} \frac{\partial^3 \hat{\psi}_{00}}{\partial \hat{\zeta}^3}\bigg|_{\zeta=0} + \frac{\hat{\gamma}}{M} \frac{\partial^3 \hat{\psi}_{01}}{\partial \hat{\zeta}^3}\bigg|_{\zeta=0} \mp \frac{5\sqrt{2}}{M} \sinh\alpha_k \frac{\partial^2 \hat{\psi}_{00}}{\partial \hat{\zeta}^2}\bigg|_{\zeta=0} \right.$$

$$\left. + \varepsilon \hat{\gamma} \frac{\partial^3 \hat{\psi}_{10}}{\partial \hat{\zeta}^3}\bigg|_{\zeta=0} + O(M^{-2}, \varepsilon \mathrm{Re}_s) \right\} d\beta = F_D^{(u),k} + F_D^{(s),k} + O(M^{-2}, \varepsilon \mathrm{Re}_s),$$

where ^ corresponds to k = 1, 2; $\hat{\zeta} = \pm(\alpha_k - \xi) M/\sqrt{2}$, $\hat{\gamma} = \cosh\alpha_k - \beta$ and $\hat{\psi}_{ij}$ (i, j = 0, 1) are the stream functions [see (13.3.11)] in the two boundary layers around the spheres and their explicit expressions could be found in Tabakova and Zapryanov (1982a).

From the analysis of the hydrodynamic interaction of the forces acting on both spheres it is established (see Fig.13.3.4.) that:

i) the drag acting on each of the particles is smaller than the drag on a single sphere at the same values of the flow parameters and this tendency is preserved with the change of the stream parameters values;

ii) if the spherical particles have equal radii then the total drag on the first sphere in the flow is greater than the total drag on the second one, while the drags due to the secondary steady streaming have equal values but opposite signs, i.e., attractive forces (see Table 13.3.1.);

iii) the drag on the first particle increases with an increase in the distance between the spheres as well as with a decrease in the second particle radius (see Table 13.3.1. and Fig.13.3.4.).

Table 13.3.1.

a=b=1, M=20, ε=0.033, Re_s=0.44

d	$F_D^1(\tau=2\pi)$	$F_D^2(\tau=2\pi)$	$F_D^1(\tau=\pi/2)$	$F_D^2(\tau=\pi/2)$	$F_D^{(s),1}$	$F_D^{(s),2}$
0.5	-0.5916	-0.5970	6.4946	6.5328	-0.0092	0.0092
1	-0.6147	-0.6268	6.6670	6.6958	-0.0043	0.0043
3	-0.6500	-0.6616	6.8816	6.8955	-0.0006	0.0006
5	-0.6581	-0.6663	6.9226	6.9312	-0.00015	0.00015

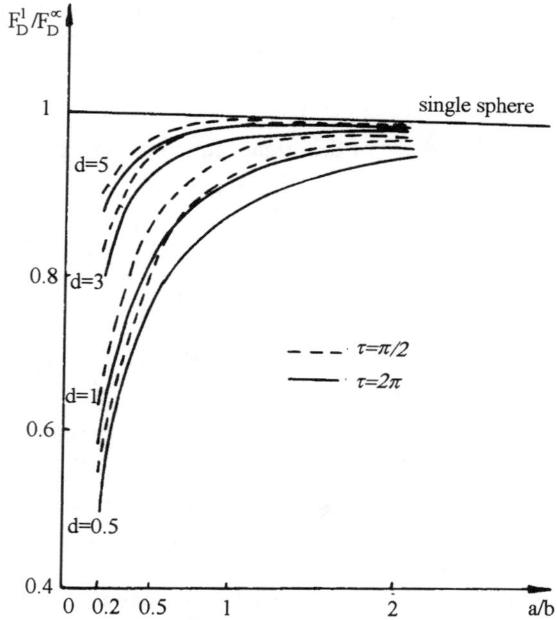

Fig.13.3.4. Ratio of the drag coefficient F_D^1 on a sphere of radius a (due to the present analysis) and the drag coefficient F_D^∞ on a single sphere (see Riley, 1966) as a function of a/b and for different d at M = 20, $\varepsilon = 0.033$, $Re_s = 0.44$, $\tau = 2\pi$ and $\tau = \pi/2$.

<u>3D Problem.</u> In the second case we suppose that the two spherical particles are motionless and a cylindrical (ρ', φ, z') and bispherical (η, ξ, φ) coordinate system is connected with them. The flow at infinity oscillates perpendicularly to the line of the spheres' centres. Since the problem is three-dimensional, the stream function concept

can not be introduced and we are going to treat this case in its u, v, w, p concept. The variables are dimensionalized in the following manner:

$$\rho' = c\rho,\ z' = cz,\ t = \omega^{-1}\tau,\ p' = U_0\omega\bar{\rho}c\,p,$$
$$v'_\rho = U_0 u,\ v'_\varphi = U_0 v,\ v'_z = U_0 w.$$

Then the Navier-Stokes equations and the continuity equation referring to the cylindrical coordinate system are given in their dimensionless form:

$$\frac{\partial u}{\partial \tau} + \varepsilon\left(u\frac{\partial u}{\partial \rho} + \frac{v}{\rho}\frac{\partial u}{\partial \varphi} - \frac{v^2}{\rho} + w\frac{\partial u}{\partial z}\right) \tag{13.3.4}$$
$$= -\frac{\partial p}{\partial \rho} + M^{-2}\left(\nabla^2 u - \frac{u}{\rho^2} - \frac{2}{\rho^2}\frac{\partial v}{\partial \varphi}\right),$$

$$\frac{\partial v}{\partial \tau} + \varepsilon\left(u\frac{\partial v}{\partial \rho} + \frac{v}{\rho}\frac{\partial v}{\partial \varphi} + \frac{uv}{\rho} + w\frac{\partial v}{\partial z}\right) \tag{13.3.5}$$
$$= -\frac{1}{\rho}\frac{\partial p}{\partial \varphi} + M^{-2}\left(\nabla^2 v + \frac{2}{\rho^2}\frac{\partial u}{\partial \varphi} - \frac{v}{\rho^2}\right),$$

$$\frac{\partial w}{\partial \tau} + \varepsilon\left(u\frac{\partial w}{\partial \rho} + \frac{v}{\rho}\frac{\partial w}{\partial \varphi} + w\frac{\partial w}{\partial z}\right) = -\frac{\partial p}{\partial z} + M^{-2}\nabla^2 w, \tag{13.3.6}$$

$$\frac{\partial}{\partial \rho}(\rho u) + \frac{\partial v}{\partial \varphi} + \frac{\partial}{\partial z}(\rho w) = 0, \tag{13.3.7}$$

where

$$\nabla^2 \equiv \frac{\partial^2}{\partial \rho^2} + \frac{1}{\rho}\frac{\partial}{\partial \rho} + \frac{1}{\rho^2}\frac{\partial^2}{\partial \varphi^2} + \frac{\partial^2}{\partial z^2}.$$

The dimensionless boundary conditions are:

i) at infinity $\eta^2 + \xi^2 \to 0$

$$u = \cos\varphi \exp(i\tau),\ v = -\sin\varphi \exp(i\tau),\ w = 0; \tag{13.3.8}$$

ii) on the spherical particles $\eta = \alpha_1$, $\eta = \alpha_2$

$$u = v = w = 0. \tag{13.3.9}$$

The formulated problem (13.3.4)-(13.3.9) will be solved at small oscillation amplitudes ε, large oscillation frequencies M and small Reynolds number of the steady flow Re_s. At these conditions the problem is singular and the method of matched asymptotic expansions is applied again.

The solution is sought in bispherical coordinates in three different regions: two Stokes boundary layers (I and III) around both spheres and an external to the boundary layers region (II). In the boundary layers the variable η is stretched as

$$\zeta = -(\eta - \alpha_1)\frac{M}{\sqrt{2}},\quad \bar{\zeta} = (\eta - \alpha_2)\frac{M}{\sqrt{2}}. \tag{13.3.10}$$

The unknown functions are expanded in double asymptotic series on the powers of the small parameters ε and M^{-1}:

$$F = F_{00} + M^{-1}F_{01} + \ ... \ + \varepsilon\left[F_{10} + M^{-1}F_{11} + \ ... \ \right] + \ ..., \tag{13.3.11}$$

where

$$F = \left(u, v, w, p, U, V, W, P, \overline{u}, \overline{v}, \overline{w}, \overline{p}\right).$$

Here u, v, w, p are the velocity components and the pressure in the region I, and U, V, W, P are the respective ones in the region II, while $\overline{u}, \overline{v}, \overline{w}, \overline{p}$ correspond to the boundary layer region III.

According to the boundary condition (13.3.8), the expansions for the first approximations of U, V, W and P are chosen to satisfy the equations (13.3.4)-(13.3.7), i.e.:

$$U_0(\rho, \varphi, z, \tau, M) = \sum_{j=0}^{\infty}\left[U_{0,0j}(\rho, z)M^{-j} + U_{2,0j}(\rho, z)M^{-j} + 1\right]\cos\varphi \ \exp(i\tau), \tag{13.3.12}$$

$$V_0(\rho, \varphi, z, \tau, M) = \sum_{j=0}^{\infty}\left[U_{2,0j}(\rho, z)M^{-j} - U_{0,0j}(\rho, z)M^{-j} - 1\right]\sin\varphi \ \exp(i\tau),$$

$$W_0(\rho, \varphi, z, \tau, M) = \sum_{j=0}^{\infty} W_{1,0j}(\rho, z)M^{-j}\cos\varphi \ \exp(i\tau),$$

$$P_0(\rho, \varphi, z, \tau, M) = \sum_{j=0}^{\infty}\left[P_{1,0j}(\rho, z)M^{-j} - i\rho\right]\cos\varphi \ \exp(i\tau).$$

Then making use of equations (13.3.7), (13.3.11) and (13.3.12) we obtain

$$i\left(U_{2,oj} + U_{0,0j}\right) = -\frac{\partial P_{1,0j}}{\partial\rho} + M^{-2}\left(L_2^2 U_{2,oj} + L_0^2 U_{0,0j}\right), \tag{13.3.13}$$

$$i\left(U_{2,oj} - U_{0,0j}\right) = \frac{P_{1,0j}}{\rho} + M^{-2}\left(L_2^2 U_{2,oj} - L_0^2 U_{0,0j}\right), \tag{13.3.14}$$

$$iW_{1,0j} = -\frac{\partial P_{1,0j}}{\partial z} + M^{-2}L_1^2 W_{1,0j}, \tag{13.3.15}$$

$$\frac{\partial U_{0,0j}}{\partial\rho} + \frac{\partial U_{2,0j}}{\partial\rho} + \frac{2U_{2,0j}}{\rho} + \frac{\partial W_{1,0j}}{\partial z} = 0, \tag{13.3.16}$$

where

$$L_m^2 = \frac{\partial^2}{\partial\rho^2} + \frac{1}{\rho}\frac{\partial}{\partial\rho} - \frac{m^2}{\rho^2} + \frac{\partial^2}{\partial z^2},$$

m=0, 1, 2, ... and j=0, 1, 2, ...

The boundary conditions (13.3.8) and (13.3.9) expressed by the new unknown quantities have the form:

$$U_{0,0j} = U_{2,oj} = W_{1,0j} = 0 \ \text{(j=0, 1, 2, ...) at } z^2 + \rho^2 \to \infty \ (\eta^2 + \xi^2 \to 0), \tag{13.3.17}$$

$$U_{0,00} = -1, \quad U_{0,0j} = 0, \quad \text{(j=1, 2, ...)}, \tag{13.3.18}$$

$$U_{2,oj} = W_{1,0j} = 0 \quad (j=0, 1, 2, \ldots) \text{ at } \eta = \alpha_1, \ \eta = \alpha_2 .$$

Equations (13.3.13)-(13.3.16) and the boundary condition (13.3.17) are appropriate for the intermediate region II, while for the regions I and III the same equations are used but written in bispherical coordinates by means of the corresponding stretched variables (13.3.10). Beside the boundary conditions (13.3.9), the unknown functions of the regions I and III must fulfil the conditions obtained after the matching with the intermediate region solution. In this way, the solutions (13.3.4) - (13.3.7) can be found as successive approximations on the small parameter M^{-1} and some of the constants can be determined in the further solution stages.

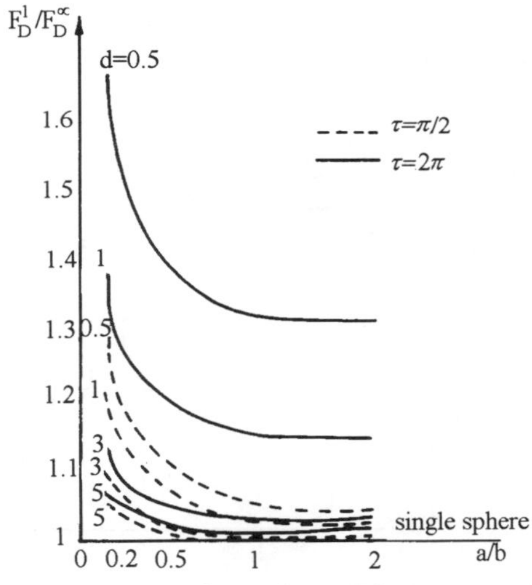

Fig.13.3.5. Ratio of the drag coefficient F_D^1 on a sphere of radius a (Tabakova and Zapryanov, 1982b) and the drag coefficient F_D^∞ on a single sphere (Riley, 1966) as a function of a/b and for different d at $M = 20$, $\varepsilon = 0.033$, $Re_s = 0.44$, $\tau = 2\pi$ and $\tau = \pi/2$.

Equating the terms in front of M^{-1} in (13.3.13) - (13.3.16), the following equations are reached:

$$L_1^2 P_{1,0j} = 0, \quad L_0^2 U_{0,0j} = 0, \quad L_2^2 U_{2,oj} = 0, \quad L_1^2 W_{1,0j} = 0, \qquad (13.3.19)$$

where j = 0. These equations and the boundary conditions (13.3.17) are satisfied if the unknown functions have the following form (Tabakova, 1981):

$$P_{1,0j} = (\cosh\eta - \beta)^{\frac{1}{2}} \sum_{n=1}^{\infty} \left[A_{n,0j} \sinh\left(n + \frac{1}{2}\right)\eta + B_{n,0j} \cosh\left(n + \frac{1}{2}\right)\eta \right] P_n^1(\beta), \quad (13.3.20)$$

$$U_{0,0j} = (\cosh\eta - \beta)^{\frac{1}{2}} \sum_{n=0}^{\infty} \left[C_{n,0j} \sinh\left(n + \frac{1}{2}\right)\eta + D_{n,0j} \cosh\left(n + \frac{1}{2}\right)\eta \right] P_n(\beta),$$

$$U_{2,0j} = (\cosh\eta - \beta)^{\frac{1}{2}} \sum_{n=2}^{\infty} \left[E_{n,0j} \sinh\left(n + \frac{1}{2}\right)\eta + F_{n,0j} \cosh\left(n + \frac{1}{2}\right)\eta \right] P_n^2(\beta),$$

$$W_{1,0j} = (\cosh\eta - \beta)^{\frac{1}{2}} \sum_{n=1}^{\infty} \left[G_{n,0j} \sinh\left(n + \frac{1}{2}\right)\eta + H_{n,0j} \cosh\left(n + \frac{1}{2}\right)\eta \right] P_n^1(\beta),$$

where $\beta = \cos\xi$, $j = 0$ and $P_n^m(\beta)$ are the associated Legendre polynomials of order n and rank m.

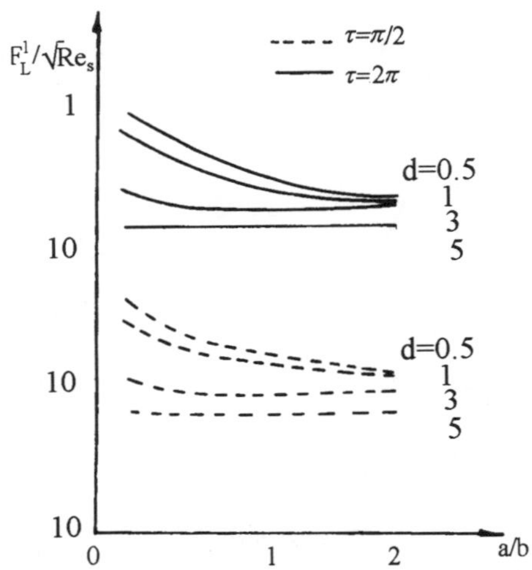

Fig.13.3.6. The total lift force coefficient F_L^1 on a sphere of radius a, as a function of a/b and for different d at $M = 20$, $\varepsilon << 1$, $Re_s << 1$, $\tau = 2\pi$ and $\tau = \pi/2$.

Since the proposed scheme applying the method of inner and outer expansions for the two bodies' case has been discussed in the previous section, here we are not going to consider the further approximations in ε and M^{-1}. The calculations are quite complicated due to the three-dimensional character of the problem (Tabakova and Zapryanov, 1982b).

From the analysis of the hydrodynamic interaction of the forces acting on the two spheres immersed in an oscillating flow, perpendicularly to their line of centres, it is established that (see Fig.13.3.5. - Fig.13.3.7.):

i) in contrast to the single particle case (Riley, 1966) and to the axisymmetrical case described in the first part of the present section, here the drag force occurs only due to the primary oscillatory flow (of order ε^0), while a lift force is induced by the secondary streaming, steady and unsteady (of order ε^1), which is due to the inertial effects;

ii) the drag acting on each of the particles is larger than the drag on a single sphere at the same values of the flow parameters and this tendency is preserved with the change of the flow parameters values;

iii) if the spherical particles have equal radii then the drags on both the spheres have equal values and signs (see Table 13.3.2.);

Table 13.3.2.

$a = b = 1,\ M = 20,\ \varepsilon = 0.033,\ Re_s = 0.44$

d	F_D^1 $\tau=2\pi$	F_D^2 $\tau=2\pi$	F_D^1 $\tau=\pi/2$	F_D^2 $\tau=\pi/2$	F_L^1 $\tau=2\pi$	F_L^2 $\tau=2\pi$	F_L^1 $\tau=\pi/2$	F_L^2 $\tau=\pi/2$	$F_L^{(s),1}$	$F_L^{(s),2}$
0.5	-0.873	-0.873	7.370	7.370	0.185	-0.185	0.01	-0.01	0.102	-0.102
1	-0.767	-0.767	7.171	7.171	0.167	-0.167	0.009	-0.009	0.089	-0.089
3	-0.692	-0.692	6.999	6.999	0.107	-0.107	0.006	-0.006	0.056	-0.056
5	-0.678	-0.678	6.970	6.970	0.077	-0.077	0.004	-0.004	0.041	-0.041

iv) the drag on the first particle decreases to unity with an increase in the distance between the spheres as well as with decrease of the second particle radius (see Table 13.3.2. and Fig.13.3.5.) and this drag behaviour is preserved when the flow parameters are changed within given limits;

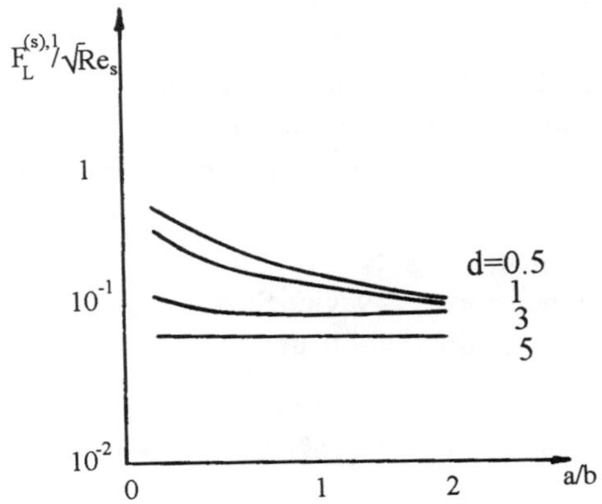

Fig.13.3.7. The steady lift force coefficient $F_L^{(s),1}$ on a sphere of radius a, as a function of a/b and for different d at $M = 20$, $\varepsilon \ll 1$, $Re_s \ll 1$.

v) the lift force on the first particle decreases monotonously to zero as the distance between the two spheres, or the radius of the second sphere, increases; a repulsive force appears between the two spheres for different time periods (see Fig. 13.3.6.) and its behaviour does not differ when the flow parameters are changed in the given limits;

vi) the steady lift force due to the secondary steady streaming has a similar character to the total lift force (see Fig.13.3.7.);

vii) for two equal spheres the lift forces are of equal value, but of opposite sign (see Table 13.3.2.), i.e., repulsive forces.

A similar analysis for the 2D problems of two parallel circular cylinders oscillating in direction parallel or perpendicular to the plane containing their axes is performed by Zapryanov et al. (1988). The observed tendencies of the drag and lift forces are the same as the corresponding forces for the spherical particles oscillation problem, presented in this section.

13.4. Viscous Flow between Two Eccentric Rotary Oscillating Spherical Particles

In this section the inner flow between two eccentric spheres, i.e., in a spherical annuli, induced by the high-frequency rotational oscillations of one of the spheres or of both of them with angular velocity $w = \overline{\Omega}\cos\omega t$, where $\overline{\Omega}$ is the amplitude of torsional oscillations and ω is their frequency. The problem is assumed axisymmetric as the oscillations are performed around the axis of symmetry. If the spheres radii are a and b, where $b > a$ or $k = b/a > 1$, then a bispherical coordinate system (7.1.1) is again suitable to be implemented and thus the spheres surfaces are defined as coordinate surfaces: $\eta = \alpha_1 > 0$ and $\eta = \alpha_2 > 0$ (see Fig.13.4.1).

Following the analysis in section 13.3 a dimensionless stream function $\Psi = \Psi'/\varepsilon c^3\omega$ is introduced. Since the spheres, or one of them, rotationally oscillate, there exists a rotational velocity $w = \rho' v_\varphi$ and its dimensionless form is $\overline{w} = w/\varepsilon c\omega$. Here $\varepsilon = \overline{\Omega}/\omega$ is the amplitude parameter, which throughout the analysis in the present section is assumed to be small, i.e., $\varepsilon \ll 1$.

Then the full dimensionless Navier-Stokes equations (1.3.15) in bispherical coordinates rewritten in Ψ - Ω formulation are:

$$\frac{\partial(E^2\Psi)}{\partial\tau} + \varepsilon\frac{2\Omega}{h_3^2h^2}\frac{\partial(\Omega,h_3)}{\partial(\eta,\xi)} - \varepsilon\frac{1}{h_3h^2}\frac{\partial(\Psi,E^2\Psi)}{\partial(\eta,\xi)} + \varepsilon\frac{2E^2\Psi}{h_3^2h^2}\frac{\partial(\Psi,h_3)}{\partial(\eta,\xi)} = M^{-2}E^4\Psi, \tag{13.4.1}$$

$$\frac{\partial\Omega}{\partial\tau} - \varepsilon\frac{1}{h_3h^2}\frac{\partial(\Psi,\Omega)}{\partial(\eta,\xi)} = M^{-2}E^2\Omega, \tag{13.4.2}$$

where $\Omega = h_3^2\overline{w}$, $\tau = \omega t$, $M^2 = \dfrac{\omega c^2}{\nu}$, the operator E^2 is given by (13.3.1′), $E^4 = E^2(E^2)$ and h_1, h_2 and h_3 by (13.3.1″). The boundary conditions for Ψ and Ω on either sphere surface and on the axis of symmetry, respectively, are:

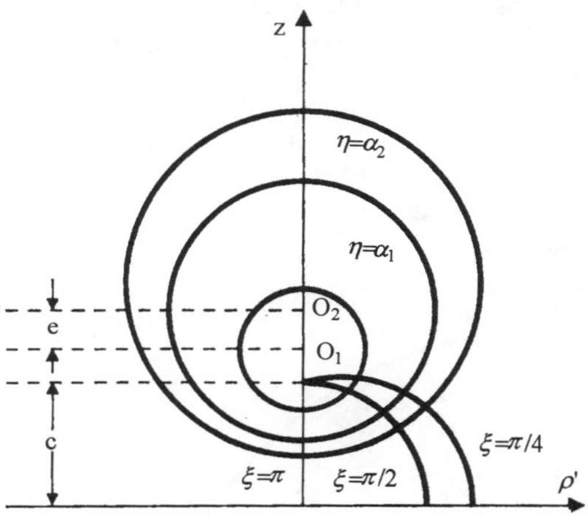

Fig.13.4.1. A sketch of two eccentric spheres geometry.

$$\Psi = \frac{\partial\Psi}{\partial\eta} = 0, \qquad \text{at} \quad \eta = \alpha_1 \ , \ \eta = \alpha_2 , \tag{13.4.3}$$

$$\Omega = \frac{\sin^2\xi}{(\cosh\alpha_1 - \cos\xi)^2}\exp(i\tau) \qquad \text{at} \quad \eta = \alpha_1 , \tag{13.4.4}$$

$$\Omega = \frac{\sin^2\xi}{(\cosh\alpha_2 - \cos\xi)^2}\exp(i\tau) \qquad \text{at} \quad \eta = \alpha_2 , \tag{13.4.5}$$

$$\Psi = 0, \qquad \Omega = 0 \qquad \text{at} \quad \xi = 0,\ \pi. \tag{13.4.6}$$

As in the previous section 13.3., it is convenient to use complex form for the unknown functions Ψ and Ω and afterwards to take only their real part.

The functions Ψ and Ω are expanded in regular power series, similar to (13.3.11) with respect to the small parameter ε. The analysis, presented here, is restricted up to $O(\varepsilon)$. It is easily shown that $\Psi_{2n} \equiv 0$ and $\Omega_{(2n+1)} \equiv 0$, while for the terms Ψ_1 and Ω_0 the following system of equations is derived from (13.4.1) and (13.4.2):

$$\frac{\partial(E^2\Psi_1)}{\partial\tau} + \varepsilon\frac{2\Omega_0}{h_3^2h^2}\frac{\partial(\Omega_0, h_3)}{\partial(\eta,\xi)} = M^{-2}E^4\Psi_1 , \tag{13.4.7}$$

$$\frac{\partial \Omega_0}{\partial \tau} = \mathrm{M}^{-2}\mathrm{E}^2\Omega_0, \tag{13.4.8}$$

with boundary conditions (13.4.3) - (13.4.6). Under the assumption of high frequencies of oscillations, M >> 1, for the system (13.4.7), (13.4.8) and (13.4.3) - (13.4.6) the method of matched asymptotic expansions is applied to find both functions Ψ_1 and Ω_0 in the core region and in the two boundary layers created around the spherical surfaces, i.e., Ψ_1 and Ω_0 are sought in singular series expansions in M^{-1} as given by (13.3.11).

Since (13.4.8) yields a linear relation, the rotational velocity Ω_0 has only unsteady part which is expressed by:

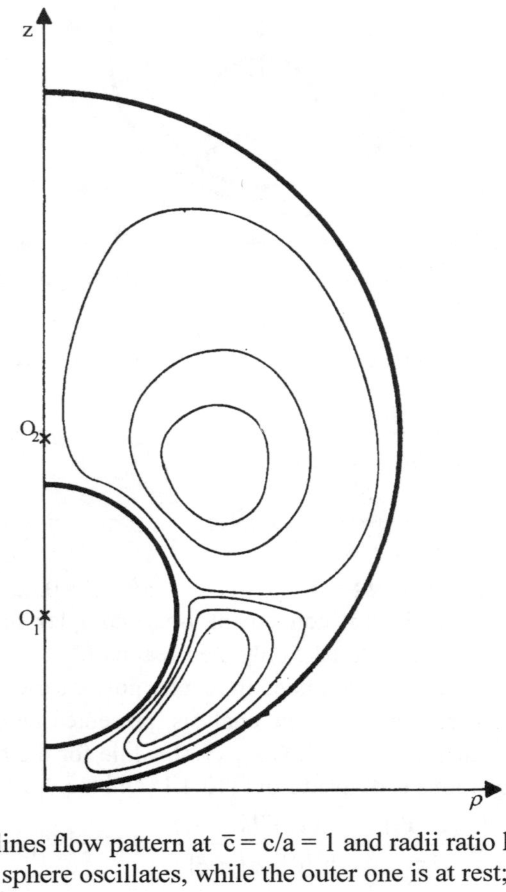

Fig.13.4.2. Streamlines flow pattern at $\overline{c} = c/a = 1$ and radii ratio k = 2.5: a) the inner sphere oscillates, while the outer one is at rest;

$$\Omega_{00} = \Omega_{01} = 0, \qquad \hat{\omega}_{00} = \frac{1-\beta^2}{\left(\cosh\alpha_k - \beta\right)^2}\exp\left(i\tau - \frac{1+i}{\cosh\alpha_k - \beta}\hat{\zeta}\right), \quad (13.4.9)$$

$$\hat{\omega}_{01} = \mp\frac{(1+i)\left(1-\beta^2\right)\sinh\alpha_k}{\sqrt{2}\left(\cosh\alpha_k - \beta\right)^4}\hat{\zeta}^2\exp\left(i\tau - \frac{1+i}{\cosh\alpha_k - \beta}\hat{\zeta}\right)$$

where ^ corresponds to k = 1, 2; $\hat{\zeta} = \pm(\alpha_k - \xi)M/\sqrt{2}$, and $\hat{\omega}_{0j}$, (j = 0, 1) are the rotational velocity functions in the two boundary layers around the spheres. However, the stream function contains an unsteady and a steady part, due to the non-linearity of (13.4.7). The explicit expressions for both parts in the boundary layer can be found in the works of Kovatcheva et al. (1985, 1988).

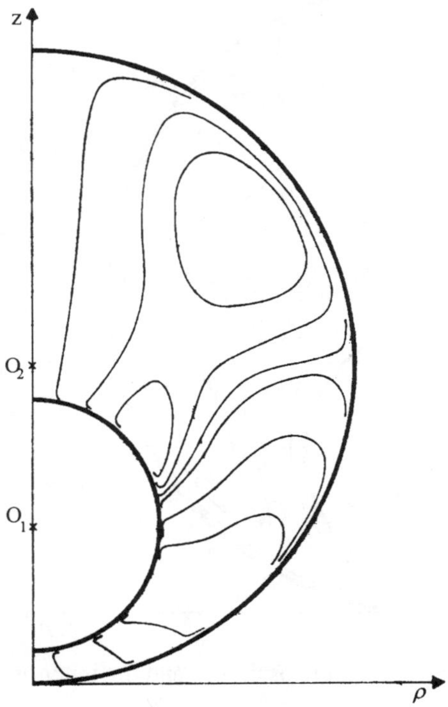

Fig.13.4.2. b) both spheres are oscillating;

Special numerical procedure is applied when tracing the streamlines in the original coordinates space (ρ, z). In order to have good accuracy for large values of eccentricity e/a, a non-uniform mesh in ξ direction is employed:

$$\xi = 2\text{arctg}\left[\frac{\sinh\alpha_2}{\cosh\alpha_2 + 1}\text{tg}\left(\frac{s.\sinh\alpha_2}{2c}\right)\right], \qquad s = (j-1)\frac{c\pi}{(N-1)\sinh\alpha_2}, \quad (13.4.10)$$

which ensures that the reference points in the original space are uniformly distributed along the meridian of the outer sphere. Then the mesh points are given:

$$\eta_i = \alpha_2 + (i-1)h_\eta, \qquad \xi_j = 0 + (j-1)h_\xi, \quad (13.4.11)$$

where h_η and h_ξ are uniform steps in the corresponding directions, and i =1, ..., M and j = 1, ..., N.

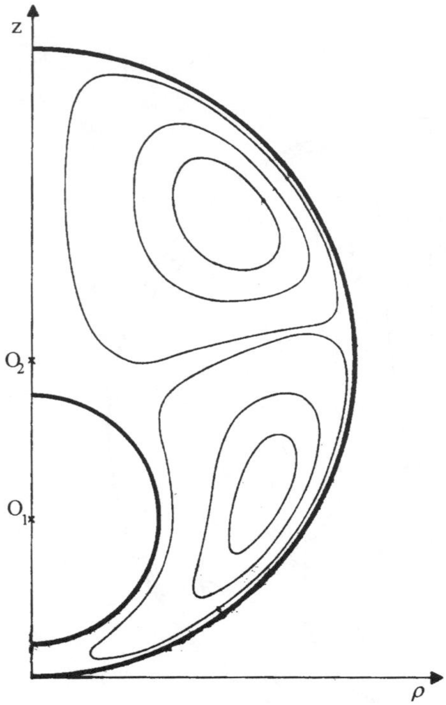

Fig.13.4.2. c) the outer sphere oscillates and the inner one is immobile.

The obtained results of small eccentricity case $\bar{c}$ = 100 are compared with the concentric spheres case (Zapryanov and Tabakova, 1979) and give a good coincidence up to fourth digit. Moreover, even for $\bar{c}$ = 10, the coincidence accuracy is within 2%.

The streamlines patterns for different eccentricities are illustrated in Fig.13.4.2. and Fig.13.4.3. Three different cases are shown in Fig.13.4.2. for $\bar{c}$ = c/a = 1 and radii ratio k = 2.5: a) the inner sphere oscillates, while the outer one is at rest; b) both spheres are oscillating; c) the outer sphere oscillates and the inner one is immobile. From Fig.13.4.2.b) it is seen that the negative streamlines coalesce with each other and form a flow known as "cat eye". However, for smaller $\overline{c} = 0.2$ and k = 1.5, the torsional oscillations of both spheres do not cause a flow of this type and only one positive vortex is observed in Fig.13.4.3. The limiting case of k >> 1 and $\overline{c} = O(1)$ approximately corresponds to the case of a sphere oscillating near a plane.

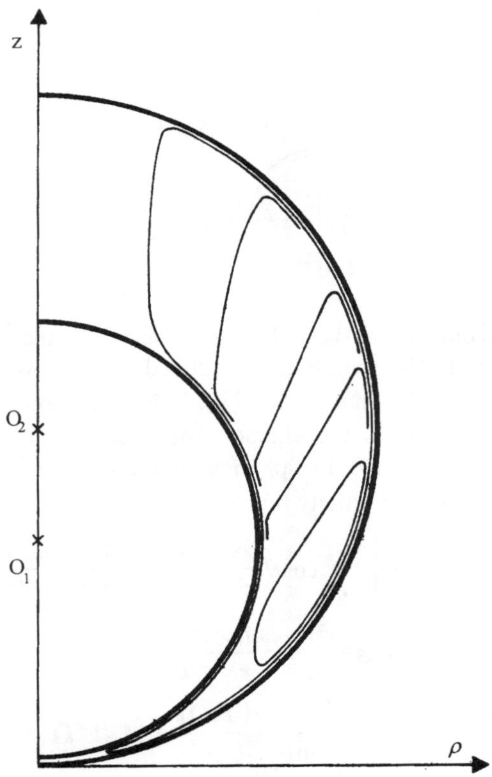

Fig.13.4.3. Streamlines flow pattern at $\overline{c} = c/a = 0.2$ and radii ratio k = 1.5 for the case of both spheres oscillations.

13.5. Numerical Modelling of the Flow Induced by the Rotary Oscillating Rigid Particle in a Spherical Container

The purpose of the present section is to solve numerically the problem of intermediate frequency oscillations of a viscous fluid in spherical annuli. Similarly to the problem studied in the previous section 13.4., the two concentric spheres of radii a and b (a > b) are considered torsionally oscillating with frequency ω and angular amplitude $\overline{\Omega}_i$ (i =1 corresponds to the inner sphere and i = 2 to the outer one), i.e., the angular velocity of each sphere is $\overline{\Omega}_i \cos(\omega t + c_i)$, where c_i is the oscillation phase. Because of the assumed axisymmetry, a spherical coordinate system (r, θ, φ) with origin coinciding with the spheres' centre and given by (1.3.13) is adopted (see Fig.13.5.1). Thus the oscillations and the induced fluid motion are independent of φ.

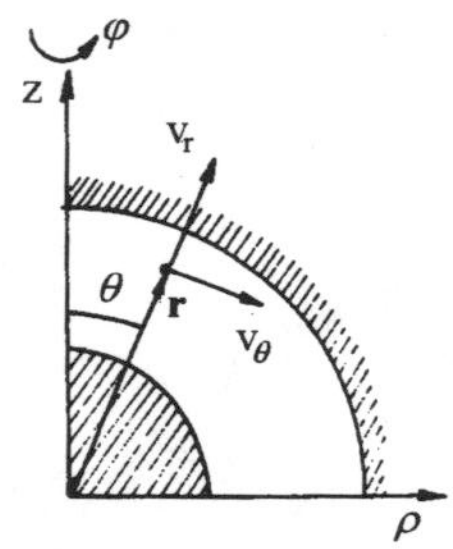

Fig.13.5.1. A sketch of 1/4 of the region between two concentric spheres

The velocity components v_r and v_θ are related to the stream function ψ by formulae (1.3.17) and (1.3.14), while the rotational velocity w is connected with the angular velocity v_φ by the relation $w = r \sin\theta \, v_\varphi$. If the outer sphere radius a is taken as a characteristic length and ω^{-1} is the characteristic time, then the full dimensionless Navier-Stokes equations (1.3.15) in spherical coordinates rewritten in Ψ - ζ - Ω formulation ($\zeta = -E^2\Psi$ is the vorticity function) are:

$$\frac{\partial \zeta}{\partial \tau} - \frac{\varepsilon}{r^2 \sin\theta}\left[2\Omega\left(\cos\theta\frac{\partial \Omega}{\partial r} - \frac{\sin\theta}{r}\frac{\partial \Omega}{\partial \theta}\right) + \frac{\partial(\Psi,\zeta)}{\partial(r,\theta)}\right.$$
$$\left. - 2\zeta\left(\cos\theta\frac{\partial \Psi}{\partial r} - \frac{\sin\theta}{r}\frac{\partial \Psi}{\partial \theta}\right)\right] = M^{-2}E^2\zeta, \tag{13.5.1}$$

$$\frac{\partial \Omega}{\partial \tau} - \varepsilon\frac{1}{r^2 \sin\theta}\frac{\partial(\Psi,\Omega)}{\partial(r,\theta)} = M^{-2}E^2\Omega, \tag{13.5.2}$$

where $\Omega = r\sin\theta\, w$, $\tau = \omega t$, $M^2 = \dfrac{\omega a^2}{\nu}$, $\varepsilon = \dfrac{\sqrt{\overline{\Omega_1^2} + \overline{\Omega_2^2}}}{\omega}$, the operator E^2 is given by $E^2 \equiv \dfrac{\partial^2}{\partial r^2} + \dfrac{\sin\theta}{r^2}\dfrac{\partial}{\partial\theta}\left(\dfrac{1}{\sin\theta}\dfrac{\partial}{\partial\theta}\right)$. Here, all the unknown functions and variables are dimensionless.

We have to note, that there is no boundary condition for the vorticity ζ, but there are two boundary conditions for the stream function Ψ on both rigid surfaces. Then the boundary conditions of the considered problem are as follows:

$$\Psi = \frac{\partial\Psi}{\partial r} = 0 \quad \text{at } r = 1 \text{ and } r = b/a = k, \tag{13.5.3}$$

$$\Omega = \alpha_1 \sin^2\theta \cos(\tau + c_1) \qquad \text{at } r = 1, \tag{13.5.4}$$

$$\Omega = \alpha_2 k^2 \sin^2\theta \cos(\tau + c_2) \quad \text{at } r = b/a = k, \tag{13.5.5}$$

$$\Omega = \Psi = E^2\Psi = 0 \quad \text{at } \theta = 0, \qquad \frac{\partial\Omega}{\partial\theta} = \Psi = E^2\Psi = 0 \quad \text{at } \theta = \pi/2. \tag{13.5.6}$$

The last condition expresses the symmetry requirement, which indicates that it is sufficient to seek the solution only in one quarter of the spherical annulus (see Fig.13.5.1). The relative amplitude parameters α_i are given by the relation $\alpha_i = \overline{\Omega}_i \big/ \sqrt{\overline{\Omega_1^2} + \overline{\Omega_2^2}}$, (i =1, 2).

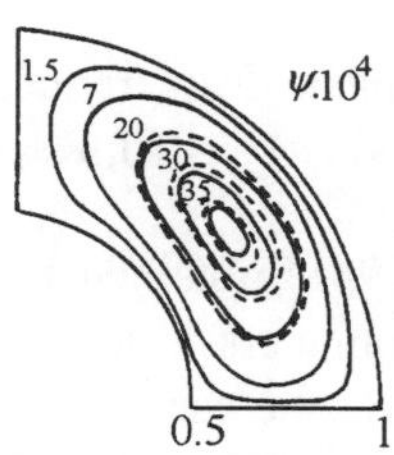

Fig.13.5.2. The streamlines flow pattern for the steady case at Re = 100 when the inner sphere is held fixed and the outer rotates, according to the results of: (——) Pearson (1967); (··········) Zapryanov and Christov (1981).

The formulated problem (13.5.1)-(13.5.6) forms a closed system for the unknown functions Ψ, ζ, Ω in the space $\boldsymbol{B} = (r, \theta, \tau)$. It is solved numerically by means of the finite-difference ADI method adapted for unsteady flow with second order accuracy in all direction of the space $\boldsymbol{B}$ (for the details of the computational procedure see Zapryanov and Christov, 1981).

If in the boundary conditions (13.5.4) and (13.5.5) the term $\cos(\tau + c_i)$ is omitted, then the problem becomes completely steady and the results of its numerical solution show a good agreement with the steady case results for the streamlines of Pearson (1967) at $Re = \varepsilon M^2 = 100$ (see Fig.13.5.2).

As stated in section 13.1. the most representative characteristics of oscillatory motion is the steady part of the flow in the cross-section plane. This secondary streaming can be obtained by averaging the solution with respect to time. The calculation of the solution can be done in the same way as for the steady flow and only the condition of convergence is different for the oscillatory case. For the latter, a convergence of two successive periods in time is required, which ensures that the periodic solution is reached. This criterion can be found in (Christov and Zapryanov, 1980).

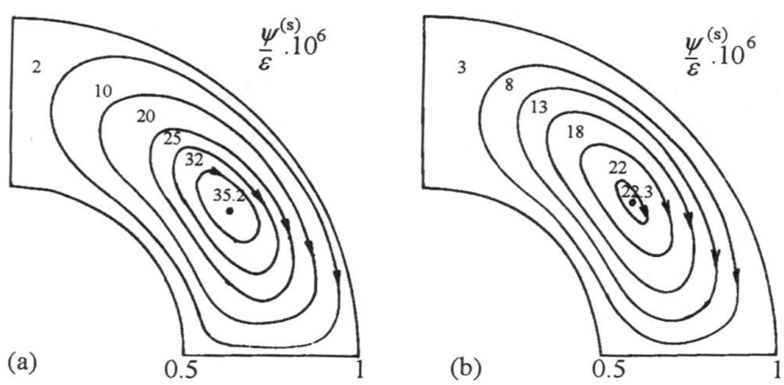

Fig.13.5.3. Steady streaming at low frequency when inner sphere held fixed, outer oscillates at M = 1, k = 0.5 and : a) $\varepsilon \leq 10$, $Re_s \leq 100$; b) $\varepsilon = 100$, $Re_s \leq 10000$.

Zapryanov and Christov (1981) obtained results of the considered problem for various values of the governing parameters, ε, M and Re_s. A comparison with the known asymptotic solutions for the two extreme cases of frequency parameter, namely $M \ll 1$ and $M \gg 1$ is performed. The low-frequencies oscillations problem permit a regular asymptotic solution as presented by Munson and Douglass (1979). Their asymptotic results for the steady streaming, when M approaches unity and ε is small, have been used by Zapryanov and Christov (1981) for comparison with their own numerical results. The agreement is quite satisfactory. Typical steady streaming flow patterns at M = 1, k = 0.5 and two different values of ε are shown in Fig.13.5.3a) and b).

The high-frequency case has been studied by Tabakova and Zapryanov (1978) and Zapryanov and Tabakova (1979) by the matched asymptotic expansions method, but as a characteristic length they have used the inner sphere radius b. Further, the comparison with that work needs to take into account the different values of the parameter M corresponding to the same physical problems.

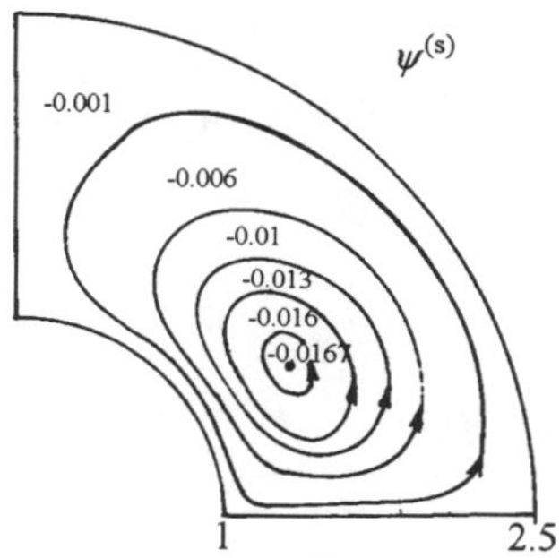

Fig.13.5.4. Steady streaming at high frequency when outer sphere held fixed, inner oscillates at M = 10 (25), k = 0.4, $\varepsilon = 1$, $Re_s = 100$.

In Fig.13.5.4. the steady flow streamlines are shown, as the parameter M of Tabakova and Zapryanov is given in brackets, i.e., M = 10 (25). The maximum of the stream function is chosen for the test. From the present results of Zapryanov and Christov, $\left|\Psi^{(s)}_{max}/\varepsilon\right| = 0.0167$ at $\varepsilon = 1$, while from the results of Tabakova and Zapryanov it is 0.0142. The maximum difference between both solutions is up to 20%, which is attributed to the Stokes boundary layer of thickness $O(M^{-1})$ developed at large M and yields a necessity of finer mesh than used in the computation.

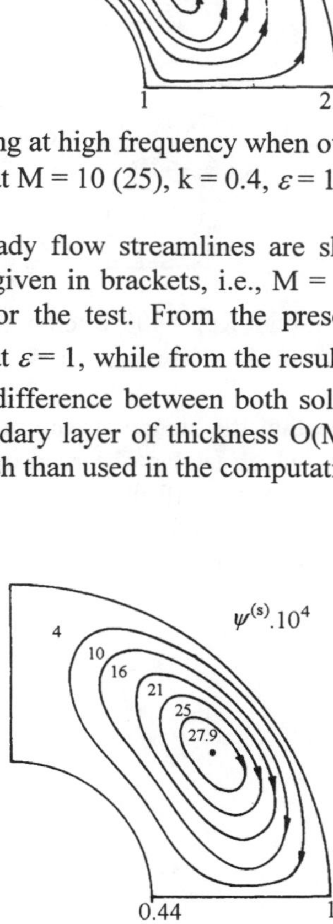

Fig.13.5.5. Steady streaming at intermediate frequency when inner sphere held fixed, outer oscillates at M = 4.4 (10), k = 0.44, $\varepsilon = 1$.

In Fig.13.5.5. the steady streaming for a truly intermediate value of the frequency parameter M = 4.4. (10) can be seen. It is evident that for the same value $Re_s = 100$ the order of $\Psi^{(s)}$ remains the same even for quite different values of the frequency parameter

M (Fig.13.5.3.b. and Fig.13.5.5). Therefore, for intermediate M the secondary streaming also depends chiefly on Re_s. This fact is confirmed also by Tabakova and Zapryanov (1978) for M >>1 and Munson and Douglass (1979) for M << 1, respectively. Fig.13.5.6. illustrates the steady secondary flow at Re_s = 400 and M = 20 (50). It is remarkable that the steady streaming exhibits a jet character near the equatorial plane.

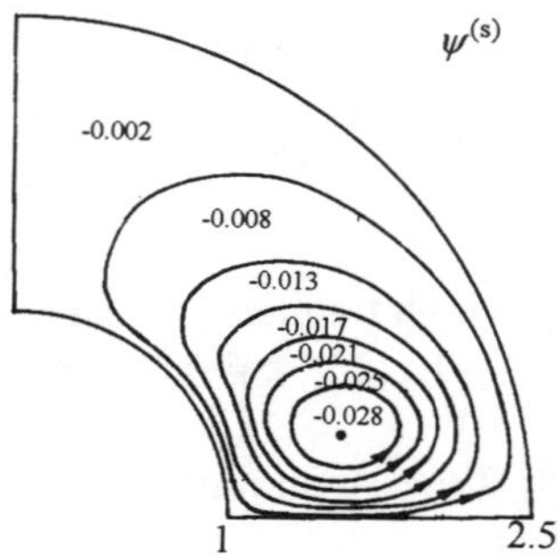

Fig.13.5.6. Steady streaming at high $Re_s = 400$ when outer sphere held fixed, inner oscillates at M = 20 (50), k = 0.4, $\varepsilon = 1$.

Finally, we can conclude that the steady streaming is governed exclusively by the steady Reynolds number Re_s, which is also observed in both asymptotic limits. The suggested numerical model permits one to fill the gap between the singular and regular asymptotic solutions of the problem, thereby securing a theoretical solution for the region of intermediate values of the parameters.

CHAPTER 14.

Finite Deformations of Drops and Bubbles at Moderate Reynolds Numbers Flows

14.1. Introduction

In the scientific literature only a small number of numerical solutions of comprising interfaces unsteady problems, based on the Euler or Navier-Stokes equations are still published, despite their importance in the explanation of a lot of hydrodynamic effects. For example, depending on the value of Weber number, the unsteady solution of the bubble rise in a quiescent viscous fluid may pass into a steady solution or may show the non-existence of a steady solution, or may be periodic in time. By means of the numerical solutions of free surfaces unsteady hydrodynamic problems, some difficulties, which are met when applying some solution algorithms to steady problems at given values of their parameters, can be explained.

The transient behaviour of a cavitation bubble in an inviscid liquid near a rigid or free surface was studied by Blake et al. (1986, 1987). In such problems the bubble volume changes with time and the evolution of the process is the subject of main interest here. The problem is modelled via a system of integral equations and solved by the boundary-integral method. A Lagrangian description of the bubble surface is employed. These techniques allow the problem to be studied in detail and produce many interesting bubble shapes. The influence of gravity is also considered. An essential advance achieved by Blake et al. is the treatment of the unsteadiness and the influence of a rigid or a free interface on the cavitation process. However, the role of surface tension has not been studied there.

For the same class problems as the one treated by Blake et al. (1987), a similar, but more sophisticated numerical method was developed by Dommermuth and Yue (1987). They used a regridding algorithm to remove the instabilities, which allows longer time simulations to be achieved than the ones obtained by the previous authors and their results are in good agreement with the experimental measurements. Two wave-body interaction problems were also treated. The results confirmed the universality and the good accuracy of the method.

The first method created for solving numerically the unsteady hydrodynamic problems with interfaces, based on the Navier-Stokes equations, is the so-called "method of marks and cells". The calculations are performed in Euler cells and a set of marked particles is exploited to trace the free surface position changes. A basic defect of this approach is neglecting of the force due to the surface tension. This method is proposed by Harlow and Welch (1965) and used by Nichols and Hirt (1971) and Foote (1973), as well.

The use of finite-difference techniques in solving partial differential equations is a well-established one. There is a substantial and rapidly growing amount of literature addressing various types of difference schemes available. The literature also abounds of fluid mechanics applications. Details of the different finite-difference methods are available from texts, such as (Richtmyer and Morton, 1967), (Roache, 1976) and Ames (1977). These methods are most suitable, or at least simplest to implement, for boundary geometry that is rectilinear. For free-surface problems the boundary usually will not intersect the mesh system at grid points that are regularly spaced. These difficulties are considerable when such intersections are time-dependent.

Various means of circumventing these difficulties exist. One of them is to use the so-called method of boundary-fitted coordinates. The basic idea of this method is to transform the physical boundaries of a given problem into coordinate lines in a mapped space, finite-difference computations can be conveniently carried out in a regular mesh system. The boundary-fitted coordinate techniques have been successfully applied by many authors [Shanks and Thompson (1977), Haussling and Coleman (1979), etc.].

At first sight there are no difficulties and unsolved problems connected with this method but, in fact, it can be applied effectively only to simple problem geometry and large computational resources. This method was exploited by Ryskin and Leal (1984a, 1984b, 1984c) and Christov and Volkov (1985) for solving the problem of a steady rise of a deformable bubble. Later it was generalised by Kang and Leal (1987) for solving the problem of unsteady rise of a deformable bubble in unbounded viscous fluid.

Another efficient way to overcome the mentioned difficulties is the application of the finite elements method (FEM). Introduced and developed in the 60s, it has established itself as a mainstream computational tool for stress analysis of complicated geometry structures and it was soon adapted to fluid flow problems.

The finite elements method offers great flexibility for solving many unsteady problems of theoretical and practical interest. Its advantages are particularly due to:

(i) The ability of representation of arbitrary complicated boundaries;

(ii) The freedom allowed in constructing grids using a priori knowledge of the sought solution, i.e., high gradient regions can be accurately represented with a fine mesh, while coarser grids are sufficient for regions of small solution variation;

(iii) Systematic rules for numerical approximation of well posed problems, with various types of boundary conditions.

As mentioned above, the application of FEM to flow analysis problems is a relatively recent development but, nevertheless, a substantial amount of literature on the topic has already emerged [Zienkiewicz, (1977); Connor and Brebbia, (1976); Giralt and Raviart, (1979); Thomasset, (1981); Cuvelier et al. (1986), etc.].

The vast majority of analytical and numerical research on fluid particle motions in a viscous fluid has been performed on the basis of four assumptions:

(i) the complete neglect of non-linear terms in Navier-Stokes equations, i.e., the Reynolds number is reduced to zero;
(ii) the shape of drops or bubbles is fixed, or their deformations are small;
(iii) quasi-steady regime of motions of the fluid particles;
(iv) the deviations from the assumptions (i), (ii) and (iii) are small.

If none of the above mentioned "standard assumptions" is fulfilled the hydrodynamic problems become very complicated.

In this chapter, some typical unsteady problems for fluid particles past by viscous flow will be solved by means of the finite elements method applied to the unsteady Navier-Stokes equations [Shopov, (1985); Zapryanov et al., (1987); Zapryanov et al., (1989); Shopov et al., (1989, 1990), Shopov and Minev (1992), etc.]. The presented calculation technology refers to the velocity-pressure formulation of the treated problems. It is characterised by a suitable choice of the finite elements, and a special assembly when solving the obtained equations system for the velocity and pressure. Since in the first stage of the solution procedure the continuity equation usage and the pressure elimination reminds to an extent the so-called procedure of static condensation, in the classical FEM this method is conventionally called the condensation method (Shopov, 1985).

In this chapter we mainly consider the following three non-steady axisymmetric problems with two deformable interfaces:
(a) the buoyancy-driven motion of a deformable bubble towards a deformable liquid/liquid interface;
(b) the buoyancy-driven motion of two deformable drops in a free or bounded quiescent viscous fluid;
(c) the buoyancy-driven motion of a deformable compound drop in a free or bounded quiescent viscous fluid.

In this connection our objectives are: to expose those factors which control the details of time evolution deformation in these three purely hydrodynamic problems; to provide an analysis, so that (a) the influence of the fluid inertia can be recognised; (b) and the time dependent deformation effects can be evaluated.

All these problems are linked with the better understanding of the basic fluid mechanics of coalescence, floatation and many phenomena of interest to technology.

As far as we are informed, the numerical solutions of the three above mentioned problems without any "standard assumptions" were obtained for the first time by Shopov (1985), Zapryanov et al. (1987, 1989) and Shopov et al. (1990).

14.2. Basic Features of the Finite Element Method

The application of the FEM starts by dividing the flow domain Ω into N non-overlapping but contiguous "finite elements" Ω_1, Ω_2, ..., Ω_N (see Fig.14.2.1), where

$\Omega = \sum_{k=1}^{N} \Omega_k$. With irregularly shaped subdivisions and no limitation, this method has great flexibility. The price is that the construction of an adequate approximation in each finite element is not possible through replacing a differential operator by a simple difference operator.

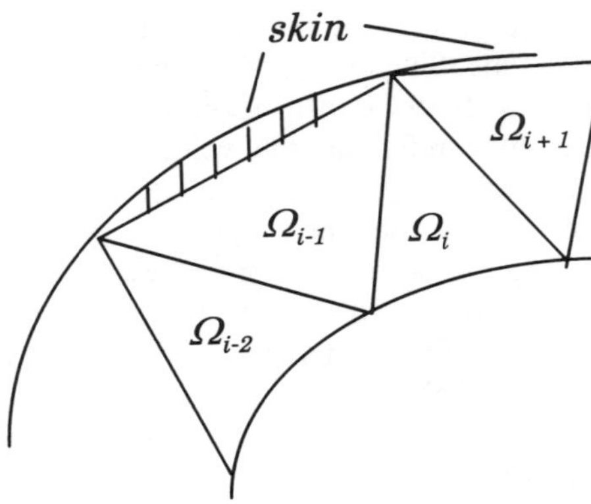

Fig.14.2.1.A part of the flow domain Ω divided into finite elements" $\ldots\Omega_{i-2}\Omega_{i-1}\Omega_i\Omega_{i+1}\ldots$.

The determination of an approximate solution in each finite element is typically done through integral constraints. Whenever possible an extremum principle is used. To fix ideas we may consider the simple case of the Laplace equation in a simply connected domain Ω with Dirichlet condition over its smooth boundary S_Ω

$$\begin{cases} \nabla^2 F = 0 & \text{in } \Omega \\ F = g & \text{given on } S_\Omega . \end{cases} \tag{14.2.1}$$

It is well known that the solution of this problem should (among all admissible functions satisfying the boundary condition) minimise the "energy functional $\Pi(F)$", defined by

$$\Pi(F) = \iiint_\Omega \frac{1}{2} (\nabla F)^2 \, d\tau , \tag{14.2.2}$$

where $d\tau$ denotes an element of the volume Ω. Taking into account the subdivision of Ω, the volume integral can be evaluated as a sum

$$\Pi(F) = \sum_{k=1}^{N} \frac{1}{2} \iiint_{\Omega_k} \left(\nabla F^{(k)}\right)^2 d\tau ,$$

where in each element Ω_k one treats $F^{(k)}$ as the solution of a Dirichlet problem with unknown boundary data along $S_{\alpha\text{-}\beta}$ (here the boundary between the elements Ω_α and Ω_β is denoted by $S_{\alpha\beta}$).

The actual construction of an approximation F_h with equation (14.2.1) as the corresponding functional to be minimised is, of course, none other than the Rayleigh-Ritz procedure. The equivalence of the Laplace differential equation and its Dirichlet condition with an extremum principle, as in the above example, cannot be expected in all problems.

It follows from the calculus of variations that a necessary condition for a functional to be extremised or made stationary is that the Euler-Lagrange equation is satisfied for the sought solution together with the boundary conditions. This Euler-Lagrange equation is precisely the partial differential equation of the problem. The variational formulation often has advantages over the partial differential formulation, especially from the numerical point of view. Unfortunately, it is sometimes impossible to find the correct functional corresponding to our partial differential equation. Here we have in view the solution of Navier-Stokes equations where no simple variational principle can be involved.

That is why we shall discuss another variational formulation based on the partial differential equation itself. This method is known as the Galerkin method and is also applicable to those PDEs for which no equivalent variational principle exists. Thus we shall treat the FEM as a particular example of the Galerkin method.

In its general form, the underlying principle of the FEM is one based on the method of weighted residuals. This kind of methods are numerical procedures for approximating the solution of a set of differential equations of the form

$$L(u_0) = f, \qquad \mathbf{r} \in \Omega \tag{14.2.3}$$

with boundary conditions

$$B(u_0) = g, \qquad \mathbf{r} \in S_\Omega, \tag{14.2.4}$$

where $\mathbf{r}$ represents the spatial coordinates x_1, x_2, x_3 and u_0 is the exact solution.

The solution u_0 is sought in a functional space V_0 consisting of sufficiently smooth functions satisfying the boundary conditions of the problem. It is assumed that the space V_0 has a countable basis Φ_1, Φ_2, ..., Φ_k, ..., which means that any function $F \in V_0$ can be expressed as an infinite linear combination of basis functions, i.e., $F = \sum_{k=1}^{\infty} \alpha_k \Phi_k$. Then the function u_0 is approximated by a set of the so called trial functions Φ_k, usually polynomials, i.e.,

$$u_0 \sim u_h = \sum_{k=1}^{M} u_k \Phi_k, \tag{14.2.5}$$

where u_k are undetermined parameters. We shall denote the vector space generated by the trial functions Φ_k ($k = 1, 2, ..., M$) with V_h and $V_h \subset V_0$.

If an arbitrary $u \in V_0$ is substituted into (14.2.3) the equality is not satisfied, i.e.

$$L(u) - f = R, \tag{14.2.6}$$

where $R = R(u)$ is called the residual, or error that results from taking u instead of the exact solution u_0. Furthermore, the error is forced to be zero, in an average sense, by setting weighted integrals of the residual equal to zero:

$$\iiint_{\Omega} [L(u) - f] w_i \, d\tau = 0, \qquad i = 1, 2, ..., M, \tag{14.2.7}$$

where w_i is a set of weighting functions, also known as test functions.

Galerkin's method is a particular weighted residual method in which the weighting (test) functions are the same as the trial functions Φ_k (k=1, 2, ..., M). So the residual is orthogonalised with respect to the trial functions Φ_k to yield:

$$\iiint_{\Omega} \left[L\left(\sum_{k=1}^{M} u_k \Phi_k \right) - f \right] \Phi_i \, d\tau = 0, \qquad i = 1, 2, ..., M, \tag{14.2.8}$$

or

$$\sum_{k=1}^{M} u_k \iiint_{\Omega} L(\Phi_k) \, d\tau = \iiint_{\Omega} f \, \Phi_i \, d\tau, \quad i = 1, 2, ..., M.$$

Since L is a linear operator, (14.2.8) produces a system of linear equations from which the coefficients u_k can be obtained. However, note that Galerkin's method is applicable to non-linear problems as well.

If we define

$$\delta u = \delta u_1 \Phi_1 + \delta u_2 \Phi_2 + ... + \delta u_M \Phi_M,$$

where δu_i terms are arbitrary increments, then (14.2.7) can be written as

$$\iiint_{\Omega} [L(u) - f] \, \delta u \, d\tau = 0 \tag{14.2.9}$$

for arbitrary δu. Furthermore, the above condition will be used when applying Galerkin's method.

Now let us consider the degree of continuity of a function. If the function u defined over a region Ω is discontinuous at discrete points of Ω, but is finite throughout Ω, then the norm of u satisfies the following condition

$$\|u\|_0 = \iiint_{\Omega} u^2 d\tau < \infty. \tag{14.2.10}$$

All functions subjected to (14.2.10), i.e., those which are square integrable, are said to belong to the $L_2(\Omega)$ functional space.

The functions whose derivatives are continuous pertain to spaces called Sobolev spaces denoted by $W_2^{(l)}(\Omega)$, where the superscript *l* corresponds to the highest derivative order. For example the Sobolev space $W_2^{(1)}$, contains all functions whose first derivative is square integrable

$$\|u\|_1 = \iiint_{\Omega} \left[u^2 + \left(\frac{du}{dx_1} \right)^2 + \left(\frac{du}{dx_2} \right)^2 + \left(\frac{du}{dx_3} \right)^2 \right] d\tau < \infty. \tag{14.2.11}$$

Let us now return to the problem of solving the differential equation (14.2.3) with the boundary condition (14.2.4) with respect to the unknown function u. We suppose that L is an n-th order differential operator and f, g $\in W_2^{(1)}$. The classical solution of this problem is a function belonging to the Sobolev space $W_2^{(n+1)}$.

If we release the continuity requirements on the sought function u and its derivatives, i.e., by lowering the order of the functional space, we obtain not the classical solution, but the so-called "weak" solution. If the weak solution is unique, it is called "generalised" solution. When the Galerkin's method is applied to solve a boundary value problem, the "weak formulation" is used, because it is better suited for the purpose. In contrast a "strong formulation" is the one based on the existence of a variational principle where a functional is made stationary. We shall note that to each variational principle it is possible to find, correspondingly, the equivalent Galerkin's procedure, but not always the vice versa.

After dividing the domain Ω into small pieces, or elements the basis (trial) functions are then defined element by element. We shall give below several examples where they are polynomials of moderate degree *l*. The elements we consider are of Lagrange type, where the degrees of freedom are the values of the unknown function. In other words the basis functions Φ_k are defined by their values at some specially chosen nodal points. The choice of the nodal points is subjected to the continuity conditions, which can be formulated for Φ_k by:

(i) Φ_k is a piecewise polynomial;

(ii) Φ_k is continuous at the inter-element faces;

(iii) $\Phi_k = 0$ on $\Gamma \subset S_\Omega$ at which $u_0 = 0$.

Further on, for the different finite elements in 2D space, we shall use the notation P_l = set of polynomials of degree $\leq l$ for triangles and Q_l = set of polynomials of degree $\leq l$ with respect to each variable for quadrangles. Thus

$$P_0 = \{1\},\ P_1 = \{1,\ x_1,\ x_2\},\ P_2 = \{1,\ x_1,\ x_2,\ x_1^2,\ x_1x_2,\ x_2^2\},\ \text{etc.}$$

$$Q_0 = \{1\}, Q_1 = \{1, x_1, x_2, x_1x_2\}, Q_2 = \{1, x_1, x_2, x_1^2, x_1x_2, x_2^2, x_1^2x_2, x_1x_2^2\}, \text{etc.}$$

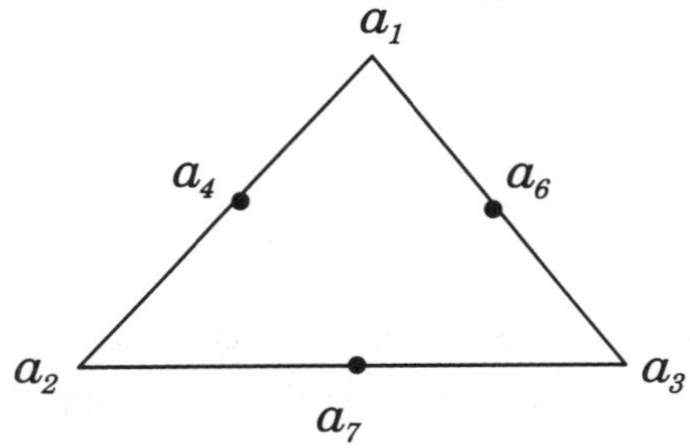

Fig.14.2.2. Triangular element T

The degrees of freedom are diminished by the value of the nodes lying on Γ. Moreover, to avoid "skins" like these in Fig14.2.1 obtained when straight triangles are used, one introduces "curved" or "isoparametric" finite elements through which the boundary is better approximated.

Let us consider the triangular element $T = (a_1, a_2, a_3)$ (see Fig.14.2.2, where the numbering is anticlockwise made), whose mid points of the sides are a_4, a_5 and a_6. In the above notation, it is a P_2-triangle. Six basis functions Φ_k (k =1, 2, ..., 6) correspond to the six nodes of this element, which must satisfy the conditions $\Phi_i(a_j) = \delta_{ij}$. If the triangular element is close to a curved boundary S (see Fig.14.2.3), then it would be better if the solution is interpolated at point a_5 on the boundary S, instead of at the midpoint $\tilde{a}_5$ on the chord a_2 a_3.

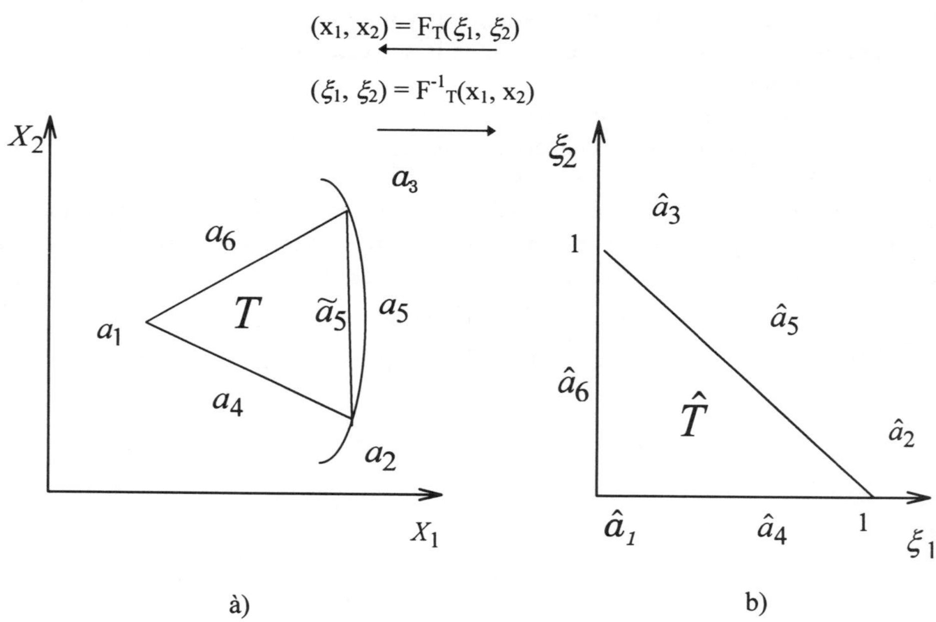

Fig.14.2.3.a) curved P_2-triangle T; b) reference straight P_2-triangle

In order to fulfil the continuity condition, the curvilinear triangle $T = (a_1, a_2, a_5, a_3)$ from the plane x_1x_2 will be transformed into a straight triangle $\hat{T} = (\hat{a}_1, \hat{a}_2, \hat{a}_3)$ from the plane $\xi_1\xi_2$. According to Fig.14.2.3b, we shall define the basis functions on $\hat{T}$ and shall use $\hat{T}$ as a reference triangle for all finite elements T in the problem to be solved. The transformation $\mathbf{F}_T$, by means of which the reference triangle $\hat{T}$ is mapped on T must satisfy the condition for the image nodes of $\hat{T}$ to be the nodes of T, i.e.,

$$\mathbf{F}_T(\hat{a}_i) = a_i, \quad i = 1, 2, ..., 6. \qquad (14.2.12)$$

In Cartesian coordinates (14.2.12) defines 6 equations for the transformation $\mathbf{F}_T$, which means that if $\mathbf{F}_T$ is a quadratic function of the two variables ξ_1 and ξ_2, then

$$\mathbf{F}_T(\xi_1, \xi_2) = \alpha_1\xi_1^2 + \alpha_2\xi_2^2 + \alpha_3\xi_1\xi_2 + \alpha_4\xi_1 + \alpha_5\xi_2 + \alpha_6,$$

where α_i are unknown coefficients. On the other hand, by use of the conditions

$$\hat{\Phi}_k(\hat{a}_i) = \begin{cases} 1, & \text{if } k = i, \\ 0, & \text{if } k \neq i, \end{cases}$$

we can represent $\mathbf{F}_T(\xi_1, \xi_2)$ in terms of the basis functions

$$\mathbf{F}_T(\xi_1, \xi_2) = \sum_{k=1}^{6} a_k \hat{\Phi}_k(\xi_1, \xi_2).$$

In this way, we get for example $\hat{\Phi}_1 = (1 - \xi_1 - \xi_2)(1 - 2\xi_1 - 2\xi_2)$ and so on.

Now the basis functions in the original plane Φ_k (x_1, x_2) are defined through the transformation function $\mathbf{F}_T$:

$$\Phi_k(x_1, x_2) = \hat{\Phi}_k(\xi_1, \xi_2) \qquad (k = 1, 2, ..., 6),$$

where $\mathbf{r} = x_1\mathbf{i} + x_2\mathbf{j} = \mathbf{F}_T(\xi_1, \xi_2)$.

We shall note also, that the integrals on T, for example

$$\iint_T \Phi_{l,i}\Phi_{m,j}\, dx_1 dx_2 = \iint_{\hat{T}} J_T \hat{\Phi}_{l,r} \hat{\Phi}_{m,s} \frac{d\xi_r}{dx_i}\frac{d\xi_s}{dx_j} d\xi_1 d\xi_2, \qquad (14.2.13)$$

transform into integrals on $\hat{T}$, as the Jacobian J_T of the transformation $\mathbf{F}_T$ is given by

$$J_T = \begin{vmatrix} \dfrac{\partial F_T^{(1)}}{\partial \xi_1} & \dfrac{\partial F_T^{(1)}}{\partial \xi_2} \\ \dfrac{\partial F_T^{(2)}}{\partial \xi_1} & \dfrac{\partial F_T^{(2)}}{\partial \xi_2} \end{vmatrix}$$

and J_T is different from zero. Then the inverse transformation $\mathbf{F}_T^{-1}$ exists which maps the triangle T into $\hat{T}$.

Clearly, the integrals involved in (14.2.13) should be computed through numerical integration using some quadrature formulae, i.e., $\iint_{\hat{T}} g(\xi_1, \xi_2)\, d\xi_1 d\xi_2$ is approximated by $\sum_{j=1}^{m} \omega_j g(\xi^j)$, where ξ^i belong to $\hat{T}$ and ω_j are weight functions.

In this way the choice of a common element $\hat{T}$ for all elements T of the solved problem allows the usage of the same formula for all finite elements independently of their being straight or curved.

Further we shall consider a convex linear quadrangle K = (a_1, a_2, a_3, a_4) as shown in Fig.14.2.4, correspondent to Q_1- quadrangle. The basis functions, which are polynomials of not higher than 2-nd degree with respect to both variables, on the reference element $\hat{K}$ are: $\hat{\Phi}_1 = (1-\xi_1)(1-\xi_2)$, $\hat{\Phi}_2 = \xi_1(1-\xi_2)$, $\hat{\Phi}_3 = \xi_2(1-\xi_1)$ and $\hat{\Phi}_4 = \xi_1\xi_2$. It is easy to show that $\hat{\Phi}_l(\hat{a}_m) = \delta_{lm}$. Here again the transformation (14.2.12) holds, mapping the quadrangle $\hat{K} = (\hat{a}_1, \hat{a}_2, \hat{a}_3, \hat{a}_4)$ into K.

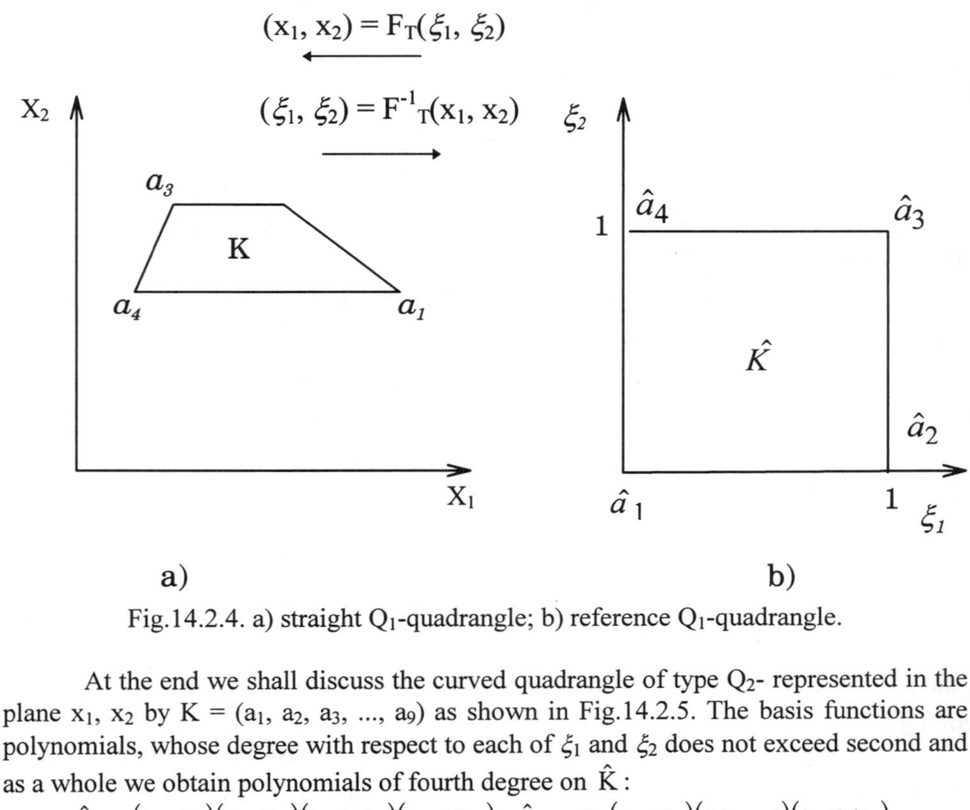

Fig.14.2.4. a) straight Q_1-quadrangle; b) reference Q_1-quadrangle.

At the end we shall discuss the curved quadrangle of type Q_2- represented in the plane x_1, x_2 by K = (a_1, a_2, a_3, ..., a_9) as shown in Fig.14.2.5. The basis functions are polynomials, whose degree with respect to each of ξ_1 and ξ_2 does not exceed second and as a whole we obtain polynomials of fourth degree on $\hat{K}$:

$$\hat{\Phi}_1 = (1-\xi_1)(1-\xi_2)(1-2\xi_1)(1-2\xi_2),\ \hat{\Phi}_2 = \xi_1(1-\xi_2)(2\xi_1-1)(1-2\xi_2),$$
$$\hat{\Phi}_3 = \xi_1\xi_2(1-2\xi_1)(1-2\xi_2),\ \hat{\Phi}_4 = \xi_2(1-\xi_1)(1-2\xi_1)(2\xi_2-1),$$
$$\hat{\Phi}_5 = 4\xi_1(1-\xi_2)(1-\xi_1)(1-2\xi_2),\ \hat{\Phi}_6 = 4\xi_1\xi_2(2\xi_1-1)(1-\xi_2),$$
$$\hat{\Phi}_7 = 4\xi_1\xi_2(1-\xi_1)(2\xi_2-1),\ \hat{\Phi}_8 = 4(1-\xi_1)\xi_2(1-2\xi_1)(1-\xi_2),$$

$$\hat{\Phi}_9 = 16\xi_1\xi_2(1-\xi_1)(1-\xi_2).$$

Here again the mappings (14.2.12) are valid, as well as the numerical integration formulae. In practice a combination between Q_2 and P_2 or Q_1 and P_1 are exploited in order to obtain better solution accuracy.

$$(x_1, x_2) = F_T(\xi_1, \xi_2)$$

$$\longleftarrow$$

$$(\xi_1, \xi_2) = F^{-1}_T(x_1, x_2)$$

$$\longrightarrow$$

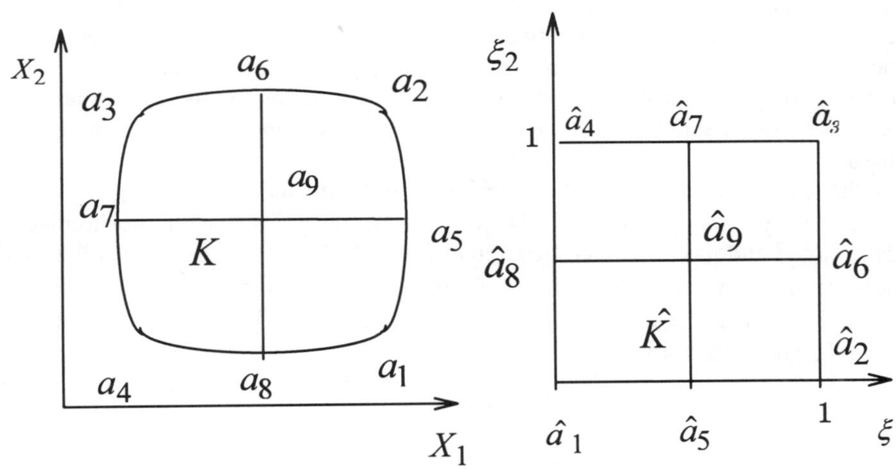

à) b)

Fig.14.2.5. a) curved Q_2-quadrangle; b) reference Q_2-quadrangle

Finally we shall summarise the general features of the finite element method, which have been outlined in this section:

(i) The method is based on a weak formulation;

(ii) The unknown quantities of the physical problem have an immediate physical meaning: $u_i = u_h(x_i)$, which are the degrees of freedom of u_h ;

(iii) The basis functions Φ_i have a small support, i.e., they do not vanish on a_i and its vicinity. Contrary to the spectral methods, there is a grid point approximation in FEM;

(iv) The matrix system is sparse, with most zero elements;

(v) The most general boundary conditions can be easily handled in a semi-automatic fashion;

(vi) The Dirichlet boundary conditions are strongly imposed, while the Neumann conditions are only approximately satisfied.

14.3. Full Formulation of the Problem of Interfaces Finite Deformations. Discretezation of the Unsteady Navier-Stokes Equations

In many recent hydrodynamic problems besides the unknown functions such as velocity components and pressure there exist additional geometrical unknowns. One defines a free boundary problem as a boundary value problem involving partial differential equations (in our case unsteady Navier-Stokes equations) on domains of which the fluid/fluid boundaries are entirely or partially unknown and must be determined as a part of the solution. One can distinguish two types of free boundary problems, according to whether the free boundary depends on time or not. In this section we shall formulate the so called moving boundary value problems which are defined to be the initial value problems for full axisymmetric Navier-Stokes equations with unknown (moving) free boundaries.

The mathematical postulation of the problems, to be treated in this chapter, will be presented here. Let the region Ω of 2D Euclidean space is filled with two immiscible liquids or liquid and gas, separated by an interface $S_{1,2}$. We shall assume that the fluids are incompressible, viscous (or inviscid, if gases), Newtonian, with constant density ρ_i and dynamic viscosity μ_i (i = 1, 2). Then the unsteady Navier-Stokes equations and (1.3.1)-(1.3.2) have the form:

$$\rho_i\left[\frac{\partial \mathbf{v}^{(i)}}{\partial t} + \mathbf{v}^{(i)}\left(\nabla . \mathbf{v}^{(i)}\right)\right] = \nabla . \mathbf{T}^{(i)} + \mathbf{F}^{(i)}, \tag{14.3.1}$$

$$\nabla . \mathbf{v}^{(i)} = 0, \tag{14.3.2}$$

$$\mathbf{T}^{(i)} = -p^{(i)}\mathbf{I} + 2\mu_i \mathbf{E}^{(i)}, \qquad (i = 1, 2), \tag{14.3.3}$$

where the accepted denotations of this book are used.

We shall consider only planar and axisymmetric problems and assume the gravity force $\mathbf{F}^{(i)}$ to have the components $(0, \rho_i g)$ in the spatial coordinate system Ox_1x_2 with radius vector $\mathbf{r} = (x_1, x_2)$.

Remark. For the sake of simplicity the problem will be formulated for the case of two fluids, but it can be easily generalised for more fluids as well.

We shall discuss three types of boundary conditions:

(i) Boundary conditions at infinity and on rigid wall S_w

$$\mathbf{v}^{(i)} = 0 \qquad \text{at } |\mathbf{r}| \to \infty, \tag{14.3.4}$$

$$\mathbf{v}^{(i)}\Big|_{S_w} = \mathbf{v}_0, \tag{14.3.5}$$

where the velocity $\mathbf{v}_0$ is prescribed.

(ii) Boundary conditions on the interface $S_{1,2}$ (t)

a) kinematic conditions

$$\mathbf{v}^{(1)}\Big|_{S_{1,2}} = \mathbf{v}^{(2)}\Big|_{S_{1,2}}, \qquad (14.3.6)$$

$$\frac{DS}{Dt} = \frac{\partial S}{\partial t} + \mathbf{v}^{(i)}.\nabla S = 0, \qquad \text{or} \qquad \frac{\partial S}{\partial t} + \mathbf{v}^{(i)}.\mathbf{n} = 0, \qquad (14.3.7)$$

where S(**r**, t) = 0 is the equation of the deformed interface $S_{1,2}$ and $\mathbf{n} = \left(\frac{\partial S}{\partial x_1}, \frac{\partial S}{\partial x_2}\right)$ is the unit normal to it, directed from fluid (1) towards fluid (2).

b) dynamic conditions

$$\mathbf{T}^{(2)}.\mathbf{n} - \mathbf{T}^{(1)}.\mathbf{n} = \sigma_{1,2}\left(\frac{1}{R_1} + \frac{1}{R_2}\right)\mathbf{n}, \qquad (14.3.8)$$

where $\sigma_{1,2}$ is the surface tension on the interface between the two fluids (1) and (2), R_1 and R_2 are the two principle curvatures of the surface $S_{1,2}$.

(iii) Boundary conditions caused by the symmetry of the solved problem. If, for example, Ox_1 is an axis of symmetry in the flow field, then on it the following conditions are imposed

$$v_1^{(i)}\Big|_{x_1=0} = 0, \quad \frac{\partial v_2^{(i)}}{\partial x_1}\Bigg|_{x_1=0} = 0, \quad \frac{\partial p^{(i)}}{\partial x_1}\Bigg|_{x_1=0} = 0. \qquad (14.3.9)$$

In order to close the unsteady Navier-Stokes problem (14.3.1) - (14.3.3), initial conditions are also needed

$$\mathbf{v}^{(i)}(\mathbf{r},0) = \mathbf{v}_0, \quad S(\mathbf{r},0) = S_0. \qquad (14.3.10)$$

For the dynamic pressure no boundary conditions are necessary (Ladyzhenskaya, 1969). In order to be determined uniquely, it must be fixed at one point of the flow domain. Usually, the pressure is fixed at infinity, i.e., it is accepted that $p(\infty, t) = 0$.

It is well established that if we fix a characteristic linear dimension L and characteristic velocity U_0, all the quantities of the problem (14.3.1) - (14.3.10) can be dimensionalized. Then on the place of μ_i the parameter Re^{-1} (Re is the Reynolds number) will stand, and $\mathbf{F} = \rho\mathbf{g}$ in (14.3.1) will be changed by $-Fr^{-1}\mathbf{k}$ (Fr is the Froude number and **k** the unit vector along the vertical axis), while in (14.3.8) We^{-1} (We is the Weber number) is found instead of $\sigma_{1,2}$. The particular dimensionalization of the different problems discussed in this chapter will be given, where necessary, in the text.

The basic physical parameters participating in the problem of determination of the deformation evolution of a given interface and the acceleration due to gravity can be combined to form the dimensionless Morton number $M = \frac{g\mu^4}{\rho\sigma^3}\lambda$, where $\lambda = \frac{\rho - \rho_2}{\rho}$.

The Morton number is a property mainly of liquids and its variations are principally due to the factor μ^4. The range of change of M varies significantly for the different liquids, i.e., for highly viscous oils $M = O(10^5)$, while for liquid metals $M = O(10^{-13})$. For bubbles in liquid $\lambda = 1$, since $\rho_2 = 0$. When the gravitational and capillary forces are taken into account, the Eötvös (Bond) number $\text{Eö} = \frac{\rho g L^2}{\sigma}\lambda$ plays an important role, for it expresses the ratio of the gravitational and capillary forces. The commonly used Reynolds, Froude and Weber numbers are respectively defined by the equations $\text{Re} = \frac{U_0 L \rho}{\mu}$, $\text{Fr} = \frac{U_0^2}{gL\lambda}$ and $\text{We} = \frac{U_0^2 L \rho}{\sigma}$.

For the problems which will be treated in this chapter: bubbles or drops moving in a viscous fluid, it is convenient to use the "equivalent" spherical radius r_c as a characteristic length scale, i.e., $L = r_c$. If the volume of the fluid particle (bubble or drop) is V and it is assumed constant, then $r_c = \sqrt[3]{\frac{3V}{4\pi}}$.

If the terminal velocity of a rising bubble in a unbounded viscous fluid is chosen as a characteristic velocity $U_0 = \frac{2}{9}\frac{\rho g r_c^2}{\mu}\lambda$, only two of the above given dimensionless numbers are independent. If we take the Reynolds number Re and Eötvös number Eö as "basic", the remaining dimensionless numbers will be expressed by means of the relations $M = \left(\frac{2}{9}\right)^2 \frac{\text{Eö}^3}{\text{Re}^2}$, $\text{We} = \frac{2}{9}\text{Re}\,\text{Eö}$ and $\text{Fr} = \frac{2}{9}\text{Re}$. Although the numbers M, We and Fr are expressed by Re and Eö, they have their own physical meaning. In particular, the Morton number characterises the fluid type, while the Weber number determines the interface deformation. The capillary number Ca defined by Re is given by the formula: $\text{Ca} = \frac{U_0 \rho}{\sigma} = \frac{2}{9}\text{Eö}$.

The virtual power principle (Connor and Brebbia, 1976) is exploited to discretize the equations (14.3.1) - (14.3.3). For the purpose (14.3.1) and (14.3.2) are multiplied by arbitrary functions $\delta \mathbf{v}^{(i)}$ and $\delta p^{(i)}$, which belong to the respective spaces, where the solution of the considered problem is sought; $\mathbf{v}^{(i)}$ and $\delta \mathbf{v}^{(i)}$ belong to $W_2^{(1)}(\Omega_j)$, the space with integrable squares of the function and its first derivative; $p^{(i)}$ and $\delta p^{(i)}$ belong to $L_2(\Omega_j)$, the space with integrable function square. The product is integrated over $\Omega_j(t)$ and the divergence theorem is applied for the integral $\iint_{\Omega_j} \nabla.\mathbf{T}^{(i)}\delta \mathbf{v}^{(i)} d\Omega_j$. In this way, in the one interface case, the obtained two integral relations are summed to obtain only one integral expression for the region $\Omega_j(t)$.

Let us consider the finite bases $\left\{\Phi_k\right\}_{k=1}^{M_v}$ and $\left\{\Pi_k\right\}_{k=1}^{M_p}$ for $\mathbf{v}$ and p, respectively, in $\Omega_j(t)$, where the FEM will be applied. Then the functions $\mathbf{v}(\mathbf{r}, t)$ and $p(\mathbf{r}, t)$ will be approximated by the series

$$\mathbf{v}_h = \sum_{k=1}^{M_v} u_k(t)\Phi_k, \qquad p_h = \sum_{l=1}^{M_p} p_l(t)\Pi_l, \tag{14.3.11}$$

where the functions $u_k(t)$ and $p_l(t)$ are unknown and must be determined. Following the Galerkin's method, we express $\delta\,\mathbf{v}^{(i)}$ and $\delta\,p^{(i)}$, respectively, by the basis functions Φ_k and Π_l, which are found by use of isoparametrical quadrilateral finite element with 9-nodes for the velocity, i.e., a biquadratic approximation is applied for the velocity and a linear approximation for the pressure with respect to each spatial coordinate (see Fig.14.2.5). In this way there are 18 parameters for the velocity and 4 for the pressure. The accuracy is $O(h^3)$ for the velocity and $O(h^2)$ for the pressure (Cuvelier et al., 1986).

On the basis of the Lagrangian approach the material derivative of velocity can be represented in the form

$$\left.\frac{D\mathbf{v}_h}{Dt}\right|_t = \frac{\mathbf{v}_h[\mathbf{r}(t),t] - \mathbf{v}_h[\mathbf{r}(t-\tau),t-\tau]}{\tau} + O(\tau).$$

Then for the discretization of the material derivative in (14.3.1) an Eulerian-Lagrangian analogue to the implicit scheme is applied with first order accuracy with respect to time.

Thus by use of the virtual powers principle and some quadrature formulae, the equations (14.3.1) - (14.3.3) are discretized with respect to the spatial coordinates and are transformed into ordinary differential equations for $u_k(t)$ and $p_l(t)$. After approximating the time derivatives, as described before, a linear system of algebraic equations for the moment t is obtained

$$\tau^{-1}\mathbf{M}.(\mathbf{U} - \mathbf{U}_L) + \mathbf{A}.\mathbf{U} + \mathbf{B}^T.\mathbf{P} = \mathbf{F}, \quad \mathbf{B}.\mathbf{U} = 0, \tag{14.3.12}$$

where $\mathbf{U} = \{u_k\}$, $\mathbf{P} = \{p_l\}$ and **M**, **A**, **B** are matrices, while $\mathbf{U}_L$ and **F** are vectors (Zapryanov et al., 1987). Here $\mathbf{U}_L$ is a vector of the velocity coefficients, corresponding to the previous position of the mesh at the moment t-τ, while the coefficients of the matrices **M**, **A**, **B** are calculated for a mesh position corresponding to the moment t, using the predictor scheme, i.e.

$$\mathbf{r}(t) = \mathbf{r}(t-\tau) + \tau\mathbf{v}(\mathbf{r},t-\tau), \quad \mathbf{v}(\mathbf{r},t-\tau) = \sum_i U_{Li}\Phi_i(t-\tau).$$

The obtained linear system (14.3.12) can be solved directly by some well known methods for linear systems or by the so-called "condensation method", proposed by Shopov (1985).

14.4. Numerical Approximation of the Surface Forces Operators. Determination of the Free Surface Position

The key to an efficient numerical solution of the problem defined in section 14.3. is to keep in mind that the governing equations are unsteady and non-linear, and that the solution domain is unknown. Thus, a general form for the solution cannot be derived and an efficient numerical procedure is generally necessary for the numerical solution of considered problems. There are two important aspects to the discretization: approximation of the variation of interfacial velocities over an element of the surface, and accurate representation of the interface shape and curvature.

Accurate representation of the interface shape is crucial in the problems considered. When a region of high curvature begins to develop, additional elements have to be added in its immediate vicinity. It is worth noting that the developed procedure is indeed efficient in dealing with the non-linear terms in the Navier-Stokes equations and in representing the time dependent boundary conditions around fluid/fluid interfaces.

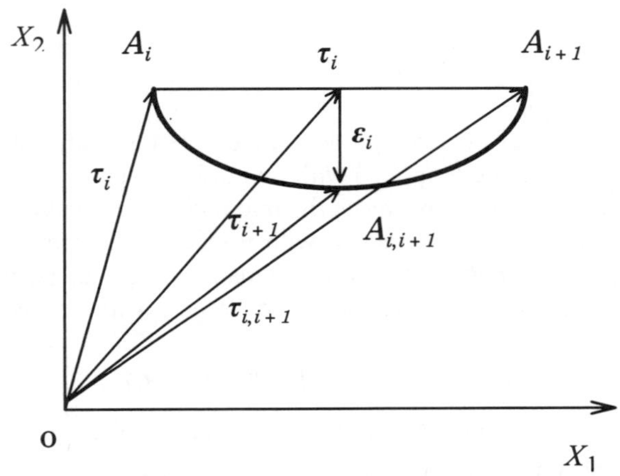

Fig.14.4.1. Local representation of each side $\overline{A_i A_{i,i+1} A_{i+1}}$ of the finite element i.

The free boundaries (interfaces) position, which is one of the objects of the solved problem, is determined by the finite elements nodes touching the boundary. The finite elements, considered in the problem are isoparametric and their sides are parabolas, constructed on three points. Locally each side $\overline{A_i A_{i,i+1} A_{i+1}}$ of the finite element i with normal vector $\mathbf{n}$ and radius vector $\mathbf{r}$ can be represented (see Fig.14.4.1) as

$$\mathbf{r} = \mathbf{r}_{i,i+1} + \boldsymbol{\tau}_i s + \boldsymbol{\varepsilon}_i s^2 . \qquad (14.4.1)$$

In the upper expression the following notations have been used: $\boldsymbol{\tau}_i = 0.5\mathbf{A}_i\mathbf{A}_{i+1} = 0.5(\mathbf{r}_{i+1} - \mathbf{r}_i)$; $\boldsymbol{\varepsilon}_i = \mathbf{r}_{i,i+1} - 0.5(\mathbf{r}_{i+1} + \mathbf{r}_i)$, where $\mathbf{r}_{i,i+1}$ is the radius vector of the point $A_{i,i+1}$, and $\mathbf{r}$ is the radius vector of an arbitrary point from the arc $\overline{A_i A_{i,i+1} A_{i+1}}$, which is a side of the considered finite element, and $-1 \leq s \leq 1$. This approximation is used by many authors, for example, by Ruschak (1980) and Keunings (1986). In this manner, every interfacial boundary Γ_d , participating in the domain boundary S_Ω of the discussed problem, is approximated with a piecewise smooth curve Γ_h, consisting of parts of parabolas.

The contribution of the surface tension forces $\mathbf{R}_{ij}$ acting on the interface Γ_d in the right-hand side of (14.3.12) is

$$\int_{\Gamma_d} \mathbf{R}_{ij} \Phi_k J\, dl = \frac{\kappa_{ij}}{We} \int_{\Gamma_d} \left(\frac{1}{R_1} + \frac{1}{R_2} \right) \mathbf{n} \Phi_k J\, dl, \tag{14.4.2}$$

where $\kappa_{ij} = \dfrac{\sigma_{ij}}{\sigma_{12}}$, We is the Weber number and R_1^{-1}, R_2^{-1} are the principle curvatures of the interface, as

$$\left| \begin{array}{l} \dfrac{1}{R_1} = \dfrac{\ddot{x}_2 \dot{x}_1 - \ddot{x}_1 \dot{x}_2}{\left(\dot{x}_1^2 + \dot{x}_2^2\right)^{3/2}}, \\ \dfrac{1}{R_2} = 0 \quad \text{– in the plane case,} \\ \dfrac{1}{R_2} = \dfrac{\dot{x}_2}{x_1 \left(\dot{x}_1^2 + \dot{x}_2^2\right)^{3/2}} \quad \text{– in the axisymmetric case,} \end{array} \right. \tag{14.4.3}$$

and the differentiation with respect to the parameter s is denoted by dot.

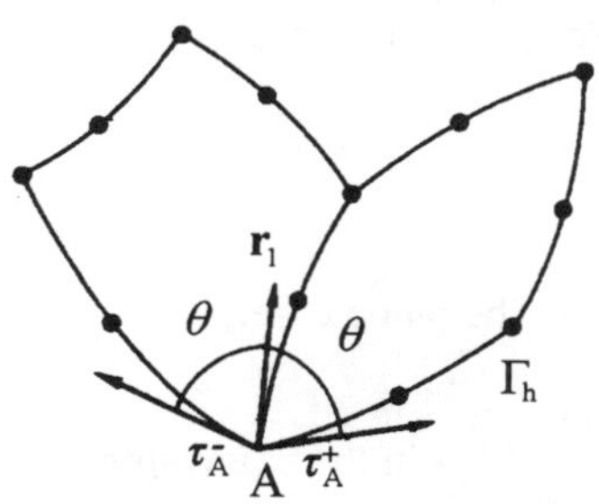

Fig.14.4.2. Sketch of the joint point A between two parabolic segments.

Since the approximation (14.4.1) is of second order, then at the common points of two parabolas Γ_h is not smooth, i.e., the tangents τ_A^- and τ_A^+ are not collinear when the nodal point A lies between two adjacent elements (see Fig.14.4.2). Therefore, $\mathbf{R}_{ij}$ tends to infinity at such points. Then the integrals connected with the surface tension are split into two parts, regular and singular. If we denote the expression $\dfrac{1}{We}\left(\dfrac{1}{R_1} + \dfrac{1}{R_2} \right)\mathbf{n}$ with $\mathbf{R}_s$, then

$$\frac{1}{\mathrm{We}}\int_{\Gamma_d}\Phi_k\left(\frac{1}{R_1}+\frac{1}{R_2}\right)\mathbf{n}J\,dl=\sum_l\left(\mathbf{R}_s,\Phi_k\right)^R+\sum_l\left(\mathbf{R}_s,\Phi_k\right)^S, \qquad (14.4.4)$$

as the summation is performed on the sides l of the finite elements, which are touching the interface. When taking into account, that

$$\sum_l\left(\mathbf{R}_s,\Phi_k\right)^R=\int_l\mathbf{R}_s\Phi_k J\,dl, \qquad (14.4.5)$$

the two principle curvatures can be calculated via formulae (14.4.3) using the approximation (14.4.1) of Γ_h and some of the well known quadrature formulae for numerical integration. However, if the regular part is calculated in this manner, the approximation conservation will not be guaranteed on the boundary with respect to one very important interfacial property of a drop or bubble immersed in an infinite, stationary fluid at weightlessness. Namely, at such conditions the drop or bubble boundary aims always to take a spherical form (Batchelor, 1967). The reason for this is, that the surface tension tends to minimise the area of the free surface. In order to guarantee the approximation conservation with respect to this physical property, the regular contribution of the surface tension will be accounted by the procedure described below.

At first, one more approximation of the interface $\widetilde{\Gamma}_h$ is introduced, as the parabolas composing Γ_h are replaced by parts of circumferences. Then the two principle curvatures are calculated on $\widetilde{\Gamma}_h$, while the integral in (14.4.5) is calculated on Γ_h by 3 or 5 nodes Gauss quadrature. Both approximations differ by $O(h^3)$ in the L_2 norm of the plane. Therefore, a special type of a quadrature for numerical calculation of (14.4.5) is obtained. At the interface approximation with arcs of circumferences, the two principle curvatures are easily obtained:

$$\left|\begin{array}{l}\dfrac{1}{R_1}=\dfrac{1}{R},\\ \dfrac{1}{R_2}=0 \quad \text{– in the plane case,}\\ \dfrac{1}{R_2}=\dfrac{1}{R}-\dfrac{x_1^c}{x_1 R} \quad \text{– in the axisymmetrical case.}\end{array}\right. \qquad (14.4.6)$$

Here x_1 is the current first coordinate of the surface point, while x_1^c and R are, correspondingly, the first coordinate of the centre and the radius of the circumference belonging to $\widetilde{\Gamma}_h$, whose curvature is being calculated.

The singular part of the integral is calculated when in the vicinity of such point A the curve Γ_d is approximated by a C^∞ curve of length ε, which is "stuck" smoothly on Γ_d at the ends of this vicinity. Furthermore, the integral $\int_l\mathbf{R}_s\Phi_k J\,dl$ is calculated and ε tends to zero. The obtained expression enters as a singular contribution from the surface tension

$$\left(\mathbf{R}_s,\Phi_k\right)^S=\frac{1}{\mathrm{We}}\Phi_k(\mathrm{A})\mathbf{n}\cos\theta\,\mathrm{J}(\mathrm{A})\,, \tag{14.4.7}$$

where $\mathbf{n}=\frac{\boldsymbol{\tau}^{+}}{\left|\boldsymbol{\tau}^{+}\right|}+\frac{\boldsymbol{\tau}^{-}}{\left|\boldsymbol{\tau}^{-}\right|}$, as $\boldsymbol{\tau}^{-}$ and $\boldsymbol{\tau}^{+}$ are the left and right tangent to Γ_d at the point A, θ is the half angle between the two tangents (see Fig.14.4.2). The expression (14.4.7) can be regarded as a penalty function at point A, which tends to straighten the angle 2θ between the tangents towards π, i.e., it smoothes the boundary at point A. When this angle is exactly equal to π the expression is zero and, therefore, its contribution to the surface tension forces is zero.

If the gravity force is not acting on the drop or bubble interface the right-hand side of (14.3.12) is equal to zero and the free boundary surface will preserve its initial shape. In this way the conservation of the numerical scheme with respect to the spherical drop or bubble form is guaranteed regardless of the mass force.

Now we shall explain how the position of the free surface is determined in time. In order to calculate the shape of an interface, a mixed Eulerian-Lagrangian scheme is used. In this manner simple integration of the Lagrange formulation is used and the computational mesh employed in the Eulerian scheme is preserved.

The interface motion naturally follows that of the mesh in the Lagrangian approach, which is determined by

$$\mathbf{r}(t+\Delta t)=\mathbf{r}(t)+\Delta t\,\mathbf{v}\left[\mathbf{r}(t),t\right],\ \mathbf{r}=x_1\,\mathbf{i}+x_2\,\mathbf{j}. \tag{14.4.8}$$

This explicit scheme written for the interface points

$$\Gamma_d(t+\Delta t)=\Gamma_d(t)+\Delta t\,\mathbf{v}\left[\mathbf{r}(t),t\right].\nabla\Gamma_d(t)\,, \tag{14.4.9}$$

is in fact an explicit discretization of the equation of motion (14.3.7).

Here we must have in mind, that the velocity **v**(**r**, t) can be divided into two parts: tangential v_τ and normal v_n . The tangential motion of the boundary points at sufficiently small Δt does not change the boundary shape, but the motion in normal direction is exactly given by the discretization (14.3.7). However, this scheme with a fixed time step may provoke disturbances on the boundary and lead to numerical instability, if Δt becomes bigger than a certain Δt_{crit}. This is due to the hyperbolic character of the equation of boundary motion. In order to "filter" the eventual numerical disturbances, cubic splines may be used (Bazhlekov, 1992). Another troublesome problem, arising when applying the FEM, is the big deformation of the finite elements in regions with high gradients of the sought fields, which requires the redefinition of the used elements (Minev, 1990).

For the boundary motion stability control the so-called special scheme of the type predictor-corrector may be used. On the predictor step the boundary, together with the whole mesh, is determined from

$$\mathbf{r}^{pr}(t+\Delta t)=\mathbf{r}(t)+\Delta t\,\mathbf{v}\left[\mathbf{r}(t),t\right], \tag{14.4.10}$$

where **v** is the velocity, which is assumed to be definitely obtained in the lower time layer.

At the initial moment **r** and **v** are the initial conditions for the boundary position and velocity. The obtained boundary $\Gamma_d^{pr}(t+\Delta t)$ is checked up for smoothness, when measuring the minimum value of the angle θ between the two tangents τ^- and τ^+ (see Fig.14.4.2) at all contact points between the finite elements and the interface. This angle is compared with a preliminary given angle θ_0 and this is in fact a boundary smoothness criterion. If $\theta < \theta_0$ then the time step is decreased and the step (14.4.10) is repeated. On the contrary, it is said that the boundary smoothness criterion is passed and the velocity field $\mathbf{v}^{cor}[\mathbf{r}^{pr}(t+\Delta t)]$ is calculated in the domain $\Omega^{pr}(t+\Delta t) = \{\mathbf{r}^{pr}(t+\Delta t)\}$ from the system (14.3.12). The corrector step for the boundary and the whole mesh is made by the scheme:

$$\mathbf{r}^{cor}(t+\Delta t) = \mathbf{r}(t) + 0.5\Delta t\,\{\mathbf{v}[\mathbf{r}(t),t] + \mathbf{v}^{cor}[\mathbf{r}^{pr}(t+\Delta t), t+\Delta t]\}. \qquad (14.4.11)$$

The obtained via this procedure boundary $\Gamma_d^{cor}(t+\Delta t)$ is checked up again for smoothness and if the criterion fails, Δt is decreased and the process is returned again to step (14.4.10). If the criterion holds, the stability control condition must be checked

$$\left| \mathbf{r}^{cor}(t+\Delta t)\Big|_{\Gamma_d^{cor}(t+\Delta t)} - \mathbf{r}^{pr}(t+\Delta t)\Big|_{\Gamma_d^{pr}(t+\Delta t)} \right| < \varepsilon\, \Delta t, \qquad (14.4.12)$$

where ε is a preliminary determined constant, controlling the calculations accuracy. This condition is equivalent to

$$\left| v_n^{cor}(t+\Delta t) - v_n^{pr}(t+\Delta t) \right| < 2\varepsilon, \qquad (14.4.13)$$

where v_n is the normal velocity component and the unit normal is taken towards $\Gamma_d^{cor}(t+\Delta t)$ and $\Gamma_d^{pr}(t+\Delta t)$, respectively. If this criterion for the normal velocity on the interface is satisfied, then the velocity and the interface position in the moment $t + \Delta t$ are assumed to be equal correspondingly to $\mathbf{v}^{cor}[\mathbf{r}^{pr}(t+\Delta t), t+\Delta t]$ and $\mathbf{r}^{cor}(t+\Delta t)$, i.e., $\Gamma_d^{cor}(t+\Delta t)$.

14.5. Interaction of a Deformable Bubble with a Rigid Wall at Moderate Reynolds Numbers

Let us consider the following two problems:

(i) the rise of a deformable gas bubble towards a plane rigid wall in an unbounded liquid at moderate Reynolds numbers;

(ii) the motion of a gas bubble in a spherical container filled with viscous liquid for moderate values of the Reynolds number.

In both problems the bubble and the liquid are initially motionless and the motion is set up by the buoyancy. The geometry of the problems is illustrated in Fig.14.5.1.a, b. The fluid in the bubble is assumed incompressible with density $\rho_0 = 0$, viscosity $\mu_0 = 0$, and the pressure, $p_0(t)$ is independent of the spatial coordinates. The ambient liquid is homogeneous, incompressible and with constant physical properties; dynamic viscosity μ and density ρ. The pressure at infinity is assumed to be constant.

The diameter $2l$ of the equivalent spherical bubble (with the same volume) is chosen to be the reference length and the so-called Stokes terminal velocity U_0 is chosen to be the characteristic velocity, $U_0 = \frac{2\rho g l^2}{9\mu}\lambda$, where $\lambda = \frac{\rho - \rho_0}{\rho}$ is the ratio of the difference in densities of the liquid and the gas to the density of the liquid and g is the gravity acceleration.

The problem depends on two geometrical parameters:

$$\gamma = \frac{H}{2l} \quad \text{and} \quad e = \frac{2R}{2l},$$

where H is the initial distance from the bubble centre to the rigid boundary and R is the radius of the container (for problem (i) R = ∞). So γ represents the dimensionless starting distance and e determines the relative importance of the wall curvature.

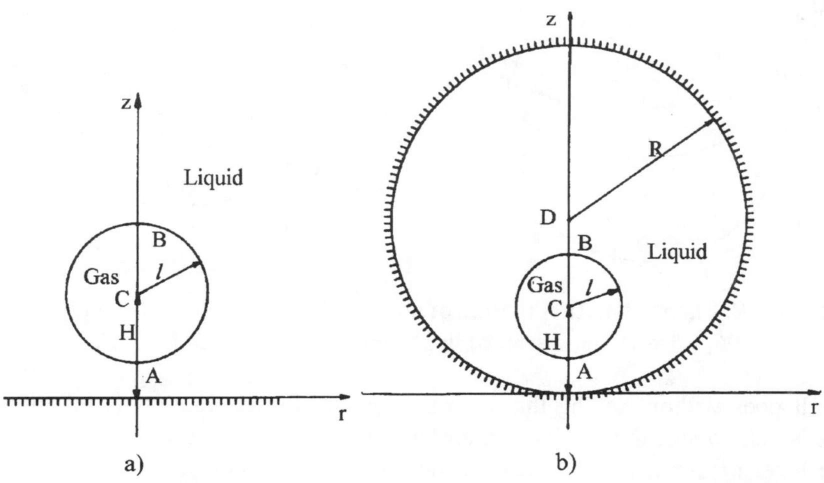

Fig.14.5.1. Geometry of a bubble in a liquid: a) near a rigid plane; b) in a spherical container

Provided the plate is horizontal and the Reynolds number is not too high, there is no asymmetry and the flow is axisymmetrical. In this way we can employ the axisymmetric Navier-Stokes and continuity equations (14.3.1)-(14.3.3). The boundary conditions at infinity and on the rigid wall are given by (14.3.4) and (14.3.5), respectively, where $\mathbf{v}_0$ = 0. The kinematic and dynamic conditions on the free surface are written with (14.3.6)-(14.3.8). From (14.3.9) and (14.3.10) we can obtain the corresponding symmetry and initial conditions.

Problems (i) and (ii) are solved for various values of the parameters. Two distinct cases are considered: when the bubble approaches the wall and when it recedes from the wall. The Reynolds number varies from 4.10^{-3} to 120, the Eötvös (Bond) number from 1 to 360. The initial distances are taken from the interval $0.55 \leq \gamma \leq 1$ ($\gamma = 0.5$ corresponds to the case when the bubble touches the wall), and the diameter of the container is allowed to vary from 2 up to ∞, the latter corresponding to the case of a planar wall. At the initial moment of time the bubble is at rest and the liquid is quiescent.

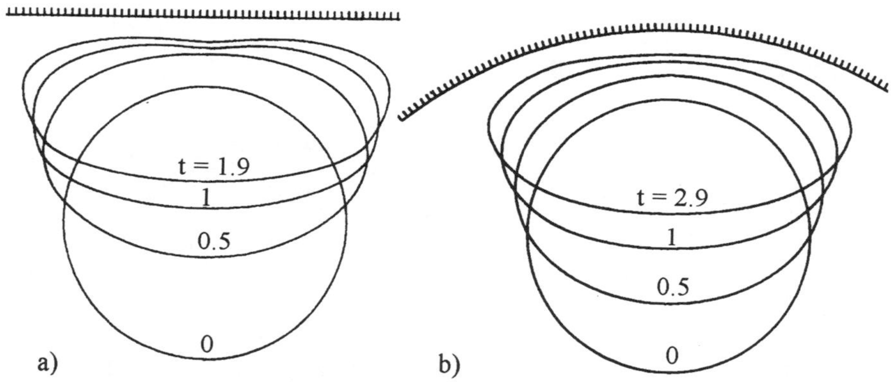

Fig.14.5.2. Shape evolution in time of an air bubble at $\gamma = 0.75$: a) in glycerol at Re = 0.396, Eö = 19.48, e = ∞; b) in glycerol at Re = 0.396, Eö = 19.48, e = 3.

It goes without saying that the question of ultimate interest is to find out the factors which control the mechanism and dynamics of the rupture of the thinning film. At the level of our analysis based on the continuum mechanics it is not possible to predict the rupture of the thin film. Indeed, because of the fact that the film rupture is associated with van der Waals and electrostatic forces or even with a purely molecular mechanism, where the film thickness is of molecular scale, we can not predict this breakthrough process. What we could find out, however, are the conditions by means of which the combination of viscous, capillary and body forces would lead to a film formation.

Following Zapryanov et al. (1987) and Shopov et al. (1990), we shall consider first gas bubbles in two different real liquid-glycerol (Fig.14.5.2.a, b) and aqueous solution (Fig.14.5.2.c). In the second case the gas and liquid are the same as those used by Bhaga and Weber (1981), see their figure 2(b). The only difference is that the results of Zapryanov et al. are obtained for a bubble with equivalent diameter twice as large as Bhaga and Weber.

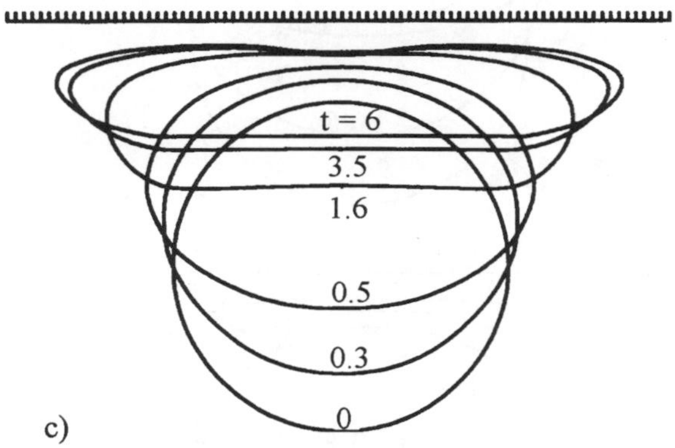

c)

Fig.14.5.2. c) in aqueous sugar solution at Re = 1.24, Eö = 70.8, e = ∝;

The dimpling phenomena is observed in both experiments for large times. A lubrication layer is formed between the wall and the gas/liquid interface. The viscous fluid is "trapped" between the bubble and the rigid boundary and causes dimpling, because the pressure drop is insufficient to overcome the viscous stress in the layer. This is evident on the axis in Fig.14.5.2.a, c and Fig.14.5.3.a.

In Fig.14.5.2.a and Fig.14.5.3.c the bubble velocity is seen to be smaller in the container [problem (ii)] than for the flow bounded by a plate [problem (i)] when remaining conditions and governing parameters are the same. This is natural and is attributed to the influence of the container. For the same instant of time the thickness of the liquid layer that is formed between the bubble and the wall is larger for the flow in the container than for the case when the flow is bounded by a plate. This is not surprising, since the non-slipping at the container wall results in a larger resistance to the outflow of the liquid from the near-wall zone than in the plate case. In fact the hydrodynamical interaction of a deformable bubble with the container wall is stronger than with a planar wall.

It is interesting to give a feeling about the influence of the four governing parameters on the properties of the scheme and algorithm as well as on the results obtained by Shopov et al. (1990).

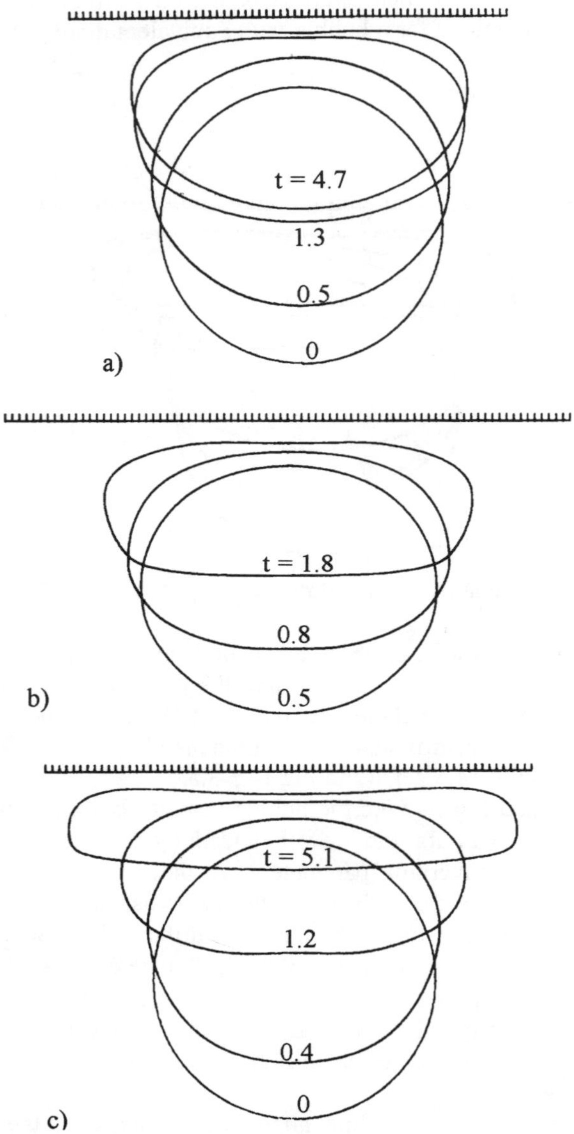

Fig.14.5.3. Gas bubble moving towards a plane at Re = 10, γ = 0.75 and: a) Eö = 7.2; b) Eö = 18; c) Eö = 36.

Fig.14.5.3. shows the results for the shape of a bubble rising towards a rigid wall (three governing parameters) for fixed Reynolds number Re = 10 and starting distance $\gamma = 0.75$ at different Eötvös numbers. It is interesting to note the occurrence of an indentation of the free surface at the rear of the bubble for comparatively large Eötvös numbers (see Fig.14.5.3.c and Fig.14.5.4.a). This effect is a consequence of the interplay between the inertia of the fluid behind the bubble and the deceleration of the particle's motion due to the presence of the wall. At the same time the pressure maximum is located at the bubble rear. The liquid at the bubble rear is accelerated due to the particle's motion (under the buoyancy, in the case). It pushes the interface and penetrates into the particle, if the surface tension is not sufficiently strong to stop it. This phenomenon could be referred to as a jet formation, by analogy with a similar effect in a different problem, collapse of a vapour cavity near a plate (see Blake et al., 1986).

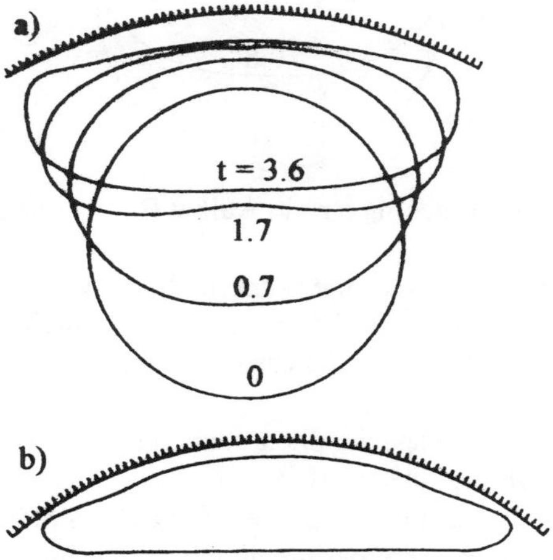

Fig.14.5.4. Gas bubble approaching a spherical container wall at $\gamma = 0.75$, e = 3: a) Re = 120, Eö = 36; b) Re = 1, Eö = 360.

In Fig.14.5.4.a another numerical experiment from this series is depicted. A slight dimpling ring is observed at the film surface for r = ±0.3125. This is a zone where the curvature of the interface changes twice its sign. That also could be considered as an analogue of the "classical" dimpling phenomenon in the case of a film between the bubble and the spherical container (see Mysels et al., 1959). This "dimple" does not grow but fades away with time. A clear pattern of a toroidal dimpling is observed for large Eötvös numbers (see fig.14.5.4.b), i.e., a dimple-ring is formed in the front part of the bubble not far from the rim. The effect is connected with the curvature of the wall and is due to the additional hydrostatic pressure gradient in the draining film because in this case the hydrostatic pressure grows in radial direction and, as a result, the pressure maximum shifts from the central zone to the film edge. It is clearly observed in the numerical experiment that the velocity in the dimpling zone remains comparatively small with respect to the velocity near the front stagnation point. A similar effect was observed experimentally by Hartland (1968) for a spherical film on a rigid sphere.

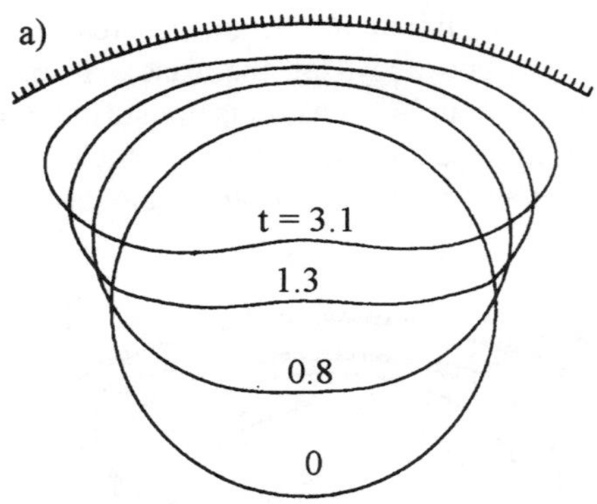

Fig.14.5.5. Gas bubble approaching a rigid wall at Eö = 18 and for: a) Re = 20, e = 3;

For the case of a curved wall the numerical results show that the film thickness will be greater at the line of symmetry than at the rim, which could be considered as a "dimpling" drainage configuration in the case of a curved wall. The film drainage will again be slow.

Ring dimpling is a more special phenomenon and only occurs at large curvatures of the wall and deformability of the interface.

In order to investigate the influence of the Reynolds number Shopov et al. (1990) have performed numerical experiments with fixed Eötvös number, Eö = 18. The first one for Re = 10 is shown in Fig.14.5.3.b. The dimpling is not very well expressed but the bubble has not yet achieved its "terminal" shape. For Re = 20 the inertial effects (the concavity in the rear part and the receding of the ends of the film area from the wall) increases (not shown on the figures). Deformations become larger, but the length of the thin liquid layer zone does not change essentially. A clear dimpling is formed at the axis of symmetry, which increases when Reynolds number increases. This confirms the important role of the particle inertia in the dimple formation. These tendencies develop for Re = 40 that can be clearly observed in Fig.14.5.5.c.

It is interesting to point out that in the case of a bubble in a container the general tendency remains the same as in the planar case (see Fig.14.5.5.a, b). The general rule that the bubble velocity in the container is smaller than near the plate holds again.

The case of a relatively large Reynolds number and Eötvös number of an intermediate value is depicted in Fig.14.5.6.a, b. For $t \leq 1.5$ the elongation effect is present, but it is to a certain degree weaker than in the previous case. Clearly its amplitude diminishes with the growth of the Reynolds number. However, one could not say that this effect is absent in the inviscid case. Something similar was observed by Blake et al. (1986) (see their figure 5) in the rather different problem of a cavitation bubble collapsing near a wall in an ideal flow for large buoyancy influence and moderate time.

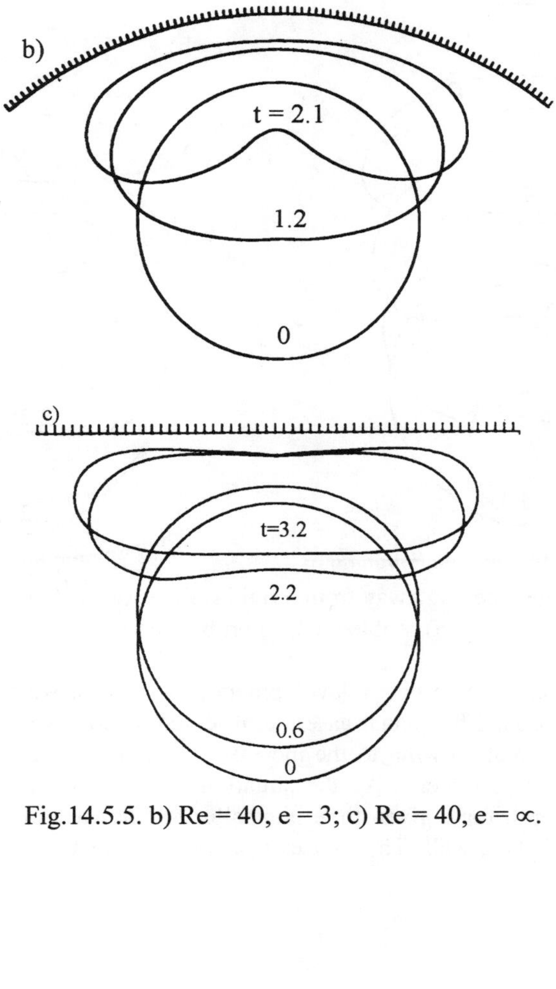

Fig.14.5.5. b) Re = 40, e = 3; c) Re = 40, e = ∝.

When the elongation of the bubble becomes sufficiently large, the surface tension rounds the bubble and after that the inertia forces come into play to preserve the somewhat higher speed of the liquid in the vicinity of the rear end, thus forming a concave shape in the rear part of the bubble. So the initial disturbance (elongation) causes an oscillation in the rear, which develops into a surface wave. This wave is fading away in all numerical experiments. Perhaps if initial disturbance were sufficiently great it could encompass the bubble and even lose stability. This effect is not observed for low Reynolds numbers.

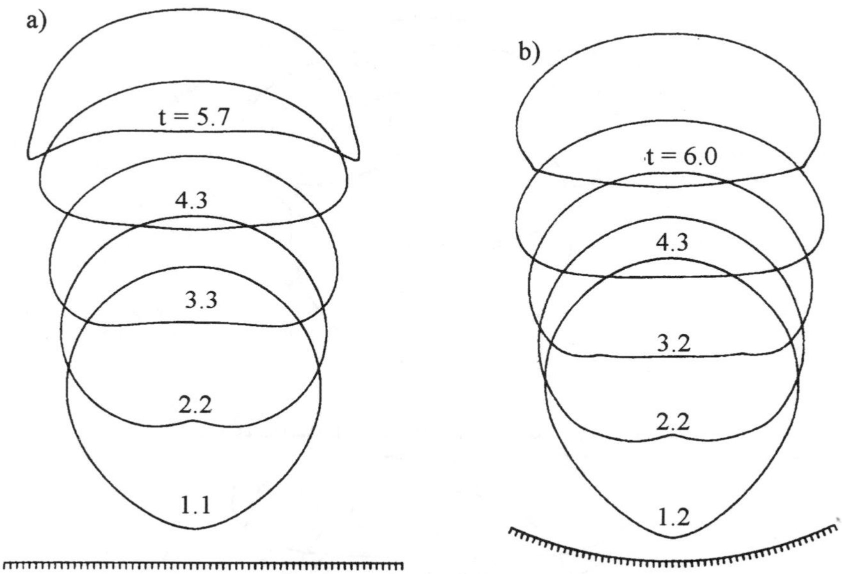

Fig.14.5.6. Bubble receding away from a rigid wall at Re =120, Eö = 12, $\gamma = 0.55$: a) problem (i); b) problem (ii)

In Fig.14.5.6.a, b the process development is shown for Re = 120 and Eö = 12. The tendencies mentioned for planar case are observed here as well. The elongation of the rear end is now greater owing to the increase of the influence of the wall, and the concavity becomes even larger. Like the situation with the overall deformations, the magnitude of the depth for a given time is smaller for the case of spherical container than for the case of plane wall. This causes a wave with greater amplitude than in the planar case.

The mentioned above results show that the proposed method in sections 14.3. and 14.4. appears to be a useful tool for solving variety of problems with boundaries of solid, gas/liquid or liquid/liquid type.

14.6. Finite Deformations of Two Viscous Drops at Moderate Reynolds Numbers

The relative motion of two droplets dispersed in another, immiscible, fluid plays an important role in a variety of industrial processes, including liquid-liquid extraction, liquid phase material processing, etc. Earlier theoretical studies of the relative motion of two drops has been based on the assumptions of Stokes flow approximation, a spherical form of fluid particles and small deformations of fluid interfaces. The resulting series solutions for the velocity and pressure show that the hydrodynamic force opposing drop motions increases without bound when the distance between the fluid particles decreases (see Chapter 7). In order to determine the nature of this singularity Barnocky and Davis (1989) and Davis et al. (1989) applied the lubrication theory as two drops become close. Finite deformations of two viscous drops in buoyancy driven motion at zero Reynolds numbers has been recently calculated by Manga and Stone (1993).

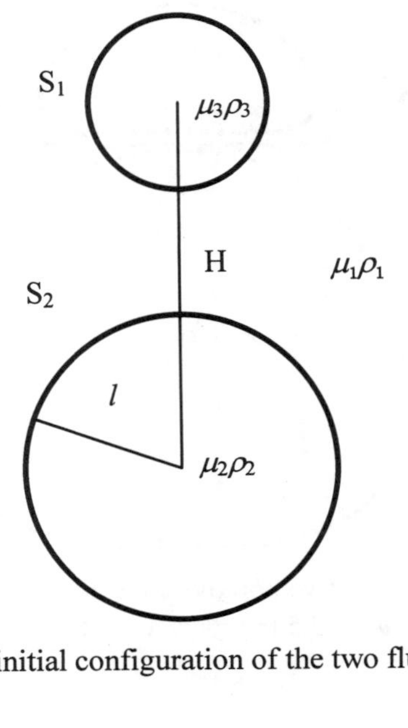

Fig.14.6.1. The initial configuration of the two fluid particles.

The first numerical solution of the problem of buoyancy driven motion of two deformable viscous drops without any ad hoc assumptions has been given by Minev (1990) and Zapryanov et al. (1998b). In this section we shall give some results from these investigations.

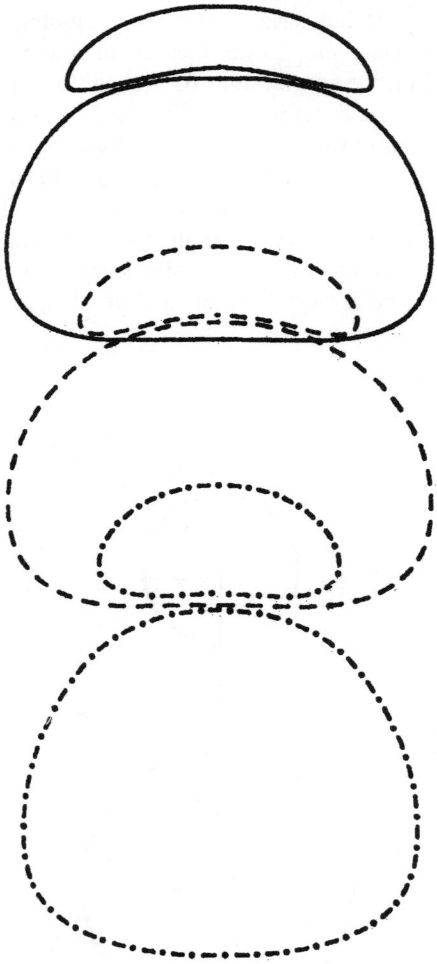

Fig.14.6.2. Coalescence of two homoviscous drops at Re = 2, We = 2.2, $\lambda = 1/3$, $\delta = 0.2$, $\gamma = 1$, d = 2 at t = 2.5 (- · — · -), t = 4.5 (··········) and t = 6.3 (———).

The interaction of two drops in parallel translation with their line of centres is modelled by considering three fluid domains, as shown in Fig.14.6.1. For convenience, the trailing drop is labelled drop 1, and the leading drop is labelled drop 2. For a characteristic length we choose the radius of one of the two drops while for a characteristic velocity, the terminal velocity of the same fluid particle.

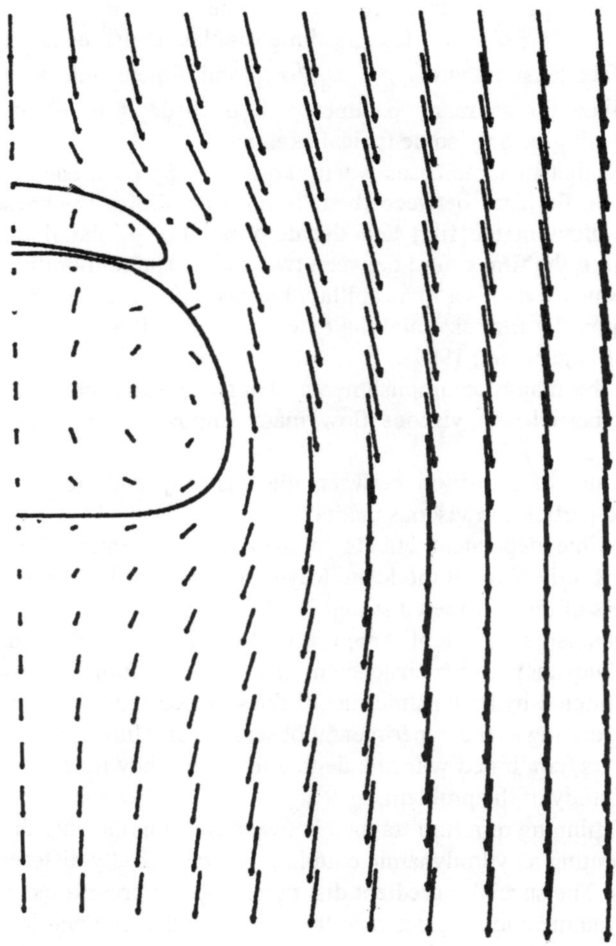

Fig. 14.6.3. The velocity field at t = 6.3 correspondent to the case given in Fig.14.6.2.

The corresponding equations for the considered problem are given in (14.3.1)-(14.3.3). We require that the velocity decays to zero far from the drops:

$$\mathbf{v}_1 \rightarrow 0, \qquad \text{as } r \rightarrow \infty \tag{14.6.1}$$

and that the velocity is continuous across all interfaces

$$\mathbf{v}_2 = \mathbf{v}_1 \text{ on } S_1 \quad \text{and} \quad \mathbf{v}_3 = \mathbf{v}_1 \text{ on } S_2, \tag{14.6.2}$$

where S_1 is the surface bounding drop 1 and S_2 is the surface bounding drop 2. Dynamic conditions on S_1 and S_2 are given in (14.3.6)-(14.3.8). From (14.3.9) and (14.3.10) one can obtain the respective symmetry and initial conditions. The time evolution of the drops shapes and drops motion depend on several parameters: Reynolds number Re, Weber number We, separation distance between the drops $d = H/l$, dimensionless viscosities ratios $\lambda_i = \mu_i/\mu_1$, (i = 1, 2, 3); dimensionless densities ratios $\delta_i = \rho_i/\rho_1$, dimensionless surface tensions ratios $\gamma = \sigma_{13}/\sigma_{12}$ and dimensionless radius of one of the two drops. Since for so many parameters it is difficult to perform parametric analysis, here we shall give only some typical examples.

It is known, that in some cases when two drops approach each other along the axis of their centres, the film between them is not of uniform thickness. Due to the radial pressure gradient in the film and the deformability of the liquid interfaces a "dimple" is formed in the film centre between two drops. The non-uniform thickness of film formed between two menisci in a capillary has been observed (Radoev et al., 1983) to be dependent upon the film size and the extent of non-uniformity increases with the film size (see also Manev et al., 1997).

In spite of the major accomplishments of many researchers in the field of the dynamics of fluid particles in viscous flow many important questions remain to be answered:

(i) What is the nature of transition between the unsteady and steady motion of the deformable fluid particles in viscous media?

(ii) What are the time dependent effects in hydrodynamic interaction between two deformable drops or bubbles at moderate Reynolds numbers and how do they depend on the parameters of the considered system?

(iii) What are mechanisms for establishment a trailing configuration or a draining film in the case of a buoyancy driven unsteady motion of two deformable viscous drops?

The used coupled hydrodynamic model does indeed make possible a variety of predictions, corresponding to the experimental observations. Unfortunately, the presence of several parameters, combined with the degree to which they range in nature hinders the full parametric study of the problem.

The rate of thinning of a film trapped between two approaching drops, according to a theory incorporating a hydrodynamic coupling, is dramatically different from earlier uncoupled theories. The special ingredient distinguishing the present theory from earlier ones is the hydrodynamic coupling between the motion in the film and in the drops. The parameter characterising that coupling is the Reynolds number, Re.

In Fig.14.6.2 we show the time evolution of deformation of the two drops moving under the action of gravity force. Reynolds number of the carrying phase is

supposed to be Re = 2, while the ratio of viscosities of the two fluids is $\lambda = \frac{\mu_2}{\mu_1} = \frac{1}{3}$.
Since the volume of the rear drop is eight times greater than the front one, the fluid particle are reached and at t = 2.5 the film with thickness Δ(2.5) = 0.086 is formed. The front drop in this moment is essentially more deformed from the rear one. This is due to the greater resistance of the carrying fluid as well as to the action of the second (rear) fluid particle, which presses the front one. Another reason for the smaller deformation of the second drop compared to the first one is because the second is moving in the wake of the first drop. Similar shapes of two moving drops are observed experimentally by Bhaga and Weber (1980).

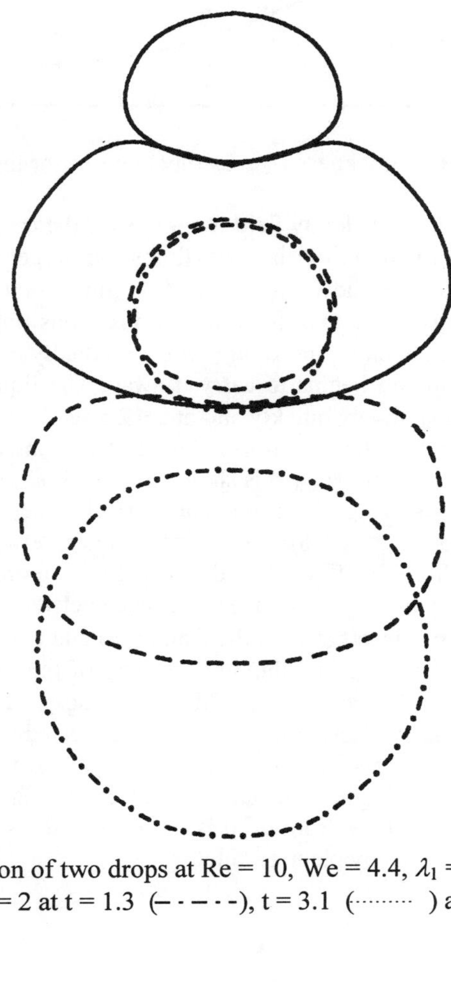

Fig.14.6.4. Interaction of two drops at Re = 10, We = 4.4, $\lambda_1 = 0.5$, $\lambda_2 = 10$, $\delta_1 = 0.2$, $\delta_2 = 2$, $\gamma = 2.2$, d = 2 at t = 1.3 (– · – · -), t = 3.1 (·········) and t = 6.4 (———).

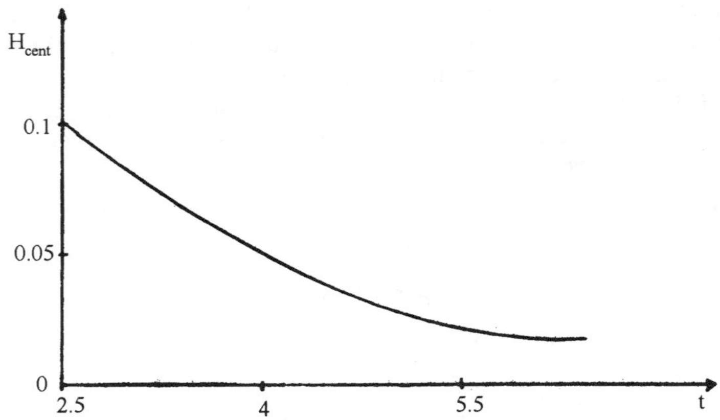

Fig.14.6.5. The film centre thickness H_{cent} as function of t for the case of Fig.14.6.4.

Fig.14.6.3. presents the velocity field at t = 6.3 for the case shown in Fig.14.6.2. The time evolution deformations of the two drops for larger Reynolds number of carrying phase, Re = 10, are shown in Fig.14.6.4. Since $\delta_1 = 0.2$ but $\delta_2 = 2$ the gravitational forces acting on the two drops are in directions opposite to one another. Due to the larger volume, the emerging fluid particle gradually reverses the direction of motion of the falling drop and begins to carry it away. The liquid layer between two drops makes thinner comparatively quickly and at $t \approx 2.5$ its thickness would be smaller than 0.1, i.e., one can conclude that the film is formed between two fluid particles. The plot of thickness at the film centre $H_{cent}(t)$ is shown in Fig.14.6.5.

The shape of the two drops is changed considerably when a rigid plane is placed at the front of their motion (Fig.14.6.6). The initial distance between the front drop and the wall is 0.5. It is seen from this figure that the shapes of two drops in the course of all time are very close to a rotational ellipsoid. The distance between the two fluid particles decreases faster than the distance between the front drop and the wall, and at $t \approx 2.1$ an uniform thin film begins to form. The thinning velocity of this film is about 2.5 times larger than in the case of a free emerging of the two drops at the same values of the parameters (see Fig.14.6.2). Therefore, it can be expected that in this case the coalescence will occur faster. The reason for it is the effect of the wall which leads to increasing pressure gradient in the film and its faster flowing. Besides this, here, a toroidal vortex motion around the front drop is not observed, as in Fig.16.6.3. and the shape of the film is approximately plane parallel.

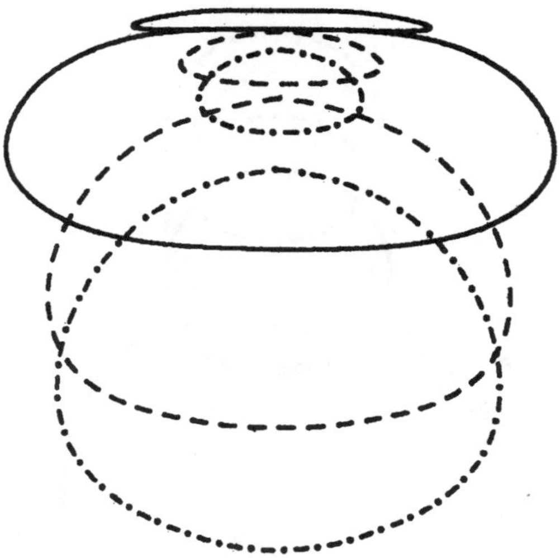

Fig.14.6.6. Two homoviscous drops approaching a wall at Re = 2, We = 2.2, $\lambda = 1/3$, $\delta = 0.2$, $\gamma = 1$, d = 1.8, t = 1.3 (– · – · ·), t = 2 (··········) and t = 3.4 (———).

The case of relatively large Reynolds number (Re = 63) and intermediate Eötvös one (Eö = 2, We = 28) is depicted in Fig.14.6.7. Since $\delta_1 = 0.8$ and $\delta_2 = 8$ the ratio of the gravitational forces acting on the particles is equal to 0.22, i.e., the buoyancy force acting on the smaller drop is greater than the buoyancy force acting on the larger one. Because of that up to t = 2.7 the larger drop is practically not moving, while the smaller one penetrates into the former and a thin film between the two particles is formed. At the end of our calculations the film profile is uneven and its minimum thickness is reduced in the ends. This is explicable since $\lambda_1 = 3.15$ and $\lambda_2 = 63$, i.e., the liquids from both sides of the boundary are more viscous from the liquid of the film and therefore they are against its rushing. In this case the slope of the plot of $H_{cent}(t)$ is much larger than in the previous considered cases. This is due to the relatively large value of the Reynolds number, Re = 63, of the carrying phase in the last case and because of the large buoyancy force acting on the larger drop.

In conclusion one can say that in the presence of finite surface tension, the main portion of the drops attain a nearly steady asymptotic configuration, and that the long-time evolution of the two drops is due to the deformation of the thin film which is trapped between them. Thus, the long-time behaviour of this film merits detailed consideration. It is seen that, if the film is more viscous phase, then the thinning is more rapid than the converse case, but aside from the case of effectively infinitely viscous surroundings, results differ quantitatively from those presuming immobile interfaces.

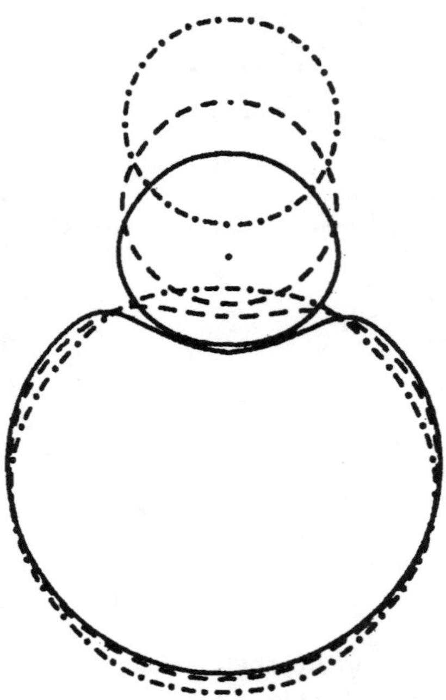

Fig.14.6.7. Interaction of two drops at Re = 63, We = 28, , $\lambda_1 = 3.15$, $\lambda_2 = 63$, $\delta_1 = 0.8$, $\delta_2 = 8$, $\gamma = 28$, d = 2 at t = 1.3 (- · — · -), t = 2.4 (··········) and t = 2.7 (———).

Following through its evolution, we can see that the flow within this film is driven by three mechanisms: the radial pressure field due to capillary forces, the gravitational field due to the discontinuity in density across the surfaces of the film and the pressure field set up to support the two drops.

The foregoing descriptions and figures manifest best the basic characteristics of thinning of a draining film, but by being abbreviated in number they fail to indicate the full spectrum of interactions of the several physical effects as they modulate the evolution shapes of the two drops.

14.7. Finite Deformations of a Compound Drop at Moderate Reynolds Numbers

Recently some interesting problems associated with the rheology of double emulsions and with the motion of encapsulated droplets in viscous liquids (in membrane separation processes) have arisen. These problems are now subjects of much research (see Chapter

7). The overall objective of the membrane separation processes is to selectively extract compounds from a liquid. The two miscible liquids (donor and receptor) are separated by a membrane through which the particles may diffuse. When the membrane phase is an immiscible liquid, there are three fluid phases. In some processes the receptor phase is dispersed as droplets which are encapsulated by the membrane phase. The transfer process using this type of membrane is carried out in a spray or diffusion column and, therefore, the performance of such apparatus will depend partly on the rise (or settling) velocity of the multiple drops. Since these cannot be calculated from either Stokes or Hadamard-Rybczynski formulae the calculation of drag coefficient and terminal settling velocity of compound multiphase drops is important.

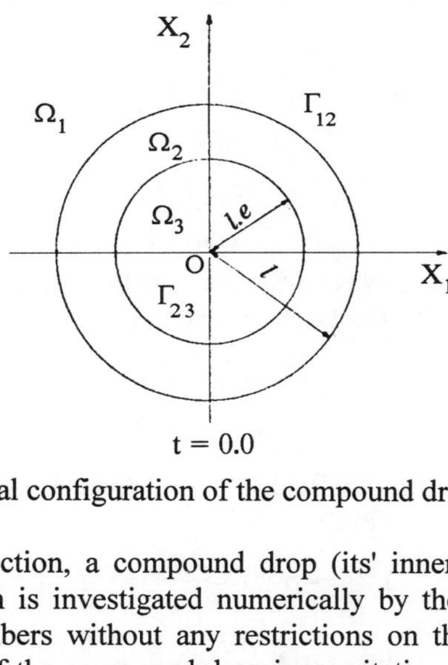

Fig.14.7.1. Initial configuration of the compound drop in the case (i_1).

In the present section, a compound drop (its' inner phase is a large drop or bubble) unsteady motion is investigated numerically by the finite-element method at moderate Reynolds numbers without any restrictions on the deformation of the free interfaces. The motion of the compound drop in gravitational field is considered in the following two cases:

(i) the free motion of a compound drop in a viscous fluid;

(ii) the rise of a compound drop toward a rigid horizontal wall in a viscous fluid.

Consider three fluids that occupy the domains Ω_i (i = 1, 2, 3): Ω_1 is the outer domain, Ω_2 is the shell phase and Ω_3 is the inner phase of the compound drop (see Fig.14.7.1 and Fig.14.7.7). The liquids are assumed to be homogenous, incompressible,

Newtonian with constant viscosities μ_i and densities ρ_i. The surface tension coefficients σ_{ij} at the interfaces Γ_{ij}, respectively, are also assumed to be constants.

The "equivalent" compound drop radius $l = (3V/4\pi)^{\frac{1}{3}}$ is chosen as the reference length l, where V is the volume of the compound drop.

The geometrical dimensionless parameters are: e is the ratio of the equivalent radius of the inner part of the compound drop phase to l ; $d = h/l$ is the dimensionless starting distance between the compound drop centre and the wall [for the case (ii)].

In case (i) two sub-cases will be discussed:

(i_1) the rise of a gas-liquid compound drop (the inner phase, Ω_3 , is a gas);

(i_2) the motion of a liquid-liquid compound drop.

The Stokes terminal velocity U_0 of a solid sphere with a density equal to the inner phase density or average compound drop density is chosen as a characteristic one, respectively: for the case (i_1) $U_0 = 2l^2\rho_1 g/9\mu_1$; for the case ($i_2$) $U_0 = 2l^2|\rho_1 - \rho_a| g/9\mu_1$, where ρ_a is the average compound drop density and $\rho_a = \rho_2(1-e^3) + \rho_3 e^3$.

Thus the problems depend on the following dimensionless hydrodynamic parameters: $\mathrm{Re} = \rho_1 l\, U_0/\mu_1$; $\mathrm{Eo} = \rho_1 l^2\, g/\sigma_{12}$; Fr = 2 Re/9 for the case (i_1), or $\mathrm{Fr} = 2\,\mathrm{Re}\big/\left\{9\left[1-\lambda_2(1-e^3)-\lambda_3 e^3\right]\right\}$ for the case (i_2); We = Fr.Eö; $\lambda_i = \rho_i/\rho_1$; $\eta_i = \mu_i/\mu_1$, (i = 1, 2, 3); $k_{jm} = \sigma_{jm}/\sigma_{12}$, (jm = 12, 23).

Then the dimensionless Navier-Stokes equations and the continuity equation describing the motion of the i-th fluid can be written in a general form as

$$\lambda_i \frac{D\mathbf{v}^{(i)}}{Dt} = \nabla.\mathbf{T}^{(i)} + \lambda_i \mathbf{F}, \tag{14.7.1}$$

$$\nabla.\mathbf{v}^{(i)} = 0. \tag{14.7.2}$$

$\mathbf{T}^{(i)}$ is the stress tensor for the i-th fluid, i = 1, 2 (and 3 when the inner part of the compound drop phase is viscous fluid),

$$\mathbf{T}^{(i)} = -p^{(i)}.\mathbf{I} + \frac{2\eta_i}{\mathrm{Re}}.\mathbf{E}^{(i)}, \tag{14.7.3}$$

$$\mathbf{E}^{(i)}\Big|_{kl} = 0.5\left(\frac{\partial v_k^{(i)}}{\partial x_l} + \frac{\partial v_l^{(i)}}{\partial x_k}\right), \tag{14.7.4}$$

where $\mathbf{I}$ is the unit tensor, $\mathbf{F}$ denotes the vector of the outer body forces $\mathbf{F} = (0, -\mathrm{Fr}^{-1})$.

The boundary conditions on the liquid/liquid interface Γ_{ij} are:

- the kinematic condition (continuity of the velocity across Γ_{ij}),

$$\left(\mathbf{v}^{(i)} - \mathbf{v}^{(j)}\right)\Big|_{\Gamma_{ij}} = 0; \tag{14.7.5}$$

- the dynamic condition (stress balance),

$$\left(\mathbf{T}^{(i)} - \mathbf{T}^{(j)}\right).\mathbf{n}\Big|_{\Gamma_{ij}} = \frac{k_{ij}}{We}\mathbf{n}.\left(\frac{1}{R_1} + \frac{1}{R_2}\right)\Bigg|_{\Gamma_{ij}}, \tag{14.7.6}$$

where **n** is the unit normal to Γ_{ij} outward to the i-th phase, R_1 and R_2 are the principle radii of curvature of Γ_{ij} [R_s (s = 1, 2) is reckoned positive if the curvature centre lies on the i-th phase].

If the inner part of the compound drop phase (Ω_3) is a gas, as in the cases (i$_1$) and (ii), it is supposed to be ideal ($\mu_3 = 0$) with a zero density ($\rho_3 = 0$) and pressure $p^{(3)} = p_0(t)$, i.e., the pressure in the gas is reckoned to be independent of the space coordinates. In this case, $\mathbf{T}^{(3)} = -p_0(t).\mathbf{I}$ and on Γ_{23} (14.7.6) has the form

$$\left(\mathbf{T}^{(2)} + p_0(t).\mathbf{I}\right).\mathbf{n}\Big|_{\Gamma_{23}} = \frac{k_{23}}{We}\mathbf{n}.\left(\frac{1}{R_1} + \frac{1}{R_2}\right)\Bigg|_{\Gamma_{23}}. \tag{14.7.7}$$

The standard boundary condition of motionlessness of the fluid at infinity is used:

$$\left|\mathbf{v}^{(1)}\right| \to 0, \qquad \text{when } |\mathbf{x}| \to \infty. \tag{14.7.8}$$

The no-slip boundary condition,

$$\mathbf{v}^{(1)}\Big|_{\Gamma_w} = 0, \tag{14.7.9}$$

is applied to the rigid wall Γ_w for the case (ii).

If Γ_{ij} is described by the equation

$$S_{ij}(\mathbf{x},t) = 0, \tag{14.7.10}$$

then the interface shape can be determined by means of the standard kinematic condition:

$$\frac{DS_{ij}}{Dt} = \frac{\partial S_{ij}}{\partial t} + \mathbf{v}.\nabla S_{ij} = 0. \tag{14.7.11}$$

The initial conditions for the velocity and the position of the interfaces are:

$$\mathbf{v}(\mathbf{x},0) = 0, \qquad S_{ij}(\mathbf{x},0) = S_{ij}^0(\mathbf{x}), \tag{14.7.12}$$

where $S_{ij}^0(\mathbf{x}) = 0$ is the equation describing the initial position of Γ_{ij}. It is assumed that at the initial moment the interfaces are concentric spheres. If there is information about the velocity field and the compound drop shape at t = 0, it would be possible to use other initial conditions.

The above described mathematical problem (14.7.1) - (14.7.12) is solved by means of a finite-element method in Galerkin formulation given in sections 14.5 and 14.6.

From mathematical point of view the problem of the free motion of compound drops in a gravitational field and that of the existence of their steady shape is considerably more difficult than the analogous problem concerning single-fluid particles. Some theoretical and experimental results (Brunn and Roden, 1985; Sadhal and Oguz, 1985; Mori, 1978; Chervenivanova and Zapryanov, 1989) show that in many cases a steady moving compound drop in the gravitational field does not exist, but some necessary conditions can be suggested.

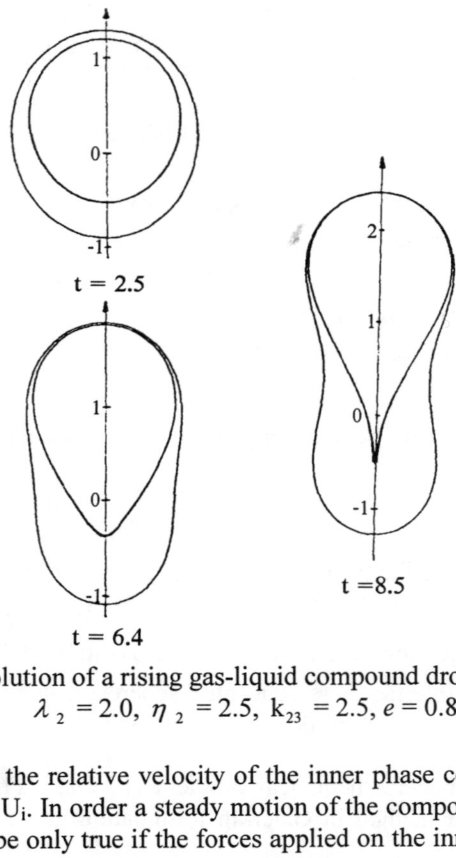

Fig.14.7.2. Shape evolution of a rising gas-liquid compound drop at Re = 1.4, Eö = 25.0, $\lambda_2 = 2.0$, $\eta_2 = 2.5$, $k_{23} = 2.5$, $e = 0.8$.

Let us denote the relative velocity of the inner phase concerning the compound drop as a whole with U_i. In order a steady motion of the compound drop to exist, U_i has to be zero. This will be only true if the forces applied on the inner phase counterbalance each other, i.e., their sum is zero. The inner phase is acted upon by the viscous forces F_v and the forces due to gravity F_b, where $F_b = (\rho_2 - \rho_3).\mathrm{vol}(\Omega_3)g$ and both F_v and F_b are the vertical components of these forces; the horizontal ones are zero. Thus, the necessary conditions for which $F_v + F_b = 0$ can be obtained. There are two possibilities for the direction of the force due to gravity, applied upon the whole compound drop $F_d = (\rho_1 - \rho_a).\text{vol}(\Omega_3 \cup \Omega_2)g$ and, therefore, for the direction of its motion:

a) $\rho_1 > \rho_a$, then $F_d > 0$ and the compound drop as a whole will move upward and F_v is positive. If $\rho_2 < \rho_3$, then $F_b < 0$ and $U_i = 0$, i.e, the inner phase will move together with the whole compound drop. However, if $\rho_2 > \rho_3$, F_b and U_i are positive, and the inner part of the compound drop will move to its outer part.

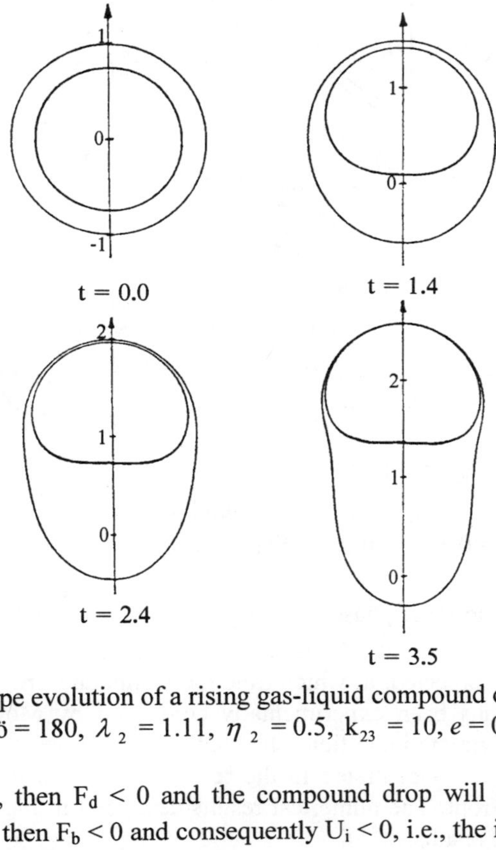

Fig.14.7.3. Shape evolution of a rising gas-liquid compound drop at Re = 1.25, Eö = 180, $\lambda_2 = 1.11$, $\eta_2 = 0.5$, $k_{23} = 10$, $e = 0.75$.

b) $\rho_1 < \rho_a$, then $F_d < 0$ and the compound drop will move downward, i.e., $F_v < 0$. If $\rho_2 < \rho_3$, then $F_b < 0$ and consequently $U_i < 0$, i.e., the inner part will move to the lower part of the compound drop. If $\rho_2 > \rho_3$, then $F_b > 0$ and since $F_v < 0$, a steady state of the compound drop will be achieved.

Further, we shall give some examples and their numerical results illustrating the shape evolution in time of the compound drop for the cases (i_1), (i_2) (see Fig.14.7.1) and (ii) (see Fig.14.7.7). More examples as well as more discussions could be found in (Bazhlekov et al., 1995).

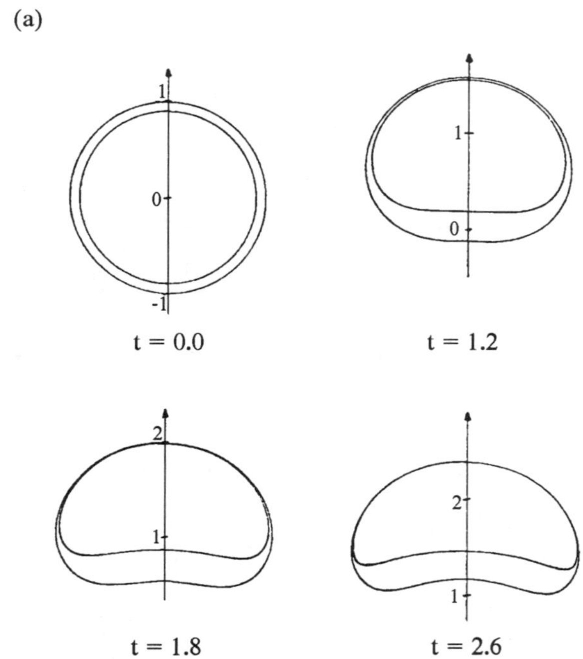

Fig.14.7.4. A free rise of a gas-liquid compound drop at Re = 5, Eö = 9, $\lambda_2 = 0.1$, $\eta_2 = 0.33$, $k_{23} = 0.5$, $e = 0.9$: (a) Shape evolution of the compound drop;

In the case (i_1) the inner phase is the lightest ($\rho_3 = 0$) and $\rho_1 > \rho_a$. Thus as a result a thin shell film in front of the compound drop has to be formed and its breakdown can also be expected, at which time the compound drop will turn into a 3S one. This conclusion has been experimentally confirmed by Mori (1978), who has pointed out that, in the gravitational field, the greatest part of the shell phase of the gas-liquid compound drop is concentrated in the rear part, while a thin film of the shell liquid is formed at its front. The numerical results described in Figs.14.7.2.-14.7.5. are in agreement with these results.

In Fig.14.7.2. a protraction of the compound drop, when the shell liquid density is twice greater than that of the ambient liquid, is observed. Moving upward the inner phase draws up a part of the shell phase and a shell liquid film is formed in the front of the compound drop. At the same time a part of the descending shell phase draws down part of the inner one, and because of this motion, the rear part of the inner phase accepts a wedge shape at t = 8.5. A qualitatively similar shapes of the inner phase has been obtained experimentally by Mori (1978). Apart from this, a "pinching" in the middle of the compound drop is also observed. Further on, the compound drop will turn into another configuration and the process can continue in three ways: (a) breakdown of the film, at which time the compound drop will turn into a 3S one; (b) separation of a certain part from the rear of the inner phase, at which time the compound drop will turn into a compound drop whose inner phase is divided into separate drops; (c) pinching of the compound drop leading to its splitting into two parts.

(b)

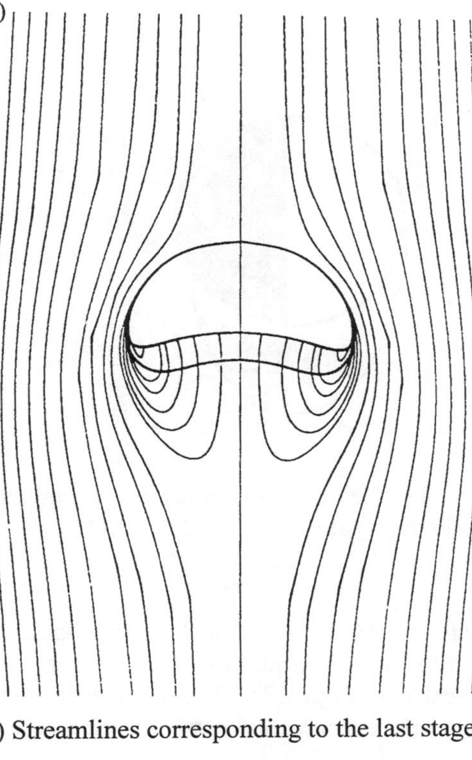

Fig. 14.7.4. (b) Streamlines corresponding to the last stage of (a) at t = 2.6.

Fig.14.7.3 displays the case when the shell phase density is greater than that of the ambient liquid, but smaller than the corresponding one in the previous case

($\lambda_2 = 1.11$). The whole shell liquid cannot be swept along at the upward motion of the inner phase because of the relatively larger shell phase volume ($e = 0.75$). A tendency of "wrapping" of Γ_{23} by Γ_{12} is observed, due to the larger surface tension coefficient of the inner interface with respect to the outside one ($k_{23} = 10$).

The shape evolution of a freely rising gas-liquid compound drop when the density of the shell liquid is considerably lower than that of the ambient liquid is shown in Fig.14.7.4a. The shell volume is relatively smaller, too. A concavity beginning to form in the rear part of the compound drop at about t = 1.2 can be seen which increases with time. The reason for the concave shape is the closed flow formed behind the drop. It is clearly displayed in Fig.14.7.4b, where the streamlines at t = 2.6 are shown. The film at the compound drop front, formed at about t = 1.5, grows thinner quickly and at time t = 2.6, its minimal thickness is 0.0005 from the "equivalent" compound drop radius. Further on, it is expected that the film in the front part of the drop will break down and the drop will turn into a 3S compound one.

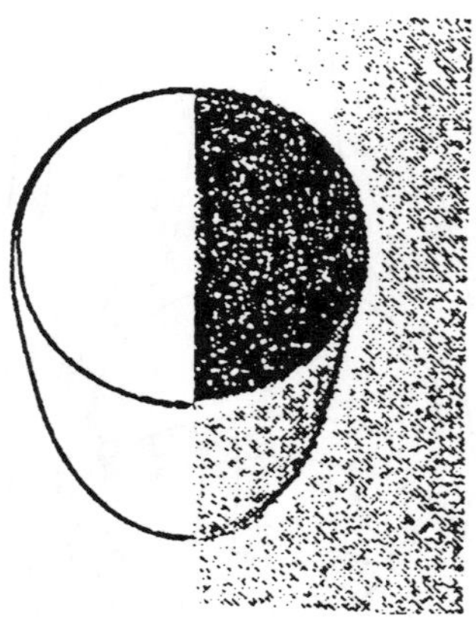

Fig.14.7.5. A comparison between the compound drop shapes obtained numerically (left) and experimentally (right) at Re = 0.016, Eö = 2.11, $\lambda_2 = 1.29$, $\eta_2 = 0.84$, $k_{23} = 3.64$, $e = 0.87$.

A comparison with the Mori's (see Mori, 1978, Fig.8a.) experimental result concerning the unsteady rise of a gas-liquid (96% aqueous glycerol) compound drop through castor oil is presented in Fig.14.7.5. The numerically obtained shape (left) is chosen to correspond best to the experimental one (right). Because the initial conditions (perfectly symmetric compound drop at t = 0) at which the numerical results are obtained are completely different from the experimental "initial conditions" and because of the good qualitative agreement, it can be concluded that in some cases the influence of the initial conditions on the compound drop shape is negligible.

The shape evolution of a freely rising compound drop of the case (i_2) is shown in Fig.14.7.6. The inner phase has a lower density than the shell liquid and therefore $F_b > 0$. The average compound drop density is lower than that of the ambient liquid, $\rho_1 > \rho_a$, thus the compound drop moves upward and consequently $F_v > 0$. Because both forces applied upon the inner phase of the compound drop are directed upward, the inner phase moves in this direction with respect to the compound drop as a whole, $U_i > 0$. Hence, the inner phase advances to the compound drop front and a thin film of the shell liquid is formed after t = 2.4. At t = 4.1 this film has uneven thickness, greater in the centre than at the periphery. Such effect is not observed when the inner phase is gas. Thus it can be concluded that this effect is a result of the action of the viscous forces from the inner liquid that retard film drainage.

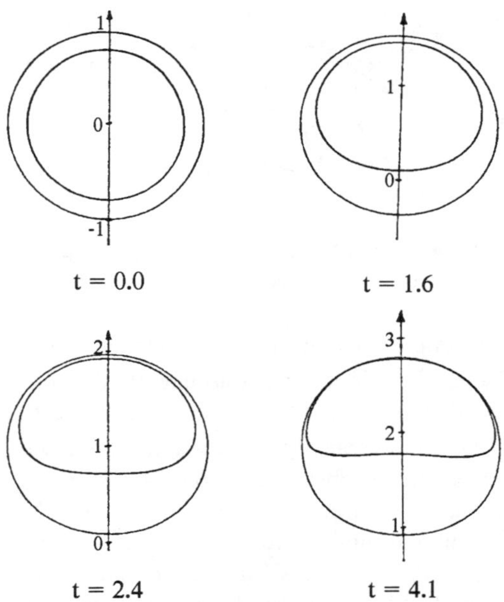

Fig.14.7.6. Shape evolution of a rising liquid-liquid compound drop at Re = 1.5, Eö = 4, $\lambda_2 = 0.5$, $\lambda_3 = 0.25$, $\eta_2 = 0.4$, $\eta_3 = 0.2$, $k_{23} = 0.4$, $e = 0.8$.

For the case (ii), the unsteady motion of a deformable compound drop under gravity in a viscous fluid toward a rigid horizontal wall is considered. Only the final stage of this process, when the compound drop is comparatively close to the wall (at a distance of about 1.5 radii), is investigated. It is assumed that at the initial moment t = 0 the compound drop is perfectly symmetrical (the interfaces Γ_{12} and Γ_{23} are concentric spheres) at a certain distance from the wall (see Fig.14.7.7) and liquids are immobile, $\mathbf{V}(\mathbf{x},0) = \mathbf{0}$. The investigations are conducted in the case when the inner phase is gas as in the case (i_1). Because of the large number of governing geometrical and dynamical parameters for the present case, a complete parametric investigation is not performed.

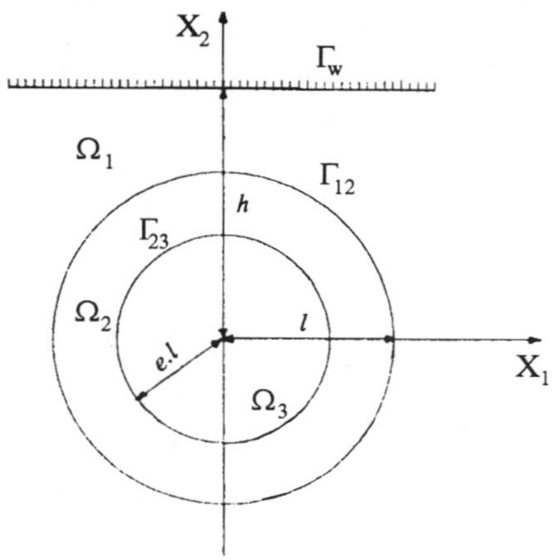

Fig.14.7.7. Initial configuration of the compound drop and the horizontal wall in the case (ii).

For most of the examples discussed here the average compound drop density ρ_a is lower than that of the ambient liquid and consequently the compound drop will move upwards (toward the wall). Since the inner phase is the lightest ($\rho_3 = 0$), most of the shell liquid is accumulated in the rear of the compound drop, while in its front a film is formed. These tendencies can be observed in Fig.14.7.8. The formation of the shell liquid film in the compound drop front part starts after about t = 0.5. At about t = 1.0, another thin film of the ambient liquid is formed between the compound drop front and the wall. The first of both films (of the shell liquid) drains more quickly because it is bounded by a viscous liquid and an ideal fluid, while the second is bounded by the solid wall and the ambient viscous liquid. Thus, it can be expected that the film of the shell liquid will break down first and the drop will turn into a 3S one before reaching the wall. In the final stage of the process the compound drop has a flattened shape, i.e., its front follows the form of the wall. It can be seen also that the inner free interface Γ_{23} is more deformed than the outer one, Γ_{12}. This can be explained by the fact that the surface tension coefficient of Γ_{23} is twice smaller than that of Γ_{12} ($k_{23} = 0.5$). A slight concavity in the rear of the inner interface can be observed which begins to disappear after t = 1.1.

The shape evolution of the compound drop at considerably higher Reynolds and Eötvos numbers (Re = 55.0, Eö = 20.0) is shown in Fig.14.7.9. In this case, the shell phase density is equal to that of the ambient liquid ($\lambda_2 = 1.0$) and consequently the compound drop moves under the gravity on the inner phase. Because of the large Eö and Re values the deformation of the compound drop as well as a inertial concavity in its rear are greater than that in the previous case. At the end of the process, after t = 14.0, the rear part of the compound drop returns to the equilibrium state at which the inertial concavity disappears.

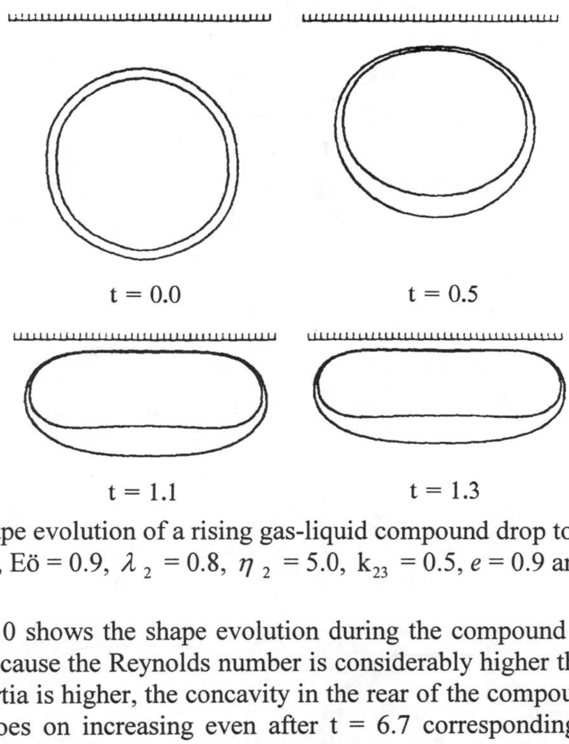

Fig.14.7.8. Shape evolution of a rising gas-liquid compound drop toward the wall at Re = 5.0, Eö = 0.9, $\lambda_2 = 0.8$, $\eta_2 = 5.0$, $k_{23} = 0.5$, $e = 0.9$ and $d = 1.5$.

Fig.14.7.10 shows the shape evolution during the compound drop's rise toward the rigid wall. Because the Reynolds number is considerably higher than in the previous case, i.e., the inertia is higher, the concavity in the rear of the compound drop is greater. The concavity goes on increasing even after t = 6.7 corresponding to the last stage illustrated on the Fig. 14.7.10. Because the Eötvos number is almost twice lower than that in the previous case, the horizontal dimension of the compound drop (flatness) in this case is smaller.

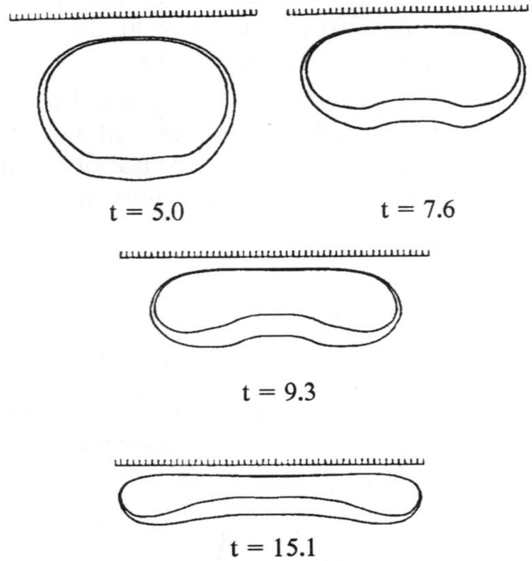

Fig.14.7.9. Shape evolution of a rising gas-liquid compound drop toward the wall at Re = 55, Eö = 20, $\lambda_2 = 1$, $\eta_2 = 11.1$, $k_{23} = 0.2$, $e = 0.9$ and $d = 1.5$.

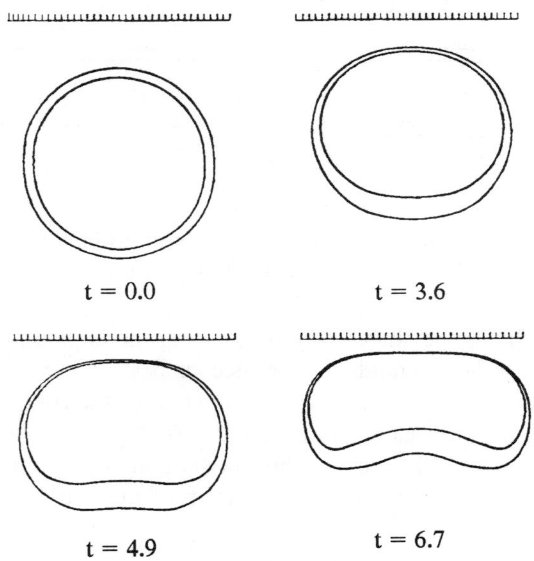

Fig.14.7.10. Shape evolution of a rising gas-liquid compound drop toward the wall at Re = 100.0, Eö = 11.25, $\lambda_2 = 0.5$, $\eta_2 = 10.0$, $k_{23} = 0.5$, $e = 0.9$ and $d = 1.5$.

Two shapes of the compound drop at t = 2.2 and t = 5.2 are shown in Fig.14.7.11. Here, the Reynolds and Eötvos numbers are considerably lower than those in the previous two cases, as the inertial concavity in the rear of the compound drop is not observed. It is noteworthy that a "dimpling" is formed in the front of the compound drop at t = 5.2, i.e., the liquid film between the wall and the compound drop has uneven thickness, greater in the centre. The shell liquid film profile is analogous, with the difference that its thickness as a whole is considerably smaller. An analogous effect has been observed (see Shopov et al., 1990) in the case of the motion of a single-phase bubble towards a rigid wall.

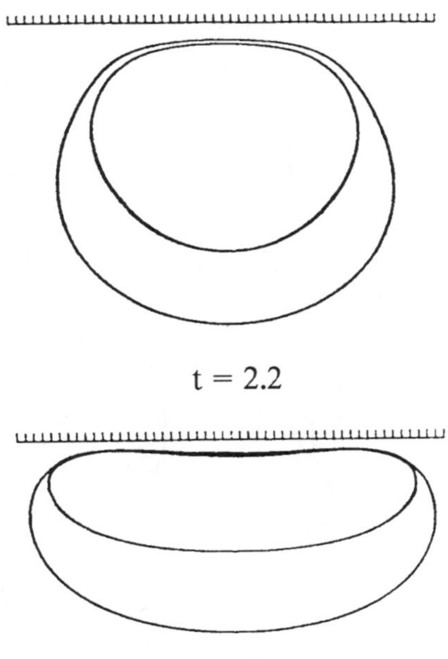

Fig.14.7.11. Two shapes of a compound drop rising toward the wall at Re = 1.2, Eö = 3.4, $\lambda_2 = 0.5$, $\eta_2 = 0.4$, $k_{23} = 0.4$, $e = 0.77$ and $d = 1.38$.

In the last case, which is considered in this chapter, the density of the shell liquid is twice as great as that of the ambient liquid ($\lambda_2 = 2.0$). The volume of the inner phase is $e^3 = 0.46$ of the volume of the whole compound drop, so the compound drop's average density ρ_a is insignificantly greater than that of the ambient liquid ($\rho_a / \rho_1 = 1.08$). Thus the compound drop moves very slightly downward as a whole (see Fig.14.7.12). At the same time the inner part of the compound drop phase moves toward the wall, while the rear of the shell phase moves downwards. Thus a protraction of the compound drop is observed at t = 5.3. Because of the surface tension coefficient σ_{23} of the inner interface being two and a half times greater than that of Γ_{12} ($k_{23} = 2.5$), Γ_{23} does not deform and practically remains spherical. The shape of the outer interface in its front is also spherical, and a film between the drop and the wall is not formed, which enables the front of the compound drop to reach the wall more quickly. Thus, at t = 5.3 the distance between the wall and the compound drop is 0.035.

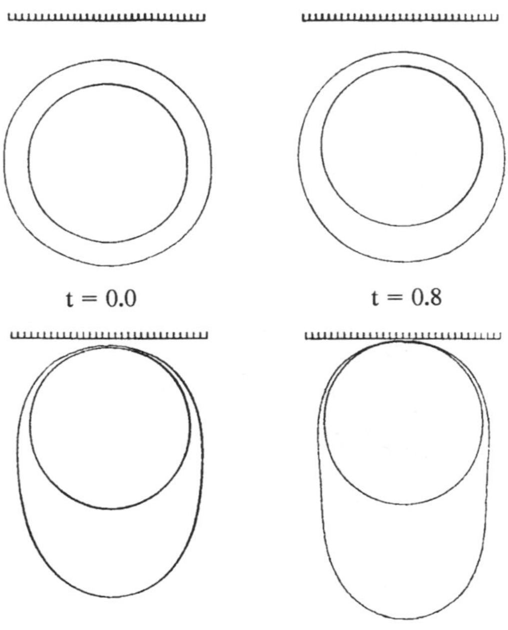

t = 0.0 t = 0.8

t = 3.5 t = 5.3

Fig.14.7.12. Shape evolution of a rising gas-liquid compound drop toward the wall at Re = 0.9, Eö = 2.1, $\lambda_2 = 2.0$, $\eta_2 = 2.5$, $k_{23} = 2.5$, $e = 0.77$ and $d = 1.38$.

Some of the results discussed in the present chapter depend on the initial conditions used and consequently cannot be related directly to real systems. However, it can be considered that the inertial and transient effects, e.g., the appearance and disappearance of a concavity in the rear of the compound drop, the formation of a "dimpling", single or double liquid film in its front, observed at the end stage of the investigated processes, are less sensitive to the initial conditions.

REFERENCES

Abramowitz, M. and Stegun, I.A. (1965), Handbook of Mathematical Functions, Dover, New York.
Acrivos, A. (1983), 4th Int. Conf. on Physicochemical Hydrodynamics, Ann. N.Y. Acad. Sci., **404**, 1.
Acrivos, A. (1985), Fluid Dynamics Transactions, **13**, 7.
Acrivos, A. and Lo, T.S. (1978), J. Fluid Mech., **86**, 641.
Acrivos, A., Batchelor, G.K., Hinch, E., Koch, D. and Mauri, R. (1992), J. Fluid Mech., **240**, 651.
Aderogba, K. (1976), J. Eng. Math., **10**, 143.
Aderogba, K. and Blake, J. (1978a), Bull. Austr. Math. Soc., **18**, 345.
Aderogba, K. and Blake, J. (1978b), Bull. Austr. Math. Soc., **19**, 309.
Adler, P.M. (1984), J. de Mécanique Theorique et Appliquée, **3**, 1.
Adler, P.M., Zuzovsky, M. and Brenner, H. (1985), Int. J. Multiphase Flow, **11**, 387.
Agrawal, S.K. and Wasan, D.T. (1979), Chem. Eng. J., **18**, 215.
Al Taweel, A.M. and Landau, J. (1976), Can. J. Chem. Eng., **54**, 532.
Alam, Md., Kawamura, T., Kuwahara, K. and Takami, H. (1981), Preprint of JSME Meeting, Tokyo, 391.
Allen, D.N. and Southwell, R.V. (1955), Quart. J. Mech. Appl. Math., **8**, 129.
Ames, W.F. (1977), Numerical Methods for Partial Differential Equations, 2nd ed., New York: XIV, 365pp.
Andrade, E.N. (1931), Proc. Roy. Soc., **A 134**, 445.
Andres, A. and Ingard, U. (1953), J. Acoust. Soc. Amer., **25**, 928.
Antanovskii, L.K. (1994), Eur. J. Mech. B/Fluids, **13**, 491.
Antanovskii, L.K. (1996), J. Fluid Mech., **327**, 325.
Aris, R. (1989), Vectors, Tensors and the Basic Equations of Fluid Mechanics, Dover.
Arp, P.A. and Mason, S.G. (1977), J. Colloid Int. Sci., **61**, 21.
Asaki, T.J. and Marston, P.L. (1994), J. Acoust. Soc. Am., **96(5)**, 3096.
Asaki, T.J. and Marston, P.L. (1995), J. Fluid Mech., **300**, 149.
Ascoli, E., Dandy, D. and Leal, L.G. (1990), J. Fluid Mech., **213**, 287.
Barker, J. and Henderson, D. (1971), Molecular Physics, **21**, 187.
Barnea, E. and Mizrahi, J. (1973), J. Fluid Mech., **92**, 435.
Barnes, H., Edwards, M. and Woodcock, L. (1987), Chem. Eng. Sci., **42**, 591.
Barnes, H., Hutton, J.F. and Walters, K. (1989), An Introduction to Rheology, Elsevier, Amsterdam.
Barnocky, G. and Davis, R.H. (1989), Int. J. Multiphase Flow, **15**, No.4, 627.
Bart, E. (1968), Chem. Eng. Sci., **23**, 193.
Barthes-Biesel, D. and Acrivos, A. (1973a), Int. J. Multiphase Flow, **1**, 1.
Barthes-Biesel, D. and Acrivos, A. (1973b), J. Fluid Mech., **61**, 1.
Basset, A.B. (1888), A Treatise on Hydrodynamics, **vol. 2, ch. 21**, 22. Cambridge: Deighton Bell.
Batchelor, G.K. (1954), Quart. J. Mech. Appl. Math., **7**, 179.
Batchelor, G.K. (1967), An Introduction to Fluid Dynamics, Cambridge Univ. Press, London.
Batchelor, G.K. (1970a), J. Fluid Mech., **41**, 545.
Batchelor, G.K. (1970b), J. Fluid Mech., **44**, 419.
Batchelor, G.K. (1971), J. Fluid Mech., **46**, 813.
Batchelor, G.K. (1972), J. Fluid Mech., **52**, 245.
Batchelor, G.K. (1974), Ann. Rev. Fluid Mech., **6**, 227.
Batchelor, G.K. (1976a), J. Fluid Mech., **74**, 1.
Batchelor, G.K. (1976b), Theoretical and Applied Mechanics, Koiter, W.T. (Ed.) North Holland, p. 33.
Batchelor, G.K. (1977), J. Fluid Mech., **83**, 97.

Batchelor, G.K. (1982), J. Fluid Mech., **119**, 379.
Batchelor, G.K. (1983), J. Fluid Mech., **131**, 155.
Batchelor, G.K. and Green, J.T. (1972a), J. Fluid Mech., **56**, 375.
Batchelor, G.K. and Green, J.T. (1972b), J. Fluid Mech., **56**, 401.
Batchelor, G.K. and Wen, C.S. (1982), J. Fluid Mech., **124**, 495.
Bauer, H.F. (1984), Mech. Res. Commun., **11(1),** 11.
Bazhlekov, I. (1992), Ph. D. Thesis, Bulgarian Academy of Sciences.
Bazhlekov, I., Shopov, P. and Zapryanov, Z. (1995), J. Colloid Int. Sci., **169**, 1.
Beard, K.V. and Pruppacher, H.R. (1969), J. Atmos. Sci., **26**, 1066.
Becker, H.A. (1959), Can. J. Chem. Eng., **37**, 185.
Beenakker, C.W. (1984a), Physica, **128A**, 349.
Beenakker, C.W. (1984b), Physica, **128A**, 48.
Beenakker, C.W. and Mazur, P. (1982), Phys. Letters, **91**, 290.
Beenakker, C.W. and Mazur, P. (1983), Physica, **120A**, 388.
Beenakker, C.W. and Mazur, P. (1984), Physica, **126A**, 349.
Beenakker, C.W., van Saarloos, W. and Mazur, P. (1984), Physica, **127A**, 451.
Belov, I.A. and Kudriavtzev, N.A. (1981), J. Phys. Eng., **XLI**, 310.
Benjamin, T.B. (1987), J. Fluid Mech., **181**, 349.
Benjamin, T.B. and Ellis, A.T. (1990), J. Fluid Mech., **212**, 65.
Benjamin, T.B. and Strasberg, M. (1958), J. Acoust. Soc. Am., **30**, 697.
Bentley, B.J. (1985), Ph.D. Thesis, California Institute of Technology.
Bentley, B.J. and Leal, L.G. (1986), J. Fluid Mech., **167**, 241.
Bentwich, M. and Miloh, T. (1978), J. Appl. Mech., **45**, 463.
Beran, M.J. (1968), Statistical Continuum Theories, New York: Interscience.
Berdan, C. II and Leal, L.G. (1982), J. Colloid Int. Sci., **87**, 62.
Berry, D. and Russel, W. (1987), J. Fluid Mech., **180**, 475.
Beshkov, V.N., Radoev, B.P. and Ivanov, I.B. (1978), Int. J. Multiphase Flow, **4**, 563.
Bestman, A.R. (1983), ZAMP, **34**, 867.
Bhaga, D. and Weber, M.E. (1981), J. Fluid Mech., **105**, 61.
Blake, J.R. (1971), Proc. Camb. Phil. Soc., **70**, 303.
Blake, J.R. (1974), J. Engng. Math., **8**, 113.
Blake, J.R. and Gibson, D.C. (1987), Ann. Rev. Fluid Mech., **19**, 99.
Blake, J.R., Taib, B.B. and Doherty, G. (1986), J. Fluid Mech., **170**, 479.
Blake, J.R., Taib, B.B. and Doherty, G. (1987), J. Fluid Mech., **181**, 197.
Blasius, H. (1908), Z. Math. Phys., **56**, 1.
Block, M.J. (1956), Nature, London, **178**, 650.
Boltze, E. (1908), Göttingen Dissertation.
Bossis, G. and Brady, J.F. (1984), J. Chem. Phys., **80**, 5141.
Bouillot, J.L., Camoin, C., Belsons, M. Blanc, R. and Guyon, E. (1982), Adv. Coll. Interface Sci., **17**, 299.
Boulton-Stone, J.M. (1995), J. Fluid Mech., **302**, 231.
Boussinesq, J. (1903), Theorie Analytique de la Chaleur en Harmonie avec la Thermodynamique et avec la Theorie Mechanique de la Lumiere, **vol. II**, Paris, Gauthier-Villars, 625 p.
Boussinesq, J. (1913a), Ann. Chim. Phys., **291**, 349.
Boussinesq, J. (1913b), C. R. Acad. Sci., **156**, 1124.
Bozzi, L.A., Feng, J.Q., Scott, T.C. and Pearlstein, A.J. (1997), J. Fluid Mech., **336**, 1.
Brabston, D.C. and Keller, H.B. (1975), J. Fluid Mech., **69**, 179.
Brady, J.F. and Acrivos, A. (1982), J. Fluid Mech., **115**, 443.
Brady, J.F. and Bossis, G. (1985), J. Fluid Mech., **155**, 105.
Brady, J.F. and Bossis, G. (1988), Ann. Rev. Fluid Mech., **20**, 111.
Brady, J.F. and Durlofsky, L. (1988), Phys. Fluids, **31**, 717.

Brady, J.F., Phillips, R., Lester, J. and Bossis, G. (1988), J. Fluid Mech., **195**, 257.
Bratuhin, Yu.K. (1975), Izv. AN SSSR, Mehanika Jidkosti i Gaza, **5**, 156. (in Russian)
Breach, D. (1961), J. Fluid Mech., **10**, 306.
Brenner, H. (1961), Chem. Eng. Sci., **16**, 242.
Brenner, H. (1962), J. Fluid Mech., **12**, 35.
Brenner, H. (1963), Chem. Eng. Sci., **18**, 1; errata, ibid. 557.
Brenner, H. (1964a), Chem. Eng. Sci., **19**, 599.
Brenner, H. (1964b), Chem. Eng. Sci., **19**, 703.
Brenner, H. (1964c), J. Fluid Mech., **18**, 144.
Brenner, H. (1966), In Advances in Chemical Engineering, **6**, 287.
Brenner, H. (1971), Ind. Engng Chem. Fund., **10**, 537.
Brenner, H. (1972a), Progress in Heat and Mass Transfer, **5**, 89.
Brenner, H. (1972b), Progress in Heat and Mass Transfer, **6**, 509.
Brenner, H. (1974), Int. J. Multiphase Flow, **1**, 195.
Brenner, H. and Cox, G. (1963), J. Fluid Mech., **17**, 561.
Brenner, H. and Happel, J. (1958), J. Fluid Mech., **4**, 195.
Brenner, H. and O'Neill, M.E. (1972), Chem. Eng. Sci., **27**, 1421.
Bretherton, F.P. (1961), J. Fluid Mech., **10**, 166.
Bretherton, F.P. (1962a), J. Fluid Mech., **12**, 591.
Bretherton, F.P. (1962b), J. Fluid Mech., **14**, 284.
Brignell, A.S. (1973), Quart. J. Mech. Appl. Math., **26**, 99.
Brinkman, H.C. (1949), Appl. Sci. Res., **A 1**, 27.
Broersma, S. (1960), J. Chem. Phys., **32**, 1626.
Brunn, P. and Roden, T. (1985), J. Fluid Mech., **160**, 211.
Buchanan, J. (1891), Proc. London Math. Soc., **22**, 181.
Buckmaster, J.D. (1972), J. Fluid Mech., **55**, 385.
Buckmaster, J.D. (1973), Trans. A.S.M.E. J. Appl. Mech., **E 40**, 18.
Bull, T.H. (1956), Br. J. Appl. Phys., **7**, 416.
Bungay, P.M. and Brenner, H. (1973), Int. J. Multiphase Flow, **1**, 25.
Burgers, J.M. (1938a), Second Report on Viscosity and Plasticity, Amsterdam, North Holland Publishing Co.
Burgers, J.M. (1938b), Kon. Ned. Akad. Wet. Verhand. (Eerste Sectie), **16**, 113.
Burgers, J.M. (1941), Proc. Koningl. Akad. Wetenschap (Amsterdam), **44**, 1045.
Burgers, J.M. (1942), Proc. Koningl. Akad. Wetenschap (Amsterdam), **45**, 9.
Burrill, K.A. and Woods, D.R. (1973), J. Colloid Interface Sci., **42**, 15.
Burton, R.A. and Mannhemer, R.J. (1967), Adv. Chem. Ser., **63**, 315.
Buscall, R., Goodwin, J.W., Ottewill, B.H. and Tadros, T.F. (1982), J. Colloid Interface Sci., **85**, 78.
Bush, M.B. and Tanner, R.I. (1983), Int. J. Numer. Meth. Fluids, **3**, 71.
Buyevich, Yu. A. (1971a), Izv. Akad. Nauk SSSR, Mekh. Zhid. Gaza, **5**, 104.
Buyevich, Yu. A. (1971b), J. Fluid Mech., **49**, 489.
Buyevich, Yu. A. (1971c), J. Fluid Mech., **52**, 345.
Buyevich, Yu. A. (1972), J. Fluid Mech., **56**, 313.
Buyevich, Yu. A. and Markov, V.G. (1972), Prikl. Matem. Mekh., **36**, 480.
Buyevich, Yu. A. and Markov, V.G. (1973a), Prikl. Matem. Mekh., **37**, 883.
Buyevich, Yu. A. and Markov, V.G. (1973b), Prikl. Matem. Mekh., **37**, 1059.
Buyevich, Yu. A. and Shchelchkova, I (1978), Progress in Aerospace Sciences, **17**, 121.
Carriere, Z. (1929), J. Phys. Radium, **10**, 198.
Carslaw, H.S. and Jaeger, J.C. (1947), Conduction of Heat in Solids, Clarendon Press, London and New York.
Cerf, R.J. (1951), J. Chim. Phys., **48**, 59.
Chaffey, C.E. (1977), Colloid Polymer Sci., **255**, 691.

Chaffey, C.E. and Brenner, H. (1967), J. Colloid Int. Sci., **24**, 258.
Chaffey, C.E., Brenner, H. and Mason, S.G. (1965), Rheol. Acta, **4**, 56.
Chaffey, C.E., Brenner, H. and Mason, S.G. (1967), Rheol. Acta, **6**, 100.
Chahine, G. (1994), Proc. Symp. Nav. Hydrodyn., 20th, Santa Barbara, CA, pp. 290-310, Washington, DC, Nat. Acad. Press.
Chakrabarti, A., Gooden, D.K. and Shail, R. (1982), J. Colloid. Interface Sci., **88**, 407.
Chambers, R. and Kopac, M. (1937), J. Cell Comp. Physiol., **9**, 331.
Chan, P. C. and Leal, L.G. (1979), J. Fluid Mech., **92**, 131.
Chandrasekhar, S. (1959), Proc. London Math. Soc. (3), **9**, 141.
Chandrasekhar, S. (1961), Hydrodynamic and Hydromagnetic Stability, Oxford, Clarendon press.
Chang, C. and Powell, R.L. (1993), J. Fluid Mech., **253**, 1.
Chang, C. and Powell, R.L. (1994a),Phys. Fluids, **A 6**, 1628.
Chang, C. and Powell, R.L. (1994b), J. Fluid Mech., **281**, 51.
Chang, H.-C. and Chen, L.H. (1986), Phys. Fluids, **29**, 3530.
Chao, B.T. (1962), Phys. Fluids, **5**, 69.
Chauveteau, G. (1982), J. Rheol., **26**, 111.
Chen, T. and Skalak, R. (1970), Appl. Sci. Res., **22**, 403.
Cheng, P.Y. and Schachman, H.K. (1955), J. Polymer. Sci., **16**, 19.
Cheng, H. and Papanicolaou, G. (1997), J. Fluid Mech., **335**, 189.
Cherukat, P. and McLaughlin, J.B. (1994), J. Fluid Mech., **263**, 1.
Cherukat, P., McLaughlin, J.B. and Graham, A.L. (1994), Int. J. Multiphase Flow, **20**, 339.
Chervenivanova, E. (1985), Ph D Thesis, University of Sofia.
Chervenivanova, E. and Zapryanov, Z. (1985), Int. J. Multiphase Flow, **11**, 721.
Chervenivanova, E. and Zapryanov, Z. (1987), PhysicoChemical Hydrodynamics, **8**, 293.
Chervenivanova, E. and Zapryanov, Z. (1988), Quart. J. Mech. Appl. Math., **41(3)** , 419.
Chervenivanova, E. and Zapryanov, Z. (1989), PhysicoChemical Hydrodynamics, **11**, 243.
Chervenivanova, E., Christov, Y. and Zapryanov, Z. (1994), Ann. Sofia Univ., **88**, 21.
Chi, B.K. and Leal, L.G. (1989), J. Fluid Mech., **201**, 123.
Childress, S. (1972), J. Chem. Phys., **56**, 2527.
Childress, W.S. (1964), J. Fluid Mech., **20**, 305.
Chisnell, R.F. (1987), J. Fluid Mech., **176**, 443.
Chong, J.S., Christiansen, E.B. and Baer, A.D. (1971), J. Appl. Polymer Sci., **15**, 2007.
Chorin, A.J. (1967), J. Comput. Physics, **2**, 12.
Chorin, A.J. (1968), Studies in Numerical Analysis 2, pp.64-70, Philadelphia, SIAM.
Christov, Ch. and Markov, K. (1985), SIAM J. Appl. Math., **45**, 289.
Christov, Ch. and Volkov, P. (1985), J. Fluid Mech., **158**, 341.
Christov, Ch. and Zapryanov, Z. (1980), Comp. Meth. in Appl. Mech. and Engng, **22**, 49.
Chwang, A.T. (1975), J. Fluid Mech., **72**, 17.
Chwang, A.T. and Wu Yao-Tsu, T. (1974), J. Fluid Mech., **63**, 607.
Chwang, A.T. and Wu Yao-Tsu, T. (1975), J. Fluid Mech., **67**, 787.
Chwang, A.T. and Wu Yao-Tsu, T. (1976), J. Fluid Mech., **75**, 677.
Clayes, I.L. and Brady, J.F. (1993a), J. Fluid Mech., **251**, 441.
Clayes, I.L. and Brady, J.F. (1993b), J. Fluid Mech., **251**, 443.
Clayes, I.L. and Brady, J.F. (1993c), J. Fluid Mech., **251**, 479.
Clift, R. and Gauvin, W.H. (1970), Proc. Chemeca '70, **1**, 14.
Clift, R., Grace, J. R. and Weber, M. E. (1978), Bubbles, Drops and Particles, Academic Press, New York, San Francisco, London.
Cole, J.D. (1968), Perturbation Methods in Applied Mathematics, Blaisedell Waltham.
Coleman, B.D., Markovitz, H. and Noll, W. (1966), Viscometric Flows of Non-Newtonian Fluids, New York: Springer.
Collins, R. (1966), J. Fluid Mech., **25**, 469.

Connor, T.T. and Brebbia, S.A. (1976), Finite Element Techniques for Fluid Flows, Newnes Butternorths, London.
Cooley M.D. and O'Neill, M.E. (1969), Proc. Cambridge Phil. Soc., **66**, 407.
Coutanceau, M. (1972), C. R. Acad. Sci., **ser. A 274**, 853.
Coutanceau, M. and Thizon, P. (1981), J. Fluid Mech., **107**, 339.
Cox, R.G. (1969), J. Fluid Mech., **37**, 601.
Cox, R.G. (1970), J. Fluid Mech., **44**, 791.
Cox, R.G. (1971), J. Fluid Mech., **45**, 625.
Cox, R.G. (1974), Int. J. Multiphase Flow, **1**, 343.
Cox, R.G. and Brenner, H. (1967), Chem. Eng. Sci., **22**, 1753.
Cox, R.G. and Brenner, H. (1968), Chem. Eng. Sci., **23**, 147.
Cox, R.G. and Hsu, S.K. (1977), Int. J. Multiphase Flow, **3**, 201.
Cox, R.G. and Mason, S.G. (1971), Ann. Rev. Fluid Mech., **3**, 291.
Curtis, S.G. and Hocking, L.M. (1970), Trans. Faraday Soc., **66**, 1381.
Cuvelier, C., Segal, A. and van Steenhoven, A.A. (1986), Finite Element Methods and Navier-Stokes Equations, D. Reidel Publishing Company, Dordrecht, Boston, Lancaster, Tokyo.
Dabros, T. A. (1985), J. Fluid Mech., **156**, 1.
Dagan, Z., Yan, Z.Y. and Shan, H. (1988), J. Fluid Mech., **190**, 299.
Dandy, D.S. and Dwyer, H.A. (1990), J. Fluid Mech., **216**, 381.
Dandy, D.S. and Leal, L.G. (1989), J. Fluid Mech., **208**, 161.
Danov, Kr., Aust, R., Durst, F. and Lange, U. (1995a), Chem. Eng. Sci., **50**, 263.
Danov, Kr., Aust, R., Durst, F. and Lange, U. (1995b), Chem. Eng. Sci., **50**, 2943.
Danov, Kr., Ivanov, I., Zapryanov, Z., Nakache, E. and Raharimalala, S. (1988), in Proceedings on Synergetics, Order and Chaos, 13-17 Oct., 1987, Madrid, Spain, (ed. M.Vilarde), World Scientific, Singapore, 1988, p. 178.
Darabaner, C.L. and Mason, S.G. (1967), Rheol. Acta, **6**, 273.
Davenport, W.G., Bradshaw, A.V. and Richardson, F.D. (1967), Chem. Eng. Sci., **22**, 1221.
Davidson, B. and Riley, N. (1971), J. Sound Vibration, **15**, 217.
Davies, J.T. and Rideal, E.K. (1961), Interfacial Phenomena, Academic Press, London.
Davies, R.M. and Taylor, G.I. (1950), Proc. Roy. Soc. London, **A 200**, 375.
Davis, A.M.J., O'Neill, M.E., Dorrepaal, J.M. and Ranger, K. (1976), J. Fluid Mech., **77**, 625.
Davis, M.H. (1971), Rep. No. NCAR-TN/STR-64, National Center for Atmospheric Research, Boulder, Colorado.
Davis, R.H. (1993), Adv. Colloid and Interf. Sci., **43**, 17.
Davis, R.H. and Acrivos, A. (1985), Ann. Rev. Fluid Mech., **17**, 91.
Davis, R.H. and Hanssen, M.A. (1988) J. Fluid Mech., **196**, 107.
Davis, R.H. and Hill, N.A. (1992), J. Fluid Mech., **236**, 513.
Davis, R.H., Schonberg, J.A. and Rallison, J.M. (1989), Phys. Fluids **A 1** (1), 77.
Day, J.T. and Genetti, W.E. (1964), B. S. Thesis, Univ. Utah. Salt Lake City, Utah.
de Bruin, R.A. (1989), Ph.D. Thesis, Technical University Eindhoven, The Netherlands.
de Bruin, R.A. (1993), Chem. Eng. Sci., **48**, 277.
de Kruif, C.G., van Iersel, E., Vrij, A. and Russel, W. (1985), J. Chem. Phys., **83**, 4717.
Dean, W.R. and O'Neill, M.E. (1963), Mathematika, **10**, 13.
Dennis, S.C.R. and Chang, Gau-Zu (1969), Phys. Fluids, **suppl.,** 88.
Dennis, S.C.R. and Walker, M.S. (1964), Aero. Res. Counc., **no. 26**, 105.
Denson, C.D., Cristiansen, E.B. and Salt, D.L. (1966), A.I.Ch.E. J., **12**, 589.
Derjaguin, B.V. and Dukhin, S.S. (1960-1961), Trans. Mining Metallurgy, **70**, 221.
Derjaguin, B.V. and Landau, L.D. (1941), Acta Physicochim., USSR, **14**, 633.
Di Prima, R.C. and Liron, N. (1976), Phys. Fluids., **19**, 1450.
Doi, M. and Edwards, S.F. (1978a), J. Chem. Soc. Faraday Trans., **74**, 560.
Doi, M. and Edwards, S.E. (1978b), J. Chem. Soc. Faraday Trans., **74**, 918.

Dommermuth, D.G. and Yue, D.K.P. (1987), J. Fluid Mech., **178**, 195.
Drew, D.A. (1988), Chem. Eng. Sci., **43**, 769.
Duineveld, P.C. (1995), J. Fluid Mech., **292**, 325.
Dukhin, S.S. and Rulev, N.N. (1977), Colloid J. USSR, **39**, 270.
Durlofsky, L., Brady, J.F. and Bossis, G. (1987), J. Fluid Mech., **180**, 21.
Dvorak, V. (1874), Ann. Phys. Lpz., **151**, 634.
Dwyer, H.A. (1989), Prog. Energy Combust. Sci., **15**, 131.
Eckstein, E., Bailey, D. and Shapiro, A. (1977), J. Fluid Mech., **79**, 191.
Edwards, D.A., Brenner, H. and Wasan, D. (1991), Interfacial Transport Processes and Rheology, Butterworth - Heinemann, Boston.
Eichhorn, R. and Small, S. (1964), J. Fluid Mech., **20**, 513.
Einstein, A. (1905), Ann. Physik, **17**, 549.
Einstein, A. (1906), Ann. Physik, **19**, 289.
Einstein, A. (1911), Ann. Physik, **34**, 591.
El Sawi, M. (1974), J. Fluid Mech., **62**, 163.
Elder, S. (1959), J. Acous. Soc. Am., **31**, 54.
Exerova, D. and Kruglykov, P.M. (1998), Foam and Foam Films. Theory, Experiment, Application., Elsevier, Amsterdam.
Falade, A. and Brenner, H. (1985), J. Fluid Mech., **154**, 145.
Famularo, J. and Happel, J. (1965), Am. Inst. Chem. Eng. J., **11**, 981.
Faraday, M. (1831), Trans. Roy. Soc. London, **121**, 229.
Faxen, H. (1921), Ph. D. Thesis, Uppsala Univ., Uppsala, Sweden.
Faxen, H. (1924), Arkiv. Mat. Astron. Fys., **18**, No. 29.
Faxen, H. (1925), Arkiv. Mat. Astron. Fys., **19A**, No. 13.
Faxen, H. and Dahl, H. (1925), Arkiv. Mat. Astron. Fys., **19A**, No. 13.
Felderhof, B.U. (1976a), Physica, **82A**, 596.
Felderhof, B.U. (1976b), Physica, **82A**, 611.
Felderhof, B.U. (1977a), Physica, **89 A**, 373.
Felderhof, B.U. (1977b), Physica, **95 A**, 572.
Felderhof, B.U. (1978), J. Phys. A: Math. Gen., **11**, 929.
Felderhof, B.U. (1979), J. Phys. C, **12**, 3121.
Felderhof, B.U. (1987), Physica, **147 A**, 203.
Felderhof, B.U. (1989), Physica, **159 A**, 1.
Felderhof, B.U. and Jones, R.B. (1987a), Physica, **146 A**, 404.
Felderhof, B.U. and Jones, R.B. (1987b), Physica, **146 A**, 417.
Feng, Z.C. and Leal, L.G. (1993), Phys. Fluids, **A 5(4)**, 826.
Feng, Z.C. and Leal, L.G. (1994), J. Fluid Mech., **266**, 209.
Feng, Z.C. and Leal, L.G. (1997), Ann. Rev. Fluid Mech., **29**, 201.
Feng, Z.C.(1992), SIAM J. Appl. Math., **52**, 1.
Feuillebois, F. (1984), J. Fluid Mech., **139**, 145.
Fischer, T.M. (1983), Thesis, Techn. Hochschule Darmstadt.
Fischer, T.M. (1986), Math. Meth. Appl. Sci., **8**, 23.
Fischer, T.M. and Rosenberger, R. (1987), ZAMP, **38**, 339.
Fitz-Gerald, J.M. (1969), Proc. Roy. Soc., **B 174**, 193.
Florence, A. and Whitehill, D. (1981), J. Colloid Int. Sci., **79**, 243.
Flumerfelt, R.W. (1980), J. Colloid Int. Sci., **76**, 330.
Flynn, H.G. (1975), J. Acoust. Soc. Am., **57**, 1379.
Foote, G.B. (1973), J. Comput. Phys., **11**, 507.
Ford, TF. (1960), J. Chem. Soc., **64**, 1168.
Fornberg, B. (1980), J. Fluid Mech., **98**, 819.
Fornberg, B. (1988), J. Fluid Mech., **190**, 471.

Forster, H. and Zuber, N. (1954), J. Appl. Phys., **25**, 474.
Francescutto, A. and Nabergoj, R. (1983), J. Acoust. Soc. Am., **73(2)**, 457.
Frankel, N.A. and Acrivos, A. (1967), Chem. Eng. Sci., **22**, 847.
Frankel, N.A. and Acrivos, A. (1970), J. Fluid Mech., **44**, 65.
Frankel, S. and Mysels, K. (1962), J. Phys. Chem., **66**, 190.
Frumkin, A. and Levich, V.G. (1947), Zh. Fiz. Khim., **21**, 1183.
Frössling, N. (1940), Lunds. Univ. Arsskr. N. F. Adv., **2**, 35.
Fuentes, Y.O., Kim, S. and Jeffrey, D.J. (1988), Phys. Fluids, **31**, 2445.
Fulford, G.R. and Blake, J.R. (1983), J. Fluid Mech., **127**, 203.
Gadala-Maria, F. and Acrivos, A. (1980), J. Rheology, **24**, 799.
Gaines, N. (1932), Physics, **3**, 209.
Gal-Or, B. and Waslo, S. (1968), Chem. Eng. Sci., **23**, 1431.
Ganatos, P., Pfeffer, R. and Weinbaum, S. (1978), J. Fluid Mech., **84**, 79.
Ganatos, P., Pfeffer, R. and Weinbaum, S. (1980a, b), J. Fluid Mech., **99**, 739; 755.
Garside,J. and Al-Dibouni, M.R. (1977), Ind. Eng. Chem. Process Des. Develop, **16**, 206.
Geller, A.S., Lee, S.H. and Leal, L.G. (1986), J. Fluid Mech., **169**, 27.
Giesenkus, H. (1962), Rheol. Acta, **2**, 50.
Giralt, V. and Raviart, P. (1979), Lect. Not. Math., Springer Verlag, 749.
Glendinning, A., and Russel, W. (1982), J. Colloid Interf. Sci., **89**, 124.
Gluckman, M.J., Pfeffer, R. and Weinbaum, S. (1971), J. Fluid Mech., **50**, 705.
Goddard, J.D. (1977), J. Non-Newt. Fluid Mech., **2**, 169.
Goddard, J.D. and Miller, C.H. (1967), J. Fluid Mech., **28**, 657.
Goldman, A.J., Cox, R.G. and Brenner, H. (1966), Chem. Eng. Sci., **21**, 1151.
Goldman, A.J., Cox, R.G. and Brenner, H. (1967a, b), Chem. Eng. Sci., **22**, 637; 653.
Goldsmith, H.L. (1968), J. Gen. Physiol., **52**, 55.
Goldsmith, H.L. (1971), Federation Proc., **30**, 1578.
Goldsmith, H.L. and Mason, S.G. (1962), J. Colloid Sci., **17**, 448.
Goldsmith, H.L. and Mason, S.G. (1963), J. Colloid Sci., **18**, 237.
Goldsmith, H.L. and Mason, S.G. (1964), Bibliotheca Anat., **4**, 462.
Goldsmith, H.L. and Mason, S.G. (1967), The Microrheology of Dispersions in Rheology (Ed. by Eirich, F.R.), Academic Press, New York, **vol.4**, chap. 2, pp. 85 - 250.
Goldstein, S. (1965), Modern Developments in Fluid Dynamics, New York, Dover Publications, Ing.
Goldstein, S. and Rosenhead, L. (1936), Proc. Cambridge Phil. Soc., **32**, 392.
Goodrich, F.C. and Allen, L.H. (1972), J. Colloid Interf. Sci., **37**, 46.
Gordon, R.J. and Schowalter, W.R. (1972), Trans. Soc. Rheol., **16**, 79.
Goren, S. (1970), J. Fluid Mech., **41, part 3**, 619.
Goren, S. and O'Neill, M.E. (1971), Chem. Eng. Sci., **26**, 325.
Grace H.P. (1971), Engng Foundation 3rd Res. Conf. on Mixing, Andover, New Hampshire.
Grace, J.R., (1983), In Handbook of Fluids in Motion, Eds. Cheremisinoff, N.P. and Gupta, R., **chap. 38**, Ann Arbor Science.
Gray, J. and Hancock, G.J. (1955), J. Exp. Biol., **32**, 802.
Green, G. (1833), Trans. Roy. Soc. Edin., Reprinted in Mathematical Papers, New York: Chelsea Publishing Co. 1970.
Green, J.T. (1971), Ph. D. Dissertation, University of Cambridge.
Greenspan, H.P. and Howard, L.N. (1963), J. Fluid Mech., **17**, 385.
Greenstein, T. (1972), J. Phys. Soc. Japan, **32**, 1398.
Greenstein, T. and Happel, J. (1968), J. Fluid Mech., **34**, 705.
Guthrie, R.I.L. and Bradshaw, A.V. (1969), Chem. Eng. Sci., **24**, 913.
Görtler, H. (1944), Ing. Arch., **14**, 286.
Haber, S. and Hetsroni, G. (1971), J. Fluid Mech., **49**, 257.
Haber, S. and Hetsroni, G. (1981), J. Colloid Interf. Sci., **79**, 56.

Haber, S., Hetsroni, G. and Solan, A. (1973), Int. J. Multiphase Flow, **1**, 57.
Haberman, W.L. and Morton, R.K. (1953), David Taylor Model Basin Rep., No. 802.
Haberman, W.L. and Morton, R.K. (1954), Proc. ASCE, **387**, 227.
Haberman, W.L. and Sayre, R.M. (1958), David Taylor Model Basin Rep., No. 1143, Washington, D.C.
Hadamard, J.S. (1911), Compt. Rend. Acad. Sci. Paris, **152**, 1735.
Halow, J.S. and Wills, G.B. (1970), AIChE J., **16**, 281.
Ham, J. and Homsy, G. (1988), Int. J. Multiphase Flow, **14**, 553.
Hamielec, A.E., Hoffman, T.Q. and Ross, L.L. (1967), AIChE J., **12**, 212.
Hancock, G.J. (1953), Proc. Roy. Soc., **A 217**, 96.
Hand, G.L. (1961), Arch. Rat. Mech. Anal., **7**, 81.
Hand, G.L. (1962), J. Fluid Mech., **13**, 33.
Hansford, R.E. (1970), Mathematika, **17**, 250.
Happel, J. (1957), J. Appl. Phys., **28**, 1288.
Happel, J. (1958), A.I.Ch.E. Journal, **4**, 197.
Happel, J. and Brenner, H. (1973), Low Reynolds Number Hydrodynamics, 2-nd ed. Noordhoff International Publishing: Leyden.
Harlow, F.H. and Welch, J.E. (1965), Phys. Fluids, **8**, 2182.
Harper, E.Y. and Chang, I-D. (1968), J. Fluid Mech., **33**, 209.
Harper, J.F. (1972), Adv. Appl. Mech., **12**, 59.
Harper, J.F. (1983), J. Austral. Math. Soc., **B25**, 217.
Harper, J.F. (1997), J. Fluid Mech., **351**, 289.
Harper, J.F. and Moore, D.W.(1968), J. Fluid Mech., **32**, 367.
Hartland, S. (1968), J. Colloid Sci., **26**, 383.
Hartunian, R.A. and Sears, W.R. (1957), J. Fluid Mech., **3**, 27.
Hashin, Z. (1964), Appl. Mech. Rev., **17**, 1.
Hashin, Z. and Shtrikman, S. (1963), J. Mech. Phys. Solids, **11**, 127.
Hasimoto, H. (1955), J. Phys. Soc. Japan, **9**, 611.
Hasimoto, H. (1959), J. Fluid Mech., **5**, 317.
Hasimoto, H. (1976), J. Phys. Soc. Japan, **41**, 2143; errata in 42, 1047.
Hassonjee, Q., Pfeffer, R. and Ganatos, P. (1992), Int. J. Multiphase Flow, **18**, 353.
Haussling, H.J. and Coleman, R.M. (1979), J. Fluid Mech. **92**, 667.
Hayakawa, T. and Shigeta, M. (1974), J. Chem. Eng. of Japan, **7**, 140.
Haywood, R.J., Renksizbulut, M. and Raithby, G.D. (1994a), Int. J. Heat Mass Transfer, **37**, 1401.
Haywood, R.J., Renksizbulut, M. and Raithby, G.D. (1994b), Numer. Heat Transfer, **A26**, 253.
He, Z., Dagan, Z. and Maldarelli, Ch. (1991), J. Fluid Mech., **222**, 1.
Hele-Shaw, H.S. (1898), Nature, **58**, 34.
Herczynski, R. and Pienkowska, I. (1980), Ann. Rev. Fluid Mech., **12**, 237.
Hershey, A.V. (1939), Phys. Rev., **56**, 204.
Hetsroni, G. (Ed.) (1982), Handbook of Multiphase Systems, Hemisphere Publishing, New York, pp. 1 - 96.
Hetsroni, G. and Haber, S. (1970), Rheol. Acta, **9**, 488.
Hetsroni, G. and Haber, S. (1978), Int. J. Multiphase Flow, **4**, 1.
Hetsroni, G., Haber, S., Brenner, H. and Greenstein, T. (1970a), Prog. Heat Mass Transfer, **6**, 591.
Hetsroni, G., Haber, S. and Wacholder, E. (1970b), J. Fluid Mech., **41**, 689.
Higdon, J.J.L. and Muldowney, G.P. (1995), J. Fluid Mech., **298**, 193.
Hill, R. and Power, G. (1956), Quart. J. Mech. Appl. Math., **9**, 313.
Hinch, E.J. (1977), J. Fluid Mech., **83**, 695.
Hinch, E.J. (1980), J. Fluid Mech., **101**, 545.
Hinch, E.J. (1988), Hydrodynamics at Low Reynolds Numbers: A Brief and Elementary Introduction, in Disorder and Mixing (Guyon, E., Nadal, J.P. and Pomeau, Y., Eds.), NATO ASI Series, Ser. E, **152**, 43 - 55.

Hinch, E.J. and Acrivos, A. (1979), J. Fluid Mech., **91**, 401.
Hinch, E.J. and Acrivos, A. (1980), J. Fluid Mech., **98**, 305.
Hinch, E.J. and Leal, L.G. (1972), J. Fluid Mech., **52**, 683.
Hirschfeld, B.R., Brenner, H. and Falade, A. (1984), PhysicoChemical Hydrodyn., **5**, 99.
Hnat, J.G. and Buckmaster, J.D. (1976), Phys. Fluids, **19**, 182.
Ho, B. P. and Leal. L.G. (1974), J. Fluid Mech., **65**, 365.
Ho, B. P. and Leal. L.G. (1975), J. Fluid Mech., **71**, 361.
Hodgson, T.D. and Woods, D.R. (1969), J. Colloid Interface Sci., **30**, 429.
Hoffman, R. (1972), Trans. Soc. Rheol., **16**, 155.
Hoffman, R. (1974), J. Colloid Interf. Sci., **46**, 491.
Holtsmark, J., Johnsen, I., Sikkeland, T. and Skavlem, S. (1954), J. Acoust. Soc. Amer., **26**, 26.
Howarth, L. (1950), Proc. Cambridge Phil. Soc., **46**, 127.
Howells, I.D. (1974), J. Fluid Mech., **64**, 449.
Hsu, R. and Ganatos, P. (1989), J. Fluid Mech., **207**, 29.
Huang, L.H. and Chwang, A.T. (1986), J. Eng. Math., **20**, 307.
Hyman, W.A. and Skalak, R. (1972a), Appl. Sci. Res., **26**, 27.
Hyman, W.A. and Skalak, R. (1972b), A.I.Ch.E. J., **18**, 149.
Ingber, M.S., Mondy, L.A. and Graham, A.L. (1989), 61-st Ann. Meeting Soc. Rheol., Montreal, Canada.
Ivanov, I.B. (Ed.) (1988), Thin Liquid Films, Marcel Dekker, New York.
Ivanov, K.P. and Rivkind, V.Yu, (1982), Izv. AN SSSR, Mehanika Jidkosti I Gaza, No.1, 167 (in Russian).
Jain, S. and Cohen, C. (1981) Macromol., **14**, 759.
Jaswon, M.A. and Symm, G.T. (1977), Integral Equation Methods in Potential Theory and Elastostatics, Academic Press.
Jefferey, R.C. (1965), J. Fluid Mech., **22**, 721.
Jeffery, G.B. (1912), Proc. Roy. Soc., **A 87**, 109.
Jeffery, G.B. (1915), London Math. Soc., **14**, 327.
Jeffery, G.B. (1922a), Proc. Roy. Soc., **A 101**, 169.
Jeffery, G.B. (1922b), Proc. Roy. Soc., **A 102**, 161.
Jeffrey, D.J. (1973), Proc. Roy. Soc., **A 335**, 355.
Jeffrey, D.J. (1974), Proc. Roy. Soc., **A 338**, 503.
Jeffrey, D.J. (1982), Mathematika, **29**, 58.
Jeffrey, D.J. and Acrivos, A. (1976), A.I.Ch.E. J., **22**, 417.
Jeffrey, D.J. and Onishi, Y. (1984a), J. Fluid Mech., **139**, 261.
Jeffrey, D.J. and Onishi, Y. (1984b), ZAMP, **35**, 634.
Jenson, V.G. (1959), Proc. Roy. Soc. Japan, **A 249**, 346.
Jeong, J.-T. and Moffatt, H.K. (1992), J. Fluid Mech., **241**, 1.
Johnson, R. (1980), J. Fluid Mech., **99**, 411.
Johnson, R.E. and Sadhal, S.S. (1983), J. Fluid Mech., **132**, 295.
Johnson, R.E. and Sadhal, S.S. (1985), Ann. Rev. Fluid Mech., **17**, 289.
Johnson, R.E. and Wu Yao-Tsu, T. (1979), J. Fluid Mech., **95**, 263.
Joseph, D.D., Nelson, J., Renardy, M. and Renardy, Y. (1991), J. Fluid Mech., **223**, 383.
Kalitzova-Kurteva, P. (1987), Ph. D. Thesis, Univ. of Sofia.
Kalitzova-Kurteva, P. and Zapryanov, Z. (1989), VI National Congress on Theoretical and Applied Mechanics, Varna, Bulgaria, Sept. 25-30, **3**, 310.
Kalitzova-Kurteva, P. and Zapryanov, Z. (1990), J. Appl. Math. Phys. (ZAMP), **41**, 20.
Kalitzova-Kurteva, P. and Zapryanov, Z. (1991), J. Eng. Math., **25**, 207.
Kaneda, Y. (1986), J. Fluid Mech., **167**, 455.
Kaneda, Y. and Ishii, K. (1982), J. Fluid Mech., **124**, 209.
Kang, I.S. and Leal, L.G. (1987), Phys. Fluids, **30**, 1929.
Kang, I.S. and Leal, L.G. (1988a), J. Fluid Mech., **187**, 231.

Kang, I.S. and Leal, L.G. (1988b), Phys. Fluids, **31(2)**, 233.
Kang, I.S. and Leal, L.G. (1989), Phys. Fluids, **A 1 (4)**, 644.
Kang, I.S. and Leal, L.G. (1990), J. Fluid Mech., **218**, 41.
Kang, I.S. and Leal, L.G. (1991), Bubble Dynamics in Quasi-time-periodic Straining Flows, in Fluid Eng.: Korea-U.S. Progress, (eds. J.H. Kim, J.M. Hyum, C.-O. Lee), New York., Hemisphere.
Kantorovich, L.V. and Krylov, V.I. (1958), Approximate Methods of Higher Analysis., Interscience.
Kanwal, R.P. (1955), Q. J. Mech. Appl. Maths, **8**, 146.
Kanwal, R.P. (1964), J. Fluid Mech., **19**, 631.
Kanwal, R.P. (1971), Linear Integral Equations, Acad. Press, New York.
Kaplun, S. (1957), J. Math. Mech., **6**, 595.
Kaplun, S. and Lagestrom, P.A. (1957), J. Math. Mech., **6**, 585.
Kapral, R. and Bedeaux, D. (1978), Physica, **91 A**, 590.
Karnis, A. and Mason, S.G. (1967), J. Colloid Int. Sci., **24**, 164.
Karnis, A., Goldsmith, H.L. and Mason, S.G. (1963), Nature, **200**, 159.
Karnis, A., Goldsmith, H.L. and Mason, S.G. (1966a), Can. J. Chem. Eng., **44**, 181.
Karnis, A., Goldsmith, H.L. and Mason, S.G. (1966b), J. Colloid Int. Sci., **22**, 531.
Katz, D.F., Blake, J.R. and Paveri-Fontana, S.L. (1975), J. Fluid Mech., **72**, 529.
Kawaguti, M. (1953), J. Phys. Soc. Japan, **8**, 747.
Keh, H.J. and Tseng, C.H. (1994), Int. J. Multiphase Flow, **20**, 185.
Keller, J.B. and Rubinow, S.I. (1976), J. Fluid Mech., **75**, 705.
Keller, J.B., Rubenfeld, A. and Molyneux, J.E. (1967), J. Fluid Mech., **30**, 97.
Kessler, D.A. and Levine, H. (1989), Phys. Rev., **A 39**, 5462.
Keunings, R. (1986), J. Comput. Phys., **62**, 199.
Khakhar, D.V. and Ottino, J.M. (1986), J. Fluid Mech., **166**, 265.
Khakhar, D.V. and Ottino, J.M. (1987), Int. J. Multiphase Flow, **13**, 147.
Kim, S. (1985a), Int. J. Multiphase Flow, **11**, 699.
Kim, S. (1985b), Int. J. Multiphase Flow, **11**, 713.
Kim, S. (1986), Int. J. Multiphase Flow, **12**, 469.
Kim, S. (1987), Phys. Fluids, **30**, 2309.
Kim, S. and Karrila, S. J. (1991), Microhydrodynamics: Principles and Selected Applications, Butterworth.
Kim, S. and Lu, S.-Y.(1987), Int. J. Multiphase Flow, **13**, 837.
Kim, S. and Mifflin, R.T. (1985), Phys. Fluids, **28**, 2033.
Kim, S. and Russel, W. (1985), J. Fluid Mech., **154**, 253.
Kirchoff, G. (1876) in Lamb, H. (1945), Hydrodynamics, Dover, New York.
Koh, C.J. and Leal, L.G. (1989), Phys. Fluids, **A 1 (8)**, 1309.
Koh, C.J. and Leal, L.G. (1990), Phys. Fluids, **A 2 (12)**, 2103.
Kojima, M., Hinch, E.J. and Acrivos, A. (1984), Phys. Fluids, **27(1)**, 19.
Kornfeld, M. and Suvorov, L. (1944), J. Appl. Phys., **15**, 495.
Kotzev, Tz. and Zapryanov, Z. (1989), VI National Congress on Theoretical and Applied Mechanics, Varna, Bulgaria, Sept. 25-30, 1989, Proceedings, p. 385.
Kotzev, Tz., Zapryanov, Z. and Toshev, E.T. (1989), Theoretical and Applied Mechanics, Bulgarian Academy of Sciences, **Year XX**, 58.
Kovatcheva, N., Christov, Ch. and Zapryanov, Z. (1985), Ann. Univ. Sofia, **7**, Livre 2, 225.
Kovatcheva, N., Christov, Ch. and Zapryanov, Z. (1988), ZAMM, **68**, 121.
Kralchevsky, P.A., Danov, K.D. and Denkov, N. (1997), Chemical Physics in Colloid Systems and Interfaces, Chapter 11, in Surface and Colloid Chemistry, Birdi (Ed.), CRC Press, Boca Raton, New York.
Kralchevsky, P.A., Danov, K.D. and Ivanov, I.B. (1995), Thin Liquid Film Physics, in Foams: Theory, Measurements and Applications, Prud'homme, R.K., (Ed.), Marcel Dekker, New York, p.1.
Krasuk, J.H. and Smith, J.M. (1963), Chem. Eng. Sci., **18**, 591.
Krieger, I. M. (1972), Adv. in Colloid Inter. Sci., **3**, 111.

Krieger, I. M. and Eguiluz, M. (1976), Trans. Soc. Rheol., **20**, 29.
Krieger, I.M. and Dougherty, T.J. (1959), Trans. Soc. Rheol., **3**, 137.
Kucaba-Pietal, A. (1986), Arch. Mech., **38**, 647.
Kuwabara, S. (1959), J. Phys. Soc. Japan, **14**, 527.
Kynch, G.J. (1956), Proc. Roy. Soc. (London), **A 237**, 90.
Kynch, G.J. (1959), J. Fluid Mech., **5**, 193.
Ladd, A.J.C. (1988), J. Chem. Phys., **88**, 5051.
Ladd, A.J.C. (1989), J. Chem. Phys., **90**, 1149.
Ladd, A.J.C. (1990), J. Chem. Phys., **93**, 3484.
Ladyzhenskaya, O.A. (1969), The Mathematical Theory of Viscous Incompressible Flow, Gordon and Breach.
Lagestrom, P.A. and Cole, J.D. (1962), J. Rat. Mech. Anal., **4**, 161.
Lai, R.Y. and Mockros, L.F. (1972), J. Fluid Mech., **52**, 1.
Lamb, H. (1945), Hydrodynamics, 6th ed. Dover: reprint of Camb. Univ. Press edition (1932).
Landau, L.D. and Lifshitz, E.M. (1953), Continuum Mechanics, Moskva, Gostehizdat (in Russian).
Landau, L.D. and Lifshitz, E.M. (1959), Fluid Mechanics, Pergamon Press, Oxford.
Landenburg, R. (1907), Ann. Phys., **23**, 447.
Lane, C.A. (1955), J. Acoust. Soc. Amer., **27**, 1082.
Langmuir, I. and Schaefer, V.J. (1937), J. Am. Chem. Soc., **59**, 2400.
Lauterborn, W. (1976), J. Acoust. Soc. Am., **59(2)**, 283.
Lauterborn, W. (1991), Int. J. Bifurcation Chaos, **1(1)**, 13.
Lauterborn, W. and Parlitz, U. (1988) , J. Acoust. Soc. Am., **84(6)**, 1975.
Lawrence, C.J. (1986), Ph. D. Thesis, The City University of New York.
Lawrence, C.J. and Weinbaum, S. (1986), J. Fluid Mech., **171**, 209.
Lawrence, C.J. and Weinbaum, S. (1988), J. Fluid Mech., **189**, 463.
Leal, L.G. (1980), Ann. Rev. Fluid Mech., **12**, 435.
Leal, L.G. (1992), Laminar Flow and Convective Transport Processes, Butherworth-Heinemann.
Leal, L.G. and Hinch, E.J. (1971), J. Fluid Mech., **46**, 685.
Leal, L.G. and Hinch, E.J. (1973), Rheol. Acta, **12**, 127.
Leal, L.G. and Lee, S.H. (1982), Adv. Colloid Interface Sci., **17**, 61.
LeClair, B.P., Hamielec, A. E. and Peuppacher, H.R. (1970), J. Atmos. Sci., **27**, 308.
LeClair, B.P., Hamielec, A. E., Peuppacher, H.R. and Hall, W.D. (1972), J. Atmos. Sci., **29**, 728.
Lee, S.H. (1979), Ph. D. Dissertation, California University of Technology.
Lee, S.H., Chadwick, R.S. and Leal, L.G. (1979), J. Fluid Mech., **93**, 705.
Lee, S.H. and Leal, L.G. (1980), J. Fluid Mech., **98**, 193.
Lee, S.H. and Leal, L.G. (1982), J. Colloid Int. Sci., **87**, 81.
Leighton, L.A. and Acrivos, A. (1985), ZAMP, **36**, 174.
Leighton, L.A. and Acrivos, A. (1987a), J. Fluid Mech., **181**, 415.
Leighton, L.A. and Acrivos, A. (1987b), J. Fluid Mech., **177**, 109.
Lerner, L. and Harper, J.F. (1991), J. Fluid Mech., **232**, 167.
Levan, M.D. (1981), J. Colloid Int. Sci., **83**, 11.
Levan, M.D. and Newman, J. (1976), A.I.Ch.E. J., **22**, 695.
Levich, V.G. (1949), Zhn. Eksp. Teor. Fiz., **19**, 18.
Levich, V.G. (1962), PhysicoChemical Hydrodynamics, Prentice-Hall, Englewood Clift, New York.
Levich, V.G. (1981), PhysicoChemical Hydrodynamics, **2**, 85.
Lewellen, P. (1982), Ph.D. Dissertation, Univ. of Wisconsin, Madison.
Li, N. and Asher, W. (1973), Int. Chem. Engng Med. Adv. Chem. Ser., **118**, 1.
Li, X. and Pozrikidis, C. (1997), J. Fluid Mech., **341**, 165.
Lighthill, M.J. (1975), Mathematical Biofluiddynamics, S.I.A.M.
Lin, C.J., Lee, K.J. and Sather, N.F. (1970b), J. Fluid Mech., **43**, 35.
Lin, C.J., Peery, J.H. and Schowalter, W.R. (1970a), J. Fluid Mech., **44**, 1.

Liron, N. (1978), J. Fluid Mech., **86**, 705.
Liron, N. and Mochon, S. (1976a), J. Eng. Math., **10**, 287.
Liron, N. and Mochon, S. (1976b), J. Fluid Mech., **75**, 593.
Liron, N. and Shahar, R. (1978), J. Fluid Mech., **86**, 727.
Loewenberg, M. and Davis, R.H. (1993), J. Fluid Mech., **256**, 107.
Loewenberg, M. and Hinch, E.J. (1996), J. Fluid Mech., **321**, 395.
Loewenberg, M. and Hinch, E.J. (1997), J. Fluid Mech., **338**, 299.
Longuet-Higgins, M.S. (1989a), J. Fluid Mech., **201**, 525.
Longuet-Higgins, M.S. (1989b), J. Fluid Mech., **201**, 543.
Longuet-Higgins, M.S. (1992), J. Acoust. Soc. Am., **91**, 1414.
Lorentz, H.A. (1907), Abhand Theor. Phys. Leipzig, **1**, 23.
Luke, J.H. (1989), SIAM J. Appl. Math., **49**, 1635.
Lundgren, T.S. (1972), J. Fluid Mech., **51**, 273.
Ma, J.T.S. and Wang, P.K.C. (1962), IBM J. Res. Dev., **6**,472.
Maneri, C.C. and Zuber, N. (1974), J. Multiphase Flow, **1**, 623.
Manev, E., Tsekov, R. and Radoev, B. (1997), J. Dispersion Sci. Tech., **18**, 769.
Manga, M. and Stone, H.A. (1993), J. Fluid Mech., **256**, 647.
Manga, M. and Stone, H.A. (1995), J. Fluid Mech., **287**, 279.
Markov, K. (1987a), SIAM J. Appl. Math., **47**, 831.
Markov, K. (1987b), SIAM J. Appl. Math., **47**, 850.
Markov, K. (1989), Math. Balkanica (New Series), **3**, 399.
Markov, K. and Christov, Ch. (1992), Mathematical Models and Methods in Applied Sciences, **2**, 249.
Marston, P.L. (1980), J. Acoust. Soc. Am., **67**, 15.
Martin, T. and Davies, G. (1976), J. Hydrometallurgy, **2**, 315.
Martinez, M.J. and Udell, K.S. (1990), J. Fluid Mech., **210**, 565.
Maru, H.C., Wasan, D.T. and Kintner, R.C. (1971), Chem. Eng. Sci., **26**, 1615.
Masliyah, J.H. (1970), Ph. D. Thesis, Univ. of British Columbia, Vancouver.
Masliyah, J.H. and Epstein, N. (1970), J. Fluid Mech., **44**, 493.
Mathon, R. and Johnston, R.L. (1977), SIAM J. Numer. Anal., **14**, 638.
Maude A.D. (1961), Br. J. Appl. Phys., **12**, 293.
Maxworthy, T. (1965), J. Fluid Mech., **23**, 369.
Maxworthy, T. (1967), J. Fluid Mech., **27**, 367.
Mazur, P. and Bedeaux, D. (1974), Physica, **76**, 235.
Mazur, P. and van Saarloos, W. (1982), Physica, **115A**, 21.
McDougald, N.K. and Leal, L.G. (1994), Proc. Symp. Nav. Hydrodyn., 20th, Santa Barbara, CA, pp. 311-27, Washington, DC, Nat. Acad. Press.
McDougald, N.K. and Leal, L.G. (1996), J. Fluid Mech., **289**, 161.
McLaughlin, J.B. (1993), J. Fluid Mech., 246, 249.
McNown, J.S. and Lin, P.N. (1952), Proc. Second Midwestern Conf. Fluid Mech., Iowa State University, p. 109.
McNown, J.S. and Malaika, J. (1950), Trans. Am. Geophys. Union, **31**, 74.
McQuarrie, D.A. (1976), Statistical Mechanics, Harper and Row.
Mei, C.C. and Zhou, X. (1991), J. Fluid Mech., **229**, 29.
Melcher, J.R. and Taylor, G.I. (1969), Ann. Rev. Fluid Mech., **1**, 111.
Melnikov, V.K. (1963), Trans. Moscow Math. Soc., **12**, 1.
Mestre, N.J. de (1973), J. Fluid Mech., **58**, 641.
Mestre, N.J. de and Russel, W.B. (1975), J. Engng Math., **9**, 81.
Mewis, J. and Metzner, A.B. (1974), J. Fluid Mech., **62**, 593.
Meyyappan, M. and Subramanian, R. R. (1984), J. Colloid Int. Sci., **97**, 291.
Meyyappan, M., Wilcox, W.R. and Subramanian, R. R. (1981a), Int. J. Multiphase Flow, **7**, 581.
Meyyappan, M., Wilcox, W.R. and Subramanian, R. R. (1981b), J. Colloid Int. Sci., **83**, 199.

Meyyappan, M., Wilcox, W.R. and Subramanian, R. R. (1983), J. Colloid Int. Sci., **94**, 243.
Mikami, T., Cox, R.G. and Mason, S.G. (1975), Int. J. Multiphase Flow, **2**, 113.
Miksis, J.M. (1981), Phys. Fluids, **24(7)**, 1229.
Miksis, J.M., Vanden-Broeck, J. and Keller, J.B. (1981), J. Fluid Mech., **108**, 89.
Miller, C.A. and Scriven, L.E. (1968), J. Fluid Mech., **32**, 417.
Milliken, W.J. and Leal, L.G. (1991), J. Non-Newtonian Fluid Mech., **40**, 355.
Milliken, W.J. and Leal, L.G. (1994), J. Colloid. Interface Sci., **166**, 275.
Milliken, W.J., Stone, H.A. and Leal, L.G. (1993), Phys. Fluids, **5**, 69.
Minev, P. (1990), Ph. D. Thesis, Bulgarian Academy of Sciences.
Miyamura, A., Iwasaki, S. and Ishii, T. (1980), Int. J. Multiphase Flow, **7**, 41.
Mok, L.S. and Kim, K. (1987), J. Fluid Mech., **176**, 521.
Mooney, M. (1951), J. Colloid Sci., **6**, 162.
Moore, D.W. (1959), J. Fluid Mech., **6**, 113.
Moore, D.W. (1963), J. Fluid Mech., **16**, 161.
Moore, D.W. (1965), J. Fluid Mech., **23**, 749.
Mori, Y. (1978), Int. J. Multiphase Flow, **4**, 383.
Munson, B.R. and Douglass, R.W. (1979), Phys. Fluids, **22**, 205.
Mysels, K., Shinoda, K. and Frankel, S. (1959), Soap Films, Pergamon Press, New York.
Nadim, A. and Stone, H.A. (1991), Studies in Appl. Math., **85**, 53.
Nakano, Y. and Tien, C. (1967), Can. J. Chem. Engng, **45**, 135.
Newman, J. (1967), Chem. Eng. Sci., **22**, 83.
Newman, J., Swinney, H.L., Berkowitz, S.A. and Day, L.A. (1974), Biochemistry, **13**, 4832.
Nguyen, T.H. (1973), Tech. Pap. Dep. Chem. Eng., McGill Univ., Montreal.
Nichols, B.D. and Hirt, C.W. (1971), J. Comput. Phys., **8**, 434.
Nir, A. and Acrivos, A. (1973), J. Fluid Mech., **59**, 209.
Nir, A. and Acrivos, A. (1990), J. Fluid Mech., **212**, 139.
Nunan, K. and Keller, J. (1984), J. Fluid Mech., **142**, 269.
Nyborg, W.L. (1953), J. Acous. Soc. Am., **25**, 68.
Nyborg, W.L. (1965), Acoustic Streaming, in Physical Acoustics (Mason, W.P., Ed.), **vol. 2B**, chap. **11**, Academic Press Inc., New York.
O'Brien, R.W. (1979), J. Fluid Mech., **91**, 17.
Ockendon, J.R. (1968), J. Fluid Mech., **34**, 229.
Odqvist, F.K.G. (1930), Math. Z., **32**, 329.
Ogura, H. (1972), I.E.E.E. Trans. Inform. Theory, **18**, 473.
Olbricht, W.L. and Kung, D.M. (1992), Phys. Fluids, **A 4**, 1347.
Olbricht, W.L. and Leal, G.L. (1982), J. Fluid Mech., **115**, 187.
Oldroyd, J.D. (1950), Proc. Roy. Soc., **A 200**, 523.
Oldroyd, J.D. (1953), Proc. Roy. Soc., **A 218**, 122.
Oldroyd, J.D. (1958), Proc. Roy. Soc., **A 245**, 278.
Oliver, D.L.R. and Chung, J.N. (1985), J. Fluid Mech., **154**, 215.
Oliver, D.L.R. and Chung, J.N. (1987), J. Fluid Mech., **177**, 1.
O'Neill, M.E. (1964), Mathematika, **11**, 67.
O'Neill, M.E. (1968), Chem. Eng. Sci., **23**, 1293.
O'Neill, M.E. (1969), Proc. Camb. Philos. Soc., **65**, 543.
O'Neill, M.E. and Majumdar, S.R. (1970), ZAMP, **21**, 164.
O'Neill, M.E. and Ranger, K.B. (1983), Phys. Fluids, **26 (8)**, 2035.
O'Neill, M.E. and Stewartson, K. (1967), J. Fluid Mech., **27**, 705.
Oseen, C.W. (1910), Ark. f. Math. Astron. og Fys., **6**, 29.
Oseen, C.W. (1911), Ark. f. Math. Astron. og Fys., **7**, 21.
Oseen, C.W. (1927), Neuere Methoden und Ergebniss in der Hydrodynamik, Akademische Verlags, Leipzig.

Pal, R. and Rhodes, E. (1985), J. Colloid Interf. Sci., **107**, 301.
Pal, R. and Rhodes, E. (1989), J. Rheology, **33**, 1021.
Pan, F.Y. and Acrivos, A. (1968), I.E.C. Fundamentals, **7(2)**, 227.
Panton, R.L. (1984), Incompressible Flow, Wiley - Interscience Publication, John Wiley and Sons.
Papazian, J.M. and Wilcox, W.R. (1976), Flight 1, Technical Report for Experiment, NASA, CR-144304.
Park, C.W. and Homsy, G.M. (1984), J. Fluid Mech., **139**, 1583.
Parlange, J.-Y. (1969), J. Fluid Mech., **37**, 257.
Parlange, J.-Y. (1970), Acta Mechanica, **9**, 323.
Parsi, F. and Gadala - Maria, F. (1987), J. Rheology, **31**, 725.
Pawar, Y. and Stebe, K.J. (1996), Phys. Fluids, **A 8**, 1738.
Payne, L.E. and Pell, W.H. (1960), . Fluid Mech., **7**, 529.
Peacemann, D.W. and Rachford, H.H. (1955), J. Soc. Indust. Appl. Maths, **3**, 28.
Pearson, C.E. (1967), J. Fluid Mech., **28**, 323.
Peterson, J.M. and Fixman, M. (1963), J. Chem. Phys., **39**, 2516.
Phan-Thien, N., Tran-Cong, T. and Graham, A. (1991), J. Fluid Mech., **228**, 275.
Phillips, R., Armstrong, R.C., Brown, R.A., Graham, A.L. and Abbott, J.R. (1992), Phys. Fluids A, **4**, 30.
Phillips, R., Brady, J. and Bossis, G. (1988a), Phys. Fluids, **31**, 3462.
Phillips, R., Brady, J. and Bossis, J. (1988b), Phys. Fluids, **31**, 3473.
Phillips, W.J., Graves, R.W. and Flumerfelt, R.W. (1980), J. Colloid Interf. Sci., **76**, 350.
Pienkowska, I. (1984), Arch. Mech., **36**, 749.
Pintar, A.J., Israel, A.B. and Wasan, D.T. (1971), J. Colloid Interf. Sci., **37**, 52.
Platikanov, D. (1964), J. Phys. Chem., **68**, 3619.
Plesset, M.S. (1949), J. Appl. Mech., **16**, 277.
Plesset, M.S. and Prosperetti, A. (1977), Ann. Rev. Fluid Mech., **9**, 145.
Pokrovskii, V.N. (1967), Kolloidnyi Zh., **29**, 576 (English translation available in Colloid Journal of the USSR, **29**, 428).
Pokrovskii, V.N.(1968)), Kolloidnyi Zh., **30**, 881 (English translation available in Colloid Journal of the USSR, **30**, 664).
Polyflow User's Manuals, (1989-1997), Polyflow s.a., Belgium.
Power, H. (1987), J. Fluid Mech., **185**, 547.
Power, H. (1993), Mathematical Methods in Applied Sciences, **16**, 61.
Pozrikidis, C. (1988), J. Fluid Mech., **188**, 275.
Pozrikidis, C. (1989a), Phys. Fluids, **A 1**, 1508.
Pozrikidis, C. (1989b), J. Fluid Mech., **202**, 17.
Pozrikidis, C. (1990a), J. Fluid Mech., **210**, 1.
Pozrikidis, C. (1990b), J. Fluid Mech., **215**, 331.
Pozrikidis, C. (1992a), Boundary Integral and Singularity Methods for Linearized Viscous Flow, Cambridge University Press.
Pozrikidis, C. (1992b), J. Fluid Mech., **237**, 627.
Pozrikidis, C. (1993), J. Fluid Mech., **246**, 301.
Pozrikidis, C. (1995), Phys. Fluids, **A6**, 3209.
Pozrikidis, C.(1997), Introduction to Theoretical and Computational Fluid Dynamics, Cambridge University Press.
Prandtl, L. (1904), Über Flüssigkeisbewgung bei sehr kleiner Reibung. Verh. Proc. III-d Int. Math. Kongr., Heidelberg, 481.
Prosperetti, A. (1976), Phys. Fluids, **19**, 195.
Prosperetti, A. (1980a), J. Fluid Mech., **19**, 149.
Prosperetti, A. (1980b), J. Fluid Mech., **20**, 333.
Prosperetti, A. (1993), Bubble Dynamics: Some Things We Did Not Know 10 Years Ago in Bubble Dynamics and Interface Phenomena (ed. J.R. Blake, J.M. Boulton-Stone, N.H. Thomas), pp. 3-16, Boston. Kluwer.

Protodiakonov, I.O., Liublinskaia, I.E. and Riijkov, A.E. (1987), Hydrodynamics and Mass-Transfer in Disperse Systems Fluid - Rigid Body, Leningrad, "Himia", Leningradskoe Otdelenie (in Russian).
Proudman, I. and Pearson J. (1957), J. Fluid Mech., **2**, 237.
Radoev, B.P., Scheludko, A. and Manev, E. (1983), J. Colloid. Interface Sci., **95**, 254.
Rahnama, M., Koch, D. and Shagfeh, S. (1995), Phys. Fluids, **A 7**, 487.
Rallison, J.M. (1977), Ph. D. dissertation, Cambridge University.
Rallison, J.M. (1978a), J. Fluid Mech., **84**, 237.
Rallison, J.M. (1978b), J. Fluid Mech., **88**, 529.
Rallison, J.M. (1981), J. Fluid Mech., **109**, 465.
Rallison, J.M. (1984), Ann. Rev. Fluid Mech., **16**, 45.
Rallison, J.M. and Acrivos, A. (1978), J. Fluid Mech., **89**, 191.
Raney, W.P., Corelli, J.C. and Westervelt, P.J. (1954), J. Acoust. Soc. Amer., **26**, 1006.
Rayleigh, Lord (1884), Phil. Trans., **A 175**, 1.
Rayleigh, Lord (1892), Phil. Mag., **34**, 177, 481.
Rayleigh, Lord (1911), Phil. Mag., **21**, 697.
Rayleigh, Lord (1917), Phil. Mag., **34**, 94.
Reed, C.C. and Anderson, J.L. (1980), A.I.Ch.E. J., **26**, 816.
Reed, L.D. and Morrison, F.A. (1974), Int. J. Multiphase Flow, **1**, 573.
Reid, W.H. (1960), Quart. J. Appl. Math., **18**, 86.
Reinelt, D.A. (1987), J. Fluid Mech., **175**, 557.
Revay, J.M. and Higdon, J.L. (1992), J. Fluid Mech., **243**, 15.
Reynolds, O. (1886), Phil. Trans. Roy. Soc., London, **A177**, 157.
Reynolds, O. (1894), Br. Assoc. Adv. Sci. Rep. 564.
Richardson, J.F. and Zaki, W.N. (1954), Trans. Inst. Chem. Eng., **32**, 35.
Richardson, P.D. (1967), Appl. Mech. Rev., **20**, 201.
Richardson, S. (1968), J. Fluid Mech., **33**, 475.
Richardson, S. (1973), J. Fluid Mech., 58, 115.
Richtmyer, R.D. and Morton, K.W. (1967), Difference Methods for Initial Value Problems, Interscience ublishers, Wiley, London.
Riley, N. (1965), Mathematika, **12**, 161.
Riley, N. (1966), Quart. J. Mech. Appl. Math., **19**, 461.
Riley, N. (1967), J. Inst. Math. Appl., **3**, 419.
Rimon, Y. and Cheng, S.I. (1969), Phys. Fluids, **12**, 949.
Rivkind, V. Ya. and Ryskin, G.M. (1976), Fluid Dyn., **11**, 5.
Rivkind, V. Ya. and Sigovtzev, G.S. (1980), Proceedings of the 6-th Heat and Mass Transfer Conference in SSSR, **5**, 77 (in Russian).
Rivkind, V. Ya., Ryskin, G.M. and Fishbein, G.A. (1976), Appl. Math. Mech., **40**, 687.
Rivlin, R.S. and Eriksen, J.L. (1955), Arch. Rat. Mech. Anal., **4**, 323.
Roache, P.J. (1976), Computational Fluid Dynamics, Hermosa Publishers, Albuquerque, New Mexico.
Roscoe, R. (1967), J. Fluid Mech., **28**, 273.
Rosenhead, L. (Ed.) (1963), Laminar Boundary Layers, Oxford, At the Clarendon Press.
Rotne, J. and Prager, S. (1969), J. Chem. Phys., **50**, 4831.
Rowe, P. N. and Henwood, G. A. (1961), Trans. Inst. Chem. Engrs., **39**, 27.
Rozin, L.A. (1957), Appl. Math. And Mech. (PMM), No.5, 21 (in Russian).
Rubenstein, J. and Keller, J. (1989), Phys. Fluids, **A 1**, 637.
Rubinow, S. and Keller, J.B. (1961), J. Fluid Mech., **11**, 447.
Rumscheidt, F.D. and Mason, S.G. (1961), J. Colloid Sci., **16**, 238.
Ruschak, K.J. (1980), Int. J. Num. Meth. Engin., **15**, 639.
Rushton, E. and Davies, G.A. (1970), Int. Fluid Dynamics Symposium, McMaster University, Canada.
Rushton, E. and Davies, G.A. (1973), Appl. Sci. Res., **28**, 37.
Rushton, E. and Davies, G.A. (1978), Int. J. Multiphase Flow, **4**, 357.

Rushton, E. and Davies, G.A. (1983), Int. J. Multiphase Flow, **9**, 337.
Russel, W.B. (1981), Ann. Rev. Fluid Mech., **13**, 425.
Russel, W.B. (1983), in Theory of Dispersed Multiphase Flow, Meyer, R.E. (Ed.), Academic Press, p.1.
Russel, W.B. and Gast, A. (1986), J. Chem. Phys., **84**, 1815.
Russel, W.B., Saville, D.A. and Schowalter, W.R. (1989), Colloidal Dispersions, University Press, Cambridge.
Rutgers, R. (1962a), Rheol. Acta, **2**, 202.
Rutgers, R. (1962b), Rheol. Acta, **2**, 305.
Rybczynski, W. (1911), Bull. Acad. Sci. Cracovie, **ser. A 40**, 257.
Ryskin, G. and Leal, L.G. (1983), J. Comput. Phys., **50**, 71.
Ryskin, G. and Leal, L.G. (1984a), J. Fluid Mech., **148**, 1.
Ryskin, G. and Leal, L.G. (1984b), J. Fluid Mech., **148**, 19.
Ryskin, G. and Leal, L.G. (1984c), J. Fluid Mech., **148**, 37.
Sadhal, S.S. and Oguz, H. (1985), J. Fluid Mech., **160**, 515.
Saffman, P.G. (1973), Stud. Appl. Math., **52**, 115.
Saffman, P.G. (1956), J. Fluid Mech., **1**, 249.
Saffman, P.G. (1965), J. Fluid Mech., **22**, 385.
Saito, N. (1952), J. Phys. Soc. Japan, **7**, 447.
Saito, S. (1913), Sci. Rep. Tohoku Imp. Univ., Sendai, Japan, **2**, 179.
Sangani, A. (1987), J. Appl. Math. Phys. (ZAMP), **38**, 542.
Sangani, A. S. and Lu, W. (1987), J. Appl. Math. Phys. (ZAMP), **38**, 557.
Sangani, A.S. and Acrivos, A. (1982), Int. J. Multiphase Flow, **8**, 343.
Sangani, A.S. and Acrivos, A. (1983), Int. J. Multiphase Flow, **9**, 181.
Sano, O. and Hasimoto, H. (1978), J. Fluid Mech., 87, 673.
Saunders, F.L. (1961), J. Colloid. Sci., **16**, 13.
Savic, P. (1953), Circulation and Distortion of Liquid Drops Falling through a Viscous Medium, Mech. Eng. Rep. MT-22, National Research Council of Canada.
Saville, D.A. (1973), Chem. Eng. J., **5**, 251.
Scheludko, A. (1967), Adv. Colloid. Interface Sci., **1**, 391.
Schlichting, H. (1932), Phys. Z., **33**, 327.
Schlichting, H. (1964), Grenzschicht - Theorie, Verlag G. Braun, Karlsruhe.
Schmitz, R. and Felderhof, B.U. (1982a), Physica, **113A**, 90.
Schmitz, R. and Felderhof, B.U. (1982b), Physica, **113A**, 163.
Schonberg, J.A. and Hinch, E.J. (1989), J. Fluid Mech., **203**, 517.
Schowalter, W.R., Chaffey, C.E. and Brenner, H. (1968), J. Colloid Sci., **26**, 152.
Schuh, H. (1953), Z. Flugwiss, **1**, 122.
Schwartz, L.W., Princen, H.M. and Kiss, A.D. (1986), J. Fluid Mech., **172**, 259.
Scriven, L.E. (1960), Chem. Eng. Sci., **12**, 98.
Segre, G. and Silberberg, A. (1961), Nature, **189**, 209.
Segre, G. and Silberberg, A. (1962a), J. Fluid Mech., **14**, 115.
Segre, G. and Silberberg, A. (1962b), J. Fluid Mech., **14**, 136.
Shah, S.T., Wasan, D.T. and Kintner, R.C. (1972), Chem. Eng. Sci., **27**, 881.
Shail, R. (1978), J. Eng. Math., **12**, 59.
Shail, R. (1979), Int. J. Multiphase Flow, **5**, 169.
Shail, R. and Gooden, D.K. (1981), Int. J. Multiphase Flow, **7**, 289.
Shankar, N., Cole, R. and Subramanian, R. R. (1981), Int. J. Multiphase Flow, **7**, 581.
Shankar, N. and Subramanian, R. R. (1983), J. Colloid Int. Sci., **94**, 258.
Shanks, S.P. and Thompson, J.E. (1977), Proc. 2nd. Int. Conf. Ship Hydrodyn. California, 202.
Shapira, M. and Haber, S. (1988), Int. J. Multiphase Flow, **14**, 483.
Shapira, M. and Haber, S. (1990), Int. J. Multiphase Flow, **16**, 305.
Shaqfeh, S. and Fredrickson, G. (1990), Phys. Fluids, **A 2**, 7.

Sherwood, J.D. (1981), Math. Proc. Camb. Phil. Soc., **90**, 529.
Shoemaker, P.D. and Marc de Chazal, L.E. (1969), Chem. Eng. Sci., **24**, 795.
Shopov, P. (1985), Ph D Thesis, Bulgarian Academy of Sciences.
Shopov, P. and Minev, P. (1992), J. Fluid Mech., **235**, 123.
Shopov, P., Minev, P., Bazhlekov, I. and Zapryanov, Z. (1989), Comptes Rendus de l'Academie Bulgare des Sciences, **42 (1)**, 43.
Shopov, P., Minev, P., Bazhlekov, I. and Zapryanov, Z. (1990), J. Fluid Mech., **219**, 241.
Sigli, D. and Coutanceau, M. (1977), J. Non-Newt. Fluid Mech., **2**, 1.
Simeonov, G. (1977), Theor. and Appl. Mech., BAS, No.**1**, 64.
Simha, R. (1949), J. Res. Nat. Bur. Stand., **42**, 409.
Simha, R. (1952), J. Appl. Phys., **23**, 1020.
Skalak, R., Chen, P.H. and Chien, S. (1972), Biorheology, **9**, 67.
Slattery, J.C. (1964), Chem. Eng. Sci., **19**, 379.
Slavchev, S. (1974/1975), Ann. Univ. Sofia, FMM, **69**, 21.
Slavchev, S. (1975), Theor. and Appl. Mech., BAS, **4**, 49.
Slavchev, S. and Simeonov, G. (1978), ZAMM, **58**, 87.
Smart, J.R. and Leighton, D.T. (1989), Phys. Fluids, **A1**, 52.
Smart, J.R. and Leighton, D.T. (1991), Phys. Fluids, **A 3**, 21.
Smoluchowski, M. (1911), Bull. Int. Acad. Polonaise Sci. Lett., **1A**, 28.
Stakgold, I. (1967, 1968), Boundary Value Problems of Mathematical Physics, 2 volumes, Macmillan.
Stewartson, K. (1957), J. Math. Phys., **36**, 173.
Stimson, M. and Jeffery, G.B. (1926), Proc. Roy. Soc., London, **Ser. A 111**, 110.
Stokes, G.G. (1851), Trans. Cambridge Phil. Soc., **9**, 8.
Stone, H.A. (1990),Phys. Fluids, **A 2 (1)**, 111.
Stone, H.A. (1994), Annu. Rev. Fluid Mech., **26**, 65.
Stone, H.A., Bentley, B.J. and Leal, L.G. (1986), J. Fluid Mech., **173**, 131.
Stone, H.A. and Leal, L.G. (1989a), J. Fluid Mech., **198**, 399.
Stone, H.A. and Leal, L.G. (1989b), J. Fluid Mech., **206**, 223.
Stone, H.A. and Leal, L.G. (1989c), J. Colloid. Interface Sci., **133**, No. 2, 340.
Stone, H.A. and Leal, L.G. (1990), J. Fluid Mech., **220**, 161.
Stoos, J.A., Yang, S.M. and Leal L.G. (1992), Int. J. Multiphase Flow, **18** , 1019.
Stover, C.A., Koch, D.L. and Cohen, C. (1992), J. Fluid Mech., **228**, 277.
Strasberg, M. and Benjamin, T.B. (1958), J. Acoust. Soc. Am., **30**, 697.
Stuart, J.T. (1966), J. Fluid Mech., **24**, 673.
Subramanian, R.R. (1981), A.I.Ch.E. J., **4**, 646.
Subramanian, R.R. (1983), Adv. Space Res., **3(5)**, 145.
Subramanian, R.R. (1985), J. Fluid Mech., **153**, 389.
Subramanian, R.S. (1987), Low-Gravity Sciences (Science and Technology Series, Ed. by J. N. Koster, J.N.), **67**, 69.
Sy, F. and Lightfoot, E.N. (1971), A.I.Ch.E. J., **17**, 177.
Sy, F., Taunston, J.W. and Lightfoot, E.N. (1970), A.I.Ch.E. J., **16**, 386.
Tabakova, S. (1981), Ph.D. Thesis, University of Sofia.
Tabakova, S. (1998), Theoretical and Applied Mechanics, Bulgarian Academy of Sciences, (in print).
Tabakova, S. and Carotenuto, L. (1994), Microgravity Quarterly, **4**, No.1, 55.
Tabakova, S. and Kolemanov, I. (1998), (in preparation).
Tabakova, S. and Zapryanov, Z. (1978), Comptes Rendus de l'Academie Bulgare des Sciences, **31**, 819.
Tabakova, S. and Zapryanov, Z. (1982a), ZAMP, **33**, 344.
Tabakova, S. and Zapryanov, Z. (1982b), ZAMP, **33**, 487.
Tabakova, S. and Zapryanov, Z. (1987), Theoretical and Applied Mechanics, Bulgarian Academy of Sciences, **Year XVIII**, No.1, 60.
Tachibana, M. (1973), Rheol. Acta, **12**, 58.

Tam, C.K.W. (1969), J. Fluid Mech., **38**, 537.
Tambe, D.E. and Sharma, M.M. (1991), J. Colloid Interface Sci., **147**, 137.
Taneda, S. (1956), J. Phys. Soc. Japan, **11**, 1104.
Taneda, S. (1979), J. Phys. Soc. Japan, **46**, 1935.
Taylor, D.T. and Acrivos, A. (1964), J. Fluid Mech., **18**, 466.
Taylor, G.I. (1932), Proc. Roy. Soc. London, **A 138**, 41.
Taylor, G.I. (1934), Proc. Roy. Soc. London, A **146**, 501.
Taylor, G.I. (1964), Proc. Int. Congr. Appl. Mech., 11th, Munich, 790.
Taylor, G.I. (1966), Proc. Roy. Soc. London, **A 291**, 159.
Taylor, G.I. and Saffman, P.G. (1981), J. Fluid Mech., **102**, 455.
Tchen, C. (1954), J. Appl. Phys., **25**, 463.
Telionis, D. (1981), Unsteady Viscous Flows, Springer Verlag.
Thom, A. (1933), Proc. Roy. Soc., **A 141**, 651.
Thomas, D.G. (1965), J. Colloid Sci., **20**, 267.
Thomasset, Fr. (1981), Implementation of Finite Element Methods for Navier-Stokes Equations, Springer Verlag, New York, Heidelberg, Berlin.
Thompson, R.L., Dewitt, K.J. and Labus, T.L. (1980), Chem. Eng. Commun., 5, 299.
Thomson, J.J. and Newall, H.F. (1885), Proc. Roy. Soc., **39**, 417.
Throop, G.J. and Bearman, R.J. (1965), J. Chem. Phys., **42**, 2408.
Tjahjadi, M.and Ottino, J.M. (1991), J. Fluid Mech., **232**, 191.
Tjahjadi, M., Stone, H.A. and Ottino, J.M. (1992), J. Fluid Mech., **243**, 297.
Tochitani, Y., Mori, Y. and Komotori, K. (1977), Wärme Stoffubertag, **10**, 71.
Tollmien, W. (1924), Göttingen Dissertation.
Tomotika, S. (1935), Proc. Roy. Soc. Lond., **A 150**, 322.
Torza, S., Cox, R.G. and Mason, S.G. (1972), J. Colloid Interf. Sci., **38**, 395.
Tozeren, H. (1982), J. Appl. Mech., **49**, 279.
Tozeren, H. (1983), J. Fluid Mech., **129**, 77.
Tozeren, H. (1984), Int. J. Numer. Meth. Fluids, **4**, 159.
Tran-Cong, T. and Phan-Thien, N. (1989), Phys. Fluids, **A 1**, 453.
Tran-Cong, T., T., Phan-Thien, N. and Graham, L. (1990), Phys. Fluids, **A 2**, 666.
Tsai, T.M. and Miksis, M.J., (1994), J. Fluid Mech., **274**, 197.
Tsai, T.M. and Miksis, M.J., (1997), J. Fluid Mech., **337**, 381.
Tsuge, H. and Hibino, S.I. (1977), J. Chem. Engng Japan, **10**, 66.
Tsukada, T., Hozawa, M., Imaishi, N. and Fujinawa, K. (1984), J. Chem. Engng. Japan, **17**, 246.
Uijttewaal, W.S.J. and Nijhof, E.-J. (1995), J. Fluid Mech., **302**, 45.
Uijttewaal, W.S.J., Nijhof, E.-J. and Heethaar, R.M. (1993), Phys. Fluids, **A 5**, 819.
Van De Ven, T.G.M. (1989), Colloid Hydrodynamics, Academic Press.
Van der Werff, J., de Kruif, C.G., Blom, C. and Mellema, J. (1989),Physical Review, **A 39**, 795.
Van Dyke, M. (1964), Perturbation Methods in Fluid Mechanics, Academic Press.
Van Dyke, M. (Ed.) (1982), An Album of Fluid Motion, The Parabolic Press, Stanford, California.
van Saarloos, W. and Mazur, P. (1983), Physica, **120A**, 77.
Vand, V. (1945), Nature, **155**, 364.
Vand, V. (1948a), J. Phys. and Colloid Chem., **52**, 277.
Vand, V. (1948b), J. Phys. Chem. Colloid., **52**, 300.
Vasseur, P. and Cox, R.G. (1976), J. Fluid Mech., **78**, 385.
Vasseur, P. and Cox, R.G. (1977), J. Fluid Mech., **80**, 561.
Verwey, E. and Oberbeek, J. (1948), Theory of Stability of Lyophobic Colloids, Elsevier.
Villat, H. (1943), Leçons sur les Fluides Visqueux, Paris, Gauthier-Villars.
Vojir, D.R. and Michaelides, E.E. (1994), Int. J. Multiphase Flow, **20**, 547.
Vuong, S.T. and Sadhal, S.S. (1989), J. Fluid Mech., **209**, 617.
Wacholder, E. (1973), Chem. Eng. Sci., **28**, 1447.

Wacholder, E. and Weihs, D. (1972), Chem. Eng. Sci., **27**, 1817.
Wakiya, S. (1953), J. Phys. Soc. Japan, **8**, 254.
Wakiya, S. (1956), Res. Rep. Fac. Eng. Niigata Univ. Japan, **5**, 1.
Wakiya, S. (1957), J. Phys. Soc. Japan, **12**, 1130.
Wakiya, S. (1959), Res. Rep. Fac. Eng. Niigata Univ. Japan, **8**, 17.
Wakiya, S. (1967), J. Phys. Soc. Japan, **22**, 1101.
Walpole, L.G. (1971), Quart. J. Mech. Appl. Math., **25**, 153.
Wang, C.Y. (1967a), J. Appl. Mech., **34**, 823.
Wang, C.Y. (1967b), J. Math. Phys., **46**, 195.
Wang, C.Y. (1968), J. Fluid Mech., **32**, 55.
Wang, C.Y. (1969), Quart. Appl. Math., **27**, 273.
Wang, H., Zinchenko, A.Z. and Davis, R.H. (1994), J. Fluid Mech., **265**, 161.
Watson, E.J. (1955), Proc. Roy. Soc. London, **A 231**, 104.
Waxman, A.M. (1984), Stud. Appl. Math., **70**, 63.
Wegener, P. and Parlange, Y. (1973), Ann. Rev. Fluid Mech., **5**, 79.
Weinbaum, S., Ganatos, P. and Yan, Z.-Y. (1990), Ann. Rev. Fluid Mech., **22**, 275.
Weinberger, C.B. and Goddard, J.D. (1974), J. Multiphase Flow, **1**, 465.
Whitehead, A.H. (1889), Quart. J. Math., **23**, 143.
Wiener, N. (1938), Amer. J. Math., **60**, 897.
Wiener, N. (1958),), Nonlinear Problems in Random Theory, Cambridge, Mass. M.I.T. Press and New York: Wiley.
Williams, M.M.R. (1997), ZAMP, **38**, 92.
Williams, W.E. (1962), ZAMP, **13**, 133.
Williams, W.E. (1966), J. Fluid Mech., **25**, 589.
Willmarth, W.W., Hawk, N.E. and Harvey, R.L. (1964), Phys. Fluids, **7**, 197.
Wills, P.R. (1979), J. Chem. Phys., **70**, 5865.
Wilson, L. and Ries, S. (1923), Am. Chem. Soc., **25**, 62.
Winnikov, S. and Chao, B.T. (1966), Phys. Fluids, **9**, 50.
Wohl, P.R. (1976), Adv. Eng. Sci. NASA CP-2001, **4**, 1493.
Wohl, P.R. and Rubinow, S.I. (1974), J. Fluid Mech., **62**, 185.
Woo, S.-W. (1971), Ph.D. Thesis, McMaster Univ., Hamilton, Ontario.
Yan, Z.Y., Weinbaum, S., Ganatos, P. and Pfeffer, R. (1987), J. Fluid Mech., **174**, 39.
Yanenko, N.N. (1971), The Method of Fractional Steps, McGraw-Hill, New York - London.
Yang, S.-M., Feng, Z.C. and Leal, L.G. (1993), J. Fluid Mech., **247**, 417.
Yang, S.-M. and Leal L.G. (1983), J. Fluid Mech., **136**, 393.
Yang, S.-M. and Leal L.G. (1984), J. Fluid Mech., **149**, 275.
Yang, S.-M. and Leal L.G. (1990), J. Multiphase Flow, **16**, 597.
Yang. Y. and Levine, H. (1992), J. Fluid Mech., **235**, 73.
Yih, C.S. (1979), Fluid Mechanics, West River Press.
Young, N.O., Goldstein, J.S. and Block, M.J. (1959), J. Fluid Mech., **6**, 350.
Youngren, G.K. and Acrivos, A. (1975), J. Fluid Mech., **69**, 377.
Youngren, G.K. and Acrivos, A. (1976), J. Fluid Mech., **76**, 433.
Yu, K.L. (1974), Ph.D. Thesis, Univ. of Houston.
Yuan, F. and Wu, W.Y. (1987), Appl. Math. Mech., SUT, Shanghai, China, **8**, 17.
Zapryanov, Z. (1974), Theoret. and Appl. Mech., BAS, **2**, 19.
Zapryanov, Z. (1975), Comptes Rendus de l'Academie Bulgare des Sciences, **28**, 1583.
Zapryanov, Z. (1977), ZAMM, **57**, 41.
Zapryanov, Z. (1982), D. Sc. Dissertation (Mathematics), Univ. Sofia.
Zapryanov, Z. (1989a), VI National Congress on Theoretical and Applied Mechanics, Varna, Bulgaria , Sept. 25-30, 1989, 380.
Zapryanov, Z. (1989b), Ann. Sofia Univ., **83**, 111.

Zapryanov, Z. (1992), Ann. Sofia Univ., **86**, 133.
Zapryanov, Z. and Chervenivanova, E. (1981), Int. J. Multiphase Flow, **7**, 261.
Zapryanov, Z. and Chervenivanova, E. (1982), Int. J. Multiphase Flow, **8**, 393.
Zapryanov, Z. and Chervenivanova, E. (1983), Euromech, 176. La Mécanique des Syst. Gaz-Liquide. Grenoble,1983, **vol.2**, 2.
Zapryanov, Z. and Christov, Ch. (1981), Comp. Meth. in Appl. Mech. and Engng, **29**, 247.
Zapryanov, Z., Dobreva, P.S. and Christov, Ya. (1998a), (in preparation).
Zapryanov, Z. and Kalitzova-Kurteva, P. (1976), University Annual, Appli. Math., **XII, book 3**, 185.
Zapryanov, Z., Kozhouharova, Zh. and Jordanova, A. (1988), ZAMP, **39**, 204.
Zapryanov, Z. and Lambova, I, (1986), ZAMP, **37**, 176.
Zapryanov, Z., Malhotra, A., Aderangi, N. and Wasan, D.(1983), International Journal of Multiphase Flow, **vol.9**, 105.
Zapryanov, Z. and Markov, N. (1994), Ann. Sofia Univ., **88**, 5.
Zapryanov, Z., Minev, P. and Shopov, P. (1998b), Proceedings of the 3D Workshop in the Field of Transport Phenomena in Two-Phase Flow, Nessebar, 1998.
Zapryanov, Z., Shopov, P., Minev, P. and Bazhlekov, I. (1987), Ann. Sofia Univ., **81**, 145.
Zapryanov, Z., Shopov, P., Minev, P. and Bazhlekov, I. (1989), in Proc. Int. Symp. On Computational Fluid Dynamics, Nagoya, Japan, 1180.
Zapryanov, Z. and Stoyanova (Tabakova), S. (1977), Theoretical and Applied Mechanics, Bulgarian Academy of Sciences, **Year VIII**, 11.
Zapryanov, Z. and Stoyanova (Tabakova), S. (1978a), Int. J. Multiphase Flow, **4**, 193.
Zapryanov, Z. and Tabakova, S. (1978b), Theoretical and Applied Mechanics, Bulgarian Academy of Sciences, **Year IX,** No.1, 65.
Zapryanov, Z. and Tabakova, S. (1979), Theoretical and Applied Mechanics, Bulgarian Academy of Sciences, **Year X**, 46.
Zapryanov, Z. and Toshev, E.T. (1985), Int. Symp. on Computational Fluid Dynamics, Tokyo, Japan, Sept. 9-12, 1985, **vol. 2**, 627.
Zapryanov, Z. and Toshev, E.T. (1986), VIII International Heat Transfer Conference, San Francisco, USA, Aug. 17-22, 1986, **vol. 5**, 2549.
Zick, A. and Homsy, G. (1982), J. Fluid Mech., **115**, 13.
Zienkiewicz, O.C. (1977), The Finite Element Method, Mc Graw-Hill, London.
Zinchenko, A.Z. (1978), PMM Applied Math. and Mech., **42**, 1046.
Zinchenko, A.Z. (1981), PMM Applied Math. and Mech., **44**, 30.
Zinchenko, A.Z. (1982), PMM Applied Math. and Mech., **45**, 564.
Zinchenko, A.Z. (1984a), PMM Applied Math. and Mech., **47**, 37.
Zinchenko, A.Z. (1984b), PMM Applied Math. and Mech., **48**, 198.
Zuzovsky, M. (1976), Ph.D. Thesis, Carnegie-Mellon Univ.
Zuzovsky, M., Adler, M. and Brenner, H. (1983), Phys. Fluids, **26**, 1714.

INDEX